Systematics and Evolution of the Sthenurine Kangaroos

Gavin J. Prideaux

Systematics and Evolution of the Sthenurine Kangaroo

Gavin J. Prideaux

UNIVERSITY OF CALIFORNIA PRESS
Berkeley • Los Angeles • London

UNIVERSITY OF CALIFORNIA PUBLICATIONS
IN GEOLOGICAL SCIENCES

Volume 146

UNIVERSITY OF CALIFORNIA PRESS
BERKELEY AND LOS ANGELES, CALIFORNIA

UNIVERSITY OF CALIFORNIA PRESS, LTD. LONDON, ENGLAND

PRINTED IN THE UNITED STATES OF AMERICA

Library of Congress Cataloging-in-Publication Data available upon request.

ISBN 978-0-520-09845-9

The paper in this publication meets the minimum requirements of
ANSI/NISO Z39.48-1992 (R 1997) (*Permanence of Paper*) {∞}

Contents

Plates, Figures and Tables

Plates

Figures

Tables

Acknowledgments

Most of the work presented here derives from a Ph.D. undertaken in the School of Biological Sciences, Flinders University of South Australia, Adelaide. I sincerely thank Rod Wells for fostering my academic interest in paleontology and for supervising the Ph.D. Rod placed many resources at my disposal and provided motivation and ideas. Dick Tedford has my gratitude for providing helpful criticism, guidance and ideas over the course of the work. Rod and Dick graciously allowed me to include a description of *"Procoptodon" williamsi*, a taxon they had originally intended to describe themselves. I am also indebted to Peter Murray, David Ride and Jim McNamara for sharing many insights into kangaroo evolution with me over the past decade.

I thank the following museum staff for extended loans of fossil and recent material: Robert Jones and Tim Flannery (Australian Museum); Betty Thompson and Tom Rich (Museum Victoria); Peter Murray and Dirk Megirian (Northern Territory Museum); Alisanne Ramsden (Queen Victoria Museum); Ralph Molnar and Joanne Wilkinson (Queensland Museum); Neville Pledge, Jim McNamara and Cath Kemper (South Australian Museum); John Long and Darryl Kitchener (Western Australian Museum); Ella Hoch (Geologisk Museum, Copenhagen); Pat Holroyd and Mark Goodwin (University of California Museum of Paleontology). I am obliged to Lindsay Hatcher not only for allowing me to examine material in his collection, but also for driving the specimens from Perth to Adelaide so that I could compare them directly with specimens from eastern Australia.

For their warm hospitality during museum visits in Australia, California and the United Kingdom, I thank Alex Baynes, Robert Jones, Noel Kemp, Judd Case and Andy Currant. Alex also kindly provided the use of personal resources for fieldwork in 1996. Seated around various campfires or their metaphorical equivalents, Grant Gully, Pyramo Marianelli and Dirk Megirian shared hours of stimulating discussions on many aspects of vertebrate life and death, although the finest ideas appear to have been consistently lost somewhere between the Alien and the hangover. I am also grateful to the many folks who helped with fieldwork, especially Linda Ayliffe, Alex Baynes, Carol Conway, Rachel Correll, Adam Cronin, Grant Gully, Lindsay Hatcher, John Long, Pyramo Marianelli, Robin McBeath, Mark Norton, Dave Rowe, Adam the Runner, Neil Taylor, James Tilly, Sue Tonkin, the Western Australian Speleological Group and Dave Wilson. Financial and logistic support for fieldwork associated with this project was provided by Flinders University, the Western Australian Department of Conservation and Land Management, the Augusta Margaret River Tourism Association, and the Royal Society of New South Wales through an Ethel Mary

Read Grant. Assistance for overseas travel was provided by Flinders University and a Visiting Student Award from the University of California Museum of Paleontology.

This monograph benefited greatly from the detailed comments provided on its penultimate incarnation by Dick Tedford and Bernie Cooke. Judd Case and Jukka Jernvall also provided some helpful feedback. I thank Mike Woodburne for his encouragement to publish the work in this series. I am indebted to Grant Gully for drawing many of the line figures and assisting with photo cropping, and to Peter Murray for the sketch reconstructions of *Hadronomas puckridgi* and *Sthenurus stirlingi*. Sheila Berg improved the manuscript with her detailed editing. Finally, I offer my deepest thanks to my wife, Rachel Correll, for tolerating my obsession with extinct boof-headed kangaroos and her amazing support during the years this monograph took to produce. My parents, Keith and Margaret Prideaux, deserve equal gratitude for their support over the duration of my academic studies. This work would never have been completed without the many sacrifices made by Rachel and my parents, and I dedicate it to them.

Abbreviations

AH	Anterior Height.
AM F	Department of Palaeontology, Australian Museum, Sydney.
AM MF	Department of Palaeontology, Australian Museum; formerly in the collection of the Mining and Geological Museum, New South Wales.
AMNH	Department of Vertebrate Paleontology, American Museum of Natural History, New York.
AW	Anterior Width.
ANU Geol.	Department of Geology, Australian National University, Canberra.
CPC	Commonwealth Palaeontological Collection, Australian Geological Survey Organisation, Canberra.
BMNH (M)	Department of Palaeontology (Mammals), Natural History Museum (formerly British Museum of Natural History), London.
cf.	L. *confer*, 'compare (with)'. When used in a binomen (e.g., *Simosthenurus* sp. cf. *maddocki*), the term denotes a specific identification that cannot be made with certainty, while highlighting an apparent affinity. This uncertainty may be a result of the specimen's incompleteness or because it lies outside of the accepted morphospace or temporal range of a species.
est.	estimated.
Fm.	Formation.
FMNH PM	Department of Paleontology (Mammals), Field Museum of Natural History, Chicago.
FU	Vertebrate Palaeontology Laboratory, School of Biological Sciences, Flinders University, Adelaide.
Gr.	Greek.
Heb.	Hebrew.
HM S	W. E. Swinton's catalog, Hunterian Museum, Glasgow University, Glasgow.
ka	kiloanna; thousands of years ago.
L	Length.
Lat.	Latin.
LF	Local Fauna.
Ma	Meganna (millions of years ago).
MCZ	Museum of Comparative Zoology, Harvard University, Cambridge, Mass.
MVZ	Museum of Vertebrate Zoology, University of California, Berkeley.
MM L	Manchester Museum, Manchester, U.K.
n	sample size.

NMV P	Department of Vertebrate Palaeontology, Museum Victoria, Melbourne (formerly National Museum of Victoria).
NSW	New South Wales.
NT	Northern Territory.
NTM P	Department of Palaeontology, Northern Territory Museums of Arts and Sciences, Darwin and Alice Springs.
PH	Posterior Height.
pers. comm.	personal communication.
pers. obs.	personal observation.
PW	Posterior Width.
QLD	Queensland.
QM F	Department of Palaeontology, Queensland Museum, Brisbane.
QVM:GFV	Department of Geology and Fossil Vertebrates, Queen Victoria Museum, Launceston.
SA	South Australia.
SADM V	Vertebrate Collection, Mines and Energy South Australia, Adelaide (formerly South Australian Department of Mines).
SAM A	Department of Anthropology, South Australian Museum, Adelaide.
SAM P	Department of Palaeontology, South Australian Museum, Adelaide.
SGM P	Spencer and Gillen Museum, Alice Springs (now Museum of Central Australia, one of the Northern Territory Museums of Arts and Sciences).
s. l.	Lat. *sensu lato*, 'in the broad sense'.
s. s.	Lat. *sensu stricto*, 'in the narrow sense'.
TAS	Tasmania.
TMM	Texas Memorial Museum, Austin.
TM Z	Department of Geology, Tasmanian Museum, Hobart.
UCMP	University of California Museum of Paleontology, Berkeley.
UCR	Department of Earth Sciences, University of California, Riverside.
unreg.	Unregistered.
USNM	United States National Museum, Washington, D.C.
VIC	Victoria.
WA	Western Australia.
WAM	Western Australian Museum, Perth.
WAM HC	Hatcher Collection, Western Australian Museum, Perth.

Abstract

The subfamily Sthenurinae (Macropodoidea, Diprotodontia) is an extinct group of robust kangaroos. The earliest sthenurine appears in the late Miocene of central Australia, but the group is most common in the Pleistocene faunas of southern and eastern Australia. Since the Sthenurinae was last reviewed over three decades ago, species diversity has more than doubled. Many species are now also represented by series of well-preserved specimens, including complete crania and skeletons. New insights generated by these discoveries provided the major impetus for this review of sthenurine systematics, functional morphology, paleoecology, biochronology and zoogeography.

Sthenurinae is diagnosed on the basis of nine craniodental synapomorphies, making it the best-defined kangaroo subfamily. Two new genera and four new species are recognized here, bringing the total to six genera and 26 species. A new tribe (Simosthenurini) is raised to include the three short-faced genera: *Archaeosimos* gen. nov., *Simosthenurus* and *Procoptodon.* A cladistic analysis of interrelationships within the Sthenurinae has prompted the dismissal of concerns previously expressed about the distinction of *Simosthenurus* and *Sthenurus.* However, *Simosthenurus (sensu lato)* is paraphyletic, because *Procoptodon* is derived from within it. Several molar crests hypertrophied in the more derived *Procoptodon* species are believed to be a result of phylogenetic character reversal, a phenomenon that may be an important but previously neglected mechanism of evolutionary change within the Macropodoidea. Compelling evidence suggests a paraphyletic origin for the Sthenurinae from within the late Oligocene to middle Miocene subfamily Bulungamayinae. The middle Miocene bulungamayine *Wanburoo* is the sister taxon of the Sthenurinae, which along with the late Miocene occurrences of *Hadronomas* in central Australia and *Archaeosimos* in southern Australia, supports a middle Miocene origin for the subfamily. No support is found for the hypothesis that *Troposodon* and *Lagostrophus* are sthenurines. These two genera have a closer affinity with the macropodines and are probably most closely allied to each other and *Protemnodon.* Craniodental similarities shared with the sthenurines are likely to reflect dietary convergence.

A fundamental step in the phylogenesis of the sthenurine lineage was the attainment of relatively large body size. The plesiomorphic *Hadronomas* exemplifies the sthenurine bauplan: it is the earliest macropodoid in the fossil record larger than a small wallaby, and several craniodental attributes indicate that it was capable of generating larger occlusal forces during the apprehension and mastication of food than any other Miocene macropodoids. The evolution of sthenurines was most likely a direct faunal response to the decline in mesic conditions from the middle Miocene. All sthenurines apart from *Hadronomas* are united by nine craniodental traits that represent a crucial circumvention of the constraints that probably restricted *Hadronomas* to only a moderately fibrous diet.

The masticatory muscles became more anteriorly oriented, the orbits became anterolaterally projected, the mandibular symphysis was rigidified, grinding became the primary function of the premolars, the cheek tooth row became laterally curved, and molar complexity increased. These adaptations improved the capacity of the craniodental system for breaking down tough vegetation and they paved the way for the cranial foreshortening characteristic of the Simosthenurini, which originated in the late Miocene.

No less than 14 of the 18 simosthenurin species recognized occur in the middle or late Pleistocene cave assemblages of southern and eastern Australia. Simosthenurins filled many of the browser niches available throughout the temperate woodlands and open forests of the Pleistocene, and the lineage may have originated in transitional areas between the retracting wetter forests and expanding sclerophyll habitats of the southeast during the late Miocene. Some Pleistocene species, such as *Simosthenurus maddocki*, were apparently highly selective feeders, while others were generalists. The evolution of *Procoptodon* is likely to have been prompted by the marked expansion of dry sclerophyll vegetation and open conditions following the intensification of aridity in the latter half of the Pliocene. *P. goliah*, the largest and most robust of all macropodoids, was the only simosthenurin widely distributed throughout the continental interior and was apparently adapted to the consumption of a tougher diet than any other sthenurine. In contrast to *Procoptodon*, three of the six Pleistocene species of *Sthenurus* had already evolved by the early late Pliocene. This lineage maintained a stronghold in the interior of the continent during the Pleistocene and is characterized by a more gracile build, longer face, broader incisors and higher-crowned molars than the Simosthenurini. *Sthenurus* species may have subsisted on the small-leafed, xeromorphic shrubs and low, dusty forbs still common throughout inland Australia today. Although sthenurine diversity and abundance reached its zenith during the middle and early late Pleistocene, all species were extinct by the end of the epoch. On the basis of data currently available, the demise of this remarkable radiation of browsing mammals appears most likely to have resulted from attrition over an extended duration due primarily to human hunting pressure.

INTRODUCTION

Because Australia has been isolated to terrestrial faunal migration for most of the Cenozoic, changes in dominant vegetation types driven by changing climate have been chiefly responsible for shaping the evolution of its marsupial radiation. From the time Australia's mammal record commences in earnest at about 24 Ma (Woodburne et al. 1994) until the present day, kangaroos (superfamily Macropodoidea) have been the continent's dominant terrestrial herbivores. They are, in effect, Australia's answer to the artiodactyl radiations of other continents. Molecular clock data have extended the genesis of kangaroos back into the Eocene (Springer and Kirsch 1991; Kirsch et al. 1997; Burk et al. 1998), but Australia's penurious Paleogene record offers no insights into the history of the group prior to the latest Oligocene. From this time through until the early middle Miocene, kangaroos were small cryptic creatures with low fiber diets (Cooke 1997a-c, 2000), not unlike modern bettongs, potoroos and the musky rat-kangaroo. As the Antarctic ice sheet expanded, sea level fell and the earth cooled following the Miocene Climatic Optimum (16–15 Ma; Shackleton and Kennett 1975; Flower and Kennett 1994), and as latitudinal temperature gradients intensified across a still northbound Australia, the central and northern regions of the continent became increasingly arid (Bowler 1982; Veevers 1984; Frakes et al. 1987). Habitats dominated by sclerophyll and xeromorphic vegetation expanded and began to reshape the late Cenozoic landscape (Kershaw et al. 1994; Martin 1998a). Macropodine and sthenurine kangaroos evolved and diversified as grasslands, shrublands, open forests and woodlands spread. By the middle Pleistocene, sthenurines had become the continent's premier lineage of browsing mammals.

SCOPE OF THE WORK

Since the previous taxonomic surveys of the Sthenurinae (Stirton and Marcus 1966; Tedford 1966), the quantity and quality of sthenurine specimens have vastly improved, courtesy of sites such as the Naracoorte Caves and Lake Callabonna. The number of species recognized has more than doubled to 26, with nigh on one-third of these brought to light over the course of this study (Prideaux and Wells 1997, 1998; Prideaux 2000; this monograph). Here, I present revised diagnoses, detailed descriptions and interspecific comparisons for all sthenurine taxa that are incompletely described in the existing literature. I also reexamine the purported sthenurine credentials of *Troposodon* and *Lagostrophus*, and consider the evidence available for the phylogenetic origin of the Sthenurinae. The cladistic analysis of interrelationships that follows is the first such investigation carried out on this group. Although the phylogeny provides stronger support for some crown clades than it does for others, the primary cladogram structure is well supported and congruent with the distribution of sthenurines through space and time. Within the systematic framework established, I then consider the major trends in sthenurine evolution from functional morphological, biochronologic, zoogeographic and paleoecological perspectives, and advance a hypothesis for their extinction.

HISTORICAL BACKGROUND

Sir Richard Owen described the first fossil marsupials from Australia in an 1838 publication of the journal of Surveyor General Sir Thomas Mitchell's expeditions. The bones were collected from Breccia Cave (now Mitchell Cave) near Wellington in New South Wales, and from them Owen described the large quadruped *Diprotodon optatum* and the extinct kangaroos *Macropus titan* and *M. atlas*. Seven years later, Owen briefly described the giant *M. goliah* in volume 1 of G. R. Waterhouse's *Natural History of the Mammalia*, but did not recognize a supraspecific distinction for either *M. atlas* or *M. goliah* until he began publishing his memoirs in the 1870s. In an abstract pertaining to a paper presented to the Royal Society of London in 1873, Owen erected *Sthenurus* as a subgenus of *Macropus* Shaw, 1790. No description was provided until the following year, however, when Owen raised *Sthenurus* to genus level, diagnosing it primarily on the basis of the distinctive, broad lower permanent premolar of the type species, *S. atlas*. Owen's concept of *Sthenurus* was not entirely clear (Lydekker 1887; Stirton 1963; Tedford 1966), and he often confused the upper dentition with that of the extinct macropodine, *Protemnodon* Owen, 1874a. Two species originally named *Sthenurus brehus* (Owen, 1874a) and *S. minor* (Owen, 1877a) have subsequently been assigned to *Protemnodon* and *Troposodon* Bartholomai, 1967, respectively. At the same time that he described *Sthenurus*, Owen (1873b, 1874b) introduced the name *Procoptodon* for *"M." goliah* and two other heavily built species, *P. rapha* and *P. pusio*, which were also collected from fluviatile deposits on the Darling Downs in southeastern Queensland. Flower (1884) synonymized both *Sthenurus* and *Procoptodon* with *Macropus*, but this amendment was never followed.

In 1895, the inaugural director of the Queensland Museum, C. W. De Vis, examined the macropodid portion of the extensive vertebrate collection from the Darling Downs. Among several new species, he described *Sthenurus pales* and *S. oreas*, avowing that: "an amalgamation of *Procoptodon* with *Sthenurus* is demanded by their verisimilitude of tooth sculpture, and by the occurrence of forms of transition between the two" (De Vis 1895:88). De Vis was led to this verdict by the dental similarities of *S. oreas* and *S. pusio* (= *S. otuel* therein). While many authorities continued to recognize *Procoptodon* as a valid genus, Glauert (1910a-b), Raven (1929), Raven and Gregory (1946) and Tate (1948) agreed with De Vis, treating *Procoptodon* as a junior synonym or subgenus of *Sthenurus*. Evolutionarily, Tate (1948) considered *Procoptodon* a likely offshoot of *Sthenurus*.

Discoveries in the early 1900s of Pleistocene vertebrates in Mammoth Cave, southwestern Australia sparked considerable enthusiasm among local authorities (e.g., Woodward 1909, 1914; Anon. 1910; Glauert 1914), and Ludwig Glauert was retained by the Western Australian Museum to excavate and report on the Mammoth Cave fossils. In 1910, he named and described *Sthenurus occidentalis*. Sixteen years later, in a footnote to a list of Western Australian fossils, Glauert raised the family Sthenuridae "to hold the genera *Sthenurus* and *Procoptodon* whose cranial characters distinguish them from all known members of the Macropodidae. The skull is extremely heavy and remarkable for its height and width; the facial portion is short, reducing the diastema so that it scarcely exceeds the length of the cutting edges of the three incisors" (Glauert 1926:71).

Although the decision to name the Sthenuridae was prompted by Glauert's excavation of the first two relatively complete crania ascribed to either genus (see Woodward 1914), no descriptive comments, beyond the familial diagnosis, were provided for either specimen until the work of Anderson (1932). He used the Mammoth Cave crania for comparison with a specimen of *Sthenurus occidentalis* collected from King Island in Bass Strait. Earlier, Scott (1906) had erroneously referred the King Island cranium to *Procoptodon rapha*, and later to *S. atlas* (Scott 1917). All subsequent authorities have opted not to abide by Glauert's recommendation, but Raven's (1929) suggestion that *Sthenurus* and *Procoptodon* be placed in their own subfamily gained widespread acceptance. Although Ride (1959), Bartholomai (1963, 1970) and Marshall et al. (1990) thought an overriding number of similarities with the Macropodinae demonstrated that *Sthenurus* and *Procoptodon* were best placed here, all agreed that these two genera formed a natural grouping among the kangaroos. Raven and Gregory (1946) and Ride (1959) were among the earliest workers to associate the characteristic sthenurine cranial morphology with a diet of tough browse.

The 1960s saw a new phase of research into sthenurines and Australian vertebrate paleontology in general. Marcus (1962) described *Sthenurus andersoni* from three Pleistocene deposits in New South Wales and recognized it as a long-jawed member of the genus, akin to *S. atlas*. Marcus categorized *S. oreas*, *S. pales* and *S. occidentalis* as short-jawed, similar to *Procoptodon*. Bartholomai (1963) reviewed the species of *Sthenurus* from Queensland and named two new species from the Chinchilla Sand in the western Darling Downs. These were the first records of sthenurines from the Pliocene. Although he had no hesitation in recognizing *S. antiquus* as a short-jawed form, Bartholomai balked at assigning *S. notabilis* to either group. The partial dentary of the only known specimen was thought to be similar to the members of the short-jawed group, but the cheek teeth were more like those of a long-jawed form. Led by the prospect of finding a closer resemblance between *Procoptodon* and the earlier *Sthenurus* species, Bartholomai (1963) suggested that the mosaic features expressed in *S. notabilis* might highlight its proximity to the phyletic divergence of the genera. Thus, Bartholomai found himself unable to justify a subgeneric distinction within *Sthenurus.*

The most comprehensive study of sthenurines prior to the present investigation was the review of *Sthenurus* published by R. H. Tedford in 1966. This work was the first of the modern era to pull all of the known morphological, geographic and chronologic data together and construct a feasible taxonomy. Tedford discussed apparent evolutionary trends within the group and laid a foundation for all subsequent research into sthenurines. Fundamentally, the review demonstrated that the geographic and evolutionary dimensions of the Pleistocene sthenurine radiation were considerably greater than previously conceived. Tedford (1966) believed that the species of *Sthenurus* had evolved along two lines, as intimated by Marcus (1962) and Bartholomai (1963). He separated the dolichocephalic and brachycephalic *Sthenurus* species into the subgenera *Sthenurus* and *Simosthenurus*, respectively. Resemblance of the *S. notabilis* dentary to *Simosthenurus* was dismissed as an artifact of the holotype's juvenile and fragmentary nature. Dental similarities between *S. notabilis* and *Procoptodon* were viewed as convergence, and *S. notabilis* was placed in the nominotypic subgenus on the basis of its high-crowned molars. Broad similarities between the species of *Sthenurus* and macropodine kangaroos were considered most likely to reflect adaptations of *Sthenurus* species to grazer or browser-grazer niches. By contrast, similarities between the species of *Simosthenurus* and *Procoptodon* were thought to reflect parallel adaptations to a browsing lifestyle. Tedford (1966) described two new Pleistocene sthenurines, *S. (Sthenurus) tindalei* from Lake Menindee and *S. (Simosthenurus) orientalis* from the Wellington Caves.

In the same year that Tedford published his *Sthenurus* work, Stirton and Marcus (1966) reviewed the taxonomy of *Procoptodon.* This was soon followed by Bartholomai's (1970) survey of *Procoptodon* in Queensland. Since Owen described the original three *Procoptodon* species, only one new form has come to light, *P. texasensis* from the Texas Caves, southeastern Queensland (Archer 1978). In a second monograph, which detailed the Pleistocene macropodid fauna from Lake Menindee in western New South Wales, Tedford (1967) described a near-complete skeleton of the largest known sthenurine, *Procoptodon goliah.* This was the first study to reveal that sthenurines could also be differentiated from other macropodids on the basis of their postcranial skeleton. The pes of *P. goliah* was essentially monodactyl due to the extreme reduction of metatarsals II, III and V. Tedford (1967) also noted that the forearm and central digits of the manus were comparatively elongate, and the scapula appeared modified for a greater degree of forelimb mobility. Presumably, these adaptations were used for reaching out to grasp and pull in browse vegetation to the mouth.

While sthenurines had been recorded in the cave fills of southeastern Australia (e.g., Tindale 1933; Gill 1957), no new species were described from the area now widely acknowledged as Australia's premier Pleistocene vertebrate-yielding region until Merrilees described the diminutive *Sthenurus gilli* in 1965. Victoria Fossil

Cave at Naracoorte in southeastern South Australia is the type locality of *Si. maddocki* Wells and Murray, 1979, and has a more diverse Pleistocene fauna than any other single Pleistocene locality in Australia (Wells et al. 1984). Among the 93 species so far identified there are 22 macropodids, including a remarkable nine sthenurines.

In an influential paper published in 1983, Tim Flannery presented the first cladistic overhaul of the Sthenurinae and proposed three crucial alterations to its systematics. Following the earlier suggestion of Pledge (1980), he raised *Simosthenurus* to generic rank, a move that gained only erratic acceptance subsequently. Flannery also hypothesized that the similarities between *Simosthenurus* and *Procoptodon* were not convergent, but reflected a common ancestry to the exclusion of *Sthenurus.* Wells and Tedford (1995) agreed and provided further evidence in support of a *Simosthenurus–Procoptodon* clade. The most novel proposition of the Flannery (1983) paper, however, was the suggestion that the Pliocene to Pleistocene genus *Troposodon* and the extant banded hare-wallaby, *Lagostrophus* Thomas, 1887, were not macropodines as previously supposed, but plesiomorphic sthenurines. Features historically believed to be definitively sthenurine in nature were thus recognized as characteristics only of the crown group. Sthenurinae *sensu* Flannery 1983 was defined on the basis of five dental traits.

The discovery of two cranial specimens of *Hadronomas puckridgi* Woodburne, 1967, in the late Miocene of central Australia prompted Murray (1989, 1991a) to infer a close relationship between *Hadronomas* and the sthenurines. Woodburne (1967), Ride (1971), Bartholomai (1972, 1978a) and Campbell (1973) had previously alluded to this possibility, but a paucity of quality cranial material prevented any resolution. Murray's (1991a) description and comparison of the first near-complete *Hadronomas* cranium provided solid support in favor of a plesiomorphic position in the Sthenurinae. This idea was later strengthened by an analysis of the postcranial skeleton (Murray 1995). Because *Hadronomas* was allied to the crown group on the basis of a set of archetypal character states conflicting with those that allegedly drew in *Troposodon* and *Lagostrophus*, serious doubts were raised about the actual relationships of these two genera.

Descriptions of 11 new species in the past decade, including the four described here, have resulted not only in a better understanding of the breadth of the crown radiation, but also provide a sound basis for resolving sthenurine origins and phylogenetic interrelationships. The recognition of two relatively plesiomorphic taxa in a latest Miocene cave assemblage from Yorke Peninsula, South Australia (Pledge 1992; this monograph), is especially useful in this regard. Two partial molars described from an early Pliocene deposit in southwestern Victoria (Flannery et al. 1992) belong to either one or two new species, but are too incomplete to describe as new species. Other taxa that shed considerable light on sthenurine phylogeny were named by Prideaux and Wells (1997, 1998) and Prideaux (2000).

METHODS

FORMAT OF THE TAXONOMIC REVIEW

The primary aim of the taxonomic review is to provide diagnostic and detailed descriptive information for each species, except in a few cases where preexisting accounts remain adequate. Because all sthenurines and almost all fossil macropodoids are diagnosed solely on craniodental features, I have restricted my investigations to this portion of the skeleton. Continuing finds of associated postcranial remains will inevitably provide additional character suites useful for improving the taxonomic understanding of many fossil macropodoid groups.

The layout of each species review takes the following form.

A comprehensive list of synonyms, namely *nomina nuda*, *nomina oblita*, junior synonyms, disused senior synonyms, misidentifications, misspellings, identifications made only to genus level, and nominal combinations different from those proposed here.

Taxonomic Authority. Authority responsible for first naming and describing the taxon.

Holotype. Element information, collection details, locality and geological age.

Paratypes. As for holotype.

Referred Specimens. These are listed by locality and registration number, followed by a basic description. Because listing every referred specimen is not feasible for species that are very abundant in particular sites, I have set a limit of 15 for each locality. If the number of specimens from a site is greater than 15, I give the locality name and the number or approximate number of specimens. Exceptions are made for species that occur in only a few localities and for specimens that are remarkably complete. I have endeavored to directly confirm the identification of all craniodental material in the majority of collections that contain sthenurines, but for a few specimens, I rely on identifications from published works. These are duly referenced where an identification could not be clearly made from plates or figures.

I have not indicated the degree of intactness with the listing of referred specimens unless a specimen consists of at least a near-complete cranium, a partial cranium and associated dentary, or a partial skeleton. Most of the thousands of referred specimens now known are fragmentary to some degree, so the extra information contributed by indicating degree of intactness would be minimal. Finally, 'juvenile' and 'adult' are terms of convenience only; they are not postulations of reproductive, behavioral or physiological development. 'Adult' is also not intended to imply that growth had terminated. If a specimen is at an ontogenetic stage where the third permanent premolar is unerupted or may be safely predicted to have been unerupted, it is deemed a juvenile. Conversely, if this tooth has erupted, the specimen is deemed an adult.

Diagnoses. Diagnostic features have been limited to autapomorphies as much as possible. These can facilitate the easiest distinction of a taxon from its relatives, but difficulties often arise when a species' distinctiveness is embodied in proportional differences or a unique combination of features known individually in other species. Therefore, many diagnoses contain comparative statements, although I have endeavored to keep them succinct. Unless otherwise indicated, diagnostic features of the cranium and dentary refer to the adult condition.

Description. For species that have been completely redescribed or described as new, I detail as many significant morphological features as I can recognize, while keeping comparisons within and between species to a minimum. Style is telegraphic; I prefer to use minimum text to convey maximum descriptive information. Descriptions generally concentrate on features consistently expressed across the type series and all of the referred specimens examined. They are also based on adult specimens, except in the case of species known from very few specimens. Deciduous premolars are also described. Juvenile and variable features are considered under *Variation*. Upper teeth are designated by uppercase abbreviations (e.g., P3, M2); lower teeth are designated by lowercase abbreviations (e.g., i1, m3).

To facilitate cross-referencing between species, the order in which aspects of the craniodental system are described is kept fairly consistent. I have not described every part of the basicranium, since I have observed very few discrete anatomical differences between sthenurine species in this region. However, proportions differences are evident and thus described. Murray (1989, 1991a) and Wells and Tedford (1995) give detailed descriptions of the sthenurine basicranium compared with other macropodoids. For some species, I compare dimensions of the cranium and dentary (see Mensuration), but mean cheek tooth dimensions are listed for all species. Raw measurement data derived from many specimens examined during this study may be obtained from the author.

Variation. Documenting the range of intraspecific variation in size and morphology helps highlight traits that are taxonomically most useful. Assessments based on a narrow sample of a geographically or temporally extensive species often lead to inaccurate systematic conclusions. Consideration of the largest possible sample reduces the likelihood of future revision. I have thus examined every substantial collection of sthenurine remains and borrowed representative material from all geographic regions and many different localities. This provided the major benefit of allowing direct comparison between material from different provenances. Occasional developmental abnormalities have been noted. Description of ontogenetic variation is kept brief, because to do this aspect justice, a more detailed study is needed than there is scope for here.

Comparison With Other Taxa. This section outlines how elements of the craniodental system of a species differ phenetically from those of any other species with which they might readily be confused. To facilitate easy cross-referencing, I have employed the same order of element comparison as that used for the descriptions. Phylogenetic commentary is excluded unless absolutely necessary.

Geographic Distribution. This is a list of regions from which a species is known. Maps of Australia showing the location of every deposit that has yielded craniodental remains for particular species are provided in The Sthenurine Radiation (figs 27-37).

Temporal Distribution. The ages or estimated ages of the deposits from which a species is known. These are given detailed consideration in The Sthenurine Radiation. Epoch and subepoch durations follow Berggren et al. (1995a-b): Late Oligocene, 28.5–23 Ma; Early Miocene, 23.8–16.4 Ma; middle Miocene, 16.4–11.2 Ma; late Miocene, 11.2–5.3 Ma; early Pliocene, 5.3–3.6 Ma; late Pliocene, 3.6–1.8 Ma; early Pleistocene, 1.8–0.78 Ma; middle Pleistocene, 0.78–0.20 Ma; late Pleistocene, 0.20–0.01 Ma.

MENSURATION

Except where indicated, tabulated craniodental measurements are based on raw data collected by the author using Mitutoyo vernier calipers fitted with metal pin extensions to the measurement arms. This maximizes accuracy in measurement, particularly of abutting or slightly overlapping teeth in cheek tooth rows. Individual dimensions (all in millimeters) are defined below. Cranial measurements are shown in fig. 1.

Cheek Dentition

Length (L): distance between anterior and posterior extremities, measured along anteroposterior axis.

Anterior Width (AW):

Premolariform teeth – maximum width of crown over center of anterior root (i.e., anterior moiety);

Molariform teeth – maximum width of protoloph or protolophid.

Posterior Width (PW):

Premolariform teeth – maximum width of crown over center of posterior root (i.e., posterior moiety);

Molariform teeth – maximum width of metaloph or hypolophid.

Anterior Height (AH):

Premolariform teeth – maximum crown height at anterior end of tooth, buccal side;

Molariform teeth – maximum crown height of protoloph or protolophid, buccal side.

Posterior Height (PH):
Premolariform teeth – maximum crown height at posterior end of tooth, buccal side;
Molariform teeth – maximum crown height of metaloph or hypolophid, buccal side.

Cranium and Dentary

Condylobasal Length: distance from anterior edge of I1 alveolus to posterior limit of occipital condyles.
Diastema Length: distance from posterior edge of I3 to anterior edge of P3 (I3–dP2 in juveniles).
Rostrum (preorbital) Length: distance from anterior edge of I1 alveolus to anterior edge of orbit.
Palatal Length: distance from posterior edge of I3 to postpalatine bar.
Cranial Width: distance between lateral extremities of left and right zygomatic arches.
Width across Frontals: distance between lateral extremities of left and right frontals.
Distance between Paroccipital Processes: distance between distal extremities of left and right paroccipital processes.
Dentary Depth: vertical depth from alveolar margin at abutment of m2 and m3 to ventral surface of ramus.
Dentary Width: maximum width of ramus measured perpendicular to dentary depth.

Regional Size Comparisons

When comparing relative sizes of crania, dentaries and teeth between representatives of a species from different geographic regions, percentage differences are derived from comparison of mean measurements for

a) condylobasal length and maximum width across zygomatic arches;
b) dentary depth and dentary width;
c) premolar length, anterior width and posterior width;
d) first molar length, anterior width and posterior width, and third molar length, anterior width and posterior width.

For example, the p3 of western *"Procoptodon" browneorum* is approximately 9% smaller than southeastern *"P." browneorum*. This is derived in the following way.

Mean length: western = 14.7; southeastern = 16.1; difference = 9%
Mean anterior width: western = 7.5; southeastern = 8.5; difference = 12%

Mean posterior width: western = 9.8; southeastern = 10.3; difference = 5%
Thus, mean difference in p3 size = 9%.

CRANIODENTAL TERMINOLOGY

Serial designation of the cheek dentition follows Flower (1867), Wilson and Hill (1897) and Luckett (1993). Marsupials replace only one cheek tooth: the third premolar (Luckett 1993; Cifelli et al. 1996; Luckett and Woolley 1996). While the marsupial first molar might be an unreplaced deciduous fourth premolar in a historic sense, convenience dictates that it be referred to as a molar (Owen 1868; Luckett 1993; Jernvall 1995). In addition, no rudimentary successor buds are discernable on the "residual" lamina of the first molar (or subsequent molars), unlike the first and second deciduous premolars (Luckett 1993; Luckett and Woolley 1996). In all macropodoids, except for hypsiprymnodontines and the propleopine *Ekaltadeta* (Wroe and Archer 1995), eruption of the third adult premolar results in the ejection of both its molariform deciduous precursor and the second deciduous premolar.

Because the cusp and crest homologies of the non-molariform premolars (dP2/dp2, P3/p3) of macropodoids are uncertain and beyond the scope of this work, elementary topographic terms are used to describe their morphology (fig. 2). Cusp and crest terminology for molariform teeth (fig. 3) are derived from classic sources (Osborn 1888; Patterson 1956; Van Valen 1966) or papers dealing with macropodoids or diprotodontians in general (Flannery and Rich 1986; Tedford and Woodburne 1987; Woodburne et al. 1987; Ride 1993; Ride et al. 1997). While I have not delved into the evolution of the macropodoid molar beyond that required to discern the relationships of the Sthenurinae, it has been necessary to use several terms not previously associated with the bilophodont macropodoid dentition. Some effort has been made to ensure consistent usage and to identify homologous structures. Interpretations of dental homologies are founded on topological congruence: similarity in structure and position, and the existence of transitional forms. While identification of homologies has not been problematic for most cristae, the absence of phylogenetically transitional forms for some structures has encouraged the occasional leap of faith on the basis of structure and position alone.

Glossary

The glossary defines morphological terms most commonly used throughout this monograph. For bones plus most external processes and foramina of the kangaroo cranium and dentary, the reader is referred to fig. 1 and Wells and Tedford (1995:fig. 9). Terms listed below are common anatomical terms or are derived from Tedford (1966, 1967), Flannery (1983, 1989), Murray (1991a) or any of the papers listed in the preceding paragraph, with the exception of three new dental terms. Many features are labeled or plainly evident on figures. Although most definitions have a wider applicability than the Sthenurinae or Macropodoidea, they are customized to make the anatomy and evolutionary morphology of sthenurines as manifest as possible. As such, my terminology occasionally differs in its application from that of some other mammal authorities.

Anterior basin. Small, often poorly demarcated basin formed immediately anterior to the longitudinal basin of the sthenurine dP2 or P3 (fig. 2).

Anterior cingulum. General term often used to denote the portion of the upper molar anterior to the protoloph and the portion of the lower molar anterior to the protolophid. I prefer to use *precingulum* for upper molariform teeth (fig. 3) and *trigonid* for lower molariform teeth (fig. 3).

Anterobuccal crest. Crest or lobe formed on the buccal side of the main crest or blade of I3 (figs 16-17). In sthenurines, it is usually only expressed as a slight vertical eminence at the anterior end of the main crest, but the crest is well developed in macropodines.

Anterolingual crest. Crest or lobe formed on the lingual side of the main crest (blade) of I3 (figs 16-17). In sthenurines, the crest is usually distinct, extending from the anterior end of I3 for half or two-thirds the length of the main crest, but is sometimes only expressed as a vertical eminence at the anterior end of the tooth.

Ascending ramus. The upright portion of the dentary posterior to the (horizontal) ramus, or tooth-bearing portion of the dentary. It bears the mandibular condyle and the areas of insertion for the three major sets of jaw-closing muscles.

Bilophodont. Molar form denoted by the presence of two prominent transverse ridges (figs 3, 12, 19).

Brachycephalic. Term describing a relatively foreshortened cranium (e.g., figs 14E-F, 15H-I).

Buccal. Term used to indicate that a dental feature is positioned on or oriented toward the cheek side of a tooth, i.e., external to its anteroposterior axis. Some prior studies of macropodoids and other diprotodontians have employed *labial* in the same sense.

Cingulum. A shelf-like structure extending along the side of a tooth crown. A cingulum may or may not be confluent with a crest (figs 3, 12, 18, 19). *Cingulid* is used in reference to structures of the lower dentition.

Crista. A crest or ridge on a cheek tooth that is associated with a cusp. A crista may connect to the apex of a cusp or be restricted to its slope (figs 3, 9, 19). *Cristid* is used in reference to structures of the lower dentition (figs 3, 12).

Cristid obliqua. Cristid directed anteriorly or anterolingually from the hypoconid. In many sthenurines, it is divided into a buccal component, which maintains contact with the hypoconid apex, and an often larger, lingual component, which is oriented more anteroposteriorly and nearer to the midline of the tooth (figs 3, 12). Prideaux (2000) referred to these as the posterior and anterior components, respectively. Colloquially, the cristid obliqua is often referred to as the 'midlink' of the lower molar.

Cusp. A raised, often tapering projection on a mammalian tooth. Cusps are frequently linked by cristae or lophs that may be so well developed that the location of a cusp can only be discriminated as the point where crests meet. See also *Stylar cusp.*

Cuspule. A diminutive cusp; often used in reference to the components of the main crest of sthenurine premolars (fig. 2).

Diastema. Space between teeth; typically between the incisors and premolars in macropodoids (figs 1, 14, 20). The part of the dentary that underlies the lower diastema in sthenurines is usually raised along the lateral side, forming a longitudinal ridge. The opposing diastema ridges enclose the median dorsal groove.

Digastric eminence. Curved expansion of the ventral border of the dentary beneath the posterior molars and the anterior part of the masseteric fossa (fig. 20).

Digastric sulcus. Shallow, often elongate fossa on the mesial surface of the digastric eminence into which the digastric muscle inserts.

Distal. Further from the origin or point of reference.

Dolichocephalic. Term applied to describe a relatively elongate cranium (e.g., fig. 15F-G).

Ectoglenoid process. Mesial expansion of the posterior extremity of the jugal bordering the glenoid fossa anteriorly (fig. 1). It represents the attachment point for the lateral mandibular ligament.

Enamel crenulation. A wrinkle on the surface of a molar that cannot confidently be recognized as a crista but nevertheless may be homologous with one (figs 3, 9). The height and thickness of certain cristae are often only slightly greater than those of the enamel crenulations. This means that while it is sometimes possible to differentiate these features by their topological relationships, at other times they cannot be distinguished.

Enamel flange. An extension of the buccal enamel layer of i1 forming an elongate rim dorsal or ventral to the dentine core (fig. 13).

Entoconid. Posterolingual cusp of the lower molar, which forms the lingual extremity of the hypolophid (fig. 3).

Genial fossa (pit). Concavity on the posterior surface of the mandibular symphysis, which represents the origin for the genioglossus and geniohyoid musculature, as well as the area whence neurovascular tissue enters the symphyseal space (fig. 10).

Hypoconid. Posterobuccal cusp of the lower molar, which forms the buccal extremity of the hypolophid (fig. 3).

Hypoloph. Term formerly applied to the posterior transverse ridge on the upper molar when the posterolingual cusp was recognized as the hypocone. This structure is now referred to as the metaloph (fig. 3).

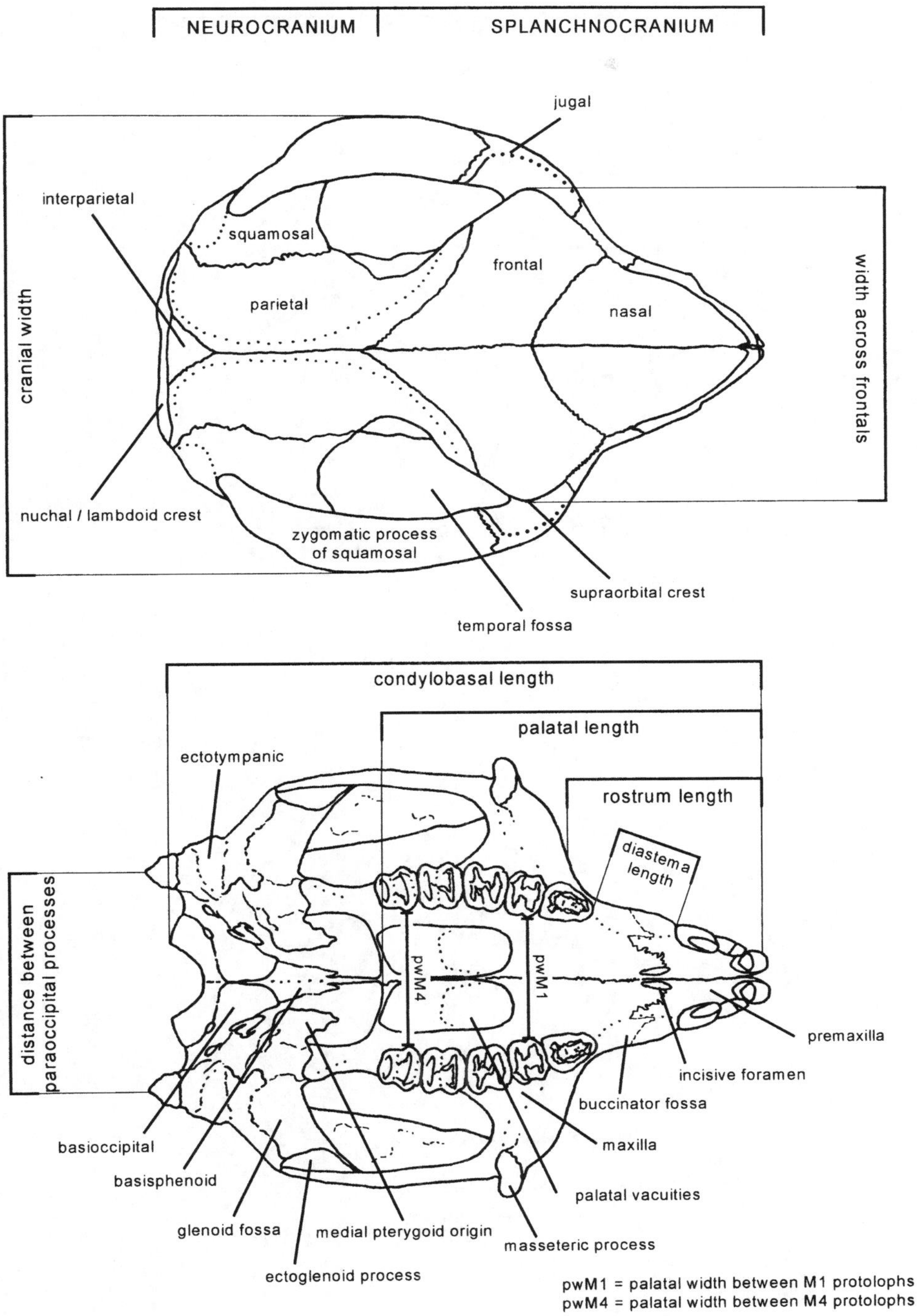

Figure 1. Key to cranial measurements and terms.

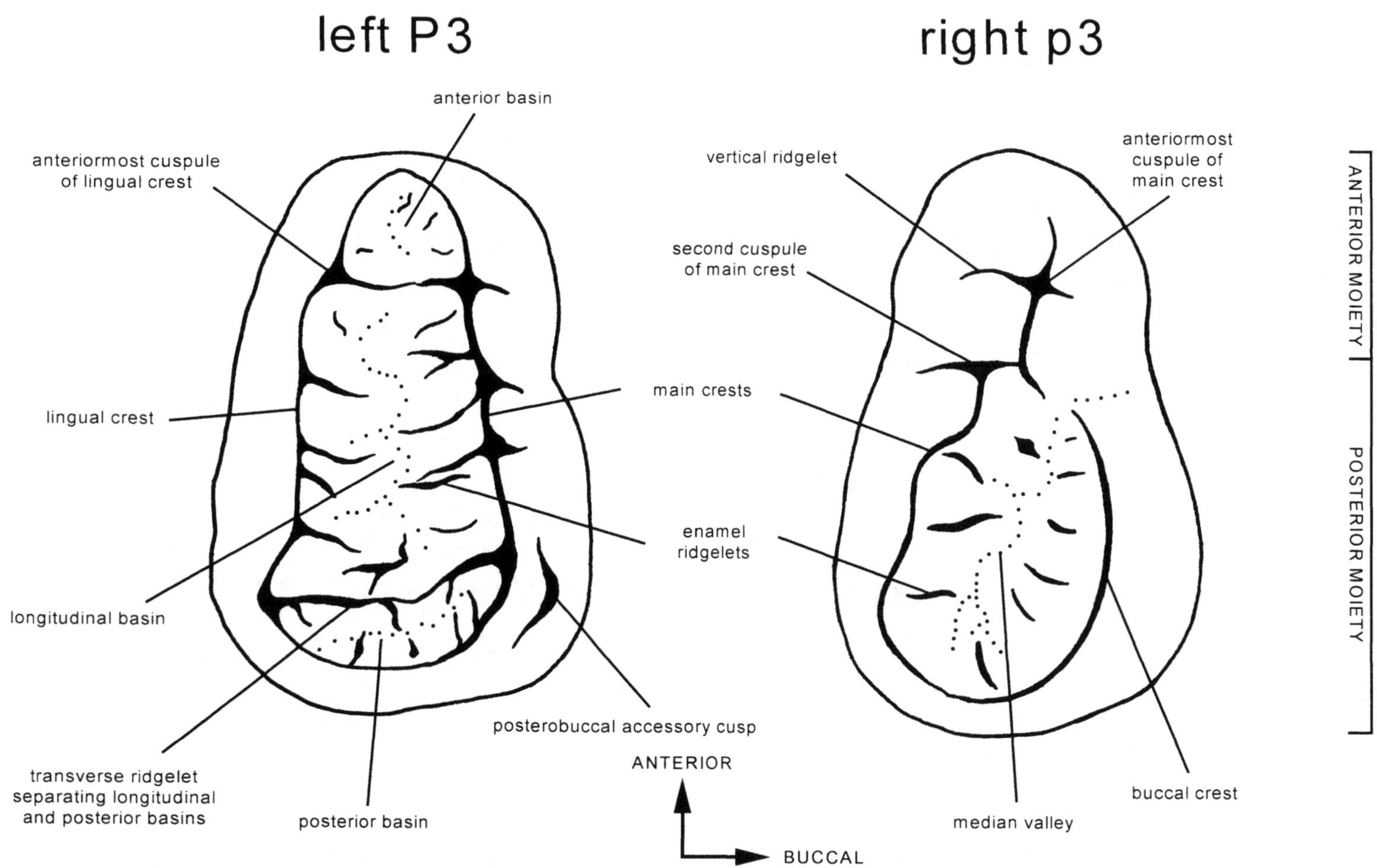

Figure 2. Key to the major features of the sthenurine upper and lower third permanent premolars.

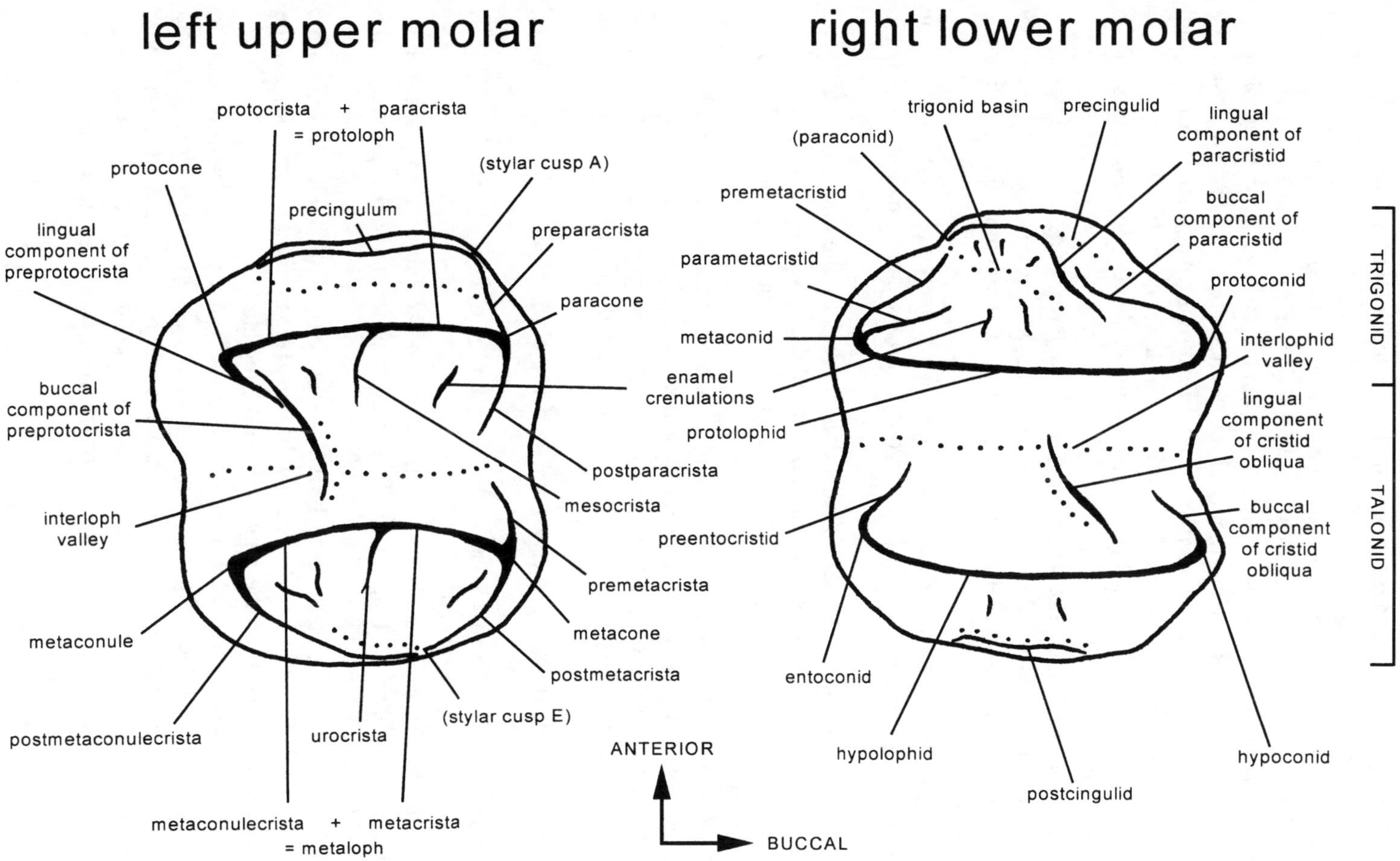

Figure 3. Key to the major features of the sthenurine upper and lower molars.

Hypolophid. High transverse ridge of the posterior half of the lower molar, linking the entoconid and hypoconid (fig. 3).

Interloph valley. Transverse valley between the protoloph and metaloph. The corresponding region on a lower molariform tooth is the *interlophid valley* (fig. 3).

Lateral. Term pertaining to a feature oriented away from the midline or anteroposterior axis of the skeleton, or positioned on a left or right outer surface.

Lingual. Term used to indicate that a dental feature is positioned on or oriented toward the tongue side of a tooth, i.e., internal to its anteroposterior axis.

Longitudinal basin. Central basin between the main and lingual crests of the dP2 or P3 of sthenurines (fig. 2).

Mandibular symphysis. The union between the anterior ends of the dentaries.

Medial pterygoid fossa. Posterior concavity on the mesial side of the dentary formed from an expansion of the angular process. The medial pterygoid muscle inserts into the fossa.

Medial pterygoid origin. Attachment area for the medial pterygoid muscle formed by the lateral surface of the pterygoid and a concavity of the alisphenoid (fig. 1).

Median dorsal groove. Longitudinal valley flanked by the diastema ridges and overlying the superior transverse tori of the mandibular symphysis.

Median valley. Valley formed between the main and buccal crests of the dp2 or p3 of sthenurines (figs 2, 18).

Mesial. Toward the midline or anteroposterior axis of the skeleton or a tooth.

Mesocrista (new). Variably developed vertical crista centrally located on the posterior face of the protoloph (figs 3, 19). It probably represents an extension of the paracrista diverted posteriorly from its union with the protocrista in the center of the protoloph.

Metacone. Posterobuccal cusp of the upper molar, which forms the buccal extremity of the metaloph (figs 3, 9).

Metaconid. Anterolingual cusp of the lower molar, which forms the lingual extremity of the protolophid (figs 3, 12).

Metaconule. Posterolingual cusp of the upper molar, which forms the lingual extremity of the metaloph (figs 3, 9). This cusp was formerly recognized as the hypocone (see Tedford and Woodburne 1987).

Metaconulecrista. Crest directed buccally from the metaconule, which forms the lingual side of the metaloph (figs 3, 9). In most bilophodont macropodoids, the crista is not discernable from the remainder of the metaloph.

Metacrista. Crista directed lingually from the metacone, which forms the buccal side of the metaloph (fig. 9). In most bilophodont macropodoids, the crista is not discernable from the remainder of the metaloph. Note that the metacrista here is not homologous with the metacrista *sensu* Van Valen (1966), which runs posteriorly from the metacone and is referred to here as the postmetacrista.

Metaloph. Posterior transverse ridge of the upper molar, linking the metacone and metaconule (fig. 2). Historically, it is composed of the metaconulecrista and metacrista.

Molariform. Having the form of a molar.

Neurocranium. Posterior region of the cranium enclosing the brain (fig. 1).

Nuchal / lambdoid crest. Posterodorsally directed crest forming the dorsal rim of the occiput (fig. 1).

Occiput. Posterior surface of the cranium bearing the occipital condyles and foramen magnum, and constituted from the supraoccipital, exoccipitals and mastoid processes of the periotic.

Palatal vacuities. Paired, elongate, elliptical voids or fenestrae within the palate (fig. 1).

Paracone. Anterobuccal cusp of the upper molar, which forms the buccal extremity of the protoloph (fig. 3).

Paraconid. Anterior cusp of the lower molar trigonid (fig. 3, 12). Where present in macropodoids, it is often located toward the anterolingual corner of the tooth, at the end of the paracristid or premetacristid.

Paracrista. Crista directed lingually from the paracone, which forms the buccal side of the protoloph (fig. 3). Note that the paracrista here is not homologous with the paracrista *sensu* Van Valen (1966), which connects the paracone with stylar cusp B. This is referred to here as the preparacrista. In macropodoids, stylar cusp B is apparently coalesced with the paracone.

Paracristid. Cristid directed anteriorly from the protoconid to the paraconid (figs 3, 12). It is often U- or L-shaped in macropodoids, with the posterior portion oriented anteroposteriorly and the anterior portion oriented more transversely. In sthenurines, the paracristid is often divided into two components. The more buccal of these (termed ?preprotostylocristid in Prideaux 2000) is shorter, maintains a direct connection with the protoconid apex, and usually terminates on the side of the lingual component, which forms the bulk of the cristid. In sthenurines, it is the lingual component of the paracristid that consists of a posterior portion and a transverse anterior portion. Colloquially, the paracristid is often referred to as the 'forelink' of the lower molar.

Parametacristid. Cristid directed anteriorly into the trigonid basin from the lingual side of the metaloph, buccal to the premetacristid when present (figs 3, 12). A distinct *parametaconid*, buccal to the metaconid, is only expressed in some plesiomorphic macropodoids (see Ride 1993; Ride et al. 1997).

Postcingulid. Shelf low on the posterior face of the hypolophid (fig. 3). The feature may be confluent with the postentocristid, oriented horizontally, A-shaped or expressed as a rounded inflation.

Postentocristid. Cristid descending posteriorly from the entoconid, which may be confluent with the postcingulid (fig. 12A).

Posterior basin. Basin formed immediately posterior to the longitudinal basin of the sthenurine dP2 or P3, which occludes with the trigonid of m1 (fig. 2).

Posterior mental foramen. Small foramen located mid-depth on the buccal surface of the dentary, usually beneath the second or third molar (fig. 20F-H).

Posterobuccal accessory cusp. Cusp buccal to the posterior end of the main crest of the dP2 or P3 in some sthenurines (fig. 2).

Posthypocristid. Cristid directed posterobuccally from the hypoconid apex to or toward a basal position on the posterolingual aspect of the lower molar (fig. 12A), where a hypoconulid is found in many marsupials. The proximal portion of the posthypocristid (i.e., closest to the hypoconid apex) may form the buccal side of the hypolophid in bilophodont macropodoids.

Postmetaconulecrista. Crista directed across the posterior face of the metaloph from the metaconule apex (figs 3, 9), which terminates basally at the position of stylar cusp E, on the posterobuccal aspect of the tooth. The term is synonymous with 'hindlink' *sensu* Stirton (1955), 'postlink' *sensu* Tedford (1966), and 'posthypocrista' *sensu* Flannery (1983). The postmetaconulecrista may be divided into distinct lingual (main) and buccal (mesial) components in sthenurines, the buccal component lying between the lingual component and the urocrista, when present.

Postmetacrista. Crista running posteriorly from the metacone apex, which terminates at or slightly dorsal of the position of stylar cusp E (figs 3, 9). The postmetacrista is sometimes divided into two distinct components in sthenurines, and the more buccal of these may represent the reemergent posterior portion of the stylar crest (i.e., between stylar cusps D and E).

Postparacrista. Crista running posteriorly from the paracone apex, which terminates at the interloph valley (fig. 3). It may or may not contact the premetacrista.

Postprotocrista. Crista running posterobuccally from the protocone apex (fig. 3). In sthenurines, the postprotocrista may be divided into two distinct components. The more lingual component retains a direct connection with the protocone apex and terminates on the side of the more mesial (buccal) component. This forms the main part of the crista and may terminate at the interloph valley or partway up the anterior face of the metaloph. Colloquially, the postprotocrista has often been referred to as the 'midlink' of the upper molar, but the postprotocrista comprises only the posterior part of the 'midlink' of certain derived *Procoptodon* species (fig. 19).

Postmetacristid. Cristid running posteriorly from the metaconid apex, which is usually small and fine where present (fig. 12A).

Postpalatine bars. Thin remnants of the palatal bone that unite and form a bridge across the palate between the posterior extremities of the maxillae. They mark the posterior border of the palatal vacuities.

Precingulid. Cingulid formed on the anterobuccal or anterior portion of the trigonid (figs 3, 12). The structure is usually most evident buccal to the paracristid, but it may extend more lingually, anterior to the transverse portion of the paracristid.

Precingulum. Cingulum on the anterior surface of an upper molar that is confluent with the preparacrista (figs 3, 19).

Preentocristid. Cristid directed anteriorly from the entoconid (figs 3, 12).

Premetacristid. Cristid directed anteriorly from the metaconid, which often terminates at the position of the paraconid (figs 3, 12).

Preparacrista. Crista directed anteriorly from the vicinity of the paracone apex, which is confluent with the precingulum (figs 3, 19). In sthenurines, it is considered to terminate at a cusp-like eminence at the anterobuccal corner of the tooth, which is probably stylar cusp A. Ancestrally, the connection between the paracone and stylar cusp A is via stylar cusp B. Because this cusp is apparently coalesced with the paracone in macropodoids, the preparacrista is, in effect, homologous with the portion of the stylar crest linking stylar cusps A and B.

Preprotocrista. Crista directed anterobuccally onto the precingulum from close to or well anterodorsal to the protocone apex. The preprotocrista is synonymous with the colloquial term 'forelink' (*sensu* Tedford 1966) and is usually diminutive in sthenurines.

Procumbent. Term describing a structure that is relatively declined. It is commonly used in reference to the anterior declination of the lower incisor or diastema region.

Protocone. Anterolingual cusp of the upper molar, which forms the lingual extremity of the protoloph (fig. 3).

Protoconid. Anterobuccal cusp of the lower molar, which forms the buccal extremity of the protolophid (figs 3, 12).

Protocrista (new). Crista directed buccally from the protocone, which forms the lingual side of the protoloph, as delimited by the mesocrista where present (fig. 3).

Protoloph. Anterior transverse ridge of the upper molar, linking the paracone and the protocone (fig. 3).

Protolophid. High transverse ridge of the anterior half of the lower molar, linking the protoconid and the metaconid (fig. 3).

Protostylid. Cusp formed buccal to the protoconid in several phalangeriform marsupials. In bilophodont macropodoids, it is either lost or united with the protoconid.

Proximal. Closer to the origin or point of reference.

Ramus (horizontal). The horizontal portion of the dentary bearing the dentition.

Ridgelet. Very small ridge on a premolar connected to a cuspule or contained within a basin or the median valley (figs 2, 18).

Rostrum. Region of the cranium anterior to the orbits (fig. 1).

Splanchnocranium. Portion of the cranium anterior or anteroventral to the neurocranium (fig. 1), which is primarily associated with ingestion.

Stylar cusp. One of five accessory cusps often expressed to varying degrees along the buccal border of the metatherian upper molar. Stylar cusps may be linked by small ridges to form a *stylar crest.* This crest may be distinct from the buccal cristae, or partially or wholly merged with them (figs 9, 19).

Supraorbital crest. Lateral edge of the frontal, which is usually positioned above the orbit. It varies substantially in its expression across the Sthenurinae. In some species the supraorbital crest is rather elongate and straight-sided, while in others the frontal tapers to a distinct, rugose *supraorbital tuberosity* (figs 1, 15).

Symphyseal boss. Rounded eminence on the anteroventral aspect of the dentary (fig. 20) formed by an enlarged inferior transverse torus of the symphyseal plate.

Symphyseal plate. Articulating surface of the dentary forming one half of the mandibular symphysis (fig. 10). In sthenurines, it is often divisible into a more anterior region (*superior transverse torus*) and a generally smaller, more posterior region (*inferior transverse torus*), which extends beneath the genial fossa (fig. 10).

Talonid. Moiety of a lower molariform tooth posterior to the interlophid valley (fig. 3).

Trigonid. Moiety of a lower molariform tooth anterior to the interlophid valley (fig. 3). Prideaux (2000) referred to this region as the precingulid / trigonid shelf.

Trigonid basin. Central portion of the trigonid enclosed posteriorly by the protolophid, anteriorly and buccally by the paracristid, and lingually by the premetacristid when present (fig. 3).

Urocrista (new). Small vertical crista that ascends the posterior face of the metaloph from the center of the unworn metaloph crest (figs 3, 9). It represents an extension of the metacrista diverted posteriorly from its union with the metaconulecrista in the center of the metaloph. The term 'postlink' has previously been applied to this feature (e.g., Flannery 1983), but that same term has also been applied to the postmetaconulecrista by R. Owen and R. H. Tedford among others. The structure also does not actually link any features in the majority of taxa.

HIGHER-LEVEL CLASSIFICATION

Since the first formal descriptions of pre-late Miocene kangaroos by Archer (1979) and Flannery, Archer and Plane (1983), family level classification of the macropodoids has seen several incarnations. Five schemes have been endorsed within the last decade:

1. an all-encompassing family Macropodidae (e.g., Case 1984; Woodburne 1984a-b; Marshall et al. 1990; Springer and Kirsch 1991; Kirsch et al. 1995);
2. families Potoroidae and Macropodidae (e.g., Archer 1981; Flannery 1989);
3. families Hypsiprymnodontidae, Potoroidae and Macropodidae (e.g., Ride 1993);
4. families Hypsiprymnodontidae and Macropodidae (e.g., Szalay 1994; Kirsch et al. 1997; Burk et al. 1998; Prideaux 1999a);
5. families Hypsiprymnodontidae, Balbaridae, Potoroidae and Macropodidae (Cooke and Kear 1999; Kear and Cooke 2001).

The protean state of family level taxonomy should be taken as a healthy sign for the field, with each of the latest reconfigurations reflecting the discovery of new fossil material, improved morphological studies or revolutionary advances in molecular analyses. The most sweeping work to date on the early radiation of macropodoids is being carried out by Bernard Cooke, primarily on the diverse range of early to middle Miocene taxa extracted from the Carl Creek Limestone at Riversleigh, northwestern Queensland (see Cooke 1997a). A conclusion of this research is likely to be a new grand classification. Wroe et al. (1998) and Kear and Cooke (2001) allude to some important implications of these ongoing studies, but a detailed phylogenetic analysis is yet to be published. Presently, the highest relatively stable taxonomic level is the subfamily, even though the Bulungamayinae is paraphyletic and the monophyly of the Macropodinae is doubtful. Throughout the monograph, I refer to Macropodoidea and the individual subfamilies, but avoid family names due to their present instability. The subfamilies and their genera are listed below.

Hypsiprymnodontinae (Collett, 1887)
- *Hypsiprymnodon* Ramsay, 1876

Propleopinae Archer and Flannery, 1985
- *Ekaltadeta* Archer and Flannery, 1985
- *Jackmahoneya* Ride, 1993
- *Propleopus* Longman, 1924

Balbarinae Flannery, Archer and Plane, 1983
- *Balbaroo* Flannery, Archer and Plane, 1983
- *Ganawamaya* Cooke, 1992
- *Nambaroo* Flannery and Rich, 1986
- *Wururoo* Cooke, 1997c
- **Galanarla* Flannery, Archer and Plane, 1983
- "macropodine genus P" (in Woodburne et al. 1994)

Potoroinae Gray, 1821
- *Aepyprymnus* Garrod, 1875
- *Bettongia* Gray, 1837
- *Borungaboodie* Prideaux, 1999a
- *Caloprymnus* Thomas, 1888
- *Milliyowi* Flannery et al., 1992
- *Potorous* Desmarest, 1804
- **Gumardee* Flannery, Archer and Plane, 1983
- **Kyeema* (*nomen nudum* in Woodburne et al. 1994)
- **Palaeopotorous* Flannery and Rich, 1986
- **Wakiewakie* Woodburne, 1984b
- "potoroine genus M" (in Woodburne et al. 1994)

Bulungamayinae Flannery, Archer and Plane, 1983
Bulungamaya Flannery, Archer and Plane, 1983
Ganguroo Cooke, 1997b
Nowidgee Cooke, 1997b
Purtia Case, 1984
Wabularoo Archer, 1979
Wanburoo Cooke, 1999
"macropodine genus W" (in Woodburne et al. 1994)
Sthenurinae (Glauert, 1926)
Archaeosimos gen. nov.
Hadronomas Woodburne, 1967
Metasthenurus gen. nov.
Procoptodon Owen, 1874b
Simosthenurus Tedford, 1966
Sthenurus Owen, 1874a
Macropodinae Gray, 1821
22 genera
* = *incertae sedis*

PHYLOGENETIC ANALYSIS

The character matrix (table 43) was assembled and saved as a data file in MacClade version 3.05 (Maddison and Maddison 1992). An equal weighting of 1.0 was applied to each character. The majority of characters were unordered (Fitch Parsimony), although 9 of the 19 multistate characters were ordered (Wagner Parsimony). The decision to order a character was made when it was considered beyond reasonable doubt that a two-state transformation could only have occurred via an intermediate state (e.g., cranium length, p3 buccal crest length, mesocrista development). Analysis of the character matrix for the 26 sthenurine and 1 outgroup taxon was executed using PAUP Version 3.1.1 (Swofford 1993). The branch-and-bound search algorithm was employed. All 37 characters used in the cladistic analysis are of craniodental origin; 18 are binary and 19 are multistate (3 or 4 states). Character polarity was determined by outgroup analysis. Polymorphic characters were scored as state unknown, unless the vast majority of specimens of a particular species expressed one particular state (e.g., state 2 for character 30 in *Sthenurus andersoni*).

As investigations into the basal relationships of the major macropodoid lineages continue (e.g., Cooke 1997a-b, 1999; Cooke and Kear 1999; Kear and Cooke 2001), it is becoming increasingly clear that macropodines and sthenurines were derived from bulungamayine stock. Because our understanding of this Miocene radiation is still very scant, the paraphyletic nature of Bulungamayinae is less of a concern than the need to verify the monophyly of both Sthenurinae and

Macropodinae. Fortunately, the complex of features exclusive to the sthenurines indicates that they arose from a common ancestry. At present, Bulungamayinae is not sufficiently represented for a delegate from this group to be selected as an outgroup for an analysis of sthenurine interrelationships. The next best option is a macropodine that shows a stage of evolution similar to the earliest sthenurine and is known from relatively complete material. On this basis, the late Miocene *Dorcopsoides fossilis* Woodburne, 1967 was chosen as the most appropriate outgroup taxon.

TAXONOMY

Order DIPROTODONTIA Owen, 1866
Suborder PHALANGERIFORMES Woodburne, 1984a
Superfamily MACROPODOIDEA Gray, 1821
Family MACROPODIDAE Gray,1821
Subfamily STHENURINAE (Glauert, 1926)

Taxonomic Authorities. Glauert, L. 1926. A list of Western Australian fossils. Geological Survey of Western Australia Bulletin 88:38-71. Raven, H. C. 1929. Kangaroo. Pp. 254-255 in Encyclopaedia Britannica, 14th ed., Vol. 13.

Type Genus. Sthenurus Owen, 1874a.

Revised Diagnosis. Splanchnocranium not markedly deflected anteroventrally; dorsal surface of neurocranium near horizontal in lateral profile, at least up to midway along frontals. Frontals inflated laterally. Posterior end of jugal deep and expanded mesially to form marked ectoglenoid process. Occiput broad, deep and oriented approximately perpendicular to dorsal cranial surface. Ectotympanic very wide and robust, with rugose surface dominated by central transverse keel; mesial end curved ventrally. Digastric sulcus and eminence well developed and posteriorly positioned. I2 crown very small and sub-cylindrical; I3 high crowned, blade-like and markedly longer than combined length of I1-2. Marked buccal cingulid or crest present on p3. Molars rather square in occlusal outline, with lophs straight and parallel.

HADRONOMAS Woodburne, 1967

Taxonomic Authority. Woodburne, M. O. 1967. The Alcoota Fauna, central Australia. An integrated palaeontological and geological study. Bureau of Mineral Resources, Geology and Geophysics, Bulletin 87:1-187.

Type species. Hadronomas puckridgi Woodburne, 1967.

Revised Diagnosis. Rostrum relatively elongate, moderately shallow and slightly tapered anteriorly. Incisive foramina very elongate, extending half of length of diastema. Frontal bears rugose posterior supraorbital tuberosity. Zygomatic arch very shallow. Jugal not widened beneath orbit; ventrolateral orbital border not expanded anterolaterally; orbit small. Masseteric process very small and rounded. Postorbital process of jugal well developed. Symphysis shallow and elongate; symphyseal plate lacks rugosity. Digastric eminence well developed, but digastric sulcus shallow. Masseteric canal penetrates body of dentary to below m4. Mandibular condyle intermediate between oval-shaped and transversely elongate. Low, distinct cingula extend along buccal and lingual sides of p3. Lophid crests oriented slightly anterobuccally relative to anteroposterior axis of lower molar. Lower molar rows straight and near parallel.

Etymology. Gr. *Hadros* 'bulky, stout, strong', *nomas* 'nomad, rover'. In reference to the hypothesized wandering habits of the type species.

Geographic Distribution. Southern Northern Territory, northern South Australia (fig. 27).

Temporal Distribution. Late Miocene to early late Pliocene.

Hadronomas puckridgi Woodburne, 1967

(pls 1-6, 108; figs 10D, 11C,G, 13E, 15D, 16F, 17C, 18E, 19D, 20G, 22B; table 1)

Taxonomic Authority. Woodburne, M. O. 1967. The Alcoota Fauna, central Australia. An integrated palaeontological and geological study. Bureau of Mineral Resources, Geology and Geophysics, Bulletin 87:1-187.

Holotype. CPC 6751, right adult dentary (missing i1, medioventral portion of ramus, ascending ramus and associated processes) collected by Michael O. Woodburne and others in 1964 from Paine Quarry, Alcoota Station, Northern Territory. Age of type locality is determined to be late Miocene by biocorrelation (Woodburne 1967; Murray and Megirian 1992; Murray et al. 1993).

Paratypes. Alcoota Station, NT: UCMP 65976, right adult dentary; UCMP 65981, left M3; UCMP 66463, right adult dentary; UCMP 69765, right i1; UCMP 69775, left i1; UCMP 70510, left P3; UCMP 70511, right P3; UCMP 70512, right P3; UCMP 70513, right P3; UCMP 70519, left M3-4; UCMP 70520, right adult maxilla; UCMP 70521, left and right adult maxillae; UCMP 70522, right m3; UCMP 70523, right m1; UCMP 70524, left adult dentary; UCMP 70525, left adult dentary; UCMP 70526, left adult dentary; UCMP 70563, left P3.

Referred Specimens. Alcoota, NT: UCMP 71031, right P3; NTM P9337, right adult dentary; NTM P9338, right i1; NTM P98132 (SGM P889), partial adult cranium; NTM P98133 (SGM P893), adult cranium; NTM P98134 (SGM P901), right adult dentary; NTM P98135, left adult dentary; NTM P98136, right adult dentary; NTM P98137, right adult maxilla; NTM P98138 right P3; NTM P98139, right adult dentary; NTM P98140, partial adult cranium; NTM P98142, right adult dentary; NTM P98143, left m4; NTM P98145, right M4; NTM P98146, right juvenile dentary; NTM P98148, right P3; NTM P98149, right M4; FU 0418, right adult maxilla; SAM P38771, right i1; SAM P38772, left m1; SAM P38773, left m2; SAM P38774, left m3; SAM P38775, left m4.

Diagnosis. As for the genus until another species is described.

Etymology. Named in honor of P. L. Puckridge, owner of Alcoota Station at the time of the initial excavations.

Description. Descriptions of the cranium, dentary and dentition provided by Woodburne (1967) and Murray (1989, 1991a) represent good accounts of the morphology of most aspects of this taxon. However, a few specimens have subsequently surfaced that improve the taxonomic understanding of *Hadronomas puckridgi* considerably and are described here.

Cranium (pls 1-3). Partial adult cranium (NTM P98140) preserves almost entire right maxilla, premaxilla and nasal, as well as right lacrimal, anterior half of right jugal and complete dentition of right side. In contrast to near-complete but crushed and distorted adult cranium (NTM P98133, pl. 1; see Murray 1991a), NTM P98140 clearly reveals elongate, moderately shallow and rather narrow nature of *H. puckridgi* rostrum (pl. 2). Marked anteroventral orientation of diastema in NTM P98140 (pl. 2) confirms minimal distortion of feature in NTM P98133. Incisor-bearing portion of premaxilla forms a rather narrow peninsula (pl. 2), but this may be overemphasized by premaxilla being broken away above I1, as well as steep angle of diastema. Anterior and posterior borders of upright portion of premaxilla are slightly sinusoidal. Premaxilla constitutes half of diastema length. Diastema moderately elongate; diastema edge rounded, but axis of edge straight (pl. 2). Narial aperture deeper than broad. Although separated from majority of specimen, associated fragment of lateral portion of frontal bears short, rugose posterior supraorbital tuberosity, which extends well beyond remainder of lateral border of expanded, tapered frontal (see reconstruction in fig. 15D). Anteorbital and buccinator fossae very shallow (pl. 2). Anterior extremity of small orbit lies directly above M2 protoloph. Small, rounded masseteric process lies directly above M3 interloph valley (pl. 2). Narrow jugal not widened into shelf beneath orbit. Ventrolateral border of orbit not expanded anterolaterally. Postorbital process of jugal distinct and pointed (pl. 2).

Upper Dentition (table 1). I1 rounded, sub-triangular in cross section, low crowned and slightly widened (pl. 2). Occlusal facet on posterior face of I1 oriented slightly anteroventrally. I2 tiny, cylindrical and very low crowned. I3 elongate and blade-like; buccal surface unornamented and slightly concave approaching occlusal surface. Although worn, anterolingual crest clearly well developed, extending about half length of main blade (pl. 2). Posterior and anterior edges of I3 slightly convex (pl. 2).

Dentary. Digastric eminence usually well developed (pls 4-5), but digastric sulcus very shallow. Posteroventral border of masseteric fossa expanded laterally. Masseteric canal penetrates body of dentary to below m4. Inferior mental foramen very small. Mandibular condyle intermediate between oval-shaped and transversely elongate, and slightly convex dorsally (pl. 4). Medial pterygoid fossa elevated such that angular process is slightly higher than level of cheek tooth row (pl. 5).

Lower Dentition (table 1). Minimally worn i1 crown (SAM P38771) rather small, narrow and slender; occlusal surface distinctly sinusoidal (pl. 6D-E; fig. 11C). Dorsal and ventral enamel flanges clearly developed. Very thin layer of enamel covers lingual surface of i1 between flanges (pl. 6E; fig. 13E).

Anterior and posterior widths of submolariform dp3 are very similar (pl. 6A-B). Hypolophid oriented anterobuccally relative to anteroposterior axis of dp3, but not as markedly as anterior lophid (pl. 6B). This is an oblique sinusoidal crest ('paralophid'), which runs anterobuccally from anterolingual extremity of interloph valley, reaches its apex at a point corresponding to protoconid, then turns more anteriorly, extending to anterior end of tooth and hooking lingually at its extremity. Adjacent to anterior extremity, just lingual of tooth midline, tiny paraconid is discernable (pl. 6B). Smooth, ventrobuccally sloping shelf formed on buccal side of trigonid crest. Low cristid obliqua intercepts base of protoconid posteriorly. Slight, low preentocristid present. Base of hypolophid posterior face inflated on buccal side.

Table 1. Cheek tooth dimensions of *Hadronomas puckridgi,* showing mean, standard deviation (in parentheses). Dimensions of holotype (CPC 6751) after Woodburne (1967). Dimensions of a P3 referred to *Hadronomas* sp. from Lake Kanunka, South Australia (SAM P32556) are also provided.

Tooth	L	AW	PW	AH	PH	n
P3	15.9 (0.68)	8.0 (0.59)	8.2 (0.33)	7.2 (0.73)	6.6 (0.76)	11
P3 Kanunka	15.1	-	8.7	-	6.0	
M1	10.2 (0.56)	9.9 (0.07)	10.0 (0.40)	4.5	4.8	4
M2	11.1 (0.54)	10.5 (0.15)	10.3 (0.35)	5.3 (0.42)	5.4 (0.35)	5
M3	11.7 (0.59)	10.6 (0.37)	10.1 (0.59)	5.0 (0.81)	5.2 (0.65)	7
M4	11.5 (0.41)	10.1 (0.42)	8.4 (0.40)	4.8 (0.31)	4.5 (0.35)	9
dp3	7.9	5.6	5.9	5.2	4.6	1
p3	14.4 (0.56)	6.7 (0.71)	7.0 (0.61)	7.2 (0.70)	7.1 (0.78)	12
p3 holotype	14.9	6.1	6.7	6.6	6.3	
m1	10.3 (0.76)	7.8 (0.37)	8.0 (0.32)	5.3 (1.11)	5.4 (1.15)	10
m2	11.5 (0.47)	8.3 (0.32)	8.4 (0.37)	5.8 (1.26)	6.1 (1.29)	11
m2 holotype	11.3	-	-	-	-	
m3	12.1 (0.36)	8.7 (0.58)	8.7 (0.80)	5.7 (0.87)	5.6 (0.71)	11
m3 holotype	11.9	8.3	7.9	-	-	
m4	11.7 (0.35)	8.7 (0.44)	7.7 (0.36)	5.2 (0.51)	5.0 (0.78)	10
m4 holotype	11.5	8.5	7.3	-	-	

Variation. Upper Dentition. Some very slight variation in size and width relative to length is observable between the nine P3 specimens, but the degree of variability is minimal relative to that seen in most sthenurine species. No intraspecific variability is evident with the upper molars.

Dentary. The adult dentaries referred to *Hadronomas puckridgi* differ in the relative development of their digastric eminence, which varies between very well developed (e.g., NTM P98134, pl. 4D), well developed (e.g., NTM P98142, pl. 5), and moderately developed (e.g., NTM P98135, pl. 4A-B). Few characters are preserved in the young juvenile ramus fragment, NTM P98146 (m2 just erupted, pl. 6A). However, it is possible to ascertain that the digastric sulcus of NTM P98146 was moderately deep, and extended anteriorly to below the m2 hypolophid, in contrast to its shallow nature in adult specimens.

Lower Dentition. Variation for p3 is similar in extent to that observed for P3, with width and development of the buccal and lingual cingula varying very slightly in proportion between specimens. I have not seen any variation with respect to the lower molar dentition that cannot be ascribed to differential preservation or wear.

Comparison With Other Taxa. Cranium. In overall cranial proportions, *Hadronomas puckridgi* most closely resembles the dolichocephalic *Sthenurus* species (figs 15, 22; table 41). However, its cranium is much shallower along its length; the orbit is more posteriorly positioned, laterally projected and thinly rimmed; the incisive foramina are very large; the zygomatic arches are very shallow; and the postorbital process is distinct. These characteristics easily separate *H. puckridgi* from all other sthenurines. A small rounded masseteric process is only seen elsewhere within the Sthenurinae in *"Procoptodon" gilli.* Lateral taper of the frontals in *H. puckridgi* is somewhat reminiscent of *Simosthenurus occidentalis* and *Si. maddocki*, but the supraorbital tuberosity is more hypertrophied and borne on a more posterior portion of the frontal (figs 15, 22B). The cranium of *H. puckridgi* is also relatively narrower than in these two *Simosthenurus* species. The marked anteroventral orientation of the *H. puckridgi* diastema is similar to that of the much larger *Sthenurus stirlingi.*

Upper Dentition. The upper incisors of *Hadronomas puckridgi* are similar to those of most sthenurines in their general form, but the combination of low crown height and a slightly broadened I1 crown is unique to this taxon. Despite the wear sustained, the anterolingual crest of I3 in NTM P98140 appears to have been as well developed as that of several other sthenurines. The P3 morphology of *Hadronomas puckridgi* is unique within the Sthenurinae, because the lingual cingulum is not raised into a crest subequal in height with the main crest (pl. 3C; fig. 23B). However, the cuspules at the ends of the coarse ridgelets directed lingually from the more central main crest cuspules, in combination with the low crest that extends anteriorly from the large posterolingual cusp, form a low rudimentary lingual crest. The upper molars of *H. puckridgi* bear some resemblance to those of *Metasthenurus newtonae* and *Archaeosimos cegsai*, particularly in their relative simplicity. However, they are intermediate between them in crown height and easily distinguished by having a squarer occlusal outline and weaker cristae, particularly relative to *M. newtonae. H. puckridgi* also differs from *A. cegsai* by lacking any sign of a mesocrista and by having a precingulum of even thickness that extends right across the anterior end of the tooth.

Dentary. Several features of the *Hadronomas puckridgi* dentary distinguish it from all other sthenurines: the extension of the masseteric canal to beneath m4; the shallow, elongate and very smooth nature of the symphyseal plate; marked development of the digastric eminence without concomitant deepening of the digastric sulcus; a mandibular condyle intermediate between the flat, oval-shaped form of most non-sthenurine macropodoids and the transverse, elongate, almost barrel-shaped condyle of all other sthenurines.

Lower Dentition. Although it is thinner laterally, the size and general morphology of the *Hadronomas puckridgi* i1 crown most closely resembles that of *Archaeosimos cegsai*, but is less robust. While the dorsal enamel flange is less pronounced in *H. puckridgi*, the two species share a very thin enamel layer on the lingual surface of i1. *A. cegsai* lacks the distinct ventral enamel flange observed in *H. puckridgi* (pl. 6E). Among non-sthenurine macropodoids, the i1 of *H. puckridgi* is very similar in almost all aspects of its morphology to that of *Dorcopsoides fossilis*, differing only by being approximately twice the size and bearing a slightly more curved ventral edge. The very much smaller i1 of *Lagostrophus fasciatus* also resembles that of *H. puckridgi* in shape of the ventral edge (figs 11, 13).

The p3 of *Hadronomas puckridgi* differs from all other sthenurines by lacking an elevated cingulid or crest on the posterobuccal aspect of the tooth. However, no other sthenurines possess low cingula that extend along the buccal and lingual sides of p3 (pl. 6C; fig. 23B). The curvature and general orientation of the main crest, the number of cuspules into which the crest is divided, and the development of the descending ridgelets are nearly identical to *Archaeosimos cegsai* and *"Simosthenurus" brachyselenis*. The lower molars of *H. puckridgi* are also similar to those of *A. cegsai* in crown height and general topographic simplicity. The paracristid and cristid obliqua are low, and there is no sign of any enamel crenulations on the lophid faces. However, compared with *A. cegsai*, the paracristid and cristid obliqua show no incipient division, the trigonid is much longer, and the paracristid is not clearly differentiable from the precingulid on the anterolingual aspect of the tooth (pls 5C, 6B). On the buccal side, the precingulid is well developed and a distinct notch is formed between the lingual extremity of the paracristid and the anterior face of the metaconid in unworn teeth. In general, the morphology of this region of the tooth most closely resembles that of *Sthenurus stirlingi* and *S. tindalei.* Both the upper and lower molar rows of *H. puckridgi* may be distinguished from all other sthenurines by their straight, essentially anteroposterior alignment and the slightly anterobuccal orientation of the lophids relative to the longitudinal axes of the molars (pls 5C, 6B).

Geographic Distribution. Southern Northern Territory (fig. 27).

Temporal Distribution. Late Miocene.

Hadronomas sp.
(pl. 3D; table 1)

Referred Specimen. Stirton Quarry, Lake Kanunka, SA: SAM P32556, left P3. Early late Pliocene.

Remarks. A P3 enamel cap collected from Stirton Quarry at Lake Kanunka (fig. 27) in 1997 represents a form close to *Hadronomas puckridgi*. SAM P32556 differs from that species by having the three cuspules at the ends of the ridgelets directed lingually from the second, third and fourth main crest cuspules coalesced into a low crest. It also lacks a strong transverse ridgelet separating the longitudinal and posterior basins (pl. 3D), although this is characteristic of some specimens of *H. puckridgi*. SAM P32556 may well lie within an acceptable range of variation for *H. puckridgi*, but because its recognition in the early late Pliocene would signify a temporal range extension of up to four million years on the basis of one tooth, I have decided to err on the side of caution until more material belonging to this form is retrieved from the Tirari Formation (see The Sthenurine Radiation).

STHENURUS Owen, 1874a

Macropus, Owen, 1838 (in part), pp. 359-363.
Macropus (Sthenurus), Owen, 1873a (*nomen nudum*), p. 128.
Protemnodon, Owen, 1874b (in part), p. 274.
Protemnodon, Owen, 1877b (in part), p. 427.
Macropus, Flower, 1884 (in part), pp. 715-717.
Sthenurus (Sthenurus), Tedford, 1966, pp. 11-33, figs 2-10.

Taxonomic Authority. Owen, R. 1874a. On the fossil mammals of Australia. Part VII. Family Macropodidae: Genera *Macropus*, *Osphranter*, *Phascolagus*, *Sthenurus* and *Protemnodon*. Philosophical Transactions of the Royal Society of London 164:245-288.

Type Species. Macropus atlas Owen, 1838.

Revised Diagnosis. Frontals not markedly inflated, except for *Sthenurus stirlingi*. I1 markedly broadened. P3 lingual crest extended anterobuccally to near base of crown. Molars high crowned.

Etymology. Gr. *sthenaros* 'strong', *ourus* 'tail'.

Remarks. This diagnosis of *Sthenurus* (*s. s.*) differs from those of Tedford (1966) and Wells and Tedford (1995) in three respects. Strong development of the paracristid, cristid obliqua and postprotocrista is not considered to be diagnostic here; these cristae are at least as well developed in *Metasthenurus* and most *Procoptodon* species. I have also found no evidence to support the view that the mandibular symphysis of *Sthenurus* is any less rigidified than in *Metasthenurus* and the simosthenurins. In addition, the lingual surface of i1 does bear a layer of enamel. Wells and Tedford (1995) flagged the absence of lingual enamel as a synapomorphy for *Sthenurus*, but did not list this feature in their generic diagnosis.

Geographic Distribution. Southwestern Australia, Nullarbor Region, northern, mid-northern, south-central and southeastern South Australia, southern Victoria, western and eastern New South Wales, southeastern Queensland (figs 27-28, 32-34, 37; table 44).

Temporal Distribution. Early late Pliocene to late Pleistocene.

Sthenurus andersoni Marcus, 1962

(pls 7-15; figs 10E, 14D, 15G, 20H, 22C; tables 2-3)

Macropus atlas, Owen, 1845b, pp. 325, 327.
Sthenurus atlas, Owen, 1874a (in part), p. 270, pl. 22, figs 5-8.
Protemnodon anak, Owen, 1874a (in part), pp. 274-275, pl. 23, figs 4-9.
Protemnodon anak, Owen, 1877b (in part), pp. 427-428, pl. 83, figs 4-9.
Macropus anak, Flower, 1884 (in part), pp. 715-716.
Sthenurus atlas, Lydekker, 1887 (in part), pp. 232-233, fig. 37.
Sthenurus atlas, De Vis, 1895 (in part), pp. 97-99, pl. 16, figs 9-10.
Sthenurus sp. cf. *andersoni*, Bartholomai, 1963, pp. 58-59.
Sthenurus (Sthenurus) andersoni, Tedford, 1966, pp. 20-25, figs 5-7.
Sthenurus (Sthenurus) andersoni, Bartholomai, 1972, pp. ix, xii, fig. 2.
Sthenurus (Sthenurus) sp., Bartholomai, 1972, pp. ix, xii, fig. 2.
Sthenurus sp. II, Marcus, 1976, p. 74, fig. 27c-d.
Sthenurus sp. cf. *oreas*, Archer, 1978, p. 82, pl. 7T.
Sthenurus atlas, Wells and Pledge, 1983, pp. 173-174.
Sthenurus atlas, Wells and Murray, 1979, p. 213.
Sthenurus sp., Williams, 1980, p. 110, site 60a.
Sthenurus atlas, Wells, Moriarty and Williams, 1984, pp. 325-326, 329.
Sthenurus (Sthenurus) andersoni, Dawson, 1985, p. 66, table 1.
Sthenurus atlas, Pledge, 1990, pp. 251, 255, 257, figs 4b, 5.
Simosthenurus sp. cf. *orientalis*, Tedford and Wells, 1990, p. 278.
Sthenurus atlas/andersoni, Pledge, 1990, p. 253, table 2.
Sthenurus sp., Wells and Tedford, 1995, pp. 19-20.
Sthenurus oreas, Dawson and Augee, 1997 (in part), p. 71.

Taxonomic Authority. Marcus, L. F. 1962. A new species of *Sthenurus* (Marsupialia, Macropodidae) from the Pleistocene of New South Wales. Records of the Australian Museum 15:299-304.

Holotype. AM MF946, left adult dentary (missing ascending ramus, i1 crown, m1 crown) collected by William Anderson in 1887 from west side of Bone Camp Gully, adjacent to Myall Creek, near Bingara, northeastern New South Wales. Age of type locality is Pleistocene (Marcus 1976).

Paratypes. Bone Camp Gully, near Bingara, NSW: AM F49661, right M2; AM F49662, right M3; AM F49663, left dP3; AM F89905 (= AM MF10), right juvenile dentary; AM MF3, left adult dentary; AM MF942, left adult dentary; AM MF1078, left adult dentary; AM MF1137, left adult dentary; UCMP 60005, right p3; UCMP 60006, right juvenile dentary; UCMP 60008, right M1-2; UCMP 60015, right adult dentary; UCMP 60016, right adult dentary.

Referred Specimens. Bone Camp Gully, near Bingara, NSW: AM F88545 (= AM MF5), left adult dentary; AM MF6, right adult dentary; AM MF1145, right dentary; AM MF1147, right juvenile dentary; UCMP 60009, right I3; UCMP 60010, right i1; UCMP 60011, right dp3; UCMP 60012, right m1; UCMP 60013, left m2; UCMP 60014, left m2.

Wellington Caves (probably mostly Mitchell Cave), NSW: AM F29557, left adult maxilla; AM F29578, right adult dentary; AM F 29590, right P3; AM F30622, left p3; AM F88610 (= AM MF21), right adult maxilla; AM F88605 (= AM MF39), right adult dentary; BMNH 42592, left juvenile maxilla; BMNH 42663Aa, one upper molar, three lower molars; BMNH 42663C, left juvenile dentary; BMNH 42663D, left maxilla; BMNH 42663E, right juvenile dentary; UCMP 57373, left adult dentary; UCMP 64967, left juvenile dentary.

Cathedral Cave, Wellington, NSW: AM F69885, left juvenile maxilla.

Cuddie Springs, near Gongolgon, NSW: repository unknown, identified in Dodson et al. (1993) and Field and Dodson (1999).

Binnia Creek, Weetaliba, NSW: FMNH PM4516, left and right adult dentaries.

Neible Station, near Weetaliba, NSW: AM F unreg., left juvenile dentary.

Toppings Creek, Monaro Region, NSW: AM F115316, left and right premaxillae.

Lake Menindee, NSW: UCMP 45673 / 55968, left and right maxillae and dentaries.

Site 50, Lake Victoria, NSW: NMV P28650, left and right adult maxillae.

The Joint, Texas Caves, southeastern QLD: QM F8239, right M1.

Gowrie, QLD: QM F2970, left adult dentary; QM F2971, left adult dentary; BMNH 50059, left adult dentary.

King's Creek, Pilton, QLD: QM F3813, left juvenile dentary; QM F9159, left adult maxilla.

Condamine River, Darling Downs, QLD: BMNH M1519, left juvenile maxilla.

Darling Downs, QLD (exact locality unspecified): QM F2976, right adult maxilla; QM F3809, left adult dentary; QM F3810, right adult dentary; QM F3811, right adult dentary; QM F3812, left adult dentary.

Chinchilla, QLD: QM F813, right adult dentary; QM F814, left juvenile dentary; UCR 23029, right lower molar; UCR 23033, left I1.

South Site, Lancefield, VIC: NMV P177755, right adult maxilla and several loose upper molars; NMV P177758, left adult maxilla; NMV P177766, right adult dentary; NMV P177768, right adult dentary; NMV P177770, left adult dentary; NMV P177771, left adult dentary; NMV P177774a, left adult dentary; NMV

P177774b, left adult dentary; NMV P unreg. A, right P3; NMV P unreg. B, left dP2, right dP2 and P3; NMV P unreg. C, right adult dentary.

Spring Creek, Minhamite, VIC: NMV P157292, right lower molar; NMV P157295, right adult dentary; NMV P157298 (in part), left m4; NMV P159934 / P198451, right juvenile dentary.

Madigan Gulf (Site 1 / 83), Lake Eyre, SA: SAM P25524, right adult dentary.

Price Peninsula, Lake Eyre, SA: AM F65476, left and right adult dentaries.

Lookout Locality, Warburton River, SA: SAM P20931, right adult dentary; SAM P20967, right adult dentary; UCMP 60867, left P3.

Green Bluff Locality, Warburton River, SA: SAM P20968, left and right adult dentaries; SAM P20969, right juvenile dentary; SAM P32552, left adult dentary.

Kalamurina, SA: SAM P40055, right adult dentary.

New Kalamurina, Warburton River, SA: UCMP 97698, right adult dentary.

Toolapinna Waterhole, Warburton River, SA: SAM P25458, left adult dentary; SAM P unreg., right adult dentary.

Warburton River ('Diamantina', unspecified), SA: SAM P24288, right adult dentary.

West Bluff, Lake Kutjitara, SA: FU 2163a, right dentary.

Malkuni Waterhole, Cooper Creek, SA: SAM P25056, left juvenile dentary; FU 2229, left adult maxilla; FU 2143, right adult dentary.

Waralamanko Waterhole, Cooper Creek, SA: SAM P25186, left adult maxilla; SAM P32551, right dentary.

Site 8, Cooper Creek, SA: UCMP 56470, right adult dentary; UCMP 60669, left adult maxilla.

Kudnampirra Waterhole (= Site 14), Cooper Creek, SA: UCMP 56472, right juvenile dentary.

Lake Callabonna, eastern SA: SAM P13672 / P13673, partial skeleton consisting of anterior portion of cranium, dentaries and most postcranial elements.

Dempsey's Lake, Port Augusta, SA: SAM P22489, right adult dentary; SAM P22494, left adult maxilla, left and right premaxillae.

Curramulka Quarry, Yorke Peninsula, SA: SAM P40062, right P3, left p3, five single loose molars.

Town Well Cave, Curramulka, Yorke Peninsula, SA: SAM P12922, right adult dentary.

Salt Creek, Normanville, SA: SAM P34820, right adult dentary.

Coorong Beach, near Murray Mouth, SA: SAM P13128, external sandstone mould of left and right juvenile dentaries.

Blanche Cave, near Naracoorte, SA: FU unreg..

Buckridge Cave, near Naracoorte, SA: Listed in Reed and Bourne (2000).

Henschke's Fossil Cave, near Naracoorte, SA: SAM P18647, adult cranium; SAM P29570, adult cranium; many SAM registered and unreg. maxillae, dentaries and single loose teeth.

Victoria Fossil Cave, near Naracoorte, SA: SAM P16457, juvenile cranium; SAM P20461, adult cranium; SAM P20499, juvenile cranium; SAM P20792, juvenile cranium; FU 0257, juvenile cranium; SAM P40056, adult cranium; 10 maxillae, premaxillae; 27 dentaries; numerous single loose teeth.

Wombat Cave, near Naracoorte, SA: FU 1802, right dP2.

Cheese and Putty Cave, near Joanna, SA: SAM P40054, partial skeleton.

Comaum Forest Cave, near Penola, SA: SAM P36626, left and right juvenile maxillae.

Unnamed swamp site, near Millicent, SA: Listed in Reed and Bourne (2000).

Goulden's Hole, near Mount Schank, SA: Listed in Reed and Bourne (2000).

Derrington St, Mount Gambier, SA: SAM P13651, right juvenile maxilla, left juvenile dentary.

Kilsby's Hole, Mount Gambier, SA: SAM P40071, right juvenile dentary, SAM P40072, right adult dentary, SAM P unreg., left M3.

Lindsay Hall Cave, Nullarbor Plain, WA: WAM 92.12.4, right P3; WAM 03.11.08, right juvenile maxilla.

Tight Entrance Cave, southwest WA: WAM HC 03.11.02, left dentary.

Last Tree Cave, Nullarbor Plain, WA: WAM unreg., several craniodental specimens, including juvenile cranium.

Leaena's Breath Cave, Nullarbor Plain, WA: WAM unreg., adult cranium.

Revised Diagnosis. Smaller than any other *Sthenurus* species. Upper molars bear few, very fine enamel crenulations. Precingulid narrow. Similar in most aspects of morphology to *S. atlas*, but differs by lacking the diagnostic characters of this species and by having proportionally shorter molars with a less well developed postparacrista.

Etymology. Named in honor of Charles W. Anderson, former director of the Australian Museum, who was studying the Bingara Fauna at the time of his death.

Description. The description of the maxilla, dentary and dentition of *Sthenurus andersoni* by Marcus (1962) and Tedford (1966, 1967) remain good accounts of this species. However, since these works were published, adult and juvenile crania have been collected from cave deposits in southeastern South Australia, while numerous dentary, maxillary and dental specimens have been collected from many sites across Australia. I describe the crania and upper incisors below, and provide an account of the variation observed in the now much wider known sample of *S. andersoni.* Several features previously considered diagnostic are evidently variable within this species.

Adult Cranium (pls 7, 8A,F, 9C, 10D; fig. 22C; table 2). Described from SAM P20461, SAM P29570 and SAM P40056. Cranium dolichocephalic; rostrum deep, elongate and rather cylindrical. Incisor-bearing portion of premaxilla elongate and moderately shallow. Upright portion of premaxilla slopes posterodorsally, with premaxilla-maxilla suture extends posterodorsally to nasal (pls 7A, 8F). Diastema elongate and deflected slightly anteroventrally relative to cheek tooth row. Anterior border of long incisive foramina lies opposite or just anterior to posterior extremity of I3 alveolus. Anteorbital and buccinator fossae on maxilla very shallow, although buccinator fossa clearly distinguishable low on maxilla, anterior to P3 (pls 7C, 10D). Nasals elongate, only narrowing markedly anterior to contact with premaxilla (pl. 9C). Masseteric process short, narrow, only very slightly flared distally (pls 7C, 8A, 10D), and composed almost entirely of maxilla. Frontal region elongate and only slightly inflated laterally, forming poorly pronounced supraorbital crest (pls 7B, 9C). Palatal vacuities extend anteriorly to opposite M1 protoloph (SAM P20461) or M2 protoloph (SAM P29570). Postpalatine bars form thin bridge across palate just posterior to M4 metaloph.

Basicranial plane moderately elevated above palatal plane (pls 7A, 8F). Medial pterygoid origin rather narrow. Basioccipital slightly flexed posterodorsally relative to basisphenoid, with median keel only moderately developed. Moderately shallow zygomatic arches parallel sagittal plane for most of length (pls 7B, 9C); jugal bears moderately narrow ectoglenoid process (pls 7C, 10D). Postglenoid process rather small. Temporal (parietal) crests convergent on sagittal suture, but low and poorly developed. Occipital region shallow, not especially broad, and oriented at 90° relative to dorsal surface of neurocranium. Vertical median occipital crest poorly developed, leading ventrally to wide, oval-shaped foramen magnum. Paroccipital processes deflected only very slightly posterior of ventral (pl. 8F). Nuchal / lambdoid crests well developed and extended posteriorly (pl. 7A-B).

Table 2. Dimensions of the adult cranium (SAM P20461, SAM P29570, SAM P40056) and dentary of eastern *Sthenurus andersoni*, showing mean, standard deviation (in parentheses) and sample size [in brackets].

Dimension	*Sthenurus andersoni*		
Condylobasal Length	218	(7.9)	[n=3]
Cranial Width	114	(1.7)	[n=3]
Palatal Length	132	(2.1)	[n=2]
Diastema Length	47.2	(4.38)	[n=2]
% Diastema Length: Palatal Length	36.0	(2.83)	[n=2]
Palatal Width between M1 Protolophs	38.8	(0.65)	[n=3]
Palatal Width between M4 Protolophs	41.9	(0.31)	[n=3]
Width across Frontals	63.3	(1.40)	[n=3]
Distance between Paroccipital Processes	63.9	(3.14)	[n=3]
Dentary Depth	28.4	(1.82)	[n=10]
Dentary Width	17.2	(2.04)	[n=10]
Dentary Depth / Width	1.65	(0.13)	[n=10]

Juvenile Cranium (pls 8B-E, 9A-B, 10A-C). Although slightly smaller, older juvenile crania (ontogenetic stage = M4 erupting; e.g., FU 0257 and SAM P20499, pls 8D, 9B, 10B) do not differ in morphology from adult crania. However, younger juvenile crania (ontogenetic stage = M3 beginning to erupt; SAM P16547, pl. 8C, and SAM P20792, pls 9A, 10A) differ by having less convergent temporal crests and less inflated frontals, with temporal crests forming continuous line with lateral edge of frontals (pls 8C, 9A); more domed dorsal surface of neurocranium (pl. 8C); proportionally larger orbit and very small masseteric process (pl. 8C); smaller nuchal / lambdoid crests and proportionally larger interparietals (pl. 9A); more anteriorly narrowed rostrum (pls 9A, 10A).

Upper Incisors (pls 8B,E-F, 10C-D). I1 angled slightly ventromesially, broadening markedly from base to occlusal surface (pl. 8B). I2 tiny, sub-cylindrical, approximately one-third size of I1. I3 large, blade-like and high crowned, with smooth buccal surface (pls 8B,E-F, 10C-D). Anterolingual crest of I3 very small and obliterated following relatively little wear (pl. 10C-D). Anterior edge of I3 straight for most of length, but curved posteriorly toward base of crown. Upper half of posterior edge largely vertical, but curves anteriorly on approach to minimally worn occlusal surface (pl. 8E).

Status of QM F813 and QM F814. Bartholomai (1963) and Wells and Tedford (1995) held the view that two juvenile dentary specimens from the Darling Downs (QM F813, QM F814; pl. 15) probably represented an undescribed species of *Sthenurus.* Although both specimens lack locality data, Bartholomai (1963) thought that their preservation was very suggestive of the early late Pliocene Chinchilla LF. While he viewed them as taxonomically close to *Sthenurus andersoni*, it was considered likely that the discovery of further material would facilitate a specific distinction. Bartholomai (1963) perceived the ramus of the Chinchilla specimens as proportionally shallower than that of *S. andersoni*, the p3 lower crowned and the sides of the lophids only very slightly convex. With regard to the p3, Bartholomai (1963:58) also observed that the "vertical median groove at the anterior limit of the postero-labial crest [is] moderately deep and extends towards the crown base." Wells and Tedford (1995:19) separated QM F813 and QM F814 from *S. andersoni* on the basis of the small size of their cheek teeth. They also noted that the p3 of QM F814 is akin to *S. andersoni* in the union observed between the buccal crest and the anteriormost cuspule of the main crest. In contrast, I am satisfied that both specimens fit into *S. andersoni*.

Contrary to Bartholomai's observations, the dentary of QM F813 (pl. 15H-I,K) is deeper, not shallower, relative to its width than in every other adult specimen of *Sthenurus andersoni* that I have examined, including material from Bingara, the Wellington Caves and southeastern Australia. Depth / width for QM F813 is 2.01, compared with the mean value of 1.65 for eastern *S. andersoni* (table 2). A value of 1.61 for QM F814 (pl. 15F-G,J) lies well within the range of variation of this sample. The other three differences noted by Bartholomai (1963) are not taxonomically significant, because each feature is variable in *S. andersoni*, even in the topotypic sample. While the cheek teeth are small relative to most eastern specimens of *S. andersoni*, their dimensions are very similar to those of QM F3813 (pl. 15A-C; table 3), a juvenile dentary from the eastern Darling Downs that Bartholomai (1963:56-57, fig. 2) was content to refer to *S. andersoni*. The sides of the lophids of this specimen are also not especially convex, but its ramus depth / width value (1.67) is very close to the mean value for *S. andersoni* (table 2). The molars of QM F3813 are very slightly narrower, but fall within the range of variation seen in the topotypic sample, as Tedford (1966) noted. Moreover, the p3 buccal crest (Bartholomai's postero-labial crest) connects anteriorly to the second cuspule of the main crest in QM F814, not the anteriormost cuspule (pl. 15J). There is no connection with the anteriormost cuspule in any specimen of *S. andersoni* that I have observed (cf. Wells and Tedford 1995:19). Overall, I find no evidence in support of a taxonomic separation of QM F813 and QM F814 from *S. andersoni*.

Variation. Cranium. Two of the four known adult *Sthenurus andersoni* crania (SAM P20461, SAM P29570) formed the basis of the description presented earlier. SAM P29570 is slightly larger and its zygomatic arch is slightly deeper than in SAM P20461, but the differences are marginal. On comparison of their lateral profiles, the dorsal surface of the cranium appears to slope slightly more anteriorly in SAM P20461, but this impression is largely created by the downwardly crushed nasal region (compare pl. 8F with pl. 7A). Otherwise, the only difference of note between the specimens is that the palatal vacuities extend anteriorly to opposite the M1 protoloph in SAM P20461 (pl. 10D), but only to the M2 protoloph in SAM P29570 (pl. 7C). The other two crania fit within this range of variation. Ontogenetic differences in the cranium have already been noted.

Upper Dentition. I have observed no morphological variation of note between the upper incisors, dP2 or dP3 specimens of *Sthenurus andersoni* (pls 8, 10-11). However, almost all of these specimens are from the deposits of Victoria and Henschke's Fossil Caves at Naracoorte in southeastern South Australia. The one specimen preserving dP3 that I have examined from another locality does vary slightly from the Naracoorte sample in both size and morphology. This is a juvenile maxilla (WAM 03.11.08) from Lindsay Hall Cave on the Nullarbor Plain (pl. 11E). Its dP3 lies outside of the smaller end of the observed size range of the Naracoorte sample for length and width dimensions (8% smaller compared with the Naracoorte average; table 3). In addition, the precingulum is thinner and extends to the anterolingual extremity of the tooth, in contrast to the Naracoorte specimens, in which the precingulum does not extend beyond the position of the rudimentary preprotocrista (pl. 11A).

As it is in the majority of sthenurine species known from large sample sizes, the P3 of *Sthenurus andersoni* is very variable in morphology (pls 10-11). Variation is observed in (a) shape of the lingual crest (straight, or central portion slightly or markedly curved buccally); (b) size and separation of the posterobuccal accessory cusp (tiny and poorly differentiated, or moderately distinct and slightly

Table 3. Cheek tooth dimensions of *Sthenurus andersoni* from eastern Australia, showing mean, standard deviation (in parentheses). Dimensions are included for the holotype (AM MF946), and specimens from southwestern WA (WAM 03.11.02), the Nullarbor Plain (WAM 03.11.08), and the Darling Downs (QM F813, QM F814 and QM F3813).

Tooth		L	AW	PW	AH	PH	n
dP2	Eastern	9.2 (0.53)	6.5 (0.37)	7.6 (0.35)	6.7 (0.75)	6.9 (0.61)	13
	Nullarbor	8.9	6.3	7.1	6.6	6.6	1
dP3	Eastern	10.4 (0.35)	9.1 (0.28)	10.4 (0.31)	5.6 (0.33)	5.9 (0.42)	13
	Nullarbor	9.5	8.6	9.7	5.8	6.2	1
P3	Eastern	15.5 (0.62)	7.9 (0.43)	9.7 (0.39)	9.6 (0.66)	9.4 (0.85)	23
	Nullarbor	14.8	7.3	8.6	8.7	8.1	
M1	Eastern	11.5 (0.47)	11.0 (0.38)	11.3 (0.37)	6.3 (0.55)	6.7 (0.54)	26
	Nullarbor	11.3	10.0	10.6	6.3	6.8	
M2	Eastern	12.6 (0.38)	11.8 (0.48)	11.6 (0.46)	7.0 (0.60)	7.5 (0.62)	26
	Nullarbor	12.0	10.6	10.8	7.0	7.5	
M3	Eastern	13.5 (0.37)	12.2 (0.56)	11.6 (0.59)	7.2 (0.58)	7.7 (0.54)	21
M4	Eastern	12.5 (0.30)	11.5 (0.34)	10.0 (0.54)	6.2 (0.22)	6.4 (0.23)	11
dp2	Eastern	8.5 (0.32)	4.7 (0.18)	6.1 (0.40)	7.6 (0.42)	6.5 (0.64)	18
	QM F3813	7.0	4.3	5.1	5.9	5.7	
dp3	Eastern	9.7 (0.54)	7.6 (0.47)	7.9 (0.35)	6.9 (0.70)	7.3 (0.67)	18
	QM F3813	9.3	6.6	6.7	5.6	6.8	
p3	Eastern	15.0 (0.68)	6.0 (0.45)	7.7 (0.52)	9.8 (0.92)	9.0 (1.04)	34
	Holotype	15.7	7.0	7.6	-	-	
	QM F814	13.2	5.4	6.5	-	-	
	QM F3813	13.0	5.5	6.7	9.1	8.6	
	Southwest	14.2	5.8	7.3	-	-	
m1	Eastern	11.3 (0.57)	8.9 (0.46)	9.1 (0.41)	8.0 (1.10)	8.7 (1.19)	43
	QM F813	9.9	7.3	7.5	-	-	
	QM F814	10.3	-	-	-	-	
	QM F3813	10.8	7.7	8.1	8.1	9.0	
	Southwest	9.7	8.1	8.3	-	-	
	Holotype	13.8	10.2	10.2	-	-	
m2	Eastern	12.6 (0.74)	10.2 (0.48)	10.2 (0.49)	8.5 (1.11)	9.4 (1.06)	48
	QM F813	10.6	8.1	8.4	6.6	7.6	
	QM F814	11.0	-	-	-	8.8	
	QM F3813	12.2	8.9	9.2	9.4	10.2	
	Southwest	10.3	9.0	8.8	7.0	7.6	

m3	Eastern	13.5 (0.70)	11.0 (0.60)	10.8 (0.64)	8.8 (0.74)	9.3 (0.70)	41
	Holotype	14.3	11.3	11.2	-	-	
	QM F813	11.5	8.8	8.5	7.3	7.1	
	QM F3813	-	10.2	-	9.7	-	
	Southwest	11.5	9.7	9.4	8.2	8.2	
m4	Eastern	12.7 (0.68)	11.0 (0.71)	9.9 (0.75)	8.0 (0.69)	7.6 (0.87)	29
	Holotype	12.7	11.1	10.4	-	-	
	Southwest	11.1	9.8	8.9	7.2	7.1	

elongated); (c) development of ridgelets in the longitudinal basin (very few, fine and numerous, or coarse); (d) thickness of the anterobuccal extension of the lingual crest (thin and short, thick and long, or intermediate); (e) development of the transverse ridgelet separating the longitudinal and posterior basins (low and poorly differentiated, moderately developed but not confluent with posterior end of the lingual crest, or strong and confluent with the posterior end of the lingual crest); (f) degree into which the main crest is differentiated into cuspules (no distinct cuspules, or five to six clear differentiated cuspules). Upper molars (pls 10-11) vary most notably in the development of (a) very fine enamel crenulations on the loph faces (absent, slight, or distinct); (b) the postparacrista (slight, moderate, or large); (c) the preprotocrista (absent, fine and low, or fine but distinct and extending halfway up anterior face of protoloph). While the dimensions of the M1 and M2 of WAM 03.11.08 (pl. 11E) lie at or just outside the smaller end of the observed size range of eastern *S. andersoni* for length dimensions, they are about 5% narrower relative to the average length for eastern *S. andersoni* (table 3).

Dentary. Minimal variation is observed between the dentary specimens of *Sthenurus andersoni* (pls 12-15), except for slight differences in size, ramus depth relative to width, and relative size of the inferior mandibular foramen. As with the majority of sthenurines for which dentary age series are known, the *S. andersoni* increases in depth relative to width with increasing ontogenetic age (pl. 12).

Lower Dentition. Morphological variation in minimally worn i1 specimens of *Sthenurus andersoni* is seen in the shape of the occlusal surface, which is quite straight in the majority of specimens (e.g., pl. 12D), but slightly more curved in AM F89905 from Bingara and QM F3813 from the Darling Downs (pl. 15A). Otherwise, only slight variation is observed in crown depth. Most variation in dp2 relates to the form of the buccal crest, which may be (a) very short and restricted to the posterobuccal corner of the tooth; (b) very short and separated from a high ridgelet directed buccally from the second cuspule of the main crest by a small notch; (c) half the length of the main crest and connected to it via a low ridgelet directed buccally from the second cuspule of the main crest; (d) half the length of the main crest and confluent with a high ridgelet directed posterobuccally from the second cuspule of the main crest. Posterior inflation of the dp2 varies slightly. In size, the dp2 of QM F8313 lies well outside of the size range covered by the other specimens examined. It is also much smaller relative to the size of its dp3 than in any other *S. andersoni* specimens (pl. 15A,C; table 3). No morphological variation and only a limited degree of size variation is apparent in dp3.

The p3 of *Sthenurus andersoni* is a highly variable tooth (pls 12-15). It may be narrow for its entire length, such that it is only slightly wider posteriorly than anteriorly, or markedly wider posteriorly than anteriorly. As such, the median valley may have the form of a narrow cleft or a wide D-shape. Main crest cuspules may be well differentiated along the entire length of the crest, discernable only for the anterior half of the crest, or poorly differentiated. In general shape, the main crest may be straight or sinusoidal. Overall, though, most variation is observed on the buccal aspect of the p3, especially in the form of the buccal crest, which may be (a) less than half the length of the main crest and crescentic (e.g., UCMP 64967); (b) half the length of the main crest and relatively straight (pl. 14D); (c) half the length of the main crest and moderately crescentic (pl. 14C); (d) two-thirds of the length of the main crest and relatively straight (e.g., SAM P20968); (e) two-thirds of the length of the main crest and moderately crescentic (pl. 14E); (f) two-thirds of the length of the main crest and crescentic (pl. 14G). The buccal crest may completely separate from the main crest anteriorly or linked to the second cuspule of the main crest via a low transverse ridgelet. In QM F3811, the anterior end of the buccal crest is met by a ridgelet directed buccally from each of the second and third main crest cuspules (pl. 15D-E). In QM F814, the buccal crest is linked only to the second cuspule of the main crest by a low, intervening ridgelet (pl. 15F,J). In size, the p3 of QM F814 and QM F3813 is approximately 14% smaller than the average of all other measured *S. andersoni* specimens from eastern Australia (table 3). Dimensions of the p3 of the *S. andersoni* dentary from Tight Entrance Cave (WAM 03.11.02, pls 12E-F, 14B) in southwestern Australia lie within the smaller end of the size spectrum for the eastern Australian specimens.

The lower molars of *Sthenurus andersoni* show very little variation in morphology. Relative size of the precingulid varies slightly between specimens, as does the thickness of the transverse anterior portion of the paracristid. Some specimens also bear very fine vertical enamel crenulations on the anterior faces of the lophids (pl. 14A), akin to those seen on the upper molars of all individuals of *S. andersoni.* Dimensions of the lower molars of QM F813, QM F814 and QM F3813 are up to 20% smaller than the average of all other measured *S. andersoni* specimens from eastern Australia (table 3), and the molars of these three specimens are slightly narrower proportionally. Lower molars of the southwestern specimen (WAM 03.11.02, Tight Entrance Cave) are approximately 13% smaller than the average of the eastern Australian specimens (table 3).

Comparison With Other Taxa. The possession of only one apparent craniodental autapomorphy (see Phylogeny) makes *Sthenurus andersoni* difficult to diagnose. Except for the presence of a few very fine enamel crenulations on the upper molars and a narrower precingulid relative to the other *Sthenurus* species, *S.*

andersoni is defined largely on the basis of its smaller size and its lack of the diagnostic characters of each of the other *Sthenurus* species.

Cranium. Among the species of *Sthenurus*, complete adult crania are known for *S. andersoni*, *S. tindalei* and *S. stirlingi*. *S. andersoni* is most easily distinguished from *S. tindalei* by its less procumbent diastema and less posterolaterally expanded frontals (adjacent to the fronto-parietal suture); and from *S. stirlingi* by its greater dolichocephaly, less inflated frontals and proportionally lower basicranial plane. *S. andersoni* also differs from *S. tindalei* and *S. stirlingi* by being much smaller and lacking an anterodorsal extension of the maxilla separating the premaxilla and nasal (Wells and Tedford 1995). The most complete adult cranium of *S. atlas* (AM F121396, pl. 16), while missing much of the rostrum, zygomatic arches, and the left maxilla, shows that this species has clearly the most similar cranium to *S. andersoni*. Relative length is not assessable, but judging from the frontals and what remains of the zygomatic arches, the *S. atlas* cranium may have been slightly wider. In addition, *S. atlas* possesses a more sinusoidal nasofrontal suture and less inflated frontals (which have a rather sinusoidal lateral border), and its temporal (parietal) crests are very low and fail to converge on the sagittal suture. Aside from this minor variation, no other interspecific differences are discernable. Comparison between juvenile crania of *S. andersoni* (SAM P16547, SAM P20792, pls 8-10) and a similarly aged, but dorsoventrally crushed partial cranium of *S. atlas* (AM F88550, pl. 17) suggests that *S. andersoni* is slightly more dolichocephalic, with a proportionally longer diastema. Despite the distortion, it also appears that the neurocranial region was slightly broader in *S. atlas*. The approximation of the palatal and basicranial planes in AM F88550 is post-depositional (cf. Tedford 1966:16). Compared with the similarly aged juvenile cranium of *Metasthenurus newtonae* (FU 0252, pl. 33), *S. andersoni* differs by having a proportionally narrower neurocranium with a more curved dorsal surface, narrower temporal fossae, and slightly less inflated frontals.

Upper Dentition. The upper incisors of *Sthenurus andersoni* most closely resemble those of *S. tindalei* and *S. murrayi*, but they are smaller, and I3 has straighter anterior and posterior edges. Relative to *S. atlas*, the *S. andersoni* I1 is proportionally narrower, while I3 is proportionally larger, with markedly less curved edges. The posterobuccal portion of the dP2 of *S. andersoni* is less inflated and the outline of the dP3 is less rounded than in *S. atlas*. Lingual termination of the dP3 precingulum at the position of the preprotocrista in *S. andersoni* is only observed elsewhere in *Metasthenurus newtonae*. The *S. andersoni* P3 differs from that of *S. atlas* by being proportionally shorter, possessing a shallower longitudinal basin and lacking a very distinct and enlarged posterobuccal accessory cusp. In all aspects of its morphology, the *S. andersoni* P3 most closely resembles that of *S. murrayi*, which may be distinguished by its larger size and the presence of a ridgelet directed anterobuccally from the anteriormost cuspule of lingual crest, which divides the anterior basin into two distinct sub-basins. While the P3 of *S. andersoni* shows some similarities in overall form to the corresponding tooth in

Simosthenurus maddocki, in the latter species the main crest is more distinctly differentiated into cuspules, the lingual crest is proportionally higher, the posterobuccal accessory cusp is larger and more distinct, and there is no anterobuccal extension of the lingual crest. The upper molars of *S. andersoni* are very similar in morphology to *S. atlas* (especially when worn), but are smaller and proportionally shorter, have a less developed postparacrista, and possess a few very fine enamel crenulations on the posterior faces of the lophs. These crenulations are finer than in *S. murrayi* and are not restricted to the posterior face of the protoloph.

Dentary. Although the dentary of *Sthenurus andersoni* is, on average, slightly smaller than that of *S. atlas*, the two species overlap considerably in dentary size and do not appear to be consistently differentiable on the basis of any proportional or morphological aspects. Where they coexist, *S. andersoni* tends to be smaller.

Lower Dentition. The elongate but robust i1 of *Sthenurus andersoni* closely resembles the i1 of *S. atlas* in its general morphology and degree of procumbency, but the tooth is slightly smaller and more slender, as Tedford (1966, 1967) noted. The same features differentiate the *S. andersoni* i1 from that of *Metasthenurus newtonae*, except that the i1 of this species also has a more sinusoidal occlusal surface. In all aspects of dp2 morphology, *S. andersoni* is extremely similar to *S. atlas* and is only consistently differentiable on the basis of its smaller size. Similarly, the dp3 and molars of *S. andersoni*, while smaller than in *S. atlas*, may only be otherwise differentiated by their proportionally narrower precingulid and less reduced anterior portion of the paracristid. However, I have never seen a specimen of *S. atlas* with fine enamel crenulations on the anterior lophid faces, so at least this may be useful for distinguishing some individuals of *S. andersoni* from *S. atlas.* Among the simosthenurins, the lower molars of *"Procoptodon" gilli* are most similar to those of *S. andersoni*, but these may be distinguished by their smaller size, more weakly developed precingulum, marked separation of the paracristid and cristid obliqua into two components, and higher crown height. Because the p3 is so variable in morphology, different variants of *S. andersoni* more closely resemble different species. Those p3 specimens that are markedly wider posteriorly than anteriorly and have a relatively long, crescentic buccal crest (e.g., pl. 14G) are most similar to *S. notabilis* or the relatively more slender form of *S. atlas.* This p3 form is observed in the holotype (see Marcus 1962:fig. 1). However, the p3 of *S. andersoni* is smaller than that of *S. atlas* and *S. notabilis*. The p3 specimens with a shorter buccal crest and well-differentiated main crest cuspules (e.g., pl. 15E) bear some resemblance to *Simosthenurus maddocki*, but the median valley and posterior portion of the tooth are relatively narrower. Specimens showing a more intermediate morphology between those that resemble *S. notabilis* or *S. atlas* and those that resemble *Si. maddocki* cannot easily be confused with any other species.

Geographic Distribution. Southwestern Western Australia, Nullarbor Region, northern, south-central and southeastern South Australia, southern Victoria, eastern and western New South Wales, southeastern Queensland (figs 27, 32; table 44).

Temporal Distribution. Early late Pliocene, middle to late Pleistocene.

Sthenurus atlas (Owen, 1838)

(pls 16-21; tables 4-5)

Macropus atlas, Owen, 1838, vol. 1, p. xix; vol. 2, p. 359, pl. 29, fig. 1.
Macropus (Sthenurus) atlas, Owen, 1873a, p. 128.
Sthenurus atlas, Owen, 1874a, pp. 265, 268-269, pl. 22, figs 3-4.
Sthenurus (Sthenurus) atlas, Tedford, 1966, pp. 11-20, figs 2-4.
Sthenurus (Sthenurus) atlas, Bartholomai, 1972, pp. ix, xii, fig. 2.
Sthenurus (Sthenurus) atlas, Dawson, 1985, p. 66, table 1.
Sthenurus sp. (nov.?), Molnar and Kurz, 1997, p. 123.

Taxonomic Authority. Owen, R. 1838. Fossil Marsupialia from the caves of Wellington Valley. Pp. 359-363 in T. L. Mitchell, Three Expeditions into the Interior of Eastern Australia, With Descriptions of the Recently Explored Region of Australia Felix, and of the Present Colony of New South Wales, 1st ed. London: T. and W. Boone.

Holotype. BMNH M10778, fragmentary right juvenile dentary (containing dp3, m1-2, erupting m4, excavated p3) collected around 1830 by Thomas L. Mitchell from Mitchell (= Breccia) Cave, Wellington, New South Wales. Age of type locality is considered to be Pleistocene (Osborne 1983, 1997).

Referred Specimens. Wellington Caves (probably mostly Mitchell Cave), NSW: AM F18875, right P3; AM F29556, left juvenile maxilla; AM F29559, left adult maxilla; AM F29560, right dP2; AM F29573, left juvenile dentary; AM F29574, right juvenile dentary; AM F29575, right juvenile dentary; AM F29579, right adult dentary; AM F29591, left P3; AM F29593, left P3; AM F29595, right P3; AM F30623, right P3; AM F47052, left and right adult dentaries; AM F47063a, left adult maxilla; AM F47063b, right adult maxilla; AM F47063c, right P3; AM F47086, adult splanchnocranium; AM F88550 (= AM MF36), juvenile cranium; AM F121396, adult cranium; BMNH 42663Ab, six lower molars; FMNH PM 1571, right adult maxilla; FMNH PM 6776, dP2; FMNH PM 6777, right P3; FMNH PM 39065, left P3.

Bone Cave, Wellington, NSW: AM F106044, left juvenile maxilla; AM F106047, right adult dentary; AM F106048, left juvenile dentary; AM F106049, left juvenile dentary; AM F106050, left juvenile dentary; AM F106052, right M2; AM F106054, right P3; F106057, left M4; UCMP 45193, right juvenile dentary.

Phosphate Mine, Wellington Caves, NSW: UCMP 52429, left juvenile dentary; UCMP 52430, left adult maxilla; UCMP 57394, right adult maxilla; UCMP 63749, right juvenile maxilla, left P3.

Bunyan Siding, near Cooma, NSW: AM F unreg., right M3, metaloph, right i1.

Teapot Creek Terrace 3, Monaro Region, NSW: AM F115370, left dp3 or m1.

Lake Menindee, western NSW: UCMP 45675, left and right juvenile dentaries; UCMP 55978, right adult maxilla.

Lake Tandou, western NSW: UCMP 115347, right adult dentary.

Mulyanna Site, Warburton River, SA: SAM P25453, left and right adult dentaries.

Lookout Locality, Warburton River, SA: QM F unreg., left juvenile dentary.

Curramulka Quarry, Yorke Peninsula, SA: SAM P40063, right I3, right P3, left p3, 11 molars.

Freestone Ck, Darling Downs, QLD: QM F5769, left and right adult dentaries.

?Darling Downs, QLD: QM F2976, right adult maxilla.

Revised Diagnosis. Most similar overall to *Sthenurus andersoni*, but slightly larger, especially dentally. P3 elongate with deep longitudinal basin and very distinct posterobuccal accessory cusp; p3 with very deep vertical cleft separating anterior end of buccal crest from main crest. Lower molars with posterior portion of paracristid oriented decidedly anterolingually, anterior portion reduced to a cusp.

Etymology. Lat. *Atlas*, mythical god who held up the heavens, apparently in reference to the large size of the species or its lower premolar, which was proportionally wider than that of any other macropodoid known at that time.

Description (pls 16-21; tables 4-5). No taxonomically enlightening specimens of *Sthenurus atlas* have come to light since the work of Tedford (1966), with the exception of one partial adult cranium (AM F121396) hitherto unrecognized in the older Wellington Caves collection. Below I describe and compare the cranium with other *Sthenurus* species. As for *S. stirlingi* and *S. tindalei*, previous descriptions and interspecific comparisons (Tedford 1966, 1967), as well as comparisons made above with *S. andersoni* together represent sufficient accounts of *S. atlas*.

Adult Cranium (AM F121396, pl. 16; table 4). Dorsal portion of cranium compressed anterior to fronto-parietal suture, and most of rostrum missing, but overall cranial form clearly dolichocephalic. Posterior portion of nasals rather narrow. Nasofrontal suture gently sinusoidal. Masseteric process short, narrow, only very slightly flared distally, and composed almost entirely of maxilla. Frontal region rather elongate, with lateral border not inflated, but slightly sinusoidal in shape. Supraorbital crest bears small rugosity toward anterior end. Palatal vacuities extend anteriorly to opposite M1 interloph valley. Postpalatine bars formed thin bridge across palate adjacent to M4 metaloph. Basicranial plane moderately elevated above palatal plane. Basioccipital aligned in same plane as basisphenoid; median keel only slightly developed. Only incomplete posterior end of zygomatic process of squamosal preserved on both sides, but these were clearly shallow, and apparently oriented in sagittal plane. Postglenoid process rather small. Temporal (parietal) crests very low and poorly developed, failing to converge on sagittal suture. Occipital region rather shallow, and oriented at 90° relative to dorsal surface of neurocranium. Vertical median occipital crest clearly developed, leading ventrally to particularly wide, oval-shaped foramen magnum. Bases of paroccipital processes suggest predominantly vertical orientation. Nuchal / lambdoid crests abraded during deposition, but were apparently well developed.

Table 4. Cranial dimensions of *Sthenurus atlas* (adult, AM F121396; juvenile, AM F88550, M2 just erupted).

Dimension	*AM F121396*	*AM F88550*
Condylobasal Length	-	150
Diastema Length	-	19.3
% Diastema Length: Palatal Length	-	
Palatal Length	-	87.6
Palatal Width between M1 Protolophs	-	-
Palatal Width between M4 Protolophs	-	-
Cranial Width	118	-
Width across Frontals	67.0	48.7
Distance between Paroccipital Processes	65.4	-

Table 5. Cheek tooth dimensions of *Sthenurus atlas*, showing mean and standard deviation (in parentheses), and dimensions of the holotype (BMNH M10778).

Tooth	L	AW	PW	AH	PH	n
dP2	10.3 (0.36)	6.8 (0.20)	8.2 (0.61)	8.8 (0.92)	8.9 (0.57)	3
dP3	11.1 (0.32)	9.3 (0.35)	10.5 (0.35)	6.0	7.2	3
P3	18.0 (0.48)	8.4 (0.35)	10.0 (0.58)	11.9 (0.59)	11.1 (0.68)	12
M1	12.7 (0.39)	11.6 (0.20)	11.5 (0.22)	7.6 (0.38)	7.7 (0.49)	9
M2	14.0 (0.51)	12.5 (0.34)	11.8 (0.31)	7.3 (0.43)	8.1 (0.54)	10
M3	14.5 (0.29)	12.4 (0.38)	11.4 (0.35)	7.7 (0.75)	8.0 (0.91)	8
M4	13.4 (0.57)	12.4 (1.21)	9.5 (1.30)	7.2 (0.17)	6.9 (0.62)	3
dp2	9.8 (0.22)	5.3 (0.65)	6.9 (0.55)	8.7 (0.39)	7.1 (0.89)	5
dp3	10.4 (0.65)	7.6 (0.42)	8.2 (0.34)	7.7 (0.00)	8.5 (0.49)	7
dp3 holotype	10.7	7.7	8.6	-	-	
p3	16.5 (0.53)	6.7 (0.63)	8.3 (0.42)	11.4 (1.13)	11.1 (0.88)	12
p3 holotype	17.0	7.0	8.3	11.8	11.4	
m1	12.4 (0.87)	9.3 (0.39)	9.6 (0.45)	9.0 (0.98)	9.8 (0.98)	13
m1 holotype	12.6	9.6	9.7	9.1	9.9	
m2	13.8 (0.88)	10.5 (0.48)	10.5 (0.45)	9.9 (0.87)	10.5 (10.5)	13
m2 holotype	15.0	10.8	10.6	10.4	10.9	
m3	14.6 (0.57)	11.6 (0.43)	11.3 (0.25)	10.3 (0.73)	10.3 (0.47)	9
m4	14.0 (0.12)	11.5 (0.50)	10.0 (0.49)	8.5 (1.28)	8.1 (1.03)	6
m4 holotype	14.2	11.9	10.4	-	-	

Comparison With Other Taxa. Cranium. *Sthenurus atlas* is very similar to *S. andersoni* in all aspects of cranial morphology and relative size. It differs from *S. andersoni* by being slightly wider, possessing a more sinusoidal nasofrontal suture, less inflated frontals (which have a rather sinusoidal lateral border as in *S. tindalei*, but lack the posterolateral frontal inflation), and very low temporal crests, which fail to meet at the sagittal suture.

Geographic Distribution. Southeastern Queensland, eastern and western New South Wales, northern, central southern and southeastern South Australia (figs 27, 33; table 44).

Temporal Distribution. Middle to late Pleistocene.

Sthenurus murrayi sp. nov.

(pl. 22; fig. 4; tables 6-7)

Sthenurus atlas, Marshall, 1973, pp. 165-166, fig. 5.

Holotype. NMV P28267, partial juvenile cranium (consisting of partial left and right maxillae and premaxillae; preserving left and right I1-3, dP2-3, M1-3, excavated P3) collected by Kenneth G. Simpson and Edmund D. Gill in October 1969 from Site 50, Lake Victoria, southwestern New South Wales. Biocorrelation suggests age of type locality is middle or late Pleistocene (Marshall 1973; pers. obs.).

Diagnosis. P3 intermediate between *Sthenurus andersoni* and *S. tindalei* in size and morphology; ridgelet directed anterobuccally from anteriormost cuspule of lingual crest divides anterior basin into two distinct sub-basins. Upper molars similar in size and outline to those of *S. atlas*, but with thicker precingulum, especially lingually. Preparacrista maintains strong connection with paracone apex on M1, but only very slight on M2-3. Posterior face of protoloph bears more distinct enamel crenulations than in *S. andersoni.* Buccal (mesial) component of postmetaconulecrista strongly developed.

Etymology. Named in honor of Peter F. Murray in recognition of his broad contributions to Australian vertebrate palaeontology and anthropology, and his masterful reconstructions of extinct vertebrates.

Description. Juvenile Cranium (pl. 22; fig. 4; table 6). Holotype consists of palatal portions of left and right premaxillae and maxillae. Incisor-bearing portion of premaxilla elongate and moderately shallow (pl. 22A; fig. 4). Only base of posterodorsally sloping upright portion of premaxilla is preserved. Rather elongate diastema deflected slightly anteroventrally relative to cheek tooth row (pl. 22A). Incisive foramina thin and fairly small; bordered laterally by distinct palatal ridges, which run from in between anterior edges of incisive foramina and I3 roots to two-thirds of way along diastema (pl. 22D). Anteorbital and buccinator fossae on maxilla appear to have been very shallow. Tips of both left and right masseteric processes have been abraded away, but clearly fine, short, composed entirely of maxilla, and positioned adjacent to M2 protoloph. Jugals broken off adjacent to anterior limit of temporal fossae. Palatal vacuities extend anteriorly to opposite M1 protoloph (pl. 22D). Maxillae broken off posterior to fully erupted M3, but right side preserves anterior portion of M4 crypt, morphology of which suggests imminent M4 eruption.

Upper Dentition (pl. 22; fig. 4; table 7). Although broken buccally and occlusally, I1 crown clearly high and broad. I2, while also fragmentary, is approximately one-third size of I1 and rather cylindrical. Large, blade-like I3 bears slightly undulating buccal surface and slight anterolingual crest (pl. 22A, 22D).

While reminiscent of P3 in morphology, dP2 is much shorter relative to width and lacks any sign of a posterobuccal accessory cusp (pl. 22A, 22D; fig. 4). Having sustained considerable wear, numerous topographical features have been obliterated, but anterior and posterior basins apparently well demarcated. Main crest markedly higher than lingual crest and divided into three cuspules, with central cuspule smaller than other two (pl. 22A). Although worn, dP3 appears similar in morphology to M1, but with lophs sloped more toward crests, relatively narrower protoloph, and precingulum terminating farther from lingual edge of tooth (pl. 22D; fig. 4).

Table 6. Palatal dimensions of the holotype of *Sthenurus murrayi* sp. nov. (NMV P28267).

Dimension	*Sthenurus murrayi* sp. nov.
Diastema Length (I3-dP2)	28.0
Palatal Length to Posterior Limit of M3	86.5
Palatal Width between M1 Protolophs	29.2
Palatal Width between M3 Metalophs	26.6
Width across Masseteric Processes	71.5

Table 7. Cheek tooth dimensions of the holotype of *Sthenurus murrayi* sp. nov. (NMV P28267).

Tooth	L	AW	PW	AH	PH
dP2	10.2	6.6	8.3	7.7	8.4
dP3	11.3	9.4	10.5	-	-
P3	16.8	8.5	10.2	11.8	11.5
M1	12.4	12.2	12.5	6.7	7.4
M2	14.1	12.4	13.0	7.6	7.9
M3	15.1	13.1	13.1	6.7	7.1

P3 longer than all molars, slightly narrower anteriorly than posteriorly, buccal and lingual sides gently undulating (pl. 22B-C; fig. 4). Main crest differentiated into five linked cuspules, with anteriormost more prominent. Each cuspule gives rise to low, rounded ridgelets on buccal side, except for more distinct ridgelet from posteriormost cuspule. This bifurcates, with anterior fork immediately terminating in small but distinct posterobuccal accessory cusp. Posterior fork curves posterolingually, uniting distally with posterior extremity of main crest, thus enclosing a small deep pocket on posterobuccal corner of P3. Lingual crest lower than main crest; oriented roughly parallel to main crest, but contorting slightly buccally toward anterior end. Anteriormost cuspule of lingual crest gives rise to three separate ridgelets, with largest directed anteriorly, then curved anterobuccally toward base of crown after intersecting with ridgelet directed anterolingually from anteriormost cuspule of main crest (pl. 22C; fig. 4). Slightly smaller ridgelet, directed anterolingually from anteriormost cuspule of lingual crest, terminates halfway along ridgelet directed anterolingually from anteriormost cuspule of main crest, effectively dividing anterior basin into two sub-basins. Third ridgelet from anteriormost cuspule of lingual crest is much smaller and directed buccally. It fails to unite with larger ridgelet directed lingually from anteriormost cuspule of main crest, resulting in incomplete separation of anterior and longitudinal basins. Deep longitudinal basin traversed by several ridgelets; complex of three contorted ridgelets separating this basin from rather large posterior basin (pl. 22C; fig. 4).

Molar row slightly curved laterally (pl. 22D; fig. 4). Upper molars high crowned; unworn lophs taper more toward crests on M1 than on M2-3. Protoloph and metaloph crests slightly convex anteriorly. Preparacrista maintains strong connection with paracone apex on M1, but not on M2-3 (pl. 22A,D; fig. 4), where very slight preparacrista links paracone apex and enlarged cusp-like portion (= ?stylar cusp A). Thick precingulum extends across entire anterior face of protoloph, supporting distinct remnant of preprotocrista just lingual of tooth midline. A parallel enamel crenulation is positioned next to preprotocrista on M3. Although worn, postprotocrista on M1 appears undivided, in contrast to incipient division on M2-3, whereupon buccal component thickens posteriorly before terminating at interloph valley. On M3, buccal component is partially bifurcated. One of three distinct enamel crenulations on posterior face of protoloph, between buccal component of postprotocrista and postparacrista, is probably mesocrista. Well-developed postparacrista bears distal kink (?reemergent stylar cusp C) successively enlarged from M1 to M3. Premetacrista fine and low, overlapping slightly with postparacrista and terminating on its buccal side. Thus, interloph valley entirely enclosed buccally. Bifurcation of postmetacrista distinct on M1, slight on M3, and absent from M2. Buccal component of postmetacrista meets dorsobuccally trending lingual component of postmetaconulecrista at position of stylar cusp E. Buccal (mesial) component of postmetaconulecrista strongly developed (pl. 22D, fig. 4), shifted one-third of way across metaloph posterior face from lingual component; conjoined dorsally with likely urocrista on M3.

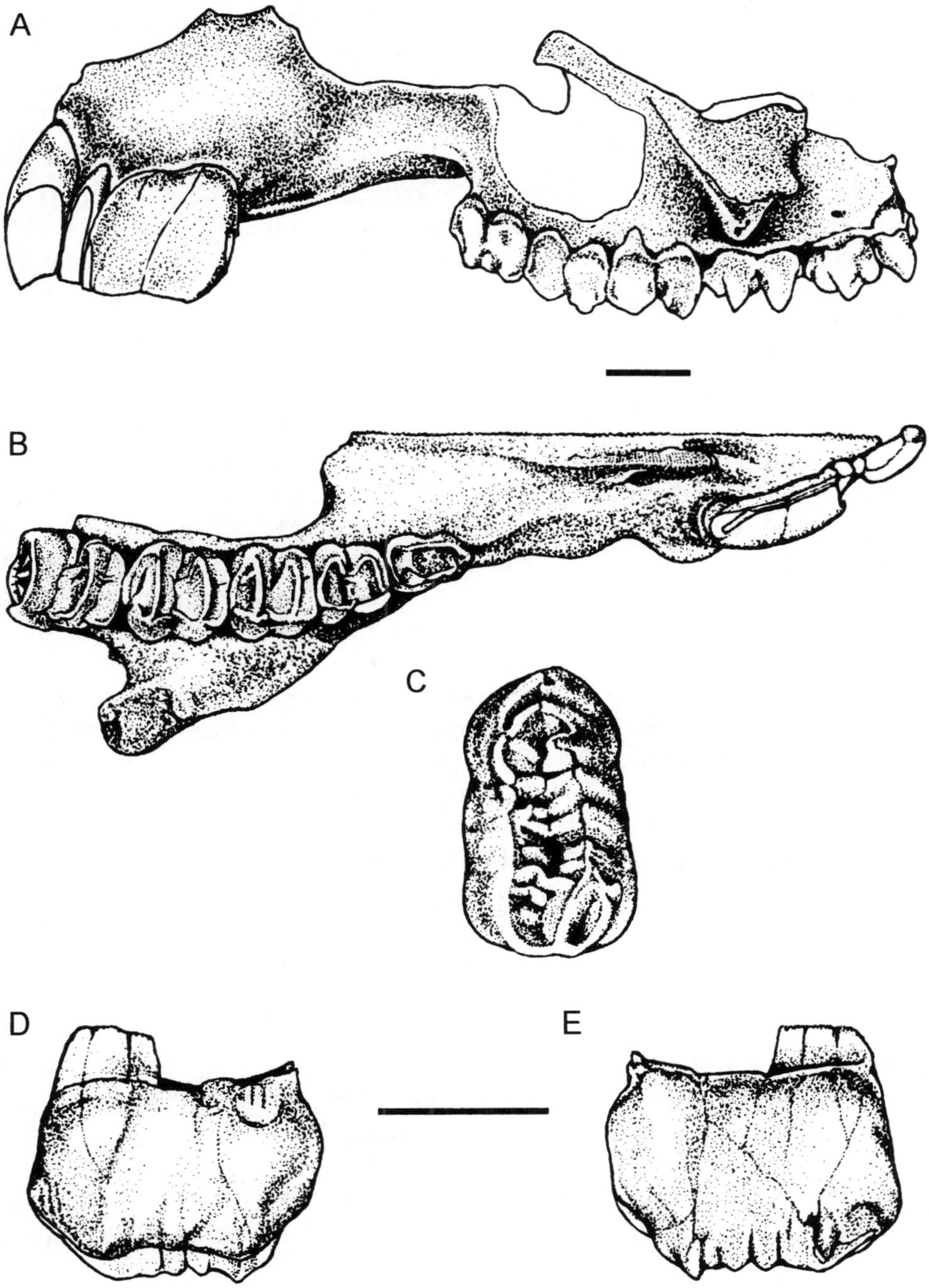

Figure 4. Partial juvenile cranium of *Sthenurus murrayi* (NMV P28267, holotype) from Lake Victoria, southwestern New South Wales. **A**, lateral view; **B**, palatal view; **C**, left P3, occlusal view; **D**, left P3, lingual view; **E**, left P3, buccal view. Modified from Marshall (1973) with permission from Museum Victoria. Scale bars = 10 mm.

Comparison With Other Taxa. Cranium. Too little of the cranium is preserved to adequately compare it with any other species, except in terms of size. In this regard, *Sthenurus murrayi* most closely resembles *S. andersoni* and *S. atlas*, rather than the much larger *S. tindalei* and *S. stirlingi. S. notabilis* is only known from a juvenile dentary fragment, but the size of its molariform teeth suggest a significantly larger cranium than for *S. murrayi.*

Upper Dentition. Both absolutely and relative to the size of the rostral region preserved, the upper incisors (especially I3) of *Sthenurus murrayi* are larger than those of the only known specimen of *S. atlas* that preserves upper incisors (AM F47086, pl. 18A-B). The anterior and posterior edges of I3 are also less curved than in AM F47086, but they are more curved than in *S. andersoni.* Overall, the *S. murrayi* I3 most closely resembles that of the slightly larger *S. tindalei* I3, and it retains a similarly slight anterolingual crest. Although larger, the P3 of *S. murrayi* is very similar in morphology to that of *S. andersoni*, and to a lesser degree, the larger *S. tindalei.* Morphologically, it may be separated from the *S. andersoni* P3 by the presence of a ridgelet directed anterobuccally from anteriormost cuspule of lingual crest, which divides the anterior basin into two distinct sub-basins. Along with this feature, the P3 of *S. murrayi* is easily distinguished from that of *S. atlas* by being relatively wider, and having a much shallower longitudinal basin and a much less distinct posterobuccal accessory cusp. The upper molars of *S. murrayi* are most similar in size and general outline to those of *S. atlas*, but they differ by having a proportionally thicker precingulum, distinct enamel crenulations on the posterior face of the protoloph, and a strongly developed buccal component of the postmetaconulecrista. The enamel crenulations on the posterior face of the protoloph are coarser than in *S. andersoni.*

Geographic Distribution. Southwestern New South Wales (fig. 34; table 44).

Temporal Distribution. Middle or late Pleistocene.

Sthenurus notabilis Bartholomai, 1963

(pl. 23; table 8)

Sthenurus (Sthenurus) notabilis, Tedford, 1966, pp. 25-26.
Sthenurus (Sthenurus) notabilis, Bartholomai, 1972, pp. ix, xii, fig. 2.

Taxonomic Authority. Bartholomai, A. 1963. Revision of the extinct macropodid genus *Sthenurus* Owen in Queensland. Memoirs of the Queensland Museum 14:51-76.

Holotype. QM F3817, partial right juvenile dentary (containing dp2-3, m1, excavated p3) collected by Jack T. Woods in April 1958 from the Chinchilla Rifle Range, Chinchilla, southeastern Queensland. Biocorrelation suggests age of type locality is early late Pliocene (e.g., Woods 1962; Bartholomai 1963, 1973a; Archer and Wade 1976; Skilbeck 1980; Tedford et al. 1992).

Revised Diagnosis. Lower permanent premolar shorter than m1. Lower molariform teeth very high crowned; paracristid markedly U-shaped anteriorly and clearly differentiable from underlying precingulid. Slight parametacristid and preentocristid evident.

Etymology. Lat. *notabilis*, 'remarkable', in reference to the relative uniqueness of the species, which Bartholomai (1963) thought bore some resemblance to the species of *Sthenurus* as well as *Procoptodon.*

Description. Sthenurus notabilis is only known from the holotype specimen, and Bartholomai's original description remains satisfactory.

Comparison With Other Taxa. Lower Dentition (pl. 23). Although *Sthenurus notabilis* is the most poorly represented member of the genus, there is no doubt about its taxonomic validity. It is the only *Sthenurus* species in which p3 is shorter than m1 (pl. 23; table 8), although it is possible that p3 length is slightly underestimated, because the tooth is not completely enameled. However, even if it were fully enameled, its length is very unlikely to have exceeded that of m1. In morphology, the *S. notabilis* p3 most closely resembles that of *S. andersoni*, from which it is distinguished by its larger size and relatively shorter length. Although dp3 and m1 are the only two molariform teeth preserved, crown height exceeds that of all other *Sthenurus* species, including *S. tindalei* (crown height index based on m1: *S. notabilis*, 0.86; *S. tindalei*, 0.84). The m1 is easily distinguished from *S. tindalei* and *S. stirlingi* by being proportionally more elongate. No other *Sthenurus* species possesses a well-developed, U-shaped paracristid that is clearly differentiable from the underlying precingulid (pl. 23C). In this respect, *S. notabilis* most closely resembles *Metasthenurus newtonae.* The presence of a slight parametacristid descending into the trigonid basin (pl. 23C) is also unique among the species of *Sthenurus.*

Geographic Distribution. Southeastern Queensland (fig. 27).

Temporal Distribution. Early late Pliocene.

Table 8. Cheek tooth dimensions of the holotype of *Sthenurus notabilis* (QM F3817).

Tooth	L	AW	PW	AH	PH
dp2	10.8	6.8	7.8	9.6	8.2
dp3	12.8	9.5	10.6	8.2	9.1
p3	16.4	7.5	9.8	10.5	10.2
m1	16.8	11.8	12.6	12.8	14.5

Sthenurus stirlingi Wells and Tedford, 1995

(pls 24-25, 109; figs 16G, 17E, 18I; table 9)

Sthenurus sp. nov., Williams, 1980, p. 105, sites 16-17.

Taxonomic Authority. Wells, R. T. and R. H. Tedford. 1995. *Sthenurus* (Macropodidae: Marsupialia) from the Pleistocene of Lake Callabonna, South Australia. Bulletin of the American Museum of Natural History 225:1-111.

Holotype. SAM P22533, near-complete adult skeleton, including cranium and dentaries, collected by a joint field party (Smithsonian Institution, American Museum of Natural History, South Australian Museum) in 1970 from Lake Callabonna, eastern South Australia. Age of the type locality is estimated to be middle Pleistocene (0.2–0.7 Ma, Wells and Tedford 1995).

Referred Specimens. Lake Callabonna, eastern SA: SAM P12882, adult cranium and dentaries; AMNH 117494, partial neurocranium and partial skeleton; AMNH 117495, partial juvenile cranium and partial skeleton; AMNH 117496, partial adult cranium, dentary and partial skeleton; AMNH 117498, left adult dentary and partial skeleton.

Billeroo Creek, eastern SA: NMV P150275 / P150276, adult cranium and dentaries.

Waralamanko Waterhole, Cooper Creek, SA: SAM P32553, left and right premaxillae.

Kudnampirra Waterhole (= Site 14), Cooper Creek, SA: UCMP 56473, right juvenile dentary.

New Kalamurina, Warburton River, SA: UCMP 97668, p3.

Bairstow's Sand Pit, Port Pirie, SA: SAM P40106, left M4.

Curramulka Quarry, Yorke Peninsula, SA: SAM P40064, right P3.

Revised Diagnosis. Very large; cranial proportions intermediate between those of dolichocephalic and brachycephalic sthenurines. Buccinator fossa deep. Frontals markedly inflated. Basicranial plane markedly elevated above palatal plane. I3 with very strongly developed anterolingual crest. Molars intermediate in crown height, very broad for length, with mesocrista moderately developed. Lower incisor upturned.

Etymology. Named in honor of Edward C. Stirling, biologist and former director of the South Australian Museum who, in the 1890s, collected and described the first Pleistocene vertebrates from Lake Callabonna with his assistant, A. H. Zietz.

Description, Variation and Comparison (pls 24-25; table 9). Wells and Tedford (1995) have provided a detailed description and comparison of *Sthenurus stirlingi*, which is most similar overall to *S. tindalei*. However, it is clearly larger and more robust than this species, resembling *"Simosthenurus" pales* in several attributes (see description of *"Si." pales*). Although crushed, the Billeroo Creek specimen (NMV P150275 / P150276) matches *S. stirlingi* in cranial proportions as well as dental dimensions and morphology. The dentaries, however, are more gracile than in the Lake Callabonna specimens, and the lower incisors (even when post-depositional distortion of the region is taken into account) appear to have been more procumbent. In these features, NMV P150275 / P150276 is more reminiscent of *S. tindalei*. At this stage, it is most parsimonious to interpret this as an intraspecifically variable trait in *S. stirlingi*.

Geographic Distribution. Northern and mid-northern South Australia (fig. 37; table 44).

Temporal Distribution. Middle to late Pleistocene.

Table 9. Cheek tooth dimensions of *Sthenurus stirlingi*, showing mean, standard deviation (in parentheses) and dimensions of holotype (SAM P22533).

Tooth	L	AW	PW	AH	PH	n
dP2	12.6	8.9	12.4	-	-	1
dP3	15.8	15.5	17.2	-	-	1
P3	20.3 (0.61)	11.2 (0.31)	14.8 (0.87)	13.8	13.3	4
P3 holotype	20.9	11.5	15.6	13.8	13.3	
M1	17.8 (1.81)	19.4 (1.03)	19.5 (0.51)	12.0	12.0	4
M1 holotype	19.0	20.6	20.2	12.0	12.0	
M2	20.4 (0.89)	20.2 (0.56)	19.7 (1.14)	12.6	12.4 (0.64)	5
M2 holotype	20.7	20.7	19.3	12.6	12.8	
M3	20.9 (0.76)	20.5 (0.51)	19.1 (0.21)	11.0 (1.06)	11.7 (0.21)	5
M3 holotype	21.0	20.8	18.9	11.7	11.8	
M4	19.7 (1.02)	18.2 (1.37)	15.4 (0.70)	9.3 (0.85)	9.2 (0.59)	6
M4 holotype	19.8	20.5	15.8	9.0	9.0	
p3	17.2 (0.25)	8.4 (0.56)	11.4 (0.85)	11.1 (0.28)	11.5 (0.00)	4
p3 holotype	17.2	8.3	11.1	11.3	11.5	
m1	15.7 (0.67)	14.4 (1.13)	15.2 (0.35)	9.6	10.2	4
m1 holotype	15.5	15.2	15.4	9.6	10.2	
m2	20.1 (0.80)	17.4	17.9 (1.20)	14.3	14.8	5
m2 holotype	21.2	17.4	17.0	14.3	14.8	
m3	20.7 (1.09)	18.7 (0.99)	17.6 (0.14)	14.5 (0.99)	14.5 (0.99)	5
m3 holotype	21.6	18.0	17.7	15.2	15.2	
m4	20.1 (0.48)	17.3 (0.64)	14.5 (0.54)	12.9 (1.41)	12.2 (0.57)	4
m4 holotype	19.7	17.6	15.0	13.9	12.6	

Sthenurus tindalei Tedford, 1966

(pls 26-29; figs 15F, 25A; table 10)

Macropus anak, Glauert, 1912, pp. 62-63.
Sthenurus atlas, Glauert, 1912, p. 64.
Sthenurus (Sthenurus) tindalei, Tedford, 1966, pp. 26-33, figs 8-10.

Taxonomic Authority. Tedford, R. H. 1966. A review of the macropodid genus *Sthenurus*. University of California Publications in Geological Sciences 57:1-72.

Holotype. SAM P13820, cranium (crushed but near complete), collected by Richard H. Tedford in the early 1950s from Unit B, northern side of Lake Menindee, western New South Wales (Tedford 1955). Age of the type locality is late Pleistocene (67±6 ka, Roberts et al. 2001). Note that the holotype registration number is incorrectly given as SAM P138201 in Tedford (1966) and Wells and Tedford (1995).

Referred Specimens. Lake Menindee, western NSW: SAM A28147, right adult dentary; UCMP 52432, right adult dentary.

Site 50, Lake Victoria, NSW: NMV P26547, left adult dentary.

Fisherman's Cliff, Moorna Station, NSW: NMV P29422, left protolophid; NMV P29423, right metaloph.

Lookout Locality, Warburton River, SA: SAM P20953, left adult dentary; SAM P20955, metatarsal IV; QM F unreg., right adult dentary.

Green Bluff Locality, Warburton River, SA: UCMP 56808, left juvenile dentary.

Cassidy Locality, Warburton River, SA: UCMP 47946, m3.

West Bluff, Lake Kutjitara, SA: FU 2161, right juvenile dentary; FU 2162b, right adult dentary; FU 2163b, right juvenile maxilla.

Katipiri Waterhole (= Site 9), Cooper Creek, SA: UCMP 56471, right juvenile dentary.

Malkuni Waterhole, Cooper Creek, SA: SAM P25058, left adult maxilla; SAM P25060, left adult maxilla.

Waralamanko Waterhole, Cooper Creek, SA: SAM P25172, right juvenile dentary.

Cooper Creek (unspecified), SA: HM S69, right juvenile dentary.

Lake Callabonna, eastern SA: AMNH 117491, partial skeleton, including cranium; AMNH 117492, partial skeleton, including cranium and right dentary; AMNH 117493, partial skeleton, including cranium and right dentary; AMNH 117499, partial skeleton, including cranium and left dentary.

Clinton, Yorke Peninsula, SA: SAM P36623, right adult dentary.

Dempsey's Lake, Port Augusta, SA: SAM P22492, left and right juvenile dentaries.

Port Augusta Sand Pit, SA: SAM P22498, left m3.

Baldina Creek, near Burra, SA: SAM P21086, right adult maxilla.

Last Tree Cave, Nullarbor Plain, WA: WAM unreg.

Balladonia Soak, WA: WAM 63.11.1 / 63.11.4 / 63.11.9 / 63.11.10, left and right adult dentaries.

Revised Diagnosis. Cranium slender and dolichocephalic. Dentary very slender, with anterior edge of ascending ramus slightly declined posteriorly; diastema and i1 markedly procumbent. Connection between p3 buccal crest and anteriormost cuspule of main crest is via a distinct, short crescentic crest. Molars very high crowned, similar to *Sthenurus notabilis*, with height of unworn lower molar crowns only slightly less than their length.

Etymology. Named in honor of Norman B. Tindale, co-discoverer of the Lake Menindee Pleistocene vertebrate locality and major contributor to the study of human prehistory and the late Cenozoic in Australia.

Description and Comparison (pls 26-29; fig. 25A; table 10). Wells and Tedford (1995) have provided a detailed description and comparison of *Sthenurus tindalei*, particularly in relation to the species that it most resembles, *S. stirlingi.* However, it is worthwhile noting here that the p3 preserved in an older juvenile specimen from Balladonia (WAM 63.11.1/-.4/-.9/-.10, pl. 29C), at the western edge of the Eucla Basin in Western Australia, differs in proportion from the remaining specimens referred to *S. tindalei.* While it falls well within the standard range for width and crown height dimensions, in length the Balladonia is 16% shorter than the average p3 length for *S. tindalei* (table 10). The lower molars fall within the standard size range for *S. tindalei* (table 10), and neither they nor the p3 differ in any topographic features (pl. 29C).

Geographic Distribution. Northern and mid-northern South Australia, western New South Wales, central and western Nullarbor Region (fig. 37; table 44).

Temporal Distribution. Middle to late Pleistocene.

Sthenurus sp. cf. *tindalei*

(pl. 27; table 10)

Sthenurus sp., Tedford, Wells and Barghoorn, 1992, p. 188.

Referred Specimen. Stirton Quarry, Lake Kanunka, SA: UCMP 56928, right juvenile maxilla. Early late Pliocene.

Remarks. UCMP 56928 is a partial juvenile maxilla from Stirton Quarry at Lake Kanunka (pl. 27B; fig. 27). It preserves what I deduce to be a slightly worn M1, based on the presence anterior to it of alveoli apparently corresponding to dP2-3, and the narrower width of the protoloph relative to the metaloph (pl. 27B; table 10). In most aspects, M1 closely resembles that of *Sthenurus tindalei.* It is similarly high crowned, has a well-developed precingulum, an absence of enamel crenulations with the exception of two on the posterior metaloph face (like *S. tindalei*), and a strongly developed postparacrista, which is distinctly incurved lingually (polymorphic for *S. tindalei*). The diminutive masseteric process is also congruent with what one might predict for a similarly aged individual of *S. tindalei*. UCMP 56928 differs from *S. tindalei* by having a much narrower M1 relative to its length and a relatively narrower protoloph relative to the metaloph. Because these features are not known to vary as substantially within species of *Sthenurus*, and because UCMP 56928 is so incomplete, I cannot allocate it to *S. tindalei* with certainty, even though *S. tindalei* occurs in the later Pliocene Fisherman's Cliff LF.

Table 10. Cheek tooth dimensions of *Sthenurus tindalei,* showing mean, standard deviation (in parentheses), and dimensions of the holotype (SAM P13820) and the Balladonia dentary (WAM 63.11.1/-.4/-.9/-.10). Dimensions of an M1 referred to *Sthenurus* sp. cf. *tindalei* from Lake Kanunka, South Australia (UCMP 56928) are also provided.

Tooth	L	AW	PW	AH	PH	n
dP2	11.0	7.6	9.9	9.8	10.2	1
dP3	13.0	12.4	14.5	-	-	1
P3	19.5 (0.74)	10.2 (0.67)	12.8 (0.75)	13.0 (0.28)	12.4 (1.13)	4
P3 holotype	19.7	10.4	12.6	13.2	13.2	
M1	15.6 (0.60)	15.9 (0.67)	16.4 (0.67)	9.9	11.1	4
M1 holotype	15.0	13.8	15.4	-	-	
Lake Kanunka	16.5	13.5	14.2	8.6	9.5	
M2	17.0 (1.02)	16.4 (0.37)	16.1 (0.37)	10.3	10.3	5
M2 holotype	17.7	15.6	16.1	10.3	10.3	
M3	17.6 (0.80)	16.6 (0.33)	15.5 (0.72)	9.6 (1.46)	10.1 (1.22)	5
M3 holotype	18.0	17.0	16.4	10.9	11.4	
M4	16.2 (0.56)	14.7 (0.61)	12.4 (1.37)	9.5 (0.28)	9.1 (0.07)	4
M4 holotype	16.5	15.2	13.6	9.7	9.0	
dp2	10.7	6.2	7.4	8.8	8.8	1
dp3	12.0	9.2	10.2	8.8	10.0	1
p3	17.5 (0.54)	7.8 (0.33)	9.3 (0.28)	10.7 (0.93)	10.6 (0.45)	7
p3 Balladonia	14.8	8.1	9.2	8.7	10.8	
m1	13.8 (1.19)	12.1 (0.34)	12.5 (0.43)	12.8	12.8	7
m2	16.4 (1.52)	13.5 (0.52)	13.7 (0.59)	13.2 (2.11)	13.7 (1.92)	9
m2 Balladonia	16.0	13.4	13.4	11.6	13.0	
m3	17.7 (1.14)	15.1 (0.97)	14.4 (0.82)	13.6 (1.16)	14.0 (1.13)	11
m3 Balladonia	16.9	14.1	13.6	13.3	14.0	
m4	17.1 (0.57)	13.9 (0.40)	12.1 (0.63)	11.2 (1.02)	10.5 (0.75)	8

METASTHENURUS gen. nov.

Sthenurus, De Vis, 1895 (in part), p. 97.
Sthenurus, Glauert, 1912 (in part), pp. 64-65.
Sthenurus, Bartholomai, 1963 (in part), p. 58, fig. 3.
Sthenurus, Lundelius, 1963, p. 77, fig. 2.
Sthenurus, Merrilees, 1965 (in part), pp. 29-30.
Sthenurus, Tedford, 1966 (in part), pp. 25, 43.
Sthenurus, Marcus, 1976 (in part), pp. 71, 74, fig. 27c-d.
Simosthenurus, Pledge, 1980 (in part), p. 137, table 3.
Sthenurus, Williams, 1980, p. 107, site 30.
Sthenurus, Williams, 1980, p. 107, site 37.
Simosthenurus, Prideaux and Wells, 1994 (in part), p. 227.
Sthenurus, McNamara, 1994, pp. 111, 115.
Sthenurus, Prideaux and Wells, 1997 (in part), pp. 191, 194.
Simosthenurus, Prideaux, 2000 (in part), pp. 1-15.

Type species. Simosthenurus newtonae Prideaux, 2000.

Generic Diagnosis. Cranium similar in size to that of *"Procoptodon" browneorum* and *Simosthenurus occidentalis*, but less brachycephalic, with a narrower, more elongate rostrum, longer diastema and less inflated frontal region. Masseteric process short, narrowed distally, twisted posteriorly. P3 short for width, crown inflated anterobuccally and thus only slightly wider posteriorly than anteriorly. Molars moderately high crowned, with very well developed cristae generally well connected to cusp apices and very few to no fine enamel crenulations. Upper molars with precingulum abruptly terminated after extending across anterior face of protoloph to position of preprotocrista. Postparacrista and premetacrista incurved lingually. Dentary with deep and rather narrow median dorsal groove, tall and narrow medial pterygoid fossa, and well-developed mylohyoid groove and processes. Lower molars bear pronounced anterior turn of lophid ends and strongly inflated hypolophid posterior face; p3 subequal in length to m2.

Etymology. Gr. *meta* 'between, near'. In reference to the evolutionarily intermediate nature of this taxon relative to *Sthenurus* and the Simosthenurini.

Geographic Distribution. Southwestern Australia, Nullarbor Region, southern and mid-northern South Australia, northwestern Tasmania, southern Victoria, eastern New South Wales, southeastern Queensland (figs 28-29; table 44).

Temporal Distribution. Middle and late Pleistocene.

Metasthenurus newtonae (Prideaux, 2000)

(pls 30-35; fig. 15E; tables 11-12)

Sthenurus oreas, De Vis, 1895 (in part), p. 97.
Sthenurus atlas, Glauert, 1912, pp. 64-65.
Sthenurus sp. cf. *oreas*, Marcus, 1962, p. 304.
Sthenurus andersoni, Bartholomai, 1963 (in part), p. 58, fig. 3.
Sthenurus sp., Lundelius, 1963, p. 77, fig. 2.
Sthenurus sp. cf. *gilli*, Merrilees, 1965, pp. 29-30.
Sthenurus andersoni, Tedford, 1966 (in part), p. 25.
Sthenurus sp. cf. *gilli*, Merrilees, 1968a, p. 65.
Sthenurus sp. II, Marcus, 1976, pp. 71, 74, fig. 27c-d.
Simosthenurus sp. II, Pledge, 1980 (in part), pp. 134, 137, table 3.
Sthenurus sp., Williams, 1980, p. 107, site 30.
Sthenurus sp. cf. *atlas*, Williams, 1980, p. 107, site 37.
Sthenurus (Simosthenurus) sp., Lundelius and Turnbull, 1989 (in part), pp. 2, 4, figs 1A, 2.
Sthenurus 'P17250', McNamara, 1994, pp. 111, 115.
Simosthenurus sp. nov., Prideaux and Wells, 1994, p. 227.
Sthenurus 'P17250', Prideaux and Wells, 1997, pp. 191, 194.
Sthenurus sp. nov., Brown and Wells, 2000, p. 101.
Sthenurus newtonae, Armand, Ride and Taylor, 2000, pp. 108-109, 112, 120.
Simosthenurus newtonae, Prideaux, 2000, pp. 1-15.
Simosthenurus newtonae, Reed and Bourne, 2000, pp. 74-76, 78, 80-82, 85-86.

Taxonomic Authority. Prideaux, G. J. 2000. *Simosthenurus newtonae* sp. nov., a widespread sthenurine kangaroo (Diprotodontia: Macropodidae) from the Pleistocene of southern and eastern Australia. Records of the South Australian Museum 33:1-15.

Holotype. SAM P17249 / P17250, partial adult cranium, and partially fused left and right dentaries, collected by Brian Brawley and others in the late 1960s from Green Waterhole Cave, near Tantanoola, South Australia. Faunal composition suggests type locality is Pleistocene in age (Pledge 1980; Newton 1988).

Paratypes. Victoria Fossil Cave, near Naracoorte, SA: FU 0227, near-complete adult cranium; FU 0252, juvenile cranium devoid of occipital region, SAM P20255 / P16632 / P16633, partial adult cranium, left and right dentaries; SAM P28969, near-complete adult cranium; FU 0179, right dentary. Age of locality is late middle Pleistocene (Ayliffe et al. 1998; Grün et al. 2000).

Referred Specimens. Victoria Fossil Cave, near Naracoorte, SA: FU 0205, right juvenile dentary; FU 0226, right m3; FU 0259, left juvenile dentary; FU 0293, left adult dentary; FU 0887, left juvenile maxilla; FU 1084, left juvenile maxilla; SAM P16550, juvenile cranium; SAM P20243, left adult maxilla; SAM P20560, right juvenile dentary; SAM P27631, left i1, m1-2; SAM P28149, left adult dentary; SAM P28478, right p3; SAM P28479, right m1, m3; SAM P28518, left i1; SAM P28671, left juvenile dentary; SAM P28996, left juvenile dentary; SAM P32533, left m3; SAM P32541, left i1; SAM P32545, right juvenile dentary.

Haystall Cave, near Naracoorte, SA: SAM P36624, right dp2.

Henschke's Fossil Cave, near Naracoorte, SA: SAM P17641, left p3; SAM P17644, right m4; SAM P17837, partial left juvenile maxilla, right dP2, right M1; SAM P17885, left P3; SAM P18554, right M1 metaloph; SAM P34807, right P3; SAM P34808, left P3; SAM P34809, right P3; SAM P34810, right p3; SAM P34811, left P3; SAM P38788, dp2-3, p3, m1-2, right p3, m1; SAM P38789 and SAM P unreg., several single loose teeth.

SOS Cave, near Naracoorte, SA: SAM P33476, right adult maxilla.

Wet (Tomato-Stick) Cave, near Naracoorte, SA: Listed in Reed and Bourne (2000).

Comaum Forest Cave, near Penola, SA: SAM P31967, left and right premaxillae and maxillae.

Goulden's Hole, near Mount Schank, SA: SAM P36620, right P3; SAM P36621, left p3; SAM P36622, left M1; SAM P38780, left M4; SAM P38781, left m4.

Kilsby's Hole, Mount Gambier, SA: SAM P38782, left P3, SAM P38783, right lower molar, SAM P38784, left upper molar, SAM P38785, right dp2.

Black Rock Gravel Pit, near Orroroo, SA: SAM P23166, left and right adult dentaries.

Baldina Creek, near Burra, SA: SAM P21035, juvenile dentary.

Curramulka Quarry, Yorke Peninsula, SA: SAM P38786, right P3, SAM P38787, right p3.

Kiana Cliff, near Mount Misery, Eyre Peninsula, SA: FU unreg., P3.

McEachern's Cave, near Nelson, VIC: SAM P17319, right adult dentary; NMV P198434 / P198438 / P198440, left and right adult maxillae; NMV P198435 / P198436, left and right adult maxilla; NMV P198439, left adult dentary; NMV P198449, left adult maxilla; NMV P198450, right P3 in maxilla fragment.

Teapot Creek Site 12, Monaro Region, NSW: AM F115319, right adult dentary.

Wellington Caves, NSW: AM F18872, left P3.

Kandos Quarry, near Rylstone, NSW: AM F73721, left adult dentary.

Darling Downs, southeastern QLD: QM F2978, right adult maxilla.

Scotchtown Cave, near Smithton, TAS: QVM:1992:GFV:232, right m3; QVM:1992: GFV:238, right M3; QVM:1992:GFV:242, right M1.

Last Tree Cave, Nullarbor Plain, WA: WAM unreg..

Leaena's Breath Cave, Nullarbor Plain, WA: WAM unreg..

Madura Cave, Roe Plain, WA: FMNH PM4356, right p3.

Lindsay Hall Cave, Nullarbor Plain, WA: WAM 92.12.3, right M1; WAM 92.12.10, right M3; WAM 00.1.1, right juvenile maxilla.

Balladonia Soak, WA: WAM 63.11.2 / 63.11.3, right adult dentary.

Tight Entrance Cave, southwestern WA: WAM 97.5.308a, left juvenile dentary; WAM 97.5.312-314/ 97.5.142, left adult dentary.

Diagnosis. As for the genus until another species is described.

Etymology. Named after Cate A. Newton, who studied the fossil fauna of Green Waterhole Cave and first recognized the morphological uniqueness of the holotype.

Description and Comparison (pls 30-35; tables 11-12). Prideaux (2000) provided a detailed description and comparison of this species, but placed it in *Simosthenurus.* Because the species is separated here at the genus level, it is germane to reproduce the comparison from Prideaux (2000) with modifications. To enable adequate cross-comparison with the dimensions of other sthenurine taxa, craniodental measurements taken from Prideaux (2000) are also provided in tables 11-12.

Table 11. Dimensions of the adult cranium and dentary of *Metasthenurus newtonae,* showing mean, standard deviation (in parentheses), sample size [in brackets], and measurements of the holotype (SAM P17249 / P17250). After Prideaux (2000).

Dimension	Mean	Holotype
Condylobasal Length	216 (3.0) [n=3]	-
Cranial Width	132 (5.6) [n=4]	136
Palatal Length	132 (1.2) [n=3]	-
Diastema Length	40.3 (1.04) [n=4]	-
% Diastema Length: Palatal Length	30.5 (0.85) [n=3]	-
Palatal Width between M1 Protolophs	37.2 (2.27) [n=5]	39.0
Palatal Width between M4 Protolophs	39.2 (4.05) [n=3]	43.1
Width across Frontals	77.5 (3.05) [n=4]	82.0
Distance between Paroccipital Processes	76.3 (5.12) [n=4]	82.0
Dentary Depth	34.8 (3.35) [n=5]	39.3
Dentary Width	20.5 (1.20) [n=5]	20.6
Dentary Depth / Width	1.70 (0.13) [n=5]	1.91

Cranium. The cranial proportions of *Metasthenurus newtonae* are intermediate between the simosthenurins, on the one hand, and *Hadronomas* and *Sthenurus*, on the other. *M. newtonae* shares with the Simosthenurini both marked inflation of the frontals and elevation of the basicranial plane above the palatal plane. Among this tribe, *Si. occidentalis* is most similar to *M. newtonae*, but it has a shorter rostrum and diastema, and a more posterodorsally flexed basioccipital region. Despite the otherwise similar occipital and basicranial proportions of these two species, the direct effect of these differences is that the portion of the cranium posterior to the end of the maxilla in *M. newtonae* is relatively longer. Inflation of the frontal region and supraorbital crest development are less pronounced in *M. newtonae* than in *Si. occidentalis* and *"Procoptodon" browneorum*, but greater than in *"Si." baileyi* and *"P." gilli.* The masseteric process is shorter and much narrower than in *Si. occidentalis* and is closest in morphology to that of *Si. maddocki*, but more twisted posteriorly. Moderate development of the temporal crests is similar to *"Si." baileyi.* The shallow anteorbital and buccinator fossae, and reduced mesial curvature of the diastema border resemble *"Si." baileyi* as well as *"P." gilli*, but both of these species have much shorter rostra.

Upper Dentition. The general shape of the *Metasthenurus newtonae* I1 is similar to that of most simosthenurins. Overall, the tooth appears quite low crowned, resembling that of *"Procoptodon" browneorum.* Although the elongate I3 alveolus suggests a moderately elongate crown, no I3 may yet be ascribed to *M. newtonae*. The dP2 of *M. newtonae* is smaller than in southeastern *"P." browneorum* and *Simosthenurus occidentalis*, but larger than in *Si. maddocki* and *Sthenurus andersoni.* Morphologically, the tooth resembles *"P." browneorum* and *"Si." baileyi*, but it is less inflated posteriorly. Although somewhat variable in form, the P3 of most *M. newtonae* individuals is quite dissimilar to those of any other sthenurine species. This is especially so because, relative to its length, the tooth is usually quite wide anteriorly as well as posteriorly. The manner in which the posterior end of the lingual crest curves buccally to partially or wholly separate the longitudinal and posterior basins is only seen elsewhere in individuals of *S. andersoni* and a P3 fragment from the early Pliocene Bow LF of central eastern New South Wales (Flannery and Archer 1984:fig. 1A). Marked inflation of the anterobuccal corner of P3 is seen elsewhere within the Sthenurinae only in occasional individuals of some simosthenurin species.

The morphology of the upper molariform teeth of *Metasthenurus newtonae* is unique among sthenurines. While similar to *"Simosthenurus" baileyi* in size, the teeth are easily distinguished by being higher crowned, lacking any noteworthy enamel crenulations, and having cristae strongly connected to cusp apices. In these features, *M. newtonae* resembles *Hadronomas puckridgi*, but this taxon is lower crowned, has more weakly developed cristae, and the molars are generally squarer in occlusal outline. Although smaller, the upper molars of *Si. maddocki* and *"Procoptodon" gilli* bear some resemblance to those of *M. newtonae* in this latter feature and in the curved nature of the postparacrista and premetacrista. However, all molar cristae of *Si. maddocki* are more weakly developed than in *M. newtonae*, while the majority are more weakly developed in *"P." gilli. Si. maddocki* may also be distinguished by the many very fine enamel crenulations that cover its loph surfaces. Overall, the molars of *M. newtonae* are most similar to *S. andersoni* and *S. atlas*, but they differ by having a stronger connection between the preparacrista

Table 12. Cheek tooth dimensions of *Metasthenurus newtonae,* showing mean, standard deviation (in parentheses), and dimensions of the holotype (SAM P17249 / P17250), Wellington Caves and Kandos specimens (AM F18872, AM F73721). After Prideaux (2000).

Tooth	L	AW	PW	AH	PH	n
dP2	10.3 (0.91)	7.7 (0.51)	9.6 (0.97)	7.1 (0.47)	7.8 (0.81)	3
dP3	11.5 (0.20)	9.8 (0.66)	10.9 (0.42)	6.2 (0.85)	7.0 (1.15)	3
P3	16.6 (0.72)	10.5 (0.82)	12.3 (0.52)	10.4 (0.61)	10.1 (0.86)	15
P3 holotype	16.9	10.6	12.9	11.0	10.8	
AM F18872	18.3	10.0	13.4	10.3	9.8	
M1	13.2 (0.32)	12.9 (0.50)	12.6 (0.38)	7.7 (0.79)	7.9 (0.71)	11
M1 holotype	13.2	13.1	12.8	8.3	8.0	
M2	14.5 (0.60)	13.9 (0.82)	13.0 (0.65)	8.2 (0.71)	8.4 (0.71)	12
M2 holotype	14.7	13.7	13.0	9.0	8.7	
M3	15.0 (0.59)	13.7 (0.53)	12.9 (0.63)	7.9 (0.48)	7.9 (0.37)	12
M3 holotype	15.0	13.8	13.3	8.3	8.3	
M4	14.1 (0.67)	12.9 (0.36)	11.1 (0.42)	6.6 (0.55)	6.2 (0.52)	8
M4 holotype	14.8	13.0	11.5	7.5	6.6	
dp2	9.5 (0.68)	6.1 (0.49)	8.1 (0.31)	7.9 (0.95)	7.1 (0.83)	6
dp3	10.4 (0.21)	8.3 (0.57)	8.8 (0.53)	7.1 (0.89)	7.2 (0.90)	6
p3	15.3 (0.74)	7.6 (0.43)	9.3 (0.33)	10.1 (0.79)	9.4 (1.16)	16
p3 holotype	15.3	7.7	9.7	10.0	10.1	
AM F73721	17.5	8.4	9.9	-	-	
m1	13.1 (0.67)	10.2 (0.37)	10.5 (0.35)	9.2 (0.98)	9.7 (1.47)	14
m1 holotype	13.1	10.1	10.7	9.1	9.7	
AM F73721	14.3	11.4	11.7	-	-	
m2	15.2 (0.61)	11.3 (0.38)	11.6 (0.34)	10.0 (1.38)	10.4 (1.35)	12
m2 holotype	15.7	11.5	11.7	9.2	9.6	
AM F73721	16.0	12.5	13.0	-	-	
m3	16.3 (0.45)	12.1 (0.29)	12.3 (0.42)	9.5 (1.43)	9.4 (1.25)	11
m3 holotype	16.6	12.1	12.4	9.5	9.5	
AM F73721	17.3	13.5	13.8	10.1	11.7	
m4	14.6 (0.79)	11.5 (0.28)	10.4 (0.44)	8.4 (0.74)	7.3 (0.55)	8
m4 holotype	15.8	11.3	10.7	8.6	7.3	
AM F73721	17.0	13.0	12.2	10.8	10.3	

and the paracone apex, a stronger premetacrista, a stronger lingual postmetaconulecrista component, and no continuation of the precingulum beyond the position of the latent preprotocrista. Although the condition where molars show this precingulum morphology is unique to *M. newtonae*, the same morphology is observed on the dP3 of *S. andersoni* and some molariform teeth of *Si. maddocki.*

Dentary. The *Metasthenurus newtonae* dentary is most similar in size and morphology to *"Procoptodon" browneorum*, but the digastric eminence of the latter species is usually larger, as is the gradient of decreasing dentary depth anteriorly. In addition, the masseteric fossa is longer in *"P." browneorum*, and the anterior root of the ascending ramus lies opposite the m3-m4 boundary or m4 protolophid. In contrast, the anterior root in *M. newtonae* leaves the ramus adjacent or just posterior to the m4 hypolophid. The slightly procumbent form of the *M. newtonae* diastema is similar only to that of *Sthenurus tindalei* and *S. stirlingi.* As in the simosthenurins and *S. stirlingi*, the symphysis of *M. newtonae* extends posteriorly below the genial fossa. The generally slender form of the symphyseal plate most resembles that of *"P." browneorum*, but the inferior transverse torus does not extend posteriorly to the same degree, and the dorsal border of the superior transverse torus is near horizontal rather than anterodorsally oriented. The narrow and deep median dorsal groove of *M. newtonae* is similar to that observed in *Simosthenurus maddocki.* Viewed posteriorly, the medial pterygoid fossa of *M. newtonae* is taller and narrower than in any other sthenurine species. Marked development of the processes overhanging the mylohyoid groove is similar to that of *"P." oreas*, but the groove in the latter species is deeper and narrower.

Lower Dentition. In size and morphology, the i1 of *Metasthenurus newtonae* is intermediate in morphology between *Simosthenurus occidentalis* and *Sthenurus andersoni.* In this sense, the tooth resembles the i1 of *"Si." pales*, but is much smaller. Size and general outline of the dp2 are comparable with those of *Si. occidentalis*, but the *M. newtonae* dp2 is relatively narrower anteriorly. The brevity of the buccal crest is similar to that observed in *Si. maddocki*, but the dp2 of this species is narrower and the main crest cuspules are more distinct. Although the protolophid base of the *Metasthenurus newtonae* dp3 is narrower than the hypolophid base, the unworn crest of the protolophid is much narrower, similar to that of *"Simosthenurus" pales. M. newtonae* appears to retain a paraconid lingual to the paracristid in the anterolingual corner of the trigonid. A similar cusp may be observed on the dp3 of other sthenurines, such as *Hadronomas puckridgi*, *Sthenurus atlas* and *"Procoptodon" gilli.* Aside also from the smoothly continuous nature of the cristid obliqua and hypolophid, the morphology of the *M. newtonae* dp3 is similar to the succeeding molars.

As with the P3, the anterior width of the *M. newtonae* p3 is not greatly exceeded by the posterior width of the tooth. Overall, the p3 of *M. newtonae* is most alike that of *"Si." brachyselenis* in size and general morphology, but it differs by being longer relative to the molars. Compared with the also similar *"Si." tirarensis*, *M. newtonae* is distinguished by being wider relative to its length and lacking the unique semicircular buccal crest form characteristic of this species. While the anterior half of the p3 in several *M. newtonae* specimens (including the holotype, pl. 30C) is especially reminiscent of *"Si." brachyselenis*, *"Si." tirarensis* and, to a lesser degree, *Si. euryskaphus*, other p3 specimens of *M. newtonae* narrow more gradually anteriorly (pls 34E, 35E).

Within the Sthenurinae, the morphology of the *Metasthenurus newtonae* lower molars is unique. In some respects, their form closely resembles that exhibited by the species of *Sthenurus*, especially *S. andersoni.* Similarities include minimal division of the paracristid and cristid obliqua, paucity of enamel crenulations on the lophid faces and, to a lesser degree, anterior turn of the lophid ends, which is more pronounced in *M. newtonae.* Incipient division of the cristid obliqua into two components is evident in most individuals of *M. newtonae* and *S. andersoni*, especially posteriorly. The more anterolingually oriented buccal component maintains a direct connection to the hypoconid apex and terminates on the buccal side of the lingual component, which is oriented more anteroposteriorly and extends into the interlophid valley. Among the simosthenurins, the lower molars of *M. newtonae* most resemble those of *"Procoptodon" gilli* in their paucity of enamel crenulations, but they are easily separated by their larger size, lower crown height, markedly inflated posterior hypolophid face, more curved anterior portion of the paracristid, and the presence of only an incipient division of the paracristid and cristid obliqua. A curved anterior portion of the paracristid is also seen in the higher-crowned *S. notabilis*, as well as in *"Si." tirarensis*, *Simosthenurus euryskaphus* and many individuals of *Si. occidentalis.* The *Simosthenurus* species are easily distinguished by the more marked division of their paracristid and cristid obliqua, distinct enamel crenulations on the lophid faces, and lophid ends that are less markedly turned anteriorly. With regard to this latter character and the marked posterior inflation of the hypolophid, *M. newtonae* is easily distinguished from all other sthenurines.

Geographic Distribution. Southwestern Australia, Nullarbor Region, southern and mid-northern South Australia, northwestern Tasmania, southern Victoria, eastern New South Wales, southeastern Queensland (fig. 29; table 44).

Temporal Distribution. Middle and late Pleistocene.

SIMOSTHENURINI tribe nov.

Type Genus. Simosthenurus Tedford, 1966.

Other Included Genera. Archaeosimos gen. nov., *Procoptodon* Owen, 1874b.

Diagnosis. Cranium brachycephalic. Postprotocrista division incipient. Mesocrista moderately developed. Lower incisor upturned, short and robust. Trigonid short. Division of paracristid and cristid obliqua incipient anteriorly grading to clear posteriorly.

ARCHAEOSIMOS gen. nov.

Simosthenurus, Pledge, 1992, pp. 133-137, fig. 19.
Sthenurus, Prideaux and Wells, 1997 (in part), pp. 181-196, fig. 3.
Sthenurus, Prideaux and Wells, 1998 (in part), pp. 4, 11-15.
Eosthenurus, Prideaux, 1999 (*nomen nudum*), p. 69A.
'Simosthenurus', Prideaux, 2000, p. 13.

Type Species. *Simosthenurus cegsai* Pledge, 1992.

Generic Diagnosis. Small simosthenurins with very low crowned, simple molars. Narrow p3 only slightly wider posteriorly than anteriorly, with very short buccal crest.

Etymology. Gr. *archaios* 'ancient, old, beginning', *simos* 'flat-nosed, pug-nosed'; in reference to the early appearance of this taxon in the fossil record and its plesiomorphic nature relative to the other simosthenurins.

Geographic Distribution. Yorke Peninsula, South Australia (fig. 27).

Temporal Distribution. Latest Miocene.

Archaeosimos cegsai (Pledge, 1992)

(pls 36-37; figs 9D-E, 12D, 18F, 19E; table 13)

Simosthenurus cegsai, Pledge, 1992, pp. 133-137, fig. 19.
Sthenurus cegsai, Prideaux and Wells, 1997, pp. 181-196, fig. 3.
Sthenurus cegsai, Prideaux and Wells, 1998, pp. 2-4, 11-15, table 1.
'Simosthenurus' cegsai, Prideaux, 2000, p. 13.

Taxonomic Authority. Pledge, N. S. 1992. The Curramulka Local Fauna: a new late Tertiary fossil assemblage from Yorke Peninsula, South Australia. The Beagle, Records of the Northern Territory Museum of Arts and Sciences 9:115-142.

Holotype. SAM P31800, left adult dentary (containing p3, m1-4, missing i1, most of ascending ramus) collected in the mid-1980s by Graham Pilkington and other members of the Cave Exploration Group of South Australia from Corra-Lynn (Correll's) Cave, near Curramulka, Yorke Peninsula, South Australia. Age of the type locality is probably latest Miocene, based on the stage of evolution and composition of the fauna (Pledge 1992; Tedford 1994; The Sthenurine Radiation).

Referred Specimens. Corra-Lynn Cave, Yorke Peninsula, SA: SAM P29872, left i1; SAM P29917, left adult maxilla; SAM P29930, right I3; SAM P30027, right adult maxilla; SAM P30153, dP3; SAM P31335, right adult dentary.

Diagnosis. Larger than *Archaeosimos correlli*, with proportionally wider ramus and shorter diastema. Compared with that of *A. correlli*, p3 is longer relative to width, has a slightly narrower median valley, and is round-ended and slightly bulbous in outline anteriorly, centrally and posteriorly.

Etymology. Named in honor of the Cave Exploration Group of South Australia (CEGSA) for discovering and collecting the fossils from the type locality.

Description. Maxilla (pl. 37A). Little of descriptive value discernable from maxillary fragments (SAM P29917, SAM P30027), except that masseteric process is stout and laterally projected.

Upper Dentition. I3 (SAM P29930, pl. 37C-D) attributed to *Archaeosimos cegsai* on basis of relatively large size and robust proportions. Crown height and length dimensions very similar. Buccal surface smooth, posterior edge concave posteriorly (pl. 37C-D). Low, posterodorsally oriented anterolingual crest evident at anterior end of lingual side of tooth (pl. 37D). Completely molariform dP3 very low crowned; quite different in general morphology from M3 (anteriormost upper molar known). Although unworn crests of both lophs are level when viewed lingually, metaloph crown is markedly higher overall due to deeper base. Vertical axis of metaloph is slightly tilted posteriorly relative to protoloph (in lingual view). Metaloph slightly wider than protoloph. Preparacrista is shifted slightly dorsolingually from paracone apex and continuous with very thin precingulum, which sweeps around from lingual side of preprotocrista fading into low vertical crest on lingual side of protocone. Protoloph crest markedly convex anteriorly, smoothly continuous lingually with postprotocrista, which ascends posterobuccally from protocone apex into interloph valley. It becomes fine and terminates before reaching metaconule apex. Postparacrista is oriented directly posteriorly from paracone apex and meets with similarly developed premetacrista. Two other cristae ascend from paracone apex toward center of posterior face of protoloph, with smaller, more buccal crista terminating before reaching interloph valley. Larger, more lingual crista represents mesocrista, and it curves posteriorly from buccal side of protoloph into interloph valley. Although flattened by wear, it appears confluent with a tiny crest on center of anterior face of metaloph. Metaloph crest much straighter than protoloph crest. Low, but distinct postmetaconulecrista ascends from metaconule apex across metaloph posterior face and terminates on buccal side, meeting with sinusoidal postmetacrista ascending from metacone apex. Small vertical crest buccal to postmetacrista on posterobuccal corner of metacone is most likely posterior portion of stylar crest. Small urocrista ascends vertically from metaloph crest in center of metaloph posterior face, terminating just inside of horizontal aspect of postmetaconulecrista, adjacent to more mesial of two postmetaconulecrista components.

Upper molars (pl. 37A; table 13) very low crowned, with metaloph slightly narrower than protoloph on M3, significantly narrower on M4. Preparacrista not smoothly connected to paracone apex; anterobuccal corner of preparacrista is expanded into cusp-like structure, probably corresponding to stylar cusp A. Precingulum smoothly confluent with preparacrista, thinning on lingual side before terminating at anterolingual corner of tooth. Postparacrista only fairly developed, but stronger than premetacrista (pl. 37A). Very small, low mesocrista descends from center of protoloph crest before terminating on side of thick, low, incipiently divided postprotocrista, which trends posterobuccally from protocone apex into interloph valley, before becoming much finer toward metaloph crest (pl. 37A). Low postmetaconulecrista terminates as short lip overhanging middle of precingulum on succeeding molar (pl. 37A). Postmetacrista is also low. Very low, barely discernable cristae or enamel crenulations evident on posterior faces of protoloph and metaloph, more central of which most likely represents urocrista.

Dentary. Described from holotype, SAM P31800 (pl. 36). Adult dentary small and short. Ramus wide relative to depth, increasingly so posteriorly. Diastema very short, slightly concave in lateral view (pl. 36A). Straight profile of ventral surface of ramus suggests a diminutive symphyseal boss. Orientation of partial alveolus indicates i1 markedly upturned (pl 37B). Profile of broken anterolingual aspect of ramus suggests partial fusion of symphysis, which extended no farther posteriorly than beneath halfway point of p3. Anterior mental foramen lies just below halfway point of diastema. Buccinator sulcus straight, narrow and shallow for most of length, but deeper at posterior end. It extends from between anterior mental foramen and p3 in direction of posterior mental foramen, terminating beneath m1 hypolophid. Posterior mental foramen is mid-depth on ramus below m2 interlophid valley. Deep digastric sulcus extends to beneath m2 interlophid valley; digastric eminence slight (pl. 37B). Anterior root of ascending ramus adjacent to m4 protolophid; anterior edge of ascending ramus angled slightly anteriorly relative to longitudinal axis of ramus. Ventral border of masseteric fossa lies just above level of posterior mental foramen, flaring slightly laterally above marked indentation. Masseteric fossa quite deep; masseteric foramen quite large for dentary size and leads into vertical masseteric canal (pl. 36C). Elliptical inferior mandibular foramen opens posteriorly.

Table 13. Cheek tooth dimensions of *Archaeosimos cegsai*.

Tooth	Specimen	L	AW	PW	AH	PH
dP3	SAM P30153	9.3	8.6	9.2	6.3	6.3
M3	SAM P29917	13.0	11.5	10.8	5.4	5.6
	SAM P30027	13.1	-	11.0	5.4	5.4
	Mean	13.1 (0.07)	11.5	10.9 (0.14)	5.4 (0.00)	5.5 (0.14)
M4	SAM P29917	12.2	11.0	9.5	5.5	5.1
p3	Holotype	14.8	6.5	7.2	8.3	7.1
m1	Holotype	10.5	-	-	-	-
m2	Holotype	12.7	10.8	10.3	5.9	5.9
m3	Holotype	12.8	11.0	10.2	7.0	6.4
	SAM P31335	13.5	10.8	10.9	7.0	7.6
	Mean	13.2 (0.49)	10.9 (0.14)	10.6 (0.49)	7.0 (0.00)	7.0 (0.85)
m4	Holotype	11.9	10.3	8.6	5.3	5.3

Lower Dentition (table 13). Short, moderately robust i1 (SAM P29872, pl. 37E-F) bears sinusoidal unworn occlusal surface and very thin enamel on lingual surface. SAM P29872 is attributed to *Archaeosimos cegsai* on basis of its size and simosthenurin morphology (contemporaneous *A. correlli* has a narrower i1 judging from partial alveolus of SAM P38801, pl. 38). Extension of buccal enamel forms distinct dorsal flange (pl. 37E). Relative to molars, p3 is elongate, round-ended and very narrow for length, with posterior third only slightly wider than anterior two-thirds (pl. 36). It is slightly bulbous anteriorly, centrally and posteriorly. Main p3 crest divided into five cuspules; apex of posteriormost cuspule twice length of any of anterior four. Low but thick vertical ridgelets descend from apices of anterior three cuspules on buccal and lingual sides. Less pronounced vertical ridgelet descends from fourth cuspule on lingual side only. Buccal crest of very short and apparently quite straight, although wear has resulted in breaching of enamel, with D-shaped facet formed atop crest (pl. 36C). Buccal crest and posterior third of main crest oriented posterolingually, toward m1 metaconid. Median valley very narrow and shallow.

Lower molars very low crowned (pl. 36A-B), with protolophid and hypolophid crests linear and close to parallel (pl. 36C). Only known m1 too worn to discern any topographic features. Paracristid and cristid obliqua of m2-4 very low and fine, with lingual cristid components shifted slightly across from protoconid and hypoconid apices. Buccal components indistinct. Trigonid very short (pl. 36C). Precingulid forms rounded eminence that extends under transverse anterior portion of paracristid. Small, right-angled notch marks lingual extremity of paracristid. Few extremely fine enamel crenulations present on protolophid anterior face, lingual to paracristid. No enamel crenulations discernable on hypolophid anterior face (pl. 36C). Hypolophid posterior face bears low, shallow inflation.

Variation. Too few specimens are known to document much intraspecific variation, but the m3 of SAM P31335 (pl. 37B) does have a slightly more pronounced posteroventral inflation of the hypolophid than in the holotype, SAM P31800 (pl. 36C). In addition, the posterior mental foramen is larger and located beneath the m3 hypolophid in SAM P31335, as opposed to beneath the m2 interlophid valley in the holotype. A tiny subsidiary foramen is present anterior to the large posterior mental foramen in SAM P31335, beneath the anterior extremity of the m3 anterior root. No such feature is present in the holotype.

Comparison With Other Taxa. Maxilla. What little is preserved of the small, stout masseteric process of *Archaeosimos cegsai* indicates that it was better developed than in *Hadronomas puckridgi*, *"Procoptodon" gilli* and most *Sthenurus* species. However, it would have been relatively smaller than that of *Metasthenurus* and most *Simosthenurus* species. Positioning of the masseteric process adjacent to M2 is typical of the simosthenurins and correlates with the postulated shortness of the cranium.

Upper Dentition. *Archaeosimos cegsai* has lower-crowned molars than any other sthenurine. In morphology, they bear some resemblance to *Hadronomas puckridgi*, from which they differ by having a precingulum that thins more toward its lingual end before terminating, an incipient mesocrista and incipiently divided postprotocrista. Overall, they are more similar to those of *Simosthenurus s. l.* Among these species, the upper molars of *A. cegsai* resemble those of *"Si." antiquus*, but the precingulum is thinner and division of the postprotocrista is slightly less distinct. In these respects, *A. cegsai* is more like *"Si." baileyi*, but the latter species has a much stronger postparacrista and better-developed enamel crenulations.

Dentary. In overall dentary size and proportion, *Archaeosimos cegsai* is most similar to *"Procoptodon" gilli*, but the *A. cegsai* ramus is proportionally wider. The dentary of *A. correlli* differs by being slightly smaller (~10%), proportionally narrower (depth / width is 1.78 versus 1.43 for *A. cegsai*), and bearing a shorter diastema. Although the *A. cegsai* dentary also bears some resemblance to that of smaller individuals of *Hadronomas puckridgi* (e.g., NTM P98135, pl. 4A-B), it is proportionally much shorter and wider (depth / width is 1.43 versus 2.18 for *H. puckridgi*). The straightness of its ventral border is reminiscent of *Sthenurus andersoni* and *Simosthenurus maddocki*, while a very short diastema typifies several other brachycephalic simosthenurins.

Lower Dentition. The degree to which the *Archaeosimos cegsai* i1 is upturned is similar to that of several simosthenurins, including *"Procoptodon" gilli*, *"Simosthenurus" brachyselenis* and *Si. occidentalis*. The size and the short, moderately robust form of the i1 crown is intermediate between that of *Hadronomas puckridgi* and other simosthenurins. Compared with that of *H. puckridgi*, the *A. cegsai* i1 has a slightly more pronounced dorsal enamel flange, a slightly less pronounced ventral enamel flange, and is proportionally wider, although the lingual enamel layer is similarly very thin.

The elongate, round-ended and slightly bulbous form (anteriorly, centrally and posteriorly) of the p3 easily distinguishes *Archaeosimos cegsai* from all other sthenurines. *A. correlli* is similar with regard to the minimal inflation of the posterobuccal portion of the crown, which results in the p3 being only slightly wider posteriorly, but the *A. cegsai* p3 is proportionally longer and the median valley narrower. While the buccal crest is also very short in *A. correlli*, *"Simosthenurus" brachyselenis*, and most individuals of *Metasthenurus newtonae*, it is shorter relative to p3 length in *A. cegsai*. The form of the main crest, including its general shape, division into cuspules and development of the vertical ridgelets from the more anterior cuspules, is extremely similar between *A. cegsai*, *Hadronomas puckridgi* and *"Si." brachyselenis*.

The lower molars of *Archaeosimos cegsai* are intermediate in size between the mean length and width dimensions for *Simosthenurus maddocki* and *"Si." baileyi*, m2-3 of the former 7% smaller and the latter 8% larger than in *A. cegsai*. The m2-3 of *A. cegsai* are 19% larger than in *A. correlli*, but in crown height and overall molar morphology, these two species closely resemble one another. *A. cegsai* differs by having a shallower inflation on the posterior face of the hypolophid, and it lacks a preentocristid, postmetacristid and well-developed parametacristid. The lower molars of *A. cegsai* are quite similar to those of *"Si." antiquus* and *"Si." baileyi* but are lower crowned, with fewer enamel crenulations, and the paracristid and cristid obliqua are only incipiently separated into two components anteriorly. *A. cegsai* also lacks the broad inflation of the posterior hypolophid face typical of *"Si." antiquus* and the distinct premetacristid and buccal component of the paracristid seen in *"Si." baileyi*.

Geographic Distribution. Yorke Peninsula, South Australia (fig. 27).

Temporal Distribution. Latest Miocene.

Archaeosimos correlli sp. nov.

(pl. 38; table 14)

Holotype. SAM P38801, partial left juvenile dentary (containing dp2-3, m1-4, excavated p3; missing i1, most of ascending ramus and pterygoid fossa) collected in 2000 by Paul Harper and Frank Hankinson of the Cave Exploration Group of South Australia from Corra-Lynn (Correll's) Cave, near Curramulka, Yorke Peninsula, South Australia. Age of the type locality is probably latest Miocene, based on the stage of evolution and composition of the fauna (Pledge 1992; Tedford 1994; The Sthenurine Radiation).

Referred Specimens. Corra-Lynn Cave, Yorke Peninsula, SA: SAM P29900, partial left adult dentary (edentulous); SAM P31340, left I3.

Diagnosis. Smaller than *Archaeosimos cegsai*, with proportionally narrower ramus and longer diastema. Compared with that of *A. cegsai*, p3 is shorter relative to width and has a slightly wider median valley. Lower molars bear well-developed parametacristid and fine, low preentocristid connected to fine, low postmetacristid.

Etymology. In reference to the original name of the type locality and in honor of Donovan H. Correll, former owner of the Corra-Lynn property, who facilitated the initial surveying by the Cave Exploration Group of South Australia that led to the discovery of the Curramulka LF.

Description. Upper Dentition. I3 (SAM P31340; pl. 39D-E) attributed to *Archaeosimos correlli* on basis of relatively small size and less robust proportions. Crown narrows toward base, with posterior edge convex and anterior edge similarly shaped, except for pointed projection at anterior end of unworn occlusal surface. Although much of lingual surface of crown is broken off, anterolingual crest is very distinct (pl. 39E). Buccal surface of I3 smooth, except for low, rounded and narrow crest, which ascends full height of crown at anterior end (pl. 39 D).

Dentary (pls 38-39). Ramus small, short and relatively gracile (depth 26.4 mm, width 14.8 mm), but given p3 remains unerupted in holotype (SAM P38801), an adult might be slightly deeper. Dentary depth gradually increases posteriorly from behind distinct symphyseal boss to estimated midpoint of digastric eminence, posterior to which most of dentary is broken off. Ramus quite narrow relative to depth (pls 38-39). Diastema straight and moderately short, as judged from referred specimen (SAM P29900) and from orientation of deepest portion of i1 alveolus in holotype. Symphyseal plate rugose, but slender dorsoventrally, with poorly developed inferior transverse torus (pl. 38B). Symphysis extends posteriorly to beneath dp3 protolophid. Anterior mental foramen lies just below diastema, well anteroventral of dp2 in holotype, but lies further below diastema in SAM P29900, with posterior border of foramen directly beneath anterior border of p3 anterior root alveolus. In holotype, buccinator sulcus is straight and shallow for entire length, extending from immediately behind anterior mental foramen to beneath m1 protolophid (pl. 38A). In SAM P29900, buccinator sulcus is deeper, especially posteriorly, and trends from between p3 anterior alveolus and anterior mental foramen to beneath anterior border of m1 posterior alveolus (pl. 39A). Posterior mental foramen positioned beneath m2 interlophid valley, just below mid-depth on ramus. Digastric sulcus moderately shallow, extending anteriorly to beneath m3 hypolophid (pl. 38B). Digastric eminence appears to have been moderately deep. Anterior root of ascending ramus adjacent to m4 protolophid. Laterally flared ventral border of masseteric fossa is aligned with level between that of posterior mental foramen and alveolar margin of cheek tooth row. Masseteric foramen moderately large, and leads into vertical masseteric canal (pl. 38C).

Lower Dentition (pl. 38; table 14). Second lower deciduous premolar (dp2) short relative to width. Main crest bears four cuspules, and runs obliquely across tooth from posterolingual corner to central position anteriorly. Buccal crest crescentic, tiny, low, and in midline of dp2 (pl. 38C). Median valley very narrow and shallow. It joins anteriorly to ridgelet directed buccally from third main crest cuspule, and posteriorly to end of main crest. Primary wear facet restricted to buccal surface of tooth, from side of large anteriormost cuspule of main crest around to buccal side of buccal crest. Protolophid of molariform dp3 narrower than hypolophid, tapering markedly toward relatively narrow crest (pl. 38C). Although wear has obliterated several features, trigonid is very short overall and precingulid very small. Paracristid apparently undivided, curving anterolingually from protoconid apex before turning lingually to terminate at anterolingual corner, where it is met by a very low, premetacristid. Distinct, although low (due to wear) parametacristid curves from metaconid apex to meet paracristid midway between protolophid apex and anterior edge of tooth. Cristid obliqua undivided, apparently terminating on buccal side of protolophid posterior face (pl. 38C). Although worn, a fine, low preentocristid appears to have been confluent with a fine, low postmetacristid on lingual side of interlophid valley. Hypolophid posterior face smooth and featureless besides attrition facet from abutting m1 trigonid. Lophid faces lack enamel crenulations.

Narrow p3 only slightly wider posteriorly than anteriorly (pl. 38D-F), and easily exceeds each of molars in length. Main crest is divided into six cuspules. Thick vertical ridgelets descend from apices of anterior three cuspules on both

Table 14. Cheek tooth dimensions of the *Archaeosimos correlli* sp. nov. holotype (SAM P38801).

Tooth	L	AW	PW	AH	PH
dp2	7.1	5.0	5.8	5.1	5.1
dp3	8.2	6.1	7.0	3.4	4.3
p3	13.5	6.4	6.9	8.2	7.0
m1	10.2	7.6	8.0	5.5	5.5
m2	11.0	8.6	8.5	5.9	6.3
m3	11.3	9.3	9.0	6.5	6.6
m4	11.3	9.4	8.7	5.3	5.6

buccal and lingual sides, while smaller ridgelets descend from each of fourth and fifth cuspules on lingual side only. Buccal crest of p3 very short, slightly curved, and extending anteriorly to buccal side of fourth main crest cuspule (pl. 38D-E). Fine low ridgelet links anterior extremity of buccal crest to base of buccal ridgelet descending from third main crest cuspule. Median valley narrow, especially posteriorly, and is filled with four enamel ridgelets, two of which descend anterolingually from buccal crest (pl. 38D).

Very low-crowned lower molars with size gradient $m1<m2<m3\approx m4$, with little variation in morphology (pl. 38A-C). Molar row slightly curved laterally; protolophid and hypolophid crests very slightly convex posteriorly. Paracristid and cristid obliqua low and moderately fine, with division into two components incipient anteriorly grading to clear posteriorly (pl. 38C). Buccal components barely visible. Trigonid very short, with precingulid little more than a rounded eminence. Small, 120° notch marks lingual extremity of paracristid. Distinct parametacristid curves anteriorly from metaconid apex then buccally, terminating one quarter of way across protolophid anterior face, directly posterior to paracristid terminus (pl. 38C). Two extremely fine and low enamel crenulations descend into trigonid basin from protolophid anterior face. Fine, low preentocristid curves into interlophid valley before meeting similarly developed postmetacristid, which descends posterobuccally from metaconid apex (pl. 38C). Several extremely fine enamel crenulations descend onto rounded inflation that extends two-thirds of way up hypolophid posterior face from crown base.

Comparison With Other Taxa. Upper Dentition. The I3 attributed to *Archaeosimos correlli* is similar to, but slightly smaller and less robust than that attributed to *A. cegsai*. It also narrows more toward the crown base, and has a slightly more concave posterior edge and a distinct pointed projection at the anterior end of the unworn occlusal surface, which *A. cegsai* lacks.

Dentary. The relatively gracile nature of the *Archaeosimos correlli* ramus is reminiscent of similarly aged individuals of *"Procoptodon" gilli* and some individuals of *Sthenurus andersoni*, but is clearly more similar to the former both in size, brevity and lateral flaring of the masseteric fossa ventral border. Compared with the ramus of *A. cegsai*, it is smaller and proportionally narrower. Although the holotype and most complete dentary specimen of *A. cegsai* is broken anteriorly, it appears, from the contour of the ventral surface, to have had a less pronounced symphyseal boss. The digastric eminence is also better developed in *A. correlli*, and its diastema is longer.

Lower Dentition. Out of those sthenurines for which the tooth is known, *Archaeosimos correlli* most closely resembles *Metasthenurus newtonae* and *"Simosthenurus" brachyselenis* in dp2 morphology, but its dp2 is relatively narrower posteriorly, the buccal crest is shorter, and the median valley is shallower and narrower. The dp3 of *A. correlli* is very similar in morphology to that of *M. newtonae*, particularly in the relative narrowness of the protolophid crest, diminutive precingulid, undivided paracristid and cristid obliqua, slight preentocristid and postmetacristid, and absence of enamel crenulations. It differs by having a distinct parametacristid, which extends from the metaconid apex to the point where the paracristid inflects from an anterolingual to more anterior direction. This is reminiscent of *Si. occidentalis* and, to a lesser degree, *"Si." pales*.

The lower molars of *Archaeosimos correlli* are smaller, but otherwise very similar to those of *A. cegsai*. They differ by bearing a well-developed parametacristid, and fine, low preentocristid confluent with a similarly developed postmetacristid, the latter arrangement unique within the Sthenurinae. A short, fine postmetacristid is evident on the *Simosthenurus maddocki* m1, but not on m2-4. In size and topographic morphology, the lower molars of *A. correlli* are also distinctly similar to molar fragments referred herein to cf. *Archaeosimos* sp. While they are indistinguishable in hypolophid form, morphology of the *A. correlli* protolophid and trigonid is intermediate between that of the Kanunka and Hamilton specimens, but both of these lack a postmetacristid. *A. correlli* is more similar to the Hamilton specimen in parametacristid form, but its cristid is shorter in the former.

Geographic Distribution. Yorke Peninsula, South Australia (fig. 27).

Temporal Distribution. Latest Miocene.

cf. *Archaeosimos* sp.

(pl. 39F-H; table 15)

Simosthenurus "Hamilton molar", Flannery, 1989, pp. 29-30.

Simosthenurus sp., Flannery, Rich, Turnbull and Lundelius, 1992, p. 12, figs 4D-E, 5C-D, table 1.

Referred Specimens. Grange Burn, Hamilton, VIC: NMV P158449, anterior half of right ?m2; NMV P158633, left ?m2 talonid.

Stirton Quarry, Lake Kanunka, SA: UCR 22656, right ?m2.

Description. Lower Dentition. NMV P158449 and NMV P158633 were described in Flannery (1989:29) and Flannery et al. (1992), so here I describe only UCR 22656 (pl. 39F-H; table 15). Low-crowned ?m2 with protolophid and hypolophid crests parallel, but oriented slightly anterolingually relative to anteroposterior axis of tooth (pl. 39H). Protolophid slightly wider than hypolophid (table 15). Buccal ends of lophid crests distinctly curved anteriorly; lingual ends only slightly so. Paracristid and cristid obliqua very low and fine. Posterior portion of lingual component of paracristid oriented anterolingually, overlapping for one-third of length with similarly oriented buccal component of paracristid (pl. 39H). Trigonid short, narrowing markedly anteriorly, and oriented rather anterolingually, such that anterior edge of molar set obliquely (posterolingually) relative to anteroposterior axis of tooth. Short anterior portion of paracristid not differentiable from precingulid, which is restricted to anterior end of molar, extending buccally only just past tooth midline (pl. 39H). Two very fine, vertical enamel crenulations wedged between paracristid and fine, but distinct parametacristid. Extremely low premetacristid terminates immediately posterolingual to position of paraconid. Posterior extremity of lingual component of cristid obliqua arises one quarter of way across anterior hypolophid face, overlapping slightly with very low buccal component, which terminates well above bottom of interlophid valley (pl. 39H). Preentocristid extremely fine and low. Enamel crenulations not discernable on hypolophid anterior face, but several very fine, vertical enamel crenulations descend onto or confluent with low, but distinct A-shaped inflation on hypolophid posterior face (pl. 39F-H).

Variation. Based on all hypolophid features, it is not possible to distinguish UCR 22656 from NMV P158633. Although there is a difference in crown height (table 15), this cannot be accurately estimated for the Hamilton specimen, because enamel has been broken off from its base. Therefore, any height discrepancy is likely to have been marginal. Protolophid height of the second Hamilton specimen (NMV P158449) is certainly lower than that of the Kanunka specimen, even though its base is also not complete. In fact, it is also lower than what one might expect to match the height of NMV P158633 if they both pertain to the same tooth position. The greater width of the lower-crowned protolophid (table 15) supports the idea that they do not, and suggests that the two Hamilton fragments may not be conspecific. UCR 22656 and NMV P158449 are similar in protolophid and trigonid morphology, but the latter has a slightly more anterobuccally expanded precingulid, slightly greater division of the paracristid, and less anteriorly turned lophid ends, especially lingually, where the parametacristid is better developed, trending first anteriorly then more buccally across the protolophid face.

Comparison With Other Taxa. The partial lower molars closely resemble *Archaeosimos correlli* in size, proportion and topographic morphology, and to a lesser degree, *Simosthenurus maddocki*. While the morphology of the entire talonid region of UCR 22656 and NMV P158633 is essentially indistinguishable from that of *A. correlli*, this tends to be the more conservative moiety of the sthenurine lower

Table 15. Lower molar dimensions of cf. *Archaeosimos* sp. Dimensions for the Hamilton specimens are from Flannery et al. (1992).

Tooth	Specimen	L	AW	PW	AH	PH
?m2	UCR 22656	12.8	9.1	8.8	7.7	8.0
?m2	NMV P158663	-	-	8.5	-	7.3
?m2	NMV P158449	-	9.1	-	5.9	-

molar. *A. correlli* is intermediate between UCR 22656 and NMV P158633 in all aspects of trigonid morphology, including that of the precingulid and paracristid. However, NMV P158633 and *A. correlli* are closer in parametacristid form. The parametacristid of NMV P158633 differs by extending slightly more anteriorly, and further into the trigonid basin. UCR 22656 and NMV P158633 also lack a postmetacristid. More material attributable to the Hamilton and Kanunka simosthenurins is required before their precise taxonomic affinities will become clearer.

Geographic Distribution. Southwest Victoria, northern South Australia (fig. 27).

Temporal Distribution. Early to early late Pliocene.

SIMOSTHENURUS Tedford, 1966

Sthenurus, De Vis, 1895 (in part), pp. 88-99, pl. 15, figs 1-9, pl. 16, figs 1-11.
Procoptodon, Scott, 1906 (in part), pp. 1-2.
Sthenurus, Glauert 1910a (in part), pp. 31-36, pl. 5, figs 6-7.
Sthenurus, Raven, 1929 (in part), p. 255.
Sthenurus, Simpson, 1930 (in part), p. 77.
Sthenurus, Anderson, 1932 (in part), pp. 383-387, figs 1, pl. 45, figs 2-3, pl. 46, figs 1-2.
Sthenurus, Raven and Gregory, 1946 (in part), pp. 6-8, fig. 2.
Sthenurus, Tate, 1948 (in part), pp. 336-338.
Procoptodon, Ride, 1959 (in part), pp. 54-56, fig. 8d.
"Sthenurus", Stirton, 1957 (in part), p. 124.
Sthenurus, Bartholomai, 1963 (in part), pp. 58-71, figs 3-10.
Sthenurus (Simosthenurus), Tedford, 1966, pp. 33-39, figs 11-13.
Sthenurus (Simosthenurus), Merrilees, 1968a, pp. 69-77, figs 3-4, 7.
Sthenurus (Simosthenurus), Wells and Murray, 1979, pp. 213-218, figs 1-6.
Sthenurus, Prideaux and Wells, 1997 (in part), pp. 181-196.
Sthenurus, Prideaux and Wells, 1998 (in part), pp. 1-15.

This list of synonyms is abbreviated, because almost all authorities prior to Tedford (1966) recognized the species now placed in *Simosthenurus* as species of *Sthenurus*. Subsequent to Tedford's work, some authorities have placed them in *Sthenurus (Simosthenurus)*, while others have recognized *Sthenurus* only.

Taxonomic Authority. Tedford, R. H. 1966. A review of the macropodid genus *Sthenurus*. University of California Publications in Geological Sciences 57:1-72.

Type species. Sthenurus occidentalis Glauert, 1910a.

Diagnoses. Simosthenurus is used in two senses in this monograph: *Simosthenurus s. l.* and *Simosthenurus s. s.* Cladistic analysis supports the holophyly of *Simosthenurus s. s.* (*Si. euryskaphus*, *Si. maddocki*, *Si. occidentalis*), but not *Simosthenurus s. l.* (see Phylogeny). Until more complete material is retrieved for *"Si." brachyselenis*, *"Si." tirarensis*, *"Si." baileyi*, *"Si." antiquus* and *"Si." pales*, character support will remain insufficient to justify separate generic designations for these clades. For now, I place them in *Simosthenurus s. l.*, but when referring to them individually I enclose the generic epithet in quotation marks to highlight the likely interim nature of this taxonomic assignment.

Simosthenurus s. s. Frontals taper laterally to rugose supraorbital tuberosity. Median valley of p3 wide. Molars bear numerous very fine enamel crenulations.

Simosthenurus s. l. Simosthenurins characterized by p3 median valley of at least intermediate width, but lacking diagnostic features of *Archaeosimos* (very low-crowned, simple molars, p3 only slightly wider posteriorly than anteriorly, with very short buccal crest) and *Procoptodon* (well-developed mesocrista, long p3 buccal crest, lophid faces bear coarse enamel crenulations).

Etymology. Gr. *Simos* 'flat- or pug-nosed'; in reference to the foreshortened cranium.

Geographic Distribution. Southwestern Australia, Nullarbor Region, northern, south-central and southeastern South Australia, central and northwestern Tasmania, southern Victoria, eastern New South Wales, southeastern Queensland (figs 28-32, 34).

Temporal Distribution. Early Pliocene to late Pleistocene.

Simosthenurus euryskaphus Prideaux and Wells, 1997

(pl. 40; table 16)

Sthenurus (Simosthenurus) oreas, Tedford, 1966 (in part), pp. 39-41, fig. 15.
Sthenurus (Simosthenurus) oreas, Marcus, 1976 (in part), pp. 69-70.
Sthenurus (Simosthenurus) euryskaphus, Prideaux and Wells, 1997, pp. 181-196, figs 4-6.

Taxonomic Authority. Prideaux, G. J. and R. T. Wells. 1997. New *Sthenurus* species (Macropodidae, Diprotodontia) from Wellington Caves and Bingara, New South Wales. Proceedings of the Linnean Society of New South Wales 117:181-196.

Holotype. AM F113131 (= AM MF2), partial left juvenile dentary, containing dp3, m1-2, excavated p3, alveoli for dp2, m3. The angular, condylar and coronoid processes are broken off; i1 has been lost (see Tedford 1966:fig. 15). Specimen collected in 1887 by William Anderson from the west side of Bone Camp Gully, adjacent to Myall Creek, near Bingara in northeastern New South Wales. Age of type locality may be middle Pleistocene (Marcus 1976). In February 2002, the author was unable to locate the holotype p3 in the Australian Museum collection. A plaster cast of the p3 (UCR 22999) is housed in the Department of Earth Sciences, University of California, Riverside.

Referred Specimen. Bone Camp Gully, near Bingara, NSW: AM F73723, right adult dentary.

Revised Diagnosis. The dp3 is similar to that of *"Simosthenurus" tirarensis*, but is smaller relative to molars, has a more tapered protolophid and a narrower, shorter trigonid. The p3 is short relative to width and shorter than m2; median valley is a broad, circular basin due to marked curvature of buccal crest and posterior end of main crest; buccal crest very weakly connected to main crest anteriorly.

Etymology. Gr. *eury-* 'broad', *skaphe* 'basin, trough', in reference to the broad, circular nature of the median valley formed between the buccal crest and posterior end of the main crest on p3.

Description. Juvenile Dentary (AM F113131, pl. 40A-C). Proportions of preserved portion of ramus suggest very short dentary. Ventral border markedly concave due to rather large digastric eminence and prominent symphyseal boss, which corresponds to large inferior transverse torus of symphysis. This extends to beneath dp3 protolophid. Symphyseal plate very rugose, but presence of numerous foramina and clean nature of separation suggest it was unfused (pl. 40B). Genial fossa rather open. Anteroventral border of dentary steep, reflecting markedly upturned nature of i1 as indicated by axis of alveolus. Digastric sulcus deep, extending anteriorly to beneath m2 protolophid (pl. 40B). Little of diastema preserved, but apparently very short. Buccinator sulcus deep anteriorly; posterior terminus below m1 protolophid shallow. Anterior mental foramen circular, and opening more dorsally than laterally, with dorsal border immediately beneath anterior end of buccinator sulcus. Posterior mental foramen relatively high on lateral dentary surface, beneath hypolophid of m1 (pl. 40A). Most of posterior region of dentary broken away, but anterior root of ascending ramus lies adjacent to m3. Masseteric canal large and near vertical. Sharp process oriented anterodorsally appears to have partially overhung mylohyoid groove (pl. 40B).

Adult Dentary (AM F73723, pl. 40E-G). Specimen preserves only portion of dentary below m4 and anterior root of ascending ramus. Although incomplete ventrally, dentary was clearly deep and robustly proportioned. Masseteric foramen apparently large. Mylohyoid groove shallow and wide.

Lower Dentition (pl. 40; table 16). Although described and illustrated by Tedford (1966:fig. 15), i1 now appears lost. A plaster cast is housed in Department of Earth Sciences, University of California, Riverside (UCR 22999). Upturned i1 moderately robust, with distinctly sinusoidal occlusal surface rising to point anteriorly. Crown curves occlusally relative to longitudinal axis of root.

Table 16. Cheek tooth dimensions of *Simosthenurus euryskaphus*.

Tooth	Specimen	L	AW	PW	AH	PH
dp3	AM F113131 (holotype)	9.4	7.8	7.9	6.5	6.5
p3	AM F113131 (holotype)	13.9	7.0	8.7	9.5	8.7
m1	AM F113131 (holotype)	12.9	9.7	9.7	8.4	8.4
m2	AM F113131 (holotype)	14.5	10.8	10.9	9.5	9.0
m3	AM F73723	-	-	12.3	-	7.7
m4	AM F73723	15.7	12.9	12.7	8.3	7.5

Alveolus of dp2 indicates tooth was slightly shorter than dp3. Completely molariform dp3 bears protolophid considerably more tapered toward crest than hypolophid, due largely to dorsolingually sloping buccal side (pl. 40C). Posterior end of paracristid possibly shifted very slightly lingually from protoconid apex, but too worn to be certain. Paracristid trends slightly anterolingually across short trigonid before turning lingually and terminating at anterolingual corner of dp3. Precingulid small but distinct. Slight parametacristid worn, but it clearly descended into center of trigonid basin, adjacent to two fine enamel crenulations. Cristid obliqua low, directed anterolingually from hypoconid and curving more anteriorly to protolophid posterior face, before trending buccally and terminating at protoconid apex (pl. 40C). Hypolophid anterior face may have borne few very fine enamel crenulations, but too worn to be certain. Directed anterobuccally from entoconid apex, very low preentocristid met in interlophid valley by extremely low, slight postmetacristid, which descends posterobuccally from metaconid apex. Hypolophid posterior face bears very low, slight postentocristid, which descends ventrobuccally from entoconid apex to about one-third of way toward crown base, then curves dorsally for a short way before again curving basally in center of tooth. Here, it becomes confluent with top of very low A-shaped inflation.

Unerupted p3 fully formed and enameled; slightly shorter than m2 (pl. 40D). In basal outline, lingual side of p3 quite straight, but curves buccally at point adjacent to anteriormost cuspule of main crest. Anterior half of p3 outline symmetrical, but posterobuccal aspect inflated relative to posterolingual aspect. Main crest divided into four linked cuspules; posterior two-fifths of crest markedly convex lingually. Anteriormost cuspule of main crest gives rise to coarse ridgelets that curve posterobuccally and posterolingually from apex, while finer ridgelet descends anteriorly. Moderately short buccal crest markedly convex buccally, restricted to posterobuccal corner of p3 (pl. 40D). Short, very fine ridgelet links anterior extremity of buccal crest to small eminence low down on buccal side of main crest, just posterior to ridgelet directed posterobuccally from anteriormost cuspule of main crest. Posterior edges of buccal and main crests form 90° notch when viewed posteriorly. Broad and rather circular median valley contains only few fine ridgelets (pl. 40D).

Lower molars low crowned, with protolophid and hypolophid crests straight and close to parallel (pl. 40A-C). Anterior turn of buccal lophid ends moderate. Moderately short trigonid only extends across central two-thirds of anterior base of protolophid. Very low buccal component of paracristid directed anterolingually from protoconid apex, terminating on buccal side of posterior portion of lingual component of paracristid. This is low, thick and shifted one quarter of way across protolophid from protoconid apex (pl. 40C). Posterior portion of lingual component slightly shorter than transverse anterior portion, which terminates at anterolingual corner of tooth. Several fine enamel crenulations descend into trigonid basin from protolophid anterior face (pl. 40C). Several finer enamel crenulations descend into trigonid basin from lingual surface of paracristid. Premetacristid and parametacristid absent. Buccal and lingual components of cristid obliqua almost identical in position, orientation and development to paracristid. Several very fine, anteroposteriorly oriented enamel crenulations on anterior hypolophid face terminate before reaching interlophid valley. Hypolophid posterior face characterized by much more pronounced A-shaped inflation or postcingulid than dp3 (pl. 40C). Very low postentocristid descends ventrobuccally and is smoothly confluent with lingual side of postcingulid inflation.

Comparison With Other Taxa. Before proceeding with the comparison, it is necessary to explain why this account of *Simosthenurus euryskaphus* differs significantly from the original diagnosis and description provided in Prideaux and Wells (1997), even though the newly referred specimen contributes only marginally to the understanding of the species. When that work was prepared, I had an inadequate grasp of the amount of intraspecific variation expressed within *Si. occidentalis*. Some features then considered diagnostic or, at least, points of difference with *Si. occidentalis* have since been recognized within the morphological spectrum of this widespread species. Moreover, the *Si. euryskaphus* holotype was compared with specimens of a slightly earlier molar eruption stage. Prideaux and Wells (1997:fig. 5) compared it with specimens of *Si. occidentalis* and *Metasthenurus newtonae* in which m3 eruption was incipient, not approaching completion. I have since become aware, from species represented by good age series, that notable changes in dentary proportions can occur during the ontogenetic period delimited by these tooth eruption stages.

Given the similarity between *Simosthenurus euryskaphus* and *Si. occidentalis*, it is also important to compare the two *Si. euryskaphus* specimens with an adult dentary from the type locality (AM F88542), which Tedford (1966) referred to *Simosthenurus orientalis* (herein a junior synonym of *Si. occidentalis*). In the review of *Si. occidentalis* provided later in this monograph, AM F88542 is shown to differ sufficiently in ramus proportions from *Si. occidentalis* for it to be excluded from that species. However, AM F88542 is just as unlikely to represent a specimen of *Si. euryskaphus.* The roots of the p3, which are all that is preserved of that tooth in AM F88542, indicate that it was considerably larger than in *Si. euryskaphus.* The digastric sulcus did not extend as far anteriorly, while the symphyseal plate is a different shape and bore a less posteriorly extensive inferior transverse torus. In other sthenurines, these features either do not vary significantly with ontogenetic age or the polarity of the proportional change is reversed. Due to its relative incompleteness and the disparate morphology of its ramus, AM F88542 cannot presently be assigned to any species (see Marcus 1976:fig. 27a-b).

Dentary. Compared with the only similarly aged specimen of *Simosthenurus occidentalis* that I have observed (SAM P16534), ramus depth to width is greater in the *Si. euryskaphus* holotype (depth / width: *Si. euryskaphus*, 1.62; *Si. occidentalis*, 1.41). Given the variability in the dentary form of adult *Si. occidentalis*, it would be unwise to deem this difference diagnostic until more specimens are recovered. Subsequent to the original description of *Si. euryskaphus*, it has also become evident that digastric eminence and sulcus development, as well as symphysis morphology, also vary within adult individuals of *Si. occidentalis.* Therefore, it is also not possible to be confident in diagnosing *Si. euryskaphus* on the basis of these features. Nevertheless, the fragmentary adult specimen (AM F73723) does appear to have had a notably deep dentary.

Lower Dentition. Since it is now clear that the lower molars of *Simosthenurus euryskaphus* fit within the morphological and size spectra of mainland southeastern *Si. occidentalis*, only the i1, dp3 and, especially, p3 are useful for distinguishing this species. While upturned to a similar degree as that of *Si. occidentalis*, the *Si. euryskaphus* i1 is more like that of *"Procoptodon" browneorum* in its comparatively slender form and the slightly sinusoidal nature of the occlusal surface. However, the i1 is considerably more upturned in *Si. euryskaphus.* Although the *Si. euryskaphus* dp3 is very similar to that of *Si. occidentalis*, it is smaller relative to the molars, the protolophid is more tapered toward its occlusal surface, and the trigonid is narrower and shorter. In each of these regards it resembles *"Si." tirarensis*, from which it differs by having a lower cristid obliqua, and by lacking the small but distinct pocket formed by the postentocristid on the posterior face of the hypolophid.

In size relative to the succeeding molars, the p3 of *Si. euryskaphus* is similar to that of *Metasthenurus newtonae*, *"Procoptodon" oreas* and, to a lesser degree, *"Si." brachyselenis. Si. euryskaphus* is also distinguished from *M. newtonae* and *"Si." brachyselenis* by the more curved posterior portion of its main crest, a longer buccal crest, and a much wider, rather circular median valley. Compared with *"P." oreas*, the *Si. euryskaphus* p3 is larger than that of the *"P." oreas* holotype (pl. 102), but is similar in dimensions to the p3 of AM F88540. Morphologically, the *Si. euryskaphus* p3 is slightly wider relative to its length, has a shorter, more crescentic buccal crest, which lacks a strong connection with the main crest anteriorly, and has a slightly wider, more circular median valley. Compared with the p3 of another *"P." oreas* specimen (AM F31922, pl. 105), the posterior portion of the *Si. euryskaphus* p3 is distinctly wider than its anterior portion, the main crest is less distinctly divided into cuspules, and it has a shorter, more crescentic buccal crest, and a wider, more circular median valley containing fewer ridgelets. In light of the inseparability of their molars, it is taxonomically pivotal that the form of the *Si. euryskaphus* p3 is markedly dissimilar to that of any *Si. occidentalis* specimen known.

Geographic Distribution. Northeastern New South Wales (fig. 34; table 44).

Temporal Distribution. ?Middle Pleistocene.

Simosthenurus maddocki Wells and Murray, 1979

(pls 41-46; fig. 12E; tables 17-18)

Sthenurus sp. I, Marcus, 1976, pp. 72-74, fig. 27e-f.
Sthenurus sp. aff. *occidentalis*, Archer, 1978, pp. 82-83, pl. 7S.
Sthenurus (Simosthenurus) maddocki, Wells and Murray, 1979, pp. 213-218, figs 1-6.
Simosthenurus sp. I, Pledge, 1980, pp. 133-134, fig. 2B.
Sthenurus (Simosthenurus) maddocki, Archer, 1984, pp. 213-218, figs 1-6.
Sthenurus maddocki, Wells, Moriarty and Williams, 1984, pp. 325-326, 329.
Sthenurus (Simosthenurus) sp., Lundelius and Turnbull, 1989, pp. 3-4, 6, fig. 1C.
Simosthenurus maddocki nullarborensis, Prideaux, 1994, pp. 133-138, table 1, figs 2-5.
Sthenurus maddocki, Prideaux and Wells, 1997, pp. 185-187, fig. 2c.
Sthenurus maddocki, Prideaux and Wells, 1998, pp. 1, 4, 11-13, 15.

Systematic Authority. Wells, R. T. and P. F. Murray. 1979. A new sthenurine kangaroo (Marsupialia, Macropodidae) from southeastern South Australia. Transactions of the Royal Society of South Australia 103:213-219.

Holotype. SAM P16999, near-complete cranium (slightly crushed, missing incisors, part of right zygomatic arch) collected in the early 1970s by Roderick T. Wells and others from Victoria Fossil Cave, near Naracoorte, South Australia. Age of type locality is late middle Pleistocene (Ayliffe et al. 1998; Grün et al. 2000).

Paratypes. Victoria Fossil Cave, near Naracoorte, SA: SAM P16513, right juvenile dentary; SAM P16548, left adult dentary; SAM P16627, left and right adult dentaries; SAM P16673, left and right adult dentaries.

Referred Specimens. Victoria Fossil Cave, near Naracoorte, SA: SAM P25608 / P25610, left and right juvenile dentaries; FU 0231, partial juvenile cranium; FU 0714, left and right adult dentaries; SAM P numerous registered dentaries, maxillae and single loose teeth.

Cathedral Cave, near Naracoorte, SA: FU unreg., adult cranium.

Crawford's Cornucopia Cave, near Naracoorte, SA: Listed in Reed and Bourne (2000).

Haystall Cave, near Naracoorte, SA: SAM P13735 + SAM P unreg., 15 single loose teeth.

Henschke's Fossil Cave, near Naracoorte, SA: SAM P29571, juvenile cranium; SAM approximately 30 registered and unreg. maxillae and dentaries, many single loose teeth.

Sand Funnel Cave, near Naracoorte, SA: SAM P16674 / P20011, left and right juvenile dentaries and maxillae.

Specimen Cave, near Naracoorte, SA: SAM P unreg., right i1.

Comaum Forest Cave, near Penola, SA: SAM P40076, left and right adult dentaries; SAM P unreg., numerous single loose teeth.

Wandilo Cave, near Mount Gambier, SA: SAM P32562, right adult maxilla.

Kilsby's Hole, Mount Gambier, SA: SAM P40070, right juvenile dentary; SAM P unreg., several single loose teeth.

Goulden's Hole, near Mount Schank, SA: SAM P24099, approximately 40 single loose teeth.

Mount Schank, SA: SAM P22518, right adult maxilla.

Green Waterhole Cave, southeastern SA: SAM P17246, cranium and dentaries; SAM P17247, adult cranium and dentaries; SAM P17252, left and right adult dentaries; SAM P17263, left and right adult dentaries; SAM P17471, partial adult cranium; SAM P18316, left adult dentary; FU 1664, left and right juvenile maxillae and dentaries; FU 1665, left and right juvenile dentaries and maxillae.

Unnamed swamp site, near Millicent, SA: FU unreg., partial maxilla.

Derrington St, Mount Gambier, SA: SAM P40077, left and right adult dentaries.

Town Well Cave, Curramulka, Yorke Peninsula, SA: SAM P12786, right adult dentary; SAM P40078, right I3.

Curramulka Quarry, Yorke Peninsula, SA: SAM P40058, left I3.

South Site, Lancefield, VIC: NMV P200789 (in part), left M4.

Dolomite quarry, near Mount Fairy, NSW: AM F72916, left adult dentary.

Wombeyan Caves, NSW: AM F54894, left adult dentary.

Bone Camp Gully, near Bingara, NSW: AM F88546 (= AM MF7), right adult dentary.

The Joint, Texas Caves, southeastern QLD: QM F8529, right dP2, dP3.

Madura Cave, Roe Plain, WA: TMM 41106-3500, left M3.

Lindsay Hall Cave, Nullarbor Plain, WA: WAM 92.9.8, left juvenile maxilla; WAM 92.12.7, left P3; WAM 03.11.09, right adult maxilla; WAM HC LH14, left adult maxilla; WAM HC LH15, right juvenile maxilla; WAM HC LH19, I1; WAM HC LH20, right juvenile maxilla; WAM HC LH21, right juvenile maxilla; WAM HC LH116, left and right adult maxillae.

Balladonia Soak, WA: WAM 61.7.2, right adult maxilla.

Revised Diagnosis. Cranium similarly proportioned to *Simosthenurus occidentalis*, but with frontals more expanded and laterally tapered. Postglenoid process very shallow. I1 much higher crowned than I2 and small I3, which is only slightly longer than combined length of I1-2. Cheek tooth rows only very slightly curved laterally. Upper molars small relative to size of cranium, bearing many fine enamel crenulations, a distinct preprotocrista, and weakly developed mesocrista. Postparacrista and premetacrista slightly incurved lingually; preparacrista strongly connected to paracone apex. Diastema region of dentary narrow; median dorsal groove deep; symphysis lacks an inferior transverse torus. Lower incisor very slender and elongate, resting wholly within upper incisor arcade. Low-crowned p3 very narrow anteriorly, wide posteriorly, with short buccal crest; median valley filled by row of cuspules.

Etymology. Named in honor of Ernest Maddock, former director of National Pleasure Resorts and president of the Cave Exploration Group of South Australia, whose efforts led to the development of the Victoria Fossil Cave tourist complex at Naracoorte.

Description. Cranium (pls 41-43; table 17). Markedly brachycephalic, with incisor-bearing portion of premaxilla moderately deep and short. Distinct premaxillary eminence absent. Upright portion of premaxilla near vertical and rather thin (pls 41A, 43A). Moderately short diastema either aligned in same horizontal plane as alveolar margin of cheek tooth row or convex dorsally. Anterior borders of moderately long incisive foramina slightly anterior to posterior border of I3 alveolus. Short rostrum tapers anteriorly, with moderately shallow anteorbital / buccinator fossae. Zygomatic arch encloses very wide temporal fossa (pls 41B, 42, 43B). Orbits quite convergent. Masseteric process small and only slightly flared (pls 41A, 42). Narial aperture slightly broader than deep. Nasals broad posteriorly but elongate, with anterior tips extended well beyond limit of premaxilla (pl. 41). Anterior border of orbit and lacrimal positioned above anterior limit of P3. Lacrimal bears two very small lacrimal foramina, separated by tuberosity.

Frontals greatly expanded laterally, tapering to rugose supraorbital tuberosity directly above orbit (pls 41, 43). Portion of jugal that forms inferior orbital rim projects anteriorly. Temporal (parietal) crests strongly developed and convergent on sagittal suture. Anteriorly, palatal vacuities terminate adjacent to protoloph or interloph valley of M1. Postpalatine bars form bridge across palate just posterior to M4. Zygomatic arch deep, with wide ectoglenoid process at posterior extremity of moderately deep jugal (pl. 42). Dorsal edge of zygomatic process of squamosal markedly sloped anteroventrally. Postglenoid process very shallow (pls 41A, 43A). Ectotympanic very wide, curved ventrally at mesial end and bears two transverse keels, with more strongly developed at anterior extreme, more weakly developed located posteriorly. Smooth, wide concavity separates keels. Small elliptical postzygomatic foramen not wholly separated from external auditory meatus. Rather large subsquamosal foramen lies just posteroventral of dorsal root of zygomatic process.

Basicranial plane well elevated above level of palatal plane (pl. 41A). Medial pterygoid origin quite shallow and not particularly wide. Basioccipital moderately flexed posterodorsally relative to basisphenoid, with well-developed median keel. Occiput broad, moderately deep, and oriented at 90° relative to dorsal surface of neurocranium (pl. 41A). Vertical median occipital crest not strongly developed, leading to wide foramen magnum bordered by broad, rather elongate occipital condyles. Nuchal / lambdoid crests slightly developed. Quite vertical paroccipital processes thick and short, not quite extending down to level of palate. Mastoid process of periotic thickly developed.

Table 17. Cranial dimensions of the holotype of *Simosthenurus maddocki* (SAM P16999) taken from Wells and Murray (1979), except for Diastema Length, which here estimates what the distance I3-P3 would have been were P3 erupted. Wells and Murray (1979) measured the distance I3-dP2.

Dimension	*Simosthenurus maddocki* holotype
Condylobasal Length	190
Diastema Length	34.7
% Diastema Length: Palatal Length	30.2
Palatal Length	115
Palatal Width between M1 Protolophs	35.4
Palatal Width between M4 protolophs	35.8
Cranial Width	128
Width across Frontals	101
Distance between Paroccipital Processes	82.7

Upper Dentition (pls 41-44; table 18). I1 crown high relative to I2-3, curved posteroventrally, and markedly over-biting i1 (pls 41A, 42B). I2 tiny by comparison, about one quarter size of I1. I3 small, with anterior edge of unworn crown quite straight, posterior edge convex. Buccal surface may bear shallow vertical undulations. Anterolingual crest very small on unworn I3.

Compared with P3, dP2 is shorter relative to width; posterobuccal accessory cusp very small if present (pl. 44B,E). Main crest straight to slightly convex buccally, but lingual crest curved to larger degree lingually and convergent toward main crest anteriorly. Main crest considerably higher than lingual crest; usually divided into three prominent cuspules, with central cuspule lowest. Anterior basin very shallow and quite vertical. Few fine crenulations in longitudinal basin, with coarsest directed lingually from central main crest cuspule. Posterior basin short, but well separated from longitudinal basin by coarse transverse ridgelet linking posteriormost cuspule of main crest and distinct posteriormost cuspule of lingual crest. Completely molariform dP3 very similar to M1, except smaller, narrower for length, with more tapered lophs and precingulum not extending or only just extending lingually past preprotocrista (pl. 44A,E).

P3 longer than all molars and usually only slightly wider posteriorly than anteriorly (pl. 44). Anterobuccal aspect of tooth slightly inflated. Posterobuccal inflation of P3 correlated with size of distinct and often large posterobuccal accessory cusp. Main crest usually divided into five very distinct cuspules, with anteriormost and posteriormost very prominent (pls 41A, 43C, 44). Anteriormost cuspule frequently bears crescentic apex, as does anterior extremity of only very slightly lower lingual crest. Buccally and lingually, all cuspules of main crest give rise to short, coarse ridgelets. Lingual crest only very slightly lower than main crest, straight for most of length, but curved anteriorly and posteriorly. Posterior edges of P3 rather straight and only very slightly convex. Bordered posteriorly by union of curved ends of main and lingual crests, anterior basin usually reduced to tiny cleft, but may occasionally be slightly larger. Narrow longitudinal basin moderately shallow, traversed by many fine ridgelets (pl. 44). Small posterior basin may contain few very fine ridgelets and usually separated from longitudinal basin by distinct transverse ridgelet.

Upper molars low crowned and small relative to size of cranium (pl. 42A). Lophs taper moderately toward unworn crests, more so on anterior teeth (pl. 44A,G). Cheek tooth row only slightly curved laterally. Size gradient along molar row is M1<M2<M3>M4, with little variation in morphology, except for much narrower metaloph relative to protoloph on M4. Protoloph and metaloph crests notably convex anteriorly. Preparacrista maintains strong connection with paracone apex (pl. 44A,G); anterobuccal aspect of crest only very slightly expanded in region of stylar cusp A. Preparacrista confluent with moderately thick precingulum, which terminates at anterolingual corner of crown. Low on anterior face of protoloph, small but distinct preprotocrista often evident on precingulum, just lingual of tooth midline (pl. 44G). Numerous fine vertical enamel crenulations on anterior face of protoloph. Lingual component of postprotocrista smoothly continuous with protoloph crest, trending dorsobuccally and terminates on side of smaller buccal component oriented more anteroposteriorly. Buccal postprotocrista component barely distinguishable from many very fine enamel crenulations coating protoloph (pl. 44A,G). Mesocrista not distinct. Postparacrista moderately developed, curving in lingually and almost mirroring relative development and orientation of preparacrista (pl. 44A,G). Premetacrista curves in similarly, but only half size of postparacrista. Postmetacrista apparently divided into two components, more buccal of which may be stylar crest. This meets postmetaconulecrista just below mid-depth on buccal side of crown in position of stylar cusp E, but no cusp is evident. Stronger, more lingual postmetacrista component curves dorsolingually, then directly dorsally, before terminating inside of corner formed by more buccal component and postmetaconulecrista. Several often bi- or trifurcated enamel crenulations dominate center of posterior face of metaloph, most central of which is probably urocrista.

Dentary (pls 41A, 42B, 45-46). Dentary short, with most of ventral surface of ramus near to straight on account of moderate to minimal development of digastric eminence and symphyseal boss. Diastema region markedly narrow and not especially short. Diastema border usually convex dorsally (pls 41A, 45B), but may be straighter and slightly upturned relative to alveolar margin of cheek tooth row. Symphysis partially fused in some specimens, but inferior transverse torus absent. Genial fossa small and deep. Median dorsal groove deep and narrow (pl. 45A). Anteroventral border of ramus steeply upturned, but i1 more procumbent. Alveolar margin of cheek tooth row curves up slightly beneath posterior two molars. Digastric sulcus moderately to very shallow, extending no farther anteriorly than beneath m3. Buccinator sulcus quite deep along entire length, but forms very deep pit posteriorly beneath m1 hypolophid or m2 protolophid (pls 41A, 45B). Anterior mental foramen slightly elongated dorsoventrally, positioned just ventral to anterior extremity of buccinator sulcus, below posterior half of diastema. Posterior mental foramen positioned low on ramus beneath m3 protolophid.

Anterior root of ascending ramus adjacent to m4 protolophid or interlophid valley, with anterior border oriented slightly anterior of vertical before inflecting to vertical. Masseteric fossa rather deep, with ventral border just above level of posterior extremity of buccinator sulcus. Posteroventral border of masseteric fossa quite flared laterally. Masseteric foramen moderately large. Medial pterygoid fossa wide, but not particularly deep, bearing well-developed angular process. Inferior mandibular foramen rather small, circular and largely opening posteriorly. Mylohyoid groove shallow and wide, with posteroventrally oriented process barely visible. Mandibular condyle moderately large. Coronoid process quite elongate.

Lower Dentition (pls 45-46; table 18). Crown of markedly slender, elongate i1 terminates in sharp point anteriorly (pls 45A-B, 46A-B). Although i1 is slightly more procumbent than anteroventral border of ramus, steep and very slightly sinusoidal unworn occlusal surface sits well above level of cheek tooth row. Mid-portion of i1 rather triangular in cross section, although buccal surface slightly convex. Lingual enamel layer thinner than buccal layer for most part, but thicker beam runs along lingual edge of unworn occlusal surface, from anterior point to base of crown. Lower incisors rest wholly within upper incisor arcade due to marked overbite of I1 (pls 41A, 42B).

Although similar to p3 in morphology, dp2 shorter relative to width and simpler in overall form (pl. 46). Anterior portion of tooth usually much narrower than posterior portion. Main crest divided into four variably distinct cuspules.

Table 18. Cheek tooth dimensions of *Simosthenurus maddocki* from eastern Australia and the Nullarbor Plain, showing mean, standard deviation (in parentheses). Dimensions of the holotype (SAM P16999), Texas Caves (QM F8539) and Balladonia (WAM 61.7.2) specimens are also given.

Tooth		L	AW	PW	AH	PH	n
dP2	Eastern	9.1 (0.29)	6.2 (0.48)	7.3 (0.47)	6.2 (0.45)	6.2 (0.53)	6
	Holotype	9.4	7.1	8.1	5.9	6.0	
	Nullarbor	7.2 (0.29)	5.0 (0.26)	5.8 (0.23)	5.9 (0.51)	5.6 (0.42)	3
	QM F8529	8.9	7.1	-	6.3	6.3	
dP3	Eastern	9.7 (0.12)	8.7 (0.32)	9.0 (0.35)	5.0 (0.31)	4.9 (0.27)	8
	Holotype	9.7	8.5	8.6	4.7	4.7	
	Nullarbor	8.4 (0.14)	7.5	8.2	4.9	4.9	2
	QM F8529	11.5	9.5	-	6.6	-	
P3	Eastern	16.0 (0.55)	8.1 (0.59)	10.3 (0.73)	8.4 (0.59)	7.8 (0.59)	11
	Holotype	16.2	8.3	10.1	8.5	8.6	
	Nullarbor	12.7 (0.68)	6.1 (0.27)	8.7 (0.21)	6.9 (0.27)	6.1 (0.44)	5
M1	Eastern	10.7 (0.33)	10.3 (0.37)	10.1 (0.40)	5.4 (0.34)	5.3 (0.46)	12
	Holotype	11.0	10.2	10.1	5.7	5.2	
	Nullarbor	9.4 (0.17)	8.4 (0.31)	8.7 (0.26)	4.6 (0.36)	4.8 (0.47)	6
	WAM 61.7.2	10.4	-	10.0	-	-	
M2	Eastern	11.1 (0.24)	10.8 (0.46)	10.3 (0.33)	5.5 (0.37)	5.7 (0.50)	10
	Holotype	11.3	11.6	10.7	5.2	5.2	
	Nullarbor	9.9 (0.25)	9.0 (0.27)	8.9 (0.16)	4.6 (0.57)	5.0 (0.71)	5
	WAM 61.7.2	11.3	10.3	9.4	-	-	
M3	Eastern	11.4 (0.26)	11.1 (0.47)	10.2 (0.43)	5.8 (0.52)	5.7 (0.56)	9
	Holotype	11.7	11.9	10.7	5.8	5.9	
	Nullarbor	10.2 (0.13)	9.3 (0.26)	8.8 (0.25)	4.6 (0.13)	4.3 (0.15)	4
M4	Eastern	10.8 (0.58)	10.8 (0.42)	9.1 (0.60)	5.3 (0.54)	4.9 (0.87)	8
	Holotype	10.6	10.8	9.2	6.0	5.9	
	Nullarbor	9.9 (0.14)	8.9 (0.14)	7.7 (0.14)	4.1 (0.07)	3.8 (0.00)	2
dp2	Eastern	8.1 (0.33)	4.8 (0.31)	5.9 (0.24)	6.5 (0.17)	5.6 (0.29)	6
dp3	Eastern	8.8 (0.18)	6.8 (0.28)	7.1 (0.29)	5.8 (0.48)	5.9 (0.51)	6
p3	Eastern	15.6 (0.60)	5.8 (0.27)	8.2 (0.30)	7.7 (0.36)	7.0 (0.65)	9
m1	Eastern	10.9 (0.52)	8.3 (0.34)	8.6 (0.31)	6.7 (0.84)	7.0 (0.73)	14
m2	Eastern	11.8 (0.52)	9.3 (0.24)	9.3 (0.23)	7.2 (0.41)	7.3 (0.48)	12
m3	Eastern	12.2 (0.58)	10.0 (0.36)	9.6 (0.31)	7.2 (0.18)	7.4 (0.20)	10
m4	Eastern	11.5 (0.53)	9.9 (0.50)	9.1 (0.38)	6.5 (0.38)	6.3 (0.47)	6

Anteriormost cuspule gives rise to buccal and lingual ridgelet, but more posterior cuspules may or may not bear short, low ridgelets. Short, crescentic buccal crest oriented rather anterobuccally, and restricted to posterobuccal corner of crown (pl. 46E). Median valley may contain few fine, one coarse or no ridgelets. Connection between anterior end of buccal crest and buccal side of main crest very fine and low, if present at all. Apart from smaller size, dp3 usually only distinguished from m1 by having slightly more tapered protolophid, slightly better developed parametacristid, and paracristid and cristid obliqua maintaining direct contact with buccal cusp apices (pl. 46E).

Low-crowned p3 (pl. 45) longer than any molar and rather narrow, although posterior portion considerably wider than anterior portion. Main crest divided into six or seven very distinct cuspules linked by fine, low connecting ridgelets. Cuspules give rise to coarse lingual and buccal ridgelets. Anteriormost cuspule of main crest much more prominent than succeeding cuspules (pl. 45B-C). Buccal ridgelets from main crest cuspules usually link distally to one another via finer ridgelets that reflect orientation of main crest. Together, they form a lower, irregular median alignment of cuspules extending through ridgelet-filled median valley (pl. 45C). Short, crescentic buccal crest restricted to inflated posterobuccal corner of p3.

Lower molars low crowned; molar row only very slightly curved laterally (pl. 45). Unworn lophid crests straight and parallel, but oriented slightly anterolingually relative to anteroposterior axis of tooth. Moderate anterior turn of lophid ends emphasized buccally. Moderately short trigonid underlain by rounded precingulid and extending to anterobuccal corner of tooth (pls 45C, 46E). Paracristid low but distinct, consisting of only one component on m1, incipiently divided into two components on m2, and clearly divided on m3-4. Anterolingually oriented posterior portion and transverse anterior portion of paracristid often meet at sharp corner (pls 45C, 46E). On m2, posterior portion of lingual component of paracristid oriented anterolingually, overlapping for one-third of length with similarly oriented buccal component of paracristid. Buccal component of paracristid lower on m3-4, but usually distinct, and terminating on buccal side of lingual component. Parametacristid usually mirrors buccal component of paracristid in orientation and development. Short, fine premetacristid occasionally present. Short, fine postmetacristid evident on m1 only. Anterior faces of lophids bear numerous very fine enamel crenulations (pls 45C, 46E). Cristid obliqua very similar in orientation and morphology to paracristid, but incipiently divided into two components on both m1 and m2. Preentocristid absent or extremely fine and low. Center of hypolophid posterior face bears several extremely fine, vertical enamel crenulations that descend from just below lophid crest onto a slight ventral inflation. Hypolophid of m4 usually only slightly narrower than protolophid.

Variation. Across its extensive range around the southern and eastern Australian periphery, *Simosthenurus maddocki* varies greatly in P3 morphology (pl. 44) and cheek tooth size. This prompted Prideaux (1994) to propose a subspecific distinction for Pleistocene specimens from Lindsay Hall Cave on the Nullarbor Plain (WAM 92.9.8, AM 92.12.7, pl. 44A-D). It is now possible to discount the validity of this separation since other material in the Hatcher Collection (now WAM) was made available for study (e.g., pl. 44G) and after closer examination of the size variation observed in *Si. occidentalis*, *"Si." pales* and *"Procoptodon" browneorum* across southern Australia. As in these species, individuals of *Si. maddocki* from southern Western Australia have molars that are smaller on average (13% in this case) than their southeastern relatives (table 18). Proportionally, the P3 of western *Si. maddocki* (pl. 44C-D,G) is 21% smaller on average than in southeastern specimens (table 18). Unfortunately, no dentary or relatively complete cranial material is known for western *Si. maddocki*, but if the distance between the anterior extremities of the left and right P3 can be taken as a relative estimate of cranial width, then it may have been about 7% smaller (based on a comparison of WAM HC LH116 and SAM P17247).

Cranium. Four crania are known for *Simosthenurus maddocki.* The two best preserved are adult specimens from Green Waterhole Cave, near Tantanoola (SAM P17246 and SAM P17247, pls 41-42). Of the two slightly younger specimens (P3 eruption imminent) from Victoria Fossil Cave, the holotype (SAM P16999, pl. 43A-B) is relatively complete but slightly crushed, while FU 0231 is missing the rostrum. Aside from the slightly larger size and deeper zygomatic arch of the Green Waterhole Cave specimens, no differences between the crania are evident.

Upper Dentition. Relatively few incisors are known for *Simosthenurus maddocki*, and no noteworthy variation has been observed. Variation in the size and morphology of cheek teeth has been mentioned above and in Prideaux (1994). In size, the cheek teeth of *Si. maddocki* are smallest on the Nullarbor Plain, intermediate in southeastern Australia and at Balladonia (WAM 61.7.2), and largest in a maxilla fragment preserving dP2-3 from the Texas Caves in southeastern Queensland (QM F8529, pl. 44E; table 18). Overall, the morphology of the upper molariform teeth is remarkably uniform; they vary only slightly in the relative development of very fine enamel crenulations that coat the loph surfaces and the width of the M4 metaloph relative to the protoloph. The posterobuccal accessory cusp is the only feature that varies appreciably on dP2. Among the specimens observed, it is best developed in QM F8529, where it forms a thin crest running half the length of the buccal side of the tooth (pl. 44E). In other individuals, it may be a distinct but small cusp (e.g., SAM P28102), or may be completely absent (e.g., SAM P20563 and WAM 92.9.8, pl. 44B). One case of a simple, peg-like supernumerary dP1 anterior to dP2 has been observed in *Si. maddocki* (FU 6723, Gresham 2000).

By contrast to the other elements of the upper dentition, P3 is highly variable in both size and morphology (pl. 44). Aside from the 21% size difference observed between the western and southeastern specimens, P3 varies in (a) length relative to width and the upper molars; (b) the degree to which it is inflated posteriorly or tapered anteriorly; (c) width and development of ridgelets in the longitudinal basin;

(d) the number of cuspules from which the main crest is constituted; (e) size and separation of the anterior and posterior basins; (f) size of the posterobuccal accessory cusp.

Dentary. Ontogenetically, the ramus remains relatively slender and enlarges quite isometrically until m4 erupts (pl. 46). At this stage, there appears to have been a growth spurt in individuals of *Simosthenurus maddocki*, with the dentary becoming much more robust and the diastema less upturned, while the medial pterygoid fossa, masseteric fossa and buccinator sulcus deepen (pl. 46). Between adult specimens, variation in the dentary is observed in overall size and robustness, the degree to which the ramus becomes shallower anteriorly, curvature of the diastema, depth of the buccinator sulcus, development of the digastric eminence (pls 41A, 45B) and depth of the digastric sulcus.

Lower Dentition. No variation not attributable to differential wear has been observed between the various lower incisors of *Simosthenurus maddocki*, with the exception of a tendency for slightly larger incisors in larger individuals. Only minor variation is seen in the dp2, with some specimens narrowing slightly less markedly anteriorly than the majority. The number of cuspules from which the main crest is constituted varies between three and four, while development of the ridgelets directed buccally into the median valley from the more posterior of the main crest cuspules also varies. Specimens of p3 may differ in (a) overall size; (b) narrowness of the anterior portion and relative inflation of the posterior portion; (c) orientation of the main crest; (d) length and curvature of the buccal crest; (e) development of the median alignment of cuspules; (f) number of ridgelets in the median valley. However, the overall magnitude of variation for p3 does not reflect that observed for its upper counterpart. Lower molars vary only slightly in width of the m4 hypolophid relative to the protolophid, development of the precingulid and buccal component of the paracristid, the degree to which the lingual component of the paracristid is shifted across the protolophid face, and the number and fineness of the enamel crenulations. Occasionally, the posterior face of the hypolophid may bear a very slight, A-shaped postcingulid, although this is usually no more than a mild inflation onto which several very fine vertical enamel crenulations descend.

I have noticed one instance of a developmental abnormality in *Simosthenurus maddocki.* In the left juvenile dentary, SAM P20630, a small area of bone on the lateral surface has been removed to expose the unerupted p3. Although not fully enameled, the tooth is topographically normal, but is rotated both anteriorly and longitudinally, such that its posterior end is dorsal and its occlusal surface lateral. The inner surface of the crypt reveals that the p3 has not merely rotated postmortem. In addition, an isolated, unenameled conical cusp was found loose in the crypt anterior to the p3.

Comparison With Other Taxa. Cranium. The cranium of *Simosthenurus maddocki* could only readily be confused with that of *Si. occidentalis.* Both are markedly brachycephalic and have greatly inflated, laterally tapered frontals. *Si. maddocki* may be distinguished by the greater development of this frontal condition and by having a rostrum that is intermediate between *Si. occidentalis* and *"Si." baileyi* in its degree of anterior taper, a much smaller masseteric process, a more anteriorly sloping dorsal zygomatic surface, and a very slight postglenoid process. In addition, the width of the temporal fossa is as great anteriorly as it is posteriorly, producing more convergent orbits than in *Si. occidentalis.*

Upper Dentition. I1 is reminiscent of that of most similarly sized simosthenurins in its general form, but differs by being higher crowned and more curved posteroventrally. Relative to the tiny I2 and small I3, I1 is much higher crowned and overbites the narrow and elongate i1, which rests wholly within the upper incisor arcade during molar occlusion. I3 is most similar to that of *Si. occidentalis* in general morphology, but is much smaller. In size, it resembles the I3 of *P. goliah*, but is neither as short nor as wide.

The dP2 of *Simosthenurus maddocki* is most alike that of *Si. occidentalis* and *Sthenurus andersoni*, but may initially be distinguished by its smaller size. It also differs from *Si. occidentalis* by having a smaller anterior basin, a better separated posterior basin, and a relatively lower lingual crest. It differs from the dP2 of *S. andersoni* by being lower crowned and more narrowed anteriorly, and having a smaller anterior basin and a shallower longitudinal basin. Although the *Si. maddocki* P3 is markedly variable in morphology, most specimens are generally intermediate between that of *Si. occidentalis* and *S. andersoni*, as with the dP2. *Si. maddocki* differs from *Si. occidentalis* by being smaller, having a very poorly developed anterior basin, more distinctly separated posterior basin, and consistently large, distinct posterobuccal accessory cusp. The P3 of *Si. maddocki* is distinguished from that of *S. andersoni* by having a main crest more distinctly differentiated into cuspules, a relatively higher lingual crest and a larger, more distinct posterobuccal accessory cusp, and by lacking an anterobuccal extension of the lingual crest toward the base of the crown.

The upper molariform teeth of *Simosthenurus maddocki* differ from the otherwise very similar teeth of *Si. occidentalis* by being smaller both absolutely and relative to the size of the cranium, as well as by maintaining a stronger union between the preparacrista and paracone apex and bearing a more curved postparacrista and premetacrista. On a number of *Si. maddocki* specimens the very fine enamel crenulations are also more numerous than in *Si. occidentalis.* The preprotocrista is also better developed in *Si. maddocki*, but is not hypertrophied to the degree observed in some individuals of *Procoptodon pusio.*

Dentary. Overall, the dentary of *Simosthenurus maddocki* is probably most similar to *"Si." baileyi*, but is not unlike that of either *Si. occidentalis* or *"Procoptodon" gilli.* Features shared with *"Si." baileyi* include a similar relative diastema length, very moderate development of the digastric eminence and sulcus, and similar flaring of the posteroventral border of the masseteric fossa. The morphology of the symphyseal plate is also quite similar in these two species. *Si. maddocki* differs from *"Si." baileyi* by having a more dorsally convex diastema, a less posteriorly inflated medial pterygoid fossa and a deeper buccinator sulcus. By comparison, the *Si. occidentalis* dentary is more robust and has a larger inferior transverse torus of the symphysis, a better-developed digastric sulcus, markedly shallower buccinator sulcus, and a shorter, flatter or more horizontally oriented diastema. The deep, narrow median dorsal groove of *Si. maddocki* resembles that of *Metasthenurus newtonae.* It is most readily distinguished from *"P." gilli* by its larger size and lack of an inferior transverse torus of the symphysis.

Lower Dentition. The i1 of *Simosthenurus maddocki* is narrower and more elongate, and has a shorter occlusal surface than that of any other sthenurine, including *Procoptodon pusio. Si. maddocki* is also the only sthenurine in which the lower incisors rest within the upper incisor arcade rather than on the posteriorly facing I1 occlusal facets. The dp2 is most similar in morphology and size to that of *"Si." brachyselenis* and *Sthenurus andersoni*, and the length of the buccal crest is intermediate between these two species. The dp2 of *Si. maddocki* differs from that of *"Si." brachyselenis* by being slightly narrower and lacking a strong continuation of the buccal crest to the posterior extremity of the main crest. From the *S. andersoni* dp2, it differs by having a shallower, ridgelet-filled median valley. The distinctness of the main crest cuspules is also much greater in *Si. maddocki* than in both *S. andersoni* and *"Si." brachyselenis.* Overall, the unique p3 morphology easily separates *Si. maddocki* from all other sthenurines, although the tooth most closely resembles the p3 form observed in many southwestern individuals of *Si. occidentalis.* In these specimens, the posterobuccal portion of the tooth is markedly expanded and underlies a crescentic, although longer buccal crest (e.g., WAM 63.2.78, pl. 57D; WAM 97.5.74, pl. 57E). The median valley of *Si. occidentalis* is also wider. Distinct division of the main crest into numerous cuspules is unique to *Si. maddocki* and some individuals of *Si. occidentalis.*

Simosthenurus maddocki also closely resembles *Si. occidentalis* in lower molar morphology, especially in the distribution of fine enamel crenulations across the lophid surfaces. *Si. maddocki* differs by having more numerous enamel crenulations, a more sharply turning, better-developed paracristid, a generally wider trigonid, and a better-developed cristid obliqua. In addition, division of the paracristid and cristid obliqua is only incipient anteriorly in *Si. maddocki* compared with *Si. occidentalis* in which it is clear on all molars.

Geographic Distribution. Nullarbor Region, southern South Australia, southern Victoria, southeastern New South Wales, southeastern Queensland (fig. 30; table 44).

Temporal Distribution. Middle to late Pleistocene.

Simosthenurus sp. cf. *maddocki*

(pl. 47; table 19)

Referred Specimens. Wellington Caves (unspecified), NSW: AM F73720, right juvenile dentary; BMNH 42663Bb, right upper molar. Pleistocene.

Town Well Cave, Curramulka, Yorke Peninsula, SA: SAM P35054, right juvenile maxilla. Pleistocene.

Remarks. SAM P35054 is partial right juvenile maxilla, which contains dP2-3, M1, and the roots of M2 (pl. 47E-F). P3 has been excavated from its crypt (pl. 47G). Very little of the maxilla is preserved, but there is enough to show a similarity to *Simosthenurus maddocki* and *Si. occidentalis*. The distal portion of the masseteric process is broken off. The dP2 is greatly narrowed anteriorly, and is close to an equilateral triangle in outline (pl. 47F). This is unique among the sthenurines for which dP2 is known, but it is worth noting that the P3 of a *Si. maddocki* specimen not preserving dP2 is also very narrow anteriorly and wide posteriorly (SAM P20011, pl. 44J). The P3 of SAM P35054 has a similar topology to most *Si. maddocki* specimens and is not nearly as narrowed anteriorly as in SAM P20011, despite the marked similarity with this specimen in posterobuccal accessory cusp development. Compared with the mean dimensions for eastern *Si. maddocki*, the SAM P35054 P3 is 15% wider and has an 18% greater crown height relative to its length. These proportional differences render the tooth quite reminiscent of *Procoptodon rapha*. While the dP3 and M1 cannot be distinguished from *Si. maddocki* in topographic morphology, they are about 22% larger on average. Although I have not measured its dimensions, the isolated upper molar, BMNH 42663Bb, is also only distinguishable from an upper molar of *Simosthenurus maddocki* by its larger size.

AM F73720 is an incomplete right juvenile dentary, which contains dp3, m1-3, and the alveolus for the partially erupted m4 (pl. 47A-C). The p3 has been excavated from its crypt (pl. 47D). The ramus resembles that of similarly aged *Simosthenurus maddocki* specimens in most aspects, but is slightly larger and slightly deeper anteriorly, and the ventral border of the masseteric fossa is level with the alveolar margin of the cheek tooth row. While the i1 crown has been broken off, the cross section of the root clearly shows that the tooth was also narrow and, judging from the orientation of its alveolus, must also have been similarly elongate. While all topographic features of the AM F73720 molars (pl. 47A-C) are very similar to those of *Si. maddocki*, everything is larger. The molars are 29% larger in length and width dimensions, and notably higher crowned (table 19). Its crown height index (see Phylogeny) of 0.63 is more comparable to *"Si." pales* (0.65) and *Metasthenurus newtonae* (0.65) than it is to *Si. maddocki* (0.51). While the dp3 dimensions are similar to those of the large *Si. maddocki* specimen from the Texas Caves (QM F8529, pl. 44E), the projected disparity in crown height (given the greater wear of AM F73720) means that these teeth would not have

Table 19. Cheek tooth dimensions of *Simosthenurus* sp. cf. *maddocki* from the Wellington Caves (AM F73720) and Yorke Peninsula (SAM P35054) compared with mean dimensions of eastern *Si. maddocki.*

Tooth		L	AW	PW	AH	PH	
dP2	SAM P35054	9.5	6.0	8.9	6.8	7.2	
	Eastern	9.1 (0.29)	6.2 (0.48)	7.3 (0.47)	6.2 (0.45)	6.2 (0.53)	6
dP3	SAM P35054	11.8	10.3	10.8	5.6	5.9	
	Eastern	9.7 (0.12)	8.7 (0.32)	9.0 (0.35)	5.0 (0.31)	4.9 (0.27)	8
P3	SAM P35054	15.9	9.0	12.1	10.0	9.1	
	Eastern	16.0 (0.55)	8.1 (0.59)	10.3 (0.73)	8.4 (0.59)	7.8 (0.59)	11
M1	SAM P35054	13.4	12.4	12.5	6.7	6.9	
	Eastern	10.7 (0.33)	10.3 (0.37)	10.1 (0.40)	5.4 (0.34)	5.3 (0.46)	12
dp3	AM F73720	10.8	8.2	9.0	-	-	
	Eastern	8.8 (0.18)	6.8 (0.28)	7.1 (0.29)	5.8 (0.48)	5.9 (0.51)	6
p3	AM F73720	16.8	7.1	9.0	11.1	10.3	
	Eastern	15.6 (0.60)	5.8 (0.27)	8.2 (0.30)	7.7 (0.36)	7.0 (0.65)	9
m1	AM F73720	14.1	10.9	10.8	8.0	8.6	
	Eastern	10.9 (0.52)	8.3 (0.34)	8.6 (0.31)	6.7 (0.84)	7.0 (0.73)	14
m2	AM F73720	14.9	12.8	12.3	10.2	10.7	
	Eastern	11.8 (0.52)	9.3 (0.24)	9.3 (0.23)	7.2 (0.41)	7.3 (0.48)	12
m3	AM F73720	16.5	13.9	13.2	10.8	11.5	
	Eastern	12.2 (0.58)	10.0 (0.36)	9.6 (0.31)	7.2 (0.18)	7.4 (0.20)	10

been theoretically able oppose one another. The p3 of AM F73720 compares well with that of larger specimens of *Si. maddocki* (e.g., SAM P16673) in length, width and the morphological details of the occlusal surface, but it is much shorter relative to the molars and much higher crowned (pl. 47D; table 19). The median valley contains several contorted enamel ridgelets, but lacks the anteroposteriorly oriented row of cuspules characteristic of *Si. maddocki.*

SAM P35054 and AM F73720 deviate from typical eastern *Simosthenurus maddocki* in the same general directions, and the disparity in size and crown height observed is not paralleled by any other wide-ranging sthenurine species. However, their incompleteness, the fact that one is a maxilla and the other a dentary, the geographic separation of the localities, and their overall similarity to *Si. maddocki* does not permit a specific separation to be made at present. Until further material is retrieved, these specimens are best referred to *Si.* sp. cf. *maddocki.*

Simosthenurus occidentalis (Glauert, 1910a)

(pls 48-57; figs 11D,H, 14E, 20G; tables 20-21)

Procoptodon rapha, Scott, 1906, pp. 1-2.
Sthenurus sp. nov., Glauert, 1909, pp. 462.
Sthenurus occidentalis, Woodward, 1909 (*nomen nudum*), pp. 210-212.
Sthenurus atlas hacketti, Le Souef, 1910 (*nomen nudum*), p. 5.
Sthenurus occidentalis, Glauert, 1910a, pp. 31-36, pl. 5, figs 6-7.
Sthenurus occidentalis, Glauert, 1910b, pp. 53-69.
Sthenurus atlas, Glauert in Scott, 1917, pp. 2-3.
Sthenurus occidentalis, Glauert, 1926, p. 71.
Sthenurus occidentalis, Simpson, 1930, p. 77.
Sthenurus occidentalis, Anderson, 1932, pp. 383-387, figs 1, pl. 45, figs 2-3, pl. 46, figs 1-2.
Sthenurus occidentalis, Raven and Gregory, 1946, p. 7, fig. 2.
Sthenurus occidentalis, Glauert, 1948 (in part), pp. 100-102, uppermost figure.
"Sthenurus" occidentalis, Stirton, 1957, p. 124.
Procoptodon sp., Ride, 1959 (in part), pp. 54-56, fig. 8d.
Sthenurus occidentalis, Marcus, 1962, p. 299.
Sthenurus occidentalis, Merrilees, 1965, pp. 22, 24-26, 29-30.
Sthenurus sp., Merrilees, 1965 (in part), pp. 22, 24-30, table 2, figs 4-6.
Sthenurus (Simosthenurus) occidentalis, Tedford, 1966, pp. 33-39, figs 11-13, 20, tables 2, 7-10.
Sthenurus (Simosthenurus) orientalis, Tedford, 1966, pp. 50-53, figs 19-20, tables 2, 7-8.
Sthenurus (Simosthenurus) occidentalis, Merrilees, 1968a, pp. 70-77, figs 3-4, 7.
Sthenurus (Simosthenurus) orientalis, Merrilees, 1968a, pp. 69, 73.
Sthenurus orientalis, Flood, 1974, p. 180, pl. 3.
Sthenurus (Simosthenurus) occidentalis, Murray and Goede, 1977, pp. 8-9, table 6, fig. 7.
Sthenurus occidentalis, Merrilees, 1979, pp. 116-118, table 1, fig. 2.
Sthenurus occidentalis, Wells and Murray, 1979, pp. 213, 215, 217-218, fig. 6C.
Simosthenurus sp. II, Pledge, 1980 (in part), p. 134, fig. 3.
Simosthenurus orientalis, Pledge, 1980, pp. 131, 134.
Sthenurus sp., Williams, 1980, p. 111, site 62.
Simosthenurus orientalis, Flannery and Hope, 1983, pp. 37, 39.
Sthenurus occidentalis, Wells, Moriarty and Williams, 1984, pp. 325-326, 329.
Simosthenurus orientalis, Flannery, 1989, pp. 29-30, fig. 10.
Sthenurus occidentalis, Lundelius and Turnbull, 1989, pp. 4-6, fig. 2.
Sthenurus orientalis, Lundelius and Turnbull, 1989, pp. 4-5, fig. 2.
Sthenurus (Simosthenurus) occidentalis, Murray, 1991a, p. 263, fig. 5B.
Simosthenurus orientalis, Turnbull, Lundelius and Tedford, 1992a, pp. 161, 163.

Sthenurus occidentalis, Prideaux and Wells, 1997, pp. 185-186, 188, 192, figs 2, 5.
Sthenurus occidentalis, Prideaux and Wells, 1998, pp. 1, 4, 10-13, 15.
Sthenurus occidentalis, Van Huet, 1999, p. 338.
Sthenurus occidentalis, Brown and Wells, 2000, pp. 100-101, table 6.
Sthenurus occidentalis, Armand, Ride and Taylor, 2000, pp. 108-109, 112, 120.

Taxonomic Authority. Glauert, L. 1910a. The Mammoth Cave. Records of the Western Australian Museum and Art Gallery 1:11-36.

Holotype. WAM 60.10.2, left and right adult dentaries (missing coronoid and condylar processes) collected by E. A. Le Souef in 1904 from Mammoth Cave, southwestern Western Australia (Glauert 1910a-b; Merrilees 1968a). Based on faunal correlation, age of type locality is estimated to be late middle or late Pleistocene (Merrilees 1968a, 1979; pers. obs.).

Referred Specimens. Mammoth Cave, southwestern WA: WAM 62.8.31, adult cranium; WAM 63.2.194, left and right adult dentaries; approximately 100 registered maxillary, premaxillary or dentary specimens registered in the WAM, BMNH and UCMP collections.

Foundation Cave, southwestern WA: WAM HC, left juvenile dentary.

Kudjal Yolgah Cave, southwestern WA: WAM HC, right maxilla, two left adult dentaries, right juvenile dentary.

Strong's Cave, southwestern WA: WAM 61.11.10, left p3.

Terrible Cave, southwestern WA: WAM HC, right juvenile maxilla, left juvenile dentary.

Tight Entrance Cave, southwestern WA: WAM 99.10.6, adult cranium; many WAM registered and unreg. maxillary and dentary specimens.

Devil's Lair, southwestern WA: WAM registered specimens.

Melaleuca Cave, near Yanchep, WA: WAM 76.6.19.

Kiana Cliff, near Mount Misery, Eyre Peninsula, SA: FU 1650, left juvenile dentary.

West end of western Brothers Island, Eyre Peninsula, SA: SAM P34930, left adult dentary.

Curramulka Quarry, Yorke Peninsula, SA: SAM P40059, right upper molar, two right upper molar fragments.

Town Well Cave, Curramulka, Yorke Peninsula, SA: SAM P35054, right juvenile maxilla; SAM P35055, right juvenile dentary.

Emu Caves (3 Hole), Kangaroo Island, SA: SAM P40065, left upper molar.

Blackford Creek, near Kingston SE, SA: SAM P13834, right adult dentary.

Alexandra Cave, near Naracoorte, SA: SAM P27296, partial juvenile cranium; SAM P27309, right adult dentary.

Cathedral Cave, near Naracoorte, SA: FU 1060, left and right adult dentaries.

Crawford's Cornucopia Cave, near Naracoorte, SA: Listed in Reed and Bourne (2000).

Fox Cave, near Naracoorte, SA: SAM P18997, right p3; SAM P19004, right juvenile dentary; SAM P19005, left juvenile dentary; SAM P19016, left adult dentary; SAM P unreg., left and right juvenile dentaries, cranial fragments; FU 0817, left juvenile dentary.

Haystall Cave, near Naracoorte, SA: SAM P13831, right p3.

Henschke's Fossil Cave, near Naracoorte, SA: SAM P17318, adult cranium; SAM P18645, adult cranium; numerous SAM registered and unreg. dentary and maxillary specimens.

Robertson Cave, near Naracoorte, SA: Listed in Reed and Bourne (2000).

Specimen Cave, near Naracoorte, SA: SAM P40066, right adult maxilla.

Victoria Fossil Cave, near Naracoorte, SA: SAM P16535, left and right juvenile dentaries; SAM P16620 / P16621 / P16648, adult cranium and dentaries; SAM P16631, left and right juvenile dentaries; SAM P16669, left and right adult dentaries; SAM P16682 / P16994, left and right adult dentaries; SAM P20489, partial juvenile cranium; SAM P20796, left and right juvenile dentaries; SAM P20798, left and right adult dentaries; SAM P20803, cranium, dentaries and partial skeleton; SAM P28673, left and right juvenile dentaries; FU 0182, left and right juvenile dentaries; FU 0215, partial juvenile cranium; FU 1092, left and right juvenile dentaries; SAM or FU registered specimens – 23 dentaries, 11 maxillae, numerous single loose teeth.

Wet (Tomato-Stick) Cave, near Naracoorte, SA: Listed in Reed and Bourne (2000).

Comaum Forest Cave, near Penola, SA: SAM P31892, right juvenile dentary; SAM P several unreg. dentaries, maxillae and numerous single loose teeth.

Wandilo Forest Cave, near Mount Gambier, SA: SAM P40068, left and right adult dentaries.

Glencoe, near Tantanoola, SA: SAM P11393, right juvenile maxilla and dentary.

Green Waterhole Cave, near Tantanoola, SA: SAM P17244, adult cranium; SAM P17245, adult cranium and dentaries; SAM P17253, adult cranium and dentaries; SAM P17498, left and right juvenile maxillae and premaxillae; SAM P20820, complete skeleton; FU 1658, left and right juvenile maxillae and premaxillae; FU 1659, left and right adult dentaries and maxillae, fragments of neurocranium; FU 1660, left and right adult dentaries, right maxilla; FU 1661, left and right adult dentaries; FU 1662, adult cranium and dentaries; FU 1663, left and right adult dentaries; FU 1666, left and right adult dentaries and fragmented cranium; several SAM and FU registered and unreg. dentary and maxillary specimens; AM F53464 / F53466, partial juvenile cranium and right dentary.

Tantanoola Cave, SA: SAM P190, right juvenile dentary.

Mount Burr Cave, SA: SAM P17454, left adult dentary.

Goulden's Hole, near Mount Schank, SA: SAM P40067, numerous single loose teeth.

Kilsby's Hole, Mount Gambier, SA: SAM P40069, right juvenile dentary; SAM P unreg., several single loose teeth.

Mount Gambier, SA (sinkhole exposed in limestone quarry): AM F unreg., right adult maxilla.

Surprise Bay, King Island, TAS: QVM:1971:39:1, partial juvenile cranium; QVM:1971:GFV:39/2, right adult dentary.

Scotchtown Cave, Smithton, TAS: QVM numerous registered dentaries and single loose teeth.

Pleisto Scene Cave, near Montagu, TAS: TM Z several registered and unreg. maxillae, dentaries and single loose teeth.

Main Cave, near Montagu, TAS: TM Z2030, upper molar.

Titan's Shelter, Florentine Valley, TAS: TM Z2024, upper molar; TM Z2025, upper molar; Z2026, dp2; Z2027, lower molar; Z2028, upper molar; TM Z3268, upper molar.

Beginner's Luck Cave, Florentine Valley, TAS: TM Z unreg., dentary.

McEachern's Cave, near Nelson, VIC: NMV P198422, P3; NMV P198425, left juvenile dentary; NMV P198427, P3; NMV P198429, P3; NMV P198441, adult maxilla; NMV P198446, M3; NMV P198447, p3.

Shire Quarry, Strathdownie, VIC: NMV numerous dentary and maxillary specimens.

Spring Creek, Minhamite, VIC: NMV P157298, left m2; NMV P157301, left adult dentary.

Foul Air Cave, Buchan, VIC: NMV P22501, right adult dentary.

Pot Luck Cave, Buchan, VIC: NMV P22505, dentary; NMV P22506, left adult maxilla.

Royal Cave, Buchan, VIC: NMV P22502, right juvenile dentary.

Trogdip Cave, Buchan, VIC: right adult dentary.

Buchan Caves, VIC: NMV P22503 / P22504, left and right adult dentaries; NMV P48989, juvenile dentary.

Honeycomb Cave, Murrindal, VIC: right adult dentary.

Puralka Caves, VIC: NMV P32267, left and right adult maxillae; NMV P172863, maxilla; NMV P172864, right p3; NMV P173000, partial neurocranium; NMV P173001, left adult dentary; NMV P173009, cranium; NMV P183002, right dentary.

Lancefield, VIC: NMV P173533, right P3; NMV P177755, left adult maxilla; NMV P177758, left adult maxilla; NMV P177771, left adult dentary; NMV P unreg., several loose molars and a right P3.

Kilcunda (28 ft shaft), VIC: NMV P1894, left adult dentary.

Werribee River valley, west of Bacchus Marsh, VIC: NMV P156983, right adult dentary.

Cow Flat Quarries, Bathurst, NSW: AM F53800, left and right adult maxillae; AM F53801, adult cranium; AM F53899, left juvenile maxilla.

Teapot Creek Site 5, Monaro Region, NSW: AM F115318, upper molar.

Mount Fairy Caves, NSW: AM F57840, left adult dentary.

Wellington Caves, NSW: AM F10201, right adult dentary; AM F73722, left juvenile dentary; AM F106046, left juvenile maxilla; MM L9302, right premaxilla.

Revised Diagnosis. Cranium robust; masseteric process widely flared. Rostrum short, as in *"Simosthenurus" baileyi*, but narrowing more anteriorly. Frontals greatly expanded and tapered laterally, although not to magnitude of *Si. maddocki.* Lower incisor short, very robust, with strongly sinusoidal occlusal surface. Lower permanent premolar distinctly divided into narrower anterior and wider posterior portions, demarcated by wide, vertical groove below anterior extremity of buccal crest.

Etymology. Lat. *occidentalis*, 'of the west', in reference to the initial recognition of the species in Western Australia.

Description. Cranium (pls 48-53; table 20). Incisor-bearing portion of premaxilla deep and short. Short, pointed premaxillary eminence may be present on anterodorsal aspect of incisor-bearing portion of premaxilla (pl. 49A-B). Upright portion of premaxilla only slightly declined, expanding very slightly from level of lacrimal upwards. Short diastema slightly sinusoidal in lateral profile and deflected anteroventrally relative to cheek tooth row. Premaxilla comprises approximately one-third length of diastema. Anterior borders of moderately long incisive foramina lie adjacent to posterior border of I3 alveolus. Short rostrum markedly tapered anteriorly, with deep buccinator and anteorbital fossae leading posteriorly onto very wide, laterally flared masseteric process (pls 48-50, 53C). Narial aperture slightly deeper than broad. One or two infraorbital foramina lie just anteroventral of lacrimal. Nasals short, broad and not quite extending to above premaxillary eminence. Anterior border of orbit and lacrimal lie dorsal to posterior half of P3. Lacrimal bears two small foramina, often with separating or more dorsally positioned tuberosity.

Frontals markedly expanded and tapered laterally, with supraorbital crest concealing mesial portion of orbit in dorsal view (pls 48B, 51). Lateral extremity of frontal usually bears distinct supraorbital tuberosity (pls 48A-B, 49B, 50A-B). Temporal (parietal) crests well developed and often, but not always, convergent on sagittal suture. Palatal vacuities terminate adjacent to M1 metaloph or interloph valley. Postpalatine bars form a bridge across palate opposite M4 metaloph. Zygomatic arch very deep, with very wide ectoglenoid process at posterior extremity of jugal (pl. 53B). Very large postglenoid process on squamosal forms borders of glenoid fossa posteriorly, becoming rigidly fused with broad, robust ectotympanic posteriorly. Ectotympanic bears rugose surface and raised ventral keel, emphasized mesially where it contacts posteroventral wing of alisphenoid. Small elliptical postzygomatic foramen incompletely separated from dorsal surface of external auditory meatus. Tiny subsquamosal foramen lies immediately posterior to dorsal root of zygomatic process of squamosal.

Table 20. Dimensions of the adult cranium and dentary of mainland southeastern *Simosthenurus occidentalis* compared with southwestern and Tasmanian *Si. occidentalis*, showing mean, standard deviation (in parentheses) and sample size [in brackets].

Dimension	Southeast	Southwest	Tasmania
Condylobasal Length	198 (4.9) [n=4]	186 (5.7) [n=2]	-
Cranial Width	143 (1.53) [n=3]	134 (9.2) [n=2]	-
Palatal Length	126 (4.1) [n=4]	117 (2.8) [n=2]	-
Diastema Length	31.8 (1.42) [n=4]	35.0 (0.1) [n=2]	-
% Diastema Length: Palatal Length	25.2 (0.01) [n=4]	29.9 (0.64) [n=2]	-
Palatal Width between M1 Protolophs	35.0 (0.46) [n=3]	35.1 [n=1]	-
Palatal Width between M4 Protolophs	41.2 (0.53) [n=3]	36.0 [n=1]	-
Width across Frontals	91.8 (3.22) [n=4]	77.3 (0.4) [n=2]	-
Distance between Paroccipital Processes	67.7 (2.67) [n=3]	62.3 [n=1]	-
Dentary Depth	37.6 (1.70) [n=13]	35.4 (1.12) [n=8]	35.5 (2.33) [n=5]
Dentary Width	23.4 (2.03) [n=13]	18.9 (1.31) [n=8]	19.7 (0.57) [n=5]
Dentary Depth / Width	1.62 (0.14) [n=13]	1.87 (0.10) [n=8]	1.80 (0.10) [n=5]

Basicranial plane markedly elevated above level of palatal plane (pl. 49). Medial pterygoid origin wide and deep. Basioccipital flexed posterodorsally relative to basisphenoid and bears well-developed median keel. Occiput broad and moderately deep, oriented at 90° relative to dorsal surface of neurocranium. Vertical median occipital crest slight to well developed, leads ventrally to moderately wide foramen magnum, bordered by large occipital condyles (pl. 50D-E). Nuchal / lambdoid crests strongly developed and extended posterodorsally. Base of long, posteroventrally oriented paroccipital process supported laterally by rigid posterior portion of alisphenoid and wide, deep mastoid process of periotic, extending out from between exoccipital and posterior extension of squamosal.

Upper Dentition (pls 48-54; table 21). Moderately high crowned I1 quite robust and elliptical in cross section (long axis of ellipse is anteroposterior, pls 48-50, 52E-F, 53). I2 very small, with rather cylindrical crown one quarter size of I1. Blade-like I3 bears smooth, featureless buccal surface, except for occasional development of very low, narrow eminence, which ascends buccal surface of tooth at anterior end (pls 49B, 50C, 52F), effectively opposite to anterolingual crest. Anterior edge of unworn I3 very slightly concave; posterior edge markedly convex.

On unworn I3, distinct anterolingual crest forms narrow, rounded eminence that ascends lingual surface of tooth at anterior end.

Although very reminiscent of P3 in morphology, dP2 much shorter relative to width (pls 53A, 54A), consistently a round-cornered rectangular shape in basal outline, and usually only slightly wider posteriorly than anteriorly. Main crest straight and parallel or very slightly convergent with lingual crest anteriorly. Main crest not clearly divided into cuspules, but anterior and posterior extremities form most prominent points, which give rise lingually to transverse ridgelets separating anterior and posterior basins, respectively. Anterior and posterior basins approximately half size of longitudinal basin, which contains numerous very fine ridgelets. Lingual crest of unworn dP2 considerably lower in height than main crest. Completely molariform dP3 essentially inseparable from M1, except by smaller size and more tapered lophs (pls 53A, 54A).

P3 longer than all molars, usually only slightly wider posteriorly than anteriorly (pls 48A,C, 49A-B, 53B, 54B-D). Individuals with wider, inflated posterior portion usually possess a distinct posterobuccal accessory cusp (e.g., SAM P20803, pl. 54D). Main crest variably differentiated into five linked cuspules giving rise to low ridgelets on buccal side and more distinct vertical ridgelets on lingual side. Main and lingual crests parallel or very slightly convergent anteriorly. Lingual crest slightly lower than main crest. Posterior and anterior edges of P3 only very slightly convex. Anterior basin small, narrow and variably separated from longitudinal basin by strong ridgelet directed lingually from prominent anteriormost cuspule of main crest. Longitudinal basin relatively narrow, moderately shallow and traversed by many fine or coarse ridgelets. Posterior basin identical to longitudinal basin in terms of contained ridgelets and may or may not be separated from it by a stronger transverse ridgelet or set of ridgelets.

Upper molars low crowned; unworn lophs taper toward crests, especially on anterior teeth (pls 48C, 53A-B, 54A-C). Size gradient along molar row M1<M2<M3>M4, with little variation between them in morphology, except for relative narrowness of M4 metaloph. Protoloph and metaloph crests slightly convex anteriorly. Preparacrista lacks or maintains very slight connection with paracone apex, but anterobuccal aspect expanded into enlarged cusp-like portion (stylar cusp A). Preparacrista forms or merges into moderately thin precingulum, which terminates at anterolingual corner of crown. Very slight preprotocrista occasionally present on precingulum in same longitudinal line as postparacrista, but usually not differentiable from numerous other fine vertical enamel crenulations on anterior face of protoloph. Postprotocrista divided into two components; low lingual component ascends posterobuccally from protocone apex and terminates alongside similarly developed, more posteriorly oriented buccal component, which extends into interloph valley. Buccal postprotocrista component and fine mesocrista often difficult to distinguish from many very fine enamel crenulations covering loph surfaces (pl. 54). Moderately to weakly developed postparacrista sometimes with distal kink or cusp (probably stylar cusp C) and only slightly thicker than

Table 21. Cheek tooth dimensions of *Simosthenurus occidentalis* from mainland southeastern Australia, Tasmania and southwestern Australia, showing mean, standard deviation (in parentheses),and dimensions of the holotype (WAM 60.10.2) and King Island specimen (QVM:1971:39:1). Dimensions of an M4 referred to *Simosthenurus* sp. cf. *occidentalis* from Batesford Quarry, Victoria (NMV P201859) are also provided.

Tooth		L	AW	PW	AH	PH	n
dP2	Southeastern	11.2 (0.55)	8.7 (0.54)	10.2 (0.55)	8.2 (0.80)	8.4 (0.79)	8
	Tasmanian	10.2	7.4	9.4	7.0	7.1	1
	King Island	9.8	7.6	9.2	7.4	7.4	
dP3	Southeastern	11.6 (0.60)	11.4 (0.57)	12.2 (0.46)	6.4 (0.50)	6.8 (0.52)	7
	King Island	10.0	9.1	10.2	5.5	5.8	
P3	Southeastern	18.2 (0.75)	10.5 (0.91)	13.7 (1.18)	11.0 (0.98)	10.7 (1.06)	19
	Tasmanian	16.2 (0.75)	9.7 (0.54)	12.7 (1.21)	9.9 (0.49)	9.5 (0.52)	5
	King Island	16.4	9.3	12.4	9.2	9.4	
	Southwestern	16.9 (0.39)	10.3 (0.68)	12.5 (0.53)	9.8 (0.38)	9.2 (0.59)	7
M1	Southeastern	13.3 (0.55)	13.2 (0.54)	13.2 (0.52)	7.1 (0.73)	7.6 (0.66)	13
	Tasmanian	11.3 (0.19)	11.0 (0.29)	11.2 (0.22)	5.7	5.7	4
	King Island	10.8	10.7	10.9	5.9	6.6	
	Southwestern	11.2 (0.22)	11.3 (0.40)	11.5 (0.64)	5.7 (0.65)	5.8 (0.36)	6
M2	Southeastern	14.6 (0.76)	14.0 (0.65)	13.6 (0.75)	7.3 (0.75)	7.8 (1.00)	15
	Tasmanian	12.0 (0.62)	11.6 (0.37)	11.2 (0.21)	6.6	7.0	4
	King Island	11.7	11.3	11.0	7.3	7.2	
	Southwestern	12.0 (0.41)	12.0 (0.27)	11.7 (0.75)	6.2 (0.52)	6.6 (0.54)	6
M3	Southeastern	15.2 (0.55)	14.4 (0.56)	13.7 (0.54)	8.0 (1.57)	8.3 (1.49)	15
	Tasmanian	12.9 (0.40)	11.9 (0.66)	11.5 (0.64)	6.9 (0.88)	6.8 (0.70)	9
	King Island	12.3	11.5	11.3	6.9	6.8	
	Southwestern	12.7 (0.68)	12.3 (0.46)	11.6 (0.92)	6.3 (0.15)	6.3 (0.52)	5
M4	Southeastern	14.3 (0.60)	13.7 (0.47)	12.1 (0.51)	7.1 (0.66)	6.9 (0.53)	14
	Tasmanian	12.2 (0.78)	11.6 (0.98)	10.6 (0.82)	6.2 (0.71)	6.1 (0.53)	6
	Southwestern	11.6 (0.55)	11.3 (0.58)	9.6 (0.92)	5.7 (0.21)	5.1 (0.17)	3
	NMV P201859	13.6	13.1	11.5	6.6	6.7	
dp2	Southeastern	9.6 (0.45)	6.6 (0.44)	8.0 (0.33)	8.1 (0.69)	6.9 (0.77)	14
	Tasmanian	9.4 (0.42)	6.7 (0.78)	7.9 (1.06)	6.7	6.7	2
	Southwestern	8.9	5.6	7.8	7.1	5.8	1
dp3	Southeastern	10.4 (0.49)	9.1 (0.39)	9.2 (0.42)	7.3 (0.81)	7.5 (0.81)	16
	Tasmanian	8.8	7.6	7.7	-	-	1
	Southwestern	9.1	8.3	8.2	-	-	1

p3	Southeastern	17.1 (0.92)	7.5 (0.44)	10.8 (0.81)	10.5 (0.88)	9.2 (0.93)	21
	Tasmanian	15.2 (0.51)	7.1 (0.77)	9.5 (0.71)	9.5 (0.70)	8.6 (0.81)	7
	Southwestern	16.7 (0.51)	7.5 (0.29)	10.0 (0.43)	10.0 (0.49)	8.1 (0.57)	7
	Holotype	16.7	7.4	9.6	10.2	9.2	
m1	Southeastern	12.9 (0.68)	10.7 (0.49)	10.9 (0.64)	8.8 (0.89)	8.9 (0.96)	27
	Tasmanian	11.0 (0.12)	9.0 (0.21)	9.2 (0.13)	6.7 (0.35)	6.8 (0.28)	7
	Southwestern	10.9 (0.60)	9.0 (0.37)	9.3 (0.49)	6.8 (0.58)	6.7 (0.44)	7
	Holotype	10.5	8.5	8.8	7.5	7.2	
m2	Southeastern	14.4 (0.99)	12.1 (0.53)	12.2 (0.59)	9.3 (1.02)	9.5 (1.04)	21
	Tasmanian	12.0 (0.19)	10.2 (0.19)	10.3 (0.33)	7.4 (0.58)	7.6 (0.71)	6
	Southwestern	12.1 (0.55)	9.9 (0.49)	10.1 (0.56)	7.3 (0.68)	7.5 (0.70)	8
	Holotype	11.8	9.1	9.3	8.2	8.5	
m3	Southeastern	15.6 (0.79)	12.8 (0.55)	12.8 (0.63)	9.6 (0.75)	9.5 (0.77)	20
	Tasmanian	13.1 (0.46)	10.7 (0.37)	10.5 (0.53)	8.3 (0.75)	8.2 (0.82)	6
	Southwestern	12.6 (0.53)	10.4 (0.53)	10.4 (0.53)	7.5 (0.39)	7.3 (0.72)	9
	Holotype	12.2	9.9	9.8	7.8	7.8	
m4	Southeastern	14.7 (0.59)	12.4 (0.56)	11.5 (0.69)	8.8 (0.40)	8.2 (0.47)	14
	Tasmanian	12.6 (0.54)	10.4 (0.57)	9.4 (0.74)	7.0 (0.91)	6.7 (0.75)	5
	Southwestern	11.9 (0.48)	10.2 (0.37)	9.4 (0.62)	6.7 (0.60)	6.1 (0.53)	8
	Holotype	11.6	9.8	8.6	7.3	6.6	

premetacrista slightly thicker than premetacrista, with which it forms low V-shaped cleft bordering interloph valley. Postmetacrista folded dorsolingually, terminating just anterior of similarly quite thin postmetaconulecrista, which curves dorsolingually across most of metaloph posterior face. Both cristae enclose several very fine enamel crenulations, plus centrally positioned urocrista and finer buccal (mesial) component of postmetaconulecrista (pl. 54).

Dentary (pls 48A, 50A-B,D, 55-57; table 21). Dentary robust and short, with short diastema oriented in same plane as alveolar margin of cheek tooth row or slightly curved (ventrally concave). Symphyseal plate very rugose, partially fused in many specimens, and underlain by moderately developed boss (pl. 57A-C). Inferior transverse torus of symphysis extends below deep genial fossa and may be up to half length of superior transverse torus. Median dorsal groove wide and rather shallow. Anteroventral border of ramus and i1 markedly upturned. Depth of ramus usually consistent along length, except for slight increase produced by symphyseal boss and moderately to well-developed digastric eminence. Digastric sulcus usually, but not always deep; may extend anteriorly to below m2. Relatively shallow buccinator sulcus extends posteriorly to beneath m1. Anterior mental foramen anteroventral to anterior end of buccinator sulcus. Posterior mental foramen positioned anywhere between mid-depth on ramus below m2 protolophid and low on ramus beneath m3 hypolophid, but occasionally absent.

Anterior root of ascending ramus adjacent to m4, with anterior border oriented slightly anterior of vertical before inflecting to vertical. Masseteric fossa shallow to moderately deep; masseteric foramen small to moderately large. Ventral border of masseteric fossa level with or just higher than posterior extremity of buccinator sulcus (pls 48A, 55E). Deep, wide medial pterygoid fossa moderately well inflated posteriorly, with very well developed angular process. Inferior mandibular foramen usually small, elliptical and opened posteromesially, but may be large, more circular in shape, and opening more mesially. Mylohyoid groove and associated processes poor to well developed. Mandibular condyle moderately large, with posterior surface of tall coronoid process only slightly anterodorsally oriented.

Lower Dentition (pls 48, 50, 55-57; table 21). Short i1 very robust and upturned, with curved ventral edge very slightly procumbent relative to anteroventral border of ramus, or continuing in same line (pls 55-57). Unworn occlusal surface markedly sinusoidal in shape and rising to distinct point anteriorly (pls 55A-D, 57F). Lingual enamel layer of i1 only slightly thinner than buccal enamel (pls 55A,C, 57B-C,F). Minimally worn occlusal surface of i1 of adult specimens level with halfway up p3.

Although similar in general morphology to p3, dp2 shorter relative to width, with main crest divided into three or four cuspules (pls 55, 56B,D, 57F). Prominent anteriormost cuspule gives rise to distinct buccal and lingual ridgelet. Buccal ridgelet contacts anterobuccal aspect of crescentic buccal crest, which is usually joined to second main crest cuspule via small ridgelet. Median valley fully enclosed and semicircular to circular in shape. Completely molariform dp3 similar in general outline to larger m1, but differs by tapering more toward lophid crests, bearing well-developed parametacristid, shorter trigonid and cristid obliqua directly contacting hypoconid apex (pl. 57D). Enamel crenulations tend also to be fewer on dp3.

Length of p3 exceeds that of each molar (pls 48A, 55E, 56, 57). Main crest divided into six distinct cuspules, each usually giving rise to coarse buccal and lingual ridgelet. Anteriormost cuspule of main crest often slightly more prominent than succeeding cuspules. Tooth narrows markedly anteriorly and is divided into smaller, narrower, often slightly inflated anterior portion and much larger, wider, buccally inflated posterior portion. Anterior and posterior portions of p3 demarcated by wide vertical groove on buccal surface, below union of buccal crest and ridgelet arising from second or third main crest cuspule (pl. 57D-E). D-shaped median valley shallow and filled with many contorted ridgelets. Posterior extremity of slightly to markedly curved buccal crest usually contacts main crest just anterior to its posterior extremity.

Although varying in size, low-crowned lower molars vary little between one another in morphology, except that m1 has slightly shorter trigonid, and m4 has slightly narrower hypolophid relative to protolophid. Lophid crests straight and parallel, with moderate anterior turn of buccal ends (pl. 57D-F). Moderately short trigonid underlain by rounded precingulid, which may extend to anterobuccal corner of tooth. Paracristid low and usually thick, especially anteriorly; may give rise to very fine enamel crenulations directed into trigonid basin. Lingual component of paracristid usually positioned one quarter of way across protolophid face from protoconid apex, but may be less so. Very low buccal component of paracristid terminates on side of lingual component. Premetacristid and parametacristid usually indistinguishable from numerous very fine enamel crenulations in trigonid basin and on protolophid anterior face (pl. 57D-F). Cristid obliqua usually arises on hypolophid slightly lingual of where paracristid arises on protolophid and trends more anteroposteriorly before terminating on protolophid posterior face. Several fine enamel crenulations dominate surface of hypolophid lingual to cristid obliqua. Preentocristid absent. Hypolophid posterior face may bear very low inflation or quite distinct A-shaped postcingulid shelf; several very fine vertical enamel crenulations consistently present.

Synonymy of Simosthenurus orientalis *Tedford, 1966. Simosthenurus orientalis* is recognized here as a junior synonym of *Si. occidentalis* (Glauert, 1910a), based on information gleaned from the many specimens collected from cave deposits in mainland southeastern Australia, Tasmania and southwestern Australia over the past 35 years. Tedford (1966) referred two specimens to *Si. orientalis*, the holotype adult dentary from the Wellington Caves (AM F10201), and another from Bingara (AM F88542).

AM F88542 is a partial ramus missing the p3 crown (see Marcus 1976:fig. 27a-b). Its extremely worn molars are reminiscent of AM F10201, but the proportions of the ramus are quite different from those of AM F10201 and from *Si. occidentalis* from anywhere across its range. Ramus depth relative to width is much less in AM F88542 than AM F10201 (depth/width: 39.0/24.7 = 1.58 compared with 38.3/19.3 = 1.98). Measurements given by Tedford (1966:50) indicated similar proportions, but depth for AM F10201 was greatly underestimated, giving a false value of 1.42. In fact, the dentary of AM F88542 is distinctly wider than that of AM F10201 and has a deeper buccinator sulcus than either AM F10201 or any specimens of *Si. occidentalis*. At present, AM F88542 cannot be allocated to the same species as AM F10201 or to any other species with confidence (e.g., *Si. euryskaphus*).

Tedford (1966) diagnosed *Simosthenurus orientalis* (AM F10201), thus separating it from *Si. occidentalis*, on the basis of its larger size, extension of the inferior transverse torus of the symphysis to behind p3, positioning of the anterior root of the ascending ramus opposite the posterior end of m4, proportionally smaller p3, and longer trigonid on the lower molars. In fact, overall dentary size is not especially useful for distinguishing AM F10201 from western or Tasmanian *Si. occidentalis*, because specimens from these regions may be just as large (e.g., TM Z3254, WAM 97.5.76). However, the dentary of southwestern *Si. occidentalis* is about 13% narrower on average than the mainland southeastern form within which AM F10201 appears to fit (table 21). While the p3 of southwestern *Si. occidentalis* is proportionally longer (26% of cheek tooth row length) than the mainland southeastern form (22% of cheek tooth row length), *Si. occidentalis* from Tasmania is intermediate (24% of cheek tooth row length). This may highlight a geographic cline in this character and strongly suggests that p3 proportion is not useful for supporting a specific separation of AM F10201. From the much larger samples now known, it is also clear that proportional dimensions of the AM F10201 trigonid fit within the range of variation seen within western and Tasmanian *Si. occidentalis*, even though the molars are larger.

Nevertheless, while the symphysis may extend to beneath the posterior extremity of p3 and the ascending ramus may take root opposite the m4 protolophid or interlophid valley, I have not observed conditions quite as extreme as expressed by AM F10201 in any individuals of *Si. occidentalis.* By themselves, though, these differences are insufficient to justify a separate species for AM F10201, especially when variation within the mainland southeastern form covers the disparity observed between AM F10201 and *Si. occidentalis* from southwestern Australia and Tasmania. I conclude that AM F10201 and the mainland southeastern form within which it comfortably fits represent a variant of *Si. occidentalis* typified by molars that are approximately 17% larger on average than in *Si. occidentalis* elsewhere. This supports the view of Merrilees (1965) that the large sthenurine specimens from the Strathdownie Quarry in southwestern Victoria were probably referable to a variant of *Si. occidentalis* and is concordant with the geographic size differences observed in other simosthenurins (e.g., *Simosthenurus maddocki*, *"Si." pales*, *"Procoptodon" browneorum*) across southern Australia.

Variation. Cranium. Complete or near-complete adult crania of *Simosthenurus occidentalis* are known from Mammoth and Tight Entrance Caves in southwestern Australia, and the Naracoorte and Green Waterhole Caves in southeastern Australia. The two western crania (WAM 62.8.31, pl. 51B; WAM 99.10.6, pls 51A, 53B) are very similar to the southeastern crania (e.g., SAM P16648, pls 49B, 50E, 51D; SAM P20803, pl. 51C; SAM P20820, pl. 48) in condylobasal length and maximum cranial width, but their temporal fossae narrow slightly more anteriorly, the frontals are slightly less inflated and the diastema is proportionally longer (table 21). Considering the similarity in overall cranial size, the longer diastema in the western crania is most likely correlated with the decreased size of the molars and I3.

The smallest cranial form is represented by an older juvenile specimen from Surprise Bay, King Island (QVM:1971:39:1, pls 52A-B). This specimen, originally described by Anderson (1932), is missing the neurocranium and M4 on both sides. M3 was fully erupted. In width across the frontals and rostral proportions, this specimen compares well in size and morphology to a partial juvenile cranium from Alexandra Cave, near Naracoorte (SAM P27296, pl. 52C-D). In this specimen, only the first molar has erupted, indicating a much younger ontogenetic age than the King Island specimen. Considering the marked increase in size and change in cranial proportions (e.g., frontal widening) that occur between the time that M1 and

M3 erupt, it seems as though the cranium of QVM:1971:39:1 was much smaller than that of *Simosthenurus occidentalis* elsewhere. Based on this limited evidence, it is conceivable that the King Island *Si. occidentalis* was an island dwarf, akin to the King Island emu, *Dromaius ater*.

Marked variation is seen within several aspects of the mainland southeastern crania. Frontals may be expanded laterally to (a) a sharp rugose point above the center of the orbit (e.g., SAM P16648, pl. 51D); (b) a more rounded rugose point above the anterior of the orbit (e.g., SAM P20820, pl. 48B); (c) a slightly more elongate, crescentic expansion above the length of the orbit (e.g., SAM P20803, pl. 51C). Frontals may also vary in their degree of dorsal inflation. In two Green Waterhole Cave specimens (SAM P20820, SAM P17247), the frontals are domed in lateral profile, angling down anteriorly such that the dorsal surface of the nasals lies at a considerably lower level than the more posterior dorsal cranial surface (pl. 48A). This may be contrasted with SAM P16648 (pl. 49B). Other points of difference include the size and shape of the masseteric process, degree of convergence of the temporal crests, shape of the nasofrontal suture, development of the lacrimal tuberosities, size of the premaxillary eminence and cranial width.

Upper Dentition. While I3 is about 20% smaller in the southwestern crania than the mainland southeastern specimens, very little variation in incisor morphology is evident across the board. I2 varies slightly in size, while I3 occasionally possesses a very low, narrow eminence on its buccal surface anteriorly. In contrast, dP2 varies in size, degree of posterior inflation and relative development of the ridgelets in the longitudinal basin. Only size variation is apparent for dP3, except for the occasional presence on the postparacrista of what is most likely a reemergent stylar cusp C (e.g., SAM P27296, pl. 54A). P3 varies markedly in size and general outline, and may be (a) a rounded, parallel-sided, rectangular shape (e.g., WAM 97.5.78, pl. 54B); (b) inflated posterolingually and slightly anterobuccally (e.g., SAM P16648, pl. 54C); (c) inflated on both sides posteriorly, but not anteriorly; or (d) inflated posteriorly and anteriorly (e.g., SAM P20803, pl. 54D). The posterobuccal accessory cusp may be small or large, conical or crescentic, tiny and barely differentiable, or completely absent. Upper molars vary slightly in the strength of the connection between the preparacrista and paracone apex, thickness of the precingulum, coarseness and number of enamel crenulations, and development of the distal kink (stylar cusp C portion) of the postparacrista. Differences in cheek tooth size between *Simosthenurus occidentalis* from southwestern Australia, Tasmania and mainland southeastern Australia are clearly evident. Both the southwestern and Tasmanian representatives possess molars that are approximately 17% smaller on average than their mainland southeastern relatives (table 21). Premolar sizes of southwestern *Si. occidentalis* fall within the lower end of the ranges for mainland southeastern *Si. occidentalis*, making the premolars proportionally longer relative to the molars. On average, the premolars of Tasmanian *Si. occidentalis* are 9% smaller than in mainland southeastern *Si. occidentalis* (table 21), which means that premolar proportional length is intermediate between representatives of the other two regions.

Dentary. Adult dentaries may vary in overall size, relative size of the digastric eminence, depth and anterior extent of the digastric sulcus, extent of the inferior transverse torus of the symphysis, shape of the diastema, size of the masseteric foramen, posterior inflation of the medial pterygoid fossa, and position of the posterior mental foramen. This latter feature may even vary between left and right dentaries of the one individual; the features may be tiny or absent on one side and moderately sized on the other. While some differences may be partially sexually dimorphic, I have not discerned any bimodality in size, proportion or discrete morphological attributes. With increasing ontogenetic age, ramus depth increases, the masseteric and mental foramina enlarge, and the digastric sulcus and inferior transverse torus of the symphysis expand (pls 55, 57A-C). Although I have not examined dentary depth and width dimensions for any Tasmanian *Simosthenurus occidentalis* specimens, the mainland southeastern dentaries are wider on average than the southwestern specimens (table 21). As with the proportionally shorter diastema, greater dentary width must be correlated with greater molar size. The Tasmanian specimens might therefore be predicted to fall closer to southwestern *Si. occidentalis* in this regard.

Lower Dentition. Only very slight variation in size and development of very shallow, longitudinal undulations on the buccal enamel surface may be observed for the i1 of *Simosthenurus occidentalis.* The dp2 varies in size and general inflation of the tooth outline. Occasionally, the anterior end of the buccal crest may be confluent with the buccally directed ridgelet from the anteriormost cuspule of the main crest rather than with that of the second cuspule. SAM P16637 shows this condition, and on its left dp2, but not the right, a very small pocket is present on the anterolingual aspect. Differences in size and slight morphological variation in dp3 and lower molar features have been alluded to above or in their description, although it is also worth noting that division of the paracristid and cristid obliqua of western specimens tends to be less marked anteriorly than in southeastern specimens.

Like its upper counterpart, the p3 of *Simosthenurus occidentalis* is a highly variable tooth. While p3 is consistently longer than any molar, the occasional specimen is aligned slightly more anterobuccally relative to the molar row. Most p3 specimens bear a main crest divided along its entire length into six distinct cuspules, but occasional specimens lack the coarse buccal and lingual ridgelets that usually arise from these cuspules. The general outline of the tooth varies in much the same way as the P3. The anterior portion of the tooth may be narrow or inflated, but is consistently much narrower than the posterior portion, which is also variably inflated buccally. The buccal crest varies in shape from a rounded L-shape (pl. 56A) to a semicircular crescent (pl. 57D-E). Contact between the buccal and main crests anteriorly is usually via the ridgelet arising from the second main crest cuspule, but is occasionally via the equivalent ridgelet from the third cuspule. On the p3 of WAM 97.5.75, the buccal crest terminates just short of contact with the third cuspule of the main crest, which lacks a buccally directed ridgelet. Posteriorly, the buccal crest most commonly curves around to contact the main crest just anterior to its posterior extremity, but the posterior ends of both crests sometimes meet in a broad V-shape. Finally, a tiny cuspule or eminence is infrequently observed on the anterior end of the tooth, halfway up the crown.

I have observed developmental abnormalities in two specimens of *Simosthenurus occidentalis*. In SAM P17245, the right p3 bears a greatly enlarged subsidiary cuspule at the front of the tooth, which is similar in height to the anteriormost cuspule of the main crest (pl. 56E; Pledge 1980:fig. 3). The subsidiary cuspule on the left p3 is much smaller, similar to that seen in other occasional *Si. occidentalis* specimens. In SAM P16535, supernumerary fifth molars are partially erupted on both the left and right dentaries. They both appear to have been turned 90° mesially, although the orientation of the left molar may only be judged from its alveolus. The right tooth is also slightly contorted, and angled anteriorly. From a developmental perspective, it is interesting to note that the hypolophid of the supernumerary molar is narrower relative to the protolophid than its counterpart in the specimen's normally formed and erupted m4.

Overall, *Simosthenurus occidentalis* is one of the most variable of all sthenurines, both within samples and across its large geographic range. Although the molars of mainland southeastern *Si. occidentalis* are clearly larger than those of southwestern and Tasmanian individuals, morphology tends to differ no more between regions than within them. Proportional differences are seen in premolar length relative to cheek tooth row length and relative dentary depth. Recovery of material referable to *Si. occidentalis* from the intervening Nullarbor Region should prove enlightening.

Comparison With Other Taxa. Cranium. In overall size and general cranial form, *Simosthenurus occidentalis* is most similar to *Metasthenurus newtonae*, *"Si." baileyi*, *Si. maddocki* and *"Procoptodon" browneorum*. It differs from *M. newtonae* by being more brachycephalic and by having more inflated frontals, deeper anteorbital and buccinator fossae and deeper zygomatic arches. *Si. occidentalis* shares a short rostrum with *"Si." baileyi*, but it narrows more markedly anteriorly. The frontals of *Si. occidentalis* are also more laterally expanded. By comparison, *"P." browneorum* has a more elongate, rather cylindrical rostrum and frontals that do not taper to any degree. *Si. maddocki* is most similar to *Si. occidentalis* in size, brachycephaly and general cranial form, but its frontals are even more expanded and tapered laterally. *Si. maddocki* also has a much smaller masseteric process, a very slight postglenoid process, a more anteriorly sloped dorsal zygomatic surface, and more convergent orbits courtesy of the great expansion of the temporal fossa anteriorly as well as posteriorly.

Upper Dentition. I1 is slightly larger than that of *"Procoptodon" gilli*, slightly larger and broader than that of *Metasthenurus newtonae* and *"P." browneorum*, and slightly larger and narrower than in *"Simosthenurus" baileyi*. While the I2 of

Si. occidentalis is very similar to that of *"Si." baileyi*, the I3 of *Si. occidentalis* has a relatively more elongate crown. Shape of the unworn I3 crown is very similar to that of *Si. maddocki*, but is much larger.

The dP2 of *Simosthenurus occidentalis* is somewhat reminiscent of the dP2 of *"Procoptodon" browneorum* and *"Si." baileyi* in size, but it is less inflated in outline, especially posteriorly. In this regard, *Si. occidentalis* most closely resembles the much smaller *Si. maddocki*, from which it differs by having a larger anterior basin and poorly demarcated posterior basin. The *Si. occidentalis* P3 is a highly variable tooth. Some specimens approach the general morphology of some individuals of *"P." browneorum*, but they are longer relative to the molars, the main and lingual crests are less convergent anteriorly, and the outline is very rarely trapezium-shaped. Others more closely resemble the form observed in some individuals of *Metasthenurus newtonae* and *"P." gilli*, but the *M. newtonae* P3 is usually shorter relative to molar length and more inflated in outline. While it would be difficult to differentiate some of the smaller P3 specimens of *Si. occidentalis* from those of *"P." gilli*, the most common outline of the tooth, a rather round-cornered rectangular shape, is unlike any *"P." gilli* P3 that I have observed.

The upper molariform teeth of *Simosthenurus occidentalis* are larger than but most reminiscent of *Si. maddocki* in morphology, especially in the extensive distribution of very fine enamel crenulations. *Si. occidentalis* differs from *Si. maddocki* by sporting a weaker connection between the preparacrista and paracone apex, and a less curved postparacrista and premetacrista. Otherwise, the upper molars of these two species are very similar.

Dentary. Among the simosthenurins, the dentary of *Simosthenurus occidentalis* is very generalized in both size and overall morphology. The overall shape of the dentary is most similar to that of *Si. maddocki*, *"Si." baileyi*, *"Procoptodon" gilli* and *"P." oreas*. It is most readily distinguished from *"P." gilli* by its greater size, better-developed digastric sulcus, and markedly less flared posteroventral border of the masseteric fossa; from *"P." oreas* by its less convex lateral ramus surface, anteroposteriorly longer masseteric fossa, and smaller masseteric foramen; from *Si. maddocki* by its greater robustness, more posteriorly extended symphysis, shorter and more horizontally oriented diastema, better-developed digastric sulcus, and markedly shallower buccinator sulcus; and from *"Si." baileyi* by its slightly greater robustness, more posteriorly extended symphysis, slightly shorter diastema, better-developed digastric sulcus, less flared posteroventral border of the masseteric fossa, and less posteriorly inflated medial pterygoid fossa.

Lower Dentition. The orientation of the *Simosthenurus occidentalis* i1 closely approximates that of the anteroventral border of the dentary and is thus similar to *"Si." baileyi*, *"Procoptodon" browneorum* and *"P." gilli*. The crown is shorter and more robust than in each of these species, and it has a more sinusoidal unworn occlusal surface. *Si. occidentalis* most closely resembles *"P." oreas* in overall i1 morphology, but its unworn occlusal surface is more sinusoidal and the lingual enamel layer is thicker. Morphology of the dp2 is intermediate between that of

"Si." baileyi and *"P." gilli*, but is closer to the former in size. The p3 is larger than but otherwise quite similar to that of *"P." gilli* and, to a lesser degree, *"P." oreas*. It differs from the p3 of *"P." gilli* by being lower crowned, and having a shallower median valley containing more ridgelets and a main crest usually divided into six distinct cuspules. The degree of development of this latter feature is variable, but usually intermediate between *"P." oreas* and *Si. maddocki.*

Lower molars of *Simosthenurus occidentalis* are inseparable from those of *Si. euryskaphus*, but the dp3 differs by being relatively larger and having a protolophid less sloped toward its crest, and a narrower, shorter trigonid. In every other aspect the molariform teeth of *Si. euryskaphus* fit within the range of variation observed within mainland southeastern *Si. occidentalis.* The fine enamel crenulations on the lophid surfaces are common to these two species and *Si. maddocki*, but those of the latter species are more numerous. The molars of *Si. occidentalis* also differ from *Si. maddocki* by having a better-developed precingulid, a lower and less sharply turning paracristid, and a lower cristid obliqua. Nevertheless, there is a tendency for the lower molars of southwestern *Si. occidentalis* to be intermediate between eastern *Si. occidentalis* and *Si. maddocki* in these features.

Geographic Distribution. Southwestern Australia, south-central and southeastern Australia, central and northwestern Tasmania, King Island, central eastern New South Wales (fig. 29; table 44).

Temporal Distribution. Late middle to late Pleistocene.

Simosthenurus sp. cf. *occidentalis*

(table 21)

Sthenurus sp., Whitelaw, 1989, p. 625.

Referred Specimen. Batesford Quarry, near Geelong, VIC: NMV P201859a, left M4. Late Pliocene.

Remarks. NMV P201859a is a left upper molar with slight wear on the loph crests and posterior face of the protoloph. All molar dimensions fall well within the range of southeastern Australian *Simosthenurus occidentalis* (table 21) and most aspects of the tooth's morphology also closely match this species, including numerous fine enamel crenulations on all loph faces. However, NMV P201859a differs by possessing a strongly developed postparacrista and well-developed premetacrista, states not observed in any other specimens of *Si. occidentalis* across its wide Pleistocene distribution. Because the variation in the form of these cristae in NMV P201859a and *Si. occidentalis* exceeds that typically observed within species of this genus, the specimen is best referred to *Si.* sp. cf. *occidentalis* pending the discovery of more complete material. A left metaloph fragment and left trigonid fragment (MNV P201859b-c) from the same locality may pertain to the same taxon, but are too incomplete to allow an adequate assessment.

"Simosthenurus" antiquus (Bartholomai, 1963)

(pls 58-59; table 22)

Sthenurus pales, De Vis, 1895 (in part), pp. 94-96, pl. 15, fig. 3.
Sthenurus oreas, De Vis, 1895 (in part), pp. 96-97, pl. 16, figs 7-8.
"Sthenurus" oreas, Stirton, 1957, p. 124.
Sthenurus antiquus, Bartholomai, 1963, pp. 67-71.
Sthenurus antiquus, Merrilees, 1965, pp. 24, 26.
Sthenurus antiquus, Merrilees, 1968a, pp. 66, 69, 73.
Sthenurus (Simosthenurus) antiquus, Tedford, 1966, pp. 49-50.
Sthenurus (Simosthenurus) antiquus, Bartholomai, 1972, pp. ix, xii, fig. 2.
Sthenurus antiquus, Archer and Bartholomai, 1978, p. 14, fig. 9C-C'.
Simosthenurus sp. cf. *occidentalis*, Williams, 1980, p. 103, site 7a.
Sthenurus antiquus, Prideaux and Wells, 1997, pp. 181, 191-194, 196, fig. 6b.
Sthenurus antiquus, Prideaux and Wells, 1998, pp. 1-4, 11-15, table 1, figs 7B, 8B, 9.

Taxonomic Authority. Bartholomai, A. 1963. Revision of the extinct macropodid genus *Sthenurus* Owen in Queensland. Memoirs of the Queensland Museum 14:51-76.

Holotype. QM F2975, fragmentary left adult maxilla (containing P3, M1-2) collected in the late 1800s from the Chinchilla Sand, Chinchilla, southeastern Queensland. On the basis of faunal correlations with deposits elsewhere, age of type locality is considered to be Pliocene (e.g., Woods 1962; Bartholomai 1963, 1973a; Archer and Wade 1976; Skilbeck 1980; Tedford et al. 1992).

Referred Specimens. Chinchilla, QLD: QM F2973, left adult dentary; QM F3816, left M2; UCR 23028, partial left upper molar.

Darling Downs (probably Chinchilla), QLD: QM F2931, left adult dentary (united with p3 originally registered as QM F2932); QM F2974, right adult dentary; QM F2976, right adult maxilla; QM F2977, left adult maxilla; QM F3387, right M3.

Revised Diagnosis. P3 very short relative to p3, markedly tapered anteriorly, with very large posterior basin relative to longitudinal basin; lingual crest markedly lower than main crest. Buccal crest of p3 oriented anterobuccally, parallel to posterior half of main crest, which folds slightly over toward buccal side anteriorly; rounded cingulid present at anterior end of crown. Hypolophid posterior face with broad inflation across basal portion.

Etymology. Lat. *antiquus*, 'old', in reference to the relative antiquity of this species, which was the first sthenurine described from a deposit unequivocally older than Pleistocene in age.

Description. Maxilla (pl. 58). Very little of maxilla is preserved even in most complete specimen (holotype, QM F2975). Nevertheless, wide root suggests large, robust masseteric process. Palatal vacuities extended anteriorly to opposite anterior edge of m1 (pl. 58C).

Upper Dentition (pl. 58; table 22). Only known P3 lacks posterolingual corner (pl. 58C), but approximate morphology may be estimated from general contours of adjacent regions (pl. 58D). Posterior extremity of P3 was clearly similar in width to protoloph of M1, but it narrows markedly anteriorly. P3 quite symmetrical in general outline, with buccal and lingual sides rather straight. Main crest markedly exceeds lingual crest in height. Anterior three cuspules of main crest accompanied by fine vertical ridgelets ascending from cuspule apices on buccal and lingual sides. Largest lingual ridgelet originates from anteriormost cuspule and is continuous with short transverse ridgelet directed buccally from anterior extremity of lingual crest. This separates narrow, shallow anterior basin from moderately deep longitudinal basin, which receives several fine lingual ridgelets from main crest cuspules and lingual crest. Large, broad posterior basin separated from longitudinal basin by coarse transverse ridgelet similar in height to lingual crest (pl. 58C-D). Several very fine ridgelets trend from coarse transverse ridgelet into posterior basin. Posterobuccal accessory cusp absent, but extremely slight eminence apparent on posterobuccal aspect of main crest.

Low-crowned upper molars (pl. 58A-C) slightly rounded in basal outline anteriorly, but becoming squarer farther back in tooth row. Protoloph subequal in width to metaloph on M1-2; protoloph slightly wider in M3; M4 is unknown. Short preparacrista does not connect to paracone apex and is reduced to cusp-like portion probably representing stylar cusp A (pl. 58C). Precingulum, which is continuous with preparacrista and thin in M1, becomes successively thicker (anteroposteriorly) in more posterior molars. Precingulum continues to lingual side of tooth and terminates near crown base, immediately anterior to protocone. Tiny hint of preprotocrista contacts precingulum just lingual of tooth midline. Postprotocrista divided into two components, with very low one connected to protocone apex and terminating on lingual side of thicker component shifted slightly buccally across protoloph posterior face. This is directed into interloph valley and onto metaloph anterior face before terminating (pl. 58C). Buccal to postprotocrista near tooth midline, moderately developed mesocrista extends to interloph valley. Postparacrista well developed, meeting weak premetacrista. Postmetaconulecrista forms a narrow shelf, sweeping across posterior face of metaloph from metaconule apex, terminating at small cusp corresponding to stylar cusp E. Postmetacrista ascends rather sinusoidally from metacone apex to stylar cusp E. Three small, low, rather vertical enamel crenulations or cristae borne on metaloph face between postmetaconulecrista and postmetacrista. Urocrista very slight on M1, but not evident on M2-3.

Dentary (pl. 59). Ramus only moderately short judging from proportions of most complete specimen (QM F2931), and becoming proportionally wider posteriorly (dentary depth / width: 1.66). Cheek teeth quite large relative to ramus size. Digastric eminence moderately developed; digastric sulcus moderately deep, extending anteriorly to beneath m3 trigonid (pl. 59B). Very small posterior mental foramen located mid-depth on ramus below m3 protolophid. Anterior root of ascending ramus adjacent to m4 protolophid. Broken anteroventral border of masseteric fossa appears to have been slightly flared laterally (pl. 59D), with ventral border probably just above level of posterior mental foramen.

Lower Dentition (pl. 59; table 22). Large, elongate p3 exceeds length of each molar and, like P3, is quite wide posteriorly, tapering smoothly and symmetrically to narrow anterior end (pl. 59D). Sinusoidal main crest runs along midline of p3 anteriorly, then curves lingually before straightening slightly and terminating at posterolingual corner of tooth. Anterior half of main crest bends over buccally relative to posterior half of crest. Base of anterior end of main crest bears rounded, cingulid-like inflation, emphasized lingually (pl. 59D). Main crest consists of five cuspules accompanied by low vertical ridgelets, which descend lingually and buccally from cuspule apices. Buccal crest approximately half length of main crest, only slightly convex buccally, and close to parallel with posterior half of main crest. Thus, it is oriented anterolingually across posterobuccal region of tooth, from center of posterior extremity of p3 to midway along tooth on buccal side. Median valley deep and open both anteriorly and posteriorly, because buccal crest does not converge toward main crest (pl. 59D). Tight complex of coarse enamel crenulations centered in median valley. Note inaccuracy of p3 drawing in Bartholomai (1963:fig. 9).

Lower molars low crowned, but slenderness of lophids (viewed laterally) may convey false impression of slightly greater crown height (pl. 59A). Lophid crests of QM F2931 relatively straight, but not quite parallel; interlophid valley slightly narrower lingually than buccally. Paracristid and cristid obliqua distinct, but quite low, with lingual component shifted about one quarter of way across lophid faces (pl. 59C-D). Buccal components indistinct. Trigonid moderately short. Lingual component of paracristid directed anteriorly and just slightly lingually, before turning rather sharply into thicker, transverse anterior portion, which extends to anterolingual corner of tooth (pl. 59C-D). Low, rounded precingulid extends from anterobuccal extremity of tooth to beneath anterior portion of paracristid. Few very fine enamel crenulations on anterior lophid faces almost completely obliterated by slight wear. Hypolophid posterior face bears marked, rounded inflation extended right across ventral aspect and overlapped by succeeding molars.

Variation. No variation is discernable between the molars and ramus fragments of this species.

Comparison With Other Taxa. Upper Dentition. The short, small nature of the P3 relative to p3 in *"Simosthenurus" antiquus* is unique among those sthenurines for which both teeth are known. Among the simosthenurins, only *"Procoptodon" williamsi*, *P. pusio*, *P. rapha* and *P. goliah* have a P3 that is shorter relative to the succeeding molars. Relative to the main crest, the lingual crest is lower relative to the main crest than in any other simosthenurins, including *"Si." pales*. The upper molars of *"Si." antiquus* are very similar to both *"Si." pales* and *"Si." baileyi.*

They differ from *"Si." pales* by having a thicker precingulum, more smoothly curved preparacrista, slightly fewer, finer enamel crenulations and are usually notably smaller in size. They differ from the similarly sized *"Si." baileyi* by having a slightly smaller postparacrista and thicker precingulum.

Dentary. In all aspects of its morphology, the ramus of *"Simosthenurus" antiquus* closely resembles *"Si." baileyi*, especially the holotype of this species (SAM P13670). The *"Si." antiquus* dentary may be distinguished by bearing relatively larger molars.

Table 22. Cheek tooth dimensions of *"Simosthenurus" antiquus*, showing mean, standard deviation (in parentheses) and dimensions of holotype (QM F2975); est. = estimated only, due to missing posterolingual corner of crown.

Tooth	Specimen	L	AW	PW	AH	PH	n
P3	Holotype	14.9	8.3	11.3 est.	8.4	10.1	
M1		13.1 (0.99)	12.2 (0.21)	12.1	6.9	-	2
	Holotype	12.4	12.0	12.1	-	-	
M2		14.9 (0.07)	13.4 (0.14)	13.0 (0.21)	7.2 (1.13)	7.4 (1.34)	2
	Holotype	14.8	13.5	12.8	6.4	6.4	
M3		16.0 (0.35)	13.6 (0.49)	12.9 (0.85)	8.3 (0.35)	8.3 (0.00)	2
M4		-	-	-	-	-	
p3	QM F2931	17.6	8.2	10.1	12.2	11.4	1
m1	QM F2931	13.8	-	11.0	8.5	8.5	1
m2		15.3 (0.85)	11.9 (0.71)	11.9 (0.61)	10.4	10.4 (0.07)	3
m3		15.8 (0.78)	12.4 (0.31)	12.2 (0.35)	10.3 (0.35)	10.3 (0.00)	3
m4		15.4 (0.49)	12.4 (0.21)	11.0	8.5 (0.35)	7.7 (0.35)	2

Lower Dentition. Despite the uniqueness of the disparity between the respective lengths of the upper and lower permanent premolars, I am confident that Bartholomai (1963) was correct in referring QM F2931 to *"Simosthenurus" antiquus*, given the manner in which the teeth occlude. A primary occlusal action in unworn premolars is the shearing of the p3 buccal crest against the lingual side of the P3 main crest, followed by the lateral excursion of the p3 buccal crest across the P3 longitudinal basin (see Functional Morphology). The coarse transverse ridgelet, which separates the longitudinal basin from the large posterior basin (which itself receives the m1 trigonid), acts as a posterior barrier to the p3 buccal crest. Because this crest is quite short, it fits perfectly into the P3 longitudinal basin, indicating that these teeth would have little trouble occluding. What is difficult to explain is why the main crest on p3 is so long, because its anterior end would not seem to occlude with anything. It remains a good possibility that, as with several sthenurine species, considerable variation in premolar size and morphology will eventually be revealed in *"Si." antiquus*. Overall, the p3 of *"Si." antiquus* is most similar in morphology to that of *"Si." baileyi*, but it differs by being larger, higher crowned and slightly narrower anteriorly, and by having a sinusoidal main crest.

The lower molars of *"Simosthenurus" antiquus* are almost indistinguishable from those of *"Si." pales* and differ only by usually being smaller and having lophids with fewer fine enamel crenulations. However, the molars of the Brisbane River *"Si." pales* specimen (QM F9428, pl. 69A-B) are small and can only be separated from those of *"Si." antiquus* by the presence of a slightly sharper postcingulid shelf on the posterior face of the hypolophid and by having the ends of the protolophid turned slightly more anteriorly. Most distinctly, QM F9428 possesses the unmistakable *"Si." pales* p3 morphology (pl. 69C), and its dentary depth as a juvenile equals that of the *"Si." antiquus* adult dentary. This suggests that if the individual had made it to adulthood, the size of its dentary would have markedly exceeded that of *"Si." antiquus*. *"Si." antiquus* also compares well with *"Si." brachyselenis* in lower molar size and general form, but it is distinguished by a better-developed anterior portion of the paracristid, slightly finer enamel crenulations on the lophid faces, and a postcingulid that is more of a low inflation than a distinct shelf. *"Si." antiquus* also appears to lack the bifurcated paracristid characteristic of the m1-2 of *"Si." brachyselenis*, although QM F2931 is slightly too worn to be certain.

Geographic Distribution. Southeastern Queensland (fig. 27).

Temporal Distribution. Early late Pliocene.

"Simosthenurus" sp. cf. *antiquus*

(pl. 60; table 23)

Simosthenurus sp. cf. *orientalis*, Tedford and Wells, 1990, p. 278.

Referred Specimens. Green Bluff Locality, Warburton River, SA: UCMP 56901, left adult maxilla. Late Pleistocene.

Cassidy Locality (Ulabarrinna), Warburton River, SA: UCMP 47947, right adult dentary. ?Late Pleistocene.

Remarks. UCMP 56901 is a left maxilla fragment preserving M3-4 (pl. 60D). Although no M4 is known for *"Simosthenurus" antiquus*, the molar morphology of UCMP 56901 closely matches that of *"Si." antiquus*. However, notwithstanding the difference in tooth positions, the size variation between M1-2 in the *"Si." antiquus* holotype (QM F2975) and M3-4 of UCMP 56901 is greater than that usually seen in one sthenurine population. Large size differences within species

Table 23. Cheek tooth dimensions of *Simosthenurus* sp. cf. *antiquus* from Pleistocene localities on the Warburton River.

Tooth	Specimen	L	AW	PW	AH	PH
M3	UCMP 56901	16.4	-	14.1	9.2	9.3
M4	UCMP 56901	15.8	14.8	12.8	8.9	8.8
m3	UCMP 47947	16.9	13.8	13.6	9.4	9.4

known from different geographic regions are commonplace, but since UCMP 56901 lacks P3, the morphology of which is diagnostic for *"Si." antiquus*, and since a record of *"Si." antiquus* in the Pleistocene of northern SA would constitute a great temporal and geographic range extension, the specific identification of UCMP 56901 must await confirmation through discovery of more complete material from the region.

UCMP 47947 (pl. 60A-C) closely matches QM F2931 in all aspects of the ramus preserved and in several aspects of m3 morphology. However, the m3 of UCMP 47947 is comparatively wider and has straighter lophid crests than in the Darling Downs *"Simosthenurus" antiquus* specimens (pl. 60C). UCMP 47947 also lacks the diagnostic premolar, akin to UCMP 56901. If UCMP 47947 is attributable to *"Si." antiquus*, then a deep symphysis that extends slightly below the genial fossa to beneath the middle of p3, a large, deep-rooted and moderately procumbent i1, a very large anterior mental foramen positioned just anterior to p3, a diastema positioned slightly lower than the alveolar margin of the cheek tooth row, and a wide, moderately deep buccinator sulcus (pl. 9E-G) could be added to the *"Si." antiquus* description. But as with UCMP 56901, the differences between the m3 preserved in UCMP 47947 and its equivalent in topotypic *"Si." antiquus*, and the disparity in age and geographic location (fig. 34), augur against a definitive specific identification for this specimen at this time.

"Simosthenurus" baileyi Prideaux and Wells, 1998

(pl. 61-63; figs 12F, 18G; table 24)

Sthenurus (Simosthenurus) occidentalis, Tedford, 1966 (in part), p.39.
Sthenurus sp. cf. *maddocki*, Williams, 1980, p. 103.
Simosthenurus sp., Flannery, 1984, pl. 6.9-1 (largest dentary in lower left photograph).
Sthenurus sp. cf. *brownei*, Patterson and Rich, 1987, p. 91.
Simosthenurus sp., Tedford and Wells, 1990, p. 277.
Sthenurus (?Simosthenurus) baileyi, Prideaux and Wells, 1998, pp. 1-15.

Taxonomic Authority. Prideaux, G. J. and R. T. Wells. 1998. *Sthenurus baileyi* sp. nov., a new fossil kangaroo from the Pleistocene of southern Australia. Transactions of the Royal Society of South Australia 122:1-15.

Holotype. SAM P13670, partial juvenile cranium with associated left and right dentaries; apparently collected by F. R. George in 1902 from a bone breccia in an eroded cave at the western end of the western Brothers Island, Coffin Bay, Eyre Peninsula, South Australia (Brown 1908). Age of type locality is Pleistocene based on faunal composition (Prideaux and Wells 1998).

Paratypes. Victoria Fossil Cave, near Naracoorte, SA: FU 0004, left and right adult dentaries; FU 0167, left P3, M1; FU 0168, right p3; FU 0294, partial right dentary; SAM P16531 / P16558, left and right adult dentaries; SAM P28282, right juvenile dentary; SAM P28659, right M2, left M3. It is possible that FU 0004, FU 0167 and SAM P28659 belong to same individual based on proximity in deposit, degree of enamel wear and occlusal fit. Specimens collected by Gavin J. Prideaux, Roderick T. Wells and others. Age of locality is late middle Pleistocene (Ayliffe et al. 1998; Grün et al. 2000).

Referred Specimens. Crawford's Cornucopia Cave, near Naracoorte, SA: Listed in Reed and Bourne (2000).

Specimen Cave, near Naracoorte, SA: SAM P36619, right adult dentary.

Curramulka Quarry, Yorke Peninsula, SA: SAM P40057, right I3, left m2.

Site 8, Cooper Creek, SA: UCMP 60674, left adult maxilla.

Lindsay Hall Cave, Nullarbor Plain, WA: WAM 92.9.10a, left and right adult premaxilla; WAM 03.11.01, right p3.

Revised Diagnosis. Cranium similar in size to *Simosthenurus occidentalis*, but with less expanded frontals, a broader rostrum and wider nasals. P3 similar to that of *"Procoptodon" browneorum*, but more rounded in outline, with narrower longitudinal basin and two cuspules anterior to prominent posterobuccal accessory cusp. Upper molars bear weak postprotocrista and very well developed postparacrista. In morphology, p3 is most similar to *"Si." antiquus*, but lower crowned, with a straighter lingual crest. Lower molars with very short trigonid; buccal component of paracristid and premetacristid well developed. Paracristid and cristid obliqua fine and low.

Etymology. Named in honor of Edwin H. Bailey for his contributions to the success of paleontological work in the Naracoorte Caves.

Description. Cranium (pl. 61). Vertical portion of premaxilla flared dorsally, providing elongate contact with nasals. Diastema short, anterior one-third comprising premaxilla and posterior two-thirds maxilla. Incisive foramina long and narrow, anterior border adjacent to posterior extremity of I3 alveolus. Rostrum short, tapered anteriorly (pl. 61B). Anteorbital and buccinator fossae rather shallow anteriorly; anteorbital fossa deepest just anterior to zygomatic arch. Masseteric processes well formed, rather narrow, distal extremities eroded off on both sides. Nasals very broad posteriorly and, although broken anteriorly, were clearly short.

Nasofrontal suture gently sinusoidal (pl. 61B). Frontals moderately inflated anteriorly; supraorbital crests only slightly developed. Temporal (parietal) crests moderately developed, not fully convergent upon sagittal suture. Large infraorbital foramen positioned anteroventral and mesial to lacrimal foramen, just below orbital center. Palatal vacuities extend to opposite anterior extremity of M1 (pl. 61C). Right lateral extremity of broken postpalatine bar lies opposite M4 interloph valley.

Upper Dentition (pls 61-62; table 24). I1 crown moderately high and slightly widened. I2 very small, sub-cylindrical and one-third size of I1. I3 high crowned but not especially long (pl. 61A,C), with a small anterolingual crest. Rounded in general outline, dP2 reminiscent of P3, especially lingually, but much shorter relative to width. Buccal and lingual crests straight, except for buccal curvature of lingual crest at posterior extremity (pls 61C, 62A). Anterior basin small, quite deep, and separated from longitudinal basin by low transverse ridgelet. Posterior basin apparently relatively large, about half size of longitudinal basin. Completely molariform dP3 similar in general outline to M1 (pls 61C, 62A), but lophs slightly oblique (not perpendicular to anteroposterior axis of tooth). Precingulum very thin, terminating before reaching lingual extremity of tooth. Although worn, premetacrista clearly well developed. Protoloph crest apparently very convex anteriorly. Interloph valley lacks enamel crenulations. Very low, barely detectable postprotocrista lies just lingual of tooth midline. Postmetaconulecrista curves dorsobuccally from metaconule to meet near-vertical postmetacrista. Elongate enamel crenulation present lingual of postmetacrista; slight urocrista positioned centrally on posterior face of metaloph.

P3 rounded in outline and tapered anteriorly (pl. 62B-C). Longitudinal basin shallow. Buccal crest barely exceeds lingual crest in height. Anterior third of lingual crest parallel to buccal crest; posterior two-thirds of crest curved lingually. Small anterior basin separated from longitudinal basin by transverse ridge descending from anteriormost buccal cuspule, which terminates adjacent to anteriormost lingual cuspule. Posterior basin short but distinct, separated from longitudinal basin by low, slightly anterobuccally oriented transverse ridgelet, which extends from posteriormost lingual cuspule to buccal crest. Posterobuccal accessory cusp well formed, slightly lower than posterior part of buccal crest. Three small, poorly separated accessory cuspules lie anterior to posterobuccal accessory cusp (pl. 62B-C).

Upper molars (pls 61-62) low crowned, with protoloph equal in width to metaloph on M1-2, but wider in M3-4. M1-2 proportionally shorter relative to width than M3-4. Precingulum thin; buccal extremity terminates at probable stylar cusp A (pls 61C, 62A,F). This is connected to paracone apex via slight preparacrista. Two to four slight, vertical enamel crenulations centered on precingulum, most lingual of which may be preprotocrista. Weak postprotocrista trends across protoloph face into interloph valley, uniting with low, posteriorly broadening mesocrista on M1-2 (pls 61C, 62A). Posterior face of M3-4 protoloph bears several fine, crenulations from which mesocrista is indistinguishable (pls 61C,

Table 24. Cheek tooth dimensions of *"Simosthenurus" baileyi*, showing mean, standard deviation (in parentheses), and dimensions of holotype (SAM P13870).

Tooth	L	AW	PW	AH	PH	n
dP2	10.5	7.5	10.0	6.2	6.1	
dP2 holotype	As above					
dP3	10.6	9.9	10.8	-	-	
dP3 holotype	As above					
P3	17.2 (0.14)	9.9 (0.28)	12.9 (0.21)	9.8 (0.28)	9.7 (1.06)	2
P3 holotype	17.1	10.1	13.0	10.0	10.4	
M1	12.3 (0.14)	12.3 (0.21)	12.3 (0.21)	6.3 (0.07)	6.6 (0.42)	2
M1 holotype	12.2	12.1	12.1	6.2	6.3	
M2	13.8 (0.21)	13.2 (0.21)	13.1 (0.14)	6.9 (0.42)	7.2 (0.35)	2
M2 holotype	13.6	13.3	13.0	7.2	7.4	
M3	14.5 (0.00)	13.7 (0.21)	12.7 (0.66)	7.0 (0.14)	6.7 (0.23)	3
M3 holotype	14.5	13.5	12.8	7.1	6.6	
M4	13.3 (0.71)	12.9 (0.71)	11.1 (0.28)	6.3 (0.57)	6.2 (0.14)	2
M4 holotype	13.8	13.4	11.3	6.7	6.1	
dp2	9.6	7.5	8.9 (0.99)	7.3	6.2 (0.71)	2
dp2 holotype	-	-	8.2	-	5.7	
dp3	9.8 (0.14)	8.9 (0.71)	9.1 (0.28)	6.0	6.0	2
dp3 holotype	9.9	8.4	8.9	-	-	
p3	16.2 (0.78)	8.0 (0.32)	9.7 (0.44)	9.5 (1.04)	9.0 (1.03)	6
p3 holotype	15.3	7.9	9.5	8.9	9.0	
m1	12.0 (0.41)	10.1 (0.22)	10.1 (0.26)	7.1 (1.01)	7.0 (0.76)	4
m1 holotype	12.1	10.0	10.2	6.4	6.5	
m2	13.3 (0.46)	11.4 (0.27)	11.2 (0.41)	8.3 (0.49)	8.4 (0.63)	4
m2 holotype	12.8	11.5	11.7	8.0	8.0	
m3	14.1 (0.64)	12.2 (0.37)	11.9 (0.42)	8.2 (0.93)	7.9 (0.71)	4
m3 holotype	13.6	12.6	12.4	7.8	7.8	
m4	13.9 (0.11)	12.1 (0.26)	11.1 (0.42)	7.4 (0.43)	7.2 (0.35)	5
m4 holotype	13.8	12.2	11.2	7.1	7.1	

62A,F). Postparacrista strongly developed, forming buccal border of interloph valley and meeting very slight premetacrista on anterior face of metaloph (pls 61A,C, 62A,D,F). Interloph valley contains few very fine to no enamel crenulations. Postmetaconulecrista sweeps across posterior face of metaloph and terminates just posterior to end of postmetacrista, except on M1, where cristae overlap slightly. Postmetaconulecrista encloses two to three small, distinct enamel crenulations or fine cristae from which urocrista is indistinguishable.

Dentary (pl. 63A-B,F). Moderately proportioned, except for lateral, shelf-like expansion of posteroventral border of masseteric fossa and posteriorly inflated medial pterygoid fossa (pl. 63A-B,F). Ramus moderately deep for width, particularly in region of symphysis, with anteroventral border steep and straight. Symphyseal plate gently tapered anteriorly and posteriorly (pl. 63B). Genial fossa small due to minimal development of inferior transverse torus, which terminates beneath middle of p3. Digastric eminence moderately developed. Digastric sulcus extends from below anterior extremity of medial pterygoid fossa to below m2 hypolophid or m3 protolophid (pl. 63B). Diastema short, with median dorsal groove deep and relatively wide. Very shallow buccinator sulcus extends from near posterior end of diastema, dorsal to large anterior mental foramen, deepening slightly posteriorly and terminating below m1 hypolophid (pl. 63A). Posterior mental foramen lies mid-depth on dentary below m2 hypolophid.

Anterior root of ascending ramus lies adjacent to m3 hypolophid. Medial pterygoid fossa inflated posteriorly, projecting well beyond border of masseteric fossa when viewed laterally. Masseteric fossa deep, largely due to laterally expanded posteroventral border (pl. 63A,F). Ventral border of masseteric fossa at level of posterior end of buccinator sulcus. Masseteric foramen moderately large and vertical in orientation. Inferior mandibular foramen rather small, opening largely posteriorly. Wide mylohyoid groove lies at anterior extremity of medial pterygoid fossa, anteromesial to inferior mandibular foramen. Groove was apparently partially overhung by anterodorsally directed process (broken off) at anteromesial border of medial pterygoid fossa and posteroventrally directed process below posterior extremity of postalveolar fossa. Mandibular condyle moderately large. Angular process well developed, rising dorsally to acute point.

Lower Dentition (pl. 63; table 24). Short, slender and upturned i1 with occlusal surface just above level of alveolar margin of cheek tooth row (pl. 63A-B,F). In length, dp2 equals dp3 (pl. 63G). In morphology, dp2 is similar to p3, but wider relative to length. As with p3, three cuspules dominate anterior half of main crest; each gives rise to ridgelet on buccal side. Ridgelets probably terminated in tiny cuspules as in p3, but due to wear sustained by only known specimens preserving dp2-3 (SAM P28282, SAM P13670, pl. 63G), these have become confluent with buccal crest, conveying impression of more elongate crest. Protolophid of completely molariform dp3 tapers more toward lophid crest than hypolophid. As with dp2, wear has removed several features. However, cristid obliqua apparently more strongly developed than on molars, curving directly from hypoconid apex into interlophid valley and terminating centrally on protolophid posterior face. Very weak preentocristid curves from entoconid into interlophid valley, terminating lingual to cristid obliqua. Enamel crenulations, similar to those on molars, were apparently present on anterior faces of dp3 lophids. Slight, rounded postcingulid on hypolophid posterior face confluent with slight postentocristid.

Considerably longer than any molar, p3 bears main crest extending from posterolingual corner of tooth to midline anteriorly (pl. 63). Three cuspules form anterior part of main crest, with each bearing pair of lateral ridgelets; one descends buccally, one lingually. Buccal ridgelets terminate at low shelf formed by three variably discernable, confluent cuspules lying immediately anterior to buccal crest. Buccal crest straight, short, equal in length to and mirroring shape of posterior part of main crest. Median valley of moderate depth and width, containing few fine ridgelets or one coarse transverse ridgelet.

Lower molars low crowned (pl. 63), with unworn protolophid and hypolophid crests straight and close to parallel. Trigonid very short. Posterior portion of paracristid progressively more separated into two components from m1 to m4. Buccal component well developed, sweeps smoothly anterolingually from protoconid apex, terminating on buccal side of very low, fine lingual component, which arises one quarter of way across protolophid (pl. 63F-G). In posterior molars, posterior portion of lingual component originates well below protolophid crest. Transverse anterior portion of paracristid similar in length and development to posterior portion. Few fine enamel crenulations extend into trigonid basin from low down on protolophid anterior face. Lingual side of trigonid bordered by well-developed premetacristid (pl. 63F-G), which terminates at anterolingual extremity of paracristid (position of paraconid). Precingulid small, extends lingually as very thin peninsula along anterior extremity of molar. Cristid obliqua morphology mirrors paracristid in all respects, except that buccal component is lower. Preentocristid very low and barely visible. Shallow interlophid valley lacks enamel crenulations. Hypolophid posterior face weakly or moderately inflated basally, or with a low shelf-like postcingulid.

Variation. Upper Dentition. P3 is similar in the holotype (SAM P13670) and FU 0167 (pl. 62B-C), with the slight occlusal wear in FU 0167 responsible for most superficial differences. However, P3 is slightly wider anteriorly in the holotype, across the whole tooth and the longitudinal basin. The lingual surface of the holotype P3 is slightly more convex and rounded than in FU 0167. The three cuspules anterior to the main posterobuccal accessory cusp are more separated in FU 0167. Only one slight difference is detectable on comparison of M1-3 of the holotype, FU 0167 (M1) and SAM P28659 (M2-3): the postparacrista is slightly larger in the holotype. While the greater wear sustained by FU 0167 and SAM P28659 might account for the difference, the manner in which teeth occlude suggests that it most likely reflects morphological variation. The M4 of UCMP 60674 (pl. 62F) is slightly smaller than in the holotype, but otherwise inseparable.

Dentary. Among the six dentary specimens known for *"Simosthenurus" baileyi*, three characters clearly vary. Depth and extent of the digastric sulcus is the most variable. Although deep and extending from the anterior extremity of the medial pterygoid fossa to below the m2 hypolophid in the holotype and SAM P16531 / P16558, the digastric sulcus is much shallower and only extends to below the m4 protolophid in FU 0004. The sulcus is even shallower in SAM P28282. The degree to which the medial pterygoid fossa is inflated posteriorly is also variable, with inflation greatest in SAM P16531 / P16558 (pl. 63A-B,F), slightly less in the holotype, FU 0004 and SAM P28282, and least in FU 0294. However, it is sufficient in the latter to mark it as a diagnostic feature for *"Si." baileyi.* Relative dentary depth relative is greater in SAM P16531 / P16558 (depth / width: 1.79) compared with FU 0004 (1.65) and SAM P28282 (1.61). The value is lowest in the holotype (1.46). Intraspecific variation in dentary depth relative to width is common among sthenurine species.

Lower Dentition. No variation that cannot be ascribed to differential wear has been noted for i1. In most specimens, p3 varies minimally in size, but those of SAM P16531 / P16558 and the holotype are slightly shorter and narrower than the majority. Morphology varies only slightly, primarily in the form of the buccal crest and width of the median valley. The anterior half of the buccal crest in SAM P28282 is slightly higher than the posterior half and curves posterolingually, becoming confluent with a transverse ridgelet crossing the median valley. This crest morphology is not seen in any other specimens, but a similar ridgelet crosses the median valley in SAM P13670. Apart from this, only the relative inflation of the anterior region of the p3 varies slightly. The Nullarbor p3 (WAM 03.11.01, pl. 63C-D) is inseparable in size and morphology from the more eastern specimens.

There is little variation in both size and morphology of the lower molars, although the development of the premetacristid, paracristid and cristid obliqua in the paratypes is slightly weaker than in the holotype, and their anterior lophid faces bear a few more enamel crenulations. The postcingulid is also more shelf-like in the paratypes than in the holotype, with the exception of FU 0294, where the ventrobuccal region of the hypolophid posterior face is more inflated.

Comparison With Other Taxa. Cranium. Although P3 is unerupted in the *"Simosthenurus" baileyi* holotype, the presence of M4 in occlusion indicates that P3 eruption was imminent. While a slight increase in size from this ontogenetic stage to full adult might have taken place, little change in morphology is likely to have occurred judging from those simosthenurins represented by both older juvenile and adult crania (e.g., *Si. maddocki*, *Si. occidentalis*, *"Procoptodon" browneorum*, *"P." gilli*). The cranium of *"Si." baileyi* is similar in size and brachycephaly to that of *Si. occidentalis.* Rostral length of the two species is similar, but the anteorbital and buccinator fossae are deeper in *Si. occidentalis.* Coupled with this, the edge of the *Si. occidentalis* diastema is mesially concave compared with the condition in *"Si." baileyi*, in which the diastema edge is more rounded, and the axis of the edge is straighter. In this regard, *"Si." baileyi* more closely resembles *"P." gilli* and *Sthenurus andersoni.*

The rostrum of *"Simosthenurus" baileyi* appears less tapered anteriorly than that of *Si. occidentalis*, largely because its frontals are less expanded and the narial aperture is relatively larger. Among those simosthenurins for which the splanchnocranium is known, frontal inflation and formation of the supraorbital crests are greatest in *Si. maddocki* and *Si. occidentalis*, and only slightly less in *"Procoptodon" browneorum.* The frontal region is quite narrow in *"P." gilli.* The proportions displayed in *"Si." baileyi* are intermediate between *"P." browneorum* and *"P." gilli*, but the nasals are very wide and constitute a greater proportion of the dorsal aspect of the rostrum than in any other simosthenurin except *Si. maddocki.* Overall, the short, broad nature of the rostrum is characteristic of *"Si." baileyi.* The anterior extent of the palatal vacuities in *"Si." baileyi* is similar to that seen in a number species, terminating close to the dP3 metaloph or what would be close to the posterior extremity of the P3 in an adult. Allowing for the damage in the holotype, the masseteric process appears to have been well developed and is intermediate in size between those of *Si. maddocki* and *"P." browneorum.*

Upper Dentition. The I1 of *"Simosthenurus" baileyi* is slightly higher crowned and slightly broader than in *"Procoptodon" browneorum.* It is relatively lower crowned than in *"P." gilli* and not as broad as in the species of *Sthenurus.* Although smaller, it is most alike the I1 of *"Si." pales.* The small, cylindrical I2 is intermediate in size between that of *"P." browneorum* and *Si. occidentalis.* I3 is most similar in size and general morphology to the i1 *"P." browneorum*, but the buccal surface is smooth and flat, lacking vertical undulations. Although slightly shorter and less inflated laterally than in *"P." browneorum*, the *"Si." baileyi* dP2 is closest in overall morphology to the dP2 of that species. Orientation of the buccal and lingual crests is also similar in *"P." browneorum*, but the posterior basin appears to have been larger in *"Si." baileyi.* Wear sustained by the only known dP3 specimen (holotype) has obliterated several characters useful for comparison, but the tooth appears to have been generally similar to that of *"P." browneorum*, but with a smaller precingulum, larger premetacrista and many fewer and finer enamel crenulations on the loph faces and in the interloph valley. The P3 of *"Si." baileyi* is most reminiscent of those of *"P." browneorum* and *"Si." antiquus* in morphology, particularly in the shape and orientation of the crests and the anterior basin. However, *"Si." baileyi* possesses a shallower and narrower longitudinal basin and a prominent posterobuccal accessory cusp with two cuspules anterior to it. Occasional premolars of *Metasthenurus newtonae* possess one small, poorly differentiated cuspule anterior to a similarly poorly differentiated posterobuccal accessory cusp. The posterior basin of the *"Si." baileyi* P3 is smaller than in either *"Si." antiquus* or *"P." browneorum.* Height of the lingual crest in *"Si." antiquus* is notably lower relative to the buccal crest than in either *"P." browneorum* or *"Si." baileyi.* The *"Si." antiquus* P3 is also smaller relative to the size of the molars.

Although similar in size to those of *Metasthenurus newtonae*, the low-crowned upper molars of *"Simosthenurus" baileyi* are much more similar in morphology to the larger *"Si." antiquus* and *"Si." pales.* They may be distinguished from those of *"Si." antiquus* by their thinner precingulum, less distinct postprotocrista and stronger postparacrista. Moreover, the postmetacrista slightly overlaps the postmetaconulecrista on the M1 of *"Si." baileyi*, in contrast to *"Si." antiquus* (see The Sthenurine Radiation). Morphologically, only the larger postparacrista, slightly weaker postprotocrista, slightly lower crown height and proportionally longer M3-4 separate the upper molars of *"Si." baileyi* from those of the larger *"Si." pales.* The postparacrista of *"Si." baileyi* is more strongly developed than in any other simosthenurin species to which it is even vaguely similar in morphology.

Dentary. In general morphology, the dentary of *"Simosthenurus" baileyi* is most reminiscent of *Si. occidentalis*, *Si. maddocki*, *"Si." antiquus* and *"Procoptodon" gilli.* The dentary differs from that of *Si. occidentalis* by having a slightly less robust ramus, with a very small inferior transverse torus of the symphysis, a shallower and less extensive digastric sulcus, a more laterally flared posteroventral border of the masseteric fossa, and a more posteriorly inflated medial pterygoid fossa; *Si. maddocki* by having a shorter diastema, a slightly more posteriorly inflated medial pterygoid fossa, and a much shallower buccinator sulcus; and *"P." gilli* by being larger and having a less flared posteroventral border of the masseteric fossa. The holotype dentary of *"Si." baileyi* is very similar overall to the most complete known dentary of *"Si." antiquus* (QM F2931), differing only in the proportionally smaller size of the dentition. Morphology of the *"Si." baileyi* symphyseal plate most resembles that of *Si. maddocki*, wherein the symphysis is gently tapered anteriorly and only extends slightly below the genial fossa. Relative to the length of the ramus, the diastema of *"Si." baileyi* is slightly longer than in *Si. occidentalis*, *"P." browneorum* and *"P." gilli.* It is most similar in length to *Si. maddocki*, but is not convex dorsally. Depth and extent of the digastric sulcus is similar to but slightly more pronounced than in *Si. maddocki.* The degree of intraspecific variation in depth and extent of the digastric sulcus is also similar between these two species. Lateral expansion of the posteroventral border of the masseteric fossa into a wide shelf is similar to *Si. maddocki*, but slightly less than in *"P." gilli.* The medial pterygoid fossa is more inflated posteriorly than in any other sthenurine species, with the exception of *P. goliah.*

Lower Dentition. Orientation of the *"Simosthenurus" baileyi* i1 closely approximates that of the anteroventral border of the dentary and in this respect is similar to *Si. occidentalis*, *"Procoptodon" browneorum* and *"P." gilli.* Procumbency of the i1 is intermediate between that displayed by *Si. occidentalis* and *"P." browneorum*, but overall morphology is closest to the latter species. Size and morphology of dp2 most resembles that of *"P." browneorum*, but it is not as narrow anteriorly relative to the posterior part of the tooth, nor is the median valley as narrow. Superficially, the dp2 buccal crest appears similar in length to that of *"P." browneorum*, but this impression is caused by the crest becoming confluent with the small cuspules anterior as a result of wear. While the *"Si." baileyi* dp3 is also similar in size and morphology to that of *"P." browneorum*, the cristid obliqua still contact the hypoconid apex, while fewer enamel crenulations are present on the lophid faces. In these respects, *"Si." baileyi* is more similar to *Si. occidentalis.*

In morphology, size relative to the molars, and orientation of the main and buccal crests, the *"Simosthenurus" baileyi* p3 is similar to that of *"Si." antiquus.* It differs by being smaller, lower crowned and slightly more inflated anteriorly, and by having a straighter main crest oriented from the posterolingual to the antero-buccal corner of the tooth. In *"Si." antiquus*, the posterior part of the main crest trends anterobuccally, then straightens down the p3 midline. The p3 of both *"Si." baileyi* and *"Si." brachyselenis* share a ridgelet traversing the median valley and the major features of the main crest, but the *"Si." baileyi* p3 is easily distinguished by its larger size, slightly greater width relative to length and longer, straighter buccal crest. In size and general outline, the *"Si." baileyi* p3 is also similar to that of *"P." browneorum*, but it is lower crowned, with a shorter buccal crest.

The lower molars of *"Simosthenurus" baileyi* are slightly lower crowned than in *"Si." antiquus*, but careful examination reveals that they do not to differ to the degree suggested by Prideaux and Wells (1998). Compared with the lower molars of *"Si." antiquus*, those of *"Si." baileyi* are smaller and have a shorter trigonid, a better-developed premetacristid and buccal component of the paracristid, and finer, lower lingual components of the paracristid and cristid obliqua, and only a slight inflation low on the hypolophid posterior face. The *"Si." baileyi* molars are thus shorter relative to width (length / width: 0.84) than those of *"Si." antiquus* (0.72). Otherwise, *"Si." baileyi* resembles *Archaeosimos cegsai* in the fine, low nature of the paracristid and cristid obliqua, short trigonid, absence of a distinct postcingulid, and relatively smooth lophid faces. The molars of *A. cegsai* differ by being slightly narrower and lacking a distinct premetacristid and buccal paracristid component.

Geographic Distribution. Southeastern, south-central and northern South Australia, Nullarbor Region (fig. 31; table 44).

Temporal Distribution. Middle to late Pleistocene.

"Simosthenurus" sp. cf. *baileyi*

(pl. 62G; table 24)

Sthenurus sp., Bartholomai, 1978b, pp. 127-128.

Referred Specimens. Bluff Downs, northeastern QLD: QM F9104, right M1 metaloph. Early Pliocene.

Remarks. QM F9104 is a partial upper molar from Bluff Downs (fig. 27) thought by Bartholomai (1978b) to belong to a species close to *"Simosthenurus" antiquus.* He distinguished it from that species on the basis of its weaker premetacrista and the fact that, instead of uniting, the postmetacrista and postmetaconulecrista overlap very slightly (pl. 62G). While the size of the Bluff Downs metaloph is closer to that of the *"Si." antiquus* M1 than the M3 (cf. Bartholomai 1963), the morphological differences still stand. In fact, QM F9104 is morphologically indistinguishable from the M1 metaloph of the *"Si." baileyi* holotype, differing only by having three slightly coarser vertical enamel crenulations on its posterior face. Given the variability of such dental attributes between individuals of well-represented species, this difference may well be taxonomically insignificant. Nevertheless, because allocation of QM F9104 to *"Si." baileyi* would represent a temporal range extension from the middle to late Pleistocene back to the early Pliocene, and because QM F9104 consists of one molar fragment, it is best referred to *"Si."* sp. cf. *baileyi* until more material is retrieved from the Bluff Downs deposit.

"Simosthenurus" brachyselenis Prideaux and Wells, 1997

(pls 64-65 table 25)

Sthenurus (Simosthenurus) oreas, Tedford, 1966 (in part), pp. 39-40.
Sthenurus oreas, Dawson, 1985, p. 66, table 1.
Sthenurus (Simosthenurus) brachyselenis, Prideaux and Wells, 1997, pp. 181-196, figs 1-3.
Sthenurus brachyselenis, Prideaux and Wells, 1998, pp. 4, 12-15, fig. 8A.

Taxonomic Authority. Prideaux, G. J. and R. T. Wells. 1997. New *Sthenurus* species (Macropodidae, Diprotodontia) from Wellington Caves and Bingara, New South Wales. Proceedings of the Linnean Society of New South Wales 117:181-196.

Holotype. AM F31026, right juvenile dentary (containing i1, dp2-3, m1 erupted, m2 in early stage of eruption, excavated p3, missing most of ascending ramus) collected in 1932 by Charles W. Anderson and W. Schevill from a phosphate mining drive, Wellington Caves, New South Wales (Dawson 1985). The stratigraphic provenance of the specimen is uncertain, but most likely a unit within the ?middle Pleistocene Mitchell Cave beds (see Osborne 1983, 1997; The Sthenurine Radiation).

Referred Specimen. Bone Cave, Wellington, NSW: AM F106043, left adult dentary.

Revised Diagnosis. Dentary short, massive and very deep relative to size of cheek teeth. Lower incisor robust, with relatively straight unworn occlusal surface, only very slightly curved dorsally at anterior extreme. Extremely short buccal crest connects to posterior extremity of main crest on dp2, which is markedly shorter than dp3. Short p3 similar in length to m1 and only inflated in posterobuccal portion; buccal crest short, crescentic, extends only one-third length of tooth.

Etymology. Gr. *brachys* 'short', *selenis* 'crescent', in reference to the short, crescentic nature of the buccal crest on dp2 and p3.

Description. Juvenile Dentary (AM F31026, m2 in initial stages of eruption, pl. 64; table 25). Ramus relatively deep and narrow for much of length. Prominent symphyseal boss projecting below ventral border of ramus corresponds to distinct inferior transverse torus of symphysis, which extends to beneath middle of dp2. Genial fossa circular. Digastric sulcus very shallow; digastric eminence small. Diastema very short and oriented in same plane as i1 occlusal surface. Anteroventral border of dentary steep and slightly curved. Buccinator sulcus narrow, deep and close to alveolar margin, extending from between anterior mental foramen and halfway point on diastema, to below hypolophid of dp3. Dorsal border of anterior mental foramen immediately below anterior end of buccinator sulcus. Small posterior mental foramen lies directly beneath protolophid of m1, just below level of ventral border of masseteric fossa. Masseteric foramen elongate, leading into slightly anteroventrally oriented masseteric canal. Inferior mandibular foramen moderately sized and elliptical.

Adult Dentary (AM F106043, pl. 65; table 25). Dentary very short and massive, with ramus depth almost twice that of width and approximately six times deeper than height of cheek teeth. Inferior transverse torus of symphysis large, producing massive boss beneath p3. Only anterior end of digastric sulcus is preserved, but depth here indicates remainder of sulcus would have been deep. Curvature of ventral border of ramus suggests well-developed digastric eminence. Along with upward curvature of alveolar margin below posterior molars, this results in increased ramus depth posteriorly. Anterior extremity of dentary not preserved, but orientation of remaining symphyseal region suggests steeply upturned i1. Buccinator sulcus shallow beneath p3, but deepens posteriorly beneath m1, terminating beneath m2 trigonid. Posterior mental foramen small and positioned beneath m3 trigonid, posteroventral to posterior extremity of buccinator sulcus. Only small portion of ascending ramus is preserved, but it shows dorsal edge of apparently large inferior mental foramen positioned at level of alveolar margin of cheek tooth row.

Lower Dentition (pls 64-65). Robust i1 upturned relative to cheek tooth row, with occlusal surface long, straight and oriented in same plane as diastema (pl. 64A-B). Anterior extremity of occlusal surface only very slightly curved dorsally. Between ventrolingual extension of buccal enamel and dorsal enamel flange, lingual enamel layer is very thin (pl. 64B). Relative to dp3 length, dp2 very short (pl. 64). Main crest consists of four prominent cuspules, with apex of posteriormost cuspule slightly longer than anterior three. Low vertical ridgelets descend lingually from anterior three cuspules; thicker buccal ridgelets descend more posteriorly from anterior two cuspules. Second of these terminates immediately anterior to buccal crest, resulting in small V-shaped notch. Transverse ridge directed buccally from posterior extremity of main crest confluent with extremely short, crescentic buccal crest, which is barely more than a swollen cusp (pl. 64C).

Relative to hypolophid, protolophid of completely molariform dp3 slightly more tapered toward crest (pl. 64C). Paracristid and cristid obliqua fine and low.

Table 25. Cheek tooth dimensions of the *"Simosthenurus" brachyselenis* holotype (AM F31026) and referred specimen (AM F106043).

Tooth	Specimen	L	AW	PW	AH	PH
dp2	AM F31026	8.1	5.1	6.1	5.7	5.5
dp3	AM F31026	10.2	7.8	8.2	6.2	7.0
p3	AM F31026	13.8	6.3	8.0	7.3	6.7
	AM F106043	13.8	7.1	8.7	-	-
	Mean (S.D.)	13.8 (0.00)	6.7 (0.57)	8.4 (0.49)	7.3	6.7
m1	AM F31026	13.9	9.9	10.3	9.3	9.4
	AM F106043	13.8	-	-	-	-
	Mean (S.D.)	13.9 (0.07)	9.9	10.3	9.3	9.4
m2	AM F106043	15.5	-	12.2	-	-
m3	AM F106043	15.9	12.6	12.9	8.1	7.7
m4	AM F106043	15.9	13.3	12.6	7.4	7.4

Trigonid rounded anteriorly and near symmetrical in outline. Rather short paracristid maintains contact with protoconid apex, and underlain by wide precingulid. Low but distinct parametacristid folds across protolophid face before thickening and terminating in center of trigonid basin, just posterior to short transverse portion of paracristid. Lophid faces lack enamel crenulations. Rounded, A-shaped inflation positioned centrally on hypolophid posterior face.

Short p3 equal in length to m1; slightly narrower in holotype due to lack of full complement of enamel (pl. 64D). Lingual side of p3 quite straight, only curving buccally at point adjacent to anteriormost cuspule of main crest. Anterior half of tooth outline almost symmetrical, but buccal side slightly more curved. Inflated posterobuccal aspect of tooth demarcated anteriorly by vertical cleft (pl. 64D). Very short, crescentic buccal crest, borne on posterobuccal inflation, extends one-third length of tooth (pl. 64D). Main crest divided into five cuspules, with posteriormost twice length of anterior four. Three anteriormost cuspules accompanied by finer lingual and coarser buccal vertical ridgelets (pl. 64D). In holotype, two transverse ridgelets cross median valley between main and buccal crests. Arising from buccal face of posteriormost cuspule on main crest, they extend to base of lingual face of buccal crest. Posterior extremities of main and buccal crests meet in wide V-shape when viewed posteriorly.

Lower molars (pls 64-65) intermediate in crown height, with unworn protolophid and hypolophid crests straight and close to parallel. Paracristid and cristid obliqua low, with former maintaining contact with protoconid apex, latter shifted slightly lingually. Short trigonid tapers symmetrically anteriorly on m1-2 due to wide precingulid, which extends to anterolingual corner of tooth beneath short, thick transverse (anterior) portion of paracristid. At point where main body of paracristid turns lingually, a finer prong continues anteriorly to edge of tooth (pl. 64C). Very low, slight premetacristid descends anterobuccally from metaconid apex of m1-2 to meet lingual extremity of paracristid (pl. 64C). Two coarse (AM F31026) or four finer enamel crenulations (AM F106043) descend into trigonid basin from low on protolophid anterior face. Anterior face of m2-4 hypolophid bears three fine, vertical enamel crenulations descending into interlophid valley, adjacent to cristid obliqua (pl. 65C). Hypolophid posterior face with postcingulid strongly developed, expressed as A-shaped inflation, with distinct shelves formed on buccal and lingual slopes.

Variation. Description of the referred adult dentary (AM F106043) places the taxonomic validity of *"Simosthenurus" brachyselenis*, hitherto known only from the holotype, beyond any reasonable doubt. It also demonstrates how difficult it can be to predict adult proportions from a young juvenile specimen. While Prideaux and Wells (1997) noted that the holotype (AM F31026) ramus was relatively deep for its age, the massive proportions of the adult ramus are remarkable. Depth of the adult dentary is approximately six times greater than the height of the cheek teeth, compared with about 2.5 times in the juvenile holotype. While the actual depth of the adult dentary is almost double that of the juvenile, its width is only about 60% greater. This is a typical allometric change observed during the ontogeny of other sthenurine species. The juvenile and adult dentaries also differ in the relative size of the inferior transverse torus of the symphysis; it is proportionally much larger in the adult. No variation is obvious with regard to the lower dentition that cannot be ascribed to differential wear.

Comparison With Other Taxa. Dentary. While the juvenile dentary of *"Simosthenurus" brachyselenis* compares well with similarly aged specimens of *Si. occidentalis* in size and general morphology, the ramus clearly differs in its greater depth. The massiveness of the adult dentary is initially reminiscent of *Procoptodon pusio* and *P. rapha*, but these species have a wider ramus and narrower diastema region. Although much shorter and more robust, the ramus of *"Si." brachyselenis* bears some resemblance to *Metasthenurus newtonae*, particularly with regard to the widened nature of the ramus immediately posterior to the posterior mental foramen and the manner in which the ramus deepens posteriorly. This latter feature is also characteristic of *"P." browneorum.*

Lower Dentition. The unworn i1 occlusal surface is rather straight in *"Simosthenurus" brachyselenis*, a profile quite reminiscent of that of *"Procoptodon" gilli.* However, the i1 of *"Si." brachyselenis* is larger and more robust. The dp2 is most similar in morphology and size to that of *Si. maddocki*, but it differs by having a slightly more inflated outline and a shorter buccal crest that continues to the posterior extremity of the main crest, and by lacking a main crest divided into distinct, rather separated cuspules.

Within the Simosthenurini, only the p3 of *P. goliah*, *P. rapha* and *P. pusio* is shorter relative to the molars than that of *"Si." brachyselenis.* However, *"Si."*

brachyselenis is most similar in p3 morphology to *Metasthenurus newtonae*, but it also resembles *"Si." tirarensis* and, to a lesser degree, *Si. euryskaphus*. The p3 differs from that of *M. newtonae* by being shorter relative to the molars and having a wider median valley. The near symmetrical anterior moiety of p3 is similar to that of *"Si." tirarensis*, *Si. euryskaphus* and some *M. newtonae* specimens. The p3 of *"Si." brachyselenis* differs from that of *"Si." tirarensis* by being shorter relative to its width as well as the length of the molars, and having an anteriorly open median valley. From *Si. euryskaphus* it differs by having a much straighter posterior portion of the main crest, a shorter buccal crest, and a much narrower median valley.

In morphology and size, the molariform teeth of *"Simosthenurus" brachyselenis* are most similar to those of *"Si." antiquus*. *"Si." brachyselenis* differs only by having a smaller anterior portion of the paracristid, slightly coarser enamel crenulations on the lophid faces, and a more distinct postcingulid shelf rather than a low inflation. *"Si." antiquus* may also lack the bifurcated paracristid on m1-2, but the only useful specimen preserving these teeth (QM F2931, pl. 59) is slightly too worn to be certain. In each of these features, bar the smaller anterior portion of the paracristid, the *"Si." brachyselenis* molars also resemble those of *"Si." pales*, but are much smaller. The only exception to this is the small-toothed Brisbane River specimen (QM F9428, pl. 69A-C), from which *"Si." brachyselenis* may be distinguished by having longer molars relative to their width.

Geographic Distribution. Central eastern New South Wales (fig. 34; table 44).

Temporal Distribution. ?Middle Pleistocene.

"Simosthenurus" pales (De Vis, 1895)

(pls 66-69; fig. 9F; table 26)

Halmaturus sp., Krefft, 1882, pl. 10, fig. 11.
Sthenurus pales, De Vis, 1895, pp. 94-96, pl. 15, figs 1, 2, 4.
Sthenurus pales, Glauert, 1910b, p. 60.
Sthenurus pales, Simpson, 1930, p. 77.
Sthenurus pales, Stirton, 1957, p. 124.
Sthenurus pales, Marcus, 1962, p. 299.
Procoptodon sp., Bartholomai, 1963, p. 66.
Sthenurus pales, Bartholomai, 1963, pp. 63-66, figs 6-8.
Sthenurus pales, Merrilees, 1965, p. 24.
Sthenurus (*Simosthenurus*) *pales*, Tedford, 1966, pp. 43-49, figs 16-18, 20, tables 2, 7-9.
Sthenurus (Sthenurus) sp. cf. *tindalei*, Tedford, 1966, pp. 33, 48.
Sthenurus (Simosthenurus) sp. nov., Tedford, 1966, p. 48.
Sthenurus pales, Merrilees, 1968a, pp. 69, 73.
Sthenurus (*Simosthenurus*) aff. *pales*, Marcus, 1976, pp. 70, 72, fig. 26.

Sthenurus sp. nov., Merrilees, 1979, p. 118.
Sthenurus (Simosthenurus) pales, Dawson, 1985, p. 66, table 1.
Sthenurus pales, Lundelius and Turnbull, 1989, pp. 4-6, fig. 2.
Sthenurus pales, Prideaux and Wells, 1997, pp. 185, 188, 191, 196.
Sthenurus pales, Prideaux and Wells, 1998, pp. 1, 11, 15.

Taxonomic Authority. De Vis, C. W. 1895. A review of the fossil jaws of the Macropodidae in the Queensland Museum. Proceedings of the Linnean Society of New South Wales 10:75-133.

Holotype. QM F815, right P3, collected in the late 1800s from an unspecified locality on the Darling Downs, southeastern Queensland. Age is probably Pleistocene. Note that the original rendering of this specimen by De Vis (1895) is inaccurate.

Referred Specimens. ?Ravensthorpe, Pilton, Darling Downs, QLD: QM F812, right adult dentary; QM F2972, left P3.

Darling Downs, QLD: QM F4931, left I3.

?Darling Downs, QLD: AM F16489, left adult maxilla.

Brisbane River, southeastern QLD: QM F9428, left and right juvenile dentaries.

Bone Camp Gully, near Bingara, NSW: AM F49665, right M4; UCMP 104918, right I3.

Reddestone Creek, near Glen Innes, NSW: material not seen and repository unknown, but specimens of four individuals were noted by Horton and Connah (1981).

Geurie, NSW: AM F41532 / F41533 / F41534, left and right adult maxillae and dentaries.

Wellington Caves (unspecified), NSW: AM F19653, left juvenile dentary; AM F29563, right adult maxilla; AM F29566, right adult dentary; AM F29594, right p3; AM F30626, right lower molar; AM F31041, left p3; AM F88539 (= AM MF22), left P3; BMNH 42663Bc, left dp3.

Macquarie River, near Gilgandra, NSW: AM F45855, right adult dentary.

Cuddie Springs, near Gongolgon, NSW: AM F29564, left adult maxilla; AM F36091, left adult dentary.

Jincumbilly, near Bombala, NSW: AM F39295, right P3.

Green Bluff Locality, Warburton River, SA: UCMP 56807, left juvenile dentary.

Henschke's Fossil Cave, near Naracoorte, SA: SAM P17586, right I1; SAM P18735, right I3; SAM P32318 / P32320, left and right adult maxilla, left I1; SAM P unreg., approximately 100 loose upper and lower incisors, premolars and molars.

Victoria Fossil Cave, near Naracoorte, SA: SAM P25587, left m2; SAM P26880, left m1; SAM P26881, left i1, left m3; SAM P26882, right m2; SAM P26883, left m4; SAM P27797, right juvenile dentary; SAM P32544, right m1; FU

0333, left p3 (may be same individual as FU1083); FU 0377, left and right adult maxillae, left I1; FU 1083, right adult dentary; FU 1631, right m2.

Comaum Forest Cave, near Penola, SA: SAM P36625, left i1, left p3.

Kilsby's Hole, Mount Gambier, SA: SAM P unreg., right M4.

Moorak, near Mount Gambier, SA: SAM P13669, right M1, right M3.

McIntyre's Gully, Gingin, WA: WAM 78.7.2, left juvenile dentary.

Tight Entrance Cave, WA: WAM 97.5.315, right adult dentary; WAM 97.5.316, left juvenile maxilla; WAM 97.5.318, left M1; WAM 97.5.319, left M3; WAM 97.5.320, left I3; WAM 97.5.321, right i1; WAM 99.11.270, right juvenile maxilla; WAM 99.11.271, left p3; WAM 99.11.272, left juvenile dentary; WAM 99.11.273, right I2-3; WAM 99.11.274, right I3; 99.11.473, right I2; WAM 99.11.574, left juvenile maxilla; WAM 99.11.490, right dP3; WAM 99.11.600, right juvenile dentary; WAM 03.11.03, left juvenile premaxilla and maxilla; WAM 03.11.04, right adult dentary; WAM 03.11.05, left juvenile dentary; WAM 03.11.06, left juvenile dentary; WAM 03.11.07, left adult dentary.

Revised Diagnosis. Very large, dentary proportions intermediate between those of brachycephalic and dolichocephalic sthenurines. I1 slightly broadened. Upper molars with postprotocrista markedly separated into two components. Dentary similar in size and morphology to *Sthenurus stirlingi*, but with diastema at level of alveolar margin of cheek tooth row. Large p3 bears very prominent anteriormost cuspule of main crest; long buccal crest terminates adjacent to short ridgelet directed posteriorly along tooth midline from anteriormost cuspule.

Etymology. Lat. *Pales*, ancient Roman god of sheep and cattle. Significance of name uncertain, but may be in reference to the grazing land of the Darling Downs in southeastern Queensland, whence the holotype was collected.

Description. Maxilla (pl. 66). Very little of adult maxilla is known, even in most complete specimen (SAM P32318 / P32320). Orbit centered above M2 protoloph. Palatal vacuities apparently extended anteriorly to adjacent to middle of P3. Wide root of masseteric process, which lies adjacent to M2-3 abutment, suggests large, robust proportions. Development of process in juvenile maxilla (WAM 03.11.03, pl. 66K) supports this idea.

Upper Dentition (pl. 66; table 26). High-crowned I1 slightly broadened and rather parallel-sided for most of crown depth, only narrowing slightly toward minimally occlusal surface (pl. 67D). Anterior face of crown bears very shallow vertical groove toward buccal side. High-crowned I2 one-third size of I1; thin and quite cylindrical near minimally worn occlusal surface, but becoming more of a flattened triangular shape in cross section toward base of crown. I3 very large and blade-like, with quite straight anterior edge and convex posterior edge (pl. 67E-F). Well-developed anterolingual crest on lingual side of main I3 blade, extends, in an elongate S-shape, to near posterolingual corner of crown base. Rounded, vertical enamel ridge runs up anterobuccal aspect of I3 blade.

Although markedly shorter relative to its width, dP2 is very reminiscent of P3 (pl. 66G,K). Very straight main crest dominated by three cuspules, with posterior cuspule twice length of anterior two. Lingual crest slightly lower in height than main crest. Slight vertical ridgelet ascends from anteriormost cuspule of main crest on buccal side; larger ridgelet on lingual side separates distinct anterior and longitudinal basins. Anterior basin bordered anteriorly by anterobuccal extension of lingual crest and contains three very fine ridgelets. Two transverse ridgelets centered in quite deep longitudinal basin. Posterior basin separated from longitudinal basin by large transverse ridge linking posterior extremity of lingual crest with main crest, adjacent to small but distinct posterobuccal accessory cusp.

Completely molariform dP3 similar in general outline to M1, but interloph valley narrows lingually (pl. 66G,K). Preparacrista shifted slightly dorsolingually from paracone apex, forming or continuous with precingulum, which terminates halfway up crown on anterolingual aspect of tooth. Preprotocrista absent. Protoloph and metaloph taper markedly toward crests. Protoloph crest smoothly continuous with low postprotocrista, which ascends posterobuccally into interloph valley, becoming fine and terminating halfway up anterior face of metaloph. Lingual to postprotocrista, small mesocrista is flanked buccally by fine enamel crenulation. Postparacrista well developed and apparently merged with stylar crest, given presence of cusp in middle of unworn crest in position of stylar cusp C. Distinct kink, adjacent to interloph valley, marks union of premetacrista and postparacrista. Fine premetacrista trends anteriorly from just lingual of metacone apex. Very slight remnant of stylar crest extends anteriorly from metacone apex, terminating on buccal side of premetacrista and suggesting coalition of metacone and stylar cusp D. Strong postmetacrista smoothly continuous with metaloph crest, curving lingually and terminating buccal to dorsal end of urocrista. Postmetaconulecrista low, extends dorsobuccally from metaconule apex to below halfway point on postmetacrista. Very slight enamel crenulation between postmetaconulecrista and urocrista probably represents buccal (mesial) component of postmetaconulecrista.

Massive P3 narrows slightly anteriorly (pl. 66C-F,H-J,L). Anterobuccal aspect slightly inflated, but buccal side relatively straighter than more convex lingual side. Main crest straight for most of length, only curving slightly lingually toward posterior end; consists of five or six connected cuspules. Each cuspule gives rise to vertical lingual and buccal ridgelet, with posteriormost buccal ridgelet variably connected to posterobuccal accessory cusp. Elongate, triangular-shaped anterior basin bordered by low continuations of main and lingual crests; encloses fine enamel ridgelets. Lingual crest lower in height than main crest and quite irregular in lingual profile. Prominent anterior cuspule of lingual crest connected by strong transverse ridgelet to anteriormost cuspule of main crest. Deep longitudinal basin traversed by numerous coarse ridgelets; poorly separated from posterior basin.

Although varying in size from M1 to M4, upper molars vary extremely little in morphology (pl. 66G,I-M), except for narrowness of M4 metaloph relative to protoloph. Crowns moderately high, with overall occlusal outline rather square.

Posterior extremity of preparacrista extremely fine posterior to enlarged stylar cusp A portion and not directly connecting to paracone apex. Preparacrista continuous with precingulum, extending to anterolingual aspect of tooth and terminating one-third of way down from crown base. Several very fine, vertical enamel crenulations high on anterior face of protoloph, but none clearly discernable as preprotocrista. Protoloph and metaloph only slightly tapered toward crests. Postprotocrista divided into two components, with one ascending posterobuccally from protocone apex and terminating alongside thicker, more buccal component, which extends into interloph valley and terminates halfway down anterior face of metaloph (pl. 66G,I-M). Finer mesocrista present buccal to postprotocrista. Distinct kink, adjacent to interloph valley, marks union of slightly buccally concave postparacrista and short, fine premetacrista, which extends anteriorly from metacone apex. Postmetacrista moderately developed, curves sinusoidally from metacone apex to terminate adjacent to buccal extremity of postmetaconulecrista. From apex of metaconule, postmetaconulecrista sweeps across metaloph posterior face, thickening more posteriorly, before terminating at position of stylar cusp E. Central of three coarse enamel crenulations between postmetaconulecrista and postmetacrista probably represents urocrista.

Dentary (pls 67-69). Ramus moderately deep and elongate. Symphyseal boss prominent, with inferior transverse torus extended under genial fossa to below middle of p3 (pls 67A-B, 68C-D). Digastric eminence moderately to well developed. Digastric sulcus extends from below anterior extremity of medial pterygoid fossa to anywhere beneath m2 hypolophid to m1 hypolophid. Diastema moderately long, continuing line of alveolar margin of cheek tooth row (pl. 68C-D). Median dorsal groove rather shallow and wide. Extremely shallow buccinator sulcus extends from buccal side of diastema ridge, dorsal to very large anterior mental foramen, to anywhere beneath m1 hypolophid and m3 hypolophid. Posterior mental foramen lies halfway between dorsal and ventral borders of ramus, anywhere beneath m2 hypolophid and m3 hypolophid.

Ascending ramus takes root adjacent or just posterior to m4. Posteroventral aspect of masseteric fossa moderately flared laterally (pls 67C, 68G). Ventral border of masseteric fossa at or just below level of alveolar margin of cheek tooth row. Large masseteric foramen leads into near-vertical masseteric canal (pls 67C, 68G). Inferior mandibular foramen moderately large and near circular in shape, with very wide mylohyoid groove anteroventral to it. In lateral profile, medial pterygoid fossa extends only slightly posteriorly beyond condylar process. Angular process rises to small point. Mandibular condyle moderately large. Coronoid process unknown.

Lower Dentition (pls 67-69; table 26). Large i1 rather slender and upturned (pl. 68). Minimally worn, sinusoidal occlusal surface rises to point anteriorly, level with point one-third of way down from p3 occlusal surface (pl. 68C-D). Thick enamel of buccal side extends onto ventrolingual aspect of i1 and dorsally to create thin dorsal enamel flange (pl. 12I). Thinner layer of enamel covers lingual surface between ventrolingual enamel extension and dorsal enamel flange (pl. 68B,D).

Although similar in morphology to p3, dp2 shorter relative to width and only slightly narrower anteriorly than posteriorly (pl. 68A-B,E, 69A-B). Main crest divided into five cuspules split such that anterior two are oriented along tooth midline, while posterior three are coalesced into crest atop posterolingual aspect of tooth. Apices of anteriormost and third cuspules of main crest each give rise to vertical ridgelet on lingual side. Crescentic buccal crest lower than main crest and extends from posterior extremity of main crest, curving around posterobuccal aspect of tooth and terminating adjacent to second cuspule of main crest, from which it is separated by small notch at top of crown. Tiny cuspule present low on anterior end of crown, slightly lingual of midline. Median valley contains few fine ridgelets.

Completely molariform dp3 similar in general outline to m1 (pls 68E, 69B), but interloph valley narrows lingually. Protolophid crest narrower than hypolophid crest due to marked taper. Trigonid short; posterior portion of paracristid maintains direct contact with protoconid apex, terminating in center of trigonid basin. Thick, transverse anterior portion overlies very wide precingulid, which extends across entire front of tooth. Parametacristid extends from metaconid apex to meet posterior portion of paracristid partway along its length (pls 68E, 69B). Premetacristid runs from metaconid apex to lingual extremity of paracristid. Low cristid obliqua maintains contact with hypoconid apex, terminating in middle of protolophid posterior face. Very low preentocristid curves from entoconid apex to terminate lingual to cristid obliqua. Cristid obliqua and preentocristid border several fine enamel crenulations. Hypolophid posterior face bears incipient postcingulid or slight inflation.

Longer than any molar and markedly narrower anteriorly than posteriorly, p3 is dominated in lateral profile by tall anteriormost cuspule of main crest (pls 67A-C, 68C-D,F-G, 69C,G-H). This is split from remainder of main crest and projects smaller crest posteriorly into median valley, in midline of tooth. Posterior four-fifths of main crest straight and divided into four cuspules, with more anterior two giving rise to vertical ridgelets lingually and buccally. As on dp2, p3 buccal crest curves across posterobuccal aspect of tooth, terminating adjacent to second cuspule of main crest, from which it is separated by small notch at top of crown (pls 67C, 68F-G, 69C,G-H). Median valley deep and wide; dominated by few very coarse ridgelets.

Table 26. Cheek tooth dimensions of *"Simosthenurus" pales* from eastern Australia and southwestern Australia, showing mean, standard deviation (in parentheses), and dimensions of holotype (QM F815).

Tooth		L	AW	PW	AH	PH	n
dP2	Eastern	11.8	9.7	11.7	-	-	1
	Western	12.2 (0.25)	8.7 (0.61)	11.0 (0.52)	10.0 (0.21)	10.4 (0.35)	3
dP3	Western	12.2 (0.21)	11.3 (0.53)	12.2 (0.56)	7.9 (0.38)	8.1 (0.32)	
P3	Eastern	20.9 (1.28)	12.5 (0.84)	14.8 (0.78)	15.2 (1.92)	14.5 (1.57)	10
	Holotype	21.9	12.0	14.1	12.6	11.8	
	Western	19.3	12.0	13.9	-	-	1
M1	Eastern	17.2 (0.67)	16.5 (0.70)	16.3 (1.15)	8.9 (0.78)	10.2 (0.85)	3
	Western	14.1 (0.78)	13.2 (0.49)	13.4 (0.10)	8.5 (0.7840	8.7 (0.32)	3
M2	Eastern	19.0 (1.45)	17.3 (1.46)	16.9 (1.41)	9.7 (0.65)	10.4 (0.96)	7
	Western	15.2 (0.42)	14.5 (0.28)	14.4 (0.00)	8.9 (0.21)	9.2 (0.64)	2
M3	Eastern	19.8 (1.10)	18.1 (1.09)	17.0 (0.92)	11.0 (0.49)	11.6 (0.42)	9
	Western	15.9 (0.12)	14.8 (0.15)	13.8 (0.12)	8.4 (0.45)	8.4 (0.45)	3
M4	Eastern	17.9 (1.60)	16.3 (1.22)	14.1 (1.00)	9.5 (1.27)	9.1 (1.04)	2
dp2	Eastern	11.1 (0.58)	7.4 (0.55)	9.0 (0.69)	9.8 (0.49)	8.9 (0.78)	5
	Western	10.7 (0.46)	7.1 (0.10)	8.4 (0.12)	10.0 (0.20)	9.7 (0.06)	3
dp3	Eastern	12.9 (0.86)	10.7 (0.56)	11.3 (0.53)	9.6 (1.55)	9.8 (1.20)	4
	Western	11.8 (0.00)	9.1 (0.00)	9.5	8.9 (0.71)	8.9 (0.64)	2
p3	Eastern	19.3 (1.01)	9.3 (0.89)	11.3 (0.82)	15.1 (1.64)	14.3 (1.05)	12
	Western	17.9 (0.25)	8.9 (0.35)	10.9 (0.78)	13.5 (0.78)	12.3 (0.45)	4
m1	Eastern	16.3 (1.37)	12.9 (1.20)	13.4 (1.03)	11.2 (1.78)	11.2 (1.46)	10
	Western	13.8 (0.45)	11.3 (0.76)	11.6 (0.61)	10.0 (0.42)	10.1 (0.57)	3
m2	Eastern	18.6 (1.60)	15.2 (1.25)	15.0 (1.09)	13.0 (1.67)	13.1 (1.45)	10
	Western	15.1 (0.26)	12.6 (0.61)	12.9 (0.64)	10.7 (0.42)	11.0 (0.00)	3
m3	Eastern	20.2 (1.57)	16.5 (0.71)	16.0 (0.66)	13.1 (1.37)	12.8 (1.71)	9
	Western	16.3 (0.31)	13.5 (0.46)	13.5 (0.62)	11.2 (0.14)	11.0	3
m4	Eastern	17.6 (1.34)	15.2 (1.43)	13.4 (1.06)	11.0 (1.12)	10.1 (1.13)	7
	Western	15.3 (0.35)	13.4 (0.57)	12.6 (0.07)	9.4	8.9 (0.07)	2

Although varying in size from m1 to m4, moderately high-crowned lower molars vary minimally in morphology, except for narrowness of m4 hypolophid relative to protolophid (pl. 67C). Molars moderately long for width; all cheek teeth large relative to size of dentary. Lophids rather slender when viewed laterally, with crests straight and very close to parallel. Trigonid quite short. Paracristid and cristid obliqua distinct, but not particularly high; clearly separated into two components, with more lingual shifted one quarter of way across from buccal cusp apices (pls 67C, 68E,G, 69B,F). Paracristid extends anteriorly and slightly lingually before thickening, turning sharply, and trending lingually. Premetacristid, parametacristid or both crests when present, very slight and low. Precingulid deep and wide, extending beneath transverse anterior portion of paracristid. Several fine to few very fine enamel crenulations on anterior faces of protolophid and hypolophid. Cristid obliqua oriented similarly to paracristid, terminating in midline of tooth, low on protolophid posterior face. Preentocristid very slight and low. Hypolophid posterior face bears marked low inflation, frequently bearing distinct shelf (postcingulid).

Variation. Upper Dentition. Significant morphological variation exists between the holotype P3 (QM F815, pl. 66A-C) and referred specimens (pl. 66D-F,H-J,L). However, they occupy a discrete morphospace, which facilitates their separation from all other taxa. Variation is displayed in (a) length relative to width; (b) relative height of the lingual crest relative to the main crest; (c) degree of curvature of the lingual crest; (d) development of the anterior extension of the lingual crest (may terminate at anterior extension of main crest or extend across to anterobuccal corner of tooth); (e) degree of posterobuccal accessory cusp development (ranging from large and giving rise to a small crest anteriorly to small but well defined); (f) degree of development of transverse ridgelets; (g) general inflation of tooth outline. By contrast, no appreciable morphological variation is detectable between the incisors or upper molars of any *"Simosthenurus" pales* specimens, except for a very low cingulum around the base of the M3-4 metaloph in some individuals. On the dP3, a distinct stylar cusp C may be absent from the postparacrista. Within samples and between regions, the cheek teeth of *"Si." pales* exhibit extensive proportional and absolute variation in size. While the molars of the southwestern (Tight Entrance Cave) specimens (pl. 66G,K) are 21% smaller on average than eastern Australian specimens, the incisors and only known P3 lie at the lower end of the range of variation seen in the east.

Dentary. Despite its extensive geographic range, relatively few dentary specimens are known for *"Simosthenurus" pales*. However, size variation is evident, with adult southwestern (pl. 67A-B) and Victoria Fossil Cave (pl. 68C-D) specimens slightly smaller than those from the Wellington Caves and the Darling Downs. Dentary depth / width values range between 1.81 for AM F29566 (Wellington Caves) to 1.47 for FU 1083 (Victoria Fossil Cave). The three Tight Entrance Cave dentaries and another Wellington Caves dentary are intermediate between these extremes, which suggests that these differences do not reflect geographic variation. This is especially evident if it is considered that m4 had not fully erupted in FU 1083, even though the specimen is designated an adult because its p3 had erupted (pl. 68C-D). The dentary probably would have deepened more, given the typical condition of increased depth with age.

Morphologically, variation is seen in the depth and anterior extent of the digastric sulcus and eminence (e.g., large in AM F29566; intermediate in WAM 97.5.315; small in FU 1083, pl. 68C-D). The posterior mental foramen may be positioned anywhere beneath the m4 protolophid and m2 hypolophid. With the exception of increased depth relative to width, a comparison of adult (e.g., pl. 68C-D) and juvenile specimens (e.g., SAM P27797, pl. 68A-B) reveals relatively little change in dentary proportions from the stages where m2 eruption is imminent to full dental eruption.

Lower Dentition. No morphological variation is detectable between dp3 specimens. In contrast, dp2 varies in posterior width relative to anterior width, degree of buccal crest curvature, degree of anterior separation of the buccal and main crests, degree to which the main crest cuspules are separated, and form of the tiny cuspule low on the anterior end of the crown. Slightly less variation in morphology is evident for p3 than P3, but differences are seen in (a) length relative to width; (b) relative height of the anteriormost cuspule of main crest; (c) length of the small crest directed posteriorly along the midline of the tooth from the anteriormost cuspule; (d) depth and width of the median valley; and (e) size of the split separating the anterior part from the larger posterior part of the main crest. A similar magnitude of size variation exists between the lower molars of southwestern and eastern *"Simosthenurus" pales* as for the upper molars. However, extensive variation in lower molar size is also observed within samples (e.g., Henschke's Fossil Cave) and between samples from eastern Australia, as attested by the large standard deviations for dental measurements (table 26). Among them, the smallest molars are observed in AM F29566 from the Wellington Caves and in QM F9428, a pair of partial juvenile dentaries from the banks of the Brisbane River (pl. 69A-B). These are intermediate in size between the other eastern and southwestern *"Si." pales* specimens. At the opposite extreme, the molars of QM F812 (and AM F16489) from the Darling Downs are decidedly beefier than all other *"Si." pales* specimens that I have observed. Lower molars may vary slightly in the number of enamel crenulations on their lophid faces and in the relative development of a postcingulid, which may be a distinct shelf or merely a posteroventral inflation.

Comparison With Other Taxa. Upper Dentition. I1 is slightly broadened in *"Simosthenurus" pales*, but not nearly to the degree seen in *Sthenurus.* Although it is much larger, *"Si." pales* resembles *"Si." baileyi* in this regard. Aside from its larger size, nothing distinguishes the I2 of *"Si." pales* from that of the majority of sthenurines. While development of a large anterolingual crest on I3 may be an artifact of large size, it does distinguish the *"Si." pales* I3 from all other sthenurines with the exception of *Sthenurus stirlingi*, in which the anterolingual crest is massively developed.

In size and general morphology, the P3 of *"Simosthenurus" pales* is most similar to that of *Sthenurus tindalei*, *S. stirlingi* and, to a lesser degree, *Metasthenurus newtonae*. It may be distinguished from *S. tindalei* and *S. stirlingi* by lacking the high transverse ridge that separates the longitudinal and posterior basins and generally weaker or less extensive development of the anterior extension of the lingual crest. The *"Si." pales* P3 differs from that of *M. newtonae* by being larger, higher crowned, bearing a strong, high transverse ridge separating the anterior and longitudinal basins, lacking the well-developed transverse ridge separating the posterior and longitudinal basins, and having coarser transverse ridgelets in the longitudinal basin.

Although the larger upper molar specimens of *"Simosthenurus" pales* are very similar in size and general morphology to those of *Sthenurus stirlingi*, they are narrower and have more enamel crenulations. The *"Si." pales* upper molars are most difficult to separate from those of *"Si." antiquus*, although they are usually larger in size and tend to have slightly coarser enamel crenulations, a less distinct stylar cusp A eminence on the preparacrista, and a slightly thinner precingulum. They can be distinguished from those of *"Si." baileyi* on the basis of their larger size, slightly greater crown height, weaker postparacrista, and slightly better-developed postprotocrista. To a lesser degree, the upper molars of *"Si." pales* resemble those of *"Procoptodon" williamsi*, but the latter species is slightly higher crowned, and the molars bear lophs that taper more toward their crests, a distinct preprotocrista, a stronger postprotocrista, and a slight postparacrista and premetacrista that curve in lingually but do not meet.

Dentary. In overall size and general morphology, the dentary of *"Simosthenurus" pales* is indistinguishable from that of *Sthenurus stirlingi*, except in the diastema region. In *"Si." pales*, the diastema continues the line of the alveolar margin of cheek tooth row, while in *S. stirlingi*, the alveolar margin curves upward anteriorly and the diastema is slightly procumbent. The dentary of each species differs from that of *S. tindalei* by retaining intermediate proportions. *S. tindalei* has a more slender dentary, a procumbent diastema, and more gently sweeping curve of the posteroventral margin. Relative to the depth of the ramus, the cheek teeth of *"Si." pales* are notably large, reminiscent of *S. stirlingi* and also *"Si." antiquus*.

Lower Dentition. The long, robust i1 of *"Simosthenurus" pales* is difficult to separate from that of *Sthenurus stirlingi*, but is slightly more slender. The i1 root of *"Si."* sp. cf. *antiquus* (UCMP 47947, pl. 60) suggests a crown similar in size and orientation to that of *"Si." pales*. It is not possible to confuse the p3 of *"Si." pales* with that of any other species. While the tooth most resembles the p3 of *"Procoptodon" browneorum* and *"P." gilli* in general form, it is much larger and has a very prominent anteriormost cuspule with a short crest directed posteriorly from it along the tooth midline, a longer buccal crest, and a marked split in the main crest immediately posterior to the anterior cuspule.

While the lower molars of *"Simosthenurus" pales* and *Sthenurus stirlingi* are similar in size and crown height, they are otherwise dissimilar. By contrast, the lower molars of *"Si." antiquus* and *"Si." pales* are barely separable on morphology alone. *"Si." antiquus* is usually smaller, and its lophids bear fewer fine enamel crenulations, but the molars of the Brisbane River *"Si." pales* specimen (QM F9428, pl. 69A-B) can be distinguished only by the slightly sharper postcingulid shelf on the posterior face of the hypolophid and a slightly more pronounced anterior turn of the protolophid ends. However, the distinctive p3 morphology clearly allies QM F9428 with *"Si." pales* (pl. 69C), and its dentary depth as a juvenile is already equivalent to that of the two known adult dentaries of *"Si." antiquus*.

Geographic Distribution. Southwestern Western Australia, northern and southeastern South Australia, eastern New South Wales, southeastern Queensland (fig. 32; table 44).

Temporal Distribution. Middle to late Pleistocene.

"Simosthenurus" tirarensis sp. nov.

(pl. 70; table 27)

Holotype. SAM P40128 (= FU 2077), partial left juvenile dentary, preserving fragment of dp2, near-complete dp3, complete, fully erupted m1-2, and p3 exposed in crypt. The i1 and most of the symphyseal region are missing, as is almost all of the dentary posterior to m2. Specimen collected by Richard H. Tedford and Roderick T. Wells in 2000 from Camel Swamp Yard (Pompapillina Member, Tirari Fm.), Warburton River, South Australia. Age of the type locality is early late Pliocene (Tedford et al. 1992).

Diagnosis. The p3 is shorter than m2 and rather narrow, with ratio of posterior relative to anterior width less than in all sthenurines except *Hadronomas* and *Archaeosimos*. Buccal crest short, semicircular, joins anteriorly to fourth cuspule of main crest, and encloses small, broadly D-shaped, but posteriorly open median valley. Molars high crowned; small, but distinct pocket formed at confluence of postentocristid and A-shaped inflation on posterior hypolophid face.

Etymology. In reference to the geological formation whence the holotype was collected.

Description. Juvenile Dentary (pl. 70A-B,F). Proportions of preserved portion of ramus suggest short dentary. Ramus wide relative to depth (depth 26.0 mm, width 19.9 mm); m3 apparently unerupted judging from manner in which ramus is broken posterior to m2 and minimal degree of wear on m2. Dentary depth gradually increases posteriorly from behind large symphyseal boss to point that would correspond to part of ramus beneath m3 hypolophid, posterior to which dentary is broken off. Diastema short and upturned; orientation of i1 alveolus suggests i1 markedly upturned. Preserved portion of symphyseal plate rugose, with small, but distinct inferior transverse torus (pl. 70B). Symphysis extends posteriorly to beneath anterior edge of dp3. Anterior mental foramen lies anteroventral of dp2, below midpoint on short, dorsally concave diastema. Straight, shallow buccinator sulcus extends posteroventrally from between anterior mental foramen and dp2 to beneath m1 trigonid (pl. 70A). Posterior mental foramen positioned mid-depth on ramus beneath m2 protolophid. Digastric sulcus distinct, extending anteriorly to beneath m1 hypolophid. Digastric eminence would appear to have been moderately deep.

Lower Dentition (pl. 70; table 27). Only part of posterobuccal side of dp2 is preserved, but tooth was evidently distinctly wider posteriorly than anteriorly. Posterior extremity of buccal crest suggests very short, crescentic form; positioned above midline of dp2 (pl. 70F). Completely molariform dp3 bears protolophid considerably more tapered toward crest than hypolophid, due largely to dorsolingually sloping buccal side (pl. 70F). Despite wear, paracristid division appears to have been incipient. Trigonid short; precingulid forms a steep slope from buccal side of paracristid to crown base. Slight premetacristid and two enamel crenulations on protolophid anterior face almost obliterated by wear. Cristid obliqua thick and undivided. Small, but distinct pocket on lingual side of hypolophid posterior face formed at point where postentocristid and postcingulid become confluent, latter not extending beyond midline of dp3.

Rather narrow p3 unerupted, but fully formed and enameled (pl. 70C-E). It is longer than m1, but considerably shorter than m2 (table 27). In basal outline, lingual side of p3 quite straight, but curves buccally at point adjacent to anteriormost cuspule of main crest. Orientation mirrored by anterior half of buccal side, making anterior moiety of p3 near symmetrical in outline (pl. 70E). Main crest bears five cuspules, all subequal in height (pl. 70D). Moderately thick ridgelets descend lingually from anterior three cuspules, but equivalent ridgelets pertaining to posterior two cuspules much less distinct. Anterior portion of main crest oriented in tooth midline, but posterior to second cuspule it inflects lingually, terminating in posterolingual corner of p3. Short, semicircular buccal crest borne on mild posterobuccal inflation of p3, and is joined anteriorly to fourth cuspule of main crest (pl. 70C,E). Median valley small and broadly D-shaped, but open posteriorly. One distinct enamel pillar sits within median valley, rising halfway to crest level from valley floor.

Lower molars high crowned, with protolophid and hypolophid crests close to parallel and straight, except for moderate anterior turn of lophid ends on buccal side (pl. 70A-B,F). Smaller m1 distinguished morphologically from m2 by showing slightly less division of paracristid and cristid obliqua, bearing a marginally more distinct premetacristid and a proportionally smaller inflation on hypolophid posterior face. Origins of thick lingual paracristid and cristid obliqua components close to buccal side of lophids (pl. 70F). Buccal components very low and rounded. Trigonid short; precingulid forms rounded shelf anterobuccal to paracristid. Distinct 90° notch marks anterolingual extremity of rather boomerang-shaped paracristid. A few fine enamel crenulations descend into trigonid basin from halfway down protolophid anterior face. Premetacristid very fine and low, terminating at anterolingual extremity of paracristid. Anterior hypolophid face bears a very fine enamel crenulation on m1, but none evident on m2. Preentocristid very low and barely visible. Compared with dp3, hypolophid posterior face has a more pronounced A-shaped inflation (postcingulid), but smaller pocket at confluence with postentocristid.

Table 27. Cheek tooth dimensions of the *"Simosthenurus" tirarensis* sp. nov. holotype (SAM P40128).

Tooth	L	AW	PW	AH	PH
dp2	-	-	-	-	6.5
dp3	10.0	8.1	-	7.2	-
p3	14.5	6.8	7.8	-	-
m1	13.5	10.7	10.8	10.1	10.3
m2	16.4	12.2	12.0	11.4	11.8

Comparison With Other Taxa. Dentary. The holotype's incompleteness and juvenile nature severely limit comparisons with the dentaries of other species, but in size and proportions, including depth relative to width, brevity of the diastema, i1 orientation and symphyseal form, it lies within the range of similarly aged individuals of *Simosthenurus occidentalis*.

Lower Dentition. The dp2 is too incomplete to facilitate adequate interspecific comparisons, but the buccal crest appears to have been very short and restricted to the posterobuccal corner of the crown, similar to that of *Metasthenurus newtonae*, *Archaeosimos correlli* and *"Simosthenurus" brachyselenis*. In size and morphology, the *"Si." tirarensis* dp3 is most like that of *Si. euryskaphus* and, to a lesser degree, *Si. occidentalis*. It differs from the former by possessing a thicker cristid obliqua and a small, distinct postentocristid pocket on the posterior face of the hypolophid. Compared with that of *Si. occidentalis*, the dp3 of *"Si." tirarensis* is longer relative its width, but smaller relative to the molars, the protolophid is more tapered toward its occlusal surface, and the trigonid is narrower and shorter.

The p3 of *"Simosthenurus" tirarensis* is shorter relative to m2 than that of *Metasthenurus newtonae*, but not as short as in *"Si." brachyselenis*. Otherwise, these three species more closely resemble one another in p3 morphology than they do any other species. *"Si." tirarensis* differs from both species in the relative narrowness of its p3 (the ratio of posterior relative to anterior width being less than in all sthenurines except *Hadronomas* and *Archaeosimos*), and in buccal crest form. Despite being short like that of *M. newtonae* and *"Si." brachyselenis*, the p3 buccal crest of *"Si." tirarensis* is semicircular and strongly connected anteriorly to the fourth main crest cuspule. The relatively symmetrical anterior p3 moiety is also similar to *Si. euryskaphus*.

"Simosthenurus" tirarensis has higher-crowned molars than any other simosthenurin except *"Procoptodon" gilli*. While m1 falls within the upper end of the southeastern *Si. occidentalis* size spectrum for length and width dimensions, the *"Si." tirarensis* m2 is 14% longer than the southeastern *Si. occidentalis* mean, and thus, more elongate. In molar topographic morphology, *"Si." tirarensis* is otherwise very similar to southeastern *Si. occidentalis* and, therefore, *Si. euryskaphus*. In addition to their greater crown height, the molars of *"Si." tirarensis* may be distinguished by the presence of a postentocristid pocket, which accompanies the pronounced A-shaped inflation on the posterior hypolophid face, a thicker paracristid and cristid obliqua, and a paucity of enamel crenulations. In the latter two traits, *"Si." tirarensis* is reminiscent of *Metasthenurus newtonae*.

Geographic Distribution. Northern South Australia (fig. 27).

Temporal Distribution. Early late Pliocene.

PROCOPTODON Owen, 1874b

Macropus, Owen, 1845a (in part), p. 59.
Halmaturus(?), Krefft, 1870, p. 9.
Halmaturotherium, Krefft, 1872 (*nomen oblitum*), p. 327.
Halmatutherium, Krefft, 1873 (*nomen oblitum*), p. 238.
Halmatutherium, Krefft, 1874 (*nomen oblitum*), p. 146.
Pachysiagon, Owen, 1873b (*nomen nudum*), p. 386.
Procoptodon, Owen, 1873b (*nomen nudum*), pp. 386-387.
Pachysiagon, Owen, 1874b, p. 784, pl. 76, figs 7-10.
Macropus, Flower, 1884 (in part), pp. 720-721.
Sthenurus, De Vis, 1895 (in part), pp. 88-99, pl. 15, figs 1-9, pl. 16, figs 1-11.
Sthenurus, Glauert, 1910b (in part), pp. 60-61.
Sthenurus, Raven, 1929 (in part), p. 255.
Sthenurus, Raven and Gregory, 1946 (in part), pp. 6-8, fig. 2.
Sthenurus (Procoptodon), Tate, 1948, p. 338.

Taxonomic Authority. Owen, R. 1874b. On the fossil mammals of Australia. Part IX. Family Macropodidae: Genera *Macropus*, *Pachysiagon*, *Leptosiagon*, *Procoptodon* and *Palorchestes*. Philosophical Transactions of the Royal Society of London 164:783-803.

Type species. Macropus goliah Owen, 1845a.

Revised Diagnosis. Robust, brachycephalic sthenurines with upper molars typified by a well-developed mesocrista. Buccal crest of p3 long. Lower molars bear coarse enamel crenulations.

Diagnoses. Procoptodon is used in two senses in this monograph: *Procoptodon s. l.* and *Procoptodon s. s.* Both are monophyletic, but *Procoptodon s. s.* comprises only the three original species (*P. goliah, P. pusio* and *P. rapha*), while *Procoptodon s. l.* also includes the five species arising pectinately between *Procoptodon s. s.* and *Simosthenurus* (see Phylogeny). When referring to these five species individually, I enclose the generic epithet in quotation marks to highlight the interim nature of this taxonomic assignment.

Procoptodon s. s. Buccinator fossa very deep. Basicranium greatly elevated, basioccipital markedly flexed. Molars with very high, coarse cristae. Postparacrista strongly developed. Mesocrista strongly developed, combined with posterior portion of postprotocrista to form 'midlink'. Dentary massively proportioned. Posterior mental foramen tiny to absent. Very slender i1.

Procoptodon s. l. Mesocrista well developed. Buccal crest of p3 long. Lophid faces bear coarse enamel crenulations.

Etymology. Gr. *pro*, 'forward'; *koptos*, 'chop', *odon*, 'tooth'; possibly in reference to the presumed action of the premolars.

Geographic Distribution. Southwestern Western Australia, Nullarbor Region, northern, mid-northern, south-central and southeastern South Australia, southern Victoria, western and eastern New South Wales, southeastern Queensland (figs 28, 30-31, 33, 35-36; table 44).

Temporal Distribution. Early middle to late Pleistocene.

Remarks. Although *Halmaturotherium* Krefft, 1872, and its unjustified emendation *Halmatutherium* Krefft, 1873, are subjective synonyms of *Procoptodon* Owen, 1874b, neither has been used as valid since 1899. As *Procoptodon* is in prevailing usage (Article 23.9.1.2, International Code of Zoological Nomenclature 1999), Davis and Ride (2000) sought conservation of the name and suppression of the disused senior synonyms. In the same submission, Davis and Ride also sought the suppression of *P. thomsonii* (Krefft, 1870) and *P. scottii* (Krefft, 1870), senior synonyms of *P. pusio* Owen, 1874b and *P. rapha* Owen, 1874b used only once since 1899 (by Dawson and Flannery 1985).

Procoptodon goliah (Owen, 1845a)

(pls 71-77, 110; figs 16H, 17D, 19F, 26; tables 28-29)

Macropus goliah, Owen, 1845a, p. 59.
Procoptodon goliah, Owen, 1873b (*nomen nudum*), p. 387.
Procoptodon goliath, Etheridge, 1878, p. 190.
Macropus goliah, Flower, 1884 (in part), pp. 720-721.
Procoptodon rapha, Lydekker, 1891, p. 572, pl. 21, fig. 1.

Sthenurus goliah, De Vis, 1895 (in part), pp. 89-93, pl. 15, figs 5-9.
Sthenurus goliah, Glauert, 1910b, p. 60.
Sthenurus (Procoptodon) goliah, Tate, 1948, p. 338.
Procoptodon rapha, Wells, 1975, p. 211, fig. top right.
Procoptodon sp. cf. *rapha*, Williams, 1980, p. 106, site 24.
Procoptodon rapha, Wells and Pledge, 1983, pp. 173-174.
Procoptodon rapha, Wells, Moriarty and Williams, 1984, pp. 325-326, 329.
Procoptodon sp. cf. *rapha*, Tedford and Wells, 1990 (in part), pp. 275, 277.
Procoptodon rapha, Tedford and Wells, 1990 (in part), p. 278.
Procoptodon goliath, Menzies and Ballard, 1994, p. 129.
Procoptodon rapha, Prideaux and Wells, 1998, p. 1.

Taxonomic Authority. Owen, R. 1845a. *Macropus goliah* (fossil). Pp. 59 in G. R. Waterhouse, A Natural History of the Mammalia. Volume 1, containing the Order Marsupiata, or Pouched Animals. London: Hippolyte Baillière.

Holotype. BMNH M1896, partial right adult maxilla (containing M1-3) collected from an unspecified locality on the Darling Downs, southeastern Queensland. Age of type locality is very likely Pleistocene.

Referred Specimens. Darling Downs (unspecified), QLD: QM F795, right adult dentary; QM F797, right adult dentary; QM F4457, left adult dentary; QM F4458, right adult dentary; BMNH numerous specimens.

King's Creek, Darling Downs, QLD: QM F1850, right adult maxilla; QM F1328, left adult maxilla.

Clifton, Darling Downs, QLD: QM F805, right adult dentary.

Bone Camp Gully, near Bingara, NSW: 42 maxillary and 85 dentary specimens registered in AM, AMNH, BMNH, MCZ, UCMP and USNM collections.

Reddestone Creek, near Glen Innes, NSW: material not seen, but specimens of three individuals noted by Horton and Connah (1981).

Castlereagh River, Binnaway, NSW: AM F106575, right adult dentary.

Governor's Hill, Lake George, NSW: Institutional and registration details not recorded by Sanson et al. (1980), left and right adult maxillae and dentaries.

Site 9, Lake Victoria, NSW: NMV P28279, left and right adult dentaries.

Lake Victoria (site unspecified), NSW: SAM P27886, cranium, left and right dentaries; SAM P14529, left M4; NMV P28277, partial skeleton.

Lake Menindee, NSW: Approximately 30 SAM and UCMP registered maxillary and dentary specimens.

Lake Tandou, NSW: NMV P13908, NMV P13912, NMV P13975, NMV P13976, as listed in Merrilees (1973).

Willandra Lakes (between Lakes Garnpung and Gogolo), NSW: AM F58038, left and right maxillae; AM F58039, right upper molar; AM F58058, molar fragments; AM F58059, left and right maxillae; AM F58060, right maxilla; AM F58062, left and right dentaries; AM F58063, right maxilla.

Willandra Lakes (between Lakes Mungo and Leaghur), NSW: AM F58061, left and right dentaries.

Tocumwal, VIC: specimen not seen, but confidently ascribed as a locality for this species by Tedford (1967) and Bartholomai (1970).

Alexandra Cave, near Naracoorte, SA: SAM P27294, left juvenile dentary; SAM P27295, left juvenile dentary.

Cathedral Cave, near Naracoorte, SA: SAM P20490, right juvenile dentary; FU 1067, left and right juvenile maxillae and dentaries.

Fox Cave, near Naracoorte, SA: Listed in Reed and Bourne (2000).

Haystall Cave, near Naracoorte, SA: Listed in Reed and Bourne (2000).

Henschke's Quarry, near Naracoorte, SA: SAM P40800, left adult dentary.

Victoria Fossil Cave, near Naracoorte, SA: SAM P16690, right adult dentary; SAM P16691, left juvenile dentary; SAM P16692, left juvenile dentary; SAM P16693, left adult dentary; SAM P16694, left and right juvenile dentaries; SAM P16695, adult cranium; SAM P16696, partial juvenile cranium; SAM P16704, left juvenile dentary; SAM P20557, left p3; SAM P20666, left juvenile dentary; SAM P25573, right M3; SAM P27293, left and right juvenile maxillae; SAM P28596, left I3; SAM P28967, right adult dentary; FU 0078, left adult dentary; FU 0147, left I3; FU 0225, left juvenile dentary; FU 0586, left m2.

Madigan Gulf (Site 1/83), Lake Eyre, SA: SAM P25523, maxillary and upper molar fragments.

Kallakoopah Creek, SA: SAM P30179, right adult maxilla.

Green Bluff Locality, Warburton River, SA: UCMP 56810, left juvenile dentary; UCMP 56831, right juvenile dentary.

Lookout Locality, Warburton River, SA: SAM P20917, right adult dentary; SAM P20956, right adult maxilla; SAM P20958, right juvenile dentary.

New Kalamurina, Warburton River, SA: UCMP 97679, dentary; UCMP 97682, left juvenile dentary.

West Bluff, Lake Kutjitara, SA: FU 2162a, left maxilla.

Malkuni Waterhole, Cooper Creek, SA: SAM P25053, left and right premaxillae; FU 2228, right juvenile maxilla.

Waralamanko Waterhole, Cooper Creek, SA: SAM P25184, right adult dentary.

Kudnampirra Waterhole (= Site 14), Cooper Creek, SA: UCMP 56541, right upper molar; SAM P unreg., molar fragment.

Site 8, Cooper Creek, SA: UCMP 60670, right m1; 60672, left lower molar.

Billeroo Creek, SA: SAM P22405, left adult dentary; SAM P22410, right M2; FU 0233, right M4.

Waupunya Creek, Terowie District, SA: SAM P10865, right adult dentary.

Orroroo, SA: SAM P13856, left adult dentary; SAM P13857, right adult maxilla.

Whydown Station, near Yunta, SA: SAM P40079, right adult dentary.

Collinsville Station, near Mt Bryan, SA: SAM P18449, left juvenile dentary.

Ulooloo, SA: SAM P11551, right adult maxilla.

Boolcunda Creek, SA: SAM P22424, left dP3.

Brady's Gully, Teetulpa, SA: SAM P23165, left adult dentary.

Pitcairn Station, near Nackara, SA: SAM P13257, left and right adult dentaries.

Bairstow's Sand Pit, Port Pirie, SA: SAM P18392, left adult dentary.

Adelaide Beach (unspecified), SA: SAM P7116, left adult dentary.

Town Well Cave, Curramulka, Yorke Peninsula, SA: SAM P12785, right m1.

Curramulka Quarry, Yorke Peninsula, SA: SAM P18930, left juvenile dentary.

Old Calca Station, Eyre Peninsula, SA: SAM P185, left and right adult maxillae and dentaries; SADM V1, m3.

Last Tree Cave, Nullarbor Plain, WA: WAM unreg., left and right juvenile maxillae and dentaries.

Revised Diagnosis. Very large, brachycephalic and robust sthenurine, with molars strongly tapered toward loph crests. I3 length equal to combined length of I1-2. Cranial proportions similar to *Procoptodon pusio*, but rostrum relatively shorter and narrower, cheek teeth relatively larger; orbit centered above P3. Dentition most similar to *P. rapha*, but with much smaller P3 posterobuccal accessory cusp, p3 median valley fully enclosed by buccal crest posteriorly and minimal enamel crenulations arising laterally from paracristid and cristid obliqua. Molars more bulbous and larger relative to size of maxilla and dentary than in *P. rapha*.

Etymology. Heb. *Goliath*, a Philistine giant (Old Testament), in reference to the very large size of this species. It was only the third fossil kangaroo described (Owen 1845a), but is still recognized as the largest and most robust species.

Description. The most complete specimen of *Procoptodon goliah* known is the skeleton documented by Tedford (1967) from Lake Menindee in western New South Wales. The only material collected subsequent to the descriptions of Stirton and Marcus (1966), Tedford (1967), Bartholomai (1970) and Marcus (1976) that sheds light on previously unknown, poorly preserved or variable features has come from the Naracoorte Caves in southeastern South Australia. The specimens are a near-complete adult cranium (SAM P16695), a partial juvenile cranium with M3 beginning to erupt (SAM P16696) and two single, little worn left I3s (SAM P28596, FU 0147).

Adult Cranium (SAM P16695, pls 71, 73C, 74C-D; table 28). Cranium massively constructed and very foreshortened. Incisor-bearing portion of premaxilla and diastema region small and narrow relative to overall cranial proportions. I1 also tiny and low crowned relative to cranium (pl. 73C). I2 one-third of I1 size and cylindrical. Diastema concave dorsally, with borders curved in from anterior extremity of P3, turning to orient very slightly anteromesially, before curving laterally again immediately posterior to I3 alveolus. Incisive foramina short, narrow, opening anteriorly (pl. 71B). Anterior border of maxilla opposite posterior border of I3 alveolus. Anteorbital and buccinator fossae extremely deep,

Table 28. Comparison between the adult crania of *Procoptodon goliah* from Naracoorte (SAM P16695) and Lake Menindee (UCMP 45475; from Tedford 1967), showing percentage difference between measurements.

Dimension	SAM P16695	UCMP 45475	% difference
Condylobasal length	220	244	10
Diastema Length	37.5	45.5	18
Cranial Width between Masseteric Processes	126	-	
Vertical Depth of Cranium above M4 Alveolus	135	163	17
Max. Cranial Width across lateral edge of orbits	165	180	8
Max. Cranial Width across anterior edge of orbits	96.5	99	3
Palatal Length	155	185	16
% Diastema Length: Palatal Length	24.2	24.6	2
Minimum Width across Diastema	23.6	23.6	0
Rostrum (preorbital) Length	56.7	64.0	11
Palatal Width between P3 Anterior Moieties	37.3	41.1	9
Palatal Width between M4 Protolophs	37.2	40.4	8
Combined Length of M1-4	85	89	4
P3 Length	16.4	16.4	0

reflecting narrowness of premaxillary and diastema regions. Upright portion of premaxilla slender, but broken off on both sides dorsally. Entire dorsal half of rostrum missing, along with ethmoturbinal bones. However, rostrum was clearly extremely short (about one quarter of overall cranium length) and probably rather cylindrical as judged from morphology of surrounding regions. Frontal region appears to have been moderately inflated and only extended above mesial edge of orbit (pl. 71A). Strongly developed temporal crests extend well above dorsal surface of neurocranium (pls 71A, 73C). Zygomatic arch very deep. Masseteric process only moderately developed for cranial size and proportions and narrows rather than flaring distally (pl. 74C-D). Maxilla extremely deep and robust, with cheek tooth row constituting just less than half of entire cranium length. Anteriorly, palatal vacuities terminate adjacent to M1 protoloph. Although missing, postpalatine bars clearly formed a narrow bridge across palate adjacent or just posterior to metaloph of not quite fully erupted M4.

Relative to orientation of cheek tooth row, entire neurocranium is greatly elevated and shifted anteriorly such that much of braincase sits directly above posterior end of maxilla (pls 71, 73C, 74C-D). Consequently, basicranial region is very short (pl. 71B) and basioccipital very strongly flexed posterodorsally relative to basisphenoid. Medial pterygoid origin extremely deep and opening as much posteriorly as ventrally. Occiput broader than deep, foramen magnum very large and rather trapezium-shaped. Postglenoid process very large and completely fused with very broad ectotympanic, although not fully extended to lateral border of zygomatic process of squamosal. Rather smooth ectotympanic fully supported posteriorly by broadened posterolateral portion of squamosal and mastoid process of periotic (pls 73C, 74D).

Juvenile Cranium (SAM P16696, pls 72, 73B, 74A-B). Cranium preserves neurocranial and maxillary regions, but lacks entire rostrum and most of basicranium. Proportional differences surprisingly negligible on comparison with SAM P16695, except unusually for greater development of nuchal / lambdoid and temporal crests in juvenile (pls 72A, 73B) relative to adult (pls 71A, 73C). This could represent intraspecific variation (e.g., sexual dimorphism: younger individual male, older adult female), but no such variation has yet been observed in other sthenurines. Most other major features, including masseteric process size and form, depth and lateral expansion of zygomatic arches, and neurocranial profile and shape are indistinguishable in proportion, suggesting remarkable isometric cranial development between ontogenetic stage corresponding to initial eruption of M3 and adult. Greater relative depth of anteorbital and buccinator fossae in adult specimen appears to represent only exception to isometric growth pattern.

I3 (FU 0147, pl. 75G-I). Tiny relative to size of cranium (e.g., SAM P16695); length equal to combined length of I1-2. Crown very short and broad, root and base of crown cylindrical. Posterior edge of I3 markedly convex posteriorly; anterior edge straight, smoothly continues line of anterior edge of root. Narrow, rounded crest ascends buccal surface of tooth at anterior end and is more distinct close to little worn occlusal surface (pl. 75G,I). Short, round, thick anterolingual crest well developed, extends to crown base at anterior end of I3 (pl. 75H-I).

A range of other *Procoptodon goliah* specimens preserving features previously described (e.g., Tedford 1967; Marcus 1976) are illustrated in pls 73-77) for comparative purposes. See also fig. 26.

Variation. On comparison with the Lake Menindee adult cranium (UCMP 45475) described by Tedford (1967), SAM P16695 is smaller in most dimensions (table 28). However, the percentage differences observed are comparable with proportional differences observed between crania of other simosthenurin species from disparate geographic regions. Also, given that the M4 of SAM P16695 is not completely erupted and that UCMP 45475 represents a very mature individual judged from extensive dental wear, SAM P16695 may have had more growing to do, which could well have brought it closer to UCMP 45475 in overall cranial dimensions.

Comparison of the posterodorsal region of the SAM P16695 cranium with Tedford's hypothetical reconstruction of this region in UCMP 45475 (Tedford 1967:pl. 3) suggests that the occiput of the latter may actually have continued along a straighter line rather than extending posterodorsally. In addition, the apex of the nuchal crests may have been no higher than the temporal crests. I am also unable to detect any differences in dental size and morphology that are clearly not well within an acceptable range of intraspecific variation. The dental measurements of both specimens fall within the observed ranges of Marcus (1976) for *Procoptodon goliah* from Bingara in northeastern New South Wales. However, Marcus noted that several mean dental measurements for the Lake Menindee specimens differ significantly from the Bingara sample, which led him to suggest that these were of sufficient magnitude to justify a subspecies distinction. I am unconvinced. There is some variation in P3 size and morphology (e.g., pls 71B, 75A,C,E), but this evident within single assemblages as much as between regions. Mean cheek tooth dimensions for the South Australian specimens are listed in table 29.

One slight but common variation in morphology observed between the South Australian specimens and those from Lake Menindee and eastern Australia relates to features within the interloph valley of the upper molars referred to by Stirton and Marcus (1966) as 'pockets'. In *P. goliah* specimens from Bingara and Lake Menindee, the strongly developed postparacrista and premetacrista curve in lingually and terminate on the lingual side of the 'midlink', which is formed by a combination of the anterior half of the mesocrista and posterior half of the postprotocrista (pl. 75A-B). This creates an anterior and a posterior 'pocket' between the respective cristae and the lophs from which they arise. By contrast, the postparacrista and premetacrista of the South Australian specimens only contact this 'midlink' irregularly and are often oriented more anteroposteriorly (pls 71B, 72B, 75D,F). While most individuals of *P. goliah* lack a strongly developed preprotocrista that directly connects to the protocone apex, this is, in fact, characteristic of the holotype M3 (BMNH M1896).

Comparison With Other Taxa. Cranium. Most parts of the cranium are known for *Procoptodon goliah*, but only one partial adult cranium is known for *P. pusio* (QM F33018, pl. 78D) and the *P. rapha* cranium is represented only by maxillary specimens (pl. 82). QM F33018 preserves the entire palatal region and the broken anterior portion of the basicranium. Despite the much smaller size of the cheek dentition, maxillary depth, cranial width (as judged by the distance between the masseteric processes) and dimensions of the incisor-bearing portion of the premaxilla are almost identical to SAM P16695. Width across the palate between the M4 protolophs is greater in QM F33018 than in *P. goliah*, but this difference is apparently due to the narrower molars. The diastema of QM F33018 is longer and narrower than in *P. goliah*, and the diastema border is more mesially concave. While the adult crania of *P. pusio* and *P. goliah* must have been generally similar in size and morphology, the palate of *P. pusio* is shorter, but the diastema is longer, constituting 29% of palatal length compared with 24%. Given their similar cranial widths and maxillary depths, these proportional differences denote a shorter rostrum and longer cheek tooth row in *P. goliah*. What remains of the lateral surface of the QM F33018 rostrum suggests that it may also have been broader than in *P. goliah*. While the crania of *Simosthenurus maddocki* and *Si. occidentalis* are similarly brachycephalic to *P. goliah*, their frontals are more inflated and laterally tapered, the rostrum is broader and the teeth are comparatively smaller.

Upper Dentition. *Procoptodon goliah* is distinguished from *P. pusio* by the much larger size of its cheek teeth (~25%), an anteriorly wider P3 that bears a posterobuccal accessory cusp, and wider molars that are more markedly sloped toward their loph crests and bear more enamel crenulations. The P3 of *P. goliah* is most similar in absolute size, size relative to the molars and overall morphology to that of *"P." williamsi*. *P. goliah* differs by lacking a small distinct cuspule on its anterior end and by possessing a posterobuccal accessory cusp, a relatively higher lingual crest and several coarse ridgelets in the longitudinal basin. The *P. rapha* P3 is easily distinguished by its relatively greater length, main crest more distinctly divided into cuspules, which give rise to better-developed buccal and lingual ridgelets, and expansion of the posterobuccal accessory cusp into a crest. While the upper molars of *P. goliah* most closely resemble those of *P. rapha* in size and morphology, *P. goliah* differs by having larger molars that taper more toward their loph crests and bear fewer enamel crenulations.

Table 29. Cheek tooth dimensions of *Procoptodon goliah* from South Australia (SA), showing mean and standard deviation (in parentheses), compared with measurements of the holotype, BMNH M1896. Large samples of *P. goliah* from Lake Menindee and Bingara in New South Wales have previously been measured by Tedford (1967) and Marcus (1976).

Tooth		L	AW	PW	AH	PH	n
dP2	SA	11.6 (0.30)	8.9 (0.14)	10.2 (0.00)	7.5 (0.57)	7.8 (0.49)	3
dP3	SA	14.4 (0.59)	13.8 (1.12)	15.1 (0.60)	7.4 (0.85)	8.9 (1.48)	4
P3	SA	16.9 (0.81)	11.4 (0.60)	13.6 (0.80)	10.5 (1.41)	10.2 (0.78)	3
M1	SA	17.6 (0.42)	17.1 (1.06)	17.5 (1.10)	8.5 (0.21)	10.0 (1.56)	6
	Holotype	17.1	17.0	17.7	-	-	
M2	SA	20.7 (1.24)	19.2 (1.20)	19.8 (1.27)	9.6 (0.92)	11.6 (1.36)	6
	Holotype	21.1	19.4	18.8	8.1	9.9	
M3	SA	23.2 (1.24)	21.2 (0.84)	20.7 (0.70)	11.1 (1.65)	12.2 (1.23)	4
	Holotype	23.2	19.5	18.8	10.0	11.0	
M4	SA	24.6 (1.37)	20.7 (0.84)	18.4 (051)	12.8 (1.18)	13.0 (0.69)	6
dp2	SA	9.3 (0.30)	7.2 (0.49)	9.2 (0.36)	6.9 (0.42)	6.7 (0.14)	5
dp3	SA	13.9 (0.50)	11.3 (0.21)	12.2 (0.26)	6.7 (0.49)	8.5 (0.64)	6
p3	SA	14.5 (0.65)	8.2 (0.53)	10.5 (0.69)	10.6 (0.65)	10.3 (1.37)	7
m1	SA	18.1 (0.98)	14.6 (0.93)	15.1 (1.04)	9.8 (1.23)	11.1 (1.24)	16
m2	SA	21.6 (1.35)	17.2 (1.21)	17.4 (1.24)	12.4 (1.13)	13.3 (0.90)	17
m3	SA	24.6 (1.17)	19.0 (0.87)	18.5 (1.27)	13.4 (1.85)	13.6 (1.52)	13
m4	SA	22.8 (1.34)	16.3 (0.92)	15.3 (0.92)	11.6	10.8	4

Dentary (pls 76-77). The extremely robust dentary of *Procoptodon goliah* is most similar to those of *P. pusio* and *P. rapha*, but may be distinguished by being relatively wider and bearing a more reduced digastric eminence. While the dentary of *"Simosthenurus" brachyselenis* is also massively constructed, it is deeper and narrower than in *P. goliah.*

Lower Dentition (pls 76-77). While I cannot distinguish *Procoptodon goliah* from *P. rapha* based on i1, the tooth is comparatively more slender and elongate in *P. pusio* and deeper crowned in both *"P." gilli* and *"P." browneorum.* The p3 of *P. goliah* differs from that of *P. rapha* by being shorter relative to the molars and having a smaller, more enclosed median valley, especially posteriorly. The p3 of *P. goliah* is so much larger than that of *P. pusio* that it would prove extremely difficult to confuse these species on the basis of this tooth. Morphologically, it is also proportionally longer and wider, and bears a longer buccal crest continuous with the ridgelets directed posterobuccally from the anteriormost cuspule and buccally from the second cuspules of the main crest. The lower molars of *P. goliah* are most similar to those of *P. rapha*, but are more tapered toward the lophid crests and have fewer distinct enamel crenulations arising from the lophid faces and the sides of the paracristid and cristid obliqua.

Geographic Distribution. Nullarbor Region, northern, mid-northern, south-central and southeastern South Australia, northern Victoria, western and northeastern New South Wales, southeastern Queensland (fig. 36; table 44).

Temporal Distribution. Middle to late Pleistocene.

Procoptodon pusio Owen, 1874b

(pls 78-81; tables 30-31)

Halmaturus(?) thomsonii, Krefft, 1870 (*nomen nudum*), p. 9.
Procoptodon pusio, Owen, 1873b (*nomen nudum*), p. 387.
Pachysiagon otuel, Owen, 1874b, pp. 784, pl. 76, figs 7-10.
Halmatutherium thomsonii, Krefft, 1882, p. 10, pl. 8, figs 1-2.
Procoptodon rapha, Lydekker, 1887 (in part), pp. 235-236.
Procoptodon otuel, Lydekker, 1887, pp. 236-237.
Sthenurus otuel, De Vis, 1895, pp. 93-94, pl. 16, figs. 1-4.
Sthenurus pusio, Glauert, 1910b, p. 60.
Procoptodon otuel, Simpson, 1930, p 75.
Procoptodon thompsonii, Dawson and Flannery, 1985, p. 474.

Taxonomic Authority. Owen, R. 1874b. On the fossil mammals of Australia. Part IX. Family Macropodidae: Genera *Macropus*, *Pachysiagon*, *Leptosiagon*, *Procoptodon* and *Palorchestes*. Philosophical Transactions of the Royal Society of London 164:783-803.

Holotype. BMNH 39996, partial left and right juvenile maxillae (containing M1-3, excavated P3 for both sides) probably collected from Gowrie, Darling Downs, southeastern Queensland (Bartholomai 1970; Mahoney and Ride 1975). Age of the type locality is very likely Pleistocene.

Referred Specimens. Gowrie, Darling Downs, QLD: QM F2979, right adult dentary; QM F2981, right adult dentary.

Clifton, Darling Downs, QLD: QM F2982, left and right adult dentaries.

King's Creek, Darling Downs, QLD: BMNH 46310, right adult dentary.

Pearson's Locality, King's Creek, Darling Downs, QLD: QM F33018, partial adult cranium.

"Armour", Macalister, Condamine River, QLD: QM F2989, left adult maxilla.

Condamine River, Darling Downs, QLD: QM F2989, left adult maxilla.

?Pilton, Darling Downs, QLD: QM F2985, right adult dentary.

Darling Downs (unspecified), QLD: QM F806, left and right adult dentaries; QM F808, right adult dentary; QM F810, right juvenile maxilla; QM F2984, right adult dentary; QM F2986, left dentary; QM F2987, right adult maxilla; QM F2988, left adult maxilla; QM F4469, left dentary; QM F4470, right dentary; QM F4471, right adult maxilla; QM F4472, left dentary.

?Darling Downs, QLD: AM F30330, left dentary; QM F2983, left juvenile dentary; QM F4467, left and right adult dentaries.

Bone Camp Gully, near Bingara, NSW: 17 maxillary, 34 dentary specimens registered in AM, UCMP and AMNH collections.

Reddestone Creek, near Glen Innes, NSW: material not seen, but specimens representing four individuals noted by Horton and Connah (1981).

Mitchell Cave, Wellington Caves, NSW: AM F19654, left juvenile dentary.

Teapot Creek Site 7, Monaro Region, NSW: AM F115310, right M1.

Revised Diagnosis. Very robust sthenurine, most similar to *Procoptodon rapha* and *P. goliah*, but with smaller cheek teeth, a shorter buccal crest on p3 and no distinct posthypocristid. Rostrum relatively longer and broader than in *P. goliah*; orbit centered above abutment of P3 and M1. Differs from all *Procoptodon* species by having small p3 which is narrower relative to molars, and an elongate i1.

Etymology. Lat. *pusio* 'little boy', probably in reference to the comparatively smaller size of the dentition of this species relative to the other two *Procoptodon* species recognized by Owen (1874b).

Comparison With Other Taxa. Cranium. Since the work of Stirton and Marcus (1966), Tedford (1967), Bartholomai (1970) and Marcus (1976), the only specimen recovered that greatly adds to the knowledge of *Procoptodon pusio* is a partial adult cranium (QM F33018) from King's Creek on the Darling Downs, southeastern Queensland (pl. 78D; tables 30-31). It consists of the premaxillae, maxillae, and a small portion of the anterior basicranial region. Subsequent to my initial examination of the specimen, it appears to have hit the deck resulting in the basicranial region being smashed off and highly fragmented.

Table 30. Dimensions of the partial adult cranium of *Procoptodon pusio* (QM F33018) from Pearson's Locality, King's Creek, Darling Downs, southeastern Queensland.

Dimension	QM F33018
Diastema Length	41
% Diastema Length: Palatal Length	29
Minimum Width across Diastema	18
Rostrum (preorbital) Length	65
Palatal Length	140
Palatal Width between M4 Protolophs	43
Depth of Maxilla above M4 Alveolus	73
Cranial Width between Masseteric Processes	124
Combined Length of M1-4	62.9
P3 Length	11.1

In overall dimensions, the partial cranium is similar in size to *P. goliah*, including maxillary depth and cranial width as assessed from the distance between masseteric processes. There are, however, several proportional differences. Palatal width between the M4 protolophs of QM F33018 is greater than in the crania referred to *P. goliah*, a difference almost certainly due to the smaller molars. The diastema of QM F33018 is longer and its diastema border is more mesially concave. The whole palate is shorter, but the diastema is longer, constituting 29% of palatal length compared with 24%. These differences reflect a longer rostrum and much shorter cheek tooth row relative to *P. goliah*. What remains of the side of the rostrum suggests that it may also have been broader as well as longer than in *P. goliah*.

Upper Dentition. I1 is the only known upper incisor of *Procoptodon pusio* and it is very similar to that of *P. goliah*. The cheek teeth most closely resemble those of *P. goliah* and *P. rapha*. They differ in several aspects of their morphology and by being smaller. P3 lacks a distinct posterobuccal accessory cusp and its posterolingual aspect (pl. 78) is more inflated than in the other species. The upper molars of *P. pusio* (pl. 78) slope less toward their loph crests, especially compared with those of *P. goliah*, and they bear fewer enamel crenulations. Although absent from the holotype, upper molars of most individuals of *P. pusio* possess a well-developed preprotocrista, which descends anterobuccally from the protocone apex or just dorsobuccal to it (pl. 78). A similarly oriented, but smaller and less extensive preprotocrista is present on the upper molars of *Simosthenurus maddocki*.
Dentary. The dentary of *Procoptodon pusio* (pls 79-81) differs from that of *P. rapha* and *P. goliah* by being smaller and slightly narrower, and by having a better-developed, more ventromesially curved digastric eminence. "*Simosthenurus*"

Table 31. Upper cheek tooth dimensions of the holotype (BMNH M39996) and partial cranium (QM F33018) of *Procoptodon pusio*. Large samples of *Procoptodon pusio* from Bingara in northeastern New South Wales and the Darling Downs in southeastern Queensland have previously been measured by Marcus (1976) and Bartholomai (1970).

Tooth	Specimen	L	AW	PW	AH	PH
P3	BMNH M39996	11.8	7.1	9.7	8.1	7.8
	QM F33018	11.7	-	9.3	-	-
M1	BMNH M39996	13.4	12.8	13.6	8.0	7.8
	QM F33018	12.5	-	-	-	-
M2	BMNH M39996	15.6	15.1	14.6	8.0	8.3
	QM F33018	16.4	14.0	14.3	-	-
M3	BMNH M39996	17.9	15.4	15.1	8.8	9.2
	QM F33018	17.7	14.6	15.0	8.3	-
M4	QM F33018	17.1	14.3	13.4	6.2	5.8

brachyselenis has a similarly massive dentary, but it is deeper relative to its width and clearly had a wider diastema region.

Lower Dentition. The i1 of *Procoptodon pusio* (pls 80-81) is more slender and elongate than in any species of *Procoptodon*, but is not as slender as the i1 of *Simosthenurus maddocki*. The lower cheek teeth of *P. pusio* (pls 79-81) resemble those of *P. goliah*, *P. rapha* and, to a slightly lesser degree, *"P." oreas*. In size they are intermediate between *P. rapha* and *"P." oreas*, but a *P. rapha* specimen from the Wellington Caves (AM F19757, pls 83E-F, 84G) does preserve molars that approach the dimensions of *P. pusio*. Morphology of the p3 also appears intermediate between *P. rapha* and *"P." oreas*, but the *P. pusio* p3 is easily distinguished by its smaller size. Compared with those of *P. rapha* and *P. goliah*, the p3 of *P. pusio* is also proportionally shorter and narrower, and it has a shorter buccal crest, which is not continuous with the high ridgelets directed posterobuccally from the anteriormost cuspule and buccally from the second cuspule of the main crest. The lower molars of *P. pusio* lack the large, distinct posthypocristid characteristic of *P. rapha* and *P. goliah*, although a very low, similarly shaped eminence is evident between a few low, fine, vertical enamel crenulations on the posterior face of the hypolophid. The paracristid is not as strongly forked as it is in *P. rapha* and *P. goliah*, and the anteriorly oriented prong is finer and lower. The transverse anterior portion of the *P. pusio* paracristid is also less extensive.

Geographic Distribution. Northeastern New South Wales, southeastern Queensland (fig. 35; table 44).

Temporal Distribution. Middle to late Pleistocene.

Procoptodon rapha Owen, 1874b

(pls 82-84; table 32)

Halmaturus(?) scottii, Krefft, 1870 (*nomen nudum*), p. 9.
Procoptodon rapha, Owen, 1873b (*nomen nudum*), p. 387.
Procoptodon goliah, Owen 1874b (in part), pp. 791-797, pl. 79, figs 2-7, 9.
?*Procoptodon goliah*, McCoy, 1879, pp. 9-11, pls 52-53.
Halmatutherium scottii, Krefft, 1882, p. 11, pl. 10, fig. 1.
Macropus rapha, Flower, 1884, p. 721.
Sthenurus rapha, Glauert, 1910b, p. 60.
Procoptodon sp. cf. *rapha*, Bartholomai, 1977, p. 47, pl. 17, fig. 3.
Procoptodon texasensis, Archer, 1978, pp. 79-82, fig. 6, pl. 8A, C-E.
Procoptodon texasensis, Archer, 1984, p. 752, fig. 258.
Procoptodon scottii, Dawson and Flannery, 1985, p. 474.
Procoptodon texasensis, Flannery, 1989, pp. 29-30, fig. 2.

Taxonomic Authority. Owen, R. 1874b. On the fossil mammals of Australia. Part IX. Family Macropodidae: Genera *Macropus*, *Pachysiagon*, *Leptosiagon*, *Procoptodon* and *Palorchestes*. Philosophical Transactions of the Royal Society of London 164:783-803.

Lectotype. BMNH 32885, partial left juvenile dentary (containing base of i1, dp2 and dp3 crowns, excavated p3) collected from the Condamine River, Darling Downs, southeastern Queensland. Owen (1874b) described *Procoptodon rapha* on the basis of two specimens, but did not designate a holotype. As a consequence, BMNH 32885 and the partial right dentary, AM F10152, became syntypes until Stirton and Marcus (1966) named BMNH 32885 the "holotype" (Mahoney and Ride 1975). According to Article 74.6 of the International Commission of Zoological Nomenclature (1999), Stirton and Marcus are thus responsible for designating BMNH 32885 the lectotype by inference of holotype status.

Referred Specimens. Condamine River, Darling Downs, QLD: AM F10152, partial right dentary.

Freestone Creek, Darling Downs, QLD: QM F782, right adult dentary.

King's Creek, Darling Downs, QLD: QM F2430, left adult dentary; QM F4460, left juvenile dentary; QM F4462, left adult dentary.

Jimbour Creek, Darling Downs, QLD: QM F4461, right adult dentary.

Glengallan Creek, near Warwick, QLD: QM F2626, left adult dentary.

Little Middle Creek, Maidenwell, QLD: QM F2636, left adult dentary.

Oakey Creek, Darling Downs, QLD: QM F9156, left and right adult dentaries.

Darling Downs (unspecified), QLD: QM F794, left and right adult dentaries; QM F796, left adult dentary; QM F798, left adult dentary; QM F799, left adult maxilla; QM F803, right adult dentary; QM F809, right juvenile dentary; QM F4463, right adult dentary; BMNH, numerous specimens.

Cement Mills, Gore, QLD: QM F4548, right adult dentary.

The Joint, Texas Caves, southeastern QLD: QM F7894, right juvenile maxilla fragment (containing dP3 fragment, excavated P3).

Bone Camp Gully, near Bingara, NSW: 24 maxillary and 51 dentary specimens registered in AM, AMNH, MCZ and UCMP collections.

Attunga Springs, near Tamworth, NSW: AM F3198, left adult dentary.

Catong Gully, near Attunga, NSW: QM F34187, right adult dentary; QM F34189, left juvenile dentary.

Mitchell Cave, Wellington, NSW: AM F19652, left juvenile dentary.

Wellington Caves (unspecified), NSW: AM F19652, left juvenile dentary; AM F19757, right juvenile dentary; AM F19758, left and right adult maxillae; AM F60967, left juvenile maxilla; AM F89906, right p3; BMNH 42663F, left lower molar; UCMP 57385, right adult dentary.

Bunyan Siding, near Cooma, NSW: ANU Geol. 48912, right adult dentary.

South Site, Lancefield, VIC: NMV P200725, left adult dentary.

Lake Colongulac, VIC: NMV P26901, adult dentary; NMV P26902, adult dentary.

Colac, VIC: NMV P1909, adult dentary.

Henschke's Fossil Cave, near Naracoorte, SA: SAM P31045, left and right adult maxillae, right adult dentary.

Katipiri Waterhole (= Site 9), Cooper Creek, SA: SAM P11543, right juvenile dentary.

Curramulka Quarry, Yorke Peninsula, SA: SAM P40060, left dP3, M2.

Cooper Creek, SA: HM S19, right juvenile maxilla; HM S62, left adult dentary.

Revised Diagnosis. Very large, robust sthenurine, most similar in morphology to *Procoptodon goliah*, but with less marked tapering of molars toward loph crests and p3 not fully enclosed by buccal crest posteriorly. Molars smaller relative to size of maxilla and dentary than in *P. goliah*. Differs from all *Procoptodon* species by possessing a P3 with a greatly developed, crest-like posterobuccal accessory cusp, relatively larger p3 with a wide median valley, and lower molars with numerous enamel crenulations arising from paracristid and cristid obliqua.

Etymology. Gr. *raphe*, 'seam', 'suture'. Significance of name uncertain, but may pertain to the apparently partially fused nature of the symphyseal union as preserved in the lectotype BMNH 32885. Alternatively, given the biblical derivation of several of Owen's (1974a-b) species names, it may pertain to Rapha, the fifth son of Benjamin (Old Testament: I Chronicles 8:2). In Hebrew, the word also means to mend or fix, so the name could be a possible double reference to the Old Testament and the symphysis.

Table 32. Cheek tooth dimensions of specimens referred to *Procoptodon rapha* from Henschke's Fossil Cave in southeastern South Australia (SAM P31045) and Wellington Caves in central eastern New South Wales (AM F19757). Large samples of *Procoptodon rapha* from Bingara in northeastern New South Wales and the Darling Downs in southeastern Queensland have been measured by Bartholomai (1970) and Marcus (1976).

Tooth	Specimen	L	AW	PW	AH	PH
P3	SAM P31045	18.1	10.5	14.2	-	-
M1	SAM P31045	16.8	17.1	17.3	-	-
M2	SAM P31045	21.4	18.3	17.4	-	-
M3	SAM P31045	22.9	18.7	17.9	-	-
M4	SAM P31045	23.6	17.5	16.9	-	-
dp2	AM F19757	9.2	6.6	8.2	7.4	7.1
dp3	AM F19757	12.3	9.0	9.9	-	-
	AM F19757	14.7	-	-	10.7	11.5
	SAM P31045	15.3	8.0	11.5	-	-
m1	SAM P31045	18.4	13.2	14.2	-	-
	AM F19757	16.2	11.9	12.0	8.6	9.5
m2	SAM P31045	20.7	10.6	11.0	-	-
	AM F19757	18.2	13.8	13.2	10.8	12.3
m3	SAM P31045	23.2	16.7	16.5	-	-
	AM F19757	20.5	14.6	-	-	-
m4	SAM P31045	23.0	15.3	15.3	-	-
	AM F19757	20.3	14.0	13.1	13.6	12.7

Remarks. No taxonomically enlightening specimens of *Procoptodon rapha* have come to light since the work of Stirton and Marcus (1966), Tedford (1967), Bartholomai (1970) and Marcus (1976). Material from Victoria Fossil Cave previously referred to *P. rapha* (Wells and Pledge 1983; Wells et al. 1984) is allocated herein to *P. goliah*. An associated pair of maxillae and a dentary from Henschke's Fossil Cave (SAM P31045, pls 82H, 84H; table 32) represent the only specimen from southeastern South Australia that belongs to this species. Bartholomai (1970) suggested that definitive placement of specimens from Lake Colongulac and Colac in southern Victoria in *P. rapha* may be premature, but for now (having not examined them) they are best left there, particularly since the Lancefield dentary (NMV P200725) is undoubtedly *P. rapha* and since *P. goliah* is only known in Victoria from a locality on its northern border. A range of other *P. rapha* specimens preserving features previously described (e.g., Tedford 1967; Marcus 1976) are illustrated in pls 82-84) for comparative purposes.

Synonymy of Procoptodon texasensis *Archer, 1978.* The holotype and only known specimen of *Procoptodon texasensis* (QM F7894) was collected from The Joint, part of the Texas Caves system in southeastern Queensland (Archer 1978). It is a fragment of right juvenile maxilla, which preserves only one readily comparable tooth, P3, but also contains a fragment of dP3. No material referable to this species has since been described. I am able to add nothing to the excellent description of the P3 provided by Archer (1978), but wish to compare the specimen again with *P. rapha*, the species most similar to *P. texasensis* in P3 morphology.

Aside from the differences that easily separate the P3 of QM F7894 from those of *Procoptodon pusio* and *P. goliah*, Archer (1978) distinguished *P. texasensis* from *P. rapha* and, as such, diagnosed it on the basis of the following features: P3 shorter, lower crowned and more oval-shaped in outline (not as proportionally wide posteriorly); lingual crest more smoothly convex; main crest more distinctly divided into four cuspules; ridgelets in the longitudinal basin less developed; coarse transverse ridgelets separating anterior and posterior basins from longitudinal basin more weakly separated; ridgelets between the enlarged, crest-like posterobuccal accessory cusp and main crest less developed. While several of these differences fall outside of the known range of variation in *P. rapha*, each one needs to be reconsidered in light of the variation that is actually observed within the large Bingara sample of *P. rapha* and the magnitude of variation noted for other simosthenurins.

The first diagnostic feature of *Procoptodon texasensis* is size related; specifically, the P3 is shorter than that of *P. rapha*. Indeed, if its length (15.9 mm) is compared with values for *P. rapha* taken from Marcus (1976), it does lie outside of the observed range, which is 16.7-18.2 mm. Thus, the tooth is 5% smaller than the most diminutive *P. rapha* P3 from Bingara. However, on considering the marked variation in P3 length common to other simosthenurins (e.g., 21% difference between western and southeastern representatives of *Simosthenurus maddocki*), I question whether this difference is significant enough to justify its listing as a diagnostic feature, especially when it is based on one specimen. While Bingara and the Texas Caves are close geographically, the ages of both deposits are unknown and other sthenurines from the region vary greatly in dental size (e.g., *Sthenurus andersoni*). In addition, the anterior width of the QM F7894 P3 (10.2 mm) is well within the observed range for Bingara *P. rapha* (9.8-11.8 mm).

In terms of size, what does differ markedly between the P3 of QM F7894 and Bingara *P. rapha* is posterior width (10.5 mm versus 13.9-16.2 mm). It is exactly this difference that provides QM F7894 with its characteristic oval shape. However, this must represent at least a partial underestimation of actual tooth width, because the lingual surface does not bear its full enamel complement (see Archer 1978:pl. 8D). This might be due to immaturity, but given that eruption was imminent (Archer 1978) this is a more unlikely explanation than abrasion by the same physical process responsible for removing the roots. Similarly, while crown height (anterior height divided by length) is less for QM F7894 (0.57) than for the mean crown height value derived from the only unworn *P. rapha* specimens for which I have obtained measurements (0.62, n = 3), an increased height of only 0.8 mm would produce for QM F7894 the equivalent of the mean crown height for *P. rapha*. Given that the rather jagged base of the crown was facing out of the breccia block when collected and that the roots have been abraded off (Archer 1978), it seems a very real possibility that actual crown height may have been greater and thus within the range of *P. rapha*.

Among the topographic features upon which *Procoptodon texasensis* was diagnosed, I do not consider the greater degree of differentiation of the four main crest cuspules a trait sufficient enough to differentiate QM F7894 from *P. rapha*, because this character clearly varies within that species (compare AM MF896, AM MF956, UCMP 60048, pl. 82). The number of main crest cuspules also varies between four and five. Moreover, the degree of difference in cuspule separation and number is less between QM F7894 and *P. rapha* than it is within other simosthenurins, e.g., *"P." gilli*, *"P." browneorum* and *"Simosthenurus" pales*. While these observations do not in themselves imply that QM F7894 belongs in *P. rapha*, especially when the degree of cuspule differentiation is greater in QM F7894 than in Bingara *P. rapha*, they do suggest that use of this character at a diagnostic level needs to be done with caution, especially with a sample size of one.

Other commonly variable characters within the Simosthenurini include coarseness and number of ridgelets in the longitudinal basin, as well as the relative development of high ridgelets separating the anterior and posterior basins from the longitudinal basin. Just how much intraspecific variability is acceptable for the development of ridgelets between an enlarged, crest-like posterobuccal accessory cusp and the main crest is uncertain. *Procoptodon rapha* and *P. texasensis* are the only species characterized by hypertrophy of this feature, although it is worthwhile noting that individuals of *Simosthenurus maddocki* and *Si. occidentalis* may express a similarly enlarged posterobuccal accessory "crest" (e.g., SAM P20011, pl. 44J; SAM P20803, pl. 54D), and that some specimens of *P. rapha* lack ridgelets in this region of the tooth (e.g., AM F60967, pl. 82C). The smoothly convex nature of the lingual crest does separate QM F7894 from all specimens of *P. rapha* that I have observed and it seems the most likely candidate for a taxonomically stable trait upon which *P. texasensis* and *P. rapha* might conceivably be separated. However, the feature is variable in other sthenurines (e.g., *"P." browneorum* and *Metasthenurus newtonae*).

Archer (1978) cited one additional piece of evidence to support his taxonomic separation of QM F7894. It relates to a dentary specimen (QM F4548) from the nearby Cement Mills fissure fill locality. Bartholomai (1977) only tentatively placed the specimen in *Procoptodon rapha*, because the molars preserved (m3-4) show less ornamentation (enamel crenulations) than Darling Downs and Bingara *P. rapha*. Given the generally reduced number of ridgelets on the P3 of QM F7894 and the overall faunal similarities between the Texas Caves and Cement Mills deposits, Archer (1978) suggested that the slightly aberrant nature of QM F4548 might support the taxonomic uniqueness of QM F7894. However, a dentary that can be confidently referred to *P. rapha* from the Wellington Caves (AM F19757, pls 83E-F, 84G) also has fewer enamel crenulations than conspecifics from Bingara and the Darling Downs. The dentary and molars are also smaller (table 32). While a geographic or temporal effect may account for the Wellington specimen's eccentricity, it does indicate that the molars of QM F4548 (see Bartholomai 1977:pl. 17, fig. 3) fall within the range of variation for *P. rapha.*

During the compilation of this monograph, it became very clear that in size, proportion and topographic morphology P3 is a highly variable tooth within simosthenurin species. Variability is highest in those species represented by large numbers of specimens from individual localities (e.g., Victoria Fossil Cave, Tight Entrance Cave) and from different regions (e.g., southwestern and southeastern Australia). I find that the balance of evidence over *Procoptodon texasensis* favors synonymy with *P. rapha*, despite the paucity of indisputably intermediate states between QM F7894 and *P. rapha* for some characters. All in all, the large morphospace occupied by the P3 of many simosthenurins forewarns against the use of one such tooth as the entire foundation of a species unless those characters in which it differs from close relatives are consistently invariable within species.

Comparison With Other Taxa. In many aspects, including dentary form, *Procoptodon rapha* is intermediate between *P. pusio* and *P. goliah* in size and morphology.

Upper Dentition (pl. 82). The P3 exceeds those of both of these species in size, but is more similar to *Procoptodon goliah* in morphology. It differs by being longer relative to the molars, having a main crest more distinctly divided into cuspules giving rise to well-developed vertical ridgelets and a posterobuccal accessory cusp greatly enlarged and effectively extended into a crest running just short of half the length of the tooth. The longitudinal basin also contains more ridgelets than that of *P. goliah*, but this is probably a correlate of the tooth's larger size. The upper molars of these two species are more difficult to distinguish, although those of *P. rapha* are slightly higher crowned and smaller on average, with lophs less tapered toward their crests and a preprotocrista that is usually coarser (see Stirton and Marcus 1966; Marcus 1976). Although the 'pockets' (*sensu* Stirton and Marcus 1966) enclosed by the markedly incurved postparacrista and premetacrista may be more common in *P. goliah* than in *P. rapha* within the Bingara sample, most South Australian specimens of *P. goliah* lack or weakly express the condition. Moreover, several specimens that Marcus (1976) referred to *P. rapha* (e.g., AM F88565, AM F88566) clearly display these 'pockets' (pl. 82I). On this basis, this feature must be abandoned as a diagnostic character for *P. rapha.*

Lower Dentition (pls 83-84). While I cannot distinguish the i1 of *Procoptodon rapha* and *P. goliah* on either absolute size or morphology, *P. pusio* has a more slender and elongate crown. The p3 of *P. rapha* may be distinguished from that of *P. goliah* by being longer relative to the molars and having a larger, less enclosed median valley, especially posteriorly. The lower molars taper less toward the loph crests than in *P. goliah* and bear slightly more enamel crenulations arising from loph faces and the sides of the paracristid and cristid obliqua. By comparison, the cheek teeth of *P. pusio* are smaller, the p3 is shorter relative to molar length and bears a relatively shorter buccal crest, the molars have a less extensive transverse anterior portion of the paracristid and lack a strongly developed prong directed anteriorly from the point at which the lingual component of the paracristid inflects. Enamel crenulations are also less extensive on the lower molars of *P. pusio*, and they lack the tall triangular-shaped eminence on the posterior face of the hypolophid, which *P. rapha* shares with *P. goliah*. The orientation of the lingual aspect of this eminence, its extension to the ventrolingual corner of the tooth and the manner in which it clearly descends from the hypolophid crest as a continuation of the buccal portion of the lophid, strongly suggest that this feature is a reemergent posthypocristid.

Geographic Distribution. Southeastern Queensland, northeastern New South Wales, southern Victoria, southeastern, central southern and northern South Australia (fig. 36; table 44).

Temporal Distribution. Middle to late Pleistocene.

"Procoptodon" browneorum (Merrilees, 1968a)

(pls 85-93; figs 5, 8E, 10F; tables 33-34)

Sthenurus atlas, Owen, 1874a (in part), pp. 271-272, pl. 22, fig. 9; pl. 24, figs 7-9.
Sthenurus atlas, Owen, 1877b (in part), pp. 423-424, pl. 84, figs 7-9.
Sthenurus atlas, Lydekker, 1887 (in part), p. 233.
Sthenurus occidentalis, Glauert, 1910a (in part), pp. 31-36.
Sthenurus occidentalis, Anderson, 1932 (in part), pp. 383-386, fig. 6; pl. 45, fig. 1.
Sthenurus occidentalis, Glauert, 1948 (in part), pp. 100-102, lowermost figure.
Sthenurus sp., Merrilees, 1965 (in part), pp. 29-31, table 4, fig. 8.
Sthenurus (Simosthenurus) orientalis, Tedford, 1966 (in part), pp. 20, 52-53.
Sthenurus (Simosthenurus) brownei, Merrilees, 1968a, pp. 65-79, figs 1-8.
Sthenurus brownei, Merrilees, 1969, pp. 86-87.
Sthenurus brownei, Merrilees, 1979, pp. 116-118, table 1, fig. 2.
Sthenurus brownei, Wells and Murray, 1979, pp. 213, 215, 217-218.
Simosthenurus brownei, Pledge, 1980, pp. 134, 137, table 3.
Sthenurus sp. cf. *brownei*, Williams, 1980, p. 108, site 48c.
Sthenurus brownei, Wells, Moriarty and Williams, 1984, pp. 325-326, 329.
Simosthenurus brownei, Flannery, 1989, pp. 29-30, fig. 10.
Sthenurus brownei, Lundelius and Turnbull, 1989, pp. 3-6, fig. 1G-H.
Sthenurus brownei, Murray, 1989, p. 128.
Simosthenurus brownei, Pledge, 1990, pp. 253, 255, 257, table 2, fig. 5.

Simosthenurus brownei, Murray, 1991a, pp. 256, 268, 273, 275, figs 9E, 11A, 12G.
Simosthenurus brownei, Wells and Tedford, 1995, pp. 30, 35.
Sthenurus brownei, Prideaux and Wells, 1997, pp. 185-186, 188, 191-192, 196, fig. 2d.
Sthenurus brownei, Prideaux and Wells, 1998, pp. 1-4, 10-13, 15, table 1, figs 7D, 8D.
Sthenurus brownei, Brown and Wells, 2000, pp. 100-101.
Simosthenurus brownei, Prideaux, 2000, pp. 5, 12-13.
Simosthenurus brownei, Reed and Bourne, 2000, pp. 74-82, 85-86.

Taxonomic Authority. Merrilees, D. 1968a. South-western Australian occurrences of *Sthenurus* (Marsupialia, Macropodidae), including *Sthenurus brownei* sp. nov. Journal of the Royal Society of Western Australia 50:65-79.

Holotype. WAM 63.2.94, right adult dentary (missing coronoid process) with small anterior portion of left dentary partially fused to it; probably collected by Ludwig Glauert prior to 1910 from the Glauert excavation, Mammoth Cave, southwestern Western Australia. Based on faunal correlation, age of type locality is estimated to be late middle or late Pleistocene (Merrilees 1968a, 1979; pers. obs.).

Paratypes. Mammoth Cave, southwestern WA: WAM 63.2.39, dentary; WAM 63.2.40 / 63.2.51, dentary; WAM 63.2.41 / 63.3.48, dentary; WAM 63.2.42 / 63.2.50, dentary; WAM 63.2.43 / 63.2.46, dentary; WAM 63.2.44, dentary; 63.2.81 / 63.2.81 / 63.2.49, dentary; WAM 63.2.82, dentary; WAM 63.2.89 / 63.2.95, dentary; WAM 63.3.16, dentary; WAM 63.9.43, dentary.

Referred Specimens. Mammoth Cave, southwestern WA: WAM 63.2.198, cranium; WAM 34 registered dentaries, maxillae and single loose teeth; BMNH M19548, left juvenile maxilla; BMNH M19549, right juvenile dentary; BMNH M19553, left i1; FMNH PM4414, left adult maxilla; FMNH PM7891, right juvenile maxilla; UCMP 45091, right juvenile maxilla.

Devil's Lair, southwestern WA: WAM 76.5.12, left adult maxilla; several WAM registered specimens.

Foundation Cave, southwestern WA: WAM HC, right squamosal, left juvenile maxilla, two right juvenile dentaries.

Labyrinth Cave, southwestern WA: WAM 69.4.3, lower molar; WAM 69.4.13, premolar.

Kudjal Yolgah Cave, southwestern WA: WAM 87.7.2, juvenile cranium; WAM 87.7.6, right adult dentary; WAM 87.7.7, juvenile cranium; WAM 87.7.8, left and right juvenile dentaries.

Moondyne Cave, southwestern WA: WAM HC, 40 dentaries, maxillae and single loose teeth.

Strong's Cave, southwestern WA: 10 WAM registered dentaries, maxillae and single loose teeth.

Tight Entrance Cave, southwestern WA: more than 200 WAM registered dentaries, maxillae and single loose teeth; WAM HC, approximately 150 dentaries, maxillae, and a partial skeleton that includes partial cranium and dentaries.

Koala Cave, Yanchep, WA: WAM unreg., several craniodental specimens.

Wanneroo, WA: WAM 61.6.2, left and right juvenile dentaries; 66.10.14, left and right juvenile dentaries.

Lindsay Hall Cave, Nullarbor Plain, WA: WAM 92.9.11, left adult maxilla; WAM HC LH1, left juvenile maxilla; WAM HC LH10, left P3; WAM HC LH11, left adult dentary; WAM HC LH12, left adult maxilla; WAM HC LH261, left adult dentary.

Kiana Cliff, near Mount Misery, Eyre Peninsula, SA: FU 1747, left adult maxilla.

Old Calca Station, Eyre Peninsula, SA: SAM P186, left adult dentary.

Emu Caves (3 Hole), Kangaroo Island, SA: SAM P13942, right adult dentary; SAM P13943, right adult maxilla; SAM P13944, left i1; SAM P13945, left adult dentary; SAM P40017, left and right adult maxillae; SAM P40018, neurocranium; SAM P40019, neurocranium.

Rocky River, Kangaroo Island, SA: SAM P22379, left juvenile dentary.

Alexandra Cave, near Naracoorte, SA: SAM P27298, left and right juvenile dentaries; SAM P27302, left adult dentary; SAM P27303, left adult dentary; SAM P27304, left juvenile dentary; SAM P27305, right adult dentary; SAM P27306, right juvenile dentary; SAM P27307, right adult dentary; SAM P27308, right adult dentary; SAM P27310, right juvenile dentary; SAM P27311, right adult dentary; FU 1796, right juvenile dentary.

Bat Cave, near Naracoorte, SA: SAM P13024, left juvenile dentary.

Cathedral Cave, near Naracoorte, SA: FU 1011, adult cranium; FU 1061, left and right adult dentaries; FU unreg., right adult dentary.

Crawford's Cornucopia Cave, near Naracoorte, SA: Listed in Reed and Bourne (2000).

Fox Cave, near Naracoorte, SA: FU 0841, left adult dentary; FU 0844, right juvenile maxilla; FU 1040, right metatarsal IV.

Haystall Cave, near Naracoorte, SA: SAM P numerous registered maxillae and dentaries.

Henschke's Fossil Cave, near Naracoorte, SA: SAM P numerous registered and unreg. maxillae and dentaries, many single loose teeth.

Possum Cave, near Naracoorte, SA: FU 1799, left adult maxilla.

Rabbit Cave, near Naracoorte, SA: Listed in Reed and Bourne (2000).

Small unnamed cave 60 m from VDC Cave, near Naracoorte, SA: SAM P40091, left and right juvenile dentaries.

Specimen Cave, near Naracoorte, SA: SAM P12924, left juvenile dentary; SAM P40081, right adult dentary; SAM P40082, right adult maxilla; SAM P40083, right juvenile maxilla; SAM P40084, left juvenile maxilla; SAM P40085, left and right premaxillae; SAM P unreg., several single loose teeth.

Victoria Fossil Cave, near Naracoorte, SA: SAM or FU registered specimens, 15 juvenile and adult crania, approximately 150 dentaries and maxillae, approximately 100 single loose teeth.

Wet (Tomato-Stick) Cave, near Naracoorte, SA: SAM P12917, right M3.

Wombat Cave, near Naracoorte, SA: SAM P unreg., left juvenile dentary, right juvenile dentary, maxillary and dentary fragments; FU unreg., right juvenile dentary and numerous loose single teeth.

Comaum Forest Cave, near Penola, SA: Listed in Reed and Bourne (2000).

Monbulla Cave, near Penola, SA: SAM P40086, left adult dentary; SAM P40087, right juvenile dentary.

Wandilo Forest Cave, near Mount Gambier, SA: SAM P40088, left adult dentary; SAM P40089, left juvenile dentary; SAM P40090, left juvenile dentary.

Morgan's Cave, Tantanoola, SA: SAM P12841, juvenile dentary.

Derrington St, Mount Gambier, SA: SAM P13656, right m4.

Gray St, Mount Gambier, SA: SAM P13668, dentary.

McEachern's Cave, southwestern VIC: NMV P198421, P3; NMV P198431 / P198432, left and right adult dentaries; NMV P198437, left juvenile dentary; NMV P198444, right juvenile dentary.

Skene St, Warrnambool, VIC: NMV P unreg., left juvenile dentary.

Gowrie, QLD: BMNH 40001, right adult dentary.

Revised Diagnosis. Cranium with frontals inflated but rather flat-sided. Rostrum quite elongate and cylindrical. P3 most similar to that of *"Simosthenurus" baileyi*, but less rounded and more of an elongate trapezium shape. While quite shallow anteriorly with a slender symphysis, dentary bears a very large, posteroventrally extended digastric eminence. Lower molars resemble those of *"Procoptodon" mccoyi*, but smaller and much narrower.

Etymology. Named in honor of Ida A. and William R. Browne, University of Sydney, for teaching and research in paleontology and the Pleistocene. Note that Merrilees (1968a) and all authors subsequently referring to this species have applied the suffix *-i* to the species name, but *-orum* is the correct ending (see Article 31.1.2, International Code of Zoological Nomenclature 1999).

Description. Cranium (pls 85-91; table 33). Incisor-bearing portion of premaxilla deep and short. Upright portion of premaxilla extends vertically to level of supraorbital crest before expanding posteriorly, then narrows again to contact nasal. Short, slightly sinusoidal diastema is deflected anteroventrally relative to cheek tooth row; maxilla contributes most of its length. Anterior borders of moderately long incisive foramina lie adjacent to posterior edge of I3 alveolus. Rostrum broad, moderately elongate and rather cylindrical, tapering only slightly anteriorly (pls 85-86, 88-91). Narial aperture slightly broader than deep (pl. 91A-B). Anteorbital and buccinator fossae on maxilla rather shallow anteriorly, but deeper posteriorly anterior to zygomatic arch. Infraorbital foramen positioned just anteroventral of lacrimal. Anterior border of orbit and lacrimal lie dorsal to P3.

Table 33. Dimensions of the adult cranium of southeastern *"Procoptodon" browneorum* (showing mean, standard deviation in parentheses and sample size in brackets) compared with the *"P." browneorum* cranium from Mammoth Cave, southwestern Australia (WAM 63.2.198). Mean dentary measurements for southeastern, Kangaroo Island and southwestern *"P." browneorum* are also shown.

Dimension	Southeast	Kangaroo Is.	Southwest
Condylobasal Length	203 (6.2) [n=5]	-	170
Cranial Width	138 (6.7) [n=5]	-	116
Palatal Length	129 (3.2) [n=5]	-	108
Diastema Length	34.5 (1.16) [n=5]	-	31
% Diastema Length: Palatal Length	26.0 (1.87) [n=5]	-	29
Palatal Width between M1 Protolophs	37.6 (1.54) [n=5]	-	38
Palatal Width between M4 Protolophs	42.2 (3.02) [n=5]	-	-
Width across Frontals	88.6 (2.90) [n=5]	-	68
Distance between Paroccipital Processes	64.8 (3.33) [n=4]	-	60
Dentary Depth	39.4 (3.33) [n=10]	34.3 (0.35) [n=2]	34.6 (3.20) [n=11]
Dentary Width	20.4 (2.02) [n=10]	18.1 (0.07) [n=2]	17.4 (1.12) [n=13]
Dentary Depth / Width	1.94 (0.10) [n=10]	1.90 (0.03) [n=2]	2.01 (0.15) [n=11]

Two small lacrimal foramina are separated by small lacrimal tuberosity. Nasals long and wide (pl. 89C). Nasofrontal suture gently sinusoidal.

Frontals broad for entire portion forward of anterior extremity of frontal-parietal suture, producing a rather long, flat-sided supraorbital crest (pls 85A-B, 86A-B, 88B-D, 89C-D). A small tuberosity is present at its anterior end. Masseteric process well formed, flared ventrally at distal extremity and composed almost entirely of maxilla (pl. 91A). Anterior limit of palatal vacuities is adjacent to anterior half of P3. Postpalatine bars form bridge across palate opposite or just posterior to M4 metaloph. Temporal (parietal) crests well developed, convergent at sagittal suture. Zygomatic arch deep, with very wide ectoglenoid process at posterior extremity of jugal. Very large postglenoid process on squamosal forms posterior border of glenoid fossa, buttressed posteriorly by broad, robust ectotympanic. Markedly rugose and tube-like ectotympanic is oriented anteromesially and ventromesially, terminating mesially at posteroventral wing of alisphenoid and anterior extension of exoccipital. Large elliptical postzygomatic foramen sits above external auditory meatus, with tiny subsquamosal foramen just below dorsal root of zygomatic process of squamosal.

Basicranial plane markedly elevated above level of palatal plane (pls 85A, 86A, 88B-D). Medial pterygoid origin wide and very deep, flanked by extensive pterygoid bone mesially and smaller alisphenoid extension anterolaterally. Basioccipital markedly flexed posterodorsally relative to basisphenoid, bearing well-developed median keel. Occiput deep, moderately wide and oriented at close to 90° relative to dorsal surface of neurocranium. Vertical median occipital crest leads ventrally to wide elliptical foramen magnum bordered by moderately large occipital condyles. Nuchal / lambdoid crests strongly developed and extended posterodorsally. Base of long, posteroventrally oriented paroccipital process bordered laterally by deep, irregular mastoid process of periotic sandwiched between exoccipital and posterior extension of squamosal.

Upper Dentition (table 34). Cross section of rather low-crowned I1 between round and slightly triangular in shape (pls 85C, 87A-C, 90, 91A). I2 tiny, round in cross section, one-third of I1 size. I3 blade-like but not especially elongate; anterior edge mostly straight, except for pointed projection at end of unworn occlusal surface (pls 85A,C, 86A,C, 87A-C). Posterior edge follows contour of anterior edge. Well-developed anterolingual crest clearly visible in unworn I3 and forms a low narrow crest which ascends lingual surface of tooth at its anterior end. Extremely low narrow crest ascends buccal surface of tooth at its anterior end, on opposite side of tooth to anterolingual crest (pl. 87A).

While generally reminiscent of P3, dP2 is far more inflated in outline and much shorter relative to width (pls 87C, 90A, 91C). Straight main crest consists of three cuspules connected by small ridges. Buccal ridgelets ascend from apices of anterior two cuspules and meet to form V-shaped fossette. Lingual crest markedly lower than main crest in height. Apices of all three main crest cuspules give rise to vertical ridgelets on lingual side. Anterior basin is a very small triangular fossette. Longitudinal basin contains coarse ridgelets. Posterior basin is very short, but well demarcated by strong ridgelets directed transversely from main crest and lingual crest.

Completely molariform dP3 very similar in morphology to M1, but it has a stronger postparacrista, a weaker postprotocrista and a postmetacrista that is clearly divided into two components (pls 87C, 90A, 91C). First component (probable stylar crest) ascends lingually from metacone apex, while second arises on lingual side of first and trends vertically to meet buccal extremity of postmetaconulecrista. All other features are essentially identical to upper molars, but urocrista is more distinct and not furcated.

P3 wide relative to its length, gradually narrowing anteriorly (pls 85C, 90B-C, 91D). General outline is a rather round-cornered, elongate trapezium shape. Main and lingual crests converge slightly anteriorly, are subequal in height, and parallel general orientation of buccal and lingual sides of tooth. Posterior edge of P3 only slightly convex, narrower anterior edge slightly more rounded. Main crest consists of three dominant cuspules linked by ridges. As with dP2, large anteriormost and central cuspules give rise to buccal ridgelets that form a V-shaped fossette, which terminates halfway up crown. V-shaped anterior basin narrow to moderately wide, bordered posteriorly by short, coarse transverse ridgelets stemming from large anteriormost cuspules of main and lingual crests. Longitudinal basin wide and moderately deep, traversed by many contorted coarse and fine enamel ridgelets. Posterior basin short, wide and well demarcated posteriorly by extensions of main and lingual crests. Generally well-formed but irregular transverse ridgelet or composite of several ridgelets separate longitudinal and posterior basins. Posterior basin encloses several parallel, longitudinal ridgelets. Posterobuccal accessory cusp usually absent, but may be present or expressed only as buccal ridgelet ascending from posteriormost cuspule of main crest.

General outline of low-crowned upper molars is particularly square (pls 85C, 86C, 87C, 90, 91C-D). Lophs taper slightly toward unworn crests, especially anteriorly. Size gradient along molar row is M1<M2<M3>M4. Upper molars vary little in morphology, except for narrowness of M4 metaloph relative to protoloph. Loph crests slightly convex anteriorly. Fine preparacrista connects to paracone apex, but stylar cusp A portion of crest is only slightly enlarged (pls 90, 91C-D). Preparacrista forms or merges into thin precingulum, which terminates at anterolingual corner of crown. Just lingual of tooth midline on precingulum, preprotocrista is usually difficult to distinguish from fine vertical enamel crenulations. Postprotocrista is clearly separated into two components. More lingual is very low and ascends posterobuccally from protocone apex, before terminating on side of thicker, more posteriorly oriented component which extends into interloph valley. Buccal postprotocrista component either stops at interloph valley or becomes very fine, terminating on anterior face of metaloph. Although slightly finer than buccal component of postprotocrista, adjacent mesocrista is well developed and centrally positioned on protoloph (pls 90, 91C-D). It extends into interloph valley and may also continue onto metaloph, becoming very fine before terminating. Well-developed postparacrista ascends posteriorly from paracone apex, meeting with short fine premetacrista. Together, they enclose interloph valley. Small eminence on postparacrista and vertical groove on buccal side insinuate a stylar crest component, probably mainly stylar cusp C. Low postmetacrista is oriented slightly dorsolingually. Postmetaconulecrista often divided into two components, one low and ascending directly from metaconule apex and terminating on side of thicker, vertically oriented component, which forms narrow shelf slightly overlapped by precingulum of succeeding molar. In addition to cristae, several fine enamel crenulations are often present on loph faces.

Dentary (pls 91-93; table 33). Ramus deep and narrow for most of length, becoming shallower anteriorly. Posteroventral aspect markedly deepened due to strongly developed digastric eminence and sulcus (pl. 92). Symphyseal plate very rugose (pl. 92); rather slender symphysis often partially fused. Symphyseal boss moderately developed. Large inferior transverse torus of symphysis extends under

Table 34. Cheek tooth dimensions of *"Procoptodon" browneorum* from southeastern Australia, Kangaroo Island and Western Australia, showing mean, standard deviation (in parentheses). Dimensions of the holotype (WAM 63.2.94) and Gowrie specimen (BMNH 40001) are also shown.

Tooth		L	AW	PW	AH	PH	n
dP2	Southeastern	10.9 (0.43)	8.7 (0.40)	10.8 (0.38)	7.0 (0.51)	7.9 (0.60)	15
	Western	10.6 (0.57)	8.1 (0.56)	10.1 (0.50)	7.3 (0.52)	7.6 (0.37)	5
dP3	Southeastern	11.3 (0.29)	10.7 (0.32)	11.0 (0.37)	5.7 (0.54)	5.9 (0.46)	15
	Western	10.1 (0.31)	9.7 (0.48)	10.2 (0.55)	5.6 (0.26)	5.8 (0.33)	7
P3	Southeastern	17.1 (0.57)	10.9 (0.68)	13.7 (0.81)	9.8 (0.76)	9.7 (0.78)	21
	Kangaroo Is.	16.0 (0.21)	9.5 (0.21)	12.1	9.9 (0.28)	10.0 (0.21)	2
	Western	15.6 (0.64)	9.6 (0.52)	12.2 (0.74)	8.7 (0.74)	8.9 (0.95)	15
M1	Southeastern	13.0 (0.42)	12.4 (0.40)	12.3 (0.34)	6.1 (0.76)	6.5 (0.62)	28
	Kangaroo Is.	12.3 (0.35)	11.6 (0.21)	11.3 (0.35)	-	-	2
	Western	11.2 (0.47)	11.1 (0.43)	11.0 (0.48)	6.5 (1.78)	6.8 (1.85)	16
M2	Southeastern	14.1 (0.37)	13.6 (0.46)	13.1 (0.43)	6.6 (0.72)	6.9 (0.57)	23
	Kangaroo Is.	12.9 (0.14)	12.1 (0.42)	11.6	6.7 (0.57)	6.8	2
	Western	11.8 (0.46)	11.4 (0.50)	11.1 (0.45)	6.2 (0.70)	6.3 (0.34)	16
M3	Southeastern	14.5 (0.39)	14.0 (0.50)	12.9 (0.54)	6.8 (0.61)	7.0 (0.50)	20
	Kangaroo Is.	13.1 (0.64)	12.0 (0.64)	11.2 (0.21)	6.8 (0.14)	6.8 (0.21)	2
	Western	12.2 (0.52)	11.4 (0.53)	10.8 (0.38)	5.9 (0.59)	6.1 (0.51)	12
M4	Southeastern	13.6 (0.57)	13.4 (0.45)	11.4 (0.45)	6.3 (0.66)	6.2 (0.53)	16
	Kangaroo Is.	12.0	10.9	9.6	6.8	5.9	1
	Western	11.7 (0.44)	11.0 (0.38)	9.7 (0.71)	5.3 (0.47)	5.3 (0.62)	6
dp2	Southeastern	9.8 (0.45)	6.5 (0.34)	9.1 (0.36)	7.5 (0.82)	7.1 (0.71)	17
	Western	8.8 (0.53)	5.9 (0.45)	8.2 (0.54)	8.0 (0.92)	7.4 (0.82)	7
dp3	Southeastern	10.4 (0.36)	9.2 (0.46)	9.3 (0.30)	7.3 (0.47)	7.2 (0.63)	17
	Western	9.6 (0.55)	8.6 (0.33)	8.6 (0.23)	7.3 (0.38)	7.5 (0.40)	7
p3	Southeastern	16.1 (0.53)	8.5 (0.34)	10.3 (0.58)	9.9 (0.85)	9.9 (0.83)	19
	BMNH 40001	17.6	8.0	9.7	11.1	8.9	
	Kangaroo Is.	14.7 (0.62)	7.8 (0.21)	9.4 (0.61)	9.3	-	3
	Western	14.7 (0.63)	7.5 (0.56)	9.8 (0.48)	8.8 (0.71)	8.1 (0.56)	13
	Holotype	14.8	7.6	9.7	9.1	8.0	

m1	Southeastern	13.1 (0.64)	10.4 (0.49)	10.6 (0.43)	8.6 (0.97)	8.8 (1.02)	29
	BMNH 40001	12.7	9.9	10.0	-	-	
	Kangaroo Is.	11.2 (0.14)	9.7	9.9	-	-	2
	Western	11.0 (0.58)	9.2 (0.37)	9.4 (0.42)	8.0 (1.09)	8.1 (1.02)	18
	Holotype	11.0	9.0	9.1	-	-	
m2	Southeastern	14.7 (0.53)	11.4 (0.39)	11.7 (0.33)	9.4 (0.94)	9.5 (0.80)	21
	BMNH 40001	13.6	11.1	11.2	7.1	7.6	
	Kangaroo Is.	12.6 (0.36)	10.7 (0.57)	10.8 (0.26)	9.7	9.7	3
	Western	12.1 (0.49)	10.0 (0.45)	10.2 (0.42)	7.7 (1.11)	8.0 (1.19)	16
	Holotype	12.0	9.7	9.7	7.6	7.6	
m3	Southeastern	14.9 (0.51)	12.0 (0.43)	12.2 (0.35)	9.3 (0.61)	9.1 (0.73)	21
	BMNH 40001	15.0	12.5	12.4	8.6	8.6	
	Kangaroo Is.	12.9 (0.47)	10.9 (0.12)	10.8 (0.32)	8.0 (1.84)	7.9 (1.70)	3
	Western	12.5 (0.42)	10.3 (0.49)	10.4 (0.39)	7.9 (0.52)	7.6 (0.55)	13
	Holotype	12.3	10.1	10.1	7.8	6.7	
m4	Southeastern	14.0 (0.43)	12.2 (0.48)	11.2 (0.40)	8.1 (0.62)	7.4 (0.78)	16
	Kangaroo Is.	12.4 (0.32)	10.7 (0.17)	9.5 (0.70)	6.6 (0.14)	6.6 (0.57)	3
	Western	11.8 (0.38)	10.3 (0.41)	9.5 (0.53)	6.5 (0.61)	6.1 (0.45)	11
	Holotype	11.8	10.2	9.5	7.2	6.5	

genial fossa to below junction of p3 and m1, with length roughly equal to half that of superior transverse torus. Moderately long diastema aligned in almost same plane as alveolar margin of cheek tooth row, but is very gently sinusoidal in lateral profile. Median dorsal groove moderately deep and wide (pls 91F, 93A,D). Anterior mental foramen located well below diastema ridge, just above mid-depth on ramus. Generally narrow, posteriorly deepening buccinator sulcus extends from diastema ridge above anterior mental foramen to below m1 hypolophid. Posterior mental foramen located just below mid-depth on ramus beneath m3 protolophid.

Anterior root of ascending ramus adjacent to m4. Anterior border inclined slightly anteriorly to midpoint before becoming vertical, then curving slightly posteriorly as ascending ramus narrows into coronoid process. In lateral profile, mandibular notch a wide, shallow W or U shape and is oriented posteroventrally. Mandibular condyle broad, but narrows mesially. Posteroventral aspect of masseteric fossa moderately flared laterally. Ventral border of masseteric fossa sits just below level of alveolar margin of cheek tooth row (pl. 92E,G). Rather elongate, elliptical masseteric foramen leads into vertical masseteric canal that does not extend into body of ramus. Inferior mandibular foramen is a rounded triangular or elliptical shape and opens posteromesially, with dorsal border roughly level with alveolar margin of cheek tooth row. Wide mylohyoid groove often partly overhung by very slight posteroventrally oriented process and well-developed anterodorsally oriented process extending from anteroventral border of medial pterygoid fossa (pls 91F, 92F,H). Elevated, deep medial pterygoid fossa lacks marked angular process.

Lower Dentition (pls 91E-F, 92-93; table 34). Small i1 not markedly robust or upturned. Minimally worn occlusal surface gently sinusoidal, rising to slight point anteriorly (pl. 92A-D). Posterior extremity of minimally worn occlusal surface level with or just above alveolar margin of cheek tooth row. Thinner layer of enamel covers lingual surface of i1 between thin dorsal enamel flange and ventrolingual extension of thicker enamel from buccal surface (pl. 92A-D).

While similar in general morphology to p3, dp2 is shorter relative to width (pl. 93C). Anteriorly, connection between buccal and main crests is either direct or via a short ridgelet arising posterobuccally from large anteriormost cuspule. Main crest divided into four cuspules, with posteriormost twice length of others. Small V-shaped fossette occasionally present on lingual side of crown below first and second cuspules. Completely enclosed median valley contains several enamel ridgelets (pl. 93C).

Completely molariform dp3 similar in general outline to m1, but differs by having slightly better developed premetacristid and protolophid crest distinctly narrower than hypolophid crest due to greater taper (pl. 93C). Trigonid very short, posterior portion of lingual paracristid component shifted from protoconid apex. Transverse anterior portion of paracristid very short, meeting with low premetacristid at position of paraconid. Trigonid underlain by well-developed precingulid, which extends three quarters of way across front of tooth from buccal side of crown. Parametacristid very low, extends into trigonid basin toward position of paraconid, just lingual of tooth midline. Cristid obliqua oriented anteroposteriorly and shifted well lingual across hypolophid face toward tooth midline; terminates on protolophid posterior face opposite posterior extremity of paracristid. Very low, fine preentocristid curves from entoconid apex to protolophid posterior face, mirroring orientation of anterior portion of cristid obliqua. Anterior faces of lophids bear two to three vertical enamel crenulations lingual to paracristid and cristid obliqua (pl. 93C). Hypolophid posterior face typically with distinct, often A-shaped postcingulid shelf.

Longer than any molar, p3 has large anteriormost cuspule of main crest shifted buccally relative to remainder of main crest (pl. 93A-B,D). Tooth narrows gently anteriorly, bearing rather shallow, wide vertical groove on buccal side descending from between anterior extremity of buccal crest and large anterior cuspule. Main crest divided into five cuspules, posteriormost twice length of more distinct anterior four. Each of anterior four cuspules bears vertical ridgelets descending from buccal and lingual sides. Large anteriormost cuspule also gives rise to small ridgelets anteriorly and posteriorly, with posterior one frequently merging with buccally directed ridgelet from second main crest cuspule and anterior extremity of buccal crest. Wide and moderately deep median valley contains numerous contorted coarse and fine enamel ridgelets, bordered buccally by curved buccal crest, which forms V-shaped notch posteriorly with posterior extremity of main crest (pls 93A-B,D).

Although varying in size from m1 to m4, low-crowned lower molars vary minimally in morphology, except for narrowness of m4 hypolophid relative to protolophid and less developed m4 postcingulid (pls 91F, 92-93). Lophid crests are straight and parallel, showing only a slight anterior turn of lophid ends on buccal side. Very short trigonid is underlain by a distinct precingulid. Lingual component of moderately high paracristid shifted well anterolingual of apex of protoconid apex, approximately one quarter of way across protolophid face. Posterior portion of lingual component is oriented anteroposteriorly before turning lingually and terminating after very short distance, where it meets a fine but distinct premetacristid extending from metaconid apex. Buccal component of paracristid lower than lingual component, curves around buccal border of lower molar from protoconid apex to inflection point on lingual component (pl. 93). Parametacristid very low, fine and extending just lingual to premetacristid into trigonid basin on m1, but is not expressed in m2-4. Two to three coarse enamel crenulations are centrally positioned on protolophid anterior face. Cristid obliqua arises on hypolophid in analogous position to paracristid on protolophid and is similarly oriented, terminating on protolophid posterior face opposite posterior extremity of paracristid. Portion of preentocristid contacting entoconid apex is very fine and low, but coarsest of two to three enamel crenulations on lingual side of hypolophid anterior face may be a buccally shifted preentocristid component. Hypolophid posterior face bears a distinct, often A-shaped postcingulid shelf overlain by several fine vertical enamel crenulations (pl. 93C-D).

Status of BMNH 40001. This partial left dentary (containing p3, m1-3) from Gowrie in southeastern Queensland has had a varied taxonomic history. Owen (1874a, 1877b) referred it to *Sthenurus atlas*, but from an examination of Owen's (reversed) figures, Bartholomai (1963:71) thought that the specimen might be referable to his new species, *Sthenurus antiquus*. Tedford (1966:52) believed the dentition of BMNH 40001 more closely resembled his new sthenurine, *Si. orientalis* (= *Si. occidentalis*). Although wear has obliterated several features of taxonomic utility on the first two molars, the balance of what is interpretable on the p3 and m3 is sufficient to refer this specimen to *"Procoptodon" browneorum* with relative confidence. This species was unknown at the time of either Bartholomai's or Tedford's review, but dimensions of the four cheek teeth preserved in BMNH 40001 fall within the ranges of southeastern *"Procoptodon" browneorum* (table 34). The p3 is clearly longer than any molar, and the large anteriormost cuspule of the main crest is shifted buccally relative to the remainder of the main crest, as is typical of *"P." browneorum*. Other *"P." browneorum* characteristics include the confluence of the ridgelet directed posteriorly from the anteriormost cuspule and the ridgelet directed buccally from the second main crest cuspule and the coarse

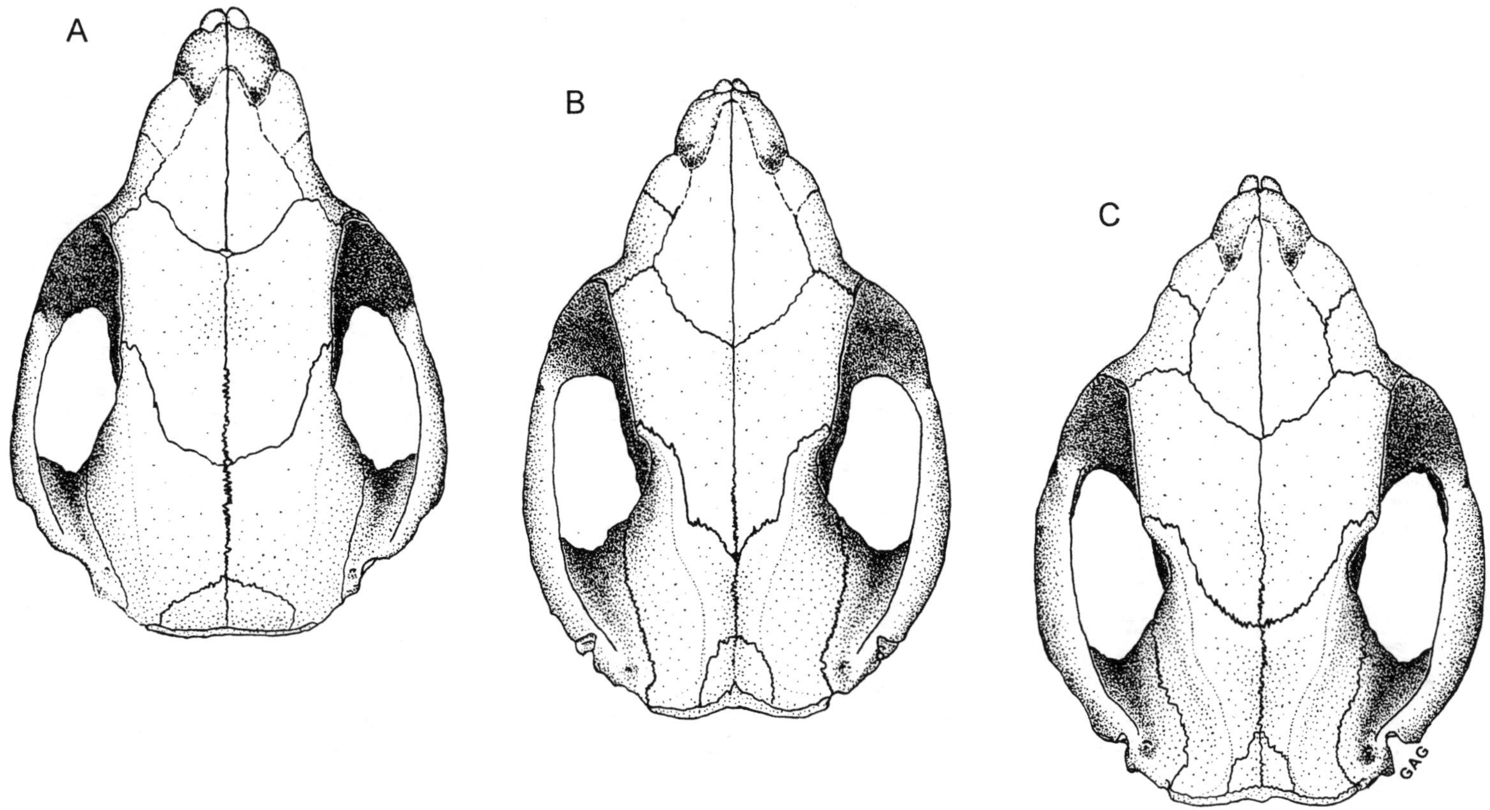

Figure 5. Age series of *"Procoptodon" browneorum* crania in dorsal view. **A**, juvenile with M1 fully erupted (SAM P20677, portion of anterior nasal and premaxillary region reconstructed); **B**, juvenile with M2 half-erupted (SAM P16542, nasals and portion of premaxilla reconstructed); **C**, juvenile with M2 fully erupted (composite of SAM P20256 and FU 0029, nasals reconstructed);.

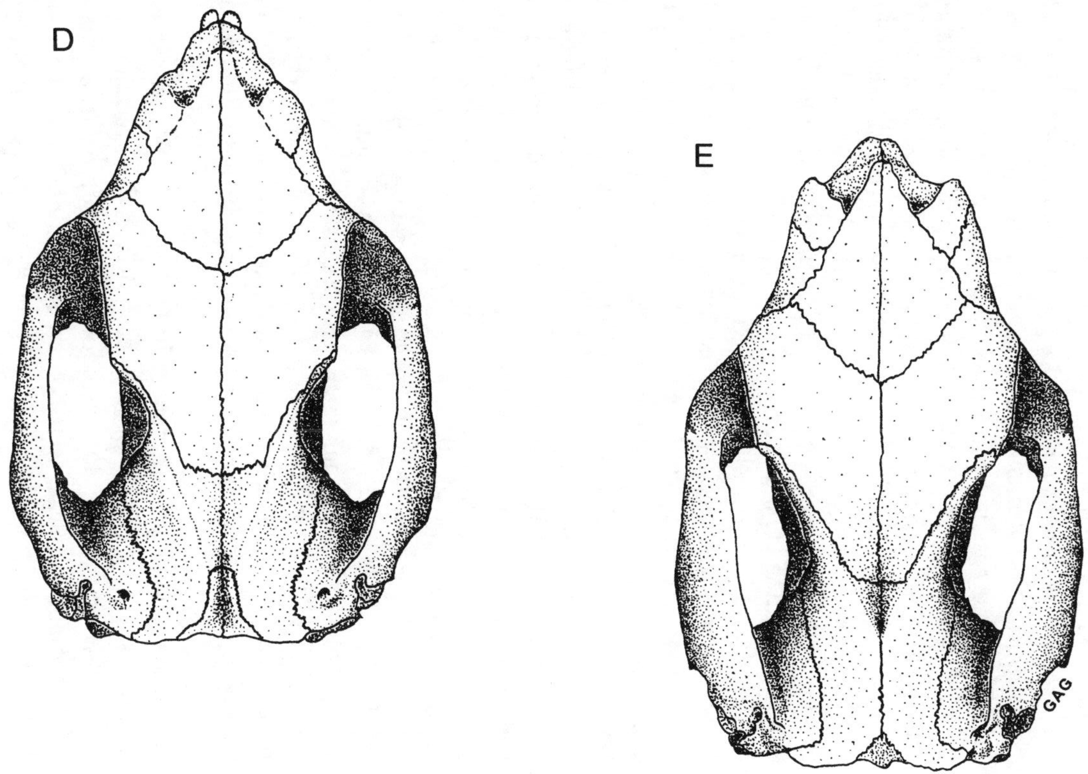

Figure 5 (continued). **D**, juvenile with M3 fully erupted (SAM P16646, frontals and nasals reconstructed); **E**, adult (FU 1011, portion of anterior nasal and premaxillary region reconstructed).

enamel crenulations in the median valley. Presence of a distinct cuspule low on the crown on the anterolingual aspect of the tooth and direct connection of the buccal crest to a small ridgelet arising from the anteriormost main crest cuspule are also observed in individuals of *"P." browneorum.* No features of the molars (especially the least worn, m3) preclude the designation of BMNH 40001 to *"P." browneorum*, including the A-shaped postcingulid and position of the lingual components of the paracristid and cristid obliqua.

Variation. "Procoptodon" browneorum is one of the most common and widely distributed sthenurine species. Not surprisingly, considerable variation in size and morphology is seen both from across its range and within individual samples (e.g., Tight Entrance Cave, Victoria Fossil Cave). While morphological variation between eastern and western representatives is no greater than that seen within individual samples, there are marked differences in size between regions according to cranial, dentary and dental dimensions (tables 33-34). Prideaux and Wells (1998:10) noted that the teeth of eastern specimens are smaller relative to the size of the dentary, but this observation is fallacious.

Cranium. Although several very well preserved crania of *"Procoptodon" browneorum* are known from the Naracoorte Caves, only one such specimen is currently known from anywhere else: WAM 63.2.198 from Mammoth Cave in southwestern Australia (pl. 86). Based on condylobasal length, width across frontals and estimated maximum cranial width, WAM 63.2.198 is 16% smaller than the crania from southeastern Australia (table 34). The only other difference between them appears to be a slightly less posterodorsally flexed basioccipital relative to the basisphenoid in the western specimen. Between the Naracoorte adult crania, slight variation is seen in overall size, neurocranial width, zygomatic arch depth, frontal inflation and rostral inflation (pls 85, 88-90). A good age series for *"P." browneorum* is preserved in the Victoria Fossil Cave sample (fig. 5), from pouch young with M1 not fully erupted (pl. 16X-AA) through to older juveniles (pl. 16I,V) to adult. Significant proportional changes occur as size increases. The most marked are associated with the massive expansion of the splanchnocranium and orbital region relative to the neurocranium, as well as frontal inflation, zygomatic arch deepening, and development and convergence of the temporal crests (fig. 5).

Upper Dentition. No morphological differences in I1-3 are evident that cannot be ascribed to differential wear. By contrast, variation in dP2 is observed in its (a) relative inflation; (b) anterior relative to posterior width; (c) anterior basin development (barely discernable to large, well bordered and containing numerous fine ridgelets); (d) lingual crest shape (irregularly curved, convex lingually, straight); (e) ridgelet development in longitudinal basin; and (f) composition of transverse enamel separation between longitudinal and posterior basins (single ridgelet, two to three combined ridgelets). No morphological variation is seen in dP3. Like its dP2, the P3 of *"Procoptodon" browneorum* is very variable in morphology. Together, though, all of the specimens occupy a discrete morphospace that permits separation from all other taxa. P3 varies in (a) relative outline inflation (especially posterolingually); (b) degree to which it is narrowed anteriorly; (c) anterior basin development (tiny valley to large V-shaped fossette); (d) distinction of the main crest cuspules; (e) lingual crest shape (irregularly curved, convex lingually, quite straight); (f) ridgelet development in the longitudinal basin; (g) posterobuccal accessory cusp development (absent, slight vertical eminence or ridgelet, discrete cusp); and (h) composition and height of separation between longitudinal and posterior basins (single ridgelet, two to three combined ridgelets, very low, subequal in height to lingual crest). Upper molars only vary slightly morphologically, most notably with regard to the relative development of the postparacrista and premetacrista. Mean length and width dimensions of the Kangaroo Island *"P." browneorum* P3 and M4 fall within the range of variation for the western population, but the means for M1-3 lie in between those of western and southeastern mainland. On average, the P3 of mainland southeastern *"P." browneorum* is 12% larger than in the western population, while the molars are 13% larger for M1, and 19% larger for M2-4.

Dentary. Marked variation is seen within and between samples of *"Procoptodon" browneorum* adult dentaries. While some differences may be partially sexually dimorphic, in none of the large samples (e.g., Tight Entrance Cave, Victoria Fossil Cave) have I been able to discern any bimodality in size, proportion or relative development of discrete morphological attributes. Dentaries vary in depth relative to both length and width. This variation is often due to differences in the development of the digastric eminence. While the digastric eminence and sulcus are large in all individuals of *"P." browneorum*, size varies between extremely large (e.g., SAM P20465, pl. 92G-H) and only moderately large (e.g., SAM P20471, pl. 91E-F; SAM P16532, pl. 92E-F). The anterior limit of the digastric sulcus may be anywhere beneath the m4 hypolophid up to the m3 protolophid, variability mirrored by the position of the anterior root of the ascending ramus. In those individuals with a very well developed digastric sulcus, the groove associated with the genial fossa tends to be well developed. Length of the masseteric fossa, and particularly size and roundness of the masseteric foramen vary noticeably between specimens of *"P." browneorum*. The pointed processes that partially overhang the mylohyoid groove also vary in size. The posterior mental foramen is positioned mid-depth or slightly below mid-depth on the ramus and may be anywhere from beneath the m2 hypolophid to the m3 interlophid valley. A smaller subsidiary foramen is sometimes present just posterior or anteroventral to the posterior mental foramen. Several tiny subsidiary anterior mental foramina may also be present below the main anterior mental foramen and p3. The buccinator sulcus varies in both depth posteriorly and posterior limit, which may be from anywhere beneath the m1 interlophid valley to the m2 protolophid. As with the cranium, proportional variation during ontogeny is marked (pl. 92). Primarily, dentary depth increases relative to width, the digastric sulcus and eminence enlarge, and length and procumbency of the diastema increase, as does the size of the masseteric foramen and height of the condyle.

Lower Dentition. Aside from the variation in size and the variable presence of two to three shallow longitudinal grooves along the buccal surface of i1, I have observed no significant morphological differences between the numerous lower incisors known for *"Procoptodon" browneorum*. The dp2 varies in (a) relative inflation of its general outline; (b) anterior relative to posterior width; (c) the anterior termination of the buccal crest (connects directly to main crest second cuspule, connects with small ridgelet arising from large anteriormost cuspule of main crest); (d) development of ridgelets in median valley; and (e) development of V-shaped fossette on lingual side of crown (absent, slight, well developed). The dp3 varies slightly in width relative to length and development of the parametacristid, coarse enamel crenulations lingual to paracristid and cristid obliqua, and postcingulid. Like P3 and dp2, p3 shows considerable morphological variation. This variation is displayed in (a) length relative to width and lower molar length; (b) presence of a vertical ridgelet arising anteriorly from the anteriormost main crest cuspule; (c) presence of a distinct cuspule low on crown on anterolingual aspect of tooth; (d) anterior termination of the buccal crest (separated from large anteriormost main crest cuspule by another small cuspule, connected directly to small ridgelet arising from anteriormost main crest cuspule); (e) degree to which cuspules of main crest are distinct (minimally, moderately); (f) relative number and coarseness of ridgelets in the median valley; and (g) curvature of the buccal crest (curved along entire length, curved posteriorly but straightening anteriorly). Lower molars vary in the same features as dp3 and in relative development of the transverse anterior portion of paracristid and the buccal components of the paracristid and premetacristid. Mean p3 and lower molar dimensions of the Kangaroo Island specimens fall near the upper end of the range for western *"P." browneorum*. On average, the p3 of mainland southeastern *"P." browneorum* is 9% larger than in the western population, while the molars are 17% larger.

Comparison With Other Taxa. Cranium. In overall proportions, *"Procoptodon" browneorum* is most similar to *Metasthenurus newtonae*, *"Simosthenurus" baileyi* and *Si. occidentalis*. It is easily distinguished from the first two species by its more inflated frontal region and more elongate, rather cylindrical rostrum. This splanchnocranial form is unique within the Simosthenurini, but does bear some resemblance to that of the larger *Sthenurus stirlingi*. The cylindrical rostrum also distinguishes *"P." browneorum* from *Si. occidentalis*, which also differs by being more brachycephalic and having more laterally tapered frontals. Although the neurocranium is proportionally narrower, the crania of southeastern *"P." browneorum*, *M. newtonae* and *Si. occidentalis* are similar in size. Western *"P." browneorum* is closer in cranium size to *"P." gilli*. Marked development of the temporal crests is most reminiscent of *Si. occidentalis* and *Si. maddocki*, while the masseteric process is intermediate in size and morphology between these two species. Intriguingly, the juvenile crania of *"P." browneorum* and *"P." gilli* are particularly similar to one another (figs 5-6), but no marked resemblances are carried on into the adult crania.

Upper Dentition. The I1 of *"Procoptodon" browneorum* is most similar to that of *Metasthenurus newtonae*, but slightly smaller and lower crowned than in other species of comparable cranial size. I2 is not dissimilar to several sthenurines in morphology and size, but the relative short I3 blade is most similar to *"Simosthenurus" baileyi* in overall morphology. It differs by having a low vertical groove on its buccal surface.

In size and morphology, the dP2 of *"Procoptodon" browneorum* is also very similar to that of *"Simosthenurus" baileyi*. It differs by being slightly more inflated laterally and having a slightly smaller posterior basin. Compared with *Si. occidentalis*, the *"P." browneorum* dP2 is shorter and has slightly coarser enamel crenulations in the longitudinal basin. While the dP3 of *"Si." baileyi* was apparently similar (it is worn in the holotype and only known example) to that of *"P." browneorum*, the precingulum is slightly thicker, the premetacrista is smaller and the loph faces bear coarse enamel crenulations. In *Si. occidentalis*, the dP3 has a weaker postparacrista, lophs that taper more toward their crests and many fine enamel crenulations. The P3 of *"P." browneorum* is also similar to that of *"Si." baileyi* in size and overall morphology, but it differs by being less rounded in general outline and having a wider longitudinal basin. While also reminiscent of the *Si. occidentalis* P3, it may be distinguished by a more trapezium-shaped outline, shorter length relative to molar length, main crest dominated by fewer cuspules and anteriorly convergent main and lingual crests.

The low-crowned upper molars of *"Procoptodon" browneorum* are easily distinguished from most sthenurines by their numerous coarse enamel crenulations, but are quite similar to those rooted into the only known maxillary specimen of *"P." oreas* (QM F3814, pl. 103A). The upper molars of *"P." browneorum* may be distinguished on the basis of (a) a stronger connection between the paracone apex and preparacrista, which bears a reduced stylar cusp A portion; (b) a slightly thinner preprotocrista; (c) a precingulum that terminates on the lingual side of the tooth; (d) a weaker postmetacrista; and (e) a postmetaconulecrista divided into two components.

Dentary. In size and morphology, the very deep dentary of *"Procoptodon" browneorum* is most similar to that of *Metasthenurus newtonae* and, to a lesser degree, the much more robust *"Simosthenurus" brachyselenis*. Although *M. newtonae* possesses a similarly large digastric eminence and sulcus, the gradient of decreasing dentary depth anteriorly is greater in *"P." browneorum*. The (anteroposterior) length of the ascending ramus and masseteric fossa is also greater in *"P." browneorum*. The sharp processes that partially overhang the mylohyoid groove can be as well developed in *"P." browneorum* as in *"P." oreas* and *M. newtonae*, but the groove itself is wider and shallower in *"P." browneorum*. The rather slender *"P." browneorum* symphysis is also similar to that of *M. newtonae*, but in this species the inferior transverse torus is smaller. The diastema of *"P." browneorum* is longer than in *Si. occidentalis*, but shorter and less procumbent than in *M. newtonae*.

Lower Dentition. The rather small size, moderately robust morphology and degree to which the i1 is upturned in *"Procoptodon" browneorum* is intermediate overall between *"Simosthenurus" baileyi* and *"P." gilli*. In size and morphology, the dp2 is most reminiscent of *"Si." baileyi*, but it has a longer buccal crest, wider median valley and narrows more markedly anteriorly. The dp3 of *"P." browneorum* is also like that of *"Si." baileyi*, but the cristid obliqua is shifted anterolingually from the hypoconid apex and there are more crenulations on the lophid faces.

The p3 of *"Procoptodon" browneorum* is similar to both *"P." gilli* and *Simosthenurus occidentalis* in size and morphology and, to a lesser degree, the smaller *"P." oreas*. In basal outline, the p3 of western and Kangaroo Island *"P." browneorum* most closely resembles that of *"P." gilli*, but they narrow less anteriorly and usually have a wider median valley with coarser transverse ridgelets. Mainland southeastern *"P." browneorum* consistently displays the latter condition and tends to have a less arcuate buccal crest. *"P." browneorum* may be distinguished from *Si. occidentalis* by having a wider, shallower vertical groove on the buccal side of the p3 crown (demarcating the anterior and posterior portions of the tooth) and a main crest much less distinctly divided into cuspules, especially posteriorly. The vertical ridgelets that descend from these cuspules on both the lingual and buccal sides are also weakly developed by comparison. In addition, the large anteriormost cuspule of the main p3 crest in *Si. occidentalis* is not separated from the remainder of the crest by a marked V-shaped notch. In this feature and in the general outline of the tooth, the *"P." browneorum* p3 is reminiscent of the much larger *"Si." pales* p3. While the tooth is also somewhat reminiscent in size and general outline to the *"Si." baileyi* p3, it may be distinguished by a more curved main crest, longer buccal crest and larger number of ridgelets in the median valley.

The low-crowned lower molars of *"Procoptodon" browneorum* are most reminiscent of those of *"P." mccoyi* and, to a slightly lesser degree, *"P." oreas* and *"P." gilli*. With *"P." mccoyi*, the lower molars of *"P." browneorum* share the same relative development, orientation and position of all major topographic features. However, even the larger southeastern form of *"P." browneorum* has smaller, narrower and higher-crowned molars than *"P." mccoyi*. Southeastern *"P." browneorum* lower molars are more similar to those of *"P." oreas* in size, and also have relatively straight lophid crests, coarse enamel crenulations on the lophid faces, lingual components of the paracristid and cristid obliqua shifted well toward the tooth midline, and a distinct postcingulid. *"P." browneorum* differs by having a shorter trigonid, a finer paracristid and cristid obliqua, a weaker transverse anterior portion of the paracristid and a slightly better developed buccal paracristid component. In these characteristics, *"P." browneorum* resembles *"Simosthenurus" baileyi*. The lower molars of the smaller western and Kangaroo Island populations of *"P." browneorum* are intermediate in size and morphology between those of southeastern *"P." browneorum* and *"P." gilli*. The lower molars of *"P." gilli* differ from those of *"P." browneorum* by having a more anterolingually oriented lingual paracristid and cristid obliqua components, a thicker transverse anterior portion of the paracristid, fewer enamel crenulations and no postcingulid.

Geographic Distribution. Southwestern Western Australia, Nullarbor Region, Kangaroo Island, southern South Australia, southwestern Victoria, southeastern Queensland (fig. 30; table 44).

Temporal Distribution. Middle to late Pleistocene.

"Procoptodon" gilli (Merrilees, 1965)

(pls 94-100; figs 6, 13F, 14F, 15I, 18H; tables 35-36)

Sthenurus gilli, Merrilees, 1965, pp. 22-31, figs 1-2, 6-7.
Sthenurus gilli, Merrilees, 1968a, pp. 66, 68-70, 73.
Sthenurus sp. cf. *gilli*, Hope et al., 1977, pp. 363, 370, 373, table 1.
Sthenurus gilli, Wells and Murray, 1979, pp. 213, 215, 217-218, fig. 6A.
Simosthenurus gilli, Pledge, 1980, pp. 131-133, 136, table 1, fig. 2A.
Sthenurus sp. cf. *gilli*, Williams, 1980, p. 109, site 48d.
Simosthenurus gilli, Flannery, 1983, pp. 22-24, fig. 6a.
Simosthenurus gilli, Flannery and Gott, 1984, pp. 404-405, tables 7-8.
Sthenurus gilli, Wells, Moriarty and Williams, 1984, pp. 325-326, 329.
Simosthenurus gilli, Flannery, 1989, pp. 29-30, fig. 10.
Sthenurus gilli, Lundelius and Turnbull, 1989, pp. 4-6, fig. 2.
Simosthenurus gilli, Pledge, 1990, pp. 253, 255, 257, table 2, fig. 5.
Sthenurus gilli, Pledge, 1992, pp. 136-137.
Simosthenurus gilli, Pledge, 1992, p. 137.
Sthenurus gilli, Prideaux, 1994, pp. 134, 137-138, table 1.
Sthenurus gilli, Prideaux and Wells, 1997, pp. 185-186, 191, 196.
Sthenurus gilli, Prideaux and Wells, 1998, pp. 1, 4, 11-12, 15.
Sthenurus gilli, Brown and Wells, 2000, pp. 100-101.
Simosthenurus gilli, Prideaux, 2000, pp. 12-13.
Simosthenurus gilli, Reed and Bourne, 2000, pp. 74-86.

Taxonomic Authority. Merrilees, D. 1965. Two species of the extinct genus *Sthenurus* Owen (Marsupialia, Macropodidae) from south-eastern South Australia, including *Sthenurus gilli* sp. nov. Journal of the Royal Society of Western Australia 48:22-32.

Holotype. NMV P21609, left and right juvenile dentaries (missing most of ascending rami) collected in 1956 or 1957 by C. Austin, W. Brooker and C. Sasse from exposed Pleistocene cave fill in Shire Quarry, Strathdownie, western Victoria (Merrilees 1965).

Paratypes. Shire Quarry, Strathdownie, VIC: 46 NMV registered dentaries; 29 NMV registered maxillae; 12 NMV registered premaxillae.

Referred Specimens. Spring Creek, Minhamite, VIC: NMV P157294, left and right dentaries; NMVP157296, right adult dentary; NMV P157297, left dp2; NMV P198423, left adult dentary; NMV P198424, left juvenile dentary; NMV P198445, left P3; NMV P198448, right P3.

Puralka Caves, VIC: NMV P172864, right p3; NMV P173005, several loose incisors; NMV P173008, three adult dentaries.

Pot Luck Cave, Buchan, VIC: NMV P22505 / P22506, left premaxilla and maxilla.

Didgeridoo Cave, Buchan, VIC: NMV P22508 / P22509 / P22510, cranium and dentaries.

Skene St, Warrnambool, VIC: NMV P four unreg. dentaries and maxillae.

South Site, Lancefield, VIC: NMV P200721, right p3.

Mount Fairy Caves, NSW: AM F unreg., left and right adult dentaries.

Derrington St, Mount Gambier, SA: SAM P13652, left juvenile dentary; SAM P unreg. (formerly SAM P13651d,f), right adult dentary, left juvenile dentary, four single loose teeth.

Gray St, Mount Gambier, SA: SAM P13668, left juvenile dentary, left adult dentary, p3.

Goulden's Hole, near Mount Schank, SA: SAM P40094, numerous single loose teeth.

Tankstand Cave, near Mount Schank, SA: Listed in Reed and Bourne (2000).

Kilsby's Hole, Mount Gambier, SA: SAM P40073, left adult dentary; SAM P40074, right adult maxilla; SAM P unreg., many single loose teeth.

Mount Burr Cave, SA: SAM P17455, right adult dentary.

Glencoe, near Tantanoola, SA: SAM P unreg., four maxillae and five dentaries.

Green Waterhole Cave, near Tantanoola, SA: SAM P17248, adult cranium and dentaries; SAM P17265, left adult maxilla; SAM P17293, adult neurocranium; SAM P17521, adult neurocranium; SAM P18326, left and right adult maxillae; FU 0194, right adult dentary; FU 1651, adult cranium; FU 1652, adult cranium (missing part of zygomatic arches, left nasal); FU 1653, adult cranium (missing jugals), left and right dentaries; FU 1654, left and right adult maxillae; FU 1655, left and right juvenile dentaries; FU 1656, left and right maxillae and dentaries; FU 1657, left and right adult dentaries; FU, several unreg. dentaries, maxillae and teeth.

Morgan's Cave, Tantanoola, SA: SAM P12841, left juvenile dentary, three single loose molars.

Tantanoola Cave "E", SA: SAM P40095, left i1; SAM P40096, left adult maxilla.

Comaum Forest Cave, near Penola, SA: SAM P unreg., many dentaries, maxillae and single loose teeth.

Unnamed cave (5L122), near Penola, SA: Listed in Reed and Bourne (2000).

Alexandra Cave, near Naracoorte, SA: SAM P13072, left and right juvenile premaxillae and left maxilla.

Bat Cave, near Naracoorte, SA: SAM P unreg., left adult dentary.

Blanche Cave, near Naracoorte, SA: SAM P unreg., left p3.

Brown Snake Cave, near Naracoorte, SA: SAM P17453, right adult dentary.

Cathedral Cave, near Naracoorte, SA: FU 1053, left and right juvenile dentaries; FU 1054, right juvenile dentary; FU 1055, right juvenile maxilla; FU 1056, left juvenile maxilla; FU 1057, left juvenile maxilla; FU 1151, cranium (missing maxillae, premaxillae); FU unreg., juvenile cranium and dentaries.

Crawford's Cornucopia Cave, near Naracoorte, SA: Listed in Reed and Bourne (2000).

Fox Cave, near Naracoorte, SA: SAM P19012, left juvenile dentary; SAM P19020, right juvenile maxilla; SAM P19026, three upper molars; SAM P19038, left and right juvenile dentaries; SAM P19047, left juvenile maxilla; SAM P unreg., three dentaries; FU 0819, left juvenile dentary; FU 0820, left m3; FU 0850, left juvenile dentary; FU 0851, right juvenile maxilla.

Haystall Cave, near Naracoorte, SA: 28 SAM P registered dentaries and maxillae.

Henschke's Fossil Cave, near Naracoorte, SA: SAM P numerous partial crania, dentaries, maxillae and single loose teeth.

James's Quarry Cave, Naracoorte, SA: SAM P12384, left p3.

Rabbit Cave, near Naracoorte, SA: Listed in Reed and Bourne (2000).

Specimen Cave, near Naracoorte, SA: SAM P12907, left P3; SAM P40099, partial cranium; SAM P40100, right juvenile maxilla; SAM P40101, right adult dentary; SAM P40102, left juvenile dentary.

Unnamed cave (5U49), near Naracoorte, SA: Listed in Reed and Bourne (2000).

Victoria Fossil Cave, near Naracoorte, SA: Numerous adult and juvenile crania, approximately 200 maxillae and dentaries registered in SAM and FU collections.

Wet (Tomato-Stick) Cave, near Naracoorte, SA: SAM P12919, right adult dentary, left upper molar.

Wombat Cave, near Naracoorte, SA: SAM P unreg., right adult maxilla, partial neurocranium; FU 1803, left dP3.

Kelly Hill Caves, Kangaroo Island, SA: SAM P12912a, right adult dentary; SAM P12912b, left juvenile dentary; SAM P12914a, right adult maxilla; SAM P34185, right adult dentary.

Rocky River, Kangaroo Island, SA: SAM P 36 registered dentaries; 12 registered maxillae; numerous single loose teeth.

Seton Rock Shelter, Kangaroo Island, SA: SAM P29773, left protoloph, right hypolophid, metaloph fragment.

Revised Diagnosis. Cranium small, similar in size to western form of *"Procoptodon" browneorum*, but with much less inflated frontals and a narrower, more anteriorly tapered rostrum. Dorsal surface of neurocranium is convex in lateral profile. Basicranial plane only slightly elevated above palatal plane. Masseteric process barely more than a rounded tumescence. Premaxilla markedly declined, becoming very thin dorsally. I3 relatively higher crowned than in any other sthenurine. Upper and lower premolars longer and higher crowned relative to molars than in any other sthenurine, with alveolar margin stepped reflecting higher crowns. P3 consistently lacks a posterobuccal accessory cusp. Upper molars high crowned with postparacrista and premetacrista curved markedly lingually, terminating at short, thick mesocrista. Lophid faces bear no or very few fine enamel crenulations.

Etymology. Named in honor of Edmund D. Gill, former curator of fossils at Museum Victoria (formerly National Museum of Victoria), for his contribution to the understanding of Australian Quaternary events.

Description. Cranium (pls 94-99; table 35). Incisor-bearing portion of premaxilla shallow and short. Upright portion of premaxilla markedly declined in lateral profile, becoming extremely narrow vertically before expanding slightly to contact nasal (pls 94B, 95). Diastema short, straight and deflected anteroventrally relative to cheek tooth row; maxilla contributes to most of length. Alveolar margin above P3 stepped relative to alveolar margin above molars Anterior borders of moderately long incisive foramina are positioned just anterior to posterior border of I3 alveolus. Moderately short rostrum tapered anteriorly and retracted dorsally due to shortened upper portions of maxilla and premaxilla. Narial aperture narrower than deep. Anteorbital and buccinator fossae rather shallow anteriorly, but deeper posteriorly anterior to zygomatic arch. Masseteric process poorly formed, barely more than a round tumescence (pls 94-96, 98-99). Infraorbital foramen and accessory foramen just anterior to anteroventral border of orbit. Anterior border of orbit and lacrimal dorsal to P3. Small posterior lacrimal foramen separated from larger, elliptical, anterior lacrimal foramen by tiny lacrimal tuberosity. Nasals short, narrowing markedly anteriorly. They project beyond contact with premaxilla for half of their length and terminate directly above point at which anterodorsal border of premaxilla inflects (boundary between incisor-bearing and upright portions). Nasofrontal sutures straight, meeting one another at or near 90° (pls 94A, 97).

Frontals rather narrow, only slightly broadened forward of anterior extremity of frontal-parietal suture, which results in minor supraorbital crest development. Anterior limit of palatal vacuities adjacent to posterior edge of P3 or just anterior to this point. Postpalatine bars form a bridge across palate opposite or just posterior to M4 metaloph. Temporal (parietal) crests weakly to moderately developed and do not meet at sagittal suture, which usually results in retention of a distinct interparietal in adult cranium. Dorsal surface of neurocranium slightly convex in lateral profile (pls 94B, 95). Zygomatic arch not especially deep; posterior extremity of jugal bears moderately developed ectoglenoid process. Zygomatic process of squamosal arises well anterior of occiput and is markedly flared laterally, resulting in greatest cranial width across posterior end of zygomatic arches (pls 97-98). Large postglenoid process forms posterior border of glenoid fossa, buttressed by only moderately wide, rugose, tube-like ectotympanic that is curved ventromesially and oriented anteromesially. Posteroventral wing of alisphenoid and anterior extension of exoccipital abut ectotympanic lingually. Small, circular postzygomatic foramen only separated dorsally from external auditory meatus by wafer-thin sliver of bone. Subsquamosal foramen barely more than a pinprick and located just below dorsal root of zygomatic process of squamosal.

Table 35. Dimensions of the adult cranium and dentary of *"Procoptodon" gilli*, showing mean, standard deviation (in parentheses) and sample size [in brackets].

Dimension	*"Procoptodon" gilli*
Condylobasal Length	165 (5.1) [n=8]
Cranial Width	111 (2.7) [n=8]
Palatal Length	100 (4.0) [n=8]
Diastema Length	23.4 (0.92) [n=8]
% Diastema Length: Palatal Length	23.4
Palatal Width between M1 Protolophs	31.2 (1.26) [n=8]
Palatal Width between M4 protolophs	32.9 (1.83) [n=7]
Width across Frontals	50.1 (2.33) [n=8]
Distance between Paroccipital Processes	51.0 (2.43) [n=8]
Dentary Depth	27.8 (1.63) [n=9]
Dentary Width	16.1 (1.11) [n=9]
Dentary Depth / Width	1.73 (0.10) [n=9]

Basicranial plane only slightly elevated above level of palatal plane (pls 94B, 95D-E). Medial pterygoid origin moderately wide and shallow, flanked mesially by pterygoid bone and anterolaterally by smaller alisphenoid extension. Basioccipital slightly flexed posterodorsally relative to basisphenoid, bears low median keel. Occiput rather shallow relative to width, vertical median crest leads ventrally to wide, elliptical foramen magnum bordered by large occipital condyles. Nuchal / lambdoid crests slight, extended posterodorsally. Basal portion of long paroccipital processes deflected slightly posteroventrally and bordered laterally by thin, deep mastoid process of periotic projected between exoccipital and posterior extension of squamosal.

Upper Dentition (pls 94-99; table 36). I1 high crowned, between round and slightly triangular in cross section (pls 94-96, 98-99A). I2 extremely small and sub-cylindrical). Blade-like I3 very high crowned, but only moderately long (pls 94B-D, 99A). Anterior edge quite straight, but projects slightly to point at anterior end of unworn occlusal surface. Posterior edge follows contour of anterior edge. Well-developed anterolingual crest clearly visible in unworn I3 (pl. 99A), forming a low, flat crest which ascends lingual surface of tooth from anterolingual corner of occlusal surface, but fades before reaching crown base. Extremely low, narrow eminence ascends buccal surface of tooth at its anterior end, immediately opposite anterolingual crest.

While reminiscent of P3, dP2 is shorter relative to its width and generally simpler in morphology (pls 98A-B, 99A). Straight main crest is divided into three or four poorly differentiated cuspules connected by high ridges. Buccal and lingual ridgelets ascend from apex of large anteriormost cuspule. Lingual ridgelet separates short, deep anterior basin from deep longitudinal basin, which contains few to no fine enamel ridgelets. In mesial profile, lingual crest rises posteriorly before terminating at large posterior cuspule. A strong transverse ridgelet connects posterior cuspule of lingual crest with posterior cuspule of main crest (pl. 99A). Short posterior basin enclosed buccally by posterior extension of main crest, lingually and posteriorly by curved posterior extension of lingual crest.

Completely molariform dP3 only separable from M1 on basis of size (pls 98A-B, 99A). P3 higher crowned, wider and much longer than succeeding molars (pls 94B-C, 95D, 98C, 99B). Outline of tooth usually rather inflated, but not necessarily. Main and lingual crests converge slightly anteriorly, following orientation of buccal and lingual sides of P3. Posterior edge of P3 only slightly convex, narrower anterior edge is slightly more rounded. Main crest divided into four cuspules, with anterior three distinct, posteriormost not distinguishable from crest. Three anterior cuspules give rise to vertical ridgelets on buccal and lingual sides. Lingual ridgelet from anteriormost crest high, separates deep V-shaped anterior basin from longitudinal basin, which is bordered by anteriorly uniting extensions of main and lingual crests. Deep longitudinal basin contains only a few fine enamel crenulations, although low lingual ridgelet from second main crest cuspule crosses basin parallel and just posterior to higher ridgelet from anterior cuspule (pls 94C, 98C, 99B). Posterolingual portion of longitudinal basin frequently contains a complex of high subsidiary ridgelets associated with high transverse ridgelet separating longitudinal basin and short posterior basin. Subsidiary ridgelets often encroach into posterior basin, which is bordered posteriorly by extensions of main and lingual crests. These form an arch shape in posterior profile. Posterobuccal accessory cusp absent.

Table 36. Mean cheek tooth dimensions of *"Procoptodon" gilli,* with standard deviation in parentheses.

Tooth	L	AW	PW	AH	PH	n
dP2	9.9 (0.43)	7.2 (0.44)	8.7 (0.46)	6.4 (0.26)	6.8 (0.42)	9
dP3	9.1 (0.32)	9.1 (0.40)	9.5 (0.42)	4.9 (0.40)	5.1 (0.27)	9
P3	16.1 (0.65)	9.1 (0.55)	11.2 (0.71)	9.7 (0.91)	9.7 (0.84)	15
M1	10.2 (0.52)	10.1 (0.36)	9.9 (0.40)	5.5 (0.56)	5.9 (0.52)	14
M2	10.9 (0.49)	10.4 (0.45)	10.1 (0.41)	6.0 (0.41)	6.6 (0.36)	14
M3	11.5 (0.47)	10.6 (0.50)	10.1 (0.54)	6.2 (0.63)	6.5 (0.65)	13
M4	10.7 (0.34)	10.1 (0.37)	9.1 (0.42)	5.1 (0.33)	5.4 (0.31)	7
dp2	8.5 (0.56)	6.4 (1.65)	8.1 (1.56)	8.9 (0.49)	8.2 (0.63)	10
dp3	8.6 (0.40)	7.7 (0.39)	7.4 (0.37)	7.0 (0.59)	6.8 (0.69)	12
p3	14.6 (0.58)	6.9 (0.41)	8.8 (0.42)	10.2 (0.64)	10.0 (0.73)	16
m1	10.0 (0.43)	8.4 (0.28)	8.3 (0.33)	7.9 (0.89)	8.3 (0.96)	19
m2	11.1 (0.59)	9.0 (0.31)	8.8 (0.46)	8.4 (0.91)	8.6 (0.82)	19
m3	11.7 (0.49)	9.6 (0.33)	9.4 (0.41)	8.1 (0.97)	8.1 (0.86)	15
m4	11.2 (0.47)	9.6 (0.33)	8.5 (0.33)	7.0 (0.59)	6.5 (0.40)	12

Upper molars high crowned and moderately narrow for length (pls 94-95, 98-99). Lophs taper slightly toward unworn crests, especially anteriorly. Size gradient along molar row is M1<M2<M3>M4. Loph crests slightly convex anteriorly. Upper molars vary slightly in morphology: M4 metaloph narrow relative to protoloph; proximity of postparacrista and premetacrista extremities increases posteriorly along tooth row; connection between preparacrista and paracone apex weakens slightly more posteriorly. Preparacrista well connected to paracone apex, with stylar cusp A portion slightly enlarged. Precingulum moderately thin, extending to anterolingual corner of crown before curving up protocone face slightly then fading out. Just lingual of tooth midline, a distinct preprotocrista is evident on precingulum; it narrows and terminates well dorsal to unworn protoloph crest. Postprotocrista clearly divided into two components; lingual component ascends posterobuccally from protocone apex and terminates on side of thick, more posteriorly oriented buccal component. This extends into interloph valley and narrows markedly before terminating just short of metaloph crest. Although rather thin initially, mesocrista trends posterobuccally in similar orientation to buccal component of postprotocrista and thickens to form a barrier to markedly curving postparacrista and postmetacrista (pls 94C, 98-99). Larger postparacrista and smaller premetacrista maintain strong connections with paracone and metacone apices, respectively. Well-developed postmetacrista folds dorsolingually from metacone apex before terminating at likely position of stylar cusp E. Postmetaconulecrista curves dorsobuccally from metaconule apex and terminates just posterior to dorsal extremity of postmetacrista. Two to three coarse enamel crenulations are enclosed by postmetaconulecrista and postmetacrista on posterior face of metaloph, one of which may be urocrista.

Dentary (pls 100-101; table 35). Dentary very short, ramus relatively narrow. Digastric eminence moderately deep, sulcus well developed and extends anteriorly to beneath m2 hypolophid or m3 protolophid. Boss below symphysis well developed, with dentary depth below middle of p3 approximating that below m4 as a result (pl. 100G-J). Anteroventral border of ramus steeply upturned, with longitudinal axis of i1 slightly more procumbent. Inferior transverse torus of symphysis extends under genial fossa from beneath p3 anterior root to below abutment of p3 and m1. Short diastema very slightly procumbent relative to plane of cheek tooth alveolar margin. Alveolar margin below p3 stepped down relative to alveolar margin of molar row. Median dorsal groove moderately deep and wide. Large anterior mental foramen borders anteroventral extremity of buccinator sulcus. Shallow for most of length, buccinator sulcus is markedly deepened at posterior extremity below m1 hypolophid. Posterior mental foramen located mid-depth on ramus beneath m3 protolophid (pl. 100G-J).

Ascending ramus takes root adjacent to m4 protolophid. Anterior border is oriented slightly anterior of vertical to halfway up, before inflecting to vertical and curving slightly posteriorly as ascending ramus narrows into coronoid process. Mandibular condyle broad, but narrows mesially. Posteroventral aspect of masseteric fossa markedly flared laterally, bordering deep attachment area for middle masseter muscle (pls 100G-J, 101D). Ventral border of masseteric fossa level with posterior extremity of buccinator sulcus or midpoint below alveolar margin of molar row and posterior mental foramen. Anteriorly positioned masseteric foramen small and elliptical leading into vertical masseteric canal. Elliptical inferior mandibular foramen opens posteromesially, with dorsal border level with alveolar margin of molar row. Wide, relatively deep mylohyoid groove slightly overhung by small, sharp dorsal and ventral processes. Medial pterygoid fossa wide, but not particularly deep. Angular process moderately well developed.

Lower Dentition (pls 100-101; table 36). Occlusal surface of slender and elongate i1 approximately level with midpoint between base and occlusal surface of p3 (pls 94B,E). In shape, minimally worn occlusal surface is very gently convex dorsally and rises to a slight point anteriorly (pl. 100C-F). Lingual enamel layer between thin dorsal enamel flange and thick ventrolingual enamel extension is much thinner than on buccal surface (pls 94E, 100C,E,G).

While generally reminiscent of p3, dp2 is shorter relative to width and has a better-developed ridgelet directed posterobuccally from anteriormost cuspule of main crest (pl. 101A-C). A tiny notch is present between this ridgelet and long buccal crest. Anterior half of main crest is either divided into two cuspules or only anteriormost cuspule backed by V-shaped fossette. Posterior half of main crest not distinctly differentiated into cuspules. Completely enclosed median valley contains only one or two small enamel ridgelets, both arising from buccal side of main crest.

Completely molariform dp3 similar in general outline to m1, but differs by having protolophid crest distinctly narrower than hypolophid crest due to more marked tapering of protolophid (pl 101A-C). Trigonid very short, with thick lingual portion of paracristid bifurcated at anterobuccal inflection. Main prong forms anterior portion of paracristid, and terminates at small paraconid distinct in unworn specimens. Minor prong continues anteriorly to anterior edge of dp3. Slight precingulid only evident beneath transverse anterior portion of paracristid, not on anterobuccal corner of crown. Premetacristid and parametacristid may be slight or moderately developed, if present. Cristid obliqua similarly developed to paracristid, curving anteriorly from hypoconid apex to buccal side of protolophid, before weakening into very slight spur terminating just below protolophid crest. Preentocristid distinct and only slightly more weakly expressed than cristid obliqua anteriorly, which it mirrors in shape. Lophid faces free of enamel crenulations, with no sign of postcingulid.

High-crowned p3 long relative to molars (pls 94B,D-E, 100I-J, 101D). Width usually much narrower anteriorly than posteriorly. Moderately deep to shallow V-shaped groove on buccal side of crown formed between anterior end of buccal crest and large anteriormost cuspule. Main crest divided into five cuspules, with posterior two usually much less distinct than anterior three. Anterior three cuspules each give rise to a vertical ridgelet on lingual side, but corresponding ridgelets on buccal side are less distinct. Slight buccal ridgelet directed from second main crest cuspule connects to anterior extremity of buccal crest, which runs approximately two-thirds of p3 length (pls 94B, 100J, 101D). Median valley moderately narrow and deep, with few fine enamel ridgelets descending into it. Union of posterior extremities of main and buccal crests forms a very shallow, broad V-shape.

Lower molars high crowned, with size gradient m1<m2<m3>m4 (pls 94B,E, 100, 101). They show little variation in morphology, except for narrowness of m4 hypolophid relative to protolophid. Lophid crests straight and parallel, with slight anterior turn of lophid ends on buccal and lingual sides. As with dp3, trigonid is very short and precingulid only underlies transverse anterior portion of rather thick paracristid. Posterior portion of lingual paracristid component is distinct, but thinner than on dp3 and shifted lingually across protolophid from protoconid apex. Premetacristid absent, parametacristid very slightly developed. Protolophid anterior face bears few very fine to no enamel crenulations. Cristid obliqua similarly oriented and developed to posterior portion of paracristid (pl. 101C-D). Preentocristid slightly developed. Hypolophid posterior face slightly inflated ventrally.

Variation. Throughout its exclusively southeastern distribution *"Procoptodon" gilli* is consistently one of the most abundant sthenurine species in cave deposits. Not surprisingly, geographic variation is not evident, but neither is variation within samples as marked as it is in many similarly well-represented species.

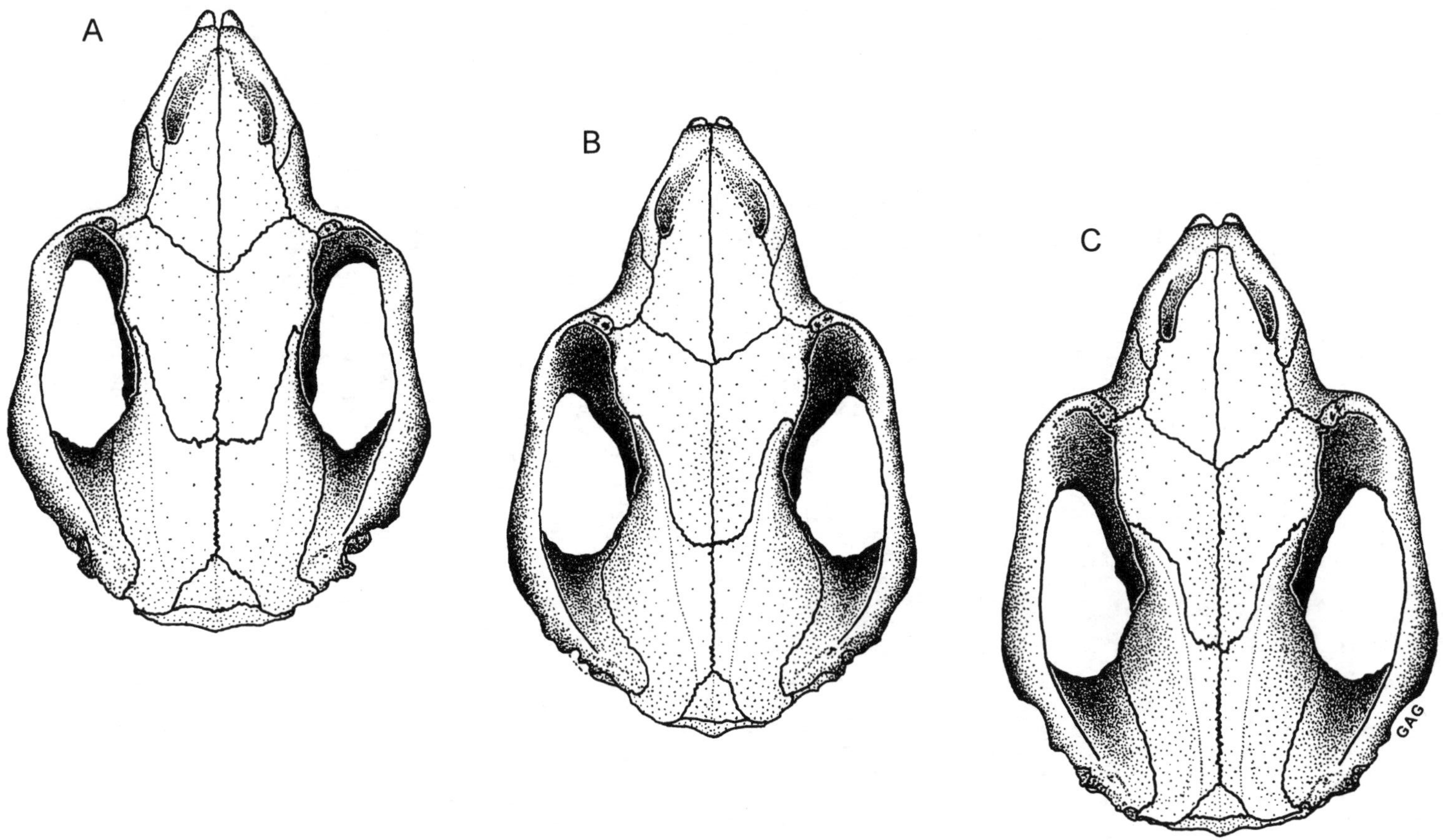

Figure 6. Age series of *"Procoptodon" gilli* crania in dorsal view. **A**, juvenile with M2 fully erupted (SAM P16504, tips of nasals reconstructed); **B**, juvenile with M3 fully erupted (SAM P20615, tips of nasals reconstructed); **C**, adult (SAM P16647).

Cranium. Surprisingly little ontogenetic variation is observed for *"Procoptodon" gilli* (pls 95-99; fig. 6), although the dorsal surface of the neurocranium is more domed in younger juvenile crania (pl. 95A). Between the adult crania within the Victoria Fossil Cave sample, slight variation is observable in overall size and cranial width, size and convergence of the temporal crests, and inflation of the frontals. Aside from these features, the crania appear remarkably uniform, and there is no obvious bimodality that might be indicative of sexual dimorphism. Among the juvenile crania expressing an identical tooth eruption stage, variation is also seen in size, degree of convergence of the temporal crests and development of a distinct supraorbital crest on the frontal (pls 95A, 97A), which is lost in adult specimens.

Upper Dentition. Variation in tooth morphology and size within *"Procoptodon" gilli* is relatively low, as is reflected in the low standard deviations for dental measurements. I1-2 vary negligibly in morphology. I3 may vary in overall size and development of the low, narrow eminence that ascends the buccal surface of the tooth at its anterior end. Minimal variation is observed for dP2, but the tooth may be slightly more inflated posteriorly in some individuals. Morphology of dP3 is remarkably uniform across the sample. It only appears to vary in relative development of the precingulum on the anterolingual corner of crown and enamel crenulations on the posterior face of the metaloph. The same features vary in the molars, but enamel crenulations on the other loph faces may also vary slightly in their expression, as does the degree to which the M4 metaloph is narrowed. P3 is the most variable element of the upper dentition. Clear differences are observed in length of the tooth relative to width and general inflation of the tooth outline. This is particularly noticeable posteriorly, but may also be evident anteriorly. P3 also varies in the degree of main crest cuspule differentiation, development of the enamel ridgelets in the longitudinal basin and the form of the complex ridgelets separating the longitudinal and posterior basins.

Dentary. Morphological variation in the dentary of *"Procoptodon" gilli* is minimal, both ontogenetically and between adult specimens. Adult dentaries may vary slightly in size, orientation of the lower incisor, depth of the buccinator and digastric sulci, and relative expansion of the posteroventral border of the masseteric fossa. Variation in all other features appears negligible.

Lower Dentition. Aside from some slight variation in size, no intraspecific variation is evident for the dp2 or p3 of *"Procoptodon" gilli.* Minimal morphological variation is also detectable for the molariform teeth. On dp3, parametacristid and enamel crenulation development is very slightly variable. On the molars, slight variation is only observed in the relative development of enamel crenulations, with some specimens possessing no crenulations and others possessing a few very fine crenulations on the anterior face of the protolophid. Relative development of the preentocristid also differs slightly between individuals. Among the species known from relatively large sample sizes, intraspecific variation is less marked in *"P." gilli* than in any other sthenurine.

From those known from at least partial age series, ontogenetic change is also least for this taxon.

Comparison With Other Taxa. Cranium. Among the sthenurine species for which the adult cranium is known, that of *"Procoptodon" gilli* may be distinguished from all others on the basis of its diminutive size, with the exception of the similarly small cranium of western *"P." browneorum* and possibly the Nullarbor form of *Simosthenurus maddocki*. *"P." gilli* is narrower across the frontals than any other simosthenurin, but is only narrower than the larger *Metasthenurus newtonae* and *"Si." baileyi*. The frontal region of *"P." gilli* is most similar to the much more dolichocephalic *Sthenurus andersoni*. The masseteric process is more slightly developed in *"P." gilli* than in any sthenurine except *Hadronomas puckridgi*. The dorsal surface of the *"P." gilli* neurocranium is also rather domed, a feature typical only of non-sthenurine macropodoids, the juvenile crania of other sthenurines and *P. goliah*. The manner in which the *"P." gilli* premaxilla is markedly declined and extremely narrowed dorsally is unique among sthenurines. In overall form, the juvenile cranium of *"P." gilli* most resembles the juvenile cranium of *"P." browneorum* (figs 5-6).

Upper Dentition. For the size of the cranium, the I1 and I3 of *"Procoptodon" gilli* are relatively and absolutely larger and higher crowned than in those of the sthenurines that they most resemble in form: *"P." browneorum*, *Simosthenurus occidentalis* and *"Si." baileyi*. While extreme reduction of the I2 is diagnostic of the Sthenurinae, in *"P." gilli* it appears even more diminutive relative to the size of I1 and I3. In absolute crown height, the I3 of *"P." gilli* only approaches that of the very large *"Si." pales* and *Sthenurus stirlingi*. For its length, though, the *"P." gilli* I3 is higher than in all sthenurines. The dP2 is larger than that of *S. andersoni*, but smaller than that of *Metasthenurus newtonae* and *"P." browneorum*, to which it is most similar in general morphology. It differs from *M. newtonae* by having a less inflated outline and from *"P." browneorum* by lacking many coarse enamel crenulations in the longitudinal basin. The P3 of *"P." gilli* is most similar to *"P." browneorum* and *"Si." baileyi*, but is absolutely smaller and narrower relative to its length, less inflated in general outline and has a deeper longitudinal basin containing fewer ridgelets. Compared with the rest of the longitudinal basin, the complex of coarse ridgelets separating it from the posterior basin, especially in the posterolingual corner, is apparently unique to *"P." gilli*. The consistent absence of any sign of a posterobuccal accessory cusp is also unique. In most aspects, the molariform teeth of *"P." gilli* are smaller than but similar to those of *M. newtonae* and to a lesser degree *S. andersoni*, although they do resemble this species more in crown height. They differ from those of *M. newtonae* by having a slightly weaker connection between the preparacrista and paracone apex, a continuation of the precingulum to the anterolingual corner of the tooth, a weaker postparacrista, a distinct mesocrista and at least one enamel crenulation on the posterior face of the metaloph.

Dentary. Although considerably smaller, the dentary of *"Procoptodon" gilli* most closely resembles that of *"P." oreas* and *Simosthenurus occidentalis* in most aspects. The dentary and diastema are short, the anteroventral border is relatively steep, and the symphysis, digastric sulcus and eminence are all similarly developed. *"P." gilli* differs from both species by possessing a greatly expanded or flared posteroventral border to the masseteric fossa, a feature expressed to a slightly lesser degree in *Si. maddocki* and *"Si." baileyi*. *"P." gilli* also differs from *"P." oreas* by having a smaller masseteric foramen, a less convex lateral ramus surface and an anteroposteriorly shorter masseteric fossa.

Lower Dentition. The i1 of *"Procoptodon" gilli* is most similar in absolute size and morphology to that of *"P." browneorum*, and is only easily distinguished by its less sinusoidal occlusal surface and slightly thinner lingual enamel layer. The dp2 is intermediate in morphology between the dp2 of *"P." browneorum* and *Sthenurus andersoni*. It differs from *"P." browneorum* by being smaller and having a less inflated outline, fewer enamel ridgelets in the median valley and a strong, direct connection of the buccal crest to the main crest anteriorly. *S. andersoni* has a slightly smaller and narrower dp2, with a shorter buccal crest. The p3 of *"P." gilli* is very similar to that of *"P." oreas*, but also resembles the p3 of *Si. occidentalis*. *"P." gilli* can be distinguished from *"P." oreas* by having the main crest of p3 less distinctly divided into cuspules anteriorly and a wider median valley with coarse enamel ridgelets. *"P." gilli* differs from *Si. occidentalis* in these features, and the posterior portion of the tooth is also not as wide relative to the anterior portion.

The lower molariform teeth of *"Procoptodon" gilli* are slightly smaller than those of the western form of *"P." browneorum*. Aspects of their morphology resemble western *"P." browneorum*, *"P." oreas*, *Sthenurus andersoni* and *Metasthenurus newtonae*. However, the molariform teeth of *"P." browneorum* are slightly wider, with a more anteriorly oriented paracristid and cristid obliqua, a thinner transverse anterior portion of the paracristid, a distinct postcingulid and coarse enamel crenulations on the lophid faces. The molars of *S. andersoni* are larger and more elongate, have a longer trigonid and an extremely weak preentocristid, and lack the small anteriorly directed paracristid prong. In each of these characteristics, *"P." gilli* is more reminiscent of *"P." oreas*, but it lacks the distinct, buccally shifted preentocristid, the anterobuccal portion of the precingulid and the two coarse enamel crenulations on the anterior face of the protolophid. A marked paucity of enamel crenulations is shared with *M. newtonae*, but the lower molars of *"P." gilli* are smaller and higher crowned, the paracristid and cristid obliqua are weaker and clearly separated into two components, the transverse portion of the paracristid is straighter and the posterior face of the hypolophid is only slightly inflated ventrally.

Geographic Distribution. Southeastern South Australia, southern Victoria, southeastern New South Wales (fig. 31; table 44).

Temporal Distribution. Middle to late Pleistocene.

"Procoptodon" mccoyi (Turnbull, Lundelius and Tedford, 1992a)

(pl. 102; tables 37-38)

Sthenurus sp. cf. *orientalis*, Whitelaw, 1991, p. 1500.
Sthenurus (Simosthenurus) orientalis, Turnbull, Lundelius and Tedford, 1992b, p. 268.

Taxonomic Authority. Turnbull, W. D., E. L. Lundelius, and R. H. Tedford. 1992a. A Pleistocene marsupial fauna from Limeburner's Point, Victoria, Australia. The Beagle, Records of the Northern Territory Museum of Arts and Sciences 9:143-172.

Holotype. NMV P23271 / P23272, left and right adult dentaries (missing lower incisors and ascending rami) collected between the mid-1800s and early 1900s from Limeburner's Point, Geelong, Victoria. Age of the type locality is either late early or middle Pleistocene based on magnetostratigraphy and faunal composition (Turnbull et al. 1992a; Whitelaw 1991, 1993).

Referred Specimens. Limeburner's Point, VIC: NMV P23238 / P23239, right adult dentary; NMV P23284, right adult maxilla.

Revised Diagnosis. Length of p3 intermediate between that of m2 and m3. Lower molars most similar to *"Procoptodon" browneorum*, but very low crowned, very wide relative to length and intermediate in size between southeastern *"P." browneorum* and southeastern *"Simosthenurus" pales*.

Etymology. Named in honor of Frederick W. McCoy, pioneer Victorian paleontologist and original authority on the vertebrate fossils from Limeburner's Point.

Description. Good descriptions of the only specimens presently referable to *"Procoptodon" mccoyi* were provided by Turnbull et al. (1992a), but I have opted to reexamine some aspects in more detail and so redescribe the holotype specimen below.

Dentary (pl. 102; table 37). Ventral surfaces of left and right holotype dentaries are abraded, but curvature of what remains indicates surface did not extend much more deeply. This suggests dentary of holotype was even shallower relative to its depth than NMV P23238-9, which has a depth to width value of 1.62. Digastric eminence and sulcus apparently only moderately developed. Symphysis may have extended only slightly below genial fossa.

Lower Dentition (pl. 102; table 38). Cheek teeth of holotype show a very unusual wear pattern probably due to malocclusion prompted either by physical deformity or an aberrant psychology. Barely more than an outer rim of enamel remains on p3 and m1 occlusal surfaces, while m2 has slightly less wear, but is still very worn. Such extreme wear would usually be taken to indicate extreme old age, but m3 shows only slightest wear, while m4 is completely unworn.

Table 37. Dentary dimensions of *"Procoptodon" mccoyi.*

Specimen	Dentary Depth	Dentary Width	Depth / Width
NMV P23271-2 (holotype)	42.0 est.	29.4	1.43
NMV P23238-9	38.4	23.7	1.62

Table 38. Lower cheek tooth dimensions of *"Procoptodon" mccoyi.*

Tooth	Specimen	L	AW	PW	AH	PH
p3	NMV P23271-2 (h'type)	16.7	9.1	11.7	-	-
m1	NMV P23271-2 (h'type)	14.0	13.4	13.8	-	-
m2	NMV P23271-2 (h'type)	15.1	15.3	14.4	-	-
	NMV P23238-9	15.6	-	-	-	-
	Mean (S.D.)	15.4 (0.35)	15.3	14.4		
m3	NMV P23271-2 (h'type)	16.5	15.7	14.7	8.8	9.6
	NMV P23238-9	17.0	15.3	-	10.0	-
	Mean (S.D.)	16.8 (0.35)	15.5 (0.28)	14.7	9.4 (0.85)	9.6
m4	NMV P23271-2 (h'type)	16.3	15.7	14.2	9.2	8.5
	NMV P23238-9	16.9	15.2	13.7	8.8	8.0
	Mean (S.D.)	16.6 (0.42)	15.4 (0.35)	14.0 (0.35)	9.0 (0.28)	8.3 (0.35)

Most topographic features have been obliterated, but p3 is clearly wide and intermediate in length between m2 and m3 (pl. 102, table 38). Wide vertical groove on buccal side separates anterior portion from more inflated posterior portion. Anteriormost cuspule of main crest was apparently prominent and shifted buccally relative to remainder of crest. Remnant of V-shaped cleft between anterior two cuspules is present on lingual side and continues toward base of crown as a narrow, shallow groove. Buccal crest probably curved, judging by shape of posterobuccal inflation which bore it, and apparently joined to second cuspule of main crest.

Lower molars very wide and very low crowned (pl. 102). First two molars too worn to discern any useful morphological features. Lophids on m3-4 very straight and parallel. Trigonid very short. Narrow precingulid extends from anterobuccal corner of crown to beneath anterolingual extremity of paracristid. Distinct posterior portion of lingual component of paracristid short, sharp and only slightly lower than protolophid crest; positioned approximately two-fifths of way across lophid face from protoconid apex. Turning at almost 90° to posterior portion, short transverse anterior portion of paracristid terminates at apparent paraconid. Premetacristid extremely low and slight. Three or four coarse enamel crenulations descend from protolophid into trigonid basin. A similar structure on buccal side of crown may be a slight buccal paracristid component. Cristid obliqua very similarly developed and oriented to posterior portion of paracristid and is paralleled lingually by three vertical enamel crenulations, finer than those on protolophid anterior face (pl. 102). Hypolophid posterior face bears low but very distinct postcingulid.

Comparison With Other Taxa. Dentary. Although the holotype dentary of *"Procoptodon" mccoyi* is fragmentary, its width appears to have been equivalent to no less than two-thirds of its depth. Notwithstanding the slightly narrower dentary of NMV P23238-9, this easily distinguishes the dentary of *"P." mccoyi* from *"P." browneorum*, wherein the average width is approximately half the depth. A dentary depth to width value of 1.50-1.62 places *"P." mccoyi* in the same range as *"P." oreas*, *Simosthenurus occidentalis*, *"Si." antiquus* and *"Si." pales*. Development of the digastric eminence and sulcus in *"P." mccoyi* appears also to have been similar to these species.

Lower Dentition. Several species of simosthenurins are known in which the p3 does not exceed the length of all molars. The relative length of the *"Procoptodon" mccoyi* p3 is reminiscent of *"P." oreas* and *"Simosthenurus" pales*, but is not as short as the p3 of *"Si." brachyselenis*, *Si. euryskaphus*, *P. pusio*, *P. rapha* or *P. goliah*. I make the width of the posterior portion of p3 to be equivalent to 70% of its length, in contrast to the 77% cited by Turnbull et al. (1992a). This falls within the range of proportional posterior widths observed for *"P." browneorum*, in which values range up to 80% in southeastern and southwestern specimens. In overall size, the p3 of *"P." mccoyi* is similar to that of southeastern *"P." browneorum*. In morphology, it resembles both southeastern and southwestern *"P." browneorum*. However, this species lacks a distinct wide groove on the buccal side of the crown, which demarcates the anterior and posterior portions of p3. Presence of this groove also distinguishes *"P." mccoyi* from *"Si." pales*, but a similar condition is present in *"P." oreas* and *Si. occidentalis*. However, a V-shaped cleft on the lingual side of the crown is also present anteriorly in *"P." browneorum* and *"Si." pales*.

The lower molars of *"Procoptodon" mccoyi* are very similar to *"P." browneorum* in their relative development, orientation and position of all topographic features. They differ by being larger than even the largest southeastern *"P." browneorum*, very low crowned and very wide. While the much smaller *Archaeosimos cegsai* is also very low crowned (tables 13, 42), no other sthenurine has lower molars that are as wide relative to their length or height as *"P." mccoyi*. In absolute size, the molars are intermediate between those of southeastern *"P." browneorum* and southeastern *"Simosthenurus" pales*. Aside from their distinction on the basis of crown height and relative width, the lower molars of *"P." mccoyi* differ from those of *"Si." pales* by having coarser enamel crenulations on the anterior face of the protolophid and a shorter transverse (anterior) portion of the paracristid despite having a higher paracristid overall.

Geographic Distribution. Central southern Victoria (fig. 28; table 44).

Temporal Distribution. Late early to middle Pleistocene.

"Procoptodon" oreas (De Vis, 1895)

(pls 103-106; table 39)

Sthenurus oreas, De Vis, 1895 (in part), pp. 96-97, pl. 16, figs 5-6.
Sthenurus oreas, Glauert, 1910b, p. 61.
Sthenurus oreas, Simpson, 1930, p. 77.
"Sthenurus" oreas, Stirton, 1957, p. 124.
Sthenurus oreas, Marcus, 1962, p. 299.
Sthenurus oreas, Bartholomai, 1963, pp. 59-63, figs 4-5.
Sthenurus oreas, Merrilees, 1965, pp. 22, 24-26, 29-30.
Sthenurus (Simosthenurus) oreas, Tedford, 1966 (in part), pp. 38-40, 42-43, figs 14, 20, tables 2, 7-8.
Sthenurus oreas, Merrilees, 1968a, pp. 65-66, 69, 73.
Sthenurus (Simosthenurus) oreas, Bartholomai, 1972, pp. ix, xii, fig. 2.
Sthenurus (Simosthenurus) oreas, Marcus, 1976, pp. 69-70, fig. 25.
Simosthenurus oreas, Pledge, 1980, p. 134.
Simosthenurus oreas, Flannery, 1989, pp. 29-30, fig. 10.
Sthenurus oreas, Lundelius and Turnbull, 1989, pp. 4-6, fig. 2.
Simosthenurus oreas, Turnbull, Lundelius and Tedford, 1992a, p. 163.
Sthenurus oreas, Prideaux and Wells, 1997, pp. 181-196, figs 3d, 6d.
Sthenurus oreas, Prideaux and Wells, 1998, p. 15.
Simosthenurus oreas, Prideaux, 2000, p. 13.

Taxonomic Authority. De Vis, C. W. 1895. A review of the fossil jaws of the Macropodidae in the Queensland Museum. Proceedings of the Linnean Society of New South Wales 10:75-133.

Holotype. QM F2923, left and right adult dentaries, collected in the late 1800s from an unspecified locality on the Darling Downs, southeastern Queensland. Age is probably Pleistocene.

Referred Specimens. Fissure fill exposed at Cement Mills, Gore, QLD: QM F3814, left adult maxilla.

?QLD (locality unknown): QM F3815, left M2.

Molong, NSW: AM F31922, left juvenile dentary.

Bone Camp Gully, near Bingara, NSW: AM F49664, left M3.

Wellington Caves (unspecified), NSW: AM F88540 (= AM MF90), right juvenile dentary; AM F88541 (= AM MF1) left adult dentary; BMNH 42663Ba, right M3.

Cathedral Cave, Wellington, NSW: AM F69880 / F69881, left and right juvenile dentaries; AM F69882, right i1; AM F69883, left p3; AM F69884, left m3.

Revised Diagnosis. Short dentary with lateral surface of ramus convex, masseteric foramen large. Lower molars with very straight lophid crests, paracristid and cristid obliqua thick but low. Buccally shifted preentocristid strongly developed, oriented parallel to cristid obliqua.

Etymology. Lat. *oreas*, 'mountain nymph'. Significance of name unknown, but may pertain to a whimsical carnal fantasy inspired by the attractive preservation of the holotype.

Description. Maxilla. Only known maxillary specimen (QM F3814, pl. 104A) preserves no features of descriptive value, except for slight curvature on mesial side opposite M2, which suggests anterior limit of palatal vacuities may have been adjacent to M1. Dawson and Augee (1997) identified AM F69885 as a partial maxilla of *"Procoptodon" oreas*, but it actually belongs to *Sthenurus andersoni.*

Upper Dentition (pl. 104A; table 39). QM F3814 preserves M1-3, which show size gradient M1<M2<M3. Crown height moderate, lophs taper only slightly toward crests. Despite greater wear, M1 appears to have differed from M2-3 by having a postmetacrista divided into two components, and a slightly better developed preprotocrista and premetacrista. First component is directed dorsally to meet postmetaconulecrista, second is oriented more dorsolingually adjacent to probable urocrista. On morphology alone, M2 and M3 are inseparable. Protoloph and metaloph crest slightly convex anteriorly. Connection of preparacrista to paracone apex very fine, but crest widens into stylar cusp A portion before becoming confluent with precingulum. This terminates just before lingual side of tooth. Positioned just lingual of tooth midline on precingulum, slight preprotocrista is only slightly coarser than adjacent enamel crenulations. Postprotocrista divided into two components, more lingual ascends posterobuccally from protocone apex and terminates on side of thicker, more posteriorly oriented crest, which extends into interloph valley (pl 104A). Here it becomes extremely fine and terminates partway up anterior face of metaloph. Distinct but slightly finer mesocrista positioned adjacent to buccal postprotocrista component and trends slightly irregularly into interloph valley, bordered buccally by smaller enamel crenulation. Ascending posteriorly from paracone apex, postparacrista joins fine premetacrista, which runs near vertically from metacone apex. Postmetacrista quite well developed, ascends rather sinusoidally from metacone apex, first trending dorsally for a short distance, then posterolingually before turning dorsally again to meet buccal extremity of postmetaconulecrista at position of stylar cusp E. Postmetaconulecrista similarly developed to postmetacrista and ascends dorsobuccally from metaconule apex, thickening and turning buccally before meeting postmetacrista. More buccal of two coarse, vertical enamel crenulations on center of metaloph probably represents urocrista. It terminates just short of metaloph crest (pl 104A).

Dentary (pls 103-106). Dentary very short, ramus rather wide relative to depth, lateral surface quite convex. Anteroventral border of ramus steep, with moderately developed boss evident below partially fused symphysis. Inferior transverse torus of symphysis extends under genial fossa from beneath p3 anterior root to below m1 protolophid (pls 103B, 104C). Genial fossa quite deep, with marked longitudinal groove on mesial side of ramus adjacent to pit. Digastric eminence not especially deep, curves ventromesially. Digastric sulcus extends from anterior extremity of medial pterygoid fossa to below m2 hypolophid (pls 103B, 104C). Short diastema parallel with but slightly lower than alveolar margin of cheek tooth row. Median dorsal groove deep and wide. Anteriorly deep buccinator sulcus extends from diastema ridge, immediately anterior to p3, to below m1 hypolophid. Anterior mental foramen borders anteroventral extremity of buccinator sulcus. Several tiny subsidiary foramina located lower on ramus beneath and posteroventral to anterior mental foramen. Posterior mental foramen positioned mid-depth on ramus below m2 hypolophid or m3 protolophid (pls 103C, 104B).

Ascending ramus takes root adjacent to m4 protolophid; anterior border is oriented vertically to slightly anterodorsally. Posteroventral aspect of masseteric fossa moderately flared laterally, with marked attachment area for middle masseter muscle evident around outer perimeter. Ventral border of masseteric fossa level with or just higher than alveolar margin of cheek tooth row. Masseteric fossa rather short anteroposteriorly and deep laterally. Very large masseteric foramen leads into near-vertical masseteric canal (pls 103A, 104D). Inferior mandibular foramen opens posteromesially, may be very large to small and elliptical to circular in shape (pls 103B, 104C). Deep, narrow mylohyoid groove overhung posteriorly by dorsal and ventral processes (pl. 103B). From just anteroventral of inferior mandibular foramen, groove extends anteriorly above digastric sulcus toward genial fossa, becoming shallower along way. Medial pterygoid fossa and other processes of ascending ramus not known.

Lower Dentition (pls 103-106; table 39). Short i1 very robust and upturned, with longitudinal axis parallel with anteroventral border of ramus. Unworn occlusal surface sinusoidal in shape and rising to point anteriorly (pl. 105). Lingual enamel layer distinctly thinner than buccal enamel. Unworn occlusal surface of i1 of juvenile dentary (AM F69881, m2 unerupted) lies just above level of diastema (pl. 105C).

Although shorter relative to width, dp2 closely resembles p3 (pl. 105). Compared with molars, dp3 has a relatively shorter trigonid, a better-developed parametacristid, which folds across protoloph face into trigonid basin, a paracristid that remains connected to protoconid apex, and a cristid obliqua with only incipient division into two components (pl. 105). Short p3 rather narrow relative to molars and only longer than m1 (pls 103, 106C-D). Tooth narrows quite smoothly anteriorly, but wide vertical groove on buccal side does serve to demarcate anterior and posterior portions of tooth. Groove is very shallow and only extends halfway to base of crown. Anterior two-fifths of main crest divided into three distinct cuspules aligned straight and anterobuccally, but posterior half of main crest is curved and oriented similarly to lingual surface of p3 (pl. 106D). Lingual surface of tooth bears small V-shaped clefts between anterior three cuspules. Median valley contains

Table 39. Cheek tooth dimensions of *"Procoptodon" oreas,* showing mean, standard deviation (in parentheses) and measurements of the holotype (QM F2923).

Tooth	L	AW	PW	AH	PH	n
M1	12.0	11.4	11.7	-	-	1
M2	13.9	13.0	12.4	7.3	8.0	1
M3	15.3 (0.30)	14.3 (0.35)	13.2 (0.46)	7.8 (0.40)	8.0 (0.60)	3
dp2	9.1 (0.49)	6.0 (0.35)	7.5 (0.35)	8.8	7.4	2
dp3	9.9 (0.85)	8.6 (0.42)	8.7 (0.21)	7.4	8.0	2
p3	13.7 (1.02)	6.5 (0.60)	8.1 (0.57)	9.4 (1.62)	8.4 (1.37)	4
p3 holotype	12.3	6.0	7.4	8.4	7.6	
m1	12.6 (0.88)	10.0 (0.74)	10.3 (0.61)	8.9 (1.25)	9.2 (0.90)	4
m1 holotype	11.7	9.2	9.5	-	-	
m2	14.4 (0.40)	10.9 (0.15)	11.1 (0.00)	8.7 (1.42)	9.7 (1.63)	3
m2 holotype	14.0	10.7	11.1	7.8	8.5	
m3	15.8 (0.80)	11.8 (0.30)	11.9 (0.10)	10.1 (0.80)	10.5 (1.10)	4
m3 holotype	14.7	12.0	12.0	9.4	9.4	
m4	14.4	11.8	10.8	8.7	7.9	
m4 holotype	As above					

coarse ridgelets, some arising buccally from cuspules of main crest. Buccal crest connects to second cuspule of main crest via high, coarse ridgelet (pls 103A,C, 106D).

Lower molars moderately low crowned, with size gradient m1<m2<m3>m4 (pls 103-106). Only differences in morphology relate to narrowness of m4 hypolophid relative to protolophid and less developed m4 postcingulid. Lophid crests straight, parallel, with very slight anterior turn of lophid ends on buccal side. Lingual component of paracristid strongly developed, extends slightly anterolingually from one quarter of way across protolophid face from protoconid, before turning lingually and thickening (pls 103A, 104D, 105). Buccal paracristid component low, extending from protoconid apex to buccal side of lingual component. Premetacristid and parametacristid absent. Two to three coarse enamel crenulations descend into trigonid basin from protolophid. Precingulid strongly developed. Thick, strongly developed cristid obliqua arises on hypolophid in position analogous to paracristid on protolophid and is oriented similarly. It terminates low on protolophid posterior face in tooth midline. Buccally shifted preentocristid strongly developed and oriented parallel to cristid obliqua (pls 103A, 104D, 105). Hypolophid posterior face bears two or three fine grooves descending onto shallow A-shaped postcingulid.

Variation. Upper Dentition. No variation of any significance is discernable in M3, the only tooth known from multiple specimens.

Dentary. The three known adult dentaries of *"Procoptodon" oreas* vary in size and ramus depth relative to width. The Wellington Caves specimen (AM F88541, pl. 104B-C) is larger and the ramus is deeper relative to its width (depth / width: 1.81) than the Darling Downs holotype (QM F2923, pl. 103; depth / width: 1.53). The holotype has larger anterior mental and inferior mandibular foramina and a deeper buccinator sulcus and genial fossa than AM F88541. In addition, the posterior mental foramen is positioned below the m2 hypolophid in the holotype, but it lies beneath the m3 protolophid in AM F88541. The right juvenile dentary, AM F88540 (Tedford 1966:fig. 14) is slightly more slender and has smaller masseteric foramen and a shallow buccinator sulcus. These differences are typical of a more immature specimen. However, its diastema is clearly longer than in the adult dentaries, which strongly suggests intraspecific variation, because the diastema of most sthenurines typically lengthens proportionally as an individual gets older. While the anterior end of the dentary is missing from the juvenile AM F31922 (pl. 106), its overall proportions vary minimally compared with the adult specimens, except for the standard condition of a shallower ramus relative to width. The dentaries of the even younger AM F69880 / F69881 (pl. 105) are shallower again and clearly lack the convex lateral surface characteristic of adult *"P." oreas* specimens.

Lower Dentition. As with the dentary, the p3 of the more southern specimens is larger than that of the Darling Downs holotype (table 39). The roots of the p3 preserved in AM F88541 from the Wellington Caves suggest that it too was also considerably larger than the holotype. In morphology, though, the p3 of the holotype, AM F31922 and AM F69883 are very similar. AM F88540 differs only by having a distinct cuspule low on the anterolingual aspect of its crown. The molars of the New South Wales specimens are also proportionally longer than those of the holotype (table 39). The only morphological difference noted with regard to the lower molars is a slightly thicker preentocristid in AM F88541 compared with the other specimens.

Comparison With Other Taxa. Upper Dentition. In morphology, the upper molars of the only known maxillary specimen of *"Procoptodon" oreas* (QM F3814) most resemble those of *"P." browneorum* and are similar in size to those of southeastern representatives of this species. Although *"P." oreas* was known only from the holotype dentaries when Bartholomai (1963) included QM F3814 in this species, I concur with his assessment based on size, occlusal fit and morphological congruence. The upper molars of *"P." oreas* may be distinguished from those of *"P." browneorum* by having a (a) weaker connection of the preparacrista to the paracone apex, which bears a more enlarged stylar cusp A portion; (b) very slightly more distinct preprotocrista; (c) precingulum that terminates before reaching the lingual side of the tooth; (d) slightly better developed postmetacrista; and (e) postmetaconulecrista that extends from the metaconule apex to the buccal side of the tooth without being intercepted by a thicker, more vertical component.

Dentary. Although the dentary of *"Procoptodon" oreas* is most similar to *Simosthenurus occidentalis* in size and morphology, it also resembles that of the smaller *"P." gilli*. The three species share a short dentary, with a short diastema and a steep anteroventral border. Morphologically, *"P." oreas* differs from both species by possessing a more convex lateral ramus surface, an anteroposteriorly shorter masseteric fossa and a larger masseteric foramen. Flaring of the posteroventral border of the masseteric fossa is also much more distinct in *"P." gilli*. The deep, narrow mylohyoid groove of *"P." oreas* is similar to that of *P. pusio*, *P. rapha* and *P. goliah*, but also resembles *Metasthenurus newtonae*.

Lower Dentition. In size and most aspects of morphology and orientation, the *"Procoptodon" oreas* i1 closely resembles that of *Simosthenurus occidentalis*. *"P." oreas* may be distinguished by its slightly less sinusoidal unworn occlusal surface and relatively thinner lingual enamel layer.

Reduction in the size of the p3 relative to the molars is a condition shared by *"Procoptodon" oreas*, *Metasthenurus newtonae*, *"Simosthenurus" brachyselenis*, *Si. euryskaphus*, *P. pusio*, *P. rapha* and *P. goliah*. However, the p3 of *"Si." brachyselenis* has a much shorter buccal crest, the p3 of *Si. euryskaphus* is much wider posteriorly, and the reduction in relative p3 size in *P. pusio*, *P. rapha* and *P. goliah* is much greater than in *"P." oreas*. Although smaller, the *"P." oreas* p3 closely resembles that of *"P." gilli* in morphology, but it is also somewhat similar to *Si. occidentalis* and *"P." browneorum*. *"P." oreas* may be distinguished from *"P." gilli* by having the anterior portion of the main crest of p3 more clearly divided into cuspules and a slightly wider median valley containing coarse enamel crenulations. In these respects that *"P." oreas* resembles *Si. occidentalis* and *"P." browneorum*.

The lower molars of *"Procoptodon" oreas* are intermediate between those of *"P." gilli* and *"P." browneorum* in morphology, but most similar to the southeastern form of the latter species in size. Intermediate morphological features include orientation of the paracristid and cristid obliqua, number of enamel crenulations on the lophid faces and postcingulid development. *"P." oreas* shares with *"P." gilli* the thick, well-developed transverse anterior portion of the paracristid, which is weak in *"P." browneorum*. *"P." oreas* has a stronger paracristid and cristid obliqua than in both *"P." gilli* and *"P." browneorum*, as well as a thicker precingulid and thicker, buccally shifted preentocristid. In these respects it approaches *P. pusio* in molar morphology.

Geographic Distribution. Southeastern Queensland; central eastern New South Wales (fig. 35; table 44).

Temporal Distribution. Middle to late Pleistocene.

"Procoptodon" williamsi sp. nov.

(pl. 107; table 40)

Holotype. SAM P25059, partial left adult maxilla (preserves P3, M1-4) collected in 1980 by Roderick T. Wells and Richard H. Tedford from the Katipiri Fm., Malkuni Waterhole, Cooper Creek, South Australia. Geological and faunal evidence suggests a late Pleistocene age, possibly last interglacial (Tedford and Wells 1990).

Referred Specimens. Phosphate Mine, Wellington Caves, AM F18903, right juvenile maxilla fragment.

Last Tree Cave, Nullarbor Plain, WA: WAM unreg., left upper molar.

Diagnosis. P3 short for width, equal in length to M1; anterior end of crown bears a small cuspule at mid-depth; deep longitudinal basin lacks transverse ridgelets. Large upper molars with bulbous base and lophs strongly tapered toward crests, similar to *Procoptodon pusio*, *P. rapha* and *P. goliah*, but with much more weakly developed cristae. M1 mesocrista slightly better developed than postprotocrista. Postparacrista fine and low.

Etymology. Named in honor of the late Dominic L. G. Williams for his contributions to the Quaternary vertebrate paleontology of South Australia.

Description of holotype. Maxilla (pl. 107A-C). Incomplete holotype maxilla (SAM P25059) broken anterior to P3, but preserves portion of palate between P3-M1 and mid-palatal suture. Shallow groove leads posteriorly to small palatal foramen just forward of anterior border of palatine vacuity, which is positioned adjacent to M1 protoloph. Although broken distally, masseteric process was clearly large and robust, with distal end apparently adjacent to abutment of M2 and M3. Portion of maxilla into which cheek teeth are set is very robust and deep.

Upper Dentition (pl. 107A-C; table 40). Cheek tooth row of holotype appears convex occlusally, because P3 and M4 are just slightly short of full eruption. P3 short for width, narrowed anteriorly and equal in length to M1 (pl. 107A-C). Main crest is divided into five cuspules, with posterior four poorly differentiated. Only first, third and fifth cuspules give rise to moderately developed vertical ridgelets on buccal and lingual sides. Small distinct cuspule present mid-depth on crown at anterior end of P3. Anterior basin small, narrow and poorly demarcated. Deep longitudinal basin lacks coarse transverse ridgelets and is well separated from anterior and posterior basins by high transverse ridges (pl. 107C). Lingual crest is lower in height than main crest and bears a large, crescentic cuspule at its posterior end. Both buccal and lingual crests are quite straight. Posterior basin enclosed by high anterior, buccal and lingual borders. Posterobuccal accessory cusp absent, but extremely slight eminence is apparent on posterobuccal aspect of main crest.

Upper molars large, moderately high crowned, with bulbous base emphasized lingually (pl. 107C). Marked taper of lophs toward crests decreases in magnitude posteriorly. Size gradient along molar row is M1<M2<M3>M4. Loph crests slightly convex anteriorly. Extensive wear on holotype M1 has obliterated many topographic features, but crown apparently very similar in morphology to M2, but protoloph is slightly wider than metaloph. Anterobuccal, stylar cusp A portion of preparacrista apparently expanded. Precingulum extends to just short of anterolingual corner of tooth. Preprotocrista distinct. Large postprotocrista connects well-worn protoloph and metaloph crests. Slight mesocrista present to buccal side of postprotocrista. Posteriormost portion of postmetaconulecrista thick, postmetacrista evidently slighter. These two cristae enclose two coarse enamel crenulations. M2 indistinguishable from M3 in morphology (pl. 107A-C). Posterior extremity of preparacrista very fine, connects to paracone apex just posterior to enlarged stylar cusp A portion. Preparacrista forms or merges into precingulum, which terminates just short of lingual side of crown, as in M1. Slight preprotocrista picks up halfway up anterior face of protoloph and contacts precingulum just lingual of midline of tooth. Several very fine vertical enamel crenulations present on anterior protolophid face between preparacrista and preprotocrista. Postprotocrista divided into two components; one ascends posterobuccally from protocone apex and terminates alongside thicker, more buccal component which extends into interloph valley, terminating just short of metaloph crest. Finer mesocrista is positioned buccal to postprotocrista in midline of tooth (pl. 107C). Very slight postparacrista and extremely low, fine premetacrista curve slightly in lingually and fail to meet. From apex of metaconule, postmetaconulecrista sweeps across metaloph posterior face, thickening more posteriorly and terminating at position of stylar cusp E. Low postmetacrista ascends very slightly lingually and fades into metaloph posterior face before meeting buccal end of postmetaconulecrista. M4 is very similar in morphology to M2-3, but metaloph is markedly narrower than protoloph, while short mesocrista is thick in region of interloph valley, but very slight anteriorly. Postparacrista and premetacrista are more lingually incurved than on M2-3, but still do not unite.

Description of AM F18903 (pl. 107D-H). Fragment of juvenile maxilla too incomplete to discern any significant features, except masseteric process lies adjacent to M1. Only slightest wear evident on protoloph M1, while metaloph lacks any sign of wear, suggesting very recent eruption (pl. 107H). Protoloph slightly wider than metaloph, in contrast to holotype M1. Buccal sides of protoloph and metaloph near vertical, with crown much lower than on lingual side (pl 107D-H). Lingual sides of lophs trend ventrobuccally from base of crown at around 105° to crests, giving crown a markedly tapered appearance when viewed anteriorly or posteriorly (pl. 107D,G). Loph crests gently convex anteriorly. Posterior extremity of preparacrista extremely slight, linking paracone apex with enlarged stylar cusp A portion of preparacrista. A small cuspule is present dorsal and anterobuccal to this portion of tooth. Preparacrista forms or is confluent with thick precingulum, similar in overall width to unworn protoloph crest. Precingulum terminates lingually at a small cuspule before reaching bulbous lingual extremity of crown base. This cuspule is evidently lacking in holotype. Short, distinct preprotocrista intersects precingulum close to tooth midline and is oriented slightly anterobuccally (pl. 107D,H). It becomes very fine and terminates well short of protocone apex. A tiny, deep, conical pit is located immediately buccal to preprotocrista.

Postparacrista and premetacrista both only slightly developed and do not contact one another, terminating just before reaching deepest part of interloph valley. Postprotocrista divided into two slightly developed components (pl. 107H). First descends posterobuccally from protocone apex and terminates just before reaching deepest part of interloph valley, but very low, fine crest on anterior face of metaloph may be posterior remnant of this component. Shorter, more buccal component is wedged between more lingual component and mesocrista, which is positioned halfway up posterior face of protoloph. Part of mesocrista adjacent to more buccal component of postprotocrista is developed similarly to that crest, but thickens markedly on approaching interloph valley and continues onto anterior face of metaloph. Thus, mesocrista is thicker over more of length than shorter, slightly higher anterior postprotocrista (pl. 107H) in contrast to holotype, in which postprotocrista development exceeds that of mesocrista. Postmetaconulecrista well developed and curves from metaconule apex, thickening buccally into a shelf (pl. 23D,F). Postmetacrista ascends vertically from metacone apex for short distance, then inflects more lingually and curves dorsally before terminating just anterior to buccal extremity of postmetaconulecrista. Two small vertical enamel crenulations centered on metaloph posterior face unite and terminate immediately lingual to postmetacrista, anterior to postmetaconulecrista shelf (pl. 107G-H). More buccal of these is probably urocrista.

Table 40. Cheek tooth dimensions of the holotype (SAM P25059) and paratype (AM F18903) of *"Procoptodon williamsi"* sp. nov.

Tooth	Specimen	L	AW	PW	AH	PH
P3	SAM P25059	16.9	10.8	13.0	12.3	12.2
M1	SAM P25059	16.9	15.4	15.7	7.9	9.7
M1	AM F18903	16.7	15.6	15.3	10.0	10.3
	Mean (St.Dev.)	16.8 (0.14)	15.5 (0.14)	15.5 (0.28)	9.0 (1.48)	10.0 (0.42)
M2	SAM P25059	18.9	17.4	16.4	10.0	11.8
M3	SAM P25059	20.0	17.9	16.8	11.3	-
M4	SAM P25059	18.6	16.8	14.5	10.7	10.1

Comparison With Other Taxa. Maxilla. No obvious characteristics separate what little is known of the adult and juvenile maxilla fragments of *"Procoptodon" williamsi* from other simosthenurins. The general size and robustness of the adult maxilla is reminiscent of *"Simosthenurus" pales*, but more similar in its brevity to *P. pusio*, *P. rapha* and *P. goliah*. Anterior extension of the palatal vacuities to adjacent to the M1 protoloph is variable within genera and, indeed, individual species. A large, robust masseteric process is also common to most of the larger, more robust sthenurines. It is also typical for the masseteric process of very young sthenurines (with M1 recently or almost erupted) to lie adjacent to M1. Interestingly, though, in the adult maxillae of most simosthenurin species, the process lies opposite M2, compared with the abutment of M2 and M3 as predicted for *"P." williamsi*. This condition is typical of the less brachycephalic *"Si." pales* and *Metasthenurus newtonae*.

Upper Dentition. Among the Simosthenurini, the P3 of *"Procoptodon" williamsi* is very short relative to the molars, but not quite as short as in *P. pusio*, *P. rapha* and *P. goliah*. Although the P3 is unknown for *"P." oreas* and *"Simosthenurus" brachyselenis*, their short p3 suggests P3 proportions similar to *"P." williamsi*. In size and morphology, the P3 of *"P." williamsi* is most similar to *P. goliah*, but it differs by lacking a posterobuccal accessory cusp and ridgelets in the longitudinal basin and by possessing a small cuspule on its anterior end and a relatively lower lingual crest. The upper molars of *"P." williamsi* are similar in size and general morphology to those of southeastern *"Si." pales*, but may be easily distinguished by their higher, more tapered lophs, distinct preprotocrista, better-developed postprotocrista and mesocrista, and a weaker postparacrista that does not contact the premetacrista. In each of these characteristics, *"P." williamsi* approaches the morphology of *P. pusio*, *P. rapha* and *P. goliah*, but is easily distinguished by its much weaker molar cristae. While the taper of the lophs is less marked than in *P. goliah*, it similar to *P. pusio* and *P. rapha*.

Geographic Distribution. Northern South Australia, central eastern New South Wales, Nullarbor Plain (fig. 33; table 44).

Temporal Distribution. Middle to late Pleistocene.

MONOPHYLY AND ORIGIN

From the time *Sthenurus* and *Procoptodon* were first isolated from the rest of the kangaroos by Glauert (1926), the agreement that they form part of a unique group within the Macropodoidea has been unanimous (Murray 1991a). Prior to the present work, the most comprehensive analyses of the features that define the Sthenurinae were carried out by Tedford (1966) and Flannery (1983). Based on a number of features that had previously been poorly studied, Flannery proposed a departure from the traditional view by arguing for the inclusion of the extant banded hare-wallaby, *Lagostrophus fasciatus*, and the Pliocene to Pleistocene genus *Troposodon* as the lowest branches of the sthenurine tree. In his 1989 review of macropodoid phylogeny, Flannery placed *Lagostrophus* in the tribe Lagostrophini, and *Troposodon, Sthenurus, Simosthenurus* and *Procoptodon* in the Sthenurini (fig. 7).

Similarities between *Lagostrophus* and *Troposodon* were first reported by Bartholomai (1967) and later by Archer (1981), but neither saw a resemblance to the sthenurines. Archer did speculate, though, on a possible relationship between sthenurines and *Hadronomas*, *Setonix* Lesson, 1842, *Dendrolagus* Müller, 1840, *Dorcopsis* Schlegel and Müller, 1845, and *Dorcopsulus* Thomas, 1922, although the only supporting evidence involved the sharing of several plesiomorphic features. Previously, Woodburne (1967), Ride (1971), Bartholomai (1972, 1978a) and Campbell (1973) had also noted similarities between the sthenurines and *Hadronomas.* New cranial and postcranial material of this late Miocene taxon allowed Murray (1991a, 1995) to argue convincingly for the inclusion of *Hadronomas* near the base of the sthenurine radiation. This move cast doubt on the proposed relationship of *Lagostrophus* and *Troposodon* to the sthenurines, because their link to the Sthenurinae is based on a set of "synapomorphies" different from those that draw in *Hadronomas.* It is appropriate here to review the evidence cited in favor of *Lagostrophus* and *Troposodon* as sthenurines.

LAGOSTROPHUS AND *TROPOSODON*

The banded hare-wallaby was one of the first Australian marsupials seen by Europeans (Prince 1995). William Dampier observed it on Dirk Hartog Island off the Western Australia coast in 1699. Peron and Le Sueur (1807) formally described the species as *Kangurus fasciatus*, but Gould (1842) later included it with the other hare-wallabies in the genus *Lagorchestes.* On the basis of several dental, cranial and pelage characters, Thomas (1887) raised the genus *Lagostrophus* for the animal, placing it within its own section of the Macropodinae. Reproductive, cytogenetic and albumin immunologic disparities (Sharman 1961; Tyndale-Biscoe 1965; Hayman and Martin 1974; Baverstock et al. 1989) have more recently been used to support a separation of *Lagostrophus* from the living macropodines. Several lines of evidence have been taken as suggestive of a potoroine allegiance. As Baverstock et al. (1989) point out, such disparities might not be unexpected if *Lagostrophus* were indeed the only extant member of the Sthenurinae. *Lagostrophus* has a very limited fossil record, but it is known from the late Pliocene Fisherman's Cliff LF in southwestern New South Wales (Marshall 1973; Whitelaw 1991).

Troposodon was proposed by Bartholomai (1967) for the fossil kangaroo that Owen (1877b) described as *Sthenurus minor*, and for which De Vis (1895) independently raised the name *Halmaturus vinceus.* Bartholomai demonstrated that *Troposodon* was not closely related to *Sthenurus* and intimated a closer alliance with *Protemnodon* and *Lagostrophus.* Since Bartholomai described *T. minor* from the Darling Downs, another four species have been recognized from the late Cenozoic of eastern Australia and the Lake Eyre Basin (Campbell 1973; Bartholomai 1978b; Flannery and Archer 1983). Campbell (1973) thought that *Troposodon* probably arose in the vicinity of *Hadronomas*, close to the divergence of *Protemnodon*, *Sthenurus* and *Macropus* Shaw, 1790. Flannery and Archer (1983) included it in the Sthenurinae.

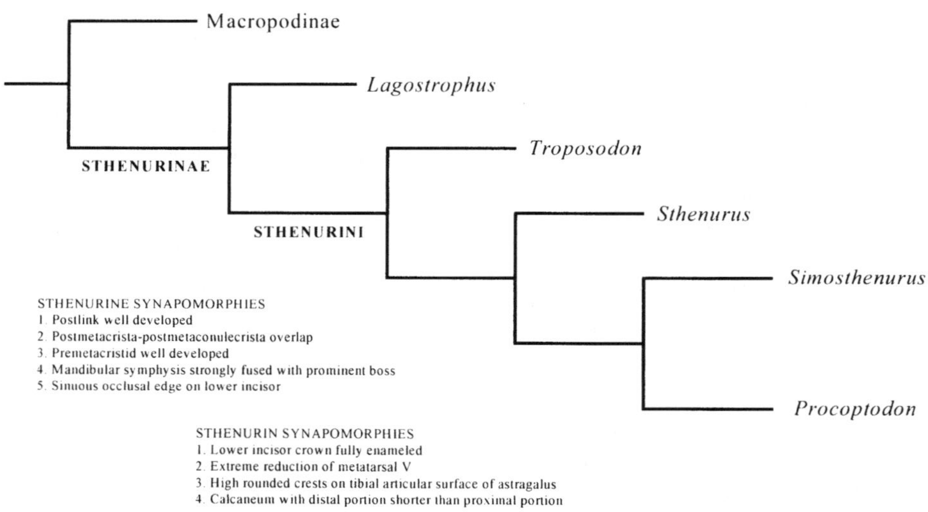

Figure 7. Hypothesis of sthenurine phylogeny according to Flannery (1983, 1989).

Character Reassessment

Flannery (1983, 1989) cited five dental synapomorphies in support of the monophyly of *Lagostrophus*, *Troposodon*, *Sthenurus* and *Simosthenurus* and *Procoptodon* (fig. 7). The nature and taxonomic distribution of each feature is reexamined in detail below. Although I later reflect on the evidence in support of *Hadronomas* as a sthenurine, it is not considered a 'sthenurine' in this section to avoid potential confusion of the issue at hand, which is the strength of the support for *Lagostrophus* and *Troposodon*.

1. Urocrista

The urocrista is a small vertical crest located in the center of the posterior face of the metaloph. It has previously been termed the postlink (see Glossary). Developmentally, it is probably an extension of the metacrista, a crest originating at the metacone and forming the buccal part of the metaloph (fig. 8). The urocrista is most clearly evident in unworn molars, where it trends posterodorsally from the point at which the metacrista meets the metaconulecrista, which is a crest that originates at the metaconule and forms the lingual part of the metaloph.

If the urocrista is a by-product of metaloph formation, the feature should be expected in at least some bunolophodont or incipiently bilophodont macropodoids. Indeed, both bulungamayine taxa for which upper teeth are known (*Nowidgee*, *Wanburoo*) bear a distinct urocrista, emphasized especially on the more anterior molariform teeth (fig. 8). Although balbarines were probably derived prior to the lineage leading to potoroines, bulungamayines, sthenurines and macropodines (Wroe et al. 1998; Cooke and Kear 1999), a urocrista is seen in the most plesiomorphic balbarine for which upper molariform teeth are known, *Nambaroo tarrinyeri* Flannery and Rich, 1986 (fig. 8). The crest is also strongly developed in the enigmatic, plesiomorphic macropodoid *Palaeopotorous priscus* Flannery and Rich, 1986. It is not surprising that no urocrista has yet been noted among the potoroines, because the rudimentary metaloph is usually formed almost entirely by the metacrista. Only the low metaloph of *Potorous* is formed by a roughly equal contribution from the metacrista and metaconulecrista.

A well-developed urocrista may also be present in these macropodines: *Dorcopsulus*, *Dendrolagus*, *Thylogale* Gray, 1837, *Wallabia* Trouessart, 1905, *Kurrabi* Flannery and Archer, 1984, *Congruus* McNamara, 1994, and *Protemnodon* (fig. 8). Therefore, this feature appears in *Palaeopotorous*, balbarines, bulungamayines, macropodines and sthenurines. It is also worth noting that a near identical structure extends posteriorly from the neometaconule in pseudocheirids (fig. 8). Such a broad distribution within the Macropodoidea argues against the idea that a well-developed urocrista represents a synapomorphy uniting *Lagostrophus* and *Troposodon* with the sthenurines.

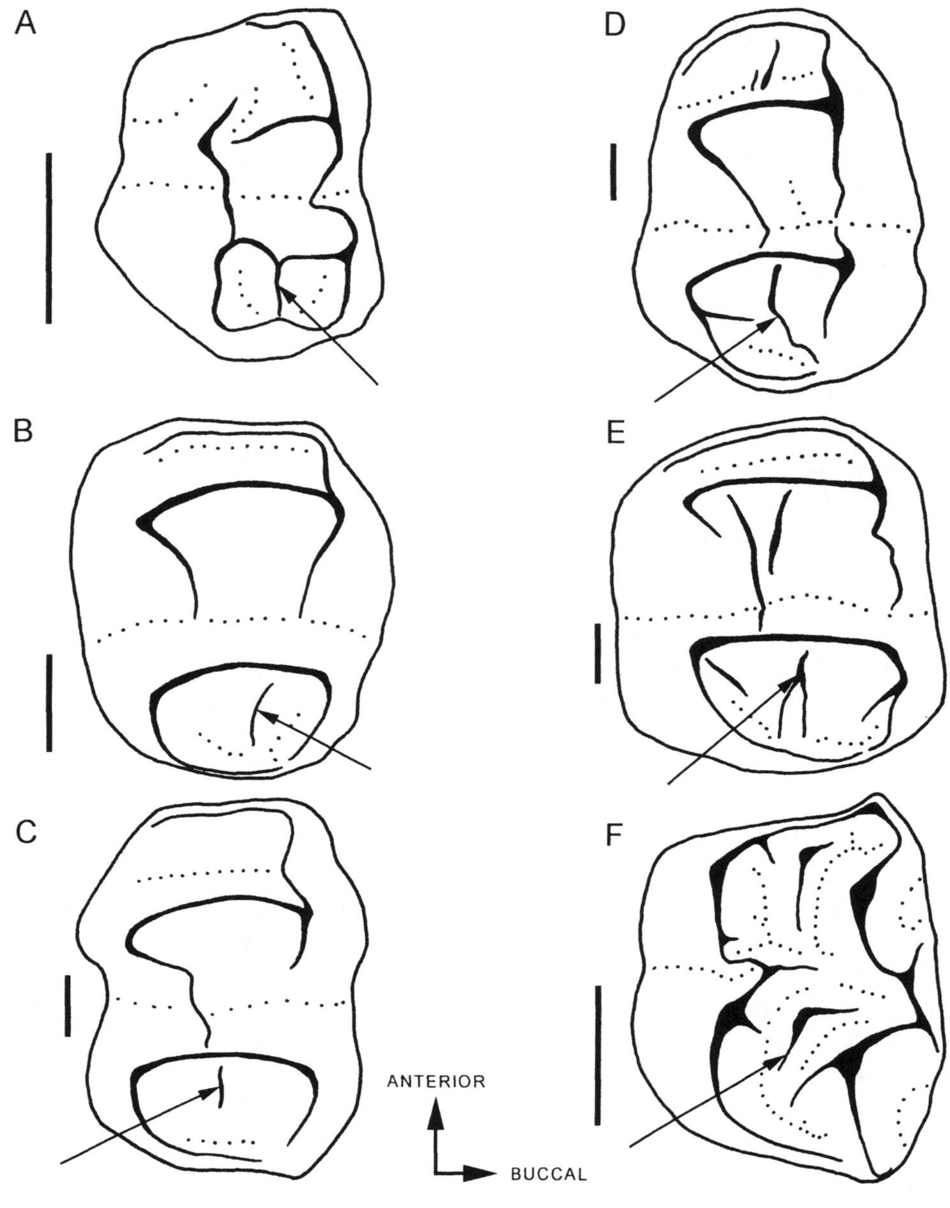

Figure 8. Left upper molars in occlusal view, showing urocrista. **A**, *Nambaroo tarrinyeri* (NMV P172992, M1, drawn from Flannery and Rich 1986, fig. 3, pl. 7); **B**, *Wanburoo hilarus* (QM F20525, M2, drawn from Cooke 1999, fig. 3); **C**, *Congruus congruus* (SAM P33475, M3, drawn from McNamara 1994, fig. 2); **D**, *Protemnodon brehus* (QM F12505, dP3); **E**, *"Procoptodon" browneorum* (SAM P20693, M2); **F**, *Pseudocheirus peregrinus* (SAM P26891, M3). In A-E, arrows point to the urocrista (bifurcated in E), and to a near identical structure extending posteriorly from the neometaconule in F. Scale bars = 2 mm.

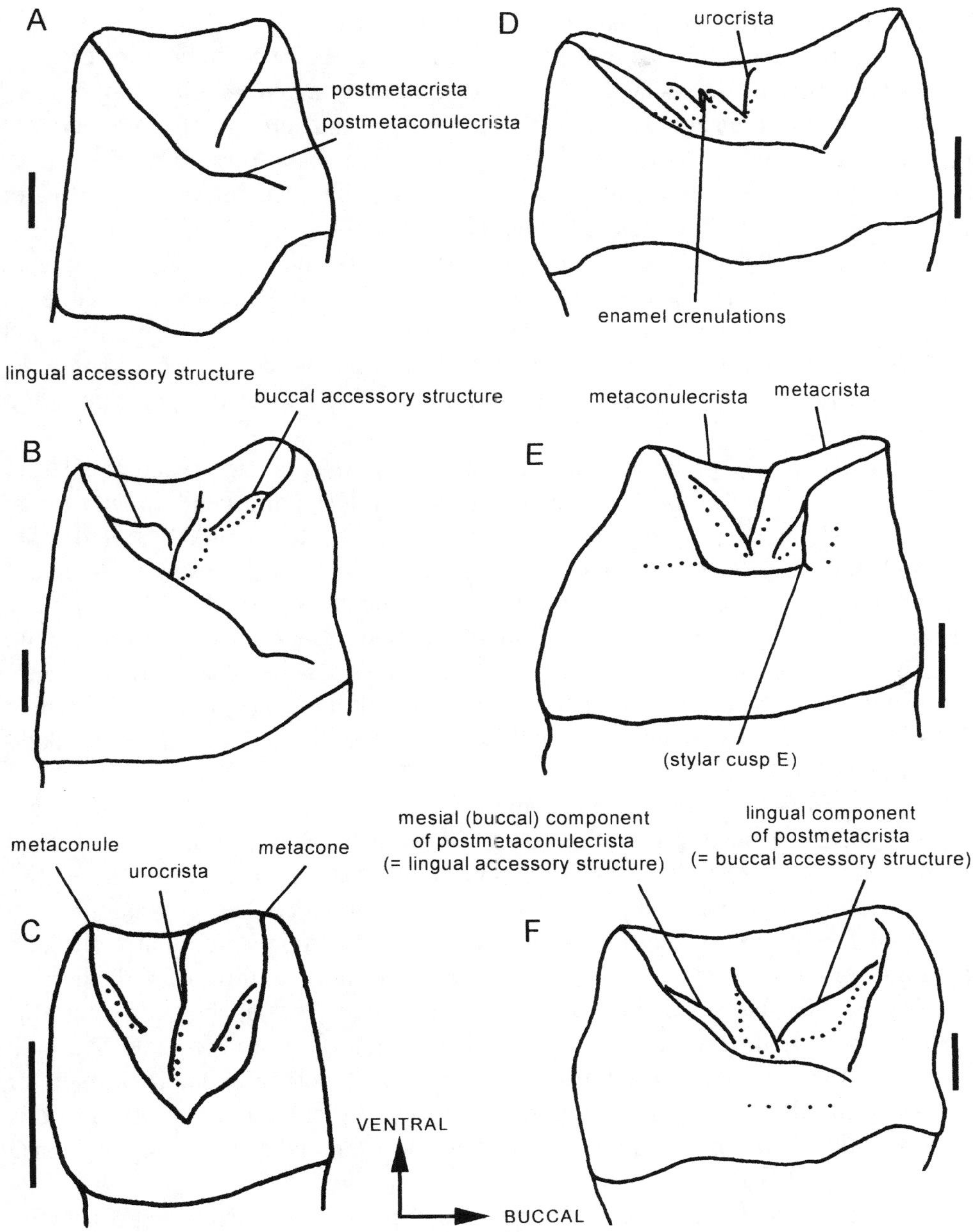

Figure 9. Left metalophs in posterior view, showing relationship of postmetacrista, postmetaconulecrista, urocrista and accessory cristae where present. **A**, *Protemnodon devisi* (QM F4712, M1); **B**, *Troposodon minor* (QM F4397, M2); **C**, *Lagostrophus fasciatus* (AM M1098, M2); **D**, *Archaeosimos cegsai* (SAM P29917, M3); **E**, *Archaeosimos cegsai* (SAM P30153, dP3); **F**, *"Simosthenurus" pales* (SAM P unreg.). Scale bars = 2 mm.

2. Overlap of postmetacrista and postmetaconulecrista

The postmetaconulecrista and postmetacrista are the two dominant cristae on the posterior face of the metaloph (figs 9-10). In many macropodoids, they ascend directly from the metacone and metaconule apices, and unite to form a dorsally convex or arch-shaped structure. Although the more distal portion of the longer postmetaconulecrista is frequently referred to as a posterior cingulum (e.g., Ride et al. 1997) and has also been termed a metacingulum (Cooke 1997b), it is not a true cingulum as such. Typically, the postmetaconulecrista and postmetacrista unite in the position of stylar cusp E, although a distinct cusp is only discernable in representatives of a few taxa (e.g., some species of *Protemnodon*, *Troposodon* and sthenurines). However, the position of stylar cusp E is marked by a sharp corner or groove in a range of different macropodoids, including balbarines, potoroines and bulungamayines.

In *Troposodon*, *Protemnodon devisi* Bartholomai, 1973a (fig. 9) and, to a lesser extent, *"Simosthenurus" baileyi*, the postmetacrista does not ascend directly to the position of stylar cusp E, but trends rather obliquely (dorsolingually) across the metaloph face, terminating near the end of the urocrista. This results in a partial overlap with the buccal end of the postmetaconulecrista. Although folded similarly across the metaloph, the postmetacrista of derived sthenurines does not extend lingually beyond the buccal extremity of the postmetaconulecrista. More importantly, this partial overlap in *Troposodon* is not the same "overlap" to which Flannery (1983) was referring when he described the condition as a likely sthenurine synapomorphy.

In *Troposodon*, simosthenurins and, to a lesser extent, *Protemnodon* and *Lagostrophus*, the postmetaconulecrista and postmetacrista are each accompanied by an accessory crest (figs 9-10). Flannery (1983) interpreted the accessory structure that is often wedged between the postmetaconulecrista and metaloph crest as a lingual extension of the postmetacrista, and it is certainly true that a virtually continuous crest is evident in *Troposodon* sp. from the Hamilton LF (Flannery et al. 1992:fig. 4B). However, in *T. minor* and the sthenurines species in which this morphology is well developed, this interpretation is difficult to reconcile; it would imply that the postmetacrista must form with the urocrista running through or across the middle of it (fig. 9). I am unsure of the developmental likelihood of this scenario, but I know of no other example in the Mammalia where molar cristae develop in such a way.

In *Lagostrophus* and in several *Protemnodon* and simosthenurin species (fig. 9), the lingual accessory structure (Flannery's overlapping portion of the postmetacrista) is slighter than it is in *Troposodon* and clearly arises from or very near to the metaconule apex. Not only does this structure usually fail to reach the urocrista, but the postmetacrista is near vertical or only slightly lingually oriented. If the above developmental scenario was possible, then it would mean that the postmetacrista in these taxa must trend dorsally from the metacone apex to the position of stylar cusp E, then disappear. Following that, the invisible postmetacrista would need to have made a sharp turn in the direction of the metaconule before reappearing in the vicinity of this cusp. On balance, it seems unlikely that the lingual accessory crest represents an overlap of the postmetacrista. A closer inspection of the upper molar of *Troposodon* sp. from Hamilton (NMV P158465) reveals a notch immediately buccal to the urocrista, which I interpret as the distal extremity of the postmetacrista. The "overlapping portion of the postmetacrista" is therefore a lingual accessory structure that terminates at the urocrista.

Occasionally in sthenurines, the lingual accessory structure is much better developed than the primary postmetaconulecrista. For this reason, and because both cristae clearly arise from the metaconule apex, I refer to them in dental descriptions as the lingual and mesial (buccal) components of the postmetaconulecrista (fig. 9). Interestingly, this situation is mirrored on the buccal side with the postmetacrista. Frequently, two cristae originate from the metacone apex or in its vicinity, and they also vary in their expression relative to each other, to other molariform teeth in the same row, and between species. Two strong cristae also arise from the region of the metacone apex in the dP3 of *Kurrabi pelchenorum* Flannery et al., 1992. In each taxon with the "double" postmetacrista, the more buccal element is probably the reemergent posterior portion of the stylar crest. A distinct postmetacrista separate from the stylar crest is clearly evident in the phalangeriforms, *Durudawiri inusitatus* Crosby and Archer, 2000, and *Priscileo roskellyae* Gillespie, 1997.

I conclude that overlap of the postmetacrista and postmetaconulecrista is not a valid synapomorphy linking *Troposodon* and *Lagostrophus* with the sthenurines. Only *Troposodon* and *Protemnodon* display any overlap between these structures and this is not the same "overlap" to which Flannery (1983) was referring when he proposed the synapomorphy. Although there are morphological similarities between the metaloph posterior faces of *Lagostrophus*, *Troposodon*, *Protemnodon* and the sthenurines, this may reflect a potential for all bilophodont macropodoids to develop these features. The second crests probably evolved as buttresses for the metaloph in response to a browsing habit, as alluded to by Murray (1991a). As with the postprotocrista, paracristid and cristid obliqua in sthenurines, the double crest morphology may have resulted from a division of the postmetacrista or postmetaconulecrista.

3. Strongly fused mandibular symphysis

Fusion of the mandibular symphysis has evolved in many different mammal groups (Beecher 1977; Lieberman and Crompton 2000). In primates, the condition appears to have arisen independently in at least 12 different lineages, five times within the Adapinae alone (Ravosa and Hylander 1994; Ravosa 1999). For this reason, Beecher (1977) cautioned against using it to support the monophyly of the Anthropoidea. In marsupials, symphyseal form and function have received minimal attention, with Ride (1959) on *Macropus* and Lieberman and Crompton (2000) on *Didelphis* two noteworthy exceptions. I comment on the function of the sthenurine symphysis later (see Functional Morphology), but it is pertinent here to discuss symphyseal morphology as it relates to the recognition of fusion as a synapomorphy uniting *Troposodon* and *Lagostrophus* with the Sthenurinae (Flannery 1983, 1989). However, even before considering whether the symphyseal condition of these taxa represents a shared derived trait in the light of Beecher's warning, it is necessary to answer two fundamental questions. What defines a fused mandibular symphysis, and did *Troposodon*, *Lagostrophus* and the sthenurines possess fused mandibular symphyses?

Although the term 'ankylosed' has been widely used to describe the nature of the symphysis in sthenurines (e.g., Glauert 1910b; Raven and Gregory 1946; Tedford 1966; Flannery 1983; Wells and Tedford 1995; Prideaux and Wells 1997; Prideaux 1999b-c), it is not strictly appropriate. 'Bony ankylosis' refers to an abnormal fusion of bones forming a joint (Dorland and Miller 1943; Melloni and Eisner 1985), not to a normal condition. 'Synostosis', which is defined as fusion of adjacent bones by means such as ossification of their connecting cartilage (Dorland and Miller 1943), is a more general, but more appropriate term to describe the non-pathological fusion of a mandibular symphysis. In many mammals, fusion is so complete that the symphyseal suture is totally obliterated (Beecher 1977, 1983). However, remnants of the suture may still be present on specimens that clearly have a well-synostosed, immovable joint (Scapino 1981:fig. 14; Ravosa and Simons 1994:fig. 4). Synostosis may occur prior to or at the time of initial dental occlusion, or very late in postnatal development (Ravosa and Simons 1994; Lieberman and Crompton 2000). In the late-fusing subfossil prosimian *Archaeolemur*, the ontogenetic sequences of increased rugosity of the symphyseal plate, synostosis from anterior to posterior, and eventually complete fusion roughly parallel the phyletic trends observed in some Eocene adapids (Beecher 1983; Ravosa and Simons 1994). With regard to the first question posed above, it is clear that a spectrum of degrees of symphyseal fusion can exist. Therefore, if symphyseal fusion is to be used at all as a phylogenetic character, it is conceivable to recognize only three discrete states: unfused, partially fused, or completely fused.

The exact nature of the symphyseal attributes of *Troposodon*, *Lagostrophus* and the sthenurines, and how they relate to supposed degrees of fusion, has never been made explicitly clear by previous authorities. Possibly in allusion to this, Murray (1991a:278) noted that there are "...difficulties ascertaining a clear distinction between the degree of mobility of the symphysis and the state of bony ankylosis. Consequently, Flannery's assertion that ankylosed dentaries are unique to sthenurines could be ambiguous." In raising symphyseal fusion as a synapomorphy, Flannery (1983) first observed that "There is . . . a tendency towards fusion of the mandibular symphysis" (p. 18); then that "the mandibular symphysis is strongly ankylosed, preventing movement between the dentaries" (p. 19). Later, this was modified to "the mandibular symphysis is usually strongly

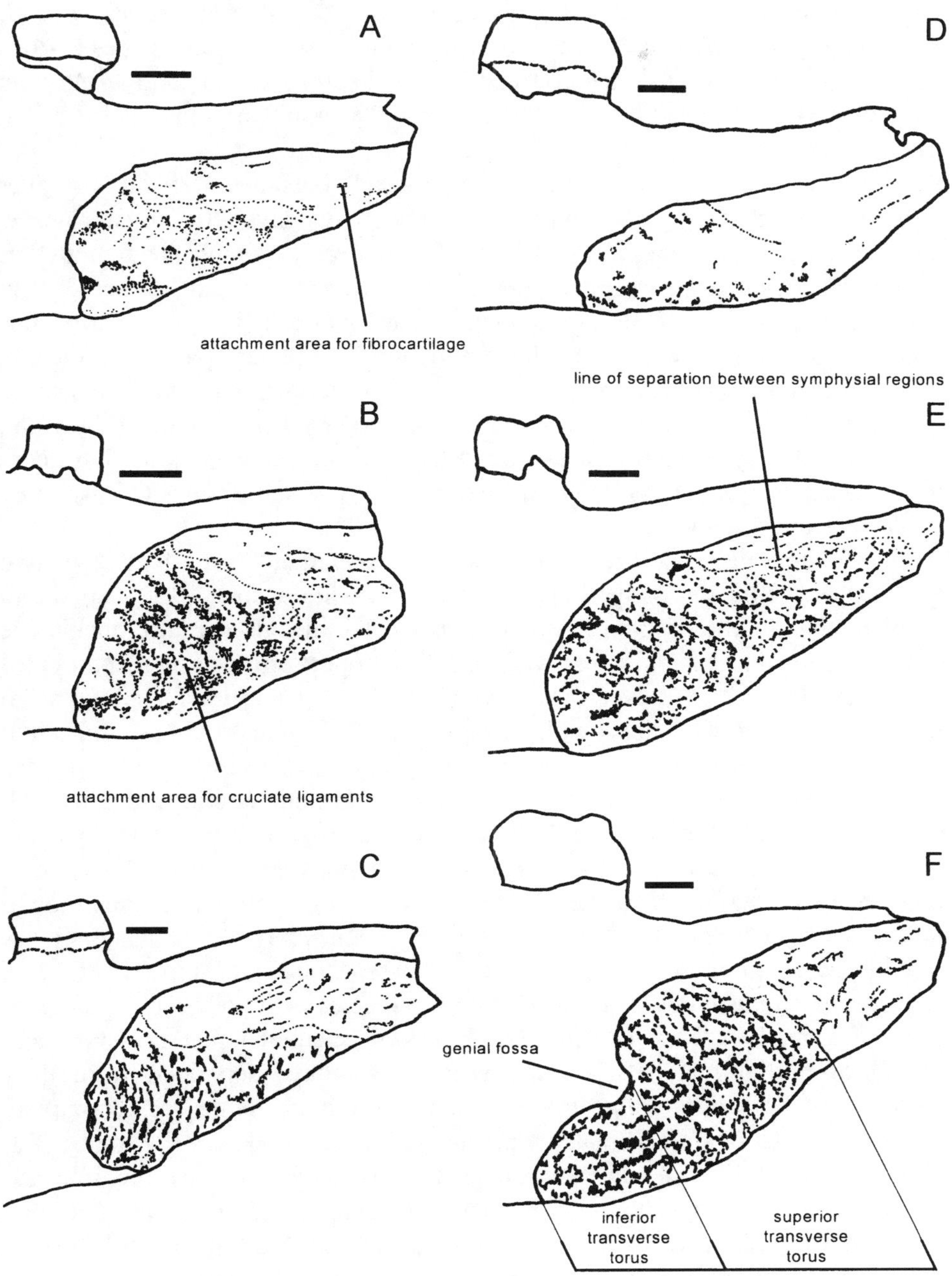

Figure 10. Left dentaries in mesial view, showing morphology of symphyseal plate. **A**, *Troposodon bowensis* (AM F60108); **B**, *Troposodon minor* (QM F4418); **C**, *Troposodon kenti* (SAM P14507); **D**, *Hadronomas puckridgi* (NTM P98142); **E**, *Sthenurus andersoni* (SAM P20494); **F**, *"Procoptodon" browneorum* (SAM P20462). Scale bars=5mm.

ankylosed" (Flannery 1989:28), perhaps because of the state expressed by the plesiomorphic *Troposodon bowensis*, for which Flannery and Archer (1983:266) noted that "ankylosis of the dentaries is weak." Tedford (1966) and Wells and Tedford (1995) also found a "weakly ankylosed" symphysis diagnostic of the genus *Sthenurus*.

How these authors envisaged "weakly ankylosed" is unclear. Uncertainty over how to interpret the degree of symphyseal fusion displayed by sthenurines is exemplified by Prideaux and Wells (1997:190) who, in describing the symphyseal plate of *Simosthenurus euryskaphus*, observed that the: "deep and rugose central and posterior regions indicate strong mandibular ankylosis." If this really does indicate strong fusion, how can it be that the symphyseal plate separated so cleanly from its counterpart? On balance, the only logical answer is that the symphysis was not fused at all. In fact, numerous foramina interspersed among the ridges and rugosities of the symphyseal plate (pl. 13B) are suggestive of a significant neurovascular supply to the symphyseal space, as Ravosa and Simons (1994) noted for juvenile *Archaeolemur*.

Many sthenurine taxa are represented by specimens in which the two symphyseal plates have separated cleanly (fig. 10), without the sort of damage to either plate that might be indicative of even partial fusion. Juvenile specimens typically fit into this category, but a significant proportion of adult dentaries of most species also had entirely unfused symphyses. Other specimens show partial fusion, often evinced when the symphyseal plate of one dentary still has attached to it (by fused bone, as opposed to sediment or glue) a portion of its counterpart. Beecher (1983) observed a similar condition among notharctine adapids. All sthenurine specimens that preserve the entirety of both halves of the symphysis, in synostosed articulation, retain at least part of the symphyseal suture. Therefore, it must be presumed that no sthenurine symphysis was completely fused, even though comprehensive, but incomplete fusion is likely to have rendered a symphysis immovable. So, although a proportion of adult specimens of most *Sthenurus*, *Metasthenurus*, *Simosthenurus* and *Procoptodon* species exhibited varying degrees of partial fusion, I have never seen a *Troposodon* specimen that preserves any indication of partial fusion. All symphyseal plates have separated from their counterparts without sustaining surface damage, although some are incomplete anteriorly due to taphonomic breakage perpendicular to the axis of the ramus (fig. 10). I have also not observed any *Lagostrophus* dentaries in which the symphyseal plates remain attached to one another by synostosis, but it is interesting to note that some modern museum specimens retain a connection via dried ligamentous tissue. Even if some individuals of *Lagostrophus* do attain partial symphyseal fusion, it cannot be considered characteristic of the taxon.

Therefore, fusion of the adult mandibular symphysis in some representatives of most sthenurines, most representatives of a few sthenurines, and no known representatives of *Troposodon* and *Lagostrophus* discounts symphyseal fusion as a synapomorphy uniting *Troposodon* and *Lagostrophus* with the sthenurines. Marked development of symphyseal rugosities must have some potential to constrain symphyseal flexibility (see Functional Morphology), as in those carnivorans with a similar morphology (symphysis class III of Scapino 1981), but they can clearly be well developed without being accompanied by synostosis. There is no question, however, that the rugose symphyseal plates of these taxa and the smoother plates of other macropodoids (e.g., *Balbaroo*, *Hadronomas*, fig. 10), are deserving of more detailed descriptions and comparisons than there is scope for in this monograph.

4. Sinuous occlusal edge on unworn i1

Morphology of the i1 varies markedly within and between macropodoid groups. Flannery (1983) considered the possession of a gently sinusoidal occlusal edge on i1 a synapomorphy uniting *Troposodon* and *Lagostrophus* with the sthenurines. However, this morphology is not exclusive to these taxa; it is also observed in *Hadronomas* and the plesiomorphic macropodine *Dorcopsoides* Woodburne, 1967 (fig. 11). Moreover, the character state pertains to the shape of the occlusal edge in an unworn state (Flannery 1983), even though no unworn i1 occlusal surface was known for *Troposodon* at the time. Since it is still unknown, a reasonable comparison can only be made between the least worn lower incisors of *Troposodon* and those of *Lagostrophus* and sthenurines that show a similar wear stage (fig. 11).

The species of *Troposodon* bear a much more slender and elongate i1 crown than *Lagostrophus*, *Dorcopsoides*, *Hadronomas* or any sthenurines except for the derived *Simosthenurus maddocki.* However, along with *Dorcopsoides*, *Hadronomas* and the other sthenurines, the occlusal wear facet of *Si. maddocki* is worn at a much greater angle to the longitudinal axis of the tooth than it is in *Troposodon* and *Lagostrophus* (fig. 11). In *T. bowensis* Flannery and Archer, 1983, *T. minor* and *T. kenti* Campbell, 1973, the wear facet is relatively long, slightly concave and worn at a very shallow angle to the i1 dorsal surface. I do not consider this a sufficient indication of a sinusoidal unworn occlusal surface in *Troposodon.* Viewed buccally, the shape of the i1 occlusal surface in *Lagostrophus* is intermediate between *Troposodon*, on the one hand, and *Hadronomas* and *Dorcopsoides*, on the other. It resembles the latter two species especially with regard to the curvature of the ventral edge of the crown (fig. 11). However, the distribution of enamel on the lingual side of the crown is markedly different between these taxa, and the i1 of *Hadronomas* and *Dorcopsoides* is deeper when scaled to the same length as that of *Lagostrophus* (fig. 11).

In summary, there is no support for linking *Troposodon* to the sthenurines on the basis of its unworn i1 occlusal surface shape. Not only is an unworn i1 of *Troposodon* unknown, but the worn occlusal surface and the overall shape of the crown are very different from those of the sthenurines. In this trait, *Lagostrophus* is more similar to *Dorcopsoides*, *Hadronomas* and the sthenurines than it is to *Troposodon*, but it shows an enamel distribution pattern more like *Troposodon* and quite unlike *Dorcopsoides*, *Hadronomas* and the sthenurines (fig. 13).

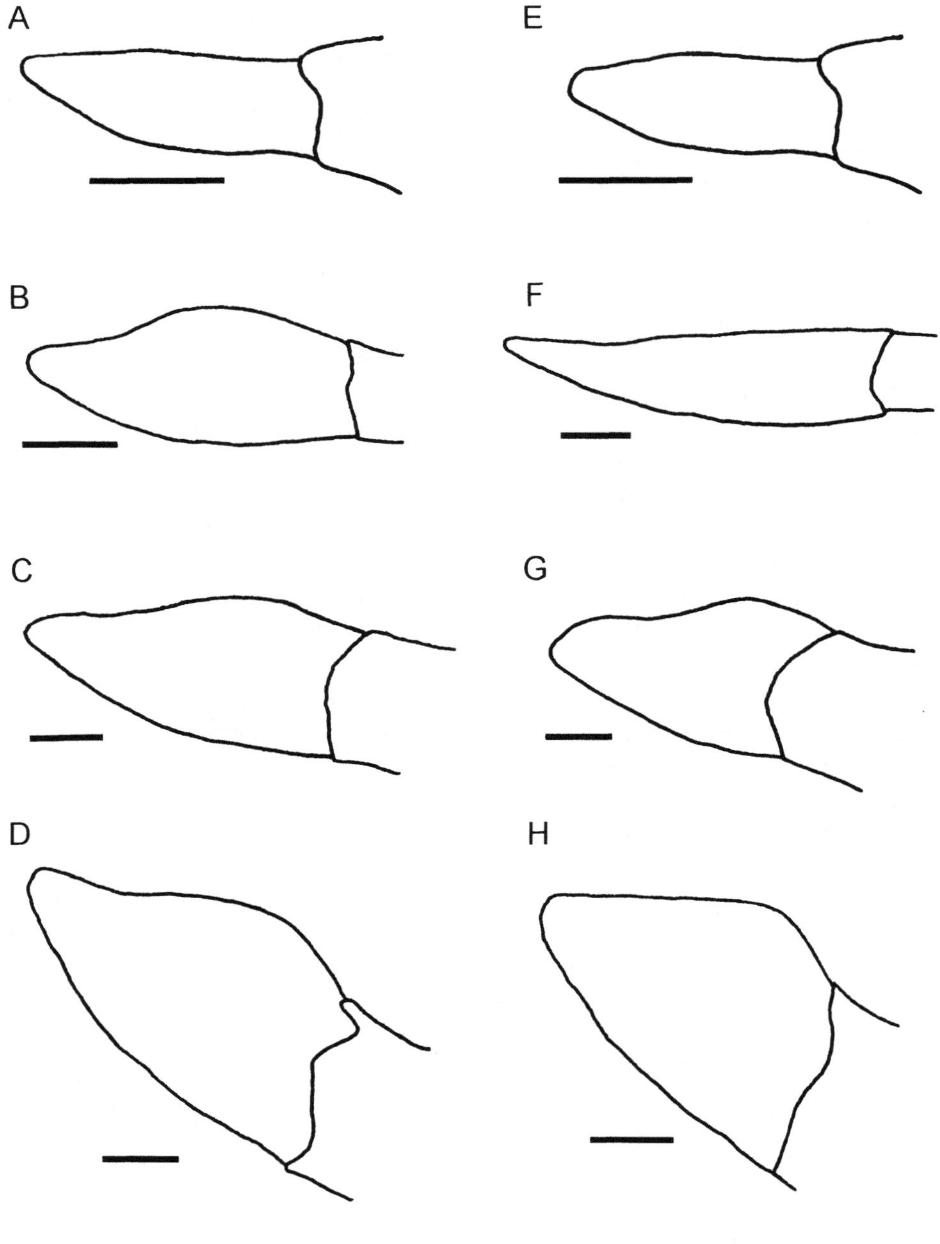

Figure 11. Left lower incisors (i1) in buccal view, showing shape of occlusal surface. UNWORN – **A**, *Lagostrophus fasciatus* (AM M1426); **B**, *Dorcopsoides fossilis* (SAM P unreg.); **C**, *Hadronomas puckridgi* (SAM P38771); **D**, *Simosthenurus occidentalis* (SAM P20480); WORN – **E**, *Lagostrophus fasciatus* (AM M1093); **F**, *Troposodon bowensis* (AM F59557); **G**, *Hadronomas puckridgi* (NTM P98142); **H**, *Simosthenurus occidentalis* (SAM P16669). Scale bars = 5 mm.

5. Premetacristid well developed, usually contacting paracristid

The paracristid is the primary crest of the archetypal mammalian trigonid, connecting the paraconid and protoconid. In most bilophodont macropodoids, the paracristid descends anterolingually from the protoconid, before turning lingually and terminating near the anterolingual corner of the tooth (fig. 12). *Lagostrophus* expresses an autapomorphic condition in which the paracristid terminates near the tooth midline, where it is met by a well-formed premetacristid (Flannery 1983:fig. 4B). The transverse anterior portion of the paracristid is usually not differentiable from the underlying precingulid in most macropodoids, but both structures are clearly evident in most sthenurines.

Mirroring the paracristid, the premetacristid runs along the anterolingual border of the trigonid and links the metaconid to the paraconid. It is extremely variable in expression across the Macropodoidea, ranging from subequal in size to the paracristid in some representatives of all groups to absent in numerous macropodines and sthenurines. The premetacristid may or may not maintain a distinct connection to the metaconid apex. Just lingual to the premetacristid, a second crest often originates from the vicinity of the metaconid. It descends into the trigonid basin and terminates anywhere between the lingual side of the paracristid and high on the lingual portion of the protolophid. In propleopines (fig. 12) and *Hypsiprymnodon*, Ride (1993) and Ride et al. (1997) referred to this crest as the parametacristid. I have not observed this structure in bulungamayines, balbarines and potoroines (except for the m1 of *Bettongia moyesi*; see Flannery and Archer 1987a:fig. 3), but several sthenurines and macropodines bear a crest in a very similar position and orientation, suggesting that the structure is, in fact, a reemergent parametacristid. Potoroines, bulungamayines and balbarines usually possess a distinct premetacristid.

The first problem with the premetacristid synapomorphy proposed by Flannery (1983) is that the premetacristid and parametacristid were not recognized as separate cristids. Although they differ quite consistently in their orientation, identification of exactly which crest is expressed can be difficult if only one is present. Fortunately, some simosthenurin and *Protemnodon* species (e.g., *"Simosthenurus" pales, "Procoptodon" browneorum, Protemnodon devisi*) show a clear premetacristid and parametacristid (fig. 12), and some individuals of *Wallabia indra* (De Vis, 1895), more faintly express both cristids. In the species of *Troposodon*, the premetacristid is completely lost, but the genus is characterized by the hypertrophy of the parametacristid (fig. 12). This parametacristid approaches and, in the more derived species (*T. minor*, *T. kenti*), connects to the anterolingually oriented, or posterior, portion of the paracristid. However, it never approaches the transverse anterior portion of the paracristid. In *P. devisi*, the parametacristid is slightly weaker than in *T. bowensis* and *T. bluffensis* Bartholomai, 1978b, but it is identical in orientation and position (fig. 12). The premetacristid of *P. devisi* is slighter than the parametacristid and descends the

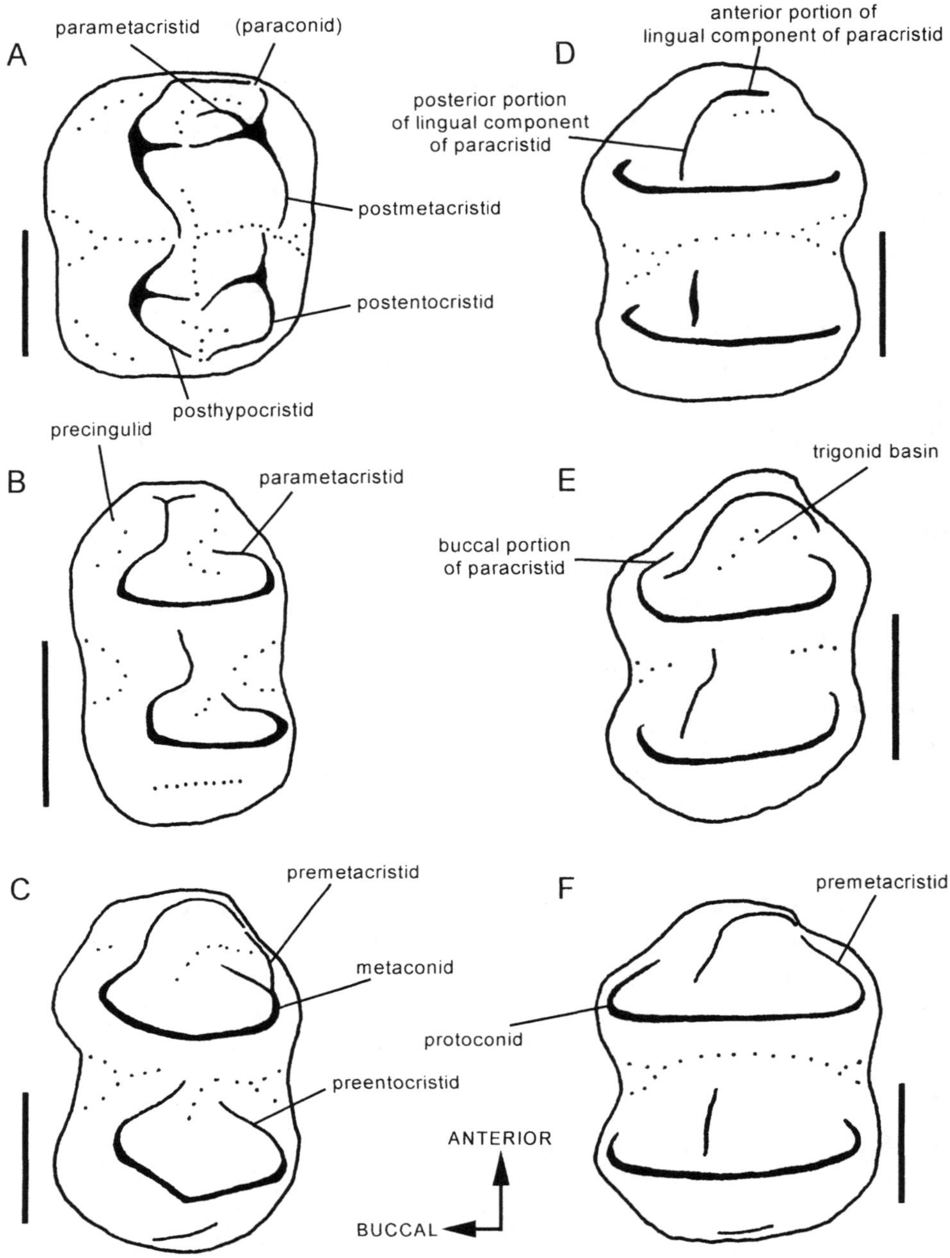

Figure 12. Left lower molars in occlusal view, showing relationship of paracristid, premetacristid and parametacristid. **A**, *Propleopus oscillans* (SAM P20815, m3); **B**, *Troposodon bowensis* (AM F60111, m2); **C**, *Protemnodon devisi* (QM F5222, m3); **D**, *Archaeosimos cegsai* (SAM P31800, m3); **E**, *Simosthenurus maddocki* (SAM P20508, m2); **F**, *"Simosthenurus" baileyi* (SAM P13670, m3). Scale bars = 5 mm.

anterior face of the metaconid, connecting with the paracristid at the position of the paraconid. *Lagostrophus* may express both a parametacristid and a premetacristid.

Certainly, the idea that the transverse anterior portion of the "paracristid" is really a strongly developed and anteriorly migrated premetacristid does not hold for *Troposodon* (cf. Flannery 1983). A premetacristid is consistently absent in members of this genus. The morphology of the *Lagostrophus* trigonid is quite different. A slight parametacristid may be evident on its molariform teeth (see Stirton 1963:fig. 11), but the crest is not nearly as well developed as in most *Troposodon* species. In contrast to most macropodoids, the *Lagostrophus* paracristid extends only to the tooth midline, where it is met by a low but distinct premetacristid. The cristids are not completely confluent, and a vertical groove demarcating their union is evident anteriorly (see Flannery 1983:fig. 4B). In sthenurines with a well-developed premetacristid (e.g., *"Simosthenurus" baileyi*, some individuals of *"Procoptodon" browneorum*), this point of contact is also clear (fig. 12). However, in contrast to *Lagostrophus*, the paracristid of these two sthenurines terminates more toward the lingual side of the trigonid, as it does in most bilophodont macropodoids.

Even if the transverse portion of the paracristid of *"Simosthenurus" baileyi* and *"Procoptodon" browneorum* actually was an anteriorly shifted premetacristid, it would be difficult to draw a phylogenetic connection with *Lagostrophus*, because these former two species are derived sthenurines. An intermediate condition in more plesiomorphic taxa would need to be demonstrable, but it is clearly not. The plesiomorphic *Metasthenurus newtonae* does possess a premetacristid on its anterior molariform teeth, but the crest is very slight and the abutment between it and the much thicker paracristid is clearly visible on the lingual side of the trigonid (pl. 34C). A well-developed premetacristid does not, therefore, represent a synapomorphy uniting *Troposodon* and *Lagostrophus* with the sthenurines.

In summary, I find very little support for the character states cited as synapomorphies uniting *Troposodon*, *Lagostrophus* and the sthenurines within the Macropodoidea. Each state is a symplesiomorphy, a structural misinterpretation or a convergent feature. A direct implication of this is that the one dental and three pedal synapomorphies (fig. 7) highlighted as evidence for the monophyly of *Troposodon* and the sthenurines (Flannery 1983, 1989) may also be questionable, and I examine them now.

1. Crown of i1 completely enameled

According to Flannery (1983, 1989), one major feature that separates *Lagostrophus* from *Troposodon* and the sthenurines is the absence of enamel on the lingual surface of i1. In combination with the three pedal characters, this difference was used to designate *Lagostrophus* the plesiomorphic sister clade to *Troposodon* and the sthenurines. I am not certain of the variability of this i1 feature in *Lagostrophus*, but in all 35 specimens that I have observed, the crown is enveloped by a relatively thick layer of enamel (fig. 13). Dentine is only exposed after

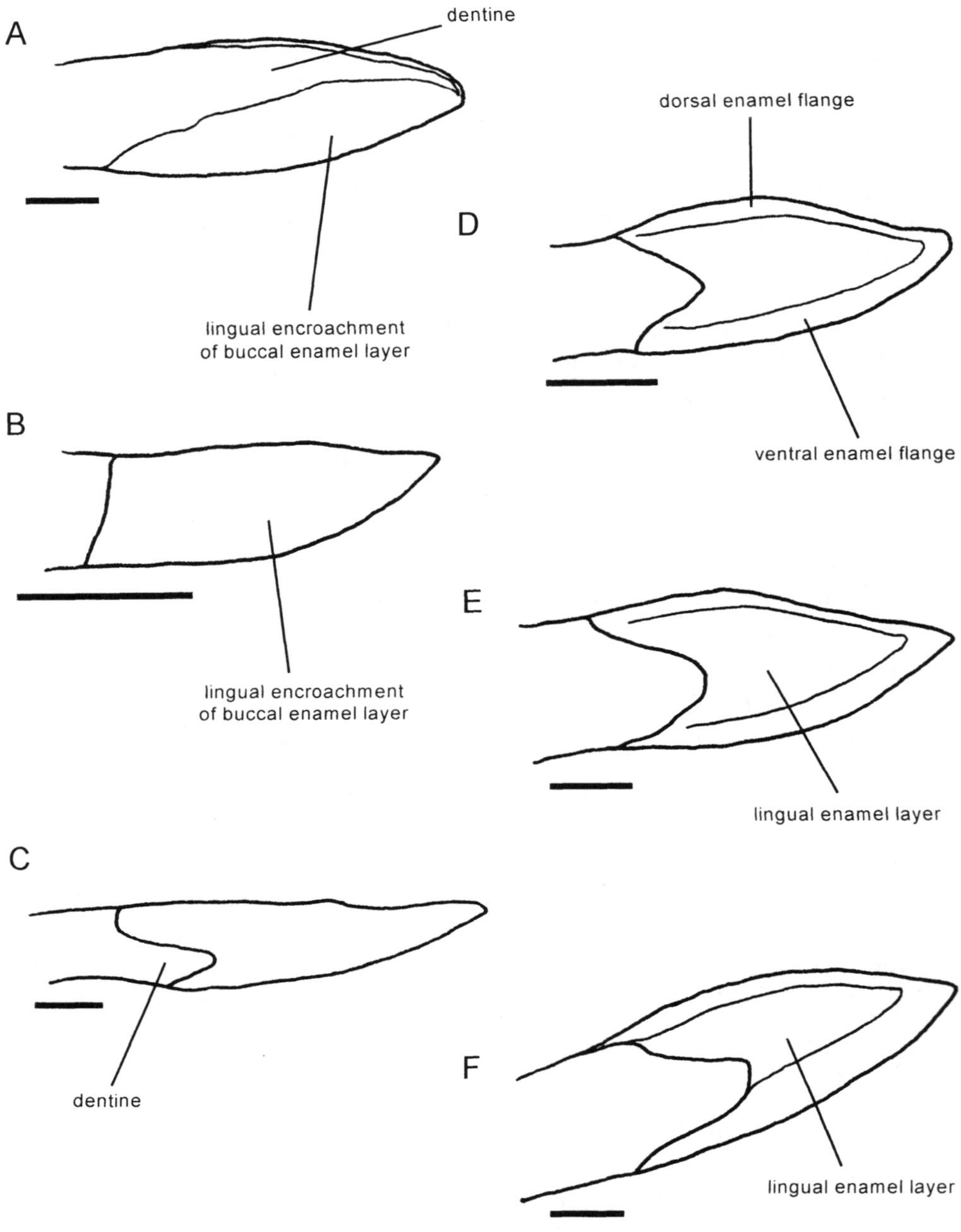

Figure 13. Left lower incisors in lingual view, showing distribution of enamel. **A**, *Macropus fuliginosus* (FU unreg.); **B**, *Lagostrophus fasciatus* (AM M1426); **C**, *Troposodon bowensis* (AM F59557); **D**, *Dorcopsoides fossilis* (SAM P unreg.); **E**, *Hadronomas puckridgi* (SAM P38771); **F**, *"Procoptodon" gilli* (SAM P28609). Scale bars = 5 mm.

considerable wear of the occlusal surface. Although this might initially seem to add weight to Flannery's hypothesis, it should be noted that several macropodines also bear enamel on the lingual i1 surface: *Dorcopsoides*, *Onychogalea unguifera* (Gould, 1840), *Dendrolagus bennettianus* De Vis, 1887, *D. lumholtzi* (Collett, 1884) and *D. matschiei* Förster and Rothschild, 1907. In addition, enamel ridges on the lingual surface of i1 are present in the balbarine *Ganawamaya.* This balbarine aside, the pattern of enamel distribution in all of these taxa can be separated into two basic forms.

The i1 of *Dorcopsoides* and *Hadronomas* has pronounced dorsal and ventral enamel flanges, which are extensions of the thick enamel layer on the buccal surface (fig. 13). On the lingual surface between the flanges, a thin layer of enamel covers the elongate dentine core, but it does not extend as far posteriorly as the buccal enamel layer. An almost identical condition is observed in *Archaeosimos cegsai* (pl. 37E). All other sthenurines differ by having a slightly thicker lingual enamel layer, but are otherwise very similar (fig. 13F, various plates). By contrast, *Lagostrophus*, *Troposodon*, *Dendrolagus* and *Onychogalea unguifera* lack pronounced dorsal and ventral enamel flanges. The dorsal and ventral surfaces of i1 are much more rounded than in *Dorcopsoides*, *Hadronomas* and the sthenurines. Moreover, the enamel on the lingual surface is approximately the same thickness as the enamel on the buccal surface, because it actually represents a continuation of the buccal enamel layer around and onto the lingual side of the tooth. In essence, the i1 is encircled by one continuous enamel layer. An intermediate condition between these taxa and those species with no lingual enamel is observed in *Macropus fuliginosus* (Desmarest, 1817) and some *Protemnodon* species (fig. 13).

The lingual enamel layer of *Dorcopsoides*, *Hadronomas* and the sthenurines appears to have formed in situ, rather than as a lingual encroachment of the thicker enamel more typically restricted to the buccal surface (see Young et al. 1990:fig. 3a). The derived bulungamayine *Wanburoo* may be an excellent structural precursor to these taxa; its i1 has no lingual enamel layer but is very similar in overall i1 shape and flange development. On balance, complete enameling of the i1 cannot be viewed as a synapomorphy allying *Troposodon* with the sthenurines, nor does it strengthen the hypothesized alliance between *Lagostrophus* and sthenurines.

2. Extreme reduction of metatarsal V

The argument that uses this element to ascribe a sthenurine affinity to *Troposodon* may be questioned on two grounds. First, one unassociated fragmentary metatarsal V (AM F63920) has been tentatively assigned to *Troposodon bowensis* by Flannery and Archer (1983). Only the proximal end of the element is preserved and that is somewhat abraded. It narrows markedly toward the broken distal extremity, suggesting that the bone was small, splint-like and could not have supported phalanges, akin to the condition only known in sthenurines. Flannery and Archer (1983) took this to imply that the metatarsal fragment most likely belonged to a sthenurine. Because *T. bowensis* was believed to be a sthenurine and more abundant than *Simosthenurus* in the Bow deposit, the specimen was thought most likely to pertain to *Troposodon.* However, given the disassociation of elements within the Bow deposit and the very small sample sizes, a *Troposodon* affinity is, at best, questionable.

Second, only a portion of the articular surface of metatarsal IV is preserved (Flannery and Archer 1983) and, unlike in *Hadronomas* and *Simosthenurus occidentalis* (see Murray 1995), it is not separated into two distinct facets angled at about 135° to one another. There is only a slight constriction in the central region of the surface and it is very nearly flat overall. According to Murray (1995), this condition typifies the majority of potoroines and macropodines. However, *Sthenurus stirlingi* and an unidentified simosthenurin from Victoria Fossil Cave also possess a relatively flat and centrally unconstricted metatarsal IV articular surface. While this might represent a return to the possibly plesiomorphic condition seen in potoroines and macropodines, the value of the double facet as a sthenurine apomorphy must be considered doubtful at present. Clearly, it cannot be used to align the Bow metatarsal with sthenurines, macropodines or *Troposodon* itself.

Interestingly, four characters of the Bow metatarsal suggest a macropodine affinity. Although much smaller, its morphology is particularly reminiscent of *Macropus rufus* (Desmarest, 1822), a species that shares several similarities with sthenurines due to the convergent reduction in relative size of metatarsal V (Murray 1995). The plantar surface is relatively narrow and gently curved posteromesially. In *Hadronomas* and the sthenurines the plantar surface is much wider proximally. On the mesial surface, a deep proximodistally oriented groove runs from the abraded proximal end to the broken distal end, between the mesial extreme of the metatarsal IV articular surface and the plantar surface. A similar groove can be seen in *Macropus*, *Protemnodon* and, to a lesser degree, *Hadronomas.* When viewed dorsally, the Bow metatarsal V narrows very gradually anterior of the metatarsal IV articular surface, in marked contrast to the rapid reduction in width visible immediately anterior to this surface in sthenurines. In *Hadronomas*, metatarsal V is not reduced (Murray 1995). The fourth similarity with macropodines relates to the shape of the lateral surface. In the Bow specimen and *M. rufus*, the surface is smooth and gently concave in contrast to a slightly uneven but generally flatter surface in sthenurines.

If these four characters do align the Bow metatarsal V with the macropodines, then the distal narrowing of the element must be convergent with sthenurines. The macropodine-like features could have arisen independently, but this does seem the less parsimonious conclusion at the moment. Regardless of whether AM F63920 belongs to a sthenurine or a macropodine, it is difficult to advocate the suggestion that it pertains to *Troposodon bowensis* just because that species is the most abundant macropodoid in the deposit. Even if the metatarsal could be allocated to *Troposodon* with confidence, the proposed synapomorphy appears very doubtful.

3. High rounded crests on tibial articular surface of astragalus

Flannery and Archer (1983) ascribed two unassociated astragali from the Bow LF to *Troposodon bowensis* and, in doing so, remarked on their similarity to *Procoptodon goliah* and *Simosthenurus occidentalis*, particularly with respect to the tibial articulation and the proximomesial process. Having examined one of these specimens (AM F63990), I fully concur with their observations. The astragalus fragment is sthenurine-like, possessing no apparent macropodine feature. Characters examined by Murray (1995) for *Hadronomas* are used in the following comparison of the Bow specimen.

General morphology of the Bow astragalus is intermediate between that of *Hadronomas* and *Simosthenurus occidentalis*, but of the two it appears closer to *Hadronomas.* Although the lateral surface and the posterolateral quadrant of the specimen are not preserved, AM F63990 is wider relative to its length than the astragalus of either *Hadronomas* or *Si. occidentalis.* The trochlear crests of AM F63990 are widely separated and the medial crest is particularly high. When viewed either anteriorly or posteriorly, the outline of the dorsal surface (the two trochlear crests and tibial articular sulcus) is almost identical in morphology to *Si. occidentalis.* In lateral and mesial profile the trochlear crests are also similarly shaped, although the anterior extremity of the medial crest is terminated by a much deeper and right-angled trochlear (bursal) notch. The position of the astragalar head is very similar to that observed in *Hadronomas*, but the neck is notably longer and dorsoventrally shallower. This means that the astragalar head of AM F63990 extends farther anteriorly from the main body of the astragalus than it does in *Hadronomas* or *Si. occidentalis*, in which it is very short and deep. The medial malleolar fossa is a rather deep and circular depression, while the distolateral process appears to have been rather deep and laterally flared, as in *Hadronomas* (see Murray 1995).

On balance, AM F63990 certainly appears to belong to a sthenurine, but assignment of the two astragali to *Troposodon bowensis* on the basis of numerical abundance alone is very contentious, especially as an indisputable sthenurine is known from the Bow LF (Flannery and Archer 1984).

4. Calcaneum with distal (lateral) portion of cuboid facet distinctly shorter than proximal portion

Flannery and Archer (1983) referred two calcanea to *Troposodon bowensis*, one nearly complete (AM F63919), the other represented by a fragment of the astragalar articular surface (AM F63918). The main feature used to align these specimens with the sthenurines, and therefore *T. bowensis*, was the form of the cuboid articulation at the anterior region of the calcaneum, which is preserved only in AM F63918. Once again, characters employed by Murray (1995) are used here to compare the two Bow calcanea with *Hadronomas*, *Simosthenurus occidentalis*, *Macropus and Protemnodon.*

Flannery and Archer (1983) suggested that the morphology of the Bow calcanea resembled sthenurines more than macropodines due to the shorter nature of the distal (lateral) portion relative to the proximal portion. This observation is ambiguous. When viewed dorsally, the cuboid articular surface of macropodoids is stepped (the lateral component projects farther anteriorly than the mesial component), but this is reduced in *Dendrolagus* and *Bohra* Flannery and Szalay, 1982. When viewed anteriorly, the lateral component is narrower than the mesial component, but it can be near to equal in width in *Protemnodon* and *Petrogale* Gray, 1837. The lateral component of the cuboid articular surface is consistently deeper (dorsoventrally) than its mesial neighbor, so "shorter" cannot be in reference to the depth of this feature. It seems, therefore, that the observation of Flannery and Archer must pertain to the degree of deflection from the horizontal of the lateral component's dorsal edge.

In the Bow calcanea the dorsal edge of the lateral component is slightly lower than the mesial component, but the degree to which this is expressed is similar in *Macropus.* The lateral dorsal edge is certainly deflected farther in sthenurines, but it is even more marked in *Protemnodon.* In other macropodines (e.g., *Dorcopsoides*) as well as *Hadronomas*, the morphology more closely approximates the potoroine condition, where the entire dorsal edge of the cuboid articular surface is horizontal. If I have not misinterpreted their observation, the proposed synapomorphy of Flannery and Archer (1983) is doubtful, because the character state is expressed in both sthenurines and macropodines. Unfortunately, no postcranial elements of bulungamayines or balbarines have yet been described. As is the case with other Bow pedal elements, there is still no convincing evidence that the calcanea are those of *Troposodon.*

A comparison of the two Bow calcanea reveals differences in size and morphology not noted by Flannery and Archer (1983). While the two specimens may represent individuals of differing ontogenetic ages, marked variation in the morphology of the medial articular facet for the astragalus might be more likely to indicate that they do not belong to the same taxon. The facet is high, rounded and, therefore, sthenurine-like in AM F63919, whereas the facet is notably lower in AM F63918, suggestive of a typical macropodine.

Overall, I find no support for any character states cited by Flannery (1983, 1989) or Flannery and Archer (1983) as evidence for a close relation between *Troposodon* and the sthenurines or between *Lagostrophus* and the sthenurines. This is consistent with the conclusions of Murray (1991a, 1995).

Macropodine Affinities

Although my primary objective here is to establish the monophyly and constituent taxa of the Sthenurinae, it is desirable to avoid leaving *Lagostrophus* and *Troposodon* in systematic limbo. Having ascertained that they are not sthenurines, it is theoretically conceivable that they may belong in one or another of the existing subfamilies or a new subfamily. Alliance to the Propleopinae, Balbarinae and Potoroinae can be immediately discounted, because *Lagostrophus* and *Troposodon* lack key diagnostic features. Unlike propleopines and potoroines, they are bilophodont and lack a deeply penetrating masseteric canal as well as a finely serrated or plagiaulacoid premolar. Unlike balbarines, the trigonid of m1 is not markedly compressed laterally, no protostylid is evident, the masseteric canal and dental canal are not well separated, and the p3 is not plagiaulacoid.

All bulungamayines are small and have elongate, gently crescentic, serrated premolars, a small buccally oriented crest arising from the entoconid and a deep masseteric canal (Flannery, Archer and Plane 1983; Cooke 1997b, 1999). Because all of these defining features might well be symplesiomorphies, the group seems most likely to represent a paraphyletic stem group from which both sthenurines and macropodines arose (Cooke 1997b, 1999). *Lagostrophus* and *Troposodon* lack these bulungamayine states, regardless of whether they are symplesiomorphies. Interestingly, it is almost as difficult to identify synapomorphies for the Macropodinae as it is for the Bulungamayinae, although perhaps this is not so remarkable if bulungamayines effectively grade into macropodines. Nevertheless, similarities to taxa within the Macropodinae do draw *Lagostrophus* and *Troposodon* toward that subfamily. Just as Bartholomai (1967) noted when describing the genus, *Troposodon* resembles both *Protemnodon* and *Lagostrophus*. It is typified by a hypertrophied parametacristid and preentocristid, and a well-developed postparacrista. None of these cristae are as well developed in *Protemnodon*, but they are better developed in representatives of this genus than in most macropodines. *Lagostrophus* differs slightly by having a well-developed premetacristid and a frequently distinct parametacristid. The metaconular accessory cusp, a small neomorphic structure at the lingual end of the interloph valley, is only observed in species of *Troposodon* and *Protemnodon*. Similarities in upper and lower premolar morphologies are also evident and the palate of both genera is completely non-fenestrate, in contrast to *Lagostrophus*.

Bartholomai (1967) also drew attention to the similarity between the L-shaped lower premolars of *Troposodon* and *Lagostrophus*, but stopped short of suggesting a close phylogenetic relationship as later alluded to by Archer (1981). Both taxa share a completely enameled i1 (fig. 13) and a similar ramus shape (fig. 20). Of particular note is the shared manner in which the ramus narrows anteriorly due to a broad concavity below the buccinator sulcus. This concavity is present, but less pronounced in *Protemnodon tumbuna* Flannery, Mountain and Aplin, 1983 (see Menzies and Ballard 1994:fig. 7). *Troposodon*, *Lagostrophus* and a large, unnamed macropodine from the Kanunka LF are also characterized by having p3 subequal in length to or shorter than m1, in contrast to *Protemnodon*. Together, all of these observations support the notion that *Troposodon* and *Lagostrophus* are best placed within the Macropodinae, but a detailed phylogenetic analysis is needed before a solid view of their relationships can be established.

Before pressing on, however, it is germane to consider these observations in the light of non-osteological studies of *Lagostrophus fasciatus*, the only extant macropodine among those just discussed. Pearson (1946, 1950) detailed the existence of two female reproductive morphologies in kangaroos. Although only a limited number of taxa were examined, potoroines were observed to differ from macropodines by possessing a short urogenital sinus, a long posterior vaginal sinus and a large anterior vaginal expansion. Intriguingly, Tyndale-Biscoe (1965) showed that the reproductive anatomy of *Lagostrophus* bore greater similarity to that of the potoroines: it possesses a large anterior vaginal expansion, a bladder attached anteriorly to the vaginal culs-de-sac and a long urethra, and lacks a permanently open median vagina. *Lagostrophus* does, however, possess a very short posterior vaginal sinus and a long urogenital sinus, features typical of macropodines. On the face of these data alone, one could suggest an intermediate phylogenetic position for *Lagostrophus* between potoroines and macropodines. However, not only are the urogenital morphologies of the taxa *Lagostrophus* most closely resembles in craniodental form unknown (they are all extinct), the urogenital systems of many apparently plesiomorphic living macropodines remain undescribed (e.g., *Dorcopsulus*, *Dorcopsis*, *Dendrolagus*). Clearly, it would be hazardous to make anything phylogenetic of the *Lagostrophus* morphology until the urogenital systems of a wider range of extant taxa are described.

Unlike the female urogenital system, the chromosomes of most extant macropodoids have been studied and vary between 2n = 10 and 2n = 32 (Hayman 1977). G-banding studies suggest that a chromosome number of 2n = 22 was probably ancestral for the macropodoids (Rofe 1979; Hayman 1990), with potoroines usually possessing either 2n = 22 or 2n = 24 chromosomes of a relatively consistent morphology. *Lagostrophus* also possesses 24 chromosomes, which has been taken as another suggestively potoroine-like feature (Sharman 1961; Hayman and Martin 1974), although they differ in morphology. However, a problem with karyotypes is that, although they can be used to imply the presence of either plesiomorphic or autapomorphic states within genera, they provide no really useful information with regard to intergeneric relations, because synapomorphic conditions cannot be identified (Clemens et al. 1989). Chromosomal number, while similar in *Lagostrophus* and the potoroines, cannot be confidently used as evidence for a closer phylogenetic proximity to that group than to the macropodines.

Using microcomplement fixation (MC'F), Baverstock et al. (1989) examined albumin immunologic relationships of the macropodoids. Their phylogenetic analysis supported the recognition of four clades: 1. *Macropus*, *Wallabia*, *Setonix*, *Onychogalea*, *Dorcopsis*, *Dorcopsulus*, *Lagorchestes*, *Dendrolagus*, *Petrogale* and *Thylogale*; 2. *Lagostrophus*; 3. *Bettongia*, *Potorous* and *Aepyprymnus*; 4. *Hypsiprymnodon*. These data also suggest that *Lagostrophus* is distinct from all other macropodines and is possibly closer to potoroines. Importantly, *Dorcopsulus*, *Dorcopsis* and *Dendrolagus* were included in the study and *Lagostrophus* did group outside of them and the other macropodines. Baverstock et al. (1989)

suggested that the MC'F data may well support Flannery's (1983) view that *Lagostrophus* is the sole extant sthenurine. Since I have demonstrated that *Lagostrophus* is not a sthenurine, this means either that it has been independently derived from bulungamayine stock or that the albumin evidence is not telling the full phylogenetic story. Baverstock et al. (1989:47) do point out that: "similarity of albumins can be due to true phylogenetic affinity, or to relatively slow rates of albumin evolution, or both."

Earlier, Kirsch (1977) examined the blood serological relationships of all marsupial groups and demonstrated a clear separation between potoroines and macropodines. However, unlike the implications from the MC'F data, *Lagostrophus* grouped with *Setonix*, *Dendrolagus* and *Dorcopsis*, which allied with the *Thylogale-Petrogale* group. Interestingly, *Onychogalea*, rather than *Lagostrophus*, took up an intermediate position between potoroines and macropodines, and *Lagorchestes* also came out as a distinct lineage. While DNA hybridization (Kirsch et al. 1995, 1997) and mitochondrial RNA (Burk and Springer 2000) studies did not examine *Lagostrophus*, Colgan's (1999) investigation of phosphoglycerate kinase DNA sequences in macropodoids and other marsupials did include data for *Lagostrophus.* Although the strict consensus of the 9602 equally most parsimonious trees placed *Lagostrophus* in a polytomy with a number of different macropodoid lineages, many trees showed it as the sister taxon of the other macropodoids, with the exception of *Dendrolagus dorianus* Ramsay, 1883, and *Hypsiprymnodon moschatus* Ramsay, 1876. However, resolution in the macropodoid portion of the tree may be questionable given that a second specimen of *D. dorianus* consistently allied with the crown group, while *Bettongia lesueur* (Quoy and Gaimard, 1824), the sole potoroine included, apparently fluctuated between the polytomy of which *Lagostrophus* is a part and a position as sister taxon of *Lagorchestes conspicillatus* (Gould, 1841).

While none of these investigations offer much more clarity than any others with regard to the immediate relationships of *Lagostrophus fasciatus*, the weight of evidence from all data sets indicates that there is something seriously odd about the animal. In all probability, it is the only living member of an otherwise entirely extinct lineage of kangaroos, one that diverged from near the base of the radiation and may include genera like *Troposodon*, *Protemnodon* and *Congruus.* The conjectural placement of *Lagostrophus* also highlights the doubts that center on the nature of the Macropodinae as currently defined. Is it monophyletic? Of the five diagnostic features listed by Flannery (1989), a broadened occlusal surface of I1 may well be the only character state not also present in at least some bulungamayines. Somewhat predictably, *Lagostrophus* is an exception. At this stage, the clarification of macropodine monophyly and the relationships of *Lagostrophus* may have to await a more comprehensive understanding of the bulungamayine radiation and improvement in the middle to late Miocene fossil record. The astute observations of Thomas (1887) and Flannery (1983) have highlighted several important craniodental features which, combined with the molecular and soft tissue data, strongly imply that the systematic position of *Lagostrophus* is far from certain. However, *Lagostrophus* is not a sthenurine.

STHENURINE SYNAPOMORPHIES

I define the Sthenurinae on the basis of nine craniodental traits that characterize no other group of macropodoids. *Hadronomas* possesses each of these nine synapomorphies, confirming the assertions on its relationships made by Murray (1991a, 1995) and first alluded to by Woodburne (1967). There are six sthenurine genera: *Hadronomas* Woodburne, 1967, *Sthenurus* Owen, 1874a, *Metasthenurus* gen. nov., *Archaeosimos* gen. nov., *Simosthenurus* Tedford, 1966, and *Procoptodon* Owen, 1874b (see Taxonomy, Phylogeny). While many of the 40 postcranial characters shared by sthenurines to the exclusion of macropodines and potoroines (Murray 1995) are probably synapomorphies, the bulungamayine conditions need to be revealed before these features may be embraced as uniquely sthenurine. More craniodental synapomorphies are also sure to be unveiled as more bulungamayine material is described.

Bulungamayine phylogeny and the basal relationships of macropodoids are currently under study by Bernard Cooke. While the dentaries and teeth of several taxa have been well documented, no bulungamayine cranium has yet been formally described. For this, my comparisons rely on the photograph of an undescribed cranium in Archer et al. (1991:122). A thorough description of the first known propleopine cranium, that of *Ekaltadeta ima* Archer and Flannery, 1985, was provided by Wroe et al. (1998). Subsequently, Cooke (2000) described the first cranium of a balbarine (*Balbaroo fangaroo*). My principal objective in the following section is to detail the derived craniodental characteristics of the Sthenurinae. This requires only the identification of character states present in sthenurines, but absent in the other macropodoids. Although this has the potential to underestimate the number of synapomorphies, because it does not differentiate homoplasies from symplesiomorphies, overestimation can only result from misinterpretation or incongruence with hitherto undescribed material.

1. Splanchnocranium not markedly deflected anteroventrally

The splanchnocranial axis of most adult macropodoids is deflected ventrally relative to the axis of the neurocranium or basicranium (fig. 14), meaning that the degree of klinorhynchy expressed is less. This is so in sthenurines largely because the rostrum and frontal region are much deeper. In most taxa, this corresponds to a dorsal cranial surface that is horizontal or very nearly horizontal, at least to midway along the frontals. The neurocrania of two specimens attributed to *Protemnodon brehus* from Victoria Fossil Cave (SAM P20819) and the Wellington Caves (AM F16490) suggest that this species has convergently attained a similar condition. Basicranial characteristics easily distinguish it from the sthenurines (see Stirton

1963:figs 3-4), but none of the cranium anterior to the posterior frontal region is preserved in either specimen to permit an assessment of the splanchnocranial profile. The dorsal neurocranial surface of all other non-sthenurine macropodoids is distinctly domed (fig. 14), as is the neurocranium of juvenile sthenurines (e.g., pls 8C, 87A). While the balbarine neurocranium is domed, the degree of klinorhynchy expressed is less than it is in many taxa and this has been interpreted as plesiomorphic within the Macropodoidea (Cooke 2000).

In *Hadronomas* and some simosthenurins, the temporal (parietal) crests extend slightly higher than the main surface of the neurocranium and can superficially convey a domed appearance. Although *Archaeosimos* is the only sthenurine genus for which the cranium is unknown, the proportions of its dentary suggest that the cranium cannot have been markedly different in form from that of other relatively plesiomorphic sthenurines. Interestingly, the slightly domed condition of the *"Procoptodon" gilli* cranium is not facilitated by temporal crest development, but rather an apparent partial reversal to the plesiomorphic macropodoid condition (fig. 14). This may well have arisen through pedomorphosis. The form of the adult cranium closely resembles that of juvenile crania of the species and the possibly ancestral *"P." browneorum* (figs 5-6). The occiput of *P. goliah* is oriented at near to 90° to the neurocranial surface (ignoring the hypertrophied temporal crests), but the cranial surface is markedly declined from the frontals anteriorly. While one could argue that this constitutes a partial reversal to the outgroup state, the condition has clearly been brought about by extreme deepening of the maxilla and the positioning of the neurocranium almost entirely above it.

2. Frontals inflated laterally

Ekaltadeta and some macropodines (e.g., *Macropus*) exhibit some dorsal inflation of the frontals, but marked lateral expansion is a sthenurine synapomorphy (fig. 15). While *Sthenurus* and *"Procoptodon" gilli* secondarily approach the plesiomorphic condition, the frontals of all other sthenurines for which the cranium is known extend laterally to some degree (fig. 15). A distinct supraorbital tuberosity is developed in a few sthenurines, and although superficially similar, the tuberosity expressed in *Hadronomas* is not homologous with that of *Simosthenurus occidentalis* and *Si. maddocki* (fig. 15).

3. Posterior end of jugal deep and expanded mesially to form marked ectoglenoid process

Murray (1991a) highlighted this feature (e.g., pls 42A 71B, 90C) as a likely synapomorphy uniting *Hadronomas* with the sthenurines and I concur with this assessment. *Hypsiprymnodon*, potoroines and macropodines have a much shallower and often narrower posterior end to the jugal. While the bulungamayine figured in Archer et al. (1991) lacks the jugal, the slender zygomatic process of the squamosal and the morphology of its ventral surface suggest that the jugal was not deep or markedly expanded posteromesially. Interestingly, the ectoglenoid process

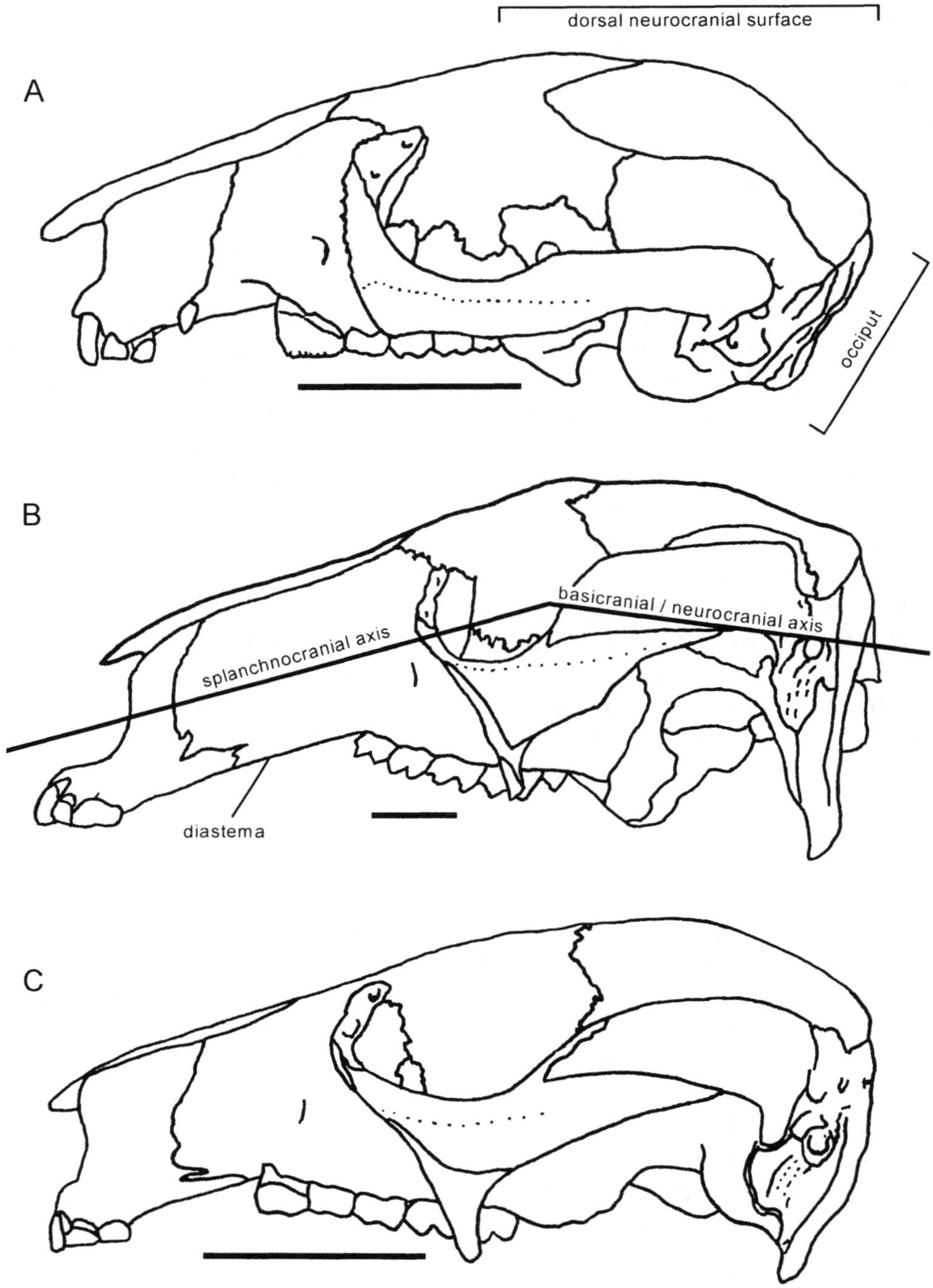

Figure 14. Crania in lateral view. **A**, *Bettongia penicillata* (FU unreg.); **B**, *Macropus giganteus* (redrawn from Wells and Tedford, 1995:fig. 9); **C**, *Lagostrophus fasciatus* (AM M1425). Scale bars = 20 mm.

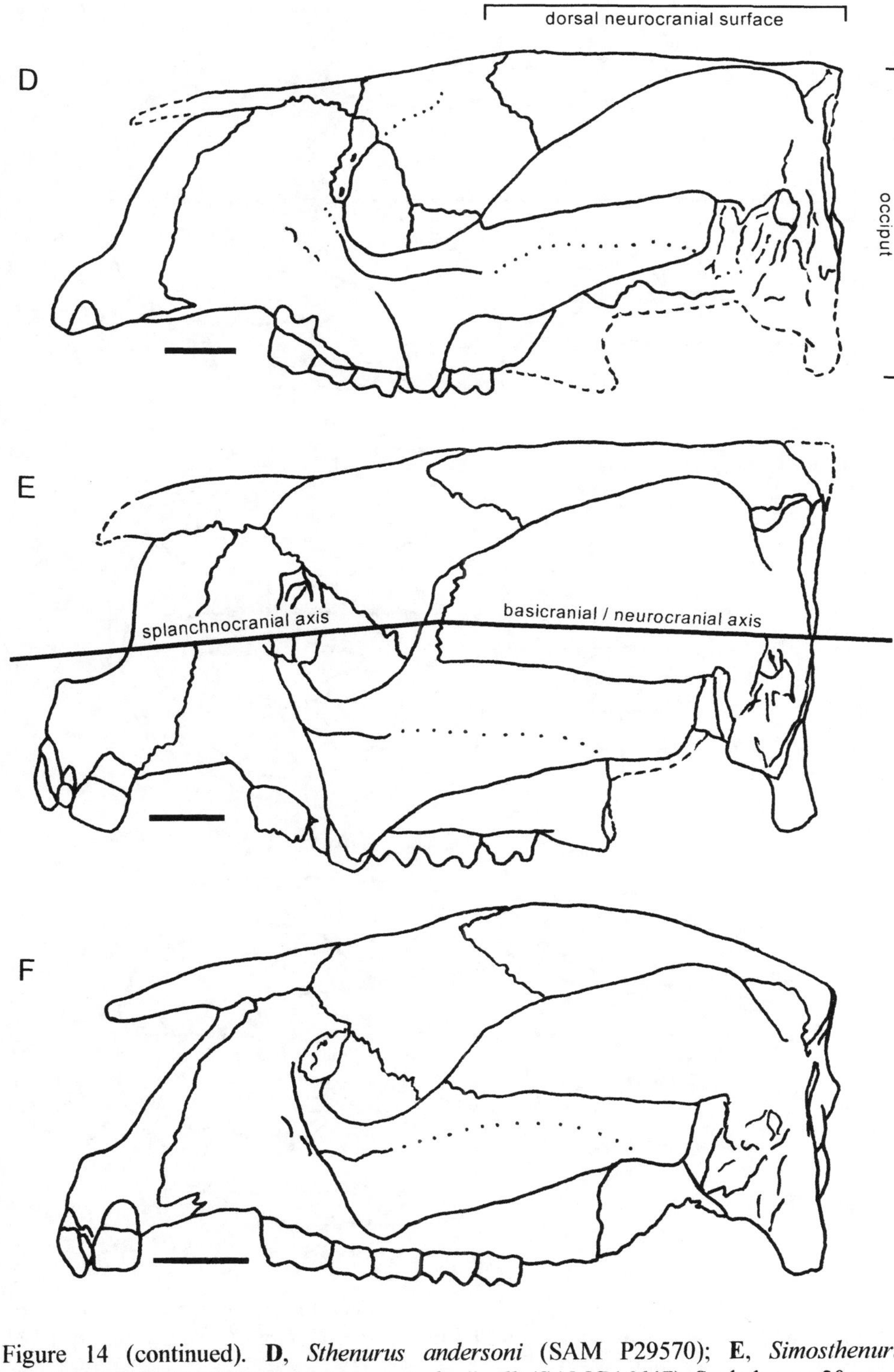

Figure 14 (continued). **D**, *Sthenurus andersoni* (SAM P29570); **E**, *Simosthenurus occidentalis* (SAM P16648); **F**, *"Procoptodon" gilli* (SAM P16647). Scale bars = 20 mm.

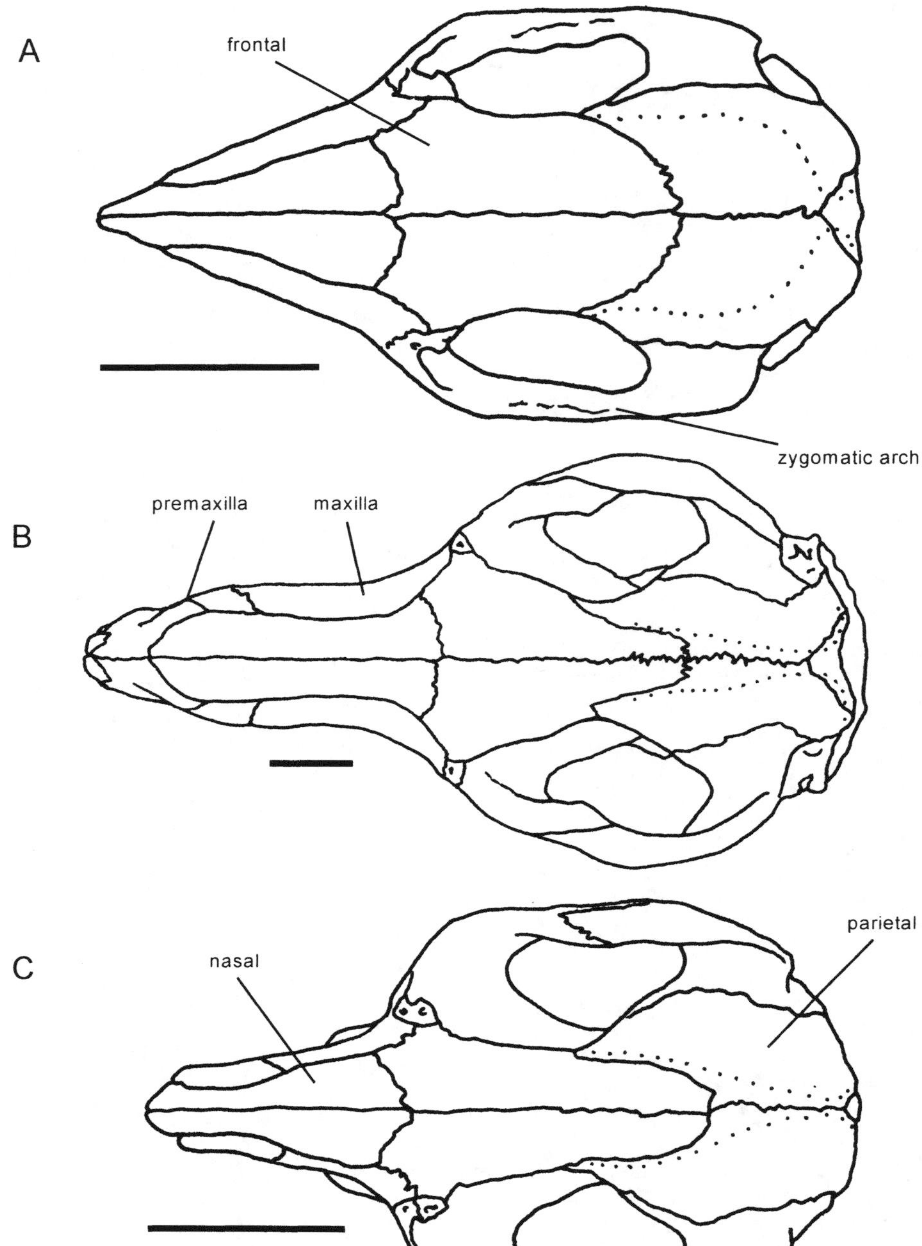

Figure 15. Crania in dorsal view. **A**, *Bettongia penicillata* (FU unreg.); **B**, *Macropus giganteus* (redrawn from Wells and Tedford 1995:fig. 9); **C**, *Lagostrophus fasciatus* (AM M1425). Scale bars = 20 mm.

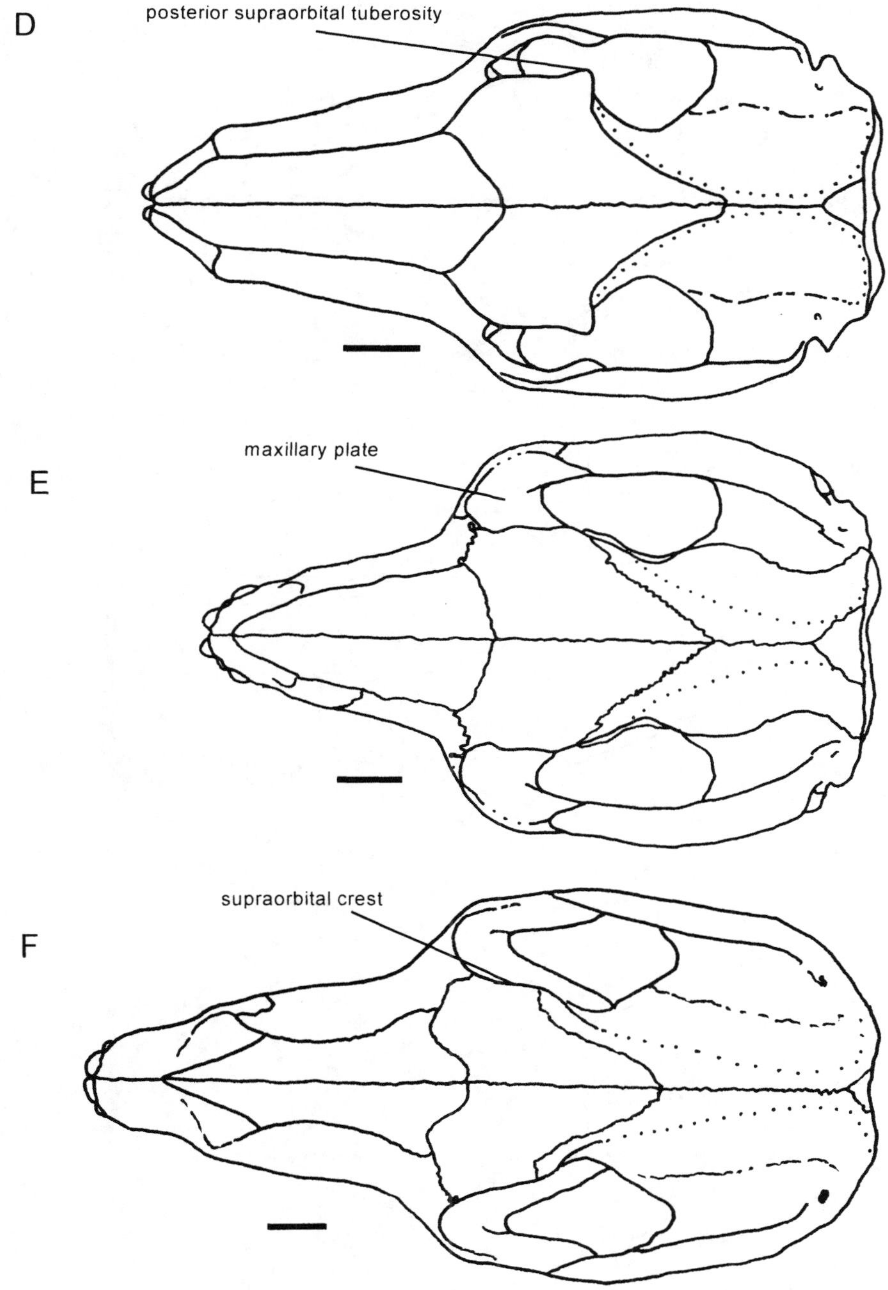

Figure 15 (continued). **D**, *Hadronomas puckridgi* (NTM P98140 + NTM P98133, reconstructed composite); **E**, *Metasthenurus newtonae* (SAM P20255, left side mirrored from right side); **F**, *Sthenurus tindalei* (redrawn from Wells and Tedford 1995:fig. 13). Scale bars = 20 mm.

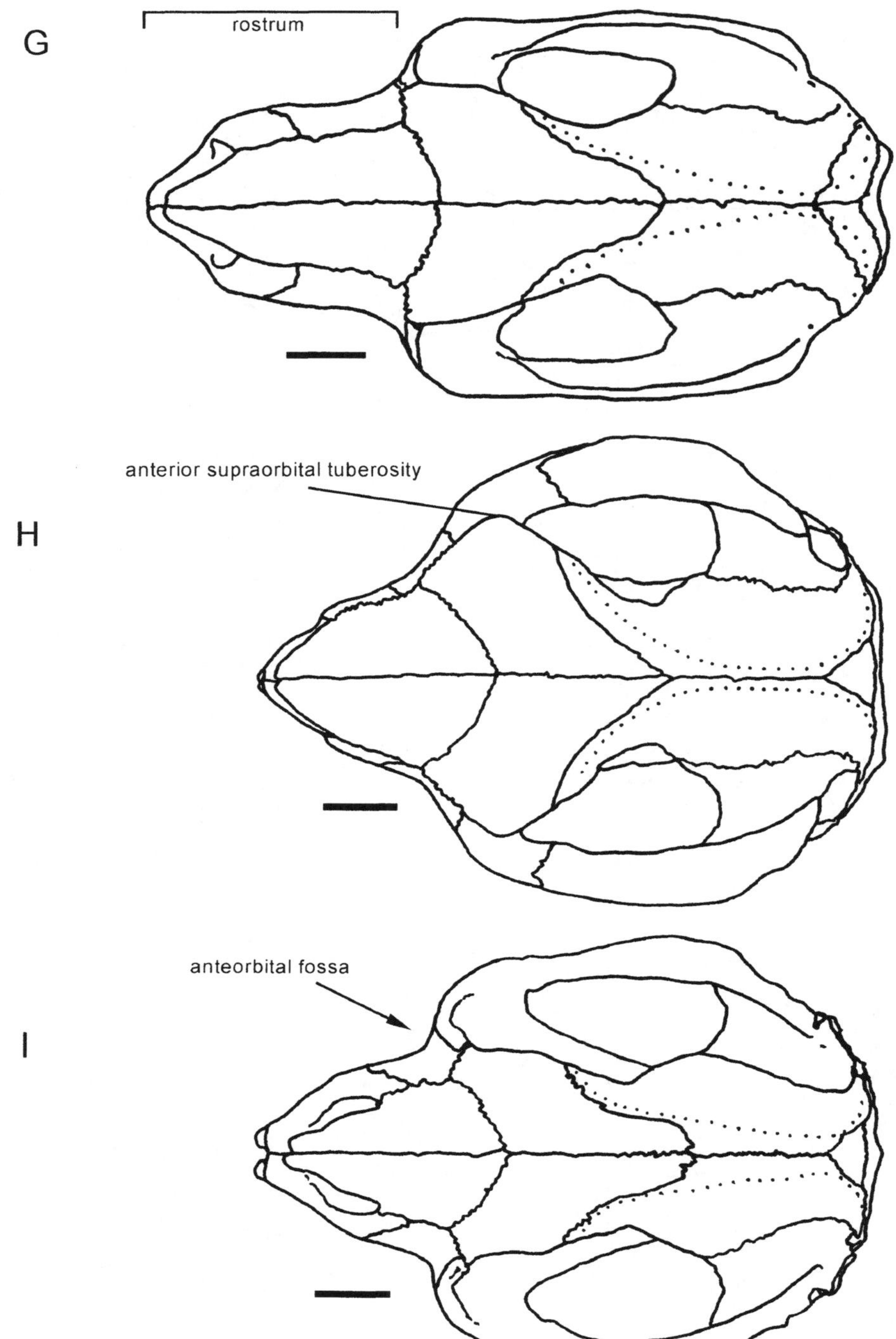

Figure 15 (continued). **G**, *Sthenurus andersoni* (SAM P29570); **H**, *Simosthenurus occidentalis* (SAM P16648); **I**, *"Procoptodon" gilli* (SAM P16647). Scale bars = 20 mm.

Ekaltadeta ima is developed similarly to that of the sthenurines. However, this would be convergent if propleopines and balbarines together form the sister taxon of the remaining macropodoids (see Wroe et al. 1998; Cooke and Kear 1999). The only balbarine cranium described (*Balbaroo fangaroo*) lacks the portion of the zygomatic arch anterior to the glenoid fossa.

4. Occiput close to vertical, broad and deep

The occiput of adult sthenurines is oriented near vertical or perpendicular to the dorsal surface of the cranium (fig. 14). This orientation and the large size of the region relative to the general proportions of the neurocranium is unique among macropodoids, with the partial exception of *Protemnodon brehus*, which has convergently evolved an occiput of similar orientation. Development of the nuchal / lambdoid crest in *Ekaltadeta* (see Wroe et al. 1998) exceeds that of the sthenurines, but its occiput is comparatively much smaller in area.

5. Ectotympanic thick, wide, rugose and ventrally keeled

Ectotympanic width in non-sthenurine macropodoids does not usually exceed its length. The bone is a smoothly surfaced, rather narrow bore ring, which lacks a marked transverse keel. By comparison, the sthenurine ectotympanic is robustly constructed and very wide, and has a rugose or ridged surface dominated by a transverse keel (Tedford 1966; Murray 1991a). The mesial end is usually curved ventrally. While the relative rugosity of the surface is slightly reduced in *Procoptodon goliah*, the overall form of the ectotympanic is a strong sthenurine synapomorphy.

6. Digastric sulcus and eminence well developed

Sthenurines are characterized by a better-developed digastric sulcus and a larger, more posteriorly positioned digastric eminence than any other macropodoid group (Murray 1991a). Although the digastric sulcus or eminence may be hypertrophied or reduced from species to species, one or the other is always better developed in the sthenurines than in the remaining macropodoids. The more centrally positioned ventral convexity of the dentaries of *Troposodon kenti*, the bulungamayine *Ganguroo bilamina* Cooke, 1997b, and some potoroines is likely to have been independently derived in each taxon.

7. I2 tiny and sub-cylindrical, I3 markedly longer than the combined length of I1-2

Extreme reduction of the I2 to a very small, sub-cylindrical tooth is exclusive to the sthenurines (Tedford 1966; Murray 1991a). Relative sizes of the upper incisors vary greatly across the Macropodoidea, but I2 is never reduced to the degree seen in sthenurines (figs 16-17). The elongate, blade-like form, absence of a well-developed anterobuccal crest, and presence of an anterolingual crest separate the sthenurine I3 from that of all other macropodoids (Tedford 1966; Murray 1991a), with the exception of the bulungamayine *Wanburoo hilarus* Cooke, 1999.

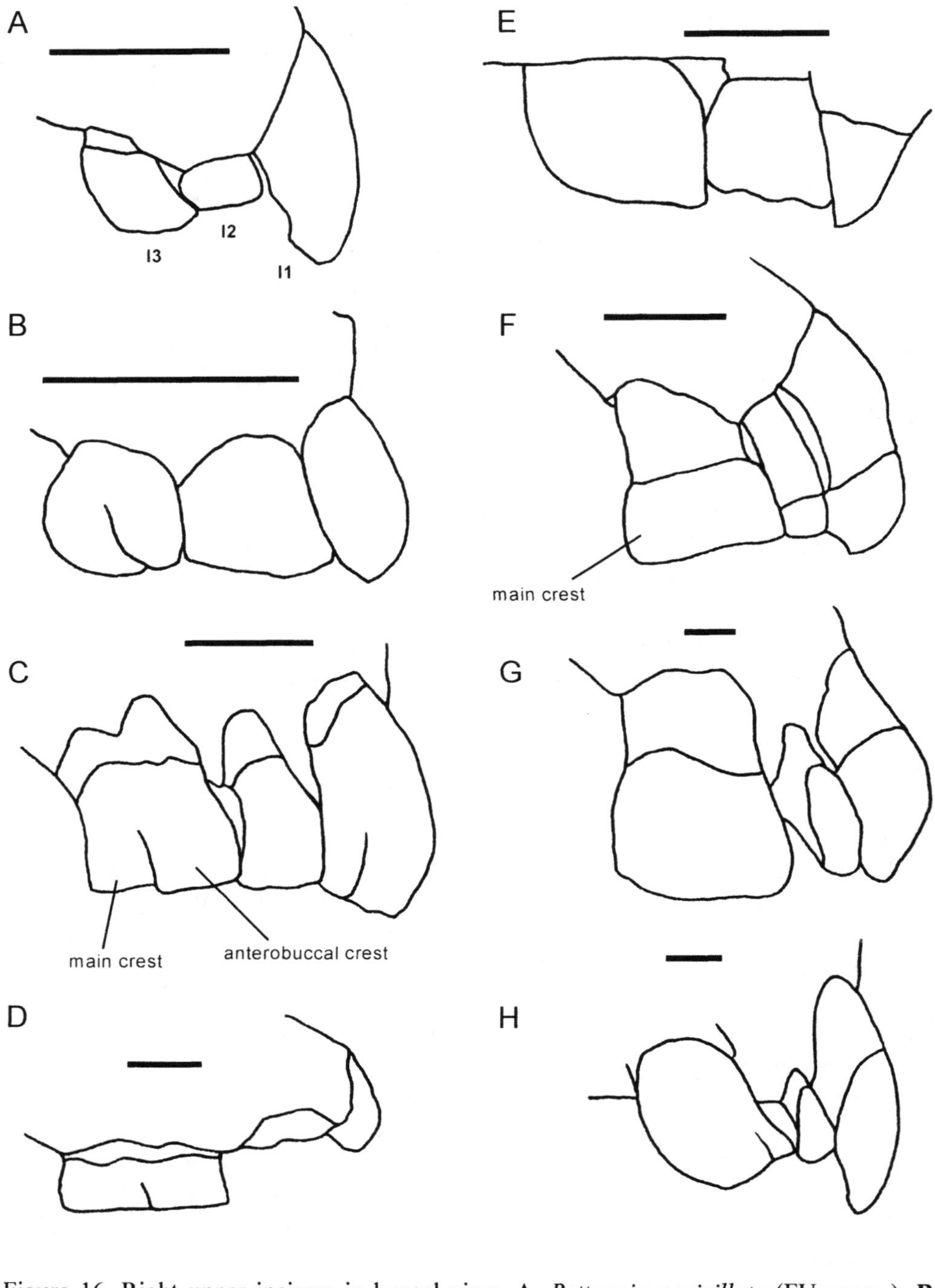

Figure 16. Right upper incisors in buccal view. **A**, *Bettongia penicillata* (FU unreg.); **B**, *Lagostrophus fasciatus* (AM M1426); **C**, *Macropus rufogriseus* (SAM P16944); **D**, ?*Troposodon bowensis* (AM F63021); **E**, *Wanburoo hilarus* (QM F20525, drawn from Cooke 1999:fig. 2); **F**, *Hadronomas puckridgi* (NTM P98140); **G**, *Sthenurus stirlingi* (SAM P22533); **H**, *Procoptodon goliah* (SAM P16695 + FU 0147). Scale bars = 5 mm.

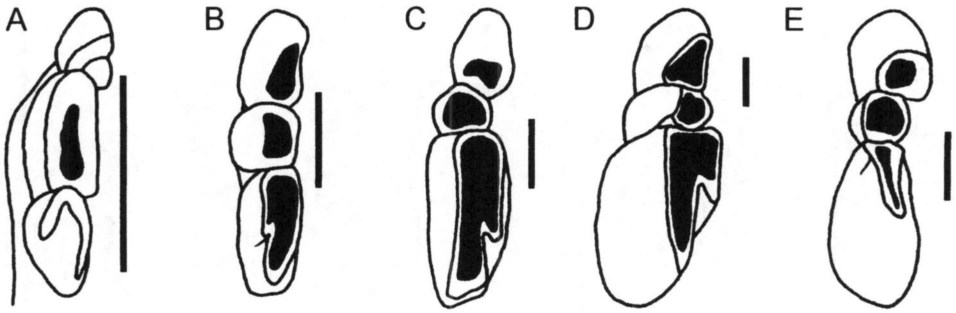

Figure 17. Right upper incisors in occlusal view. **A**, *Lagostrophus fasciatus* (AM M1426); **B**, *Macropus rufogriseus* (SAM P16944); **C**, *Hadronomas puckridgi* (NTM P98140); **D**, *Sthenurus stirlingi* (SAM P22533); **E**, *Procoptodon goliah* (composite of SAM P16695 + FU 0147). Scale bars = 5 mm.

By comparison, the I2 of *Wanburoo* is not as reduced, and it has a distinct groove on its buccal surface (Cooke 1999:fig. 2) that is absent in sthenurines (fig. 16). Interestingly, several sthenurines do retain a low, vertical eminence that ascends the anterobuccal aspect of I3 and is most likely the homologue of the anterobuccal crest in other macropodoids. Among sthenurines, the structure is most evident in *Procoptodon goliah* (pl. 75G,I; figs 16-17). While an elongate I3 has arisen independently in *Macropus (Macropus)*, the anterobuccal crest is not only distinct in these species, but is accompanied by a second, neomorphic accessory crest near identical in form to the anterobuccal crest itself (fig. 16).

A partial premaxilla (AM F63921) from the early Pliocene Bow LF "tentatively referred to *Troposodon bowensis*" (Flannery and Archer 1984:265) also possesses an I3 that exceeds the combined length of I1-2 (fig. 16D). I2 is not preserved, but the root indicates a crown length around two-thirds that of I3 and twice that of I1, which is represented by the broken crown base. AM F63921 clearly does not belong to a sthenurine, because of the relative size of I2, and the presence of a well-developed anterobuccal crest on the buccal side of I3, which is elaborated in a manner similar to that of *Macropus (Macropus)* species (see Flannery and Archer 1984:265, fig. 3A-B).

8. p3 bears marked buccal cingulid or crest

Presence of a distinct cingulid or crest on the buccal side of p3 is unique to the sthenurines. Running the length of the tooth, the cingulid is low in *Hadronomas*, but is hypertrophied into a short posterobuccal or elongate buccal crest in the other five genera (fig. 18). While the development of vertical ridgelets on the buccal and lingual surface of the main crest may be strong in some taxa (e.g., propleopines), all non-sthenurine macropodoids lack any sign of a buccal cingulid (fig. 18).

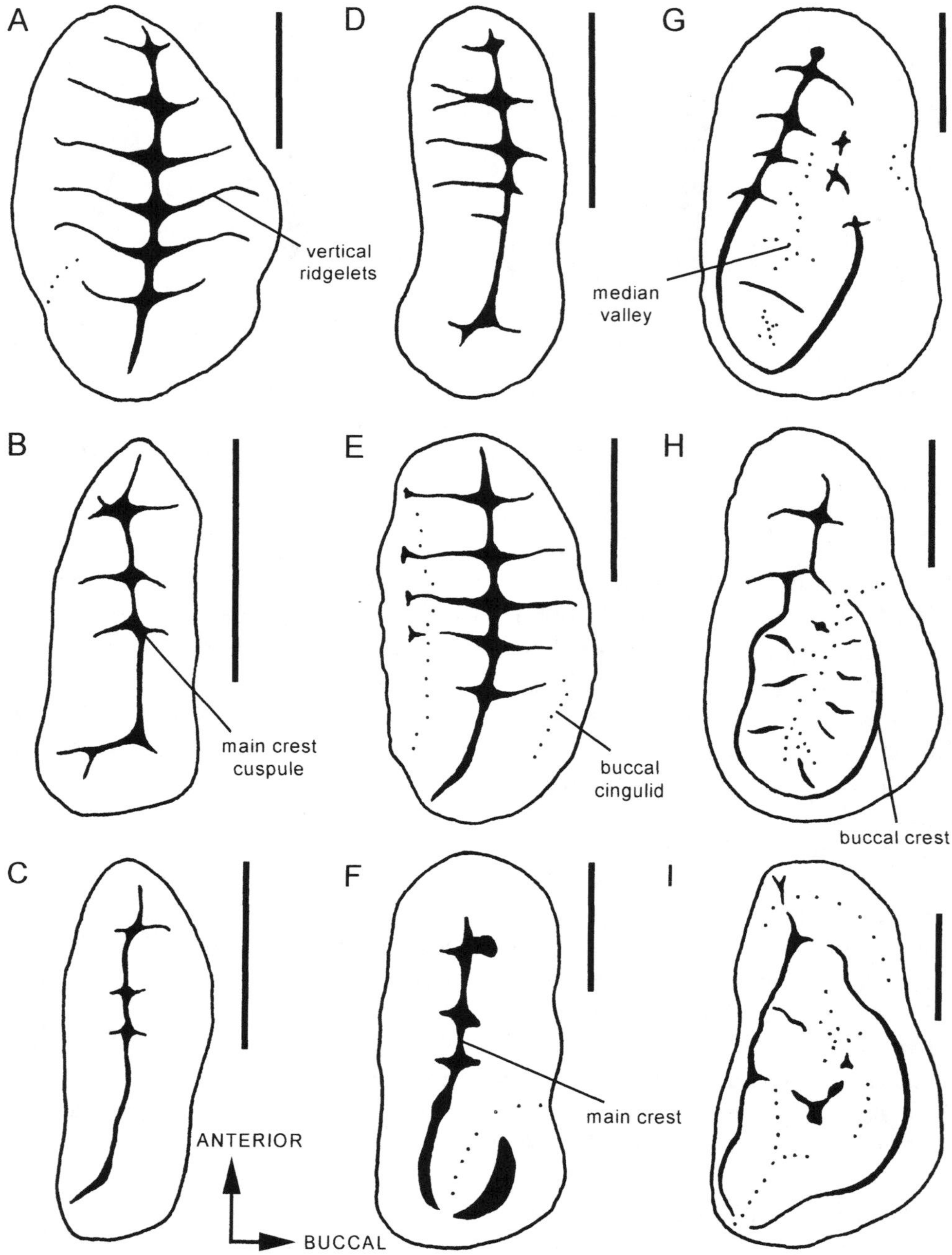

Figure 18. Right p3 in occlusal view. **A**, *Propleopus oscillans* (SAM P20815); **B**, *Wallabia bicolor* (SAM P32457); **C**, *Dorcopsoides fossilis* (NTM P unreg.); **D**, *Wanburoo hilarus* (QM F20525, drawn from Cooke 1999:fig. 5); **E**, *Hadronomas puckridgi* (NTM P98146); **F**, *Archaeosimos cegsai* (SAM P31800); **G**, *"Simosthenurus" baileyi* (SAM P13670); **H**, *"Procoptodon" gilli* (FU 1653); **I**, *Sthenurus stirlingi* (SAM P22533). Scale bars = 5 mm.

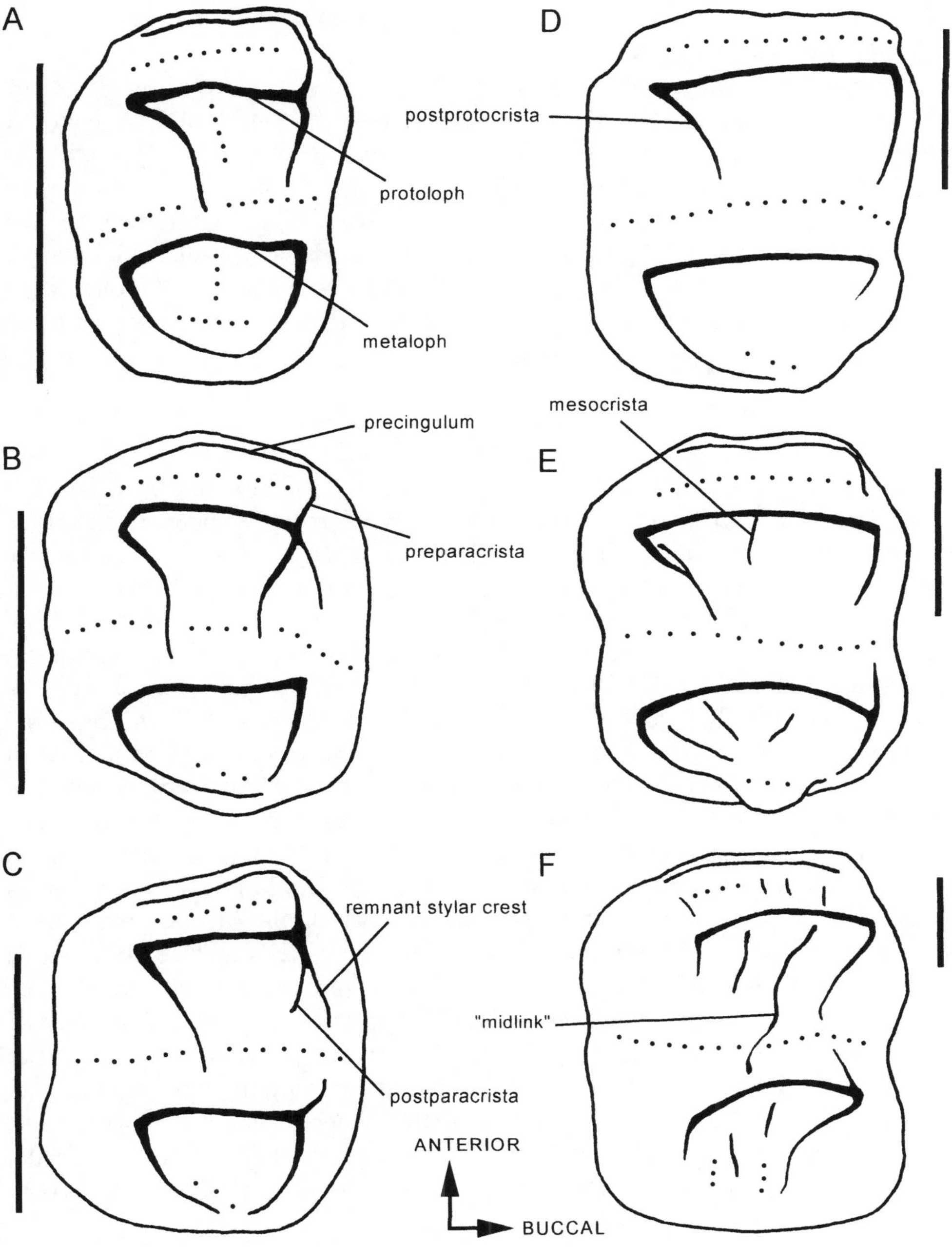

Figure 19. Left upper molars in occlusal view. **A**, *Lagorchestes leporides* (SAM P25552, M3); **B**, *Dorcopsoides fossilis* (NTM P98150, M2); **C**, *Wanburoo hilarus* (QM F20525, M2, drawn from Cooke 1999:fig. 3); **D**, *Hadronomas puckridgi* (NTM P98140, M3); **E**, *Archaeosimos cegsai* (SAM P29917, M3); **F**, *Procoptodon goliah* (SAM P16695, M3). Scale bars = 5 mm.

9. Molars rather square in occlusal view, lophs close to parallel and oriented near perpendicular to anteroposterior axis

Although this synapomorphy is probably the most subjective of all of the character states that unite the sthenurines, it has the widest utility given the taphonomic biases favoring preservation of individual teeth or the central portions of maxillae and dentaries. While plesiomorphic sthenurines (e.g., *Hadronomas*) and plesiomorphic macropodines (e.g., *Dorcopsoides*) resemble derived bulungamayines (e.g., *Wanburoo*) in the relative simplicity of their molars, sthenurines are distinguishable as a whole by their combination of molar traits. Sthenurine molars are squarer in occlusal view than those of other macropodoids, which tend to be more elongate or rounded (fig. 19).

Omitted Features

With nine craniodental synapomorphies for the Sthenurinae now established, it is necessary to discuss why several features previously considered sthenurine synapomorphies have not been included here. To a large degree it is due to the improved understanding of the pre-Pliocene record for the Macropodoidea.

Large palatal vacuities were viewed as diagnostic of the sthenurines by Tedford (1966) and Wells and Tedford (1995), and it is true that all sthenurines possess them. However, whether they represent a sthenurine synapomorphy is arguable, because the palate is fenestrated in members of all macropodoid subfamilies. While the palatal vacuities in most of these species are smaller than in the sthenurines, an almost identical condition is observed in *Hypsiprymnodon, Caloprymnus*, *Lagostrophus*, *Onychogalea unguifera* and, to a slightly lesser degree, the bulungamayine in Archer et al. (1991). Until the distribution of states among Miocene macropodoids becomes better known, large palatal vacuities delimited posteriorly by a narrow postpalatine bar might just as likely be a symplesiomorphy for sthenurines. Tedford (1966) and Wells and Tedford (1995) also listed a V-shaped incisor arcade as a synapomorphy for the Sthenurinae. Likewise, Murray (1991a) noted the anteriorly pointed nature of the *Hadronomas* incisor arcade and premaxilla, and considered them reminiscent of other sthenurines. However, he avoided citing this as a synapomorphy, viewing it instead as a direct correlate of the unique form of the upper incisors. Perhaps just as important, numerous other macropodoids have a V-shaped incisor arcade, including *Hypsiprymnodon*, potoroines, *Wanburoo*, *Lagostrophus* and some species of *Dendrolagus.*

Presence of a posterior mental foramen has often been regarded as an exclusively sthenurine character (e.g., Tedford 1966; Murray 1991a; Wells and Tedford 1995) due to its absence in *Hypsiprymnodon*, most potoroines and the macropodines. However, a posterior mental foramen may be present in balbarines, bulungamayines, *Dorcopsulus* and *Dorcopsis*. A large, unnamed macropodine from the Kanunka LF also has a posterior mental foramen (pers. obs.). This suggests that it may be either a sthenurine symplesiomorphy or a feature common to some browsing mammals (e.g., koala, camelids). Tedford (1966) also considered the relatively late eruption of the permanent premolar (synchronous with or following the eruption of M4) a sthenurine synapomorphy, but species of *Dendrolagus*, *Dorcopsis*, *Setonix*, *Thylogale*, *Petrogale* and *Lagorchestes* also display this state (Tate 1948; pers. obs.). While the eruption sequences of Miocene macropodoids are as yet undocumented, wide distribution of the typical sthenurine eruption state within the Macropodinae implies that the character needs to be used with caution.

An elevated medial pterygoid fossa is an outstanding feature of the sthenurine dentary and has been included as a synapomorphy by Murray (1991a) and Prideaux (1999b-c). Most macropodoids have an ascending ramus that is distinctly reclined and a medial pterygoid fossa set only slightly higher than the level of the ramus, with the angular process typically no higher than the level of the cheek tooth row (fig. 20). However, the posterior region of the dentary of *Troposodon kenti*, *T. minor*, *Protemnodon tumbuna*, *P. otibandus* Plane, 1967, and some *Dendrolagus* species is very similar to that of the sthenurines. This morphology may well have arisen independently in these taxa, perhaps due to dietary similarities, but the possibly exists that it is symplesiomorphic. I have opted here to omit this feature from consideration until the states of a wider array of bulungamayines become known. Even if it is shown to be derived for the Sthenurinae, consideration needs to be given to how tightly correlated a vertical ascending ramus and elevated medial pterygoid fossa are to the restructuring of the posterior portion of the cranium (see Functional Morphology), something already taken into account by the occipital synapomorphy.

Finally, the 18 purported sthenurine synapomorphies mentioned by Prideaux (1996) included not only the preceding features, but also four others subsequently identified as symplesiomorphies.

PHYLOGENETIC DERIVATION

Despite more than half a century of speculation, sthenurine origins have remained unclear, largely because the pre-Pliocene fossil record was so poorly known until recently. Not surprisingly, pioneering work by Flannery, Cooke and others is now showing that the evolutionary history of macropodoids, prior to the radiation of sthenurines and macropodines, was more complex than had been imagined. Here, I briefly review previous ideas on the origin of sthenurines and consider the implications of the latest evidence from the Miocene.

Similarities in premolar morphology and brachycephaly led Raven and Gregory (1946) to suggest that sthenurines may have been derived from a *Dendrolagus*-like ancestor. Their classic tree shows *Sthenurus* departing from the main stem after the potoroines, but before the macropodines. Tate (1948) positioned sthenurines similarly, as a floating sister lineage to the macropodines.

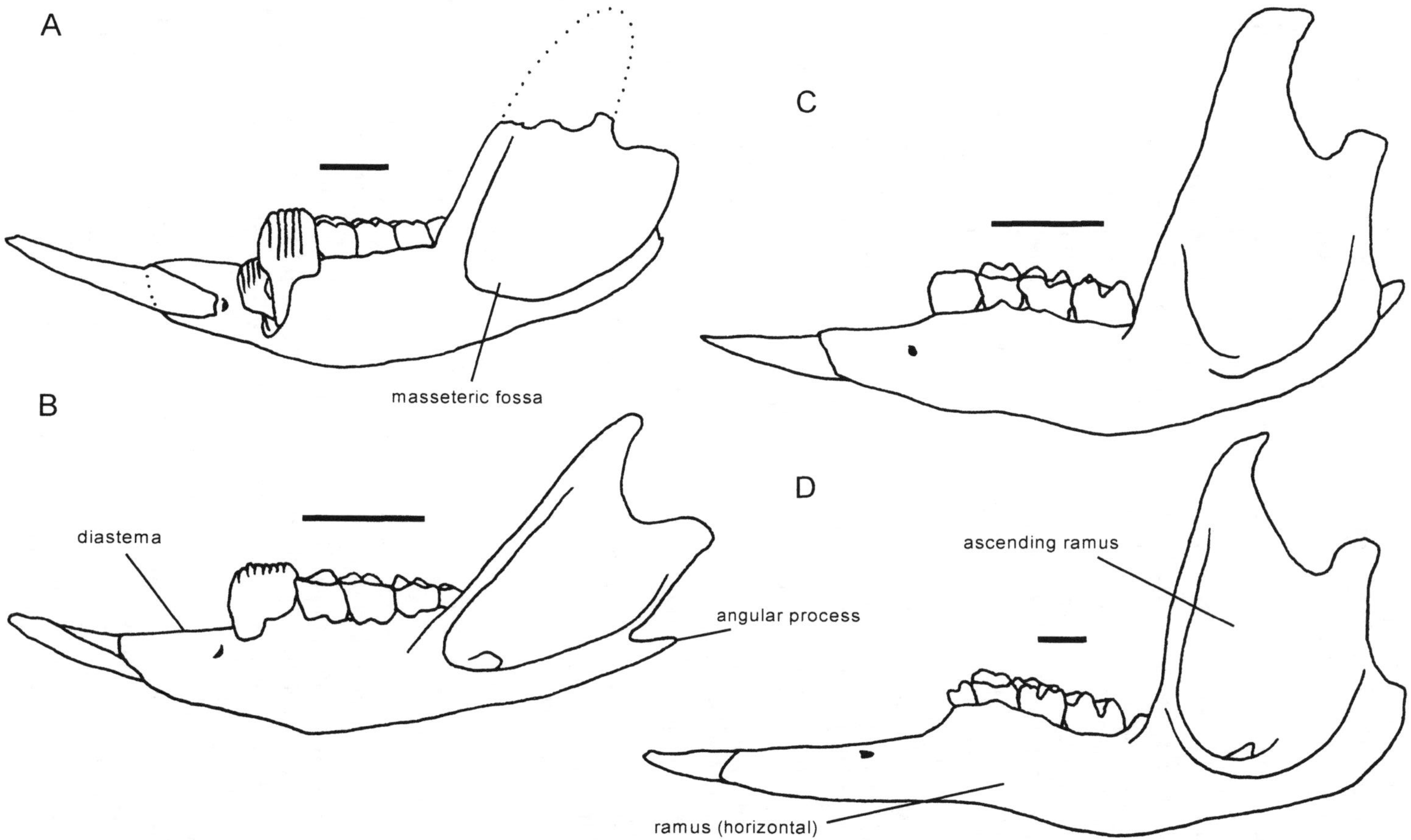

Figure 20. Left dentaries in lateral view. **A**, *Ekaltadeta ima* (QM F12435, drawn from Wroe 1996:fig. 1, pl. 2); **B**, *Bettongia penicillata* (FU unreg.); **C**, *Lagostrophus fasciatus* (AM M1425); **D**, *Macropus giganteus* (drawn from Wells and Tedford 1995:fig. 9). Scale bars = 10 mm.

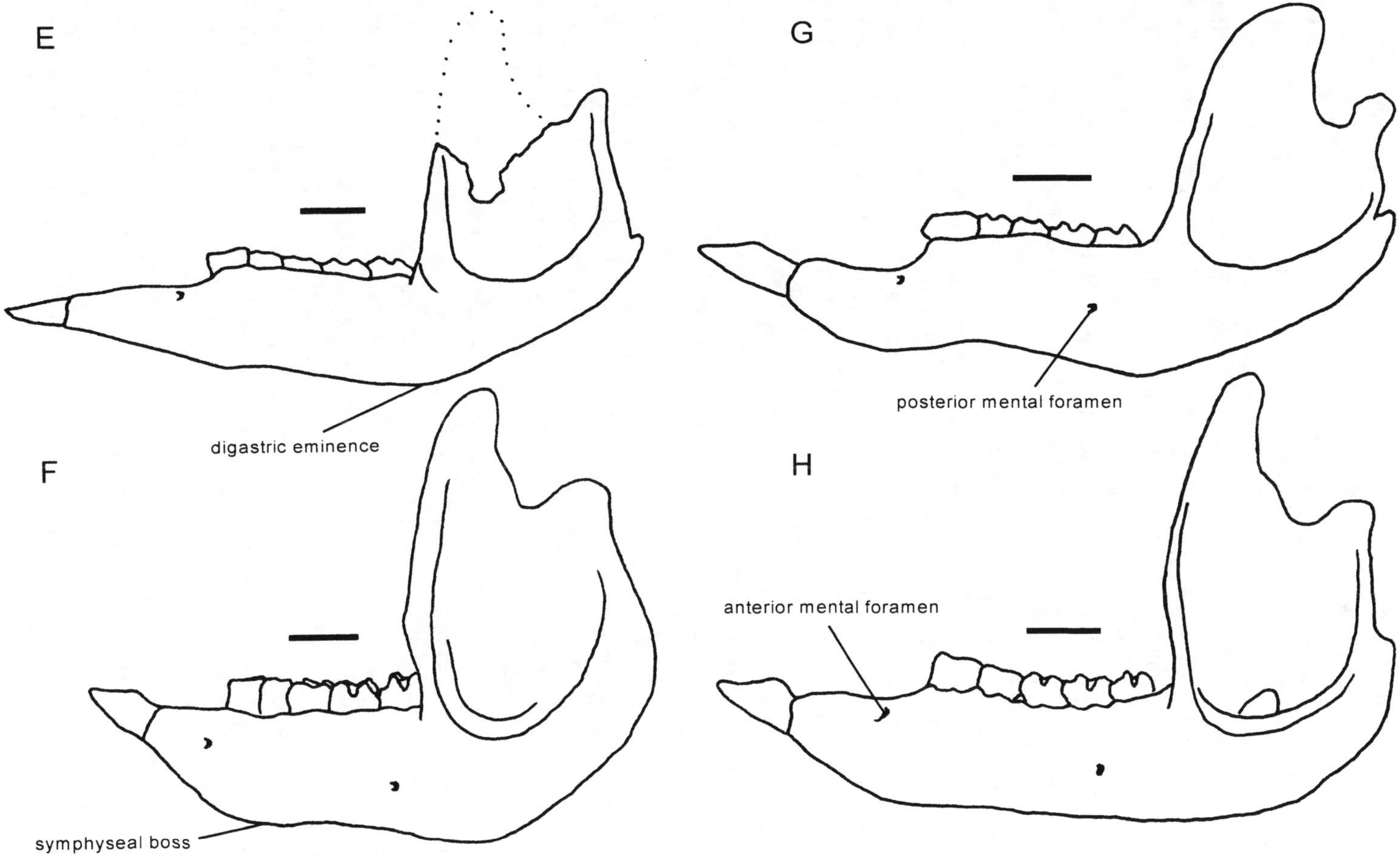

Figure 20 (continued). **E**, *Troposodon kenti* (SAM P14507); **F**, *Hadronomas puckridgi* (NTM P98142; note symphyseal region distorted post-depositionally); **G**, *Simosthenurus occidentalis* (SAM P20803); **H**, *Sthenurus andersoni* (SAM P20496). Scale bars = 20 mm.

While Stirton and Marcus (1966) and Tedford (1966) suggest a considerable antiquity for the sthenurine lineage, they balked at an estimate of the time of divergence due to the paucity of pre-Quaternary fossils known at the time. Bartholomai (1970) thought that this rarity existed because the group did not originate until the late Pliocene, but he later acknowledged Woodburne's (1967) now upheld claim that *Hadronomas* might reside near the base of the sthenurine radiation (Bartholomai 1972, 1978a). At the time, though, Bartholomai thought that *Hadronomas* might just as feasibly be a precursor to the macropodine *Protemnodon*, as did Campbell (1973). Ride (1971) also believed that a close relationship between *Hadronomas* and sthenurines was likely due to similarities in I3 morphology, with the *Protemnodon* I3 typically macropodine-like in comparison. He took this to imply a divergence between sthenurines and macropodines some time prior to the late Miocene.

Woodburne et al. (1994) reported the presence of several different macropodoids in the Etadunna and Wipajiri Fms of northern South Australia (late Oligocene to ?early Miocene), but only *Purtia mosaicus* Case, 1984, and *Wakiewakie lawsoni* Woodburne, 1984b, have been described. Flannery, Archer and Plane (1983) raised two new subfamilies, Bulungamayinae and Balbarinae, to include new forms from the early to middle Miocene of Riversleigh in northwestern Queensland and the middle Miocene of Bullock Creek in the Northern Territory. Most late Oligocene to middle Miocene macropodoids described subsequently have been placed into either of these groups. Potoroines, propleopines and hypsiprymnodontines were evidently much rarer elements in these faunas (Cooke 1997a). While no sthenurines or macropodines have been positively identified in any deposits older than late Miocene, Flannery, Archer and Plane (1983) and Flannery (1989) considered balbarines their most likely ancestors. Bulungamayines, potoroines, propleopines and hypsiprymnodontines were grouped together in the Potoroidae.

Balbarines and bulungamayines are the only bilophodont macropodoids known prior to the late Miocene, but only one specimen of a possible derived bulungamayine is known from the late Miocene (see The Sthenurine Radiation). The gap in the fossil record between the middle and late Miocene remains the primary factor clouding the origin of sthenurines and macropodines, but recent insights have shown that balbarines have a decidedly more plesiomorphic craniodental morphology than do bulungamayines, probably became lophodont much earlier, and retained a quadrupedal gait and posture (Cooke 1997b-c, 2000; Cooke and Kear 1999). Proximity to the base of the Macropodoidea is suggested by retention of a markedly compressed trigonid and distinct protostylid on m1, a buccally averted, rather plagiaulacoid premolar, and a frontal-squamosal contact. Some balbarines also retain a hypoconulid and stylar cusps C and D. Wroe et al. (1998), Cooke and Kear (1999), and Kear and Cooke (2001) have intimated that balbarines and propleopines may constitute a monophyletic group that arose near the base of the macropodoid radiation and that this may eventually warrant their distinction at the family level.

Retention of these phalangeroid-like features in balbarines is distinctive, but it is their unique pattern of hypolophid evolution that negates a phyletic affinity with bulungamayines (Cooke 1997b-c). Another balbarine synapomorphy may also be the rounded, heart-shaped I3 morphology that is characteristic of *Balbaroo fangaroo* (see Cooke 2000) and several undescribed balbarines from the Etadunna and Wipajiri Fms (pers. obs.). Sthenurines, macropodines, bulungamayines and potoroines lack these two balbarine features. Plesiomorphic members of these four subfamilies share an elongate, blade-like premolar. Cooke (1999) noted that *Hadronomas* and the macropodines were more similar in premolar morphology to bulungamayines than to balbarines, but this is also true of potoroines. An elongate premolar must therefore be considered a potential synapomorphy uniting potoroines, bulungamayines, sthenurines and macropodines as a monophyletic lineage within the Macropodoidea, i.e., family Macropodidae.

If this holds true, then a clade consisting of bulungamayines, sthenurines and macropodines is also likely. Bulungamayines portray a clear structural progression from potoroine-like bunolophodonty to bilophodonty (Cooke 1997b), and undescribed crania of *Bulungamaya*, *Ganguroo* and ?*Wabularoo* share an alisphenoid-parietal contact with sthenurines and macropodines (Cooke 2000). The phylogenetic significance of the alisphenoid-parietal and frontal-squamosal conditions has long been deliberated (Pearson 1950; Case 1984; Flannery and Archer 1987a-b; Burk et al. 1998; Cooke 1999). Now it would seem that if the frontal-squamosal contact seen in propleopines, balbarines and potoroines, except *Bettongia moyesi* and variably *B. lesueur* (Flannery and Archer 1987), is derived relative to the hypsiprymnodontine and phalangeroid states, then the expression of the opposing state in bulungamayines, sthenurines and macropodines is likely to represent a synapomorphic reversion to the plesiomorphic phalangeriform condition. Bulungamayinae clearly looks like the only candidate for the ancestral group whence Sthenurinae and Macropodinae were derived, as Cooke (1999) suggested.

Indeed, the relatively large Riversleigh bulungamayine for which Cooke (1999) raised the name *Wanburoo hilarus* appears to represent a stage of evolution close to the origin of the sthenurine lineage. Flannery (1989:32) referred to this taxon as the Gag Site macropodine, because its attributes implied a more derived phylogenetic position than those of any of its bilophodont contemporaries. Although Cooke (1999) noted some resemblance to the plesiomorphic macropodine *Dorcopsoides*, he thought that an antecedence to *Hadronomas* might be more likely. Compared with more plesiomorphic bulungamayines, traits suggestive of their phylogenetic proximity include minimal convexity of the ventral border of the ramus, a deep i1 crown, a ventrally inflated hypolophid posterior face, a relatively longer trigonid, robust, coarsely ridged premolars, and the fact that *Wanburoo* is larger than any other bulungamayine (Cooke 1999).

Cooke's observations are strongly supported by the morphology of the upper incisor arcade (figs 16-17). I3 is easily the longest crowned upper incisor in *Wanburoo*, which is a feature typical of sthenurines (e.g., Ride 1959; Tedford 1966). However, it is slightly shorter than the combined length of I1-2, unlike all sthenurines, except for the highly divergent *Procoptodon goliah.* Moreover, the *Wanburoo* I2 is about twice the length of I1, while the I2 of sthenurines is tiny. Significantly, though, the characteristic morphology of sthenurine I3 is shared by *Wanburoo.* The crown of most macropodoids has a vertical groove on the buccal surface, which may be distinct to very shallow. This groove marks the end of a distinct anterobuccal lobe or crest. Because the macropodine I3 is usually oriented slightly anterolingually, Ride (1957, 1959) referred to the anterobuccal crest as the anterior lobe and to the main crest as the posterior lobe. Most *Thylogale* species (e.g., see Archer 1984:fig. 270) and *Lagorchestes leporides* (Gould, 1840) are exceptions to the typical macropodine (i.e., plesiomorphic) design: both crests extend the length of the I3. The main crest of sthenurines and *Wanburoo* is also blade-like, but the buccal surface is primarily flat (figs 16-17). In sthenurines, a thin eminence runs up the anterior edge of the buccal surface, a feature that I believe, as apparently did Ride (1959), to be the homologue of the anterobuccal crest of balbarines, macropodines, at least two undescribed bulungamayines from the Lake Eyre Basin, and some potoroines. It is incipient in *Hypsiprymnodon.* Among the sthenurines, the homologous eminence is best developed in the short, wide I3 of *Procoptodon goliah* (pl. 75G,I; figs 16-17). Any such eminence is lacking from the buccal surface of the *Wanburoo* I3.

On the lingual surface of the main I3 crest of sthenurines and *Wanburoo*, a shallow posterodorsally oriented groove extends from near the anterior edge of the occlusal surface to the base of the crown (e.g., pls 2B, 37D, 75H; Cooke 1999:fig. 2B). This groove marks the posterior limit of an apparently neomorphic, anterolingual crest. However, only in the very large *Sthenurus stirlingi* is it hypertrophied to a degree similar to that of the anterobuccal crest of many macropodines (fig. 17D). The sthenurine anterolingual crest is often barely more than a low eminence, but it is distinct on the minimally worn I3 of plesiomorphic sthenurines, such as *Hadronomas*, *Archaeosimos* and *Sthenurus andersoni.* I consider the extreme reduction or absence of the anterobuccal crest and presence of the anterolingual crest to be a synapomorphy for *Wanburoo* and the sthenurines. It will prove interesting to see if a slightly less derived bulungamayine (e.g., *Ganguroo*) expresses an antecedent I3 morphology.

If *Wanburoo* is the sister taxon of the Sthenurinae, then loss of C1 is another synapomorphy for the sthenurines. C1 is present in all other macropodoid groups and is symplesiomorphic. Loss of C1 has occurred in many macropodines, but its status as the plesiomorphic condition in that subfamily is attested by its presence in *Dorcopsulus, Dendrolagus, Dorcopsis, Onychogalea and Lagorchestes.* Increased mandibular condyle breadth might be a sthenurine synapomorphy, even though that of *Hadronomas* is intermediate in form between the oval-shaped condyle of

Wanburoo and the transversely elongate condyle of all other sthenurines. However, this will not be demonstrable until the appropriate outgroup is identified, because transverse elongation of the condyle is known in propleopines, balbarines, potoroines and, most importantly, *Ganguroo bilamina.* Because macropodines are typified by an oval-shaped mandibular condyle, the *Wanburoo* morphology could potentially be antecedent to them and the sthenurines.

Wanburoo also resembles the sthenurines in p3 morphology, being intermediate in overall p3 form between *Hadronomas* and several of the more plesiomorphic of the other sthenurines (fig. 18). In *Metasthenurus*, *Archaeosimos* and *"Simosthenurus" brachyselenis*, the anterior two-thirds of the p3 is rather parallel-sided and the posterior third is slightly inflated. The anterior two-thirds of the main crest is divided into cuspules that each give rise to buccal and lingual ridgelets, but the posterior third of the crest lacks these features and is curved slightly posterolingually. *Wanburoo* shares these other traits, but it lacks the very distinctive crest on the posterobuccal aspect of p3. By comparison, the p3 of *Hadronomas* is wide centrally and narrows both anteriorly and posteriorly. *Hadronomas* resembles *Ganguroo* in this respect, but the tooth is not widened to the same degree and is, therefore, not quite so reminiscent of a Cornish pasty. Interestingly, though, the lingual side of the p3 crown of *Ganguroo* does have a bulbous base, a feature that Cooke (1997b) interpreted as a poorly defined lingual cingulid. If this morphology is antecedent to the distinct lingual cingulid of *Hadronomas*, then perhaps the loss of C1 and broadening of the mandibular condyle occurred independently in *Hadronomas* and the other sthenurines. If so, *Wanburoo* may have diverged from the main sthenurine line between *Hadronomas* and the clade from which the *Sthenurus*, *Metasthenurus*, *Archaeosimos*, *Simosthenurus* and *Procoptodon* arose. On balance, though, it is presently more parsimonious to consider the similarities between *Wanburoo* and these sthenurines convergent. After all, *Wanburoo* lacks any sign of the low buccal cingulid present in *Hadronomas* that could be a precursor to the buccal crest of other sthenurines. The description of a *Wanburoo* cranium would almost certainly put such speculation to rest one way or the other.

In summary, all evidence strongly suggests that sthenurines and macropodines arose from bulungamayines. While this renders the Bulungamayinae paraphyletic, the group is crucial evolutionarily. Within it, the long-sought transition from bunolophodonty to bilophodonty is manifest (Cooke 1997b). I agree with Cooke (1999) that the derived middle Miocene bulungamayine *Wanburoo hilarus* is a more than credible precursor to the sthenurine radiation, and the incisor similarities highlighted here only strengthen this notion. Based on the evidence now available, *Wanburoo* is the likely sister taxon of the Sthenurinae. If this holds true, then sthenurines must have arisen prior to the late Miocene, which accords well with the suggestions of both Ride (1971) and Flannery (1989). The idea of a middle Miocene origin is supported by the presence of *Hadronomas* in the late Miocene at Alcoota and sthenurine-like postcranial remains in the middle Miocene at Bullock

Creek (Murray and Megirian 1992). Accounts of the *Wanburoo* cranium and bulungamayine postcranial material are eagerly awaited to confirm the craniodental synapomorphies proposed here and the possible postcranial synapomorphies suggested by Murray (1995).

PHYLOGENY

This section comprises the first comprehensive cladistic analysis of the Sthenurinae. Collections of high-quality material from sites such as the caves of southeastern Australia have greatly augmented the number of characters from which several species are known, although the 12 new species described since 1992 have been pivotal in determining the phylogenetic outcome. An increase in the number of characters usually strengthens an analysis, but the effect on tree structure may be weaker if the best-known taxa are highly derived, if several taxa are very incompletely known, or if there are major gaps in the fossil record. Taxa occupying intermediate or lower positions on a tree can be highly effective in breaking up long branches (Graybeal 1998). That is, they facilitate a finer-grained view of character transformations. Several recently described sthenurines are especially informative in this regard (e.g., *Metasthenurus newtonae*).

CHARACTERS

Despite the analytical advances presented by the advent of cladistic methods, character selection and state scoring are fundamentally subjective exercises. Investigators cannot avoid introducing their own biases into a phylogenetic analysis, but need to ensure that their analyses are, at the very least, repeatable. As discussed earlier (Methods), all 37 characters used (table 43) here pertain to the craniodental system, and 22 of them are dental features. This is a major bias in itself, prompted as much by preferential preservation and historical precedence as character diversity. Shifts in dietary adaptations reflect only one dimension of taxic evolution, but it is a crucial one, and thus a sound basis for the first cladistic study of sthenurine interrelationships. As more skeletal associations are discovered, postcranial characters (especially pedal) will provide excellent means for testing hypotheses based on craniodental data alone. Below I provide a description of the 37 characters and their states and refer the reader to illustrations of taxa that aptly depict particular states.

1. Cranium length

Marcus (1962) and Tedford (1966) recognized two broad categories of cranium and dentary length in sthenurines. In adult cranial proportions, dolichocephalic sthenurines resemble other macropodoids more than the brachycephalic sthenurines do. The strong correlation between cranium and dentary length means that species known only from dentaries can be readily scored for this character. While most taxa possess either of the extreme states, a third, intermediate state has been introduced to accommodate a condition seen in some species. To reduce scope for investigator error in state scoring, a mean length to width value, or cephalic index, was calculated for each species represented by crania, and these were then ranked (table 41). Species for which no crania are known, but which are represented by at least moderately complete dentaries were slotted into a category based on proportional dentary similarities with species of known cranial proportions.

States: 0, dolichocephalic; 1, intermediate; 2, brachycephalic. Ordered.

Table 41. Cephalic indices for the 26 sthenurine species calculated by dividing condylobasal length by maximum width measured from the lateral surface of one zygomatic arch to the other, then ranked highest to lowest; est. = estimated from proportions of dentary or incomplete crania; ? = presently inestimable.

State	Species	Cephalic Index
0	*Sthenurus tindalei*	2.03
	Sthenurus andersoni	1.94
	Hadronomas puckridgi	1.93
	Sthenurus atlas	est.
	Sthenurus murrayi	est.
1	*Sthenurus stirlingi*	1.67
	Metasthenurus newtonae	1.64
	"Simosthenurus" antiquus	est.
	"Simosthenurus" pales	est.
2	*"Procoptodon" gilli*	1.49
	Simosthenurus maddocki	1.48
	"Procoptodon" browneorum	1.47
	Simosthenurus occidentalis	1.38
	Procoptodon goliah	1.30
	Archaeosimos cegsai	est.
	Archaeosimos correlli	est.
	"Simosthenurus" brachyselenis	est.
	"Simosthenurus" baileyi	est.
	Simosthenurus euryskaphus	est.
	"Procoptodon" mccoyi	est.
	"Procoptodon" oreas	est.
	Procoptodon pusio	est.
	Procoptodon rapha	est.
?	*Sthenurus notabilis*	-
	"Simosthenurus" tirarensis	-
	"Procoptodon" williamsi	-

2. Neurocranial curvature

In most macropodoids, the dorsal surface of the cranium is domed, especially in the neurocranial region. By contrast, the dorsal profile of the cranium of most adult sthenurines is near to horizontal, at least to the nasofrontal suture. Two sthenurines for which the neurocranium is known, *"Procoptodon" gilli* and *P. goliah*, show a partial reversal to the outgroup or ancestral condition (see fig. 14).

States: 0, domed; 1, not domed.

3. Anterolateral projection of orbit

In all adult sthenurines, with the exception of *Hadronomas*, the orbital capsule (enclosed by the jugal) is projected anterolaterally and floored by an expanded maxillary plate. This results in a relatively enlarged anteorbital fossa and a greater degree of orbital convergence than is observed in most other macropodoids (see fig. 15).

States: 0, absent; 1, present.

4. Anterodorsal extension of maxilla

In all non-sthenurine macropodoids and the majority of sthenurines, the suture between the premaxilla and the maxilla ascends relatively smoothly dorsally or posterodorsally to contact the nasal. In *Sthenurus tindalei* and *S. stirlingi*, a peninsula of maxilla is extended anterodorsally between the dorsal extremity of the premaxilla and nasal, such that the latter elements no longer abut (pl. 24A; Wells and Tedford 1995). In *S. andersoni*, post-depositional compression fractures may form in an almost identical position to the anteroposteriorly oriented portion of the premaxilla-maxilla suture in *S. tindalei* and *S. stirlingi*, but the suture actually continues to the nasal (pls 7A, 8F). The only partial cranium of *S. atlas* that preserves this region is too crushed to allow assessment of this character.

States: 0, absent; 1, present.

5. Buccinator fossa

The buccinator fossa is immediately adjacent to the diastema on the lateral surface of the maxilla. Its depth in adult specimens varies considerably across the Sthenurinae; compare pls 7C, 48C, 71B.

States: 0, shallow; 1, deep; 2, very deep. Ordered.

6. Nasofrontal suture

The suture formed between the nasal and frontal bones varies in shape between sthenurine species. The outgroup condition is a straight nasofrontal suture; compare pls 94A, 61B, 24B (and Wells and Tedford 1995:fig. 13).

States: 0, straight; 1, gently sinusoidal; 2, strongly sinusoidal (S-shaped). Unordered.

7. Frontal inflation

The frontals of most macropodoids are only very slightly inflated laterally, if at all. In many adult sthenurines, the frontals are markedly expanded laterally, although some taxa secondarily approach the outgroup condition. Among those species in which the frontals are inflated, most have quite straight lateral edges. In some species, the frontals taper laterally and bear a rugose supraorbital tuberosity. See fig. 15.

States: 0, not marked; 1, marked; 2, tapered laterally. Unordered.

8. Basicranial elevation

Tedford (1966) considered the level of the basicranium relative to the palate a feature of importance in the distinction of adult dolichocephalic and brachycephalic sthenurines. Two states were recognized: basicranial and palatal planes nearly approximating one another and basicranial plane elevated well above palatal plane. I have expanded the number of character states to three and find that the basicranial and palatal planes do not nearly approximate one another in any sthenurine. Tedford's view may have been conceived on the basis of the crushed crania referred to *Sthenurus atlas* and *S. tindalei.* The outgroup condition and the character state retained by at least three *Sthenurus* species may be more accurately described as slightly elevated (see fig. 14). In many sthenurines, the basicranial plane is considerably higher than the palatal plane (e.g., pl. 49) and, relative to the basisphenoid, the basioccipital is flexed dorsally. In *Procoptodon goliah* and *P. pusio*, the basicranial plane is higher again, and the basioccipital is more strongly flexed, reflecting lifting and anterior rotation of the neurocranium (e.g., pls 73C, 74D).

States: 0, slight; 1, marked, basioccipital moderately flexed; 2, great, basioccipital markedly flexed. Ordered.

9. I1 Shape

Increased breadth of the I1 has arisen convergently in the Macropodinae and some members of the Sthenurinae. Sub-cylindrical (not broadened) represents the ancestral macropodoid condition as evidenced by hypsiprymnodontids, balbarines, potoroines, plesiomorphic macropodines and most sthenurines. Two derived states are recognized; compare pls 67D, 8B.

States: 0, sub-cylindrical; 1, slightly broadened; 2, markedly broadened. Unordered.

10. Relative premolar length

In most macropodoids, the third permanent premolar exceeds each of the molars in length, although shortening of the premolar is observed in some taxa. The two derived sthenurine character states are formulated relative to the third molar (largest), and the first molar (shortest). Premolar and molar lengths are mean measurements for each species. If premolar and molar dimensions fell within ± 0.2

mm of one another, they were considered equal. If a discrepancy between upper and lower cheek teeth was revealed, the state derived from the lower teeth was preferentially chosen.

States: 0, longer than all molars; 1, equal to or shorter than third molar; 2, equal to or shorter than first molar. Unordered.

11. Anterior extension of P3 lingual crest

A pronounced anterior extension of the P3 lingual crest is characteristic of the species of *Sthenurus.* Instead of terminating at the anterior extremity of the anterior basin, the lingual crest continues toward the anterobuccal base of the crown (e.g., pl. 18; fig. 4).

States: 0, absent; 1, present.

12. Molar crown height

Tedford (1966) recognized two broad categories of crown height in sthenurines (high crowned, low crowned) and placed considerable taxonomic weight on the character. As the number of recognized species has increased, it has become clear that molar crown heights cover a wide spectrum. Molars of different sthenurines also cover quite a size range, which can make visual assessment of relative crown height rather perilous. For these reasons, a quantification of this character was introduced: the crown height index. This was calculated as

$$\frac{(\text{m2 posterior height})^2}{\text{m2 length x m2 posterior width}}$$

Mean values for each species were derived after calculating height / length and height / width values for each individual measured, ranking them highest to lowest, then eliminating the lowest 20% of values to help avoid the problem of having the more worn individuals lower the mean value (for species represented by values for at least five individuals). Mean height/length and height/width values were then multiplied to give the crown height index for each species (table 42). These were ranked lowest to highest, with two large gaps in the rankings (between 0.34-0.43 and 0.65-0.71) chosen as arbitrary dividers between character states. Species for which no m2 is known were allocated a state by comparing the molars from which they are known with teeth in the same positions in other species.

States: 0, low to intermediate; 1, very low; 2, high. Unordered.

13. Very high and coarse molar cristae

The cristae of most sthenurine molars are much lower than the lophs, with the exception of those of *Procoptodon pusio*, *P. rapha* and *P. goliah.* In these species, the cristae are especially coarse and approach the height of the lophs (e.g., pls 78, 82).

States: 0, absent; 1, present.

Table 42. Crown height indices derived from m2 for the 23 sthenurine species ranked lowest to highest, showing posterior height / length, and posterior height / posterior width values. * calculated from m1 posterior height; ** calculated from m3 anterior height; est., estimated from upper molars.

State	Species	PH / L	PH / PW	Crown Height Index	n
0	*Hadronomas puckridgi*	0.56	0.77	0.43	4
	Archaeosimos correlli	0.57	0.77	0.44	1
	Simosthenurus euryskaphus	0.59	0.78	0.46	1
	Procoptodon pusio	0.61	0.79	0.48	1
	Procoptodon goliah	0.62	0.79	0.49	8
	"Simosthenurus" baileyi	0.64	0.78	0.50	4
	Simosthenurus maddocki	0.63	0.81	0.51	7
	Simosthenurus occidentalis	0.67	0.80	0.54	23
	"Simosthenurus" antiquus	0.65	0.86	0.56	3
	"Procoptodon" browneorum	0.68	0.85	0.58	27
	"Procoptodon" oreas	0.68	0.87	0.59	3
	Sthenurus stirlingi	0.70	0.87	0.61	1
	*"Simosthenurus" brachyselenis**	0.68	0.91	0.62	1
	Procoptodon rapha	0.68	0.93	0.63	2
	"Simosthenurus" pales	0.73	0.89	0.65	9
	Metasthenurus newtonae	0.70	0.93	0.65	6
	"Procoptodon" williamsi	-	-	est.	-
1	*Archaeosimos cegsai*	0.53	0.57	0.30	1
	*"Procoptodon" mccoyi***	0.56	0.60	0.34	1
2	*"Simosthenurus" tirarensis*	0.72	0.98	0.71	1
	Sthenurus andersoni	0.77	0.95	0.73	30
	Sthenurus atlas	0.78	0.96	0.75	10
	Sthenurus tindalei	0.76	1.00	0.76	4
	"Procoptodon" gilli	0.81	1.02	0.83	12
	*Sthenurus notabilis**	0.86	1.15	0.99	1
	Sthenurus murrayi	-	-	est.	-

14. Lophs taper toward unworn crest

The molar lophs of most macropodoids taper only slightly toward their crests. In most sthenurines, the buccal and lingual sides of the lophs are near parallel, but often slightly convex in anterior or posterior view (see fig. 9). In a few sthenurines, the unworn loph crests are markedly narrower than their base. This is primarily produced by the angling of the lingual side on upper molars (e.g., pls 75, 107; fig. 19F). The taper is due largely to the angling of the buccal side on lower molars.

States: 0, slight; 1, marked.

15. Preprotocrista development

Although the preprotocrista of most macropodoids is not as well developed as it is in potoroines and plesiomorphic bulungamayines, a distinct preprotocrista is evident in some sthenurines (e.g., pls 44, 78).

States: 0, absent or indistinct; 1, distinct.

16. Postparacrista development

This character assesses the relative size (height and length) of the postparacrista, which varies considerably across the Sthenurinae; compare pls 2C, 66, 27A.

States: 0, moderately developed; 1, well developed; 2, strongly developed. Unordered.

17. Postparacrista and premetacrista orientation

In the outgroup and the majority of sthenurines, the postparacrista and premetacrista are oriented close to anteroposteriorly. However, in some sthenurine species, both cristae curve markedly toward the midline of the tooth; compare pls 54, 99.

States: 0, oriented essentially anteroposteriorly; 1, incurved lingually.

18. Postprotocrista division

Across the Macropodoidea, the postprotocrista is typically expressed as a single crest (see figs 8, 19). In many sthenurines, it is partially or wholly divided into a lingual component, which maintains contact with the protocone apex, and a more buccal (mesial) component, which is oriented more anteroposteriorly; compare pls 2A, 37A, 91C-D.

States: 0, absent; 1, incipient; 2, marked. Ordered.

19. Mesocrista development

The mesocrista is absent in most non-sthenurine macropodoids, but is distinct in numerous sthenurines (e.g., pls 62A, 91C-D; fig. 19). *Procoptodon pusio*, *P. rapha* and *P. goliah* express the most derived mesocrista state; they combine it with the posterior portion of the postprotocrista and this becomes the dominant anteroposterior crista. In these three species, the buccally shifted anterior portion of the postprotocrista remains as an isolated spur on the buccal side of the protoloph posterior face (pls 75, 78, 82; fig. 19).

States: 0, absent or weakly developed; 1, moderately developed; 2, well developed; 3, strongly developed. Ordered.

20. Dentary proportion

Relative to the size of the cheek teeth, the adult dentaries of *"Simosthenurus" brachyselenis*, *Procoptodon pusio*, *P. rapha* and *P. goliah* are much more massive than in all other macropodoids; compare pls 65, 76-77, 80-81, 83-84 with, e.g., pls 36, 63.

States: 0, moderate; 1, massive.

21. Symphyseal plate rugosity

The marked development of sub-parallel ridges and irregular rugosities on the adult symphyseal plate is typical of all sthenurines, except *Hadronomas.* A similar but less distinctive condition has arisen convergently in a few macropodine genera (fig. 10). In most macropodoids, the anterior region of the dentary is shallower relative to the main body of the ramus than it is in most sthenurines. This increased depth, which results from greater symphyseal depth and development of a prominent boss, is almost certainly correlated with increased stiffness of the symphyseal union. Partial fusion is widespread within all genera, except *Hadronomas.*

States: 0, absent; 1, present.

22. Inferior transverse torus of symphysis

A well-developed inferior transverse torus, or posterior extension of the symphysis (below the genial fossa), is characteristic of the adults of many but not all sthenurines (fig. 10). It is observed in no other macropodoids.

States: 0, absent or poorly developed; 1, well developed.

23. Posterior mental foramen

Tedford (1966) considered the presence of a posterior mental foramen diagnostic of the Sthenurinae, but the feature has since been observed in representatives of most macropodoid groups. However, the foramen is larger in most sthenurines than it is in other macropodoids. In *Procoptodon pusio*, *P. rapha* and *P. goliah*, the foramen is either secondarily reduced or lost (e.g., pls 80, 83).

States: 0, tiny or absent; 1, well developed.

24. Masseteric foramen

Invasion of the body of the dentary by the deep masseter muscle is a macropodoid characteristic, but in the adults of most taxa, the masseteric foramen remains small to moderate in size and is restricted to the anteroventral portion of the masseteric fossa. Sthenurines tend to have a moderately sized masseteric foramen, but in *"Procoptodon" oreas*, *P. pusio*, *P. rapha* and *P. goliah*, the feature is markedly enlarged and expanded posteriorly, reducing the size of the posteroventral shelf of the masseteric fossa (e.g., pls 81B, 104D).

States: 0, moderate; 1, large.

25. Masseteric canal

Although the masseteric foramen of most macropodoids is relatively small, the canal into which the foramen leads often penetrates well into the ramus. However, in all sthenurines except *Hadronomas*, the masseteric canal is much shorter and essentially vertical (see various plates; figs 20, 22).

States: 0, extends at least as far as beneath m4; 1, vertical.

26. Mylohyoid groove (and associated processes)

Positioned anterior to the inferior mental foramen, the mylohyoid groove of most macropodoids is broad, very shallow and not overhung by any sharp processes (cf. lingula in humans). Adult dentaries of some sthenurines have a narrower, deeper mylohyoid groove that is partially overhung by two small processes (see pls 34B, 103B). Their function may have been to protect the mylohyoid nerves and blood vessels from pressure applied by contraction of the adjacent digastric and hyoid musculature.

States: 0, shallow, rather wide; 1, deep, rather narrow.

27. Orientation of i1

The i1 occlusal surface of most macropodoids and several sthenurines is aligned with or set just slightly above the level of the diastema (fig. 20). In *Sthenurus stirlingi* and the Simosthenurini, the i1 is upturned, such that its occlusal surface is above the level of the diastema (e.g., pls 25, 55).

States: 0, procumbent; 1, upturned.

28. Morphology of i1

Morphology of the i1 crown varies markedly across the Macropodoidea (figs 11, 13, 20) and within the sthenurines; compare pls 6D-E, 68, 55, 45.

States: 0, short, slender; 1, short, robust; 2, long, robust; 3, very slender. Unordered.

29. p3 anterior to posterior width

Posterior p3 width is consistently greater than anterior width across the Sthenurinae, but the degree to which it is wider varies significantly. Like molar crown height, it is a continuous character, which makes straight visual state assessment difficult. The character was therefore quantified, with mean values for each species calculating simply by dividing posterior width by anterior width. Species were then ranked highest to lowest, with two gaps in the rankings (at 1.37 and 1.14 within the range 1.44-1.04) chosen as arbitrary dividers between character states.

States: 0, slightly wider posteriorly; 1, distinctly wider posteriorly; 2, much wider posteriorly. Unordered.

30. p3 buccal crest length

A buccal crest on p3 is characteristic of all sthenurines except *Hadronomas*, but its relative length varies greatly across the Sthenurinae. State 1 here takes into account those taxa with a short crest restricted to the posterobuccal corner of the tooth (e.g., pl. 30C). In sthenurines expressing the intermediate condition, the buccal crest is usually no more than half the length of the main crest and may have a distinct vertical cleft anterior to it (e.g., pl. 59). If the buccal crest contacts or terminates adjacent to the second cuspule of the main crest, it is defined as long (e.g., pl. 23F). A very long buccal crest runs the entire length of p3, contacting the anteriormost cuspule of the main crest via a short intermediary crest (e.g., pl. 29C) or the ridgelet directed buccally from the anteriormost cuspule (e.g., pl. 77).

States: 0, absent; 1, short; 2, intermediate; 3, long; 4, very long. Ordered.

31. Median valley width on p3

While its formation is quite obviously directly correlated with the advent of a buccal crest, width of the p3 median valley is not proportional to buccal crest length. State 0 is expressed in those taxa with a narrow or no median valley (e.g., pl. 36C). In sthenurines expressing state 2, length of the median valley is subequal to its width (e.g., pl. 40D).

States: 0, narrow (or absent); 1, intermediate; 2, wide. Ordered.

32. Trigonid size

In most sthenurines the trigonid is very short relative to the rest of the tooth. The outgroup and plesiomorphic condition is intermediate between this and the long, wide dimensions observed in *Sthenurus stirlingi* and *S. tindalei*, and independently derived in some macropodines (e.g., *Protemnodon*, *Macropus*); compare pls 34, 63, 25.

States: 0, intermediate; 1, short; 2, long, wide. Unordered.

33. Paracristid and cristid obliqua division

The paracristid and cristid obliqua of most macropodoids each typically consist of a single crest oriented anteriorly or anterolingually, which maintains a direct connection to either the protoconid or the hypoconid apex (see fig. 12). In many sthenurines, these cristids are incipiently or wholly divided into a buccal component, which maintains contact with the cusp apex, and a larger, lingually shifted component. In those taxa expressing state 2, the more anterior molars bear a paracristid and cristid obliqua less distinctly separated into components than the more posterior teeth; compare pls 5C, 34C-D, 38C, 67C.

States: 0, no division; 1, incipient division; 2, incipient division anteriorly grading to clear division posteriorly; 3, clear division on all molars. Ordered.

34. Anterior portion of paracristid

The anterior portion of the lingual component of the paracristid is transversely oriented or rather U-shaped in most macropodoids and usually extends to the anterolingual corner of the crown (fig. 12). Within the Sthenurinae, particularly within *Sthenurus*, the anterior portion of the paracristid may be reduced or not differentiable from the underlying precingulid (e.g., pls 19, 28).

States: 0, very distinct; 1, markedly reduced; 2, not distinct. Ordered.

35. Enamel crenulations on lophids

Distinct enamel crenulations on the lophid faces are seen among macropodoids only within the Sthenurinae. The lophid faces of those species expressing state 1 are dominated by numerous very fine crenulations, a condition reflected on the upper molars. Those expressing state 2 have fewer, coarser crenulations that are generally oriented anteroposteriorly; compare pls 38C, 57F, 93C-D.

States: 0, absent or few very fine; 1, numerous very fine; 2, coarse. Unordered.

36. Posthypocristid

Possession of a distinct posthypocristid is a plesiomorphic condition within the Macropodoidea. The recession of the feature is correlated directly with the evolution of bilophodonty. Among the sthenurines, *Procoptodon rapha* and *P. goliah* bear a coarse, ventrolingually oriented crest, which descends toward the base of the crown from the hypolophid crest (pls 77, 84). Congruence in morphology and position with plesiomorphic macropodoids strongly suggests that the feature is a reemergent posthypocristid (see Dental Character Reversal).

States: 0, absent; 1, present.

37. Molar row orientation

The molar row of the majority of macropodoids is oriented relatively straight anteroposteriorly. In all sthenurines, with the exception of *Hadronomas*, the molar row is curved laterally; compare pls 5C, 14E.

States: 0, straight; 1, curved laterally.

CLADISTIC ANALYSIS

Branch and bound searches revealed 15 equally most parsimonious trees for the data matrix of 27 species and 37 characters (table 43). Tree length is 115 steps, the combined or ensemble consistency index (CI) is 0.53 and the ensemble retention index (RI) is 0.76. CI is the minimum number of changes a character may show on a particular tree divided by the number of changes that it does show (Kluge and Farris 1969). An ensemble CI of 0.53 is the mean of the 37 individual consistency indices. This implies that the optimal tree has just under twice as many steps than it would if there was no apparent homoplasy. However, as an estimator of the

Table 43. Character matrix for the cladistic analysis of the Sthenurinae. 1. Cranium length; 2. Neurocranial curvature; 3. Orbital anterolateral projection; 4. Premaxilla-nasal contact; 5. Buccinator fossa; 6. Nasofrontal suture; 7. Frontal inflation; 8. Basicranial elevation; 9. I1 shape; 10. Relative premolar length; 11. Anterior extension of P3 lingual crest; 12. Molar crown height; 13. Very high and coarse molar cristae; 14. Lophs taper toward unworn crest; 15. Preprotocrista development; 16. Postparacrista development; 17. Postparacrista and premetacrista orientation; 18. Postprotocrista division; 19. Mesocrista development; 20. Dentary proportion; 21. Symphyseal plate rugosity; 22. Inferior transverse torus of symphysis; 23. Posterior mental foramen; 24. Masseteric foramen; 25. Masseteric canal; 26. Mylohyoid groove (and associated processes); 27. i1 orientation; 28. i1 morphology; 29. p3 anterior to posterior width; 30. p3 buccal crest length; 31. p3 median valley width; 32. Trigonid size; 33. Paracristid and cristid obliqua division; 34. Anterior portion of paracristid; 35. Enamel crenulations on lophids; 36. Posthypocristid; 37. Molar row orientation.

Taxon	1	2	3	4	5	6	7	8	9	10	11	12	13	14	15	16	17	18	19	20	21	22	23	24	25	26	27	28	29	30	31	32	33	34	35	36	37
Hadronomas puckridgi	0	1	0	0	0	0	2	0	1	0	0	1	0	0	0	0	0	0	0	0	0	0	1	0	0	0	0	0	0	0	0	0	0	0	0	0	0
Sthenurus andersoni	0	1	1	0	0	0	0	0	2	0	1	2	0	0	?	0	?	0	0	0	1	0	1	0	1	0	0	2	1	?	1	0	1	0	0	0	1
Sthenurus atlas	0	1	1	?	0	1	0	0	2	0	1	2	0	0	0	1	0	1	0	0	1	0	1	0	1	0	0	2	1	3	1	0	1	1	0	0	1
Sthenurus murrayi	0	?	?	?	0	?	?	?	2	0	1	2	0	0	0	1	0	1	0	?	?	?	?	?	?	?	?	?	?	?	?	?	?	?	?	?	1
Sthenurus notabilis	?	?	?	?	?	?	?	?	?	2	?	2	0	0	?	?	?	?	?	0	?	?	?	?	?	?	0	?	1	3	1	0	1	0	0	0	?
Sthenurus stirlingi	1	1	1	1	1	2	1	1	2	1	1	0	0	0	0	2	0	1	1	0	1	1	1	0	1	0	1	2	1	4	1	2	3	2	0	0	1
Sthenurus tindalei	0	1	1	1	0	2	0	0	2	1	1	2	0	0	0	2	?	1	0	0	1	1	1	0	1	0	0	2	1	4	1	2	3	2	0	0	1
Metasthenurus newtonae	1	1	1	0	0	0	1	1	0	1	0	0	0	0	0	1	1	0	0	0	1	1	1	0	1	1	0	2	1	1	0	0	1	0	0	0	1
Archaeosimos cegsai	2	?	?	?	?	?	?	?	?	0	?	1	0	0	0	0	0	1	1	0	1	?	1	0	1	0	1	1	0	1	0	1	2	0	0	0	1
Archaeosimos correlli	2	?	?	?	?	?	?	?	?	0	?	1	0	0	?	?	?	?	?	0	1	0	1	0	1	0	1	?	0	1	0	1	2	0	0	0	1
Simosthenurus euryskaphus	2	?	?	?	?	?	?	?	?	1	?	0	0	0	?	?	?	?	?	0	1	1	1	?	1	0	1	1	1	2	2	1	3	0	1	0	1
Simosthenurus maddocki	2	1	1	0	1	1	2	1	0	0	0	0	0	0	1	0	1	2	0	0	1	0	1	0	1	0	1	3	2	1	2	1	2	0	1	0	1
Simosthenurus occidentalis	2	1	1	0	1	1	2	1	0	0	0	0	0	0	0	0	0	2	1	0	1	1	1	0	1	0	1	1	2	2	2	1	3	0	1	0	1
"Simosthenurus" antiquus	1	?	?	?	?	?	?	?	?	0	?	0	0	0	0	1	0	1	1	0	1	1	1	?	1	?	1	?	1	2	1	1	3	0	0	0	1
"Simosthenurus" baileyi	2	1	1	0	1	1	1	1	1	0	0	0	0	0	0	1	0	1	1	0	1	1	1	0	1	0	1	1	1	2	1	1	3	0	0	0	1
"Simosthenurus" brachyselenis	2	?	?	?	?	?	?	?	?	2	?	0	0	0	?	?	?	?	?	1	1	1	1	0	1	0	1	1	1	1	1	1	2	0	0	0	1
"Simosthenurus" pales	1	?	1	?	?	?	?	?	1	?	0	0	0	0	0	1	0	2	1	0	1	1	1	0	1	0	1	2	1	3	1	1	3	0	0	0	1
"Simosthenurus" tirarensis	?	?	?	?	?	?	?	?	?	1	?	2	0	0	?	?	?	?	?	?	1	1	1	?	?	?	1	?	1	1	1	1	2	0	0	0	1

Table 43 (continued).

Taxon	1	2	3	4	5	6	7	8	9	10	11	12	13	14	15	16	17	18	19	20	21	22	23	24	25	26	27	28	29	30	31	32	33	34	35	36	37
Procoptodon goliah	2	0	1	0	2	?	1	2	0	2	0	0	1	1	1	1	1	2	3	1	1	1	0	1	1	1	1	3	1	4	1	1	3	0	2	1	1
Procoptodon pusio	2	?	1	?	2	?	1	2	0	2	0	0	1	1	1	1	1	2	3	1	1	1	0	1	1	1	1	3	1	3	1	1	3	0	2	0	1
Procoptodon rapha	2	?	?	?	2	?	?	?	?	2	0	0	1	1	1	1	1	2	3	1	1	1	0	1	1	1	1	3	1	4	2	1	3	0	2	1	1
"*Procoptodon*" *browneorum*	2	1	1	0	1	1	1	1	0	0	0	0	0	0	0	0	0	2	2	0	1	1	1	0	1	0	1	1	1	3	1	1	3	0	2	0	1
"*Procoptodon*" *gilli*	2	0	1	0	1	0	0	0	0	0	0	2	0	0	1	0	1	2	2	0	1	1	1	0	1	0	1	1	1	3	1	1	3	0	0	0	1
"*Procoptodon*" *mccoyi*	2	?	?	?	?	?	?	?	?	1	?	1	0	0	?	?	?	?	?	0	1	1	?	?	1	0	?	?	1	3	?	1	3	0	2	0	1
"*Procoptodon*" *oreas*	2	?	?	?	?	?	?	?	?	1	?	0	0	0	1	0	0	2	2	0	1	1	1	1	1	1	1	1	1	3	1	1	3	0	2	0	1
"*Procoptodon*" *williamsi*	?	?	?	?	1	?	?	?	?	2	0	0	0	1	1	0	1	2	2	?	?	?	?	?	?	?	?	?	?	?	?	?	?	?	?	?	1
Dorcopsoides fossilis (outgroup)	0	0	0	0	0	0	0	0	?	0	0	0	0	0	0	0	0	0	0	0	0	0	0	0	0	0	0	0	0	0	0	0	0	0	0	0	0

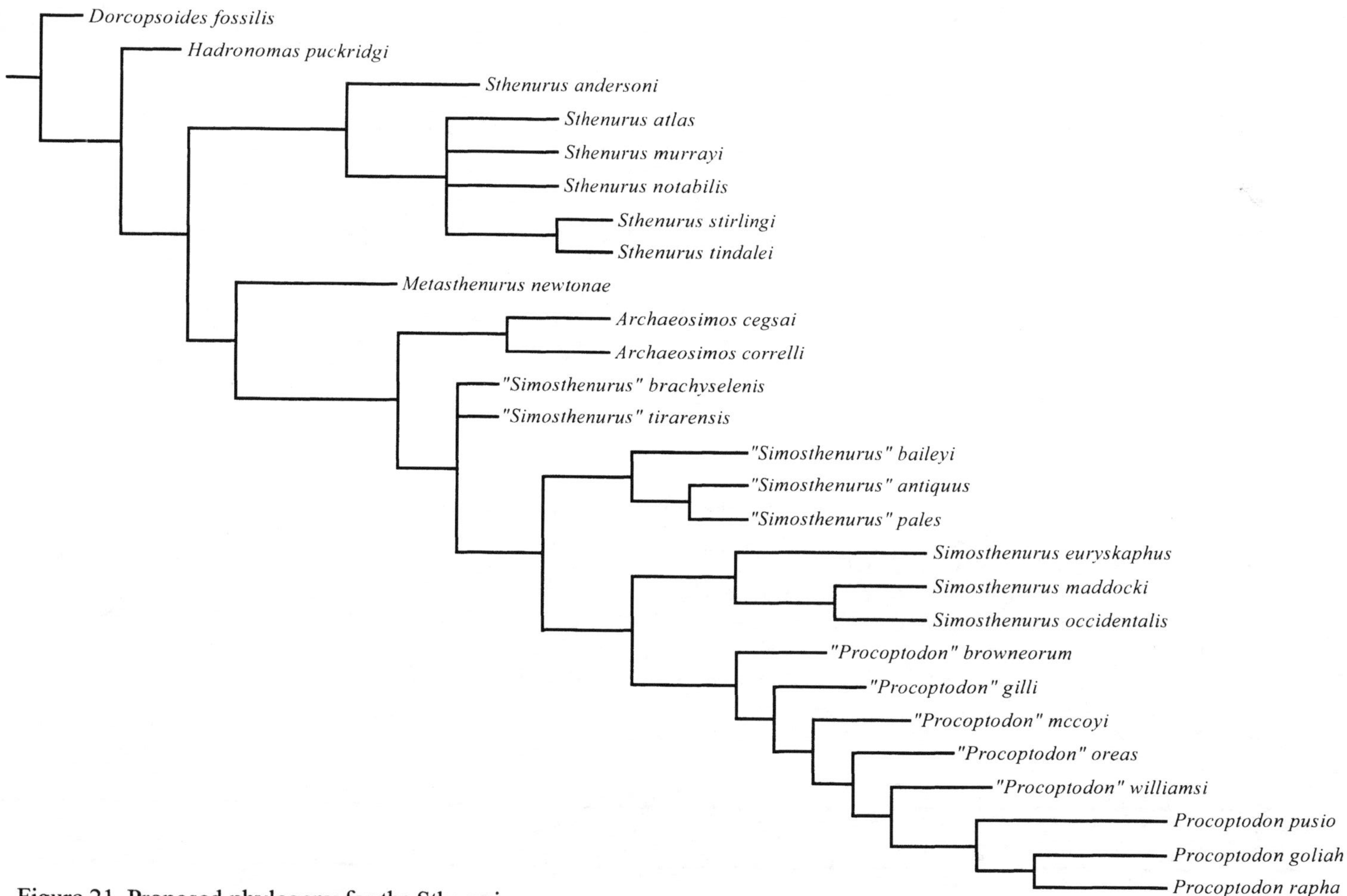

Figure 21. Proposed phylogeny for the Sthenurinae.

goodness-of-fit of character data to cladistic reconstructions, it must be borne in mind that CI is inversely proportional to the number of both species and characters in the analysis (Archie 1989a-b; Farris 1989). Thus, at the character or ensemble level, CI overestimates homoplasy in larger data sets and is questionable as the best measure for comparing between trees. Because it takes into account tree length, RI is a good comparative statistic for individual characters (Farris 1989; Archie 1990), but the ensemble RI is less useful, because it consistently underestimates homoplasy with increased taxa and characters (Archie 1989b, 1990). Perhaps the best tests of any phylogeny are to see how well topology is maintained as more data are eventually added to the matrix, or how well analyses of alternative data sets corroborate the original hypothesis.

The strict consensus of 15 equally most parsimonious trees depicting relationships within the Sthenurinae (fig. 21) reveals a decidedly stable overall topology. The poorly represented *Sthenurus notabilis* and *S. murrayi* are responsible for the polytomies within *Sthenurus*, and thus most of the uncertainty within the phylogeny as it stands. Uncertainty within the simosthenurin arm of the phylogeny is manifested as a trichotomy between *"Simosthenurus" brachyselenis*, *"Si." tirarensis* and a clade consisting of the remainder of *Simosthenurus s. l.* and *Procoptodon*. Exclusion of *S. murrayi* and either *"Si." brachyselenis* or *"Si." tirarensis* from the analysis results in one most parsimonious tree (115 steps). Collection of more material pertaining to each of these poorly known species is required to further clarify sthenurine interrelationships.

To determine the effects of different clades on tree shape, I ran a series of heuristic searches with particular taxa excluded. Strict consensus trees produced after the removal of the following taxa prompted no other topological changes relative to the original 115 step strict consensus tree:

1. *Hadronomas* (112 steps, 15 trees);
2. *Sthenurus* (88 steps, 3 trees);
3. All *Sthenurus* species except *S. andersoni* (93 steps, 3 trees);
4. *Sthenurus notabilis* (114 steps, 9 trees);
5. *Sthenurus murrayi* (115 steps, 3 trees);
6. *Sthenurus notabilis* and *S. murrayi* (114 steps, 3 trees);
7. *Archaeosimos* (112 steps, 15 trees);
8. *Simosthenurus s. l.* (92 steps, 5 trees);
9. *Simosthenurus s. s.* (103 steps, 15 trees);
10. *"Simosthenurus" brachyselenis* (113 steps, 5 trees);
11. *"Simosthenurus" tirarensis* (113 steps, 5 trees);
12. *"Simosthenurus" brachyselenis* and *"Si." tirarensis* (113 steps, 5 trees);
13. *"Simosthenurus" baileyi*, *"Si." antiquus* and *"Si." pales* (109 steps, 15 trees);
14. *Simosthenurus maddocki* (108 steps, 15 trees).
15. *Procoptodon s. l.* (86 steps, 15 trees);
16. *"Procoptodon" gilli* (109 steps, 15 trees).

By excluding the following taxa, strict consensus trees markedly different in topology to the original 115 step tree resulted.

1. *Sthenurus andersoni* (115 steps, 60 trees).
2. *Sthenurus andersoni* and *Metasthenurus* (108 steps, 90 trees);
3. *Sthenurus andersoni*, *Metasthenurus* and *Archaeosimos* (104 steps, 90 trees).
4. *Sthenurus* and *Metasthenurus* (81 steps, 45 trees);
5. *Metasthenurus* (109 steps, 81 trees);
6. *Metasthenurus* and *Archaeosimos* (105 steps, 117 trees);
7. *Metasthenurus*, *Archaeosimos*, *"Simosthenurus" brachyselenis* and *"Si." tirarensis* (99 steps, 27 trees).
8. *"Simosthenurus" brachyselenis*, *"Si." tirarensis*, *"Si." baileyi*, *"Si." antiquus* and *"Si." pales* (105 steps, 20 trees);
9. *Simosthenurus occidentalis* (115 steps, 390 trees).
10. *"Procoptodon" browneorum* (114 steps, 45 trees).

Strict consensus trees 4 and 8-10 retain the same general topology as the original tree (fig. 21), but collapse all or some of the species of *Simosthenurus s. l.* into polytomies with *Procoptodon s. l.*, or with *"P." browneorum*, *"P." gilli* and the remainder of the *Procoptodon* clade. *Si. occidentalis* and *"P." browneorum* are therefore pivotal in resolving relationships within the middle part of the Simosthenurin arm of the tree, which is hardly surprising given that they represent three of the five most completely known simosthenurins (along with the highly autapomorphic *"P." gilli* and highly derived *P. goliah*). Strict consensus trees 1-3 and 5-6 differ markedly from the original tree. Removal of *Sthenurus andersoni* and *Metasthenurus* together (tree 2) collapses the phylogeny into a picket fence, although *Sthenurus* and *Procoptodon* are retained as end posts. Removal of *Metasthenurus* in concert with *Archaeosimos* (tree 6) produces a similar tree, except that *Sthenurus* is maintained as a sister taxon of the simosthenurins. Trees 5 and 7 differ by breaking up the *"Si." baileyi* – *"Si." antiquus* – *"Si." pales* clade, placing *"Si." pales* as sister taxon of the remainder of the simosthenurins, followed successively by *"Si." antiquus*, then *"Si." baileyi* as sister taxon of a *Simosthenurus s. s.* – *Procoptodon* clade. Removal of *S. andersoni* alone (tree 1) produces a similar strict consensus result, except that *Metasthenurus* is reconstructed as the sister taxon of a *Sthenurus* – Simosthenurini clade.

Two general conclusions can be drawn from these manipulations. First, *Sthenurus andersoni* and *Metasthenurus newtonae* exert a very significant influence on tree topology. As the earliest derivatives of each major sthenurine clade (i.e., *Sthenurus* and its sister taxon), they stabilize the phylogeny by breaking up the long branch lengths between the very plesiomorphic *Hadronomas* and the crown groups. Second, the most completely known species play crucial roles in shaping a phylogeny that includes nine species for which only one or two incomplete specimens are known. Character support for all clades and taxa in the strict consensus of the 115 equally most parsimonious trees are considered individually below.

APOMORPHIES

With a phylogenetic hypothesis in place, it is necessary to scrutinize the character state changes for each branch of the tree. While it might be more parsimonious to interpret only unambiguous state changes as apomorphies, missing data often renders it impossible to reveal on which branch a particular change took place. This may occur even if a character clearly changes only once. Two ancillary optimization procedures are frequently employed to reveal the uppermost or lowermost branches on which a state change might have occurred if it is ambiguous. Accelerated Transformation (ACCTRAN) pushes change down the tree, favoring state change nearer the base and subsequent reversal. Delayed Transformation (DELTRAN) pushes change up the tree, favoring the parallel acquirement of a state (Swofford and Maddison 1987; Maddison and Maddison 1992; Swofford 1993). On occasion, data not included in the analysis may suggest which change is more likely, allowing it to be cited as an apomorphy for a particular clade with relative confidence.

In addition to reviewing apparent synapomorphies for various clades, I review the autapomorphies for each species as derived from both the taxonomic descriptions and the cladistic analysis. Because autapomorphies recognized beforehand were excluded from the analysis, those revealed by it are homoplastic with apomorphies elsewhere on the tree. Note that I employ the terms synapomorphy and autapomorphy in a broad sense, as character states that define a lineage whether or not those states also appear elsewhere on the tree. Although some branches are only moderately supported at best, the merits of the tree lie in the good character support for several critical branches. The nine craniodental synapomorphies shared by the six sthenurine genera provide very strong support for the monophyly of the group (see Monophyly and Origin). Most features have been charted previously (e.g., Tedford 1966; Murray 1991a) and together represent a unique functional character complex (see Functional Morphology).

Hadronomas puckridgi

1. Frontals taper to rugose posterior supraorbital tuberosity.
2. I1 slightly broadened.
3. Freestanding postorbital process of jugal.
4. Mandibular condyle intermediate between oval-shaped and transversely elongate.
5. p3 bears low, distinct cingula along length on buccal and lingual sides.
6. Molars very low crowned.

Six craniodental autapomorphies are recognized for *Hadronomas puckridgi.* Lateral taper of the frontals to a rugose supraorbital tuberosity is also observed in the much more derived *Simosthenurus maddocki* and *Si. occidentalis*. Significantly, though, the shape of the frontal in these two species is quite different from that of *Hadronomas* and the supraorbital tuberosity is located above rather than posterior to the eye orbit. The arrangement has clearly arisen independently in each genus. A slightly broadened I1 has also been independently derived in *"Si." baileyi* and *"Si." pales*.

The nature of the freestanding postorbital process of the jugal in *Hadronomas* was highlighted by Murray (1991a). Although he considered it a macropodoid plesiomorphy, given its presence in *Potorous* and many phalangeroids, he reasoned that the expression of the process in *Hadronomas* was more likely to represent a reappearance of that feature and, hence, a sthenurine synapomorphy. However, in no other sthenurine for which the zygomatic arch is known does the postorbital process project above the squamosal. For this reason and because *Potorous* is the only other macropodoid with a freestanding postorbital process, I regard this feature as a *Hadronomas* autapomorphy.

Morphology of the mandibular condyle in *Hadronomas* is intermediate between the circular or oval shape seen in macropodines (e.g., *Dorcopsoides*, *Protemnodon*) and the transversely elongate form typical of all other sthenurines. I know of no other macropodoid with a condylar morphology approximating that of *Hadronomas.* Presence of a low cingulid along both the buccal and lingual sides of p3 is also unique to *Hadronomas.* The posterior portion of the buccal cingulid is probably homologous with the buccal crest characteristic of all other sthenurines, but its morphology and the presence of a lingual cingulid (similarly unknown in any other macropodoid) constitute a solid autapomorphy for *Hadronomas.* As more bulungamayine and early sthenurine material is discovered, more *Hadronomas* autapomorphies will be revealed.

Sthenurinae, excluding *Hadronomas*

1. Orbit projected anterolaterally.
2. Neurocranium further elevated and anteriorly rotated; zygomatic arch oriented anteroventrally.
3. Symphyseal plate rugose, often partially fused to opposing plate.
4. Masseteric canal vertical.
5. Lingual cingulum of P3 and posterobuccal cingulid of p3 raised into crests.
6. i1 crown long and robust.
7. p3 buccal crest short.
8. Incipient division of paracristid and cristid obliqua.
9. Molar row curved laterally.

The uniqueness of these nine features strongly unites *Sthenurus*, *Metasthenurus*, *Archaeosimos*, *Simosthenurus* and *Procoptodon* as a monophyletic clade within the Sthenurinae. Some *Protemnodon* species have a long and robust i1, but the crown is much broader. The combined adaptive significance of these traits rendered the evolution of this lineage one of the most important events in the history of the macropodoid radiation (see Functional Morphology).

Sthenurus

1. I1 markedly broadened.
2. P3 lingual crest extends anterobuccally to near base of crown.
3. Molars high crowned.
4. p3 buccal crest length intermediate.
5. p3 median valley width intermediate.

These five features provide strong support for the monophyly of *Sthenurus*. Anterior extension of the P3 lingual crest is present in some individuals of *"Simosthenurus" pales* and *"Procoptodon" browneorum*, but it is much lower and extends around to the buccal side of the crown. Although absence of marked frontal inflation is taxonomically diagnostic of *Sthenurus* within the Sthenurinae (the feature having clearly originated independently in the highly autapomorphic *"Procoptodon" gilli*), it cannot presently be supported as a synapomorphy. There is a distinct possibility that the condition represents a macropodoid plesiomorphy retained in *Sthenurus*, since the character is unordered and *Hadronomas* expresses state 2, the *Metasthenurus*-Simosthenurini clade state 1, and the outgroup state 0. Theoretically, therefore, any state is as likely as the other for the clade giving rise to all sthenurines except *Hadronomas*.

Two of the three features cited by Wells and Tedford (1995) in support of the monophyly of *Sthenurus* are not included here. I find that the 'links' (postprotocrista, paracristid, cristid obliqua) are no better developed in *Sthenurus* than they are in several other taxa, including *Metasthenurus newtonae* and some *Procoptodon* species. In the past, I also have considered well-developed 'links' a *Sthenurus* synapomorphy (e.g., Prideaux and Wells 1998), but I now believe that an illusory perception of 'link' size is conveyed by the relatively high lophs and by the absence of adjacent enamel crenulations. The second feature omitted here is purported absence of enamel from the lingual face of i1. Although the lingual enamel is not as thick in *Sthenurus* as it is in some derived simosthenurins, it is present in all *Sthenurus* species.

Sthenurus andersoni

1. Upper molars bear a few, very fine enamel crenulations.

Sthenurus andersoni is the most widespread and best represented species of *Sthenurus*, but I can identify only one craniodental autapomorphy. *S. andersoni* possesses every feature that characterizes the genus, but is defined only by proportional differences, its smaller size, and the presence of a few, very fine enamel crenulations on the posterior loph faces. The paucity of craniodental autapomorphies suggests that *S. andersoni* must be very similar in structure to the ancestral taxon whence *Sthenurus* was derived.

Sthenurus, excluding *Sthenurus andersoni*

1. Postparacrista well developed.
2. Postprotocrista incipiently divided into two components.
3. p3 buccal crest long.

While three synapomorphies support the monophyly of *Sthenurus* to the exclusion of *S. andersoni*, it is worth recognizing that *S. murrayi* is known only from a palate, and *S. notabilis* from a partial dentary. PAUP reconstructs the unknown states for *S. murrayi* and *S. notabilis* on the basis of parsimony alone. Until these two species becomes better known, it is unlikely that the precise relationships within this portion of the tree will be resolvable.

Sthenurus atlas

1. P3 elongate with deep longitudinal basin and very distinct posterobuccal accessory cusp.
2. Anterior portion of paracristid reduced to a cusp-like structure.

This species is defined by two dental autapomorphies. The validity of the second feature may be open to question, because it has already been cited as an apomorphy for the *Sthenurus atlas–S. stirlingi–S. tindalei* clade (the character is ordered). However, regardless of cladistic regulations, the fact is that, among the Sthenurinae, only in *S. atlas* is the anterior portion of the paracristid reduced to a cusp-like structure, which may or may not represent the paraconid.

Sthenurus murrayi

1. Ridgelet directed anterobuccally from anteriormost cuspule of P3 lingual crest divides anterior basin into two distinct sub-basins.
2. Preparacrista has strong connection with paracone apex on M1 but not on M2-3.
3. Posterior face of protoloph bears relatively distinct enamel crenulations.
4. Buccal (mesial) component of postmetaconulecrista strongly developed.

Four autapomorphies are presently identifiable for *Sthenurus murrayi.*

Sthenurus notabilis

1. p3 shorter than m1.
2. Paracristid markedly U-shaped anteriorly and clearly differentiable from underlying precingulid.

3. Slight parametacristid and preentocristid evident.

Sthenurus notabilis is defined by three autapomorphies, but is known from only one juvenile dentary with four teeth. One upper molar would be enlightening.

Sthenurus tindalei–Sthenurus stirlingi

1. Anterodorsal extension of maxilla.
2. P3/p3 intermediate in length between first and third molars.
3. Postparacrista strongly developed.
4. Inferior transverse torus well developed.
5. p3 buccal crest very long.
6. Trigonid long and wide.
7. Paracristid and cristid obliqua clearly separated into two components.
8. Anterior portion of paracristid not distinct at all.

As previously emphasized by Wells and Tedford (1995), the phylogenetic allegiance of these two taxa is beyond reproach. Only DELTRAN reconstructs an anterodorsal extension of maxilla as a synapomorphy for this clade, since the states for *S. atlas*, *S. murrayi* and *S. notabilis* are unknown.

Sthenurus tindalei

1. Cranium markedly dolichocephalic.
2. Dentary very slender.
3. Diastema markedly procumbent.
4. Connection of p3 buccal crest to anteriormost cuspule of main crest is via a short, crescentic crest.

Sthenurus tindalei is well defined by four autapomorphies.

Sthenurus stirlingi

1. Cranium length intermediate.
2. Buccinator fossa deep.
3. Frontals markedly inflated.
4. Basicranial plane markedly elevated above palatal plane.
5. I3 with very strongly developed anterolingual crest.
6. Molar crown height intermediate.
7. Molars very wide for length.
8. Mesocrista moderately developed.
9. i1 upturned.

Sthenurus stirlingi is one of the most autapomorphic sthenurines. Remarkably, of the nine autapomorphies, only strong development of the I3 anterolingual crest and molar breadth were not revealed by the phylogenetic analysis. This means that these features originated independently elsewhere within the Sthenurinae or that something is alarmingly awry with the analysis. Clearly, the former is the more likely given the strong synapomorphies allying *S. stirlingi* with *Sthenurus*. Although it was never explicitly stated, reasons for questioning the taxonomic validity of *Simosthenurus* (Prideaux and Wells 1994, 1997, 1998) revolved not only around the enigmatic nature of the taxon since named *Metasthenurus newtonae*, but the marked similarities in craniodental morphology between *S. stirlingi* and *"Si." pales*. It is now clear that *S. stirlingi* independently evolved an entire character complex, probably in response to a diet tougher than that of its closest relatives (see Functional Morphology).

Metasthenurus–Simosthenurini

1. Cranium length intermediate.
2. Frontals markedly inflated.
3. Basicranial plane markedly elevated above palatal plane.
4. Inferior transverse torus well developed.

These features provide firm evidence for the monophyly of *Metasthenurus* and the simosthenurins. While it is possible that an unbroadened I1 could be a synapomorphy for this lineage, it may well be a symplesiomorphy given that the outgroup also expresses this condition.

Metasthenurus newtonae

1. Masseteric process short, narrowing distally and twisted posteriorly.
2. Precingulum abruptly terminated after extending lingually across anterior face of protoloph to position of preprotocrista.
3. Postparacrista well developed.
4. Postparacrista and premetacrista incurved lingually.
5. Median dorsal groove of dentary deep and rather narrow.
6. Mylohyoid groove and processes well developed.
7. p3 subequal in length to m2.
8. Lower molars bear pronounced anterior turn of lophid ends.
9. Posterior face of hypolophid strongly inflated.

During the course of this study, *Metasthenurus newtonae* proved the most intriguing and problematic of all the sthenurines, despite the fact that it is one of the better represented species. Its role in shaping the proposed phylogeny is critical. *M. newtonae* possesses a suite of features suggestive of *Sthenurus* (e.g., *S. andersoni*) or the Simosthenurini (e.g., *Simosthenurus occidentalis*, *"Procoptodon" browneorum*), which are counterbalanced by nine characteristics typical only of *M. newtonae*, plus a number of retained plesiomorphies. Initially, it was allocated to *Simosthenurus*, with its origin proposed to have occurred soon after the putative split between *Simosthenurus* and *Sthenurus* (Prideaux and Wells

1994). After considering a number of options, including the possibility that it may be related to *Si. euryskaphus* and *Si. occidentalis* (Prideaux and Wells 1997), I now believe that the weight of evidence favors *M. newtonae* as a plesiomorphic sister taxon of the diverse brachycephalic lineage (tribe Simosthenurini). On balance, a distinction at the generic level is clearly justifiable (cf. Prideaux and Wells 1997; Prideaux 1999b, 2000).

By examining the similarities between *Metasthenurus newtonae* and the species of *Sthenurus* more critically, it is evident that they share no conceivable synapomorphies. Because *S. andersoni* is the earliest derivative of the dolichocephalic lineage, it is hardly surprising that some similarities between this species and *M. newtonae* are evident. These symplesiomorphies include a shallow buccinator fossa, molar cristae with strong, direct connections to cusp apices and very few or no fine enamel crenulations, and a similar i1 morphology. There are also similarities with *S. notabilis*, which shares with *M. newtonae* a paracristid that is U-shaped anteriorly. Similarities between the juvenile crania of *M. newtonae* and *S. andersoni* also support the idea that neither is very far derived from the common ancestor of their respective parent lineages.

Phylogenetically and evolutionarily, *Metasthenurus newtonae* is an excellent intermediary between the Simosthenurini and the plesiomorphic *Hadronomas*. The nine autapomorphies that characterize it and the retained plesiomorphies upon which they are superimposed suggest a long period of phylogenetic separation. A late Miocene derivation for *Metasthenurus* is likely.

Tribe Simosthenurini

1. Cranium brachycephalic.
2. Postprotocrista incipiently divided into two components.
3. Mesocrista moderately developed.
4. i1 upturned.
5. i1 short, robust.
6. Trigonid short.
7. Division of paracristid and cristid obliqua incipient anteriorly grading to clear posteriorly.

The monophyly of tribe Simosthenurini is strongly supported by seven synapomorphies, which together reflect a decisive dietary shift within the subfamily (see Functional Morphology). While the cranium of several simosthenurin species is unknown, there is a good correlation between dentary and cranium length, so it is only for *"Procoptodon" williamsi* that a state for this character cannot be determined. With the exception of *"Simosthenurus" antiquus* and *"Si." pales*, which, from their dentary proportions, appear to have reverted to the intermediate condition, all simosthenurins are brachycephalic.

Archaeosimos

1. p3 only slightly wider posteriorly than anteriorly.
2. Molars very low crowned.

Only two synapomorphies are recognized for this genus at present, which is represented by just the two poorly known species. Future discoveries may unveil a few more synapomorphies for *Archaeosimos*, but it is likely to remain defined more by it possession of the basal simosthenurin traits and lack of the more derived features of this tribe. *Archaeosimos* can thus be viewed as an excellent structural precursor to the rest of the Simosthenurini. Note that although p3 in *Hadronomas* and the bulungamayines is also only slightly wider posteriorly than anteriorly, they are typically much wider centrally.

Prior to this analysis, *Archaeosimos cegsai* and *"Simosthenurus" brachyselenis* were thought to be the two most plesiomorphic sthenurines, with the exception of *Hadronomas* (Prideaux and Wells 1997, 1998). However, upon reexamining the evidence for the rather slapdash phylogenetic appraisal presented in the 1998 paper, it became clear that that analysis incorporated a number of plesiomorphies masquerading as apomorphies. Perhaps most important, it did not consider the information provided by the then undescribed *Metasthenurus newtonae*. Anyone unlucky enough to stumble across a copy of my Ph.D. thesis (Prideaux 1999c) will discover a different phylogenetic arrangement again, but as this work is unpublished and only half-baked, I do not feel a relentless urge to address it here.

Archaeosimos cegsai

1. p3 rather elongate, round-ended, and slightly bulbous in outline anteriorly, centrally and posteriorly.

One autapomorphy is presently identifiable for this plesiomorphic simosthenurin.

Archaeosimos correlli

1. Parametacristid short, but well developed.
2. Fine, low, but distinct preentocristid confluent with similarly developed postmetacristid.

Two distinctive dental features distinguish *Archaeosimos correlli* from all other sthenurines.

Simosthenurus s. l.–Procoptodon

1. p3 median valley width intermediate.

Only one synapomorphy supports the monophyly of the clade whence all simosthenurins except *Archaeosimos* are derived. When more remains attributable to *Archaeosimos* and the more plesiomorphic species of *Simosthenurus s. l.* are described, character support for this clade should be reexamined.

"Simosthenurus" brachyselenis

1. Dentary massive, especially in depth relative to cheek teeth size.
2. p3 subequal in length to m1.

Two autapomorphies are recognized for *"Simosthenurus" brachyselenis*, although the second arises convergently on other parts of the tree. While the dentary of *Procoptodon pusio*, *P. rapha* and *P. goliah* is also massively constructed, *"Si." brachyselenis* is easily distinguished from all sthenurines by the great depth of the ramus relative to the size of the cheek teeth. Indeed, the dentary depth of the one adult specimen known is approximately six times that of the height of the molars. In *P. pusio*, *P. rapha* and *P. goliah*, the massiveness of the dentary is largely a result of greatly increased ramus width rather than depth. In addition, the upper and lower premolars of *P. pusio*, *P. rapha* and *P. goliah* are dwarfed by the size of the molars. Cranial specimens of *"Si." brachyselenis* promise to shed more light on its exact phylogenetic relationships. If the eccentricity of its cranial traits mirror that of the dentary, separate generic status may eventually be warranted.

"Simosthenurus" tirarensis

1. p3 shorter than m2.
2. p3 buccal crest short, but joined anteriorly to fourth cuspule of main crest.
3. Molars high crowned.
4. Small, but distinct pocket formed at confluence of postentocristid and A-shaped inflation on posterior hypolophid face.

"Simosthenurus" tirarensis is characterized by three autapomorphies, the first two of which represent traits that arise elsewhere within the tribe. *"Procoptodon" gilli* is the only other high-crowned simosthenurin, while *Si. euryskaphus* and *"P." oreas* are the only other simosthenurins wherein p3 is intermediate in length between m1 and m2. Each of these state shifts is very likely to be convergent, but more specimens of *"Si." tirarensis* must be described before its precise relationships will become manifest.

Simosthenurus s. l.–Procoptodon, excluding *"Simosthenurus" brachyselenis* and *"Simosthenurus" tirarensis*

1. p3 buccal crest length intermediate.
2. Paracristid and cristid obliqua clearly separated into two components on all molars.

The two character states interpreted as synapomorphies for this clade appeared independently in *Sthenurus*, each on a different branch within that lineage. Adaptively, this is significant (see Functional Morphology), but the marked phylogenetic separation between these groups engenders substantial confidence in the level of support that they provide for this simosthenurin lineage. Nevertheless, future discoveries of more specimens of *"Simosthenurus" brachyselenis* and *"Si." tirarensis* will be crucial in assessing the relative phylogenetic strength of this clade.

"Simosthenurus" baileyi–"Simosthenurus" antiquus–"Simosthenurus" pales

1. I1 slightly broadened.
2. Postparacrista well developed.

Two synapomorphies support the monophyly of this clade, but the I1 state cannot be assessed in *"Simosthenurus" antiquus*, because no upper incisors are known for this species. Moreover, the postparacrista state of neither *"Si." brachyselenis* nor *"Si." tirarensis* can be evaluated, as no upper teeth are known. This means that although the postparacrista of these three species is better developed than in *Archaeosimos*, a question mark still resides over this shared state. For the validity of this clade to receive more rigorous support, corroborating evidence from hitherto undescribed elements of *"Si." antiquus*, *"Si." pales*, *"Si." brachyselenis* or *"Si." tirarensis* is a necessity. Until then, recognition of a new genus for these three simosthenurins would be taxonomically perilous.

"Simosthenurus" baileyi

1. Rostrum very broad and short, with very wide nasals.
2. P3 bears two cuspules anterior to posterobuccal accessory cusp.
3. Premetacristid and buccal component of paracristid very well developed.

These features presently appear to be solid autapomorphies for *"Simosthenurus" baileyi*, but no upper craniodental elements are known for *"Si." brachyselenis* or *"Si." tirarensis*. Although the cranium is also yet to be described for *"Si." antiquus* and *"Si." pales*, the rostrum of either species is unlikely to have been as stout as in *"Si." baileyi*, as their respective dentaries suggest intermediate cranial lengths. Presence of a posterobuccal accessory cusp, buccal to the posterior extremity of the main P3 crest, is a common feature in sthenurines, but separate smaller cuspules anterior to this feature have not been noted in any other species except *Metasthenurus newtonae*, in which one poorly differentiated cuspule is observed in rare individuals.

It is worth noting here that the molars of *"Simosthenurus" baileyi* were originally described as very low crowned (Prideaux and Wells 1998). However, following the development of the crown height index (see Character 12, table 42), the state "very low crowned" is now considered to apply only to *Archaeosimos*,

Hadronomas and *"Procoptodon" mccoyi*, which are notably lower crowned than *"Si." baileyi.* The phylogeny presented here also refutes the idea that *"Si." baileyi* might be the sister taxon of the remaining *Simosthenurus* species (cf. Prideaux and Wells 1998), due chiefly to the reconsideration of *"Si." brachyselenis* and *Metasthenurus newtonae.*

"Simosthenurus" antiquus–"Simosthenurus" pales

1. Cranium length intermediate.

As with the lower branch, the cladistic union of *"Simosthenurus" antiquus–"Si." pales* could benefit from more rigorous support. If my extrapolation of cranium length based on dentary proportion is valid for these two species, then reversion to the intermediate state is a solid synapomorphy.

"Simosthenurus" antiquus

1. P3 with lingual crest markedly lower than main crest.
2. P3 with very large posterior basin relative to longitudinal basin.
3. P3 short relative to p3.
4. p3 with top of anterior portion of sinusoidal main crest flexed over to buccal side.
5. Posterior face of hypolophid with broad inflation extended right across basal portion.

Four of the five autapomorphies for *"Simosthenurus" antiquus* relate to the third permanent premolars, and unfortunately, only one specimen of each tooth is known. Nevertheless, these features all vary minimally within species for which large samples are known, supporting the likelihood that they are unique to the species.

Prideaux and Wells (1998) misinterpreted several aspects of the dental morphology of *"Simosthenurus" antiquus* in light of the phylogeny presented here. They put *"Si." antiquus* forward as the least derived species of *Sthenurus*, based partly on an overestimation of molar crown height and partly on the disparity between the height of the main and lingual crests of P3. In fact, molar crown height in *"Si." antiquus* is much lower than in most *Sthenurus* species (table 42), and the lingual crest of the P3 is lower compared with the main crest than it is in *all* other sthenurines, with the perhaps quite significant exception of the *"Si." pales* holotype P3 (QM F815).

"Simosthenurus" pales

1. Postprotocrista markedly separated into two components.
2. p3 buccal crest long.
3. Anteriormost cuspule of p3 main crest very prominent.
4. Very large.

Although four autapomorphies are identified for *"Simosthenurus" pales*, it is the least phylogenetically potent character of size that is foremost in the practical recognition of this species. Otherwise, the prominent nature of the anteriormost cuspule of the p3 main crest easily distinguishes *"Si." pales* from all other sthenurines, especially in combination with the elongate buccal crest. As mentioned earlier, many aspects of the dentary and dentition of *"Si." pales* are highly convergent with *Sthenurus stirlingi*. When an *"Si." pales* cranium is found, it will prove interesting to see whether more convergent similarities with *S. stirlingi* are revealed.

Simosthenurus s. s.–Procoptodon

1. Postprotocrista divided into two clearly separate components.

Although this synapomorphy represents a critical state shift, one would ideally like to see more character support for this clade, particularly given that *Si." pales* is characterized by the same state shift. Again, the retrieval of more specimens attributable to *"Si." brachyselenis*, *"Si." tirarensis*, *"Si." antiquus*, *"Si." pales* or *Si. euryskaphus* may prove enlightening. While I am confident that the phylogeny presented here is a feasible hypothesis for relationships within the Simosthenurini, it is a concern that two species of *Simosthenurus s. s.* and the two most plesiomorphic *Procoptodon* species constitute four of the best known taxa, but can only weakly resolve this portion of the tree. Alternative data sets, like pedal morphology, may help clarify relationships in the future. Articulated pedal remains are known for each of these species (*Simosthenurus occidentalis*, *Si. maddocki*, *"Procoptodon" browneorum*, *"P." gilli*), but have not yet been examined from a phylogenetic perspective. Bishop (1997) believed that the variation in pedal morphology between sthenurines might have considerable phylogenetic utility.

Simosthenurus s. s.

1. Frontals taper laterally to rugose anterior supraorbital tuberosity.
2. p3 median valley wide.
3. Lophid faces bear numerous very fine enamel crenulations.

The monophyly of *Simosthenurus s. s.* is quite well supported by three synapomorphies, although the cranium is known only for *Simosthenurus occidentalis* and *Si. maddocki*. However, because the molars of the poorly known *Si. euryskaphus* fit wholly within the range of size and morphological variation observed for *Si. occidentalis*, there is little doubt that this species fits within this clade.

Simosthenurus euryskaphus

1. p3 shorter than m2.
2. p3 median valley is a wide, circular basin due to marked curvature of buccal crest and posterior end of main crest.

Two solid p3 autapomorphies are identified for this poorly represented species.

Simosthenurus maddocki–Simosthenurus occidentalis

1. p3 much wider posteriorly than anteriorly.

Support for this clade is very weak, particularly given that the lower molars of *Simosthenurus euryskaphus* fall within the range for *Si. occidentalis*. More material of this poorly known species is needed to confidently resolve relationships within *Simosthenurus s. s.* Retrieval of more specimens of the possible new species referred to herein as *Si.* sp. cf. *maddocki* may also be enlightening.

Simosthenurus occidentalis

1. Masseteric process widely flared.
2. Lower incisor short and very robust, with strongly sinusoidal occlusal surface.
3. p3 distinctly divided into narrow anterior and wide posterior portions, demarcated by wide, vertical groove below anterior extremity of buccal crest.

Despite being one of the most completely known of all sthenurine species, it is difficult to identify autapomorphies for *Simosthenurus occidentalis*. This is largely due to the marked variability observed within and between populations of this widespread species, and its rather generalized simosthenurin form.

Simosthenurus maddocki

1. Postglenoid process very shallow.
2. I1 much higher crowned than I2-3.
3. I3 only slightly longer than combined length of I1-2.
4. Cheek tooth rows only very slightly curved laterally.
5. Upper molars small relative to size of cranium.
6. Upper molar cristae directly connect to cusp apices.
7. Preprotocrista distinct.
8. Postparacrista and premetacrista incurved lingually.
9. Mesocrista weakly developed.
10. Inferior transverse torus absent.
11. i1 very slender and elongate.
12. p3 buccal crest length intermediate.
13. p3 very low crowned, slender anteriorly, with a row of cuspules between buccal and main crests.

In stark contrast to *Simosthenurus occidentalis*, more autapomorphies are recognized for *Si. maddocki* than for any other sthenurine species. *Si. maddocki* is the only sthenurine known for which the length of I3 only slightly exceeds the combined I1-2 length. In *Procoptodon goliah*, I3 is equal to I1-2 in length, but the I3 of all other sthenurines is markedly longer than I1-2. Compared with *Si. occidentalis*, the frontals of *Si. maddocki* are inflated and laterally tapered to a slightly greater (and less variable) degree, and the very fine enamel crenulations coating the molar surfaces are more numerous. These provide two more qualitative characteristics by which *Si. maddocki* may be distinguished from all other sthenurines. Together, these 13 traits form a unique functional complex related to the dietary adaptations of *Si. maddocki* (see Functional Morphology).

Procoptodon s. l.

1. Mesocrista well developed.
2. p3 buccal crest long.
3. Lophid faces bear coarse enamel crenulations.

Three synapomorphies provide solid support for the monophyly of *Procoptodon s. l.* (see Taxonomic Implications). Because the lophids of the sister group (*Simosthenurus s. s.*) bear numerous very fine enamel crenulations, and the remainder of the sthenurines plus *"P." gilli* retain state 0, only ACCTRAN reconstructs coarse enamel crenulations as a *Procoptodon* synapomorphy. However, given that all species of *Procoptodon s. l.* except *"P." gilli* show this condition, I consider it highly likely that the paucity of crenulations in *"P." gilli* is a reversal correlated with several similarly homoplastic autapomorphies in other craniodental characters.

"Procoptodon" browneorum

1. Frontals very inflated, but rather flat-sided.
2. Rostrum rather cylindrical.
3. Digastric eminence very large and posteroventrally extended.

While *"Procoptodon" browneorum* is one of the best known sthenurines, it is characterized by only three autapomorphies, two features of the splanchnocranium and strong development of the digastric eminence. A paucity of autapomorphies commonly seems to typify taxa that have diverged only slightly from a clade ancestor (e.g., *Sthenurus andersoni*, *Archaeosimos cegsai*), which is congruent with the phylogenetic placement of *"P." browneorum.*

Procoptodon s. l., excluding *"Procoptodon" browneorum*

1. Dorsal surface of neurocranium domed.
2. Preprotocrista distinct.
3. Postparacrista and premetacrista incurved lingually.

The monophyly of this lineage is only fairly supported at best. The second synapomorphy is known for *"Procoptodon" gilli* and the four crown species, while the first is known only in *"P." gilli*, *P. pusio* and *P. goliah.* If these features are eventually shown to be present in *"P." mccoyi*, *"P." oreas* and *"P." williamsi*, then this lineage will be characterized by two reversals to plesiomorphic macropodoid conditions. Aside from its convergent appearance in *Simosthenurus maddocki*, a distinct preprotocrista has only been noted elsewhere within the Macropodoidea in potoroines and bunolophodont bulungamayines. The third feature is open to some question, because only ACCTRAN reconstructs it as a synapomorphy for this lineage. Like the distinct preprotocrista, it is shared by *"P." gilli*, *"P." williamsi*, *P. pusio*, *P. rapha* and *P. goliah*, but the change is not unambiguous, because of the possibility of an independent origin in *"P." gilli*, given that no upper molars are known for *"P." mccoyi.* At present, I view this as the less likely scenario.

"Procoptodon" gilli

1. Frontal inflation not marked.
2. Basicranial plane only slightly elevated above palatal plane.
3. Masseteric process barely more than a rounded tumescence.
4. Premaxilla markedly declined and becoming very thin dorsally.
5. I3 very high crowned.
6. Premolars large, with crown height markedly greater than molars.
7. Molars high crowned.
8. Very few or no enamel crenulations on lophid faces.

"Procoptodon" gilli is one of the most divergent sthenurine species and is typified by eight autapomorphies. Several of these, particularly the first three and possibly the last two, suggest reversal to the plesiomorphic macropodoid or outgroup conditions. Marked similarities between the juvenile crania of *"P." browneorum* and both juvenile and adult crania of *"P." gilli* (figs 5-6) suggest that this diminutive species may have arisen pedomorphically from an ancestor very close to *"P." browneorum.* Dental similarities are most evident in *"P." gilli* and smaller (Western Australian or Kangaroo Island) individuals of *"P." browneorum* than they are in *"P." gilli* and the larger (mainland southeastern Australian) specimens. In some aspects of molar morphology, the smaller *"P." browneorum* specimens are intermediate between eastern *"P." browneorum* and *"P." gilli*, most notably in terms of the expression of enamel crenulations on the lophid faces. An ancestor-descendant relationship between a population of *"P." browneorum* and *"P." gilli* is a conceivable evolutionary scenario.

Procoptodon s. l., excluding *"Procoptodon" browneorum* and *"Procoptodon" gilli*

1. Premolar equal to or shorter than third molar.

This lineage is upheld by only one synapomorphy. Corroboration of its monophyly will rely on the discovery of more material attributable to *"Procoptodon" mccoyi.* Evolutionarily, just as with *"P." gilli*, the marked similarities between *"P." browneorum* and *"P." mccoyi* suggest that a population of the more plesiomorphic species may have given rise to the latter.

"Procoptodon" mccoyi

1. Lower molars very wide relative to length.
2. Lower molars very low crowned.

Two dental autapomorphies are identifiable for *"Procoptodon" mccoyi.* Its lower molars are wider relative to their length than in any other sthenurine and lower crowned than in any species, except *Archaeosimos cegsai.* Otherwise, *"P." mccoyi* is mainly characterized by proportional differences from *"P." browneorum*, the species it most closely resembles. Until a little-worn p3 or any moderately well preserved cranial or upper dental elements of *"P." mccoyi* are discovered, its current phylogenetic position should be considered rather tenuous.

"Procoptodon" oreas–"Procoptodon" williamsi–Procoptodon pusio–Procoptodon rapha–Procoptodon goliah

1. Mylohyoid groove deep and rather narrow.
2. Masseteric foramen large.

Practically, both of these features only unite *"Procoptodon" oreas*, *P. pusio*, *P. rapha* and *P. goliah*, because only one maxillary specimen of *"P." williamsi* is known. However, the latter species is drawn into this part of the tree on the basis of upper dental characters that firmly ally it with *P. pusio*, *P. rapha* and *P. goliah.* Further support for the monophyly of these five *Procoptodon* species is provided by the increased coarseness of the molar cristae in *"P." oreas*, which may be antecedent to the very high, coarse cristae characteristic of *P. pusio*, *P. rapha* and *P. goliah* (Prideaux and Wells 1997). Similarly, the lateral surface of the ramus of *"P." oreas* is slightly more convex than in most sthenurines, which also provides an appropriate phylogenetic precursor to the wide, massive dentaries of the three crown species. In his description of *"P." oreas*, De Vis (1895) appears to have used the same similarities between these four species as the basis for synonymizing *Procoptodon* with what was then an all-encompassing genus *Sthenurus.*

"Procoptodon" oreas

1. Postparacrista and premetacrista oriented anteroposteriorly.
2. Paracristid and cristid obliqua thick but low.
3. Buccally shifted preentocristid strongly developed and parallel to cristid obliqua.

Three autapomorphies are recognized for *"Procoptodon" oreas.* The second and third typify no other sthenurines, while the first is an ACCTRAN prediction, i.e., it assumes that lingual curvature of the postparacrista and premetacrista is actually a synapomorphy for the species of *Procoptodon s. l.*, excluding *"P." browneorum.* More autapomorphies are bound to be identified when *"P." oreas* becomes better represented, especially by upper dental and cranial elements.

"Procoptodon" williamsi–Procoptodon pusio–Procoptodon rapha–Procoptodon goliah

1. Premolar shorter than or equal in length to first molar.
2. Lophs taper toward crests.

These two dental features are solid synapomorphies for *"Procoptodon" williamsi*, *P. pusio*, *P. rapha* and *P. goliah.* Extreme shortening of the premolar relative to the molars evolved independently in both *Sthenurus notabilis* and *"Simosthenurus" brachyselenis.*

"Procoptodon" williamsi

1. P3 has no coarse transverse ridgelets in deep longitudinal basin.
2. P3 bears a small distinct cusp mid-depth on crown at anterior end of tooth.
3. M1 mesocrista slightly better developed than postprotocrista.
4. Postparacrista fine and low.

Four autapomorphies are identifiable for *"Procoptodon" williamsi*, even though it is only known from the holotype maxilla and paratype M1. Collection of more *"P." williamsi* material will enable the various synapomorphies cited for this portion of the tree to be better assessed. Positioning of *"P." williamsi* as the sister taxon of the three most derived *Procoptodon* species confirms the astuteness of the observations made by Tedford and Wells (1990:277).

Procoptodon s. s.

1. Buccinator fossa very deep.
2. Basicranium greatly elevated, basioccipital markedly flexed.
3. Molars with very high, coarse cristae.
4. Postparacrista strongly developed.
5. Mesocrista strongly developed, combined with posterior portion of postprotocrista to form 'midlink'.

6. Dentary massively proportioned.
7. Posterior mental foramen tiny to absent.
8. i1 very slender.

While I recognize eight synapomorphies for *Procoptodon s. s.*, some may also characterize *"Procoptodon" williamsi*, and possibly also *"P." oreas* and *"P." mccoyi*. However, until more fossils referable to these species are described, they are best considered synapomorphies for *P. pusio*, *P. rapha* and *P. goliah*. It is also important to note that the basicranium of *P. rapha* is unknown, so the second shared trait is estimated from *P. pusio* and *P. goliah* alone. Together, these eight synapomorphies represent a critical functional complex (see Functional Morphology) and highlight why these three species have been placed in their own genus for the past 130 years. As I discuss shortly (Taxonomic Implications), the concept of *Procoptodon s. l.* presented here is phylogenetically informative when the structure of the overall phylogeny is considered, even if it is only an interim measure.

Procoptodon pusio

1. Preprotocrista well developed, contacts protocone apex.
2. p3 small and narrow relative to molars.
3. i1 elongate.

These three features are solid autapomorphies for *Procoptodon pusio*.

Procoptodon rapha - Procoptodon goliah

1. p3 buccal crest very long.
2. Posthypocristid present.

Considering the general similarity between these two species and the difficulties often encountered in distinguishing them (see Marcus 1976), it was initially surprising that only two synapomorphies were revealed by the analysis. However, the manner in which *Procoptodon rapha* and *P. goliah* differ from *P. pusio* is based largely on proportional rather than discrete features. Extreme anterior extension of the buccal crest of p3 has evolved in parallel in these two species and the large, derived *Sthenurus stirlingi* and *S. tindalei*. In contrast, a strong posthypocristid is unique to *P. rapha* and *P. goliah* among the sthenurines, even though the A-shaped postcingulid often observed on the posterior hypolophid face of other sthenurines is a partial expression of the same trait. If the structure is correctly interpreted as a posthypocristid, then it constitutes a reemergence of a plesiomorphic macropodoid feature last observed in early bulungamayines, but independently derived in the largest *Macropus* species (see Dental Character Reversal).

Procoptodon rapha

1. P3 with greatly developed, crest-like posterobuccal accessory cusp.
2. p3 median valley wide.
3. Lower molar enamel crenulations well developed, with numerous arising laterally from paracristid and cristid obliqua.

Three solid autapomorphies are identified for *Procoptodon rapha.*

Procoptodon goliah

1. Molars strongly tapered toward loph crests.
2. p3 with small median valley fully enclosed by buccal crest posteriorly.
3. I3 length equal to combined I1-2 length.

Three autapomorphies are recognized for *Procoptodon goliah.* Reduction in length of the I3, which is a partial reversal to the outgroup condition, may eventually be revealed in more *Procoptodon* species than *P. goliah*, but the upper incisors of *P. rapha*, *P. pusio*, *"P." williamsi*, *"P." oreas* and *"P." mccoyi* are unknown.

TAXONOMIC IMPLICATIONS

One of the major objectives of modern systematics is to ensure that the taxonomy of a group adequately reflects relationships based on common descent. The phylogeny presented here provides strong support for the monophyly of *Hadronomas*, *Sthenurus*, *Simosthenurus s. s.* and *Procoptodon.* Concerns previously held over the distinction of *Simosthenurus s. l.* and *Sthenurus* (e.g., Prideaux 1996; Prideaux and Wells 1997) are dismissed, although *Simosthenurus s. l.* is paraphyletic, since *Procoptodon* arises from within it. This possibility was highlighted by Flannery (1989). In time, *"Si." brachyselenis*, *"Si." tirarensis*, *"Si." baileyi*, *"Si." antiquus* and *"Si." pales* will be more assuredly placed, but better material is needed for each species.

Since Owen raised *Procoptodon* in 1874, the taxon has proved remarkably stable. Archer (1978) named *P. texasensis* on the basis of two partial teeth, but I recommend that this species be treated as a synonym of *P. rapha.* While the phylogeny strongly supports the monophyly of the three traditional species (*Procoptodon s. s.*), I also employ the generic name in a broader sense (*Procoptodon s. l.*) to include five other species arising pectinately between Owen's three species and *Simosthenurus s. s.* This concept will undoubtedly have its critics, but as an interim measure I believe it to be more informative than including these species in *Simosthenurus s. l.*, placing each of the five species into a new genus, or lumping all sthenurines arising from the sister clade of *Sthenurus* into *Simosthenurus*, as proposed by Prideaux (1999b-c). At this stage, the monophyly of *Procoptodon s. l.* is supported by three good dental synapomorphies. Until more useful specimens of *"P." mccoyi*, *"P." oreas* and *"P." williamsi* come to light or alternative data sets are incorporated into phylogenetic analyses, it is most systematically informative, if only temporary, to recognize these taxa plus *"P." browneorum* and *"P." gilli* as species of *"Procoptodon"*.

Another important taxonomic change prompted by this analysis is the emplacement of two species arising between *Sthenurus* and *Simosthenurus* into two new genera, *Metasthenurus* and *Archaeosimos.* Evolutionarily, they highlight two crucial stages in the development of the brachycephalic lineage distinguished here as tribe Simosthenurini. *Metasthenurus* is the sister taxon of the Simosthenurini; *Archaeosimos* is the most plesiomorphic genus within it. Contrary to the proposal of Prideaux (1996), I do not now believe that the separation of *Hadronomas* or, for that matter, *Sthenurus* or *Metasthenurus* into tribes is yet warranted. While each taxon is as distinctive as the Simosthenurini, raising separate tribes would fail to augment the proposed taxonomy in any informative way until such time as new taxa closely allied with *Hadronomas*, *Sthenurus* or *Metasthenurus* are described.

DENTAL CHARACTER REVERSAL

Homoplasy pervades most cladistic data sets and can be a tormenting hindrance to phylogenetic interpretations. Systematists often equate homoplasy with convergence, but this need not be the case. Phylogenetic character reversal has been documented in a number of different lineages and may be an important but underestimated mechanism of evolutionary change (Van Valen 1978; Stiassny 1992 and papers cited therein). It has been suggested that spontaneous atavisms in individuals and character reversals in taxa stem from a reactivation of conserved morphogenic pathways (Hall 1984; McCune 1990; Stiassny 1992), although opinions differ on their role in rapid evolutionary change (cf. Gould 1980; Levinton 1986). Subtle reversals, however, need not invoke a "hopeful monster" scenario. It is estimated that only a small selective advantage is required for an allele to become widespread in a population after only a few tens of generations (Strickberger 1976; Marshall et al. 1994; Raff 1996). Over millions of years, the potential for widespread character reversal must be quite significant, even if it is only occasionally recognized. Phocine seals are an excellent example (Wyss 1988; Berta and Wyss 1994). Here, I consider examples of phylogenetic character reversal within the sthenurine dentition and consider possible developmental mechanisms that might have led to character re-expression.

The mesocrista, urocrista, preprotocrista and posthypocristid are lacking from the molars of the three most plesiomorphic sthenurine taxa, *Hadronomas*, *Metasthenurus* and *Sthenurus andersoni*, although a fine, low mesocrista and urocrista are visible on the dP3 of *Metasthenurus* (pl. 35A). These two crests appear on the molars of the more plesiomorphic simosthenurins (e.g., *Archaeosimos*, pl. 37A), where they are also fine and low. Their progressive enlargement takes place within successively more derived taxa. A posthypocristid is conspicuous only in the highly divergent *Procoptodon rapha* and *P. goliah* (pls 77, 84), while a well-developed preprotocrista is distinct only in *P. goliah*, *P. pusio* and *Simosthenurus maddocki* (e.g., pls 44, 78). Antecedent morphologies, however, are often evident in other simosthenurins, where they take the form of an A-shaped postcingulid (e.g., pl. 93) or a small eminence just lingual of the tooth midline on the precingulum (pl. 99), respectively.

These four cristae are interesting from a developmental evolutionary perspective, because they are not unique to sthenurines. The plesiomorphic bulungamayines *Nowidgee* (see Cooke 1997b) and *Purtia* (see Case 1984) express them and they also occur in balbarines (e.g., *Nambaroo*, see Flannery and Rich 1986), potoroines, propleopines and hypsiprymnodontines. A urocrista is present in several macropodines (see Monophyly and Origin) and a large posthypocristid typifies the large species of *Macropus*, but is lacking from the more diminutive species. The structure is most outstanding in *M. pearsoni* (Bartholomai, 1973b) where it often terminates at a distinct, reemergent hypoconulid. Despite the 15- to 20-million-year time gap between the last expression of these four cristae (each disappeared with the parallel development of true bilophodonty in both bulungamayines and balbarines; see Cooke 1997b-c), I believe that the mesocrista, urocrista and posthypocristid of sthenurines are the reemergent homologues of these ancient structures. Their topological relationships are essentially identical to the plesiomorphic configurations, a phylogenetic "fade-out" is seen within the Bulungamayinae and a clearly traceable reemergence is evident within the Sthenurinae.

Marshall et al. (1994) proposed that silenced genes and developmental pathways may be able to lie inactive for up to six million years before the chances of re-expressing a lost phenotypic character become overwhelming small. Re-expression may occur over shorter spans of time if, for example, a reverse mutation in control circuitry reactivates genetic and developmental expression above a threshold level (Marshall et al. 1994; Raff 1996). Structures that reappear after much longer periods of phenotypic dormancy can presumably do so because the genes controlling their expression have remained active during the intervening period. Keränen et al. (1998) have demonstrated that a close association exists between the expression domains of the developmental regulatory gene *Fgf-4* and secondary enamel knots, epithelial control centers for the formation of cusps (Jernvall 1995; Jernvall and Thesleff 2000). As folding of the epithelial sheet appears to be partly regulated by differences in rates of cell proliferation caused by secondary knot activity (Keränen et al. 1998), it is conceivable that subtle heterochronic variations in gene activity might determine the degree of expression of a crest. Perhaps evolutionary fluctuations in the activities of cusp activators and inhibitors, which Jernvall and Thesleff (2000) believe might determine cusp patterns, can also control the formation and size of a crest. At present, too little is known about these underlying interactions to guess at what the fundamental genetic mechanism for the return of a lost crest might be.

However, it is known that the duration over which a cusp develops is proportional to its size and that this is dictated by exactly when its formation is initiated (Jernvall 1995). It would seem reasonable, therefore, to assume that the relative size of a crest is controlled in a similar manner. For example, in the plesiomorphic simosthenurin *Archaeosimos*, the mesocrista is very small (pl. 37A; fig. 19E), so it can only have had a short developmental period. In *Procoptodon goliah*, the mesocrista is subequal in size to the postprotocrista and postparacrista (pl. 75; fig. 19F), which implies a much earlier initiation and a long developmental period. The genes and developmental pathways responsible for enlarging a crest must also account for its reduction by controlling the timing of its onset. Evolutionarily, this type of heterochrony is known as predisplacement if formation is initiated earlier than in the ancestor or as postdisplacement if initiation in the descendant is relatively later (Alberch et al. 1979; Jernvall 1995). With this in mind, there may be two feasible explanations for the evolutionary reversal patterns seen in sthenurine molars, which neither rely on an explicit understanding of the complexities of the underlying genetic interactions nor contradict the six-million-year limit postulated by Marshall et al. (1994).

Because an enamel crown is formed from a folded, continuous epithelial sheet, not from discrete elements, the phylogenetic "fade-out" of a crest might be simply a reduction in its size relative to that of the surrounding enamel, or a relative increase in the thickness of the surrounding enamel. This suggests that the epithelial region that would form the crest still *does* form a sliver of enamel homologous with the crest, but is indistinguishable with the naked eye. The evolutionary implication would be that the reappearing sthenurine crests were not actually "lost", but their development was postdisplaced to such a degree that they did not exceed the thickness of the surrounding enamel. Support for this idea is provided by the fluctuating degree of expression shown by other sthenurine molar crests (e.g., premetacrista, buccal component of paracristid, premetacristid), which postdisplace (fade-out) and then predisplace (reemerge) over much shorter periods of evolutionary time. Sometimes crests are so fine and low that they may be observed only under magnification.

An alternative explanation for dental character reversal in sthenurines could be the one postulated to account for the phylogenetic reappearance of m2 and the m1 metaconid / trigonid complex in the European lynx (Kurtén 1963; Werdelin 1987) and M3 in the bat-eared fox (Van Valen 1964): developmental field expansion. Because molars are serially homologous, an expansion of the molarization field and re-execution of the developmental program for one molar may stimulate the reemergence of its successor (Jernvall 1995). Supernumerary molars, as observed in manatees (Domning and Hayek 1984) and the little rock wallaby or nabarlek (Sanson 1989), owe their existence to the same phenomenon, as most likely do the spontaneous atavistic occurrences of supernumerary teeth (e.g., P1 in roe deer, Kierdorf and Kierdorf 1990; dP1 in Lumholtz's tree kangaroo and *Simosthenurus maddocki*, and m5 in *Simosthenurus occidentalis*, see Taxonomy). It is also believed that gradients in cusp or crest expression along a molar row, such as with the division of the paracristid in *"Simosthenurus" brachyselenis*, are produced by developmental field gradients (see Butler 1956; Van Valen 1970). Jernvall (2000) has suggested that the basis for these gradients may be the interplay between the spacing of cusp initiation and the termination of crown development, which determines potential cusp pattern and degree of pattern realization. Even if a crest is lost from all bar the first of the molars, the potential for re-expression on the more posterior molars exists, because the developmental field can expand posteriorly or its gradient can shift. However, in the plesiomorphic sthenurines, none of the reemergent crests are visible on any of the molars. In the putative ancestor of the European lynx, m1 is, in fact, the only molar in the lower dentition.

While Kurtén (1963), Werdelin (1987) and Jernvall (1995) put forward a developmental explanation for the reappearance of m2, they did not explicitly address the issue of how the metaconid / trigonid complex on m1 might have originated.

Under the expanded developmental field hypothesis, a solution to the lynx problem and to the reemergence of the sthenurine molar crests probably lies with the molariform deciduous premolar. Like molars, deciduous premolars are first-generation cheek teeth. The hindmost deciduous premolars are uniformly molariform across the Mammalia (Jernvall 1995). They are not recognized as molars, because they are replaced by the third adult premolars. In sthenurines, the third deciduous premolars closely resemble the first molars. Clearly, they are not only part of the same eruption series, but represent the anterior limit of the same developmental series. Intriguingly, the dP3 and dp3 often express topographic features that are not distinct on the molars. The complex dP3 attributed to *Archaeosimos cegsai*, which preserves a distinct mesocrista and urocrista among other unnamed crests, is an excellent case in point. Indeed, the dP3 of both *Metasthenurus newtonae* (pl. 35A) and the likely sister taxon of the Sthenurinae, *Wanburoo hilarus* (Cooke 1999:fig. 3), possess a very fine, low mesocrista and urocrista not evident at all on their upper molars. Although the dP3 of *Hadronomas puckridgi* is unknown, one might predict the presence of these crests in this plesiomorphic sthenurine. Unfortunately, the protocone apex of the only known *Wanburoo* dP3 has been broken off, so I am unable to assess the state of the preprotocrista in that taxon or *Hadronomas.*

Critically for the developmental field hypothesis, there is no sign of a posthypocristid on the posterior face of the dp3 hypolophid in *Wanburoo* or any plesiomorphic sthenurines. In fact, if Cooke's (1997b) idea that the hypolophid of derived bulungamayines is largely constituted of the posthypocristid holds true, then my interpretation of cristid reemergence in *Procoptodon rapha*, *P. goliah* and *Macropus* (and hypoconulid in *M. pearsoni*) would be highly questionable. There is a possible alternative to the mechanism proposed by Cooke (1997b), at least to account for hypolophid formation in the bulungamayines whence sthenurines and macropodines were derived. That is, the hypolophid is not formed exclusively of the posthypocristid (cf. Cooke 1997b:fig. 4), but also from the crest directed buccally from the entoconid. This crest clearly extends to the buccal side of the tooth in the plesiomorphic bulungamayine *Nowidgee* (Cooke 1997b:fig. 2A), but it, like the posthypocristid, is not discernable within the body of the hypolophid of derived bulungamayines (e.g., *Ganguroo*, Cooke 1997b:fig. 3C). Perhaps in sthenurines and macropodines, the crest directed buccally from the entoconid makes up the lingual part of the hypolophid crest, while the proximal portion of the posthypocristid makes up the buccal part. This situation is clearly evident on the unworn lower molars of *P. goliah* (pl. 77) and *M. pearsoni*, but whether this arrangement characterizes all sthenurines and macropodines or evolved convergently and cryptically under the cover of an outwardly simple hypolophid needs closer examination. While interpretations of phylogenetic character reversal need to be made with great care, this posthypocristid example intimates that they may have the potential to illustrate something of the original mode of formation of a structure, at least in a general sense, by replaying part of its evolutionary development in reverse.

There are other examples of apparent evolutionary reversals in the macropodoid cheek dentition, such as the apparent reemergence of the parametacristid, the posterior portion of the stylar crest (buccal component of the postmetacrista), and stylar cusps A, C and E. It is beyond the scope of this work to discuss them in detail; the main objective has been to highlight the potential significance of character reversals as an influence on directions of macropodoid evolution. Not all homoplasy is convergence. The phylogenetic line from plesiomorphic bulungamayines to derived simosthenurins reveals a trend toward simplification, followed by a trend toward increased complexity of the molars. Relative complexity of phenotypic expression has fluctuated over time, but the genes and developmental pathways must have been preserved. Functions are different upon reemergence, but forms and topologies are similar. I have put forward two hypotheses to explain the patterns observed, but neither would appear to explain everything. A combination of the two might be feasible, but much more thorough work is needed. From a phylogenetic perspective, it might be fruitful to carry out a wide investigation of the relative expression and distribution of particular features through time, through space, between taxa, within species, and within individuals. One of the many questions that could be addressed is do third deciduous premolars have the potential to act as character reservoirs? Perhaps selection to eliminate certain features is relatively less on deciduous molariform teeth than it is on molars. From a developmental perspective, it might be interesting to map the activities of the secondary enamel knots associated with the posterior face of the hypolophid in *Macropus*, given the variation in posthypocristid expression across the genus. Do crests that arise from a particular cusp lie within the extended domain of the cusp-forming secondary knot or are they formed through the interaction of different knots? If the former is true, then given that only subtle variations in the sequential activation of cusp-forming secondary enamel knots can result in large variations in cusp expression (Jernvall 2000), even subtler variations in these enamel knots activity may control the relative expression of crests.

As a concluding remark it is worth noting the apparent prescience of G. H. H. Tate (1948:254) who observed in relation to the Macropodinae that: "Numerous characters and trends seem to play genetic 'hide-and-seek' in this subfamily."

FUNCTIONAL MORPHOLOGY

The primary aim of this section is to document functional trends evident in the evolution of the sthenurine craniodental system and to speculate on their possible significance in terms of dietary ecology. Before doing so, I briefly review the major evolutionary trends evident in bulungamayines, the ancestral group whence sthenurines were derived.

BULUNGAMAYINES

All of the macropodoids in the late Oligocene Etadunna, Namba and Riversleigh "System" A assemblages are characterized by small body sizes and molars of low to intermediate crown height (e.g., Flannery and Rich 1986; Woodburne et al. 1994; Cooke 1997a). Like their modern representatives, bunodont potoroines and hypsiprymnodontids were probably omnivorous, frugivorous or mycophagous. Bunolophodont bulungamayines (e.g., *Nowidgee*, *Purtia*) may well have been capable of consuming a higher proportion of vegetative material. Dentally, their closest modern analogue is the rufus bettong, *Aepyprymnus rufescens.* This taxon has higher-crowned molars and better-developed sub-lophs than other modern potoroines, a morphology which corresponds to a diet consisting largely of herbs, grass shoots, succulent roots and tubers, as well as the potoroine staple of truffles when abundant (Dennis and Johnson 1995). The fully bilophodont morphology of balbarines and derived bulungamayines suggests that they were more committed browsers. Enamel microwear studies should prove enlightening.

Significantly, while bunolophodont bulungamayines are present and balbarines are abundant and relatively diverse in Riversleigh "Systems" A and B, both are much rarer in "System" C, in contrast to bilophodont bulungamayine diversity (Cooke 1997a). Indeed, there is no evidence to suggest that the balbarine lineage outlasted the middle Miocene. In the Lake Eyre Basin, bilophodont bulungamayines (*Bulungamaya*) first appear in the Ngama LF alongside the bunolophodont *Purtia.* Both continue into the Treasure LF, but *Purtia* is absent from the latest Oligocene or earliest Miocene Kutjamarpu LF, which contains three lophodont bulungamayine species (Woodburne et al. 1994). Whether the absence of bunolophodont species from the Wipajiri Fm. indicates an earlier onset than the similar trend observed at Riversleigh may not become evident until the many new Lake Eyre Basin macropodoid taxa are eventually described.

The evolutionary acceleration of bilophodont bulungamayines and concomitant deceleration of balbarines is intriguing, especially as balbarines had clearly evolved full bilophodonty by the latest Oligocene (Cooke 1997c, 1999). Indeed, balbarines would appear to have already filled feeding niches similar to those subsequently occupied by derived bulungamayines and their earliest macropodine descendants. Moreover, Riversleigh balbarines were slightly larger on average than bulungamayines, which is consistent with data from the Lake Eyre and Tarkarooloo Basins (Flannery and Rich 1986; pers. obs.). Only the derived "System" C bulungamayine *Wanburoo* would appear to have attained a size similar to the larger balbarines (Cooke 1999). Therefore, it might be presumed that balbarines were the first macropodoid group to exploit predominantly leaf-eating niches. It is also interesting to note that two "System" A balbarines, *Nambaroo couperi* and *Wururoo dayamayi* (Cooke 1997c), show signs of having been adapted to quite resistant plant material. This is evident especially in *Wururoo*, which has a stout ramus, relatively thick lower incisor base, and large plagiaulacoid premolar. I suspect that the solution to why this crucial faunal change took place may relate more to differences in the locomotory systems of balbarines and bulungamayines.

Several different lines of faunal evidence favor increasingly dry, seasonal conditions during the deposition of "System" C, with a significant reduction in the extent of rainforest (Archer et al. 1995, 1997). For this "system", the faunal evidence would seem to agree with regional climatic interpretations derived from geological and oceanic evidence (Stein and Robert 1986; Locker and Martini 1986, 1989; Megirian 1992; Flower and Kennett 1994). Morphological changes observed in late Oligocene to middle Miocene bulungamayines, along with increased diversity and body size, more likely reflect a shift into open wet sclerophyll niches than dry sclerophyll habitats. While the Riversleigh "System" C and Bullock Creek LFs do provide the earliest conclusive evidence of drier, more seasonal conditions in central and northern Australia, there is no clear evidence of any grazing adaptations (Archer et al. 1995). The oft-mentioned, but still undescribed molar of a "high-crowned" macropodoid from Bullock Creek and the faunal composition of the Riversleigh "System" C assemblages provide clues that may forecast an incipient expansion of abrasive vegetation during the middle Miocene (Archer et al. 1989, 1994, 1995; Murray and Megirian 1992), although no coeval floral assemblages within cooee of Bullock Creek or Riversleigh are known. However, the trend toward a drier climate with seasonal rainfall in the Murray Basin is indicated by a reduction in *Nothofagus* and an overall increase in Araucariaceae, Casuarinaceae, *Eucalyptus* and grass pollen (Truswell et al. 1985; Martin 1993; Kershaw et al. 1994).

Quite apart from its potential as a paleoenvironmental indicator, the high-crowned molar from the Bullock Creek LF may constitute the earliest sign of a bulungamayine-macropodine transition, if there was in fact a single phyletic transition. Unless new sites are discovered or Encore Site at Riversleigh yields some proto-macropodines, the collection of more material from Bullock Creek or as yet undiscovered middle Miocene sites may be the only avenue by which the monophyly of the Macropodinae might be confirmed, given that the closest relatives of this group are extinct. With the possible exception of Encore Site and the upper part of "System" C, no deposits fall within the 7-8 million year interval between the accumulation of the Bullock Creek and Alcoota LFs. The Alcoota macropodine (*Dorcopsoides*) shows a greater similarity to later forms than to bulungamayines, which is not so surprising considering the magnitude of the time gap. However, the presence of a *Wanburoo*-like taxon at Alcoota highlights the potential for future discoveries to contribute to the elucidation of the bulungamayine to sthenurine-macropodine transition.

With this in mind, it is fortunate that sthenurines are so solidly united as a monophyletic group. The craniodental character complex, from which nine synapomorphies have been identified, more than likely reflects a considerable antiquity for the lineage. This supposition would be confirmed if the relatively large macropodoid in the Bullock Creek LF (Murray and Megirian 1992) is eventually shown to lie near the base of the sthenurine lineage. When bulungamayine and balbarine postcranial remains are described, many of the distinctive sthenurine characteristics described by Murray (1995) will undoubtedly be revealed as sthenurine synapomorphies.

More support for a middle Miocene origin for sthenurines is signaled by the similarities shared by *Hadronomas puckridgi* and *Wanburoo hilarus* (Cooke 1999). While *Wanburoo* is certainly less derived than *Hadronomas*, it appears to share three synapomorphies with sthenurines, based on the limited material from which it is known (Cooke 1999; see Monophyly and Origin). If *Wanburoo* is ancestral to the sthenurine lineage, then sthenurines may well have originated around the time Riversleigh "System" C was accumulating. If *Wanburoo* is the sister taxon of the sthenurines, then they must have arisen by the early middle Miocene.

Craniodental morphology indicates that *Wanburoo* is also adaptively intermediate between other bilophodont bulungamayines (e.g., *Ganguroo*) and *Hadronomas puckridgi.* It is much larger than any other bilophodont bulungamayine, its ventral dentary margin is less convex, and its lower incisor is broader and deeper (Cooke 1999). In each of these features and in its greater overall robustness, *Wanburoo* bears more resemblance to *Hadronomas* and *Dorcopsoides* (Cooke 1999). Upper incisor morphology also closely resembles that of the sthenurines (fig. 16). All of these attributes suggest that, among the Bulungamayinae, *Wanburoo* was adapted to a relatively tougher browse diet than any other taxon.

EVOLUTION OF THE STHENURINE BAUPLAN

As a whole, the Sthenurinae is easily separated from all other macropodoid groups on the basis of several unique aspects of cranial architecture (see Monophyly and Origin). Combined, these constitute what might be termed the sthenurine bauplan. Most obvious is the general form of the splanchnocranium. In adult sthenurines, it is not markedly deflected anteroventrally relative to the neurocranial or basicranial axis (figs 14, 22). That is, the degree of klinorhynchy expressed is less, a characteristic that has resulted primarily through deepening and incipient foreshortening of the rostrum. One result is a flatter dorsal cranial surface compared with other macropodoids, especially in the neurocranial region. The occiput is near perpendicular to the neurocranial surface, a morphology brought about by the relative elevation and anterior rotation of the neurocranium (figs 14, 22). The neurocranium is also deepened by the expansion of its dorsal aspect, upon which distinct temporal and nuchal (or lambdoid) crests are borne. The region is broadened by the expansion of the posterior extremity of the squamosal and the mastoid process of the periotic. Elevation and anterior rotation of the back of the neurocranium results in a reduction in the proportion of temporalis muscle fibers pulling more posteriorly and an increase in those pulling more vertically (fig. 22).

Frontal inflation is another sthenurine characteristic, although the degree of inflation expressed varies across the group. Lateral expansion of the frontals and development of the supraorbital crest appear to be both geometrical and functional correlates of the reconfiguration of the temporalis musculature, which is also reflected by a more vertical orientation of the ascending ramus (Murray 1991a; figs 20, 22). Proportional differences between sthenurine taxa must reflect variation in masticatory capabilities. No detailed soft tissue, biomechanical or in vivo bone strain data have been published for any taxon that might be a half-reasonable modern analogue, with the possible exception of the koala, *Phascolarctos* (Davison and Young 1990), but it is still worthwhile contemplating the likely functional and adaptive significance of the morphology of this region.

The plesiomorphic sthenurine *Hadronomas* differs from the other members of the group in several important ways. The cranium is shallower along its entire length and the relatively smaller orbit is positioned more posteriorly such that its posterior border is aligned above the edentulous posterior end of the maxilla. The ascending ramus is more reclined and the coronoid process is wider when viewed laterally, extending anteriorly to just behind the eye orbit (fig. 22A). An important implication of these differences is that the proportion of temporalis muscle fibers pulling vertically was relatively less in *Hadronomas*, even though it was greater in this taxon than in all other macropodoids. While a more vertical temporalis might theoretically have conveyed a greater orthal bite force component during the mastication of resistant food items, *Hadronomas* seems to have been evolutionarily constrained by the morphology and position of its orbit. Any further anterior tilting of the coronoid process or anterior encroachment of the temporalis muscle would have placed excessive pressure on the back of the orbit or reduced visual acuity during forceful mastication. To counter such effects, several different mammal groups have ossified the postorbital ligament, which runs from the lateral portion of the frontal to the zygomatic arch (Noble et al. 1999). Although less marked, the presence of a distinct, freestanding postorbital process on the jugal and an

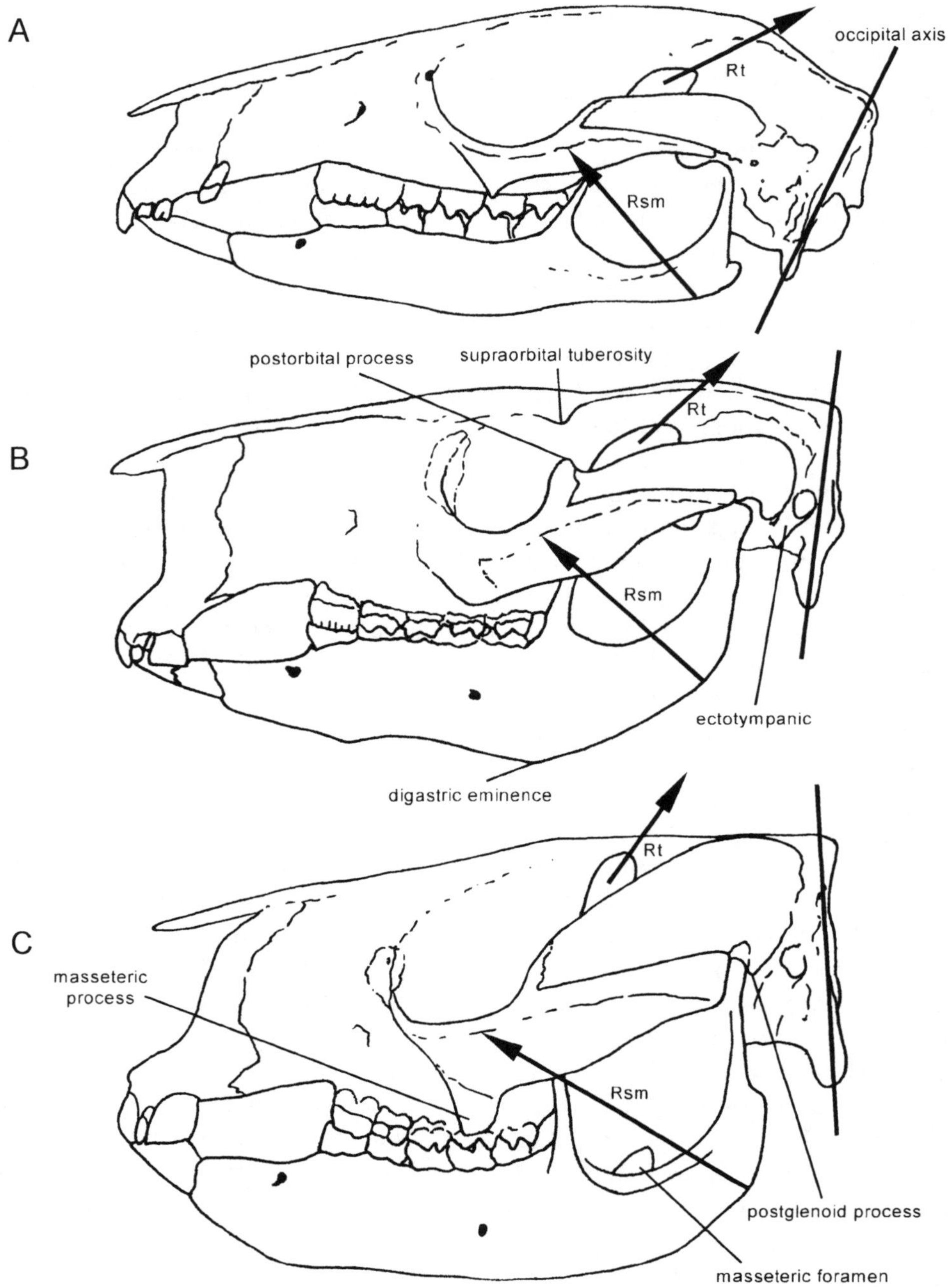

Figure 22. Articulated cranium and dentary in lateral view. **A**, *Dorcopsis hageni* (MVZ 109392); **B**, *Hadronomas puckridgi* (composite utilizing NTM P98133, P98134, P98140, P98142); **C**, *Sthenurus andersoni* (composite utilizing SAM P20496, P29570, P31877). Rsm, Rt = estimated resultant direction of pull of superficial masseter, temporalis.

expanded frontal bearing a rugose supraorbital tuberosity in *Hadronomas* attest to a similar selective pressure and adaptive response (figs 15, 22A). Whether the rugosity of the supraorbital tuberosity indicates that the *Hadronomas* postorbital ligament had become chondrified, at least mesially, is difficult to determine.

Judging from the relatively horizontal zygomatic arch and small, rounded masseteric process, the anteriorly oriented component of the masseter musculature in *Hadronomas* was much less emphasized than in other sthenurines (fig. 22). With the noteworthy exception of rodents, masseteric processes are well developed in mammals with a large and more anteriorly oriented external superficial masseter (e.g., simosthenurins, zygomaturines, sloths, glyptodonts). Typically, the jugal is deep, so the shallow nature of the *Hadronomas* jugal relative to other sthenurines suggests that the internal layer of the superficial masseter was also not as enlarged. Overall, the form of the *Hadronomas* zygomatic arch is not easily differentiable from that of most non-sthenurine macropodoids, with the exception of the postorbital process and the wide ectoglenoid process. Termination of the masseteric canal below m4 resembles the condition expressed by many non-sthenurine macropodoids, presumably because it corresponds to a similar zygomatic arch and masseter configuration, and the blade-like morphology of the premolars. For a discussion of the functional significance of deep masseter insertion into the dentary, refer to Sanson (1989).

On the mesial side of the *Hadronomas* dentary, the medial pterygoid fossa is elevated such that the angular process is set above the level of the cheek tooth row. However, the fossa is slightly shorter and the angular process is slightly lower than in other sthenurines. Still, even this degree of elevation would have resulted in a greater proportion of medial pterygoid muscle fibers pulling in an anterior direction relative to most non-sthenurine macropodoids, wherein the fibers are oriented almost entirely vertically. Nevertheless, elevation of the medial pterygoid fossa may largely have been an inevitable geometric consequence of a more vertical ascending ramus. It is also likely that any further elevation of the fossa in *Hadronomas* would have been constrained by the need for the medial pterygoid muscle to still counter the ventrolateral rotation of the dentary caused by contraction of the more vertically oriented masseter fibers, as *Hadronomas* retained a flexible symphysis.

Much of the cranial restructuring in the ancestral sthenurine lineage appears to have centered on the more vertical orientation of the jaw adductor muscles. This would have promoted greater orthal masticatory force (Murray 1991a), an action compatible with the simple, low-crowned molars of *Hadronomas.* The relatively straight alignment of the molar row must have impeded mesial rotation of the dentary, but the slightly oblique setting of the loph and lophid crests clearly allowed some lateral translation (pls 2C, 5C). Certainly, the first phase of molar occlusion in *Hadronomas* was loph-lophid shearing (*sensu* Sanson 1980). This action was initiated at the sharp points of the upper buccal cusps and lower lingual cusps. Murray (1989) noted that these points become more emphasized with increasing wear in *Hadronomas*, and this is typical of most sthenurines (Wells and Tedford 1995). During the power phase of the occlusal cycle, which involves the interlocking of lophs and lophids with their receiving valleys (phase 2, *sensu* Sanson 1980), simple crushing may have been at least as important as grinding and more important than shearing. This is suggested by the absence of well-developed cristae and enamel crenulations, as well as the apparent restriction of transverse motion to a fairly straight lateral excursion (cf. most simosthenurins, plus *Sthenurus stirlingi*, Wells and Tedford 1995).

A more orthal masticatory stroke also relates to a shift in premolar morphology and function that is incipient in *Hadronomas* and fully developed in derived sthenurines. The plesiomorphic macropodoid p3 consists of one crest that is either straight or curved posterolingually. An unworn crown often looks like a line of cuspules connected by short crests, together constituting the main crest. In general, vertical ridgelets descend from the buccal and lingual sides of each cuspule (fig. 18). The relative coarseness of the vertical ridgelets varies as much between taxa as the distinctiveness and number of cuspule apices and the overall length of the main crest. The cuspules and ridgelets greatly enhance the shearing capacity of the main crest when it occludes with the lingual side of its upper counterpart, the morphology of which essentially mirrors the p3 main crest. A range of macropodoids (*Hadronomas*, *Protemnodon*, *Troposodon*, *Lagostrophus*, *Congruus*, *Wallabia*, *Setonix*, *Dendrolagus*, *Wanburoo*, the bulungamayine in Archer et al. 1991) have enhanced premolar occlusion by developing a broad cingulum on the lingual side of P3. The lingual periphery is often raised into a low crest (especially posteriorly), which encloses a shallow longitudinal basin (fig. 23A). Function is related to the consumption of resistant vegetative material, such as petioles or twigs, which often do not sever fully when the blades of the main crests shear past one another during the first bite. When this circumstance eventuates, continuation of the occlusal cycle would result in the strong compression of the resistant item between the p3 main crest and the shelf or longitudinal basin of the P3. Repetition of the premolar occlusal cycle would eventually soften the twig to a stage where it is severed by crest-on-crest shearing. This is an important line of evidence attesting to the likelihood that extinct macropodoids with this P3 morphology were browsers, just like their modern compatriots.

In contrast to all other taxa expressing this P3 morphology, *Hadronomas* markedly widened the base of its p3 by developing a distinct cingulid on both the buccal and lingual sides of the tooth (fig. 23B). This gives the tooth a broad triangular form in cross section. Importantly, the lingual border of the *Hadronomas* P3 shelf is not raised into an elongate crest (fig. 23B). Instead, a series of three or four large cuspules are formed at the ends of the ridgelets directed lingually from the main crest. In a relatively unworn state, the main crests of the *Hadronomas* premolars would have occluded in the same scissor-like manner as in the other taxa, with the p3 crest propping on the inside of the lingual row of P3 cuspules during the exertion of maximum compressive force. However, in specimens showing more wear, the lingual row of cuspules is obliterated. The p3 could no longer have been restricted in its excursion across the lingual side of the P3, meaning that occlusion resulted not only in wearing of the crest of p3, but the entire buccal slope of the tooth. The greater the wear, the more horizontal and wider the occlusal surface (compare pls 6C, 5C, 4C). Although the molars do not appear to have been especially well adapted to lateral translation under compression, the shift in premolar morphology must have improved the grinding capacity of the cheek tooth row.

When the functional implications of these changes are considered, it is not surprising to find compatible adjustments in the morphology of the craniomandibular joint and incisor arcade. While the condyle of *Hadronomas* is narrower than that of the other sthenurines, it is broader than in *Wanburoo* and the macropodines. Directly related to this increased breadth is the morphology of the ectotympanic. Although the tympanic region is poorly preserved in both known *Hadronomas* basicrania, all indications are that it was similarly developed, even if not quite as wide as in other sthenurines (Murray 1991a). The wide, robust ectotympanic is completely fused with the postglenoid process, which would have given rigid support to the craniomandibular joint. Murray (1991a) suggested that the expanded ectoglenoid process functioned as both an anterior guide for the condyle and an attachment area for the joint-stabilizing lateral craniomandibular ligament. The glenoid morphology of *Hadronomas* indicates that large forces generated during mastication must have resulted in a tendency for the condyle to be disarticulated as well as rammed back against the postglenoid process.

The increased size of the digastric eminence (figs 20, 22) and sulcus in sthenurines suggests an enlarged digastric muscle. Proximity of the angular and paroccipital processes suggests sthenurines were not capable of attaining a notably wider gapes than other macropodoids, so the posterior shift and enlargement of the digastric insertion was probably most important for reseating the condyle before the commencement of the next occlusal phase. This muscle performs a similar function in elephants (Maglio 1972). As the masseter musculature became more anteriorly oriented in more derived sthenurines, the digastric muscle may have been an important antagonist, given its roughly 90° alignment to the mean orientation of the masseter and medial pterygoid musculature.

Aspects of sthenurine upper incisor morphology reflect an efficient mechanism for apprehending resistant items. Relative enlargement of the I3 (figs 16-17) means that the i1 occluded mainly against this tooth. The tip of i1 occludes against the posterior surface of the sub-cylindrical I1, producing a vertical wear facet. I1 may have been as important in preventing the dentary from being thrust anteriorly during mastication of tough items (Wells and Tedford 1995) as it was in providing a nipping mechanism. I2 is so reduced that one may wonder why the tooth was retained in sthenurines at all. Perhaps its main job was as a spacer between the posteriorly concave I1 and the I3. Because I3 is the incisor closest to the condyle, it is better placed mechanically in terms of bite force transmission. If its surface area was less, greater incision force per unit area might have been possible, but this probably needed to be balanced against the reduction in the quantity of herbage that could be grasped per bite. Enlargement of I3 and reduction in the size of I2 might be a functional compromise between maximizing bite area and bite force. While a reduction in rostral length in plesiomorphic sthenurines might have achieved a similar end to altering incisor proportions, splanchnocranial restructuring is a much more major evolutionary step, and it was exactly this type of cranial restructuring that became a hallmark of the Simosthenurini.

Having considered possible functional implications of the upper incisor form that sets sthenurines apart from all other macropodoids, it is interesting to recall the incisor morphology of the derived bulungamayine, *Wanburoo hilarus*, because it is clearly antecedent to the sthenurine design. Both sthenurines and *Wanburoo* have diminished the well-developed anterobuccal crest on I3 typical of other macropodoids (figs 16-17), instead developing a clear anterolingual crest. In effect, it is not possible to separate the I1 and I3 of *Wanburoo* and the sthenurines on morphology, with the possible exception of a sharper unworn occlusal edge on the *Wanburoo* I3 (see Cooke 1999:fig. 2). The crown of the *Wanburoo* i1 also more closely resembles those of *Hadronomas*, *Archaeosimos* and *Dorcopsoides* than it does any other derived bulungamayine. The only noteworthy change in incisor morphology from *Wanburoo* to sthenurines is a reduction in the size of I2, an evolutionary change that may be correlated with a reduction in overall incisor surface area and a more forceful bite.

All available evidence suggests that *Wanburoo* represents the bulungamayine taxon best adapted to the consumption of tougher browse material, but the earliest of the sthenurines patently took the trend to another level. One change very likely to have set the evolutionary stage for the sthenurine lineage was increased body size. *Hadronomas* is the earliest macropodoid known that exceeds the size of a small wallaby. Because body size is inversely proportional to metabolic rate, larger mammals have lower energy requirements per unit volume (Demment and Van Soest 1985). Larger size facilitates a greater absolute throughput of feed, meaning that under drier conditions, where lower quality (higher fiber) vegetation is abundant, larger body size is often favored. Increased body size and the re-gearing of the craniodental complex to the consumption of tougher material during the earlier phases of sthenurine evolution was most likely driven by the expansion of sclerophyll browse vegetation during the latter half of the Miocene. The appearance in the late Miocene or early Pliocene of the large browsing macropodines *Protemnodon* and *Troposodon* supports the idea that relatively higher fiber browsing niches were expanding at around this time. Both of these genera show such a marked level of convergence with plesiomorphic sthenurines that it is easy to envisage why they have often been viewed as derivatives from the same phylogenetic vicinity (e.g., Bartholomai 1967; Campbell 1973; Archer 1981; Flannery 1983). Contrary to one of several citations incorrectly attributed to Prideaux (1999c) by Burk and Springer (2000:232), there is currently no evidence to suggest that "The appearance of larger kangaroos (better able to exploit grasses as a food resource) in the middle to late Miocene fossil record is correlated with the appearance of grasslands in the middle Miocene."

Hadronomas is the oldest and most plesiomorphic member of the Sthenurinae. The nine synapomorphies that unite *Sthenurus*, *Metasthenurus*, *Archaeosimos*, *Simosthenurus* and *Procoptodon* as a monophyletic clade represent a crucial circumvention of the evolutionary constraints that probably restricted *Hadronomas* to only a moderately resistant diet. In combination with the postcranial adaptations discussed by Tedford (1967), Adnams-Hodges (1988), Murray (1995) and Wells and Tedford (1995), these advances would have been critical in facilitating the diversification of sthenurines from the late Miocene onward.

SPECIALIZATIONS OF THE CRANIODENTAL SYSTEM

While the elevation and anterior rotation of the neurocranium is a characteristic that also sets *Hadronomas* apart from all other macropodoids, it is more marked in the other sthenurine genera (fig. 22). The anteroventral orientation and more posterior termination of the zygomatic arch, as well as the slightly more anteriorly tilted ascending ramus, highlight this additional restructuring (fig. 22). The main driving force behind this continued cranial reconfiguration was probably toward the enlargement and more anterior orientation of the masseter musculature, which would have produced an increase in the lateral masticatory component. Although at present there is no way to be sure whether the ancestor of the main sthenurine lineage (i.e., *Sthenurus*-*Metasthenurus*-Simosthenurini) possessed a distinct supraorbital tuberosity and postorbital process like *Hadronomas*, the likelihood is that it did not. The lineage seems to have bypassed potential problems associated with protecting and maintaining the stability of the orbit during mastication by projecting the orbital capsule anterolaterally and flooring it with an expanded maxillary plate (fig. 15), in addition to retaining a laterally narrower, taller coronoid process (figs 20, 22). These changes resulted not only in an enlarged anteorbital fossa for the attachment of rostral muscles and increased stereoscopic vision through orbital convergence (fig. 15), but also anterior and lateral expansion of the temporalis musculature. While enlargement of the temporalis in an unmistakably herbivorous lineage is the reverse of what occurs in most herbivores (e.g., Turnbull 1970; Gillette and Ray 1981; MacFadden 1992), a similar condition characterizes the deinothere proboscideans, since they also had bilophodont molars that functioned similarly. An enlarged temporalis certainly would have enhanced orthal masticatory force by increasing the total number of muscle fibers pulling vertically, while that portion pulling more posteriorly would have ensured precise occlusion between lophs and lophids. Perhaps both orbital convergence and temporalis enlargement were exaptive (*sensu* Gould and Vrba 1982) rather than adaptive (i.e., part of the restructuring process that allowed the sthenurine neurocranium to become more elevated and anteriorly rotated), but both clearly had directly useful effects.

Lateral curvature of a molar row may be viewed as one way to maintain its overall length in a shortened skull, and thus a consequence of brachycephaly rather than an adaptation, but the presence of a curved molar row in the species of *Sthenurus* (e.g., pls 11, 22, 25) suggests that this cannot account for the advent of the feature. Powered by the more anteriorly oriented masseter musculature, the curved molar row and perpendicular lophs probably facilitated more mesial rotation and lateral translation during occlusal phase 2 than in *Hadronomas*, regardless of the restrictions imposed by the bilophodont molar condition. Even slightly more translation and rotation of lophs and lophids within their valleys would have improved the grinding capacity of the molars during compression. Because lines projected lingually from the loph crests in *Sthenurus stirlingi* converge beneath the masseteric process on the contralateral side, Wells and Tedford (1995) suggested that the superficial masseter muscle layer may have acted as the antagonist to mesial rotation of the lower jaw during occlusal phase 2 (although I suspect this is purely coincidental). The tendency for condylar disarticulation on the occluding side, as the bite point became a secondary fulcrum under strong compression, would have been especially marked considering the large anterior component of all three sets of masticatory muscles. Contraction of the digastric muscle, locking of the incisors, and propping of the contralateral condyle against its robustly supported postglenoid process may also have exerted considerable control over both mesial and posterior rotation of the dentary.

The adaptive significance of the incipient division of the paracristid and cristid obliqua in the basal members of this group is not immediately obvious, although it would seem to have resulted in decreased shearing capabilities during occlusal phase 2. The independent origins of this particular morphology in both *S. stirlingi* and the Simosthenurini, in which division of these cristids is both marked and supplemented by a similar division of the postprotocrista and emergence of the mesocrista, must related to enhanced phase 2 grinding.

If the *Sthenurus–Metasthenurus*–Simosthenurini lineage could be characterized by but one dental adaptation, it would be the transformation of the premolars from shearing to predominantly crushing and grinding teeth. The lineage is unique among macropodoids through its possession of a neomorphic buccal crest on p3, which is subequal in height to the main crest (figs 18, 23). In *Archaeosimos* and numerous individuals of *M. newtonae*, the buccal crest is very short and only separated from the posterior extremity of the main crest by a shallow, narrow median valley. A distinct wear facet is produced in precisely this posterobuccal position on the p3 of most other macropodoids, but none have evolved a distinct enamel eminence. However, a low cusp or crescent has arisen convergently in this position in phascolarctids, although the P3 longitudinal basin, which receives this structure, is shallower than in the sthenurines due to a much lower lingual crest. The same applies to macropodines like *Protemnodon*, *Troposodon*, *Dendrolagus*

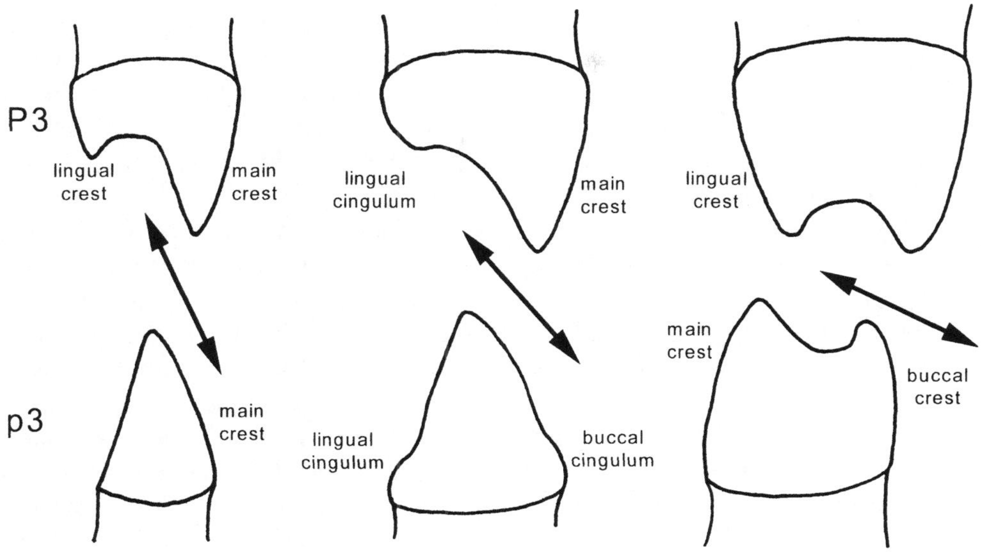

Figure 23. Diagrammatic transverse sections showing premolar occlusion in three browsing kangaroos. **A**, *Wallabia bicolor*; **B**, *Hadronomas puckridgi*; **C**, *Simosthenurus occidentalis*.

and *Wallabia* (although they have no posterobuccal eminence on p3). By contrast, the P3 lingual crest in sthenurines is usually either slightly lower than or equal in height to the main crest (fig. 23C). Although the *Archaeosimos* P3 is unknown, the manner in which the premolars occluded in the other genera, including those with a short buccal crest on p3 (especially *Metasthenurus*), strongly suggests that a similar occlusal mechanism also must have been present in *Archaeosimos.* Compared with that of *Archaeosimos* and *Metasthenurus*, the p3 buccal crest of *Sthenurus*, *Simosthenurus* and *Procoptodon* is longer and the median valley wider.

A major effect of the advent of the p3 buccal and P3 lingual crests is that the ancestral blades (main crests) no longer meet (fig. 23C). Instead of being aligned down the tooth midline, most of the p3 main crest is oriented posterolingually. Due to the inflation of its posterobuccal aspect, the main crest never approaches its traditional counterpart on P3, which is itself shifted buccally from the tooth midline (fig. 23C). Manipulation of articulated specimens reveals that the first phase of the premolar occlusal cycle is the shearing of the curved p3 buccal crest against the lingual surface of the main crest and the high transverse ridgelet separating the longitudinal and posterior basins (the latter receiving the trigonid of the m1). At the same time, the p3 main crest shears past the lingual side of the P3 lingual crest. Continuation of the occlusal cycle into a second phase sees the dentary trend more mesially, with the p3 buccal crest grinding food between the often corrugated surface of the longitudinal basin and against the inside of the P3 lingual crest. Evidently, use of the p3 buccal and P3 lingual crests in phase 2–equivalent grinding after phase 1–equivalent shearing resulted in rapid rounding of the enamel relative to both upper and lower main crests. Although *Sthenurus* enhanced phase 1–equivalent shearing by increasing crown height and length of the p3 buccal crest, grinding capacity would have increased as shearing capacity decreased, due to the increased area of the uneven enamel surface. Therefore, for most of an individual's adult life, the primary function of the premolars would have been crushing and grinding, not shearing.

Permanent premolars were often the latest permanent teeth to erupt in sthenurines, but the first molars were the earliest. By the time the p3 buccal and P3 lingual crests had worn down, large areas of dentine were already exposed on the first molars and their shearing efficiency had diminished. Functionally, they became a continuation of the premolar grinding surface. The account given by Wells and Tedford (1995) of the likely shifts in occlusal function along the cheek tooth row of *Sthenurus* applies equally well to *Metasthenurus* and the simosthenurins. Based on wear patterns, the main role of the more posterior molars would have been shearing, the more intermediate molars both shearing and grinding, and the anterior cheek teeth grinding and crushing. Wells and Tedford (1995) suggested that plant material would have been fed initially to the back of the cheek tooth row for dicing, then moved successively more anteriorly until the bolus reached the grinding surface for rupturing of cell walls. In no other macropodoids, including *Hadronomas*, was grinding the primary function of the permanent premolars, but a remarkably similar functional division characterizes the cheek dentition of deinotheres (Harris 1975).

This major shift in premolar form and function is correlated with another characteristic of the *Sthenurus*–*Metasthenurus*–Simosthenurini clade: shortening of the masseteric canal to a vertical cul-de-sac. Ride (1959:56) observed that the "almost complete reduction of the canal probably took place as the ascending ramus became tilted forwards." While this is very likely true, derived macropodines with a reclined ascending ramus (e.g., *Macropus giganteus* Shaw, 1790) can also have a similarly abbreviated masseteric canal. Sanson (1989) noted a positive correlation between the size of the sectorial premolars and the development of the deep masseter muscle, which is reflected in the anterior extent of the masseteric canal. As the large *Macropus* species and these sthenurines have both moved away evolutionarily from a sectorial premolar, reduced size of the masseteric canal cannot be envisaged as an adaptation, but rather as a result of premolar modification. However, a vertical masseteric canal was probably exaptive in *Sthenurus*, *Metasthenurus* and the simosthenurins, providing an increased proportion of masseter fibers pulling vertically, rather analogous to the anterolateral expansion of the temporalis. One of the defining features of the clade containing the five most derived *Procoptodon* species is the relative enlargement of the masseteric foramen and the vertical canal into which it opens (see Phylogeny).

Increased symphyseal rigidity, in addition to the enlargement of the inferior transverse torus, must have been important in enabling force generated by muscle sets on both sides of the head to be involved in phase 2 grinding, as well as resisting wishboning and dorsoventral shear forces (e.g., Ravosa and Hylander 1994). Increased thickness of the lingual enamel on i1 would certainly have assisted in the apprehension of resistant food items, but it would also have strengthened the anterior region of the craniodental complex if incisors became forcibly occluded during grinding. This is particularly interesting in view of the fact that most grinding actually occurred at the anterior end of the sthenurine cheek tooth row and involved the highly specialized, non-sectorial premolars.

Aspects of the craniodental system of sthenurines resemble those of other mammal groups (e.g., xenarthrans), but their macropodoid heritage placed considerable evolutionary constraints on feasible modifications. The bilophodont molar battery and the precise occlusal mechanism of the incisors appear to have been the two most paramount constraints, just as they were with grazing macropodines, which were driven toward molar progression as their primary means of improving dental durability (Janis 1990). In many respects, restructuring of the sthenurine craniodental complex has its closest analogue in the zygomaturine diprotodontids, which were also constrained by their bilophodont molars and diprotodont incisor condition. Murray (1992) showed that the morphological shifts from *Neohelos*-grade to *Zygomaturus*-grade zygomaturines largely centered on increased anterior flexion of the neurocranium and ascending ramus, more anteriorly oriented masticatory muscles, shortening of the rostrum, greater curvature of the cheek tooth row, and increased grinding capacity of the premolars (fig. 24). The changes in zygomaturine masticatory function correspond well to those drawn here for the sthenurines, but by comparison the premolars may have played a more important masticatory role in sthenurines.

Sthenurus

Based on their relatively more macropodine-like features, Tedford (1966) suggested that the species of *Sthenurus* may have been mixed feeders (browser-grazers). They predominated in the drier, open conditions of the continental interior and are united by their dolichocephalic cranial form, procumbent lower incisors, broadened upper first incisors and high-crowned cheek teeth (fig. 25A), although the very derived *S. stirlingi* secondarily evolved a shorter cranium (pls 24, 109). While the craniodental anatomy of *Sthenurus* suggests a diet not requiring great masticatory force, there was clearly a need for a long diastema for manipulating herbage prior to mastication, a large incisor surface area for a larger yield per bite, and a prolonging of the life of the molars as a response to increased wear (fig. 25A). These are characteristics typical of grazers, but the anteroposterior shearing crests of *Sthenurus* molars are not nearly as well developed as they are in the large macropodine grazers. If the evolution of higher cristae was not constrained in some way by other components of the masticatory system, then shearing in occlusal phase 2 cannot have been an evolutionary priority for *Sthenurus*.

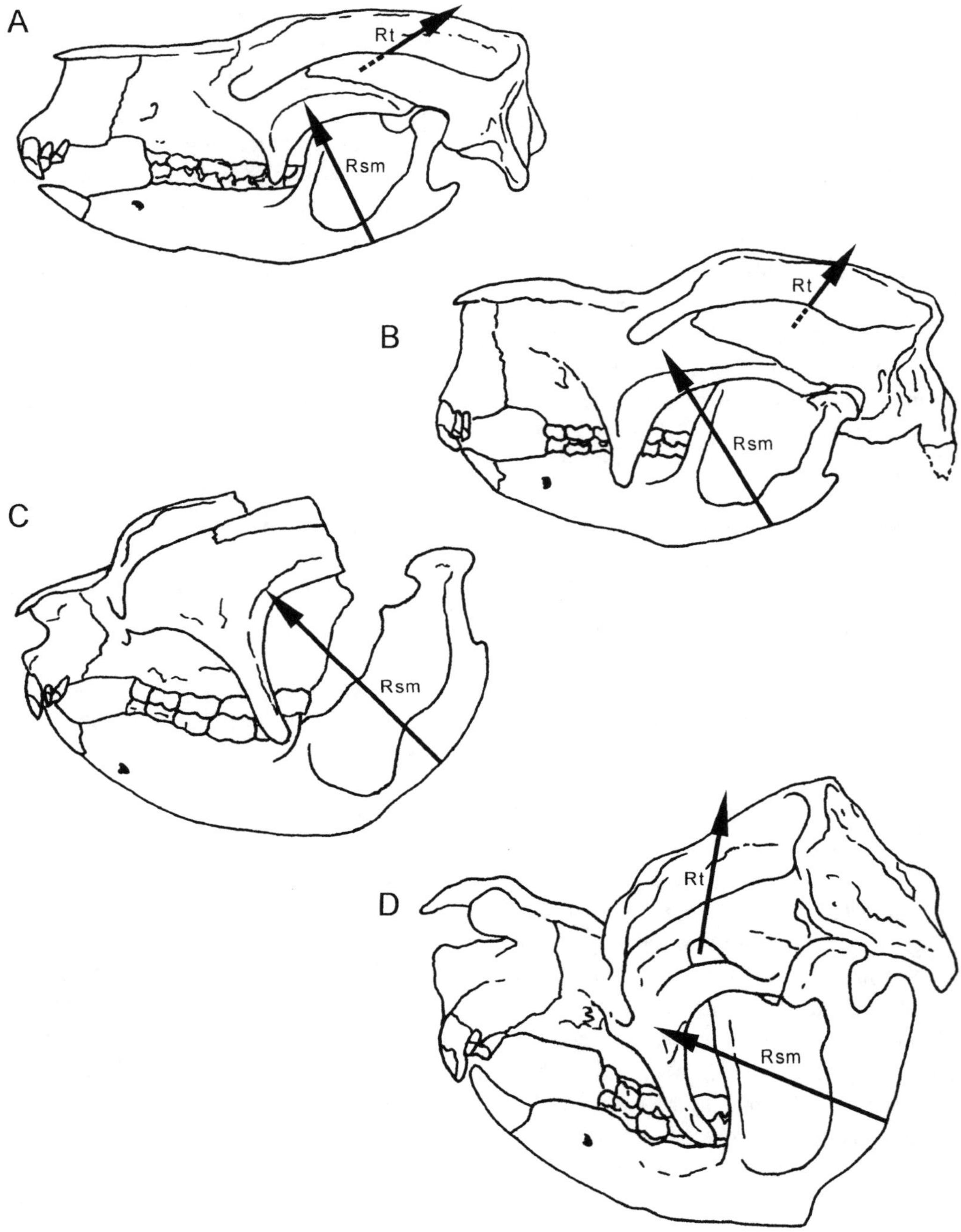

Figure 24. Articulated cranium and dentary of zygomaturine diprotodontids in lateral view (adapted from Murray 1992). **A**, *Neohelos stirtoni* (early Miocene); **B**, *Kolopsis torus* (late Miocene); **C**, *Maokopia ronaldi* (Pleistocene); **D**, *Zygomaturus trilobus* (Pleistocene). Rsm, Rt, estimated resultant direction of pull of superficial masseter, temporalis.

In all probability, the craniodental bauplan constrained most sthenurines to variations on the browsing theme. Grazing macropodines maximized shearing efficiency and molar durability by evolving strong anteroposterior molar cristae, a dorsoventrally curved molar row, and molar progression (Janis 1990). While they certainly have higher-crowned molars than other macropodoids, the degree of hypsodonty expressed is much less than in grazing ungulates, because the need to maintain a precise occlusion of the shearing blades in three planes means that even moderate wear results in a significant decrease in shearing efficiency (Sanson 1980; Fortelius 1985; Janis and Fortelius 1988; Janis 1990). Still, if lineage longevity, taxonomic diversity, geographic ubiquity (across the isolated continent on which they evolved) and numerical abundance are ways to measure evolutionary success, then the grazing macropodine design is difficult to fault. Macropodines are, however, heavily reliant on independent dentary movement facilitated by a very flexible symphysis. In sthenurines, the massive premolar, rigid symphysis and mode of incisor occlusion probably constrained the development of any of these adaptive options for grazing, which arose independently in at least two lines of macropodines (*Petrogale*, *Macropus*). If the species of *Sthenurus* consumed any amount of grass, then it may have been during the appearance of new growth, which in the southern interior today follows the generally infrequent rains.

Although Wells and Tedford's (1995) detailed anatomical analysis centered on *Sthenurus stirlingi* (pl. 109), the most divergent of the *Sthenurus* species, they closely compared the species with *S. andersoni* and *S. tindalei.* They concluded that all three species were predominantly stand-up browsers with highly mobile, elongate grasping forearms easily capable of reaching high, leaf-laden branches. Many features of the locomotory system, vertebral column and forelimbs in *Sthenurus* are also typical of simosthenurins, e.g., *Procoptodon goliah* (Tedford 1967) and *Simosthenurus occidentalis* (Adnams-Hodges 1988). The incipient development of similar modifications in *Hadronomas puckridgi* (Murray 1995) shows that these features are ancestral rather than homoplastic (Wells and Tedford 1995). When eventually studied in detail, it is likely that subtle differences in postcranial anatomy between sthenurines will reflect preferences for different terrains, and possibly, different food procurement techniques. To date, the work of Bishop (1997) on the Naracoorte sthenurines is the only study to have considered the functional significance of interspecific differences in postcranial morphology, although Bishop dealt only with the pedal system. For the moment, comparative craniodental morphology, size and geographic distribution are the chief avenues by which the habits of different sthenurines may be inferred.

If the broader nature of the incisors of *Sthenurus* (pl. 25A) relative to the simosthenurins (pl. 25B) is adaptive, and not an ontogenetic consequence of splanchnocranial elongation (I am not aware of any reason why it should be), then these teeth were designed for maximizing bite area rather than bite force. Many extant mammalian browsers do have a broad incisor array, but this tends to be either a retained ancestral morphology that has not been selected against or, presumably, an adaptation to strip off many small items at once. If the species of *Sthenurus* were browsers as opposed to mixed feeders, then the high-crowned lophs must suggest a more abrasive dicotyledonous diet than that of most other sthenurines. Independent paleodietary assessment through enamel microwear and carbon isotope analyses are needed to test these morphology-based postulations. However, it is feasible that *Sthenurus* was primarily adapted to the consumption of small-leafed shrubs (e.g., chenopods) or low dusty forbs common in the more open habitats throughout which the species of this genus were distributed (see The Sthenurine Radiation). Such a diet would explain the need for minimal anteroposterior shearing crest development and minimal crenulation of the enamel surface for grinding larger, thicker leaves. Before Eurasians introduced rabbits and ungulates (e.g., camels, sheep) to the Australian continent, shrubs were under-exploited by mammals as a food resource. Forbs also form a low proportion of the diet of modern herbivorous marsupials, such as the large extant *Macropus* species and wombats. On the assumption that only minor shifts in the diets of these animals have taken place since the late Pleistocene, it would be extraordinary if no marsupials had been exploiting the substantial forb and shrub resources prior to the close of the Pleistocene. The species of *Sthenurus* are likely candidates.

Sthenurus stirlingi is an unusual, highly derived member of the genus (pl. 109). It is closely related to *S. tindalei* and quite possibly derived from it, but *S. stirlingi* has a craniodental system reminiscent of the simosthenurins in several regards. The species is particularly similar to *"Simosthenurus" pales*. Both were very large, had apparently very similar cranial proportions, and possessed an upturned i1, high-crowned premolars, a long buccal crest on p3, a similarly divided postprotocrista, and a similarly developed mesocrista (pls 24-25, 66-69). *"Si." pales* shows incipient broadening of the I1, increased development of the anterolingual crest on I3, and increased crown height, but none of these approach the degree of development seen in *S. stirlingi*. Just how much dietary convergence there might have been between these species is hazardous to predict faced with the current data, but it may be important that *"Si." pales* is one of the few simosthenurin species with a Pleistocene distribution that extended into the continental interior. The relatively more brachycephalic condition expressed by *S. stirlingi* does suggest a capacity for greater masticatory force production than its congenerics. Larger, tougher leaves from low tree branches may have formed a substantial proportion of its diet (Wells and Tedford 1995), but the very broad upper incisor arcade and high-crowned molars might also indicate that shrubby and somewhat abrasive materials remained important dietary constituents. Perhaps *"Si." pales* filled a very similar dietary niche: no deposits are known to preserve the remains of both species.

Simosthenurini

Four craniodental specializations unite *Metasthenurus* and the Simosthenurini to the exclusion of *Hadronomas* and *Sthenurus*, while an additional seven unite the

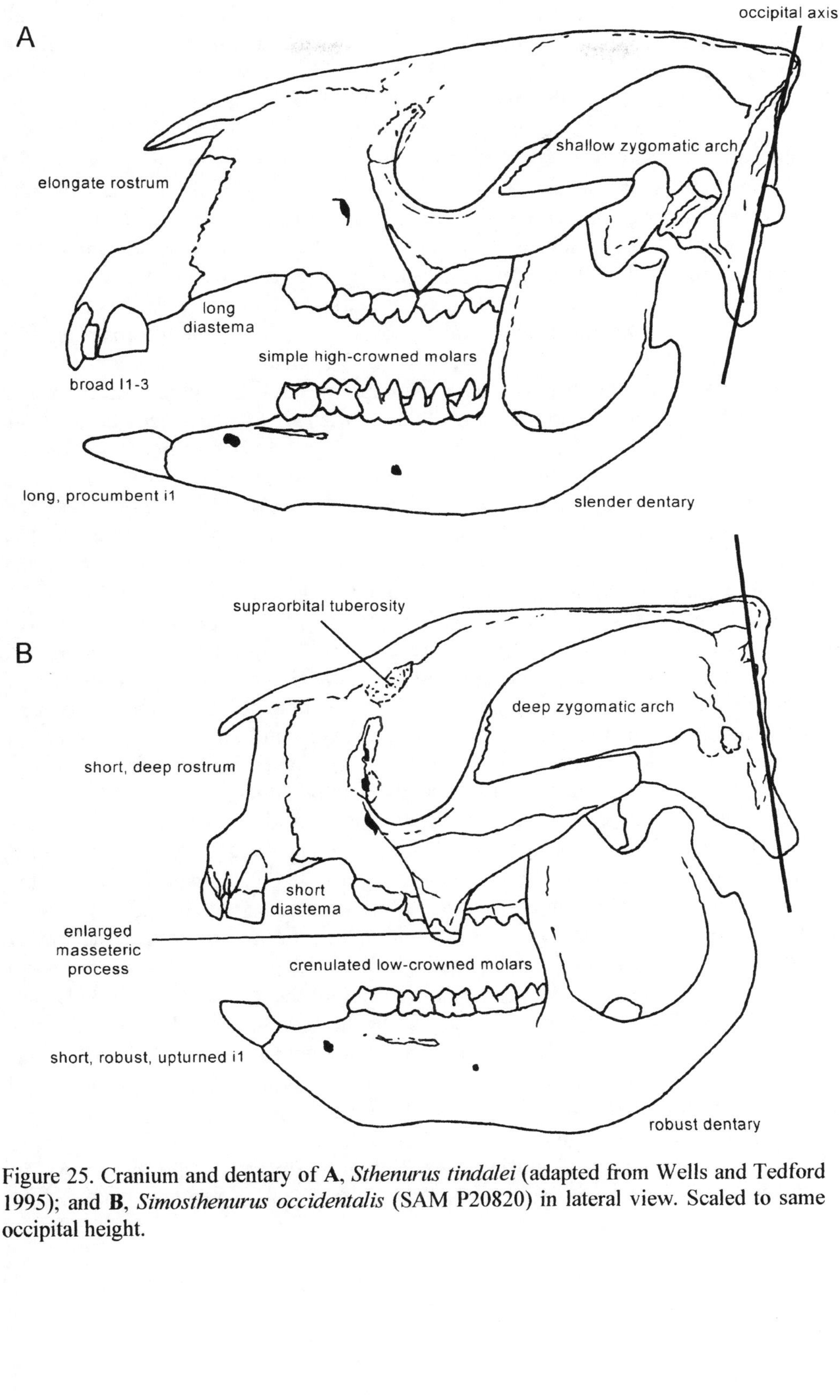

Figure 25. Cranium and dentary of **A**, *Sthenurus tindalei* (adapted from Wells and Tedford 1995); and **B**, *Simosthenurus occidentalis* (SAM P20820) in lateral view. Scaled to same occipital height.

simosthenurins to the exclusion of *Metasthenurus* (see Phylogeny). In essence, all of these features represent a continuation of the general adaptive course set early in sthenurine evolution toward the consumption of increasingly tough vegetation. Cranial restructuring in the form of further elevation and deepening of the neurocranium, as well as deepening of the zygomatic arch and ascending ramus, resulted in an even more vertical temporalis and larger attachments for all of the major jaw adductor muscles (fig. 25B). As the superficial masseter became enlarged and anteriorly oriented, facilitated in significant part by the lateral flaring of the masseteric fossa, the masseteric process became extended and thickened. The degree to which this is emphasized, however, is less than that seen in some diprotodontoids (fig. 24) and browsing xenarthrans (Gillette and Ray 1981).

Brachycephaly, brought about chiefly by the shortening of the rostrum, is a characteristic of the Simosthenurini (fig. 25B). *Metasthenurus* retains cranial proportions intermediate between the plesiomorphic, dolichocephalic *Hadronomas* and the simosthenurins. Biomechanically, a shortened rostrum places the incisors closer to the craniomandibular joint, enabling more forceful incision. Simosthenurins also have short, robust, upturned lower incisors (fig. 25B). These provided a reduced surface area for the apprehension of food items, but increased bite force per unit area. It is also more mechanically advantageous for the occlusal strain placed on the lower incisors to be directed at an angle closer to the longitudinal axis of the tooth. In all likelihood, only a limited proportion of the simosthenurin diet was collected by the incisors. Much of the diet was probably fed straight onto the premolars using the forepaws, as it is in the koala (Davison and Young 1990). The greater robustness of the splanchnocranial region and large, posteriorly extended inferior transverse torus of the mandibular symphysis are two further indications of the increased capacity to withstand large occlusal forces at the anterior end of the dentition in simosthenurins. The entire cranium was clearly well adapted for the mastication of resistant vegetative matter.

Simosthenurins, as well as *Metasthenurus*, are also typified by markedly inflated frontals. To what degree this feature was a consequence of neurocranial restructuring is difficult to say, but a correlation is likely. I suspect that with the lateral expansion of the temporal fossa, expansion of the frontals was essential to maintain a relatively short, robust postorbital ligament and, therefore, visual acuity during forceful mastication. A freestanding postorbital process of the form seen in *Hadronomas* is not present in these sthenurines, but the zygomatic arch is deeper, and there is a distinct step or inflection point on its anterodorsal aspect, directly posterior to the orbit (fig. 25B). The postorbital ligament apparently attached to this point. Lateral expansion of the frontal region would also have afforded an expanded origin for the anterior portion of the temporalis, possibly an exaptive trait. *Simosthenurus s. s.* may have exploited this to maximum effect by greatly inflating the frontals and partially extending them above the orbit. As in *Hadronomas*, the presence of a distinct supraorbital tuberosity (fig. 25B) might indicate that the postorbital ligament had become partially chondrified in these species.

If the principal evolutionary explanation for inflated frontals in simosthenurins is to maintain a short, robust postorbital ligament to protect the orbit, then the secondarily derived condition of narrower frontals in *"Procoptodon" gilli* (pl. 97) requires explanation. In this species, the frontals are no more inflated laterally than in the dolichocephalic *Sthenurus* species (fig. 15). Moreover, the anterodorsal aspect of the zygomatic arch lacks a distinct step. These features suggest that the need for a strong postorbital ligament had diminished relative to that of most simosthenurins, implying that force magnitudes produced during mastication were comparatively less in *"P." gilli.* The morphology of the dentition is consistent with this idea: *"P." gilli* has higher-crowned incisors, premolars and molars than any other simosthenurin species, and its molars are characterized by a distinct lack of enamel crenulations. In all of these regards, *"P." gilli* is convergent with *Sthenurus*, perhaps indicating a relatively softer but more abrasive diet. While its pes was not especially well adapted to a thick understorey (Bishop 1997), its craniodental morphology, small adult body size (est. 30 kg), and abundance in southeastern Australian cave faunas suggest that *"P." gilli* might be fairly reconstructed as a low-level shrub or forb browser, at home in the heathland or *Banksia* scrub habitats common in these regions today (Specht 1994). Intriguingly, the marked semblance between the cranial form of adult *"Procoptodon" gilli* and juvenile *"P." browneorum*, and the minimal change in overall cranial shape from juvenile to adult in *"P." gilli* (figs 5-6) suggests a pedomorphic origin for this species from the larger *"P." browneorum.* This hypothesis is supported by similarities between the respective cheek teeth of *"P." gilli* and the smaller morphs of *"P." browneorum* (see Taxonomy). Similar evolutionary transitions have been reported in other mammals (e.g., phocid seals, King 1972; Wyss 1994).

In many regards, *"P." browneorum* is the quintessential simosthenurin. The species is second only to *"P." gilli* in abundance in the diverse sthenurine faunas of southeastern Australia and is common in most Pleistocene cave deposits of southern Australia in which it is found. Its craniodental morphology suggests a capability for processing particularly tough dietary items. The premolars are markedly widened and the molar surfaces are covered in coarse enamel crenulations, which could have been especially useful for grinding fibrous leaves, twigs and shrubby material. The very large digastric eminence may highlight the importance of reseating of the condyle following the power occlusal stroke (phase 2), a stage during which the bite point would have acted as a secondary fulcrum. While *"P." browneorum* is commonly confused with *Simosthenurus occidentalis*, its slightly larger, less brachycephalic cranium and larger, more slender calcaneum and metatarsal IV suggest a larger body size (est. 60 kg) and a greater adeptness in more open habitats.

Most sthenurine species are estimated to have fallen within a 40-60 kg adult weight range (Murray 1984; pers. obs.). Among them, *Simosthenurus maddocki* possesses the most specialized craniodental morphology. In contrast to all other simosthenurins, it has a small postglenoid process (pls 41A, 43A), which suggests a relative reduction in masticatory forces tending to drive the condyle back into the glenoid fossa. This is consistent with the very reduced digastric sulcus and eminence. The elongate and very slender lower incisors (pl. 45A-B), which are so reduced that they rest within the upper incisor arcade (pl. 42B), bespeak highly selective browsing. Presence of a long manipulative tongue may be reflected by the deep median dorsal groove. Relative to the overall size of the cranium and dentary, the cheek teeth of *Si. maddocki* are smaller than in any other sthenurine and are covered in many very fine enamel crenulations. In this regard, the premolars reflect the morphology of the molars. Indeed, the wide median valley, which is penetrated by an unusual row of small cuspules (pl. 45C), may have been used as a knurled platform for gripping rounded objects (e.g., seeds) during premolar occlusion. The cheek tooth rows are also less curved in *Si. maddocki*, which suggests limited lateral grinding movements in occlusal phase 2 and a greater emphasis on crushing. Together, these features might suggest that *Si. maddocki* was more frugivorous, or at least specialized for selecting softer, newer growth, than its similarly sized contemporaries.

In light of this dietary hypothesis, it is interesting to contemplate the likely functional significance of its hyper-inflated frontals, which terminate above the orbit at a very rugose supraorbital tuberosity (pl. 41A-B). If I am correct in positing that lateral frontal inflation is correlated with the need for a short, robust postorbital ligament to protect the eye from the larger, more anteriorly extending temporalis musculature, why might the condition be so hypertrophied in *Simosthenurus maddocki*, and to only a slightly lesser degree in *Si. occidentalis* (pls 51D)? Perhaps the supraorbital portion of the frontal was co-opted to help protect the eye from sharp vegetative protrusions encountered during selective browsing. The more anteriorly projected nature of the inferior orbital rim of *Si. maddocki* (pl. 41A) may well have served the same function.

At a more general level, one of the most conspicuous evolutionary trends in the *Metasthenurus*–Simosthenurini lineage relates to successively more marked division of both the paracristid and the cristid obliqua into two separate components (compare pls 34C-D, 38C, 67C). This unusual developmental phenomenon, which is mirrored in the upper molars by the division of the postprotocrista (compare pls 35A, 37A, 66), also took place in *Sthenurus*. To my knowledge, no such condition has been reported in any other group of bilophodont mammals. *M. newtonae* expresses an intermediate condition between *Hadronomas* and the Simosthenurini, as *S. andersoni* does between *Hadronomas* and the more derived *Sthenurus* species (compare pls 14, 28). In *M. newtonae*, the paracristid has a kink in it halfway along its length, but remains undivided in all teeth. The cristid obliqua of dp3 is undivided, but a slight division is evident on m1. This division becomes slightly more distinct farther back in the tooth row (pl. 34C-D). In the more plesiomorphic simosthenurins (e.g., *"Simosthenurus" brachyselenis*), division of the paracristid and cristid obliqua is only incipient on the more anterior teeth, but the lingual components of both cristids are clearly shifted across the lophid anterior faces more posteriorly. In most simosthenurins, the lingual component of each cristid approaches the tooth midline. A parallel shift took place in the lineage whence *S. stirlingi* and *S. tindalei* were derived.

Mesial displacement of the main components of the paracristid, cristid obliqua and postprotocrista highlights a de-emphasis on the continuation into occlusal phase 2 of the shear produced in phase 1 (see Sanson 1980 for a detailed account of this process in macropodines), but I am unable to hazard a reasonable guess about the initial adaptive significance of anteroposterior crest division. The grinding efficiency of the molars would seem only to have improved beyond that provided by smooth enamel surfaces once the divisions had become distinct and the mesocrista, urocrista and enamel crenulations had emerged. In simosthenurins, it is generally the case that the greater the mesial shift in these three crests, the better the development of the enamel crenulations, especially on the posterior faces of the lophs and the anterior faces of the lophids. The paracristid, cristid obliqua and postprotocrista are often only slightly higher and thicker than the enamel crenulations, such that it may only be possible to differentiate them by their topological relationships. Combined, these features indicate that the chief function of the molars during occlusal phase 2 was pulverization. The roughened surfaces would have been especially well designed for rupturing the integrity of leaves with waxy cuticles and thick cell walls. Widening of the premolars combined with increased p3 buccal crest length and the enhancement of ridgelets clearly increased the overall area and unevenness of the premolar surface in simosthenurins, providing a morphological shift complementary to that of the molars.

The most eccentric craniodental morphology seen across the entire Macropodoidea characterizes the most derived of the *Procoptodon* species: *P. pusio*, *P. rapha* and *P. goliah.* The best known of these is *P. goliah*, an excellent full reconstruction of which is presented in plate 110. *P. goliah* was a large sthenurine with an extremely brachycephalic cranium, greatly enlarged masticatory muscles, near-cylindrical dentaries, deep maxillae and bulbous molars (pls 71-77; fig. 26). While the molar cristae are undoubtedly homologous with those of other simosthenurins, they are hypertrophied to such a degree that they approach the height of the unworn loph crests. In addition, the buccal (mesial) component of the postprotocrista is split into anterior and posterior moieties, the latter becoming confluent with the enlarged mesocrista (pl. 75; fig. 19F). The 'midlink' is thus composed anteriorly of mesocrista and posteriorly of postprotocrista. I am unsure of the functional importance of the extreme buccal shift of the 'midlink', but it is important to note that a phylogenetically antecedent condition is seen in *"P." williamsi* (especially on the M1 of AM F18903, pl. 107H), which shares a tapered loph condition with *P. pusio*, *P. rapha* and *P. goliah.* In numerous individuals of both *P. pusio* and *P. goliah*, the preprotocrista is also well developed and often directly contacts the protocone apex.

The most important functional consequence of the hypertrophying of all cristae was that occlusion in the typical interlocking manner became impossible. Interlocking of molars is reduced in grazing *Macropus* species due to the enlargement of the anteroposterior crests (Sanson 1980, 1991), but the condition is not nearly as hypertrophied as in *Procoptodon rapha* and *P. goliah.* These species appear to have circumvented the evolutionary constraints imposed by bilophodonty in a somewhat analogous fashion to many different ungulates early in their evolution (e.g., see Janis et al. 1998). The occlusal surface of a moderately worn molar row of *P. goliah* bears considerable resemblance to that of a hippopotamus and some pigs; the molar surfaces become a complex of different sized enamel loops in a variety of orientations ("a flattened clover-leaf pattern" *sensu* Janis and Fortelius 1988). Microwear studies will prove informative in reconstructing the occlusal motion of *P. goliah,* but the resemblance between a worn *P. goliah* cheek tooth row profile (e.g., pl. 76I) and those of elephants (see Maglio 1972:fig. 4) and some glyptodonts suggests that a more fore-and-aft masticatory motion may have become important in this derived sthenurine lineage.

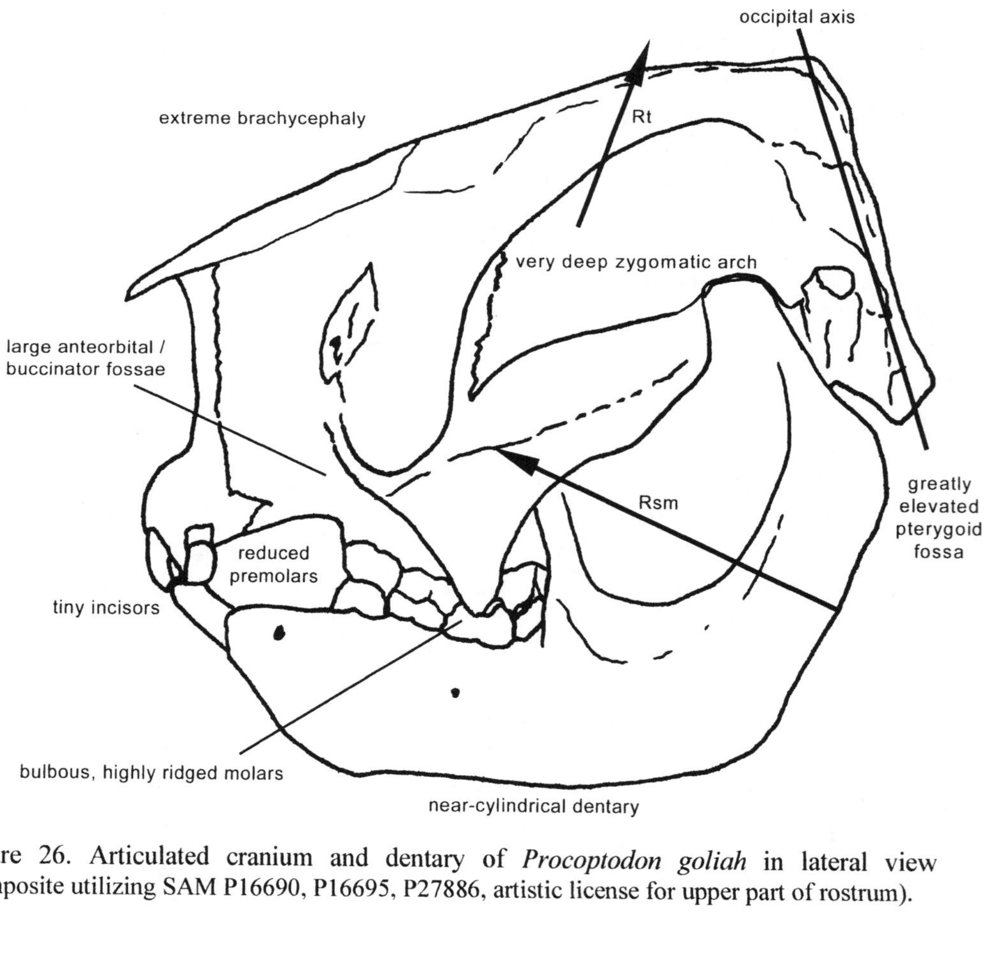

Figure 26. Articulated cranium and dentary of *Procoptodon goliah* in lateral view (composite utilizing SAM P16690, P16695, P27886, artistic license for upper part of rostrum).

Extreme incisor reduction is another intriguing feature of *Procoptodon goliah* (pls 71, 73-75; fig. 26). As in the koala, small upturned incisors have the mechanical potential to facilitate transmission of large bite forces for tearing tough leaves from branches (Wells and Tedford 1995). However, for a 200 kg kangaroo (Murray 1984), dietary yield per incisor bite could not have been satisfactory as the primary means of apprehending food (consider pl. 110). Browse items would also have been fed straight onto the anterior cheek teeth with the assistance of the manipulative forearms, similar to the koala (Davison and Young 1990) and several modern macropodoids (e.g., Menzies 1989; Sanson 1989; Prideaux 1990). It is also possible that the very large anteorbital and buccinator fossae in *P. goliah* not only reflect more robust cheek muscles important for keeping very resistant browse material centered on the cheek teeth during mastication, but also larger, more manipulative snout, analogous to that of many ungulate browsers and probably also *Zygomaturus trilobus* (Murray 1992; figs 24, 26).

In a thought-provoking departure from the orthodox view of *Procoptodon goliah* as a browser on very tough vegetation (e.g., Ride 1959; Tedford 1967), Sanson suggested that its elaborate molar morphology might indicate a grazing habit (Sanson et al. 1980; Sanson 1991). Its large body size and wide distribution throughout the more open, drier regions are not incongruous with this idea, and the molars do resemble those of the large, grazing *Macropus* species in several regards. De-emphasis of the premolars is another similarity. Theoretically, one can quite readily imagine that, as in elephants and mammoths, the molars of *P. goliah* could have been very effective at shearing grass. In fact, if the number of high crests provides a general indication of shearing capacity, then the *P. goliah* shearing system may have been more efficient than the *Macropus* design. How, though, does this relate to all other aspects of the craniodental system, which suggest that a capacity to generate and withstand increasingly large masticatory forces was a major selective force in the evolution of the more derived *Procoptodon* species?

In terms of its marked brachycephaly and rather cylindrical dentaries (pls 76-77; fig. 26), *Procoptodon goliah* bears some proportional resemblance to the elephantids. Modern elephants are primarily browsers, but the diets of Pliocene and Pleistocene *Loxodonta* and *Elephas* were dominated by grasses (Cerling et al. 1999). Mammoths are known to have taken some browse, but they were primarily grazers (e.g., Agenbroad and Mead 1996; Lister 1996). While elephantids evolved from a long line of brachycephalic proboscideans, their emergence late in the Miocene is thought to have been correlated with adaptations including increased body size, brachycephaly, a more vertical temporalis, loss of the permanent premolars, and enlarged molars that no longer interlocked, greatly emphasizing poly-shearing (Maglio 1972; Todd and Roth 1996). The parallels with the most divergent of the simosthenurins are undeniable and only support Sanson's idea. Is it possible that *P. pusio* and particularly *P. rapha* and *P. goliah* were the macropodoid adaptive equivalents of the elephants – consumers of coarse, fibrous herbage, monocots and dicots alike? Enamel isotope and microwear analyses combined with more detailed morphological investigations promise to shed much light on this problem. If the results from this work eventually corroborate Sanson's grazer hypothesis, the most difficult question to solve will be, with such small incisors and presumably no trunk (the anterodorsal aspect of the cranium is not known for these species), by what mechanism might they have been able to grasp and transport swards of grass to the cheek tooth battery?

Overall, many aspects of an animal's anatomy and environment influence diet, not just body size and the mechanical capabilities of the craniodental system. Body size is important, but there is no simple inverse correlation between it and diet quality. Competition and digestive adaptations are very important (e.g., Janis 1976; Hofmann 1989; Bodmer 1991). Thick cuticles and cell walls composed of cellulose and lignin make sclerophyll leaves tough to chew and difficult to digest. Plants also produce a range of toxic secondary metabolites that place considerable metabolic constraints on consumers (e.g., King et al. 1978; Cooper and Owen-Smith 1985; Freeland 1991; Lawler et al. 1998). Many browsers are often highly selective for this reason: adaptation to a narrower range of feeding deterrents can mean greater metabolic and digestive efficiency (Freeland 1991; Chiarello 1998). Similar factors must also have been crucial determinants of sthenurine diets. We know, for example, that arid-adapted chenopods produce such toxic compounds in response to constraints imposed by their nutrient-poor environment (Leigh 1994). The only large feral herbivore that thrives on this vegetation today is the camel (Newman 1983). So, while particular craniodental morphologies might have allowed each sthenurine species to exploit a range of tough browse items in terms of craniodental mechanics, metabolic and digestive specializations probably constrained them to a narrower realized range. This presumable holds for the most diverse Pleistocene sthenurine community known: the Victoria Fossil Cave assemblage at Naracoorte. Nine sthenurines occur in the site and even the rarest species (*"Simosthenurus" baileyi*, *"Si." pales*) are present in the top, middle and bottom layers of the thick sedimentary sequence (pers. obs.). Unless the stratigraphy is homogenized, which is apparently not the case (e.g., Wells et al. 1984; Grün et al. 2000), only some major paleoecological differences can have facilitated their coexistence.

While low, the odds of finding preserved sthenurine gut contents are not infinitesimal. At Lake Callabonna, plant material has been found associated with *Diprotodon* remains (Wells and Tedford 1995), and in the Morwell fire-holes, gut contents found in direct association with remains of *Protemnodon anak* confirm it as a browser (Flannery 1984). However, the two most promising avenues by which morphology-based hypotheses of sthenurine dietary ecologies may be tested are isotopic analysis of enamel, which enables inferences on the proportions of C_3 and C_4 plants in diets (e.g., Wang et al. 1994; MacFadden 1998; Cerling et al. 1999), and microwear analysis of enamel, which facilitates inferences on the predominance of dicots and grasses in the diets, as judged from differences in abrasion patterns (e.g., Solounias et al. 1988; Solounias and Moelleken 1992; MacFadden et al. 1999).

THE STHENURINE RADIATION

In response to a progressively drier, more seasonal Australian environment, sthenurine and macropodine macropodoids radiated more conspicuously during the late Cenozoic than any other lineage of marsupials. Because terrestrial records are very incomplete and the geochronology is poorly resolved, it is difficult to generate a clear picture of Tertiary environmental change across Australia. This notwithstanding, there have been some dramatic improvements in our understanding of Australian marsupial evolution and late Cenozoic biochronology over the past two decades. Modern paleontologists are in a much stronger position than their predecessors to map the radiation of several common groups. One of these is the Sthenurinae. Below I review the climatic and vegetational regimes prevailing immediately prior to and during sthenurine evolution, and consider the distribution of sthenurines through time and space in light of phylogenetic relations, functional morphology and paleoenvironmental data.

LATE OLIGOCENE TO MIDDLE MIOCENE

By the late Oligocene, global cooling and rainfall seasonality had passed the thresholds necessary to induce evolutionary responses in the southeastern Australian vegetation, which included declining *Nothofagus* (Martin 1994) and expanding dry, araucariaceous rainforest taxa (Macphail et al. 1994), *Eucalyptus* (Pole et al. 1993), open vegetation and scrub (N. F. Alley, pers. comm. in Martin 1997a). Inland, there is solid paleobotanical evidence for a significant decrease in rainfall by the time the upper units of the latest Oligocene Etadunna Fm. were deposited (Truswell and Harris 1982; Martin 1990). A carbonaceous shale underlying faunal zone D lacks *Nothofagus*, but *Acacia* is present and *Casuarina* is abundant (Truswell and Harris 1982; W. K. Harris, pers. comm. in Pledge 1984). The mere presence of *Acacia* is significant, because its pollen scores are consistently low wherever recorded (Sluiter and Kershaw 1982). Fossil plant material associated with the overlying Wipajiri Fm. is still unstudied (Archer et al. 1994). In the Tarkarooloo Basin to the southeast, pollen from the upper part of the Namba Fm. includes *Acacia*, *Eucalyptus* and proteaceous species, but lacks gymnosperms except for the xeric-adapted *Callitris* (Martin 1990). Despite the lack of palynological evidence for grasslands (Martin 1990; cf. Callen and Tedford 1976), this pollen spectrum suggests the presence of widespread dry sclerophyll forest. The age of the upper part of the Namba Fm. is poorly constrained, but an early Miocene pollen assemblage from Lake Menindee, to the southeast, is also suggestive of drier conditions (Martin 1988).

Even though the predominance of low-crowned browsers and the distributions of the Etadunna and Namba Fms advocate the idea of widespread permanent and ephemeral lakes in central into the earliest Miocene, some evidence suggests that the clays from which these formations are largely composed may be associated with periodically drying conditions (Norrish and Pickering 1983; Woodburne and Case 1996; see also Stein and Robert 1986). However, no doubt exists about the paleoenvironmental implications of the dolomitic units within these formations: they unequivocally point to evaporitic, seasonal conditions (Stirton et al. 1961; Wells and Callen 1986; Woodburne et al. 1994). Relatively more xeric conditions in the middle Miocene of northern and central Australia are indicated by the extensive distribution of palygorskite clays and limestone facies (Megirian 1992), the formation of duricrusts in inland Australia (Wopfner 1997), and the marked increase in illite and aeolian-derived terrigenous sediments carried across by westerlies and deposited in the southwestern Pacific (Stein and Robert 1986). The appearance of opal phytoliths at 14.4 Ma in cores from the same area may or may not be the earliest sign of grasslands in Australia (Locker and Martini 1986, 1989).

Paleoenvironmental interpretations of the poorly stratified, richly fossiliferous Carl Creek Limestone of northwestern Queensland have been strongly debated over recent years (e.g., Megirian 1992; Archer et al. 1995), and as yet, there has been no resolution between the "habitat mosaic" and "hyper-rainforest" hypotheses. McGowran and Li (1994) suggested that temporal fluctuations in prevailing climates may account for the high vertebrate diversity and differences in paleoenvironmental interpretation for the older parts of the sequence, but there appears to be consensus that by the later stages of vertebrate accumulation (e.g., ?late Miocene Encore Site), a distinct mesic-adapted to xeric-adapted shift in the fauna had taken place. Included in these strata are the earliest known wombats with ever-growing teeth (Archer et al. 1995), an apparent grazing adaptation, and two species of *Wanburoo* (Cooke 1997a), the likely sister taxon of the Sthenurinae.

LATE MIOCENE

By the time the central Australian mammal record commences again with the Alcoota LF in the late Miocene (Woodburne 1967; fig. 27), at least eight million years seem to have passed since the Bullock Creek LF accumulated (Murray and Megirian 1992; pers. obs.). Sthenurine and probably macropodine evolution had been underway for some time, even though *Hadronomas puckridgi* and *Dorcopsoides fossilis* are the earliest recognized members of these subfamilies. A *Wanburoo*-like taxon and a possible medium-sized propleopine also occur in the Alcoota LF, but each is known from a single specimen too incomplete to identify to genus (pers. obs.). If more material eventually confirms the presence of these two taxa, it will be interesting from a biocorrelative perspective to compare them with later Riversleigh propleopines and bulungamayines. Within the Waite Fm., the unit containing the Alcoota LF is overlain by a fluviatile deposit containing the Ongeva LF (Murray et al. 1993; Megirian et al. 1996). Faunal comparisons and stratigraphic evidence suggest that the age difference between the fossiliferous units may be one to two million years, placing the Ongeva LF close to the Miocene-Pliocene boundary. No macropodoid material identifiable to genus is known from the Ongeva LF, but the presence of a *Dorcopsoides*-sized taxon is revealed by several postcranial elements (Megirian et al. 1996).

Hadronomas is the only sthenurine of unequivocal late Miocene age, but four of the other five genera occur in the early Pliocene (*Metasthenurus* by phylogenetic inference). At least two of them, *Sthenurus* and *Simosthenurus*, were widespread, so

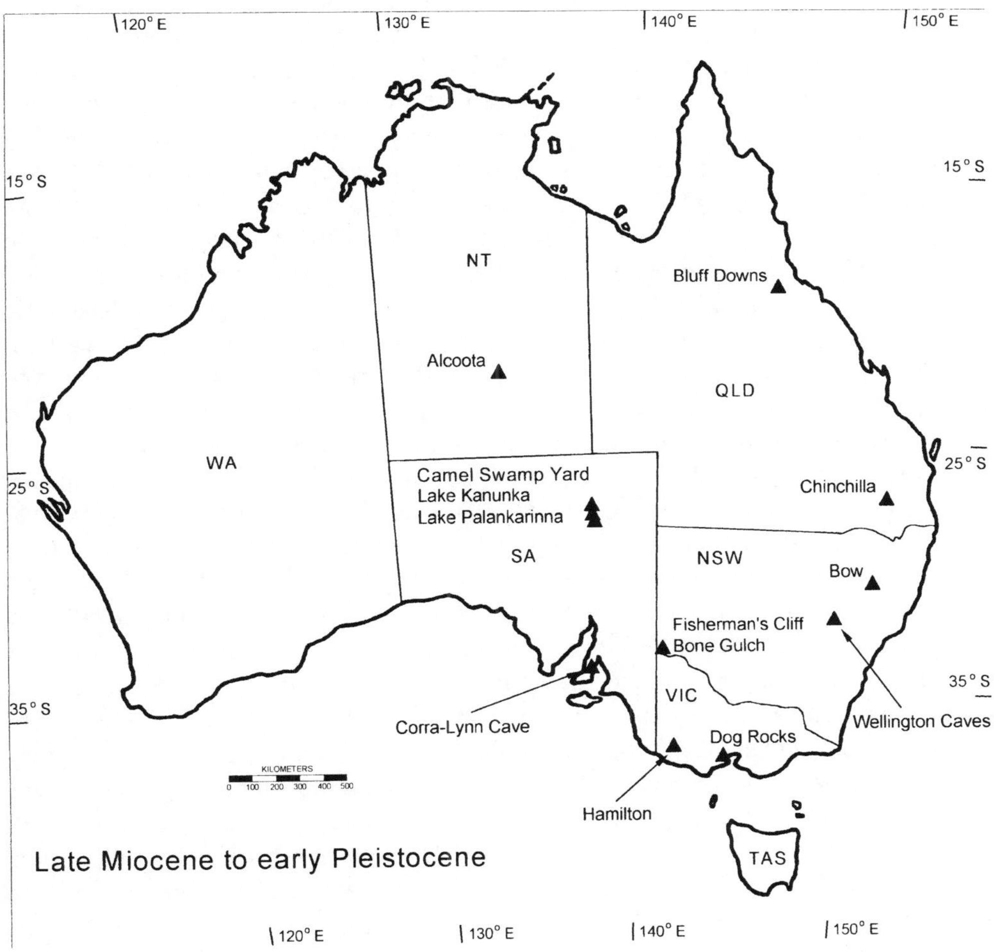

Figure 27. Map showing late Miocene to early Pleistocene localities yielding sthenurines.

divergence of the major sthenurine clades had to have been a late Miocene phenomenon. No paleovegetational data are directly associated with the Alcoota faunas, although bores in the region have yielded similarly aged pollen assemblages. At Tempe Downs, 250 km southwest of Alcoota, Macphail (1996a) reported on a ?late Miocene to Pliocene palynoflora dominated by Casuarinaceae and Cyperaceae, which suggests relatively dry and open conditions (Macphail 1997). These pollen data are congruent with inferences made from the composition of the Alcoota LF and the nature of the accumulation. The Alcoota LF seems very likely to represent a thanatocoenosis biased toward large animals as a result of waterhole tethering and an element of fluvial concentration (Murray and Megirian 1992). Seasonal and drying conditions are implied by the taphonomy of the site and supported by the very large body size and browsing adaptations of the dominant herbivores (diprotodontoids, dromornithids). Functional inferences made earlier for *Hadronomas* are congruent with these observations.

The only other assemblage with sthenurines that is probably late Miocene is the Curramulka LF of central Yorke Peninsula, South Australia (Pledge 1992; Tedford 1994; fig. 27). Its age is somewhat equivocal at present, because the fauna contains a mix of taxa with apparent late Miocene, early Pliocene or early Pleistocene affinities, and was collected from six different passages within Corra-Lynn Cave (Pledge 1992). It is possible, therefore, that all elements of the Curramulka LF are not contemporaneous, although it will prove difficult to verify this with absolute dating methods. Nevertheless, rodents are absent from each individual sample and the presence of other small vertebrates argues against the idea of taphonomic sorting. Rodents make their earliest appearance in the Parwan LF of southern Victoria (4.12–4.20 Ma, Whitelaw 1992), so the Curramulka LF can be assumed to extend at least as far back as the very early Pliocene. The fauna contains *Archaeosimos cegsai* and *A. correlli*, the two most plesiomorphic members of the Simosthenurini. A single lower incisor (SAM P40061) from a fissure fill exposed in the nearby Curramulka Quarry is very similar to that of *Archaeosimos cegsai*, but the deposit also contains eight other sthenurines with predominantly middle to late Pleistocene distributions. Several rather plesiomorphic macropodines have been collected from the same fissure fill (J. A. McNamara, pers. comm. 1998) and it would seem likely that the fauna is of mixed age.

The most temporally and geographically proximate palynoflora to the Curramulka LF is an early Pliocene pollen assemblage from northern Eyre Peninsula, 200 km to the northwest. It is dominated by *Casuarina*, Asteraceae, Chenopodiaceae, Cyperaceae, Poaceae and Myrtaceae, including *Eucalyptus* (Truswell and Harris 1982). These point to drier, more open conditions, although small amounts of *Nothofagus*, *Dacrydium*, *Lagarostrobus* and podocarpaceous pollen suggest relictual rainforest. Benbow et al. (1995) proposed that, apart from these rainforest pockets, southern South Australia was largely covered with open woodland or savanna from the late Miocene to the early Pliocene. However, before we can confidently postulate what paleohabitats the Curramulka LF might sample, the precise systematic relationships of its various elements need to be better understood and the contemporaneity issue should be more effectively addressed. Although the craniodental features that ally the two *Archaeosimos* species with the Simosthenurini do point to a tougher diet than that of *Hadronomas* or any earlier macropodoid, they are still among the lowest-crowned sthenurines known and maintain very short p3 buccal crests, so their diet clearly cannot have been especially fibrous or abrasive.

PLIOCENE

While data from the late Miocene to earliest Pliocene is very limited, sthenurines are better represented in the Alcoota and Curramulka LFs than they are in any of the eastern Australian early Pliocene deposits. From Bluff Downs in far northern Queensland to Hamilton in the southeast, sthenurines are known from a paltry total of five tooth fragments. The oldest of these localities, a paleosol overlying the marine Grange Burn Fm. in southwestern Victoria, yields the Hamilton LF (Gill 1957; Turnbull and Lundelius 1970; fig. 27). It is, in fact, Australia's oldest firmly dated Neogene assemblage; potassium-argon dating of an overlying basalt and magnetostratigraphic assessment of the paleosol have bracketed the age of the fauna to between 4.35 and 4.42 Ma (Whitelaw 1991; Flannery et al. 1992). A range of macropodoids occur in the Hamilton LF, but only two partial molars referable to the Sthenurinae have been recognized. Flannery (1989) and Flannery et al. (1992) viewed the Hamilton taxon as the earliest known and most plesiomorphic *Simosthenurus* species, a prescient assessment given similarities with *Archaeosimos correlli*, the plesiomorphic simosthenurin described herein. The Hamilton fragments may well be referable to *Archaeosimos*, but are simply too incomplete to place in this genus with confidence, so I have referred them to cf. *Archaeosimos* sp. (see Taxonomy).

A palynoflora directly associated with the Hamilton LF suggests that dry (araucariaceous) rainforest was the main vegetation in the surrounds, but some open sclerophyll forest or woodland also appears to have been present (Macphail 1996b). The pollen data strongly support the paleoenvironmental interpretation deduced previously from the composition of the fauna (Turnbull and Lundelius 1970; Flannery et al. 1992), which is dominated by small to medium-sized terrestrial browsers and arboreal forms whose closest extant relatives are rainforest dwellers. Although it is possible that the small molar size of the Hamilton simosthenurin reflects small body size, dimensions actually lie within the known range of the similarly low-crowned *Simosthenurus maddocki* (tables 15, 18), which probably weighed in the 40-60 kg range predicted for most sthenurines.

Farther to the north, the Bow LF of central eastern New South Wales (fig. 27) contains a similarly diverse array of marsupials and a similar paucity of sthenurines to the Hamilton LF. Stage-of-evolution and genus-level correlations suggest that the assemblages are roughly coeval (Flannery and Archer 1984; Flannery et al. 1992; pers. obs.), despite sharing only one species (*Jackmahoneya toxoniensis* Ride, 1993). Fifteen macropodoid taxa have been recognized from each fauna, with both dominated by macropodines. Hypsiprymnodontines, propleopines, potoroines and sthenurines are all rare. One upper molar (AM F60670), a partial P3 (AM F64002) and a partial astragalus (AM F63990) constitute the record for sthenurines at Bow, since I do not consider *Troposodon* a sthenurine (see Monophyly and Origin). AM F60670 is now lost (Flannery and Archer 1984), but the original description (low crowned with enamel crenulations) is suggestive of a simosthenurin; indeed, this prompted the authors to refer the tooth fragments to *Simosthenurus*. However, it is worth bearing in mind that *Archaeosimos correlli* (late Miocene) is relatively low crowned and *Sthenurus andersoni* (late Pliocene to late Pleistocene) typically possesses a few fine enamel crenulations on its upper molars. Despite missing the buccal enamel surface and the anterior end of the crown, the P3 specimen (AM F64002) appears to have been slightly more elongate than is typical of the Simosthenurini. Perhaps even more indicative is the manner in which the posterior extremity of the lingual crest curls back into the longitudinal basin, making a U shape and not connecting to the main crest. This is often observed in specimens of *Metasthenurus newtonae* and some individuals of *S. andersoni.* As *M. newtonae* and *S. andersoni* are plesiomorphic members of the two lineages that diverged after the separation of *Hadronomas*, it should not be surprising that the P3 of a taxon from the early Pliocene Bow LF resembles the P3 of either of these two taxa.

Unlike Hamilton, no palynoflora is directly associated with the Bow LF. However, Pliocene pollen assemblages drilled from the nearby valleys of the Mooki, Namoi and Castlereagh Rivers do reveal a marked expansion of dry rainforest and more open vegetation compared with the Miocene; the appearance of grass and composite (Asteraceae) pollen is especially portentous (Martin 1979, 1980, 1981; Kershaw et al. 1994; Macphail 1997). Differences in the apparent adaptations of the macropodoids suggest that the Bow LF samples different paleohabitats, including wetter forest (e.g., *Dendrolagus*, *Kurrabi*, *Jackmahoneya*), and open woodland or grassland (e.g., *Macropus dryas* De Vis, 1895, *M. pavana* Bartholomai, 1978b). The assemblage probably accumulated in an area enclosed by riparian forest and surrounded by open vegetation (Skilbeck 1980; Flannery and Archer 1984). Perhaps the paucity of sthenurines in the fauna relates to a lack of suitable paleohabitats, but it is equally likely that they were uncommon in this region of eastern Australia at the time, as they apparently were at Hamilton and Bluff Downs.

The Bluff Downs LF has been retrieved from the Allingham Fm., which crops out near Charters Towers in the state's northeast (Archer and Wade 1976; fig. 27). Recent radiometric dating of an overlying basalt places a lower age limit of 3.6 Ma on the Allingham Fm. (Mackness et al. 2000), implying that the 4.0 Ma age estimate for the Bluff Downs LF (Archer and Wade 1976) may be close to the mark. Taxonomic composition of the fauna resembles that of the approximately 0.5 million year younger Chinchilla LF and the two assemblages share several species (Archer and Wade 1976; Bartholomai 1978b; Tedford et al. 1992). Expectedly though, some of the Bluff Downs marsupials express slightly earlier stages of evolution than their Chinchilla counterparts (Archer and Wade 1976). While 18 different marsupials have been reported from the Bluff Downs LF (Archer and Wade 1976; Bartholomai 1978b; Archer and Dawson 1982; Mackness 1995a), recent excavations have apparently led to "a dramatic increase in the number of taxa represented" (Mackness 1995b:15). The diversity of waterbirds in the fauna has been taken to indicate a predominantly fluvio-lacustrine environment, while the bias among the mammals toward large browsing diprotodontoids and macropodines suggests surrounds of sclerophyll forest and woodland (Boles and Mackness 1994; Archer et al. 1995; Mackness et al. 2000).

Among the eight different Bluff Downs macropodoids identified, only one metaloph (QM F9104, pl. 62G) is attributable to the Sthenurinae. Again, this rarity is puzzling in view of the postulated paleohabitat associations of simosthenurins in the Pleistocene of eastern and southern Australia. Although QM F9104 agrees well in size and morphology with *"Si." baileyi*, recognition of this otherwise middle to late Pleistocene taxon in the early Pliocene would represent an extraordinary temporal range extension on the basis of one molar fragment, so I have referred it to *"Si."* sp. cf. *baileyi* pending retrieval of more material from the site. While character support is admittedly weak, it is interesting to note that the strict consensus phylogeny places *"Si." baileyi* as the plesiomorphic sister taxon of a *"Si." antiquus–"Si." pales* clade (fig. 21). Given a possible phylogenetic derivation just prior to that of *"Si." antiquus*, which appears about 0.5 million years later at Chinchilla in southeastern Queensland, the presence of a species close to *"Si." baileyi* at Bluff Downs is reasonable from a stage-of-evolution and zoogeographic perspective. By phylogenetic inference, the presence of *"Si."* sp. cf. *baileyi* at Bluff Downs might also be taken to imply that the lineage whence *"Si." brachyselenis* was derived had diverged prior to 4.0 Ma. This would also be congruent with a putative late Miocene age for the more plesiomorphic *Archaeosimos*.

The last of the early Pliocene sites containing sthenurines is a series of adjacent, coeval channel fills (Tirari Fm, Mampuwordu Member) that crop out along the edge of Lake Palankarinna in northern South Australia and contain the oldest inland Pliocene fauna, the Palankarinna LF (Stirton 1955; Stirton et al. 1961; Stirton 1967; Tedford et al. 1992; fig. 27). An age of 3.9 Ma has been ascribed to these deposits by magnetostratigraphy (Tedford et al. 1992), which is in keeping with the original estimate based on the stage of evolution of the diprotodontids (Stirton 1955). Diversity in the assemblage is low and the high-crowned, presumed grazing macropodine *Prionotemnus palankarinnicus* Stirton, 1955, is the dominant taxon. These factors, plus the occurrence of Australia's oldest known bilby, suggest that open habitats were well established in central Australia by 3.9 Ma (Tedford et al. 1992; Archer et al. 1995). Among the five marsupials recognized in the Palankarinna LF, two have been tentatively reported from Pliocene sites elsewhere: *P. palankarinnicus* and *Zygomaturus keanei* in the early late Pliocene Kanunka LF to the north (Tedford et al. 1992) and *P. palankarinnicus* in the early late Pliocene Chinchilla LF of southeastern Queensland (Bartholomai 1975). However, after comparing the Chinchilla specimen with the least worn teeth (QM F3589) to topotypic examples of *P. palankarinnicus*, I believe that the Chinchilla form belongs in a new species differing from *P. palankarinnicus* in size, crown height and premolar proportion.

At least one relatively large sthenurine occurs in the Palankarinna LF, as recognized by Tedford (1966) on the basis of a partial i1 (UCMP 45420). An associated I1 and I2 (UCMP 57233) also pertain to the Sthenurinae, but several large postcranial elements are more suggestive of *Protemnodon* (Tedford et al. 1992). I am unable to refer UCMP 57233 to a genus, but the unbroadened nature of the I1 indicates that it does not pertain to *Sthenurus*. The i1, however, probably does not belong to the same taxon; indeed, the dorsal and ventral contours of the broken crown base are suggestive of an elongate, *Sthenurus*-like crown. The relative size of this tooth also suggests that the largest early Pliocene sthenurine so far identified may be represented in the Palankarinna LF. Although at least two of the main fossil-yielding quarries in the Mampuwordu Member have been worked out, every effort should be made to probe for more channel-fills incised into the Etadunna Fm. The fauna may be especially illustrative with regard to the early stages of evolution of the relatively more arid adapted sthenurines, at least two of which occur in the 0.5 million year younger Kanunka LF.

Species diversity in the 3.4 Ma Pompapillina Member of the Tirari Fm., which yields the Toolapinna and Kanunka LFs (Tedford et al. 1992), contrasts starkly with that of the Mampuwordu Member. At least 35 different marsupials and two conilurine rodents occur in the Kanunka LF (Stirton et al. 1961; Tedford et al. 1992; Prideaux 1997; fig. 27), making it the most diverse Pliocene assemblage on the continent. Most of this diversity is provided by the macropodoids. At least 23 species are present, which easily exceeds the 15 species from Bow and Hamilton, and the 11 species from Chinchilla. Such a diversity of macropodoids is only equaled in a single assemblage by the late middle Pleistocene deposit in Victoria Fossil Cave in southeastern South Australia (Wells et al. 1984). If the Kanunka LF does represent a fairly contemporaneous assemblage, then the array of variously adapted macropodoids may reflect the sampling of a mosaic of habitats. Notably, the taxa apparently suited to mesic habitats (e.g., *Dendrolagus*, *Thylogale*) tend to be more plesiomorphic and have Miocene or early Pliocene records elsewhere. Those that probably inhabited more open areas (e.g., *Troposodon kenti*, *Macropus pearsoni*) tend to be more derived (Prideaux 1997).

One sthenurine specimen has thus far been identified from the Toolapinna LF from the Camel Swamp Yard locality on the Warburton River (fig. 27), the holotype of *"Simosthenurus" tirarensis* (pl. 70). It represents the earliest unequivocal record of the Simosthenurini in inland Australia. The Kanunka LF contains three sthenurines. The first is represented by an unworn P3 enamel cap that belongs to a form near to the late Miocene *Hadronomas puckridgi* (SAM P32556, pl. 3D). The recognition of *Hadronomas* in a late Pliocene fauna seems somewhat anachronistic, and the possibility of reworking from older strata does need to be considered. However, the only fossiliferous unit through which the Pompapillina Member incises is part of the Etadunna Fm., which is too old to be a source for sthenurine remains. *Sthenurus* sp. cf. *tindalei* is represented by a partial maxilla (UCMP 56928, pl. 27B) and possibly a proximal pedal phalanx (UCMP 60753). The presence of a derived *Sthenurus* species in the Kanunka LF, and both *S. andersoni* and *S. notabilis* in the approximately coeval Chinchilla LF, suggests that much of the cladogenesis within *Sthenurus* may have taken place by 3.4 Ma. An unworn right m2 (UCR 22656, pl. 39F-H) is identical in hypolophid size and morphology to NMV P158633 from the early Pliocene Hamilton LF, but like the Hamilton simosthenurin, can only tentatively be referred to *Archaeosimos* (see Taxonomy). No direct evidence of paleovegetation is yet known from the Tirari Fm. or any similarly aged strata in the area, but the presence of a taxon close to the Hamilton simosthenurin (and *Archaeosimos correlli*) further suggests persistence of well-forested habitats in the region at 3.4 Ma. On the other hand, the large, high-crowned *S.* sp. cf. *tindalei* intimates more open conditions, supporting the idea of substantial habitat heterogeneity deduced from the diversity of macropodines (Prideaux 1997).

The Bluff Downs LF and the early late Pliocene Chinchilla LF from the western Darling Downs in Queensland (Woods 1960, 1962; fig. 27) are approximate temporal equivalents of the Palankarinna and Kanunka LFs (Tedford et al. 1992). Taxonomic composition and stage of evolution of the macropodoids and diprotodontoids correlate well with those of the Kanunka LF, and the two sites share at least four macropodoids (Tedford et al. 1992). Three sthenurines are known from the Chinchilla LF (Bartholomai 1963): *"Simosthenurus" antiquus*, which is slightly more derived than *"Si." baileyi*, and two plesiomorphic *Sthenurus* species, *S. andersoni* and *S. notabilis*. *S. notabilis* is known only from Chinchilla, *"Si." antiquus* is also apparently known only from Chinchilla (although a related form, *"Si."* sp. cf. *antiquus*, occurs in the Pleistocene of northern South Australia), while *S. andersoni* is the most widespread Pleistocene sthenurine. The presence of *Sthenurus* in the Chinchilla LF constitutes the earliest irrefutable record of the genus in eastern Australia. If the optically stimulated luminescence age of 75±9 ka (Roberts et al. 2001) for the Lake Callabonna assemblage in northeastern South Australia is both accurate (cf. 0.2–0.7 Ma, Wells and Tedford 1995) and applicable to the in situ remains of *S. andersoni*, then this species may well have survived for 3.5 million years. Although no paleovegetational data exist for southeastern Queensland in the early late Pliocene, several large browsing diprotodontoids and grazing macropodine species suggest the presence of open forest, savanna and grassland (Archer et al. 1995). Farther south, late Pliocene pollen records from northeastern New South Wales suggest the onset of relatively drier, more open conditions (Martin 1979, 1980, 1981; Kershaw et al. 1994).

Opening of forests and the spread of grasslands and shrublands took place in the eastern Highlands at 2.5–2.9 Ma (McEwan Mason 1991) and replaced what was most likely the last area of rainforest in New South Wales well after similar transitions occurred farther inland (Martin 1998a). The peak reached in phytolith frequency at 2.5 Ma in the southwestern Pacific suggests that continental aridity reached a maximum at this time (Locker and Martini 1989). It is no surprise, therefore, that the Fisherman's Cliff LF from the Lake Victoria region in southwestern New South Wales (fig. 27), which is dated by magnetostratigraphy to 2.4–2.9 Ma (Whitelaw 1991), includes the earliest records of the relatively arid-adapted *Dasyuroides*, *Lagostrophus* and *Lasiorhinus* (Marshall 1973; Crabb 1982; Tedford 1994). The only sthenurine thus far recognized from the Fisherman's Cliff LF is *Sthenurus tindalei*, which has a widespread Pleistocene distribution through inland Australia and a close relative in the early late Pliocene Kanunka LF. A fragment of the posterior face of a metaloph (NMV P26882) from the nearby, but slightly younger Bone Gulch LF (Marshall 1973; fig. 27) may also belong to *S. tindalei*, but it is too incomplete to assign to a species. Using magnetostratigraphy, Whitelaw (1991) constrained the age of the Bone Gulch LF to 1.9–2.4 Ma, making it one of only two dated latest Pliocene faunas that have produced sthenurines. The other record is *Simosthenurus* sp. cf. *occidentalis* from the 2.03–2.48 Ma Dog Rocks LF of central southern Victoria (Whitelaw 1989; fig. 27).

PLEISTOCENE

The complex underground system of the Wellington Caves in central eastern New South Wales (fig. 28) contains sediments that range from Holocene back at least to early Pliocene age, and many contain vertebrate fossils (Osborne 1983, 1997). Sthenurine remains have been collected from the Wellington Caves for 170 years, but stratigraphic details for much of the older material were not recorded (Dawson 1985). The holotype of *"Simosthenurus" brachyselenis* and the paratype of *"Procoptodon" williamsi* were collected from phosphate mining drives that intercept strata of ?middle Pleistocene (Mitchell Cave beds) and early Pliocene (Big Sink beds) ages (Osborne 1983, 1997). While their stratigraphic provenances may never be known with certainty, their preservation is unlike that of the material described by Dawson et al. (1999) from the Big Sink Local Fauna (pers. obs.), so the likelihood is that they came from the Mitchell Cave beds. Thus far, the early Pliocene Big Sink LF has produced no sthenurines (Dawson et al. 1999) and no other indisputable Pliocene localities are known within the Wellington Caves. There have been suggestions of a late Pliocene or early Pleistocene age for components of the Bone Cave fauna based on biocorrelation of macropodines, dasyurids and rodents (Dawson 1995; Dawson and Augee 1997). *Sthenurus atlas* and *"Si." brachyselenis* have been collected from Bone Cave, but their age is uncertain, as the assemblage is probably mixed (L. Dawson, pers. comm., 1999).

Stratigraphic correlation and normal magnetic polarity constrain the age of the Limeburner's Point LF from Geelong in central southern Victoria (fig. 28) to either the Jaramillo Subchron (1.07–0.99 Ma) or Brunhes Chron (0.78 Ma-present; Whitelaw 1993). An age within the late early to middle Pleistocene interval seems most likely (Turnbull et al. 1992a). If the fauna is early Pleistocene, then its sthenurine, *"Procoptodon" mccoyi*, is the earliest known record for *Procoptodon s. l.* Tedford (1994) suggested that the occurrence of *"P." mccoyi* in the Limeburner's Point LF might highlight an initial phase of Pleistocene speciation in *Simosthenurus* (= Simosthenurini herein), but the realization that some "late Pleistocene" faunas are actually of middle Pleistocene age (e.g., Naracoorte Caves assemblages, Ayliffe et al. 1998; Brown and Wells 2000; Grün et al. 2000) implies that most sthenurines had probably originated by the early middle Pleistocene.

The understanding of Australia's Pleistocene mammal stratigraphy and chronology is weak compared with that of other parts of the world. Until relatively recently, the default age of "late Pleistocene" has typically been applied to any late Cenozoic locality lacking taxa that look like phylogenetic precursors to later forms or containing any extinct marsupials believed not to have persisted into the Holocene. The lack of resolution in the Pleistocene is due to several factors, which include absence of widespread ice sheets and volcanism, isolation to intercontinental faunal interchange (with the exception of bats, rodents and humans), a relative paucity of localities, limitations of conventional dating techniques, and too few rigorous stratigraphic studies. So, not only may responses of the mammal fauna to Pleistocene climate changes have been more subtle in Australia than in Europe and North America (given the lack of both immigration and extreme climatic fluctuations), but there are fewer means by which to obtain data that can reveal what happened where and when. Pleistocene researchers interested in faunal changes through space and time have to be extremely judicious in selecting deposits for detailed study; sites need to be conducive to applicable geochronologic techniques and the revelation of subtle biostratigraphic patterns.

Spatial Distributions

Notwithstanding our mediocre understanding of Pleistocene mammal stratigraphy and chronology, sthenurines were plainly never more diverse or abundant than during this epoch. The majority of sthenurine species have Pleistocene records in either the cave deposits of Australia's southern and eastern periphery or the fluvio-lacustrine localities of inland South Australia and New South Wales (fig. 28, table 44). Whether their abundance was a purely middle and late Pleistocene phenomenon is not possible to determine; too few late Pliocene and early Pleistocene assemblages are known or recognized. Therefore, although it might entail time-averaging of up to two million years, I have no choice but to treat all of the Pleistocene localities yielding sthenurines as coeval for the purposes of investigating zoogeographic patterns. This undoubtedly means that many potentially important signals will be concealed by noise, but a time-averaged appraisal of Pleistocene species distributions still reveals several general patterns. As a final caveat, it is worth noting that the concentration of sthenurine localities in temperate southern Australia (fig. 28; table 44) may not only reflect actual taxon distributions, but also the greater populations of potential discoverers inhabiting these areas compared with much of western and northern Australia.

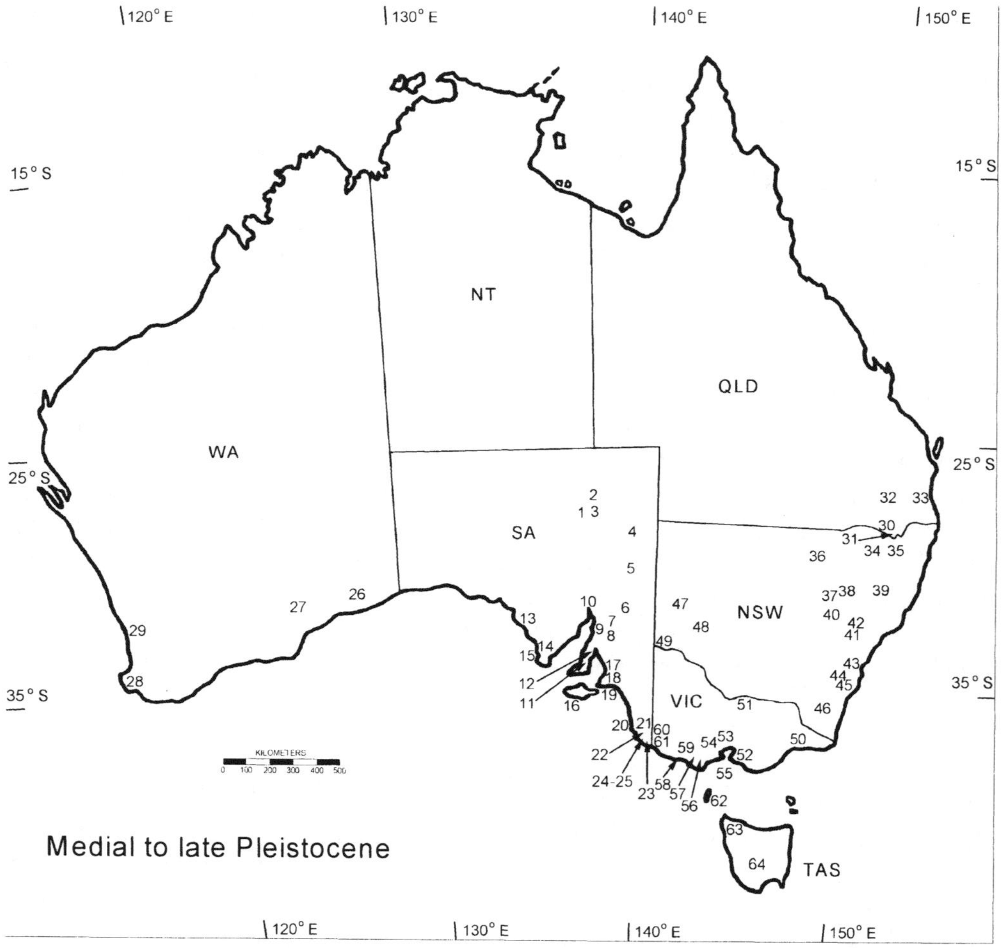

Figure 28. Map showing middle to late Pleistocene localities yielding sthenurine craniodental remains identifiable to species. Refer to table 44 for locality numbers.

Table 44. Pleistocene localities and sthenurine species occurrences.

Locality / Species	*Sthenurus andersoni*	*Sthenurus atlas*	*Sthenurus murrayi*	*Sthenurus stirlingi*	*Sthenurus tindalei*	*Metasthenurus newtonae*	*Simosthenurus euryskaphus*	*Simosthenurus maddocki*	*Simosthenurus occidentalis*	"*Simosthenurus*" cf. *antiquus*	"*Simosthenurus*" *baileyi*	"*Simosthenurus*" *brachyselenis*	"*Simosthenurus*" *pales*	*Procoptodon goliah*	*Procoptodon pusio*	*Procoptodon rapha*	"*Procoptodon*" *browneorum*	"*Procoptodon*" *gilli*	"*Procoptodon*" *mccoyi*	"*Procoptodon*" *oreas*	"*Procoptodon*" *williamsi*
SOUTH AUSTRALIA																					
1. Lake Eyre:																					
Madigan Gulf (Site 1 / 83)	x													x							
Price Peninsula	x																				
2. Warburton River:																					
Cassidy Locality					x					x											
Green Bluff Locality	x				x					x			x	x							
Kalamurina	x																				
Kallakoopah Creek														x							
Lookout Locality	x	x			x									x							
Mulyanna Site		x																			
New Kalamurina	x			x										x							
Toolapinna Waterhole	x																				
West Bluff, Lake Kutjitara	x				x									x							
Warburton River (unspecified)	x																				
3. Cooper Creek:																					
Katipiri Waterhole					x																
Kudnampirra Waterhole	x			x										x							
Malkuni Waterhole	x				x									x							x
Site 8	x										x			x							
Waralamanko Waterhole	x			x	x									x							
Cooper Creek (unspecified)					x											x					
4. Lake Callabonna	x			x	x																
5. Billeroo Creek				x										x							
6. Yunta District:																					
Brady's Gully, Teetulpa														x							
Whydown Station														x							
Manunda Creek														x							
7. Peterborough District:																					
Black Rock Gravel Pit						x															
Boolcunda Creek														x							
Orroroo														x							
Pitcairn Station, near Nackara														x							
Waupunya Creek, near Terowie														x							
8. Burra District:																					
Baldina Creek					x	x															
Ulooloo														x							
Collinsville Station														x							

Locality / Species	*Sthenurus andersoni*	*Sthenurus atlas*	*Sthenurus murrayi*	*Sthenurus stirlingi*	*Sthenurus tindalei*	*Metasthenurus newtonae*	*Simosthenurus euryskaphus*	*Simosthenurus maddocki*	*Simosthenurus occidentalis*	"*Simosthenurus*" cf. *antiquus*	"*Simosthenurus*" *baileyi*	"*Simosthenurus*" *brachyselenis*	"*Simosthenurus*" *pales*	*Procoptodon goliah*	*Procoptodon pusio*	*Procoptodon rapha*	"*Procoptodon*" *browneorum*	"*Procoptodon*" *gilli*	"*Procoptodon*" *mccoyi*	"*Procoptodon*" *oreas*	"*Procoptodon*" *williamsi*
9. Bairstow's Sand Pit, Port Pirie				x										x							
10. Port Augusta District:																					
Dempsey's Lake	x				x																
Port Augusta Sand Pit					x																
11. Curramulka:																					
Curramulka Quarry	x	x		x		x		x	x	x				x		x					
Town Well Cave	x							x	x					x				x			
12. Clinton					x																
13. Old Calca Station														x			x				
14. Kiana Cliff, near Mt Misery						x			x								x				
15. West Brothers Is., Coffin Bay									x	x											
16. Kangaroo Island:																					
Emu Caves (3 Hole)									x								x				
Kelly Hill Caves																		x			
Rocky River																	x	x			
Seton Rock Shelter																		x			
17. Adelaide Beach (unspecified)														x							
18. Salt Creek, Normanville	x																				
19. Coorong Beach, near Murray Mouth	x																				
20. Blackford Creek, near Kingston SE									x												
21. Naracoorte Caves:																					
Alexandra Cave									x					x			x	x			
Bat Cave																	x	x			
Blanche Cave	x																x	x			
Brown Snake Cave																		x			
Buckridge Cave	x																				
Cathedral Cave								x	x					x			x	x			
Cheese and Putty Cave	x																				
Crawford's Cornucopia Cave								x	x		x						x	x			
Fox Cave									x					x			x	x			
Haystall Cave						x		x	x					x			x	x			
Henschke's Fossil Cave	x					x		x	x				x			x	x	x			
James's Quarry Cave																		x			
Possum Cave																	x				
Rabbit Cave																	x	x			
Robertson's Cave									x												
Sand Funnel Cave								x													
SOS Cave						x															
Specimen Cave								x	x		x						x	x			

Locality / Species	*Sthenurus andersoni*	*Sthenurus atlas*	*Sthenurus murrayi*	*Sthenurus stirlingi*	*Sthenurus tindalei*	*Metasthenurus newtonae*	*Simosthenurus euryskaphus*	*Simosthenurus maddocki*	*Simosthenurus occidentalis*	"*Simosthenurus*" cf. *antiquus*	"*Simosthenurus*" *baileyi*	"*Simosthenurus*" *brachyselenis*	"*Simosthenurus*" *pales*	*Procoptodon goliah*	*Procoptodon pusio*	*Procoptodon rapha*	"*Procoptodon*" *browneorum*	"*Procoptodon*" *gilli*	"*Procoptodon*" *mccoyi*	"*Procoptodon*" *oreas*	"*Procoptodon*" *williamsi*
Small cave 60m from VDC Cave																	x				
Unnamed cave (5U49)																		x			
Victoria Fossil Cave	x					x		x	x		x		x	x			x	x			
Wet (Tomato-Stick) Cave						x			x								x	x			
Wombat Cave	x																x	x			
22. Penola District:																					
Comaum Forest Cave	x					x		x	x				x				x	x			
Monbulla Cave																	x				
Unnamed cave (5L122)																		x			
23. Mt Gambier District:																					
Derrington St	x							x									x	x			
Gray St																	x	x			
Kilsby's Hole	x					x		x	x				x					x			
Moorak													x								
Wandilo Cave								x	x												
Wandilo Forest Cave																	x				
Goulden's Hole, near Mt Schank	x					x		x	x									x			
Tankstand Cave, near Mt Schank																		x			
Mt Gambier (site unspecified)									x												
Mt Schank (site unspecified)								x													
24. Tantanoola District:																					
Glencoe									x									x			
Green Waterhole Cave						x		x	x									x			
Morgan's Cave																	x	x			
Mt Burr Cave									x									x			
Tantanoola Cave									x									x			
Tantanoola Cave "E"																		x			
25. Unnamed swamp site, Millicent	x							x													
WESTERN AUSTRALIA																					
26. Nullarbor Region:																					
Lindsay Hall Cave	x					x		x			x						x				
Madura Cave						x		x													
Last Tree Cave	x				x	x								x							x
Leaena's Breath Cave	x					x															
27. Balladonia Soak					x	x		x													
28. Leeuwin-Naturaliste Region:																					
Devil's Lair									x								x				
Foundation Cave									x								x				
Kudjal Yolgah Cave									x								x				

Locality	Species	*Sthenurus andersoni*	*Sthenurus atlas*	*Sthenurus murrayi*	*Sthenurus stirlingi*	*Sthenurus tindalei*	*Metasthenurus newtonae*	*Simosthenurus euryskaphus*	*Simosthenurus maddocki*	*Simosthenurus occidentalis*	"*Simosthenurus*" cf. *antiquus*	"*Simosthenurus*" *baileyi*	"*Simosthenurus*" *brachyselenis*	"*Simosthenurus*" *pales*	*Procoptodon goliah*	*Procoptodon pusio*	*Procoptodon rapha*	"*Procoptodon*" *browneorum*	"*Procoptodon*" *gilli*	"*Procoptodon*" *mccoyi*	"*Procoptodon*" *oreas*	"*Procoptodon*" *williamsi*
Mammoth Cave										x								x				
Moondyne Cave																		x				
Labyrinth Cave																		x				
Strong's Cave										x								x				
Terrible Cave										x												
Tight Entrance Cave		x					x			x				x				x				
29. Yanchep District:																						
Koala Cave																		x				
McIntyre's Gully, Gingin														x								
Melaleuca Cave										x												
Wanneroo																		x				
QUEENSLAND																						
30. The Joint, Texas Caves		x							x								x					
31. Cement Mills fissure fill, Gore																	x				x	
32. Darling Downs:																						
Clifton															x	x						
Macalister, Condamine River		x														x						
Condamine River (unspecified)		x														x	x					
Freestone Creek			x														x					
Glengallan Creek, near Warwick																	x					
Gowrie		x														x		x				
Jimbour Creek																	x					
King's Creek, Pilton		x														x						
King's Creek															x	x	x					
Little Middle Creek, Maidenwell																	x					
Oakey Creek																	x					
Pearson's Locality, King's Creek																x						
?Ravensthorpe, Pilton														x								
Darling Downs (unspecified)		x	x				x							x	x	x	x				x	
33. Brisbane River														x								
NEW SOUTH WALES																						
34. Bone Camp Gully, near Bingara		x						x	x					x	x	x	x				x	
35. Reddestone Creek, near Glen Innes														x	x	x						
36. Cuddie Springs		x												x								
37. Macquarie River, near Gilgandra														x								

Locality / Species	*Sthenurus andersoni*	*Sthenurus atlas*	*Sthenurus murrayi*	*Sthenurus stirlingi*	*Sthenurus tindalei*	*Metasthenurus newtonae*	*Simosthenurus euryskaphus*	*Simosthenurus maddocki*	*Simosthenurus occidentalis*	*"Simosthenurus" cf. antiquus*	*"Simosthenurus" baileyi*	*"Simosthenurus" brachyselenis*	*"Simosthenurus" pales*	*Procoptodon goliah*	*Procoptodon pusio*	*Procoptodon rapha*	*"Procoptodon" browneorum*	*"Procoptodon" gilli*	*"Procoptodon" mccoyi*	*"Procoptodon" oreas*	*"Procoptodon" williamsi*
38. Binnaway-Weetaliba:																					
Binnia Creek	x																				
Castlereagh River														x							
Neible Station	x																				
39. Attunga District:																					
Attunga Springs																x					
Catong Gully																x					
40. Wellington District:																					
Bone Cave		x										x									
Cathedral Cave	x																			x	
Geurie													x								
Mitchell Cave	x	x													x	x					
Molong																				x	
Phosphate Mine		x										x									x
Wellington Caves (unspecified)						x		x	x				x			x				x	
41. Cow Flat Quarries, Bathurst									x												
42. Kandos						x															
43. Wombeyan Caves								x													
44. Governor's Hill, Lake George														x							
45. Mt Fairy:																					
Dolomite quarry								x													
Mt Fairy Caves									x									x			
46. Monaro Region:																					
Bunyan Siding, near Cooma		x														x					
Jincumbilly, near Bombala													x								
Teapot Creek Site 5									x												
Teapot Creek Site 7															x						
Teapot Creek Site 12						x															
Teapot Creek Terrace 3	x																				
Toppings Creek	x																				
47. Menindee Region:																					
Lake Menindee	x	x			x									x							
Lake Tandou		x												x							
48. Willandra Lakes:																					
Between Lakes Garnpung and Gogolo														x							
Between Lakes Mungo and Leaghur														x							
49. Lake Victoria:																					
Site 9														x							
Site 50			x		x																
Lake Victoria (unspecified)	x													x							

Locality / Species	*Sthenurus andersoni*	*Sthenurus atlas*	*Sthenurus murrayi*	*Sthenurus stirlingi*	*Sthenurus tindalei*	*Metasthenurus newtonae*	*Simosthenurus euryskaphus*	*Simosthenurus maddocki*	*Simosthenurus occidentalis*	"*Simosthenurus*" cf. *antiquus*	"*Simosthenurus*" *baileyi*	"*Simosthenurus*" *brachyselenis*	"*Simosthenurus*" *pales*	*Procoptodon goliah*	*Procoptodon pusio*	*Procoptodon rapha*	"*Procoptodon*" *browneorum*	"*Procoptodon*" *gilli*	"*Procoptodon*" *mccoyi*	"*Procoptodon*" *oreas*	"*Procoptodon*" *williamsi*
VICTORIA																					
50. Buchan Caves:																					
Didgeridoo Cave																		x			
Foul Air Cave									x												
Honeycomb Cave									x												
Pot Luck Cave									x									x			
Royal Cave									x												
Trogdip Cave									x												
Buchan Caves (unspecified)									x												
51. Tocumwal														x							
52. Kilcunda (28 ft shaft)									x												
53. Lancefield:																					
South Site	x							x								x		x			
Lancefield (unspecified)									x												
54. Werribee River valley, west of Bacchus Marsh									x												
55. Limeburner's Point, Geelong																			x		
56. Colac																x					
57. Lake Colongulac																x					
58. Skene St, Warrnambool																	x	x			
59. Spring Creek, Minhamite	x								x									x			
60. Strathdownie District:									x									x			
Shire Quarry																					
Puralka Caves									x									x			
61. McEachern's Cave, near Nelson						x			x								x				
TASMANIA																					
62. Surprise Bay, King Is.									x												
63. Smithton District:																					
Main Cave, near Montagu									x												
Pleisto Scene Cave, near Montagu									x												
Scotchtown Cave						x			x												
64. Florentine Valley:																					
Beginner's Luck Cave									x												
Titan's Shelter									x												

Metasthenurus newtonae and *Simosthenurus occidentalis* had extremely similar Pleistocene distributions (Prideaux 2000; fig. 29). *Si. occidentalis* is known from several cave deposits in southwestern Australia (table 44) and both species occur in the Tight Entrance Cave assemblage (Gully 1997), which extends back into the late middle Pleistocene (Prideaux et al. 2000). Farther east, only *M. newtonae* is known from Balladonia and the Nullarbor Region, but both species are known from the western side of Eyre Peninsula. Remains of *M. newtonae* and *Si. occidentalis* are common in the cave faunas of southeastern South Australia (e.g., Pledge 1980; Wells et al. 1984; table 44), although the latter is much better known in Victoria. Farther north, both have records on the eastern side of New South Wales, but only *M. newtonae* is known to have extended into southeastern Queensland (Prideaux 2000). They are also the only sthenurines recorded in Tasmania (fig. 29), where *Si. occidentalis* is by far the more abundant (Murray and Goede 1977; Goede and Murray 1979). *M. newtonae* is represented by only a few teeth from Scotchtown Cave. Pleistocene deposits on King Island in Bass Strait (Scott 1906; Anderson 1932) have also produced an apparent dwarf form of *Si. occidentalis.* Although *M. newtonae* is uncommon throughout its entire range, *Si. occidentalis* becomes rarer with decreasing latitude and is known from no farther north than Wellington (fig. 29). In molar size, *Si. occidentalis* varies across its range, being small in southwestern Australia and Tasmania, and largest in mainland southeastern Australia. *M. newtonae* does not vary in size between regions more than it does within regions with one exception: a large-toothed form in the Kandos and Wellington Caves deposits.

The distribution of *Simosthenurus maddocki* is similar to that of *Si. occidentalis* and *Metasthenurus newtonae* (figs 29-30; table 44). A single, worn maxillary fragment from Balladonia on the western edge of the Eucla Basin represents the westernmost occurrence of *Si. maddocki*, but it is the best represented sthenurine in the Lindsay Hall Cave on the Nullarbor Plain. By contrast, *Si. maddocki* is usually an uncommon species wherever it occurs in eastern Australia. The species is also known from three sites in eastern New South Wales (Flannery and Hope 1983; table 44) and, like *M. newtonae*, extended into southeastern Queensland. *"Procoptodon" browneorum* had a Pleistocene distribution similar to those of *M. newtonae*, *Si. occidentalis* and *Si. maddocki* (figs 29-30), but it has no record yet in either New South Wales or Tasmania. However, it must have existed in New South Wales, since it is known from Gowrie on the Darling Downs (see Taxonomy). *"P." browneorum* is a common member of many southern Australian cave assemblages (table 44). It is smallest in southern Western Australia and largest in mainland southeastern Australia. Dimensions for Kangaroo Island specimens fall near or just beyond the upper end of the range of variation for western *"P." browneorum*.

"Procoptodon" gilli overlaps in distribution with several sthenurine species that include southeastern Australia within their range, but it is the only Pleistocene sthenurine that appears to have been restricted to this region (fig. 31; table 44).

Figure 29. Middle to late Pleistocene distributions of *Metasthenurus newtonae* and *Simosthenurus occidentalis*.

Figure 30. Middle to late Pleistocene distributions of *Simosthenurus maddocki* and *"Procoptodon" browneorum.*

Figure 31. Middle to late Pleistocene distributions of *"Procoptodon" gilli* and *"Simosthenurus" baileyi*.

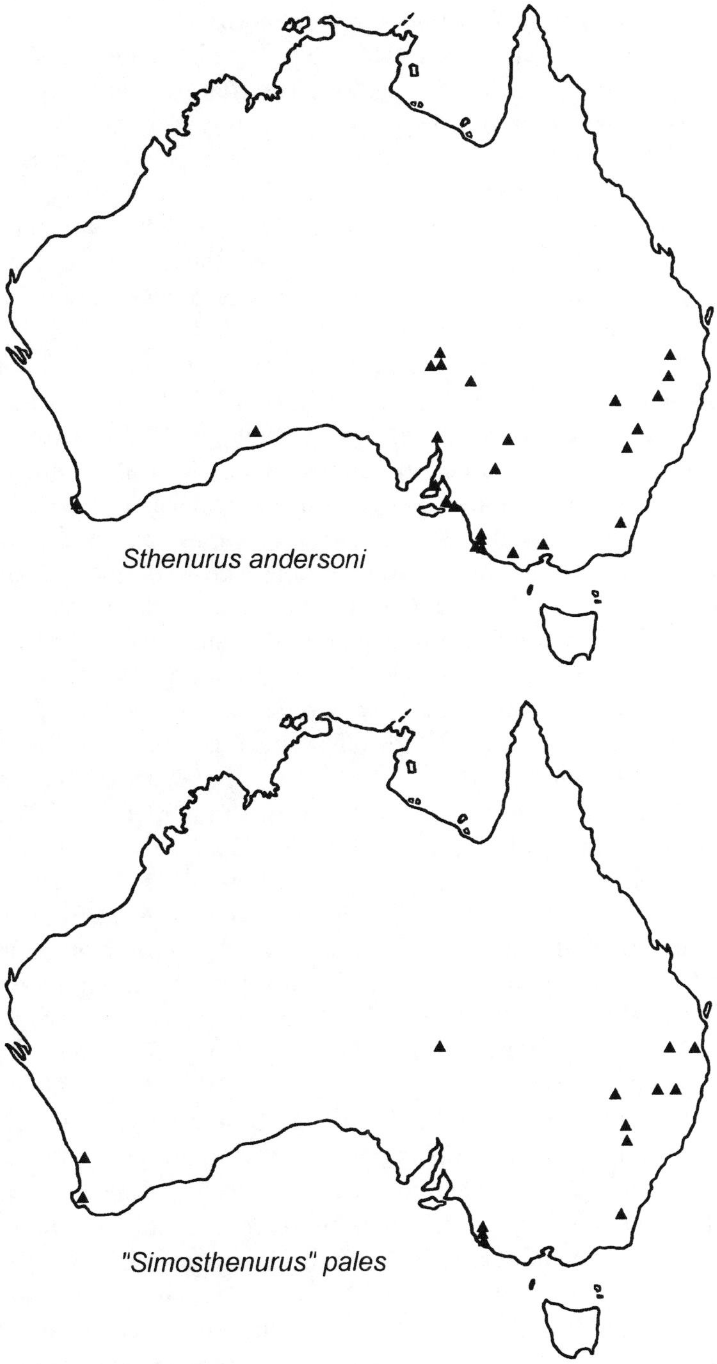

Figure 32. Middle to late Pleistocene distributions of *Sthenurus andersoni* and *"Simosthenurus" pales*.

The record of *"P." gilli* from the Nullarbor Region (Merrilees 1965; Horton 1984) is erroneous; the p3 upon which this record was based (Lundelius 1963:fig. 2) is referable to *Metasthenurus newtonae* (fig. 29). The Black Swamp deposit at Rocky River on Kangaroo Island is the westernmost locality from which *"P." gilli* is known. Mount Fairy Caves in southeastern New South Wales marks its northeastern limit. On Kangaroo Island and in the many cave deposits of southeastern Australia, *"P." gilli* is consistently the most abundant sthenurine species in what are often very diverse sthenurine faunas (table 44). With the exception of *"P." mccoyi*, which is known from only one site in central southern Victoria (fig. 28; table 44), no other sthenurine has an exclusively southern distribution. *"Simosthenurus" baileyi* is known from only one site distal to the southern periphery of the continent between Lindsay Hall Cave on the Nullarbor Plain and the Naracoorte Caves (Prideaux and Wells 1998; fig. 31; table 44), but this is a site along the Cooper Creek in northern South Australia, which suggests a distribution quite unlike that of the other Pleistocene species.

Sthenurus andersoni and *"Simosthenurus" pales* are the most widely distributed of all Pleistocene sthenurines, although *S. andersoni* is known from more individual sites (fig. 32; table 44). In fact, it occurs in a greater number of localities and has a wider distribution than any other sthenurine. It is also the only *Sthenurus* species with an extensive distribution around the southern and eastern flanks of Australia. The ranges of *S. andersoni* and *"Si." pales* appear to have encompassed the combined ranges of all other species, apart from the two species that extend into Tasmania. *S. andersoni* and *"Si." pales* are both present in the Tight Entrance Cave deposit in southwestern Australia (Gully 1997), but only *"Si." pales* is recorded from farther north at McIntyre's Gully, near Gingin (identified as *Sthenurus* sp. nov. in Merrilees 1979; table 44). *S. andersoni* is known from caves on the Nullarbor Plain and is often recovered from sites around the South Australian gulfs (e.g., Curramulka, Port Augusta, Normanville). Both species occur in deposits in southeastern and northern South Australia (e.g., Pledge 1990; Tedford and Wells 1990), eastern New South Wales (e.g., Dawson 1985) and southeastern Queensland (Bartholomai 1963), but only *S. andersoni* has been collected from southern Victoria (Flannery and Gott 1984) and southwestern New South Wales (Tedford 1967; Marshall 1973; fig. 32; table 44). *"Si." pales* is rare wherever recorded, and although it is not nearly as uncommon, *S. andersoni* is also usually not abundant in any site relative to other kangaroo species present.

In inland and eastern Australia, the Pleistocene range of *Sthenurus atlas* appears to have overlapped extensively with *S. andersoni* (figs 32-33). Most deposits that have yielded remains of *S. atlas* have also produced its more plesiomorphic congeneric (table 44), suggesting that these two outwardly similar species occupied the same depositional areas and perhaps very similar habitats (Tedford 1967). Unlike *S. andersoni*, however, *S. atlas* occurs in no southern localities (cf. Wells and Pledge 1983; Wells et al. 1984; Pledge 1990), except for a fissure fill in Curramulka Quarry on Yorke Peninsula. Marshall (1973) recorded

Figure 33. Middle to late Pleistocene distributions of *Sthenurus atlas* and *"Procoptodon" williamsi*.

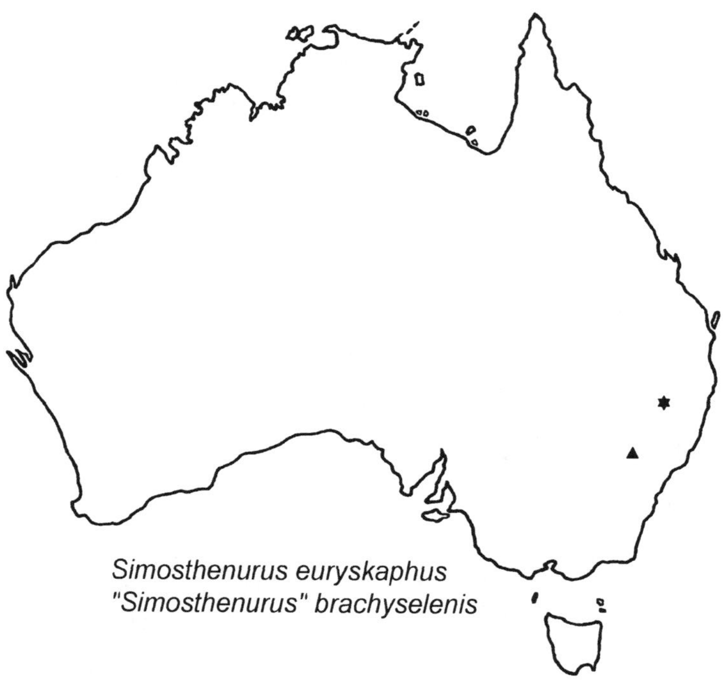

Figure 34. Middle to late Pleistocene distributions of *Sthenurus murrayi* (star), *"Simosthenurus"* sp. cf. *antiquus* (triangle), *Simosthenurus euryskaphus* (star) and *"Simosthenurus" brachyselenis* (triangle).

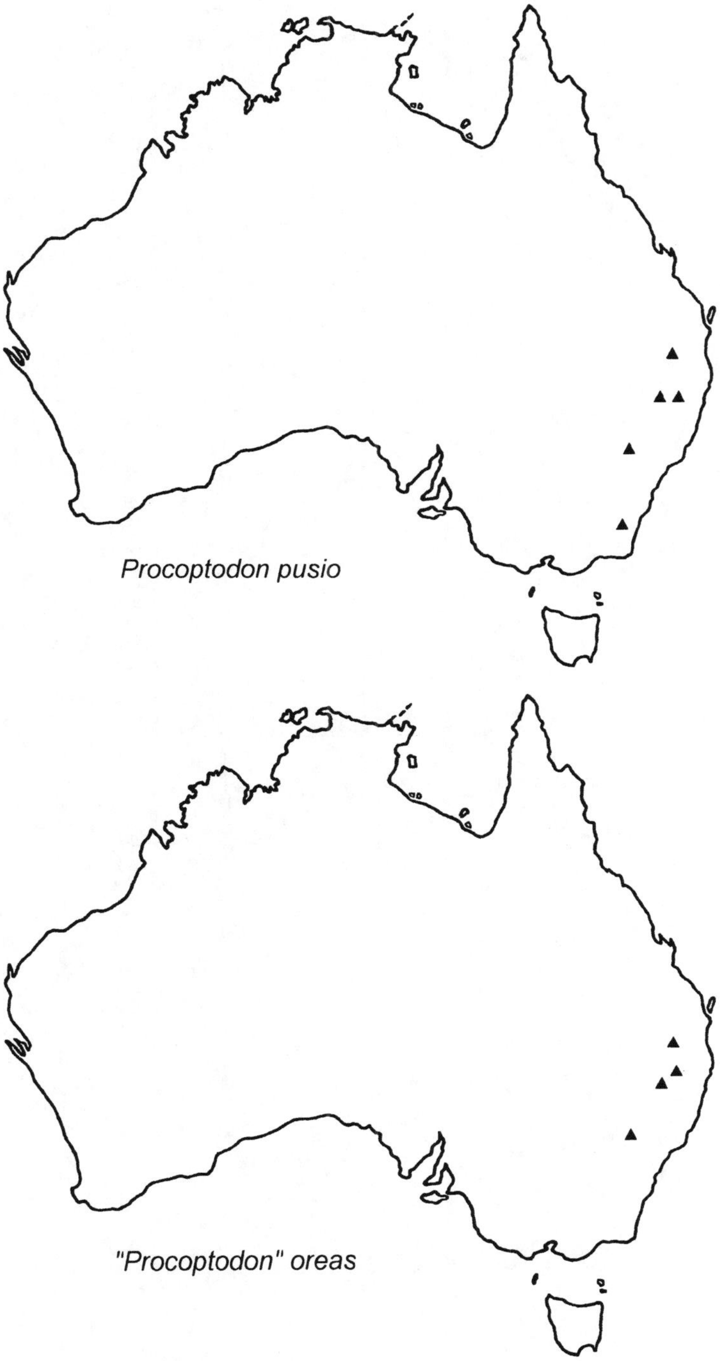

Figure 35. Middle to late Pleistocene distributions of *Procoptodon pusio* and *"Procoptodon" oreas*.

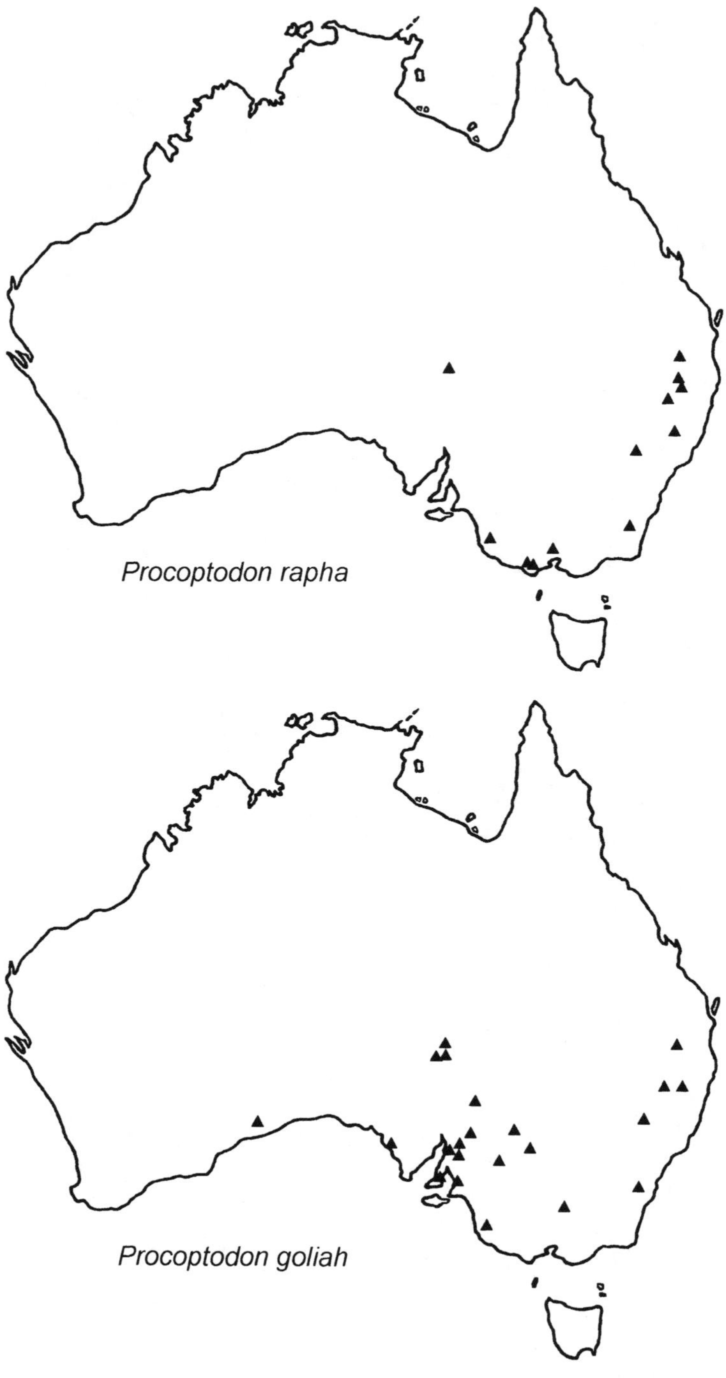

Figure 36. Middle to late Pleistocene distributions of *Procoptodon rapha* and *Procoptodon goliah*.

S. atlas from Lake Victoria in southwestern New South Wales on the basis of a juvenile palate designated herein as the holotype of *Sthenurus murrayi* (pl. 22; figs 4, 34). Other species that may have had a similar, mainly inland and eastern Australian distribution include *"Procoptodon" williamsi* and perhaps *"Simosthenurus" antiquus*. *"P." williamsi* has been collected from Pleistocene deposits in the eastern Lake Eyre Basin, Nullarbor Plain and Wellington Caves (fig. 33; table 44), but *"Si." antiquus* is only known with certainty from its early late Pliocene type locality of Chinchilla on the Darling Downs. If two Warburton River specimens (UCMP 47947, UCMP 56901) can eventually be referred to *"Si." antiquus* with confidence, then this species may also have had a similar distribution (fig. 34).

Procoptodon pusio and *"P." oreas* share an almost identical Pleistocene distribution to one another on the eastern flank of Australia, and are not known outside of this region (fig. 35; table 44). Their northernmost occurrence is on the eastern Darling Downs (Bartholomai 1963, 1970) and they have been found no farther south than central eastern New South Wales (e.g., Dawson 1985; Dawson and Augee 1997). Although *P. pusio* is common in the Darling Downs and Bingara deposits (Bartholomai 1970; Marcus 1976), it is rarer in the Wellington Caves. *"P." oreas* is represented by more than one individual in only one site: Cathedral Cave at Wellington (Dawson and Augee 1997). Marcus (1976) recorded *"P." oreas* from Bingara on the basis of a single upper molar (AM F49664) and a dentary since designated as the holotype of *Simosthenurus euryskaphus* (Prideaux and Wells 1997; pl. 40; fig. 34). A Wellington Caves specimen placed in *"P." oreas* by Tedford (1966) was designated as the holotype of *"Si." brachyselenis* by Prideaux and Wells (1997; pl. 64; fig. 34). As mentioned earlier, the age of this specimen and the Bone Cave dentary referred to *"Si." brachyselenis* (pl. 65) may never be known due to uncertainties in their stratigraphic provenances, although they are probably no older than late Pliocene.

Procoptodon rapha occurs in the same major eastern localities in about the same proportion as *P. pusio* (Bartholomai 1970; Marcus 1976; Dawson 1985; figs 35-36; table 44), but it has also been recorded in southern Victoria (Tedford 1967; Bartholomai 1970; Van Huet 1999), southeastern South Australia (Wells et al. 1984; Pledge 1990) and northern South Australia (Tedford and Wells 1990; fig. 36). After reexamining all of the South Australian material referred to *P. rapha*, it has become evident that only SAM P31045 from Henschke's Fossil Cave and three Cooper Creek specimens (SAM P11543, HM S19, HM S62) are not better referred to *P. goliah*. SAM P31045 fits within *P. rapha* in all attributes, but the molars of the Cooper Creek and Curramulka specimens taper slightly more toward their crests than is typical of *P. rapha*. Clearly, more material is desirable to affirm the presence *P. rapha* in inland Australia.

P. goliah is the only Pleistocene simosthenurin with a wide inland Australian distribution (fig. 36). Indeed, it is recorded from more individual sites through South Australia and western New South Wales than any other sthenurine (table 44).

In eastern Australia, *P. goliah* is abundant in deposits of the eastern Darling Downs and Bingara (Bartholomai 1970; Marcus 1976). It also occurs in the Naracoorte Caves (Brown and Wells 2000; Reed and Bourne 2000). Until recently, *P. goliah* was known from no farther west than Calca at the top of Eyre Peninsula (Merrilees and Ride 1965), but recognition of the species in an assemblage in Last Tree Cave on the Nullarbor Plain (pers. obs.) extends its Pleistocene range approximately 400 km to the west (fig. 36).

Of the nine sthenurine species with Pleistocene records west of Spencer Gulf, *Sthenurus tindalei*, *S. andersoni*, *"Simosthenurus" baileyi*, *"Si." pales* and *Procoptodon goliah* are also recorded in northern South Australia (table 44). *Metasthenurus newtonae*, *Si. maddocki*, *Si. occidentalis* and *"Procoptodon" browneorum* are common in cave deposits around the southern periphery, but are rare in eastern New South Wales and southeastern Queensland (table 44). Species usually encountered in this eastern region are more likely to turn up in inland rather than southern sites, with the exception of the very widely distributed *S. andersoni* and *"Si." pales*. One possible explanation for this geographic pattern may relate to temporal variations in dominant vegetation types associated with Pleistocene climatic oscillations. The central Australian arid zone expanded during drier periods and contracted during wetter times (e.g., Bowler 1982; Dodson 1989; Belperio 1995). The predominantly southern species probably became restricted to the mesic southwestern and southeastern parts of the continent during relatively drier periods, but may have expanded their range up the east coast to areas around Bingara and the Darling Downs, or west across the Great Dividing Range to the Wellington region, during phases of higher rainfall.

Hope (1982) made some interesting observations regarding the distributions of *Diprotodon optatum* and *Zygomaturus trilobus*, the two common but infrequently co-occurring Pleistocene diprotodontids. By overlaying their distributions on a map showing the climatic zones for 18 ka deduced by Bowler (1982), the Wellington, Bingara and Darling Downs localities were all shown to fall close to the boundary between the semi-arid zone and the more humid, peripheral zone. While the inference of such a young age for these megafaunal assemblages would now be refuted (see Baynes 1997; Field and Dodson 1999; Roberts et al. 2001), the pattern itself is important. It either outlines a transitional zone between suitable *Diprotodon* and *Zygomaturus* habitats or has resulted from time-averaging of more mesic-adapted and more xeric-adapted chronofaunas, since the chronology and stratigraphy of these sites is so poorly understood. Hope (1982) also noted that the distribution of *Sthenurus* was generally more similar to that of *Diprotodon*, while *Simosthenurus* had a similar distribution to *Zygomaturus*. The systematic reappraisal of the Sthenurinae presented here means that eastern versus inland patterns can no longer be polarized in precisely the manner envisaged by Hope, but it is clear that detailed stratigraphic and chronologic investigations of the Wellington Caves and Darling Downs localities could offer significant potential to improve our understanding of the responses of the Pleistocene fauna to climatic fluctuations.

The fact remains, though, that those sthenurines apparently adapted to more mesic, temperate habitats tend to be rare in the east. If the different deposits at Wellington and the Darling Downs accumulated sthenurine remains during both wetter and drier phases, or sampled a range of different habitats at one time, why do those species with wide inland distributions or adaptations to consuming higher-fiber vegetation easily outweigh in abundance species such as *Metasthenurus newtonae*, *Simosthenurus maddocki* and *Si. occidentalis*? Perhaps, even during the wetter phases of the Pleistocene, the Darling Downs was too far north and Wellington too far inland (west of the Great Dividing Range) to regularly sample the more mesic-adapted, typically more southern sthenurines. Viewed another way, it is probably relevant that most of the common eastern species (*"Procoptodon" oreas*, *P. pusio*, *P. rapha*, *P. goliah*) arose from one derived clade (fig. 21). Morphological trends within this cluster were geared toward greater robustness and a capacity for processing tougher and possibly more abrasive vegetation than in any other sthenurines (see Functional Morphology). Perhaps this eastern region of Australia was the center of evolution for this *Procoptodon* clade, although I acknowledge that the region of highest diversity for a group need not imply its center of origin.

The late early or middle Pleistocene Limeburner's Point LF of southern Victoria may be the oldest deposit containing *Procoptodon s. l.* (in the form of *"P." mccoyi*), but much of the cladogenesis within the genus is likely to have occurred prior to this time. Pollen records from the Mooki, Namoi, Castlereagh and Gwydir Rivers of northeastern New South Wales continue through the Pliocene and into the early Pleistocene (Martin 1979, 1980, 1981; Kershaw et al. 1994). They indicate replacement of dry rainforest in the late Pliocene to early Pleistocene and a marked expansion of open vegetation as evinced by a large increase in grass and composite (Asteraceae) pollen. *Eucalyptus* also increased relative to Casuarinaceae and a substantial reduction in the canopy cover over the more open vegetation types has been suggested (Kershaw et al. 1994). Farther inland, in the vicinity of Bourke, late Pliocene to early Pleistocene palynofloras also indicate open areas dominated by composites, grasses and chenopods (Martin 1997b). Rainfall was below the lower limit for the maintenance of wet sclerophyll, and Casuarinaceae and *Eucalyptus* were the main trees in a sparse cover (Martin 1997b). The pollen record from this area unfortunately lacks an early late Pliocene component, but by extrapolation, a vegetative structure similar to that implied by the late Pliocene and early Pleistocene record from farther east might be predicted.

These environmental conditions are consistent with what would be expected if this region was a center of diversification for the lineage of more robust *Procoptodon* species. Stepping farther back in time, similar factors may account for the presence of *Sthenurus andersoni* and *S. notabilis* in the early late Pliocene at Chinchilla (fig. 27), along with likely open terrain taxa that also occur in the

Kanunka LF: *Diprotodon*, *Prionotemnus*, *Macropus woodsi*, *Troposodon minor*, *T. kenti* and *S.* sp. cf. *tindalei*. From their craniodental morphology, however, *S. andersoni* and *"Si." pales* may have been more closely bound to riparian areas than the other taxa, which is in keeping with their extensive distribution around the southern periphery of the continent.

The final Pleistocene sthenurines to consider are the two species with an almost exclusively inland distribution, *Sthenurus stirlingi* and *S. tindalei* (fig. 37; table 44). *S. stirlingi* has been recorded from four South Australian sites, extending from New Kalamurina on the Warburton River in the north to Port Pirie in the south. *S. tindalei* overlaps with *S. stirlingi* for much of its range, but occurs over a wider geographic extent and in a greater number of localities (fig. 37; table 44). The species reaches into southwestern New South Wales (Tedford 1967; Marshall 1973) and is also known from the Nullarbor Plain and western edge of the Eucla Basin in Western Australia (identified as *S. atlas* and *Macropus anak* in Glauert 1912). As a form very close to *S. tindalei* is recorded in the early late Pliocene Kanunka LF (fig. 27), it may be quite reasonable to infer that the preferred habitat of this lineage changed minimally right through until the late Pleistocene. Because pollen is preferentially preserved in anaerobic conditions of permanent lacustrine environments, palynological sequences for the Pliocene and Pleistocene of inland Australia are very rare (Martin 1998b). However, Pliocene or Pleistocene pollen from a core drilled north of Lake Eyre is generally suggestive of the extant arid shrublands that exist in the Lake Callabonna region today, including abundant grasses, chenopods and composites, and some Casuarinaceae and Myrtaceae (Martin 1998b). A more direct indication of *S. stirlingi* and *S. tindalei* habitat is provided by a preliminary analysis of a Lake Callabonna pollen sample. From this, Wells and Tedford (1995:14) suggested that the surrounding vegetation may have been "a saltbush dominated savanna with scattered wattle, she-oak, and native pine crossed by eucalypt-lined water courses." This implies perhaps a slightly higher precipitation than in the same area today.

While *Sthenurus stirlingi* certainly appears to have been adapted to the consumption of tougher vegetative material than *S. tindalei* (Wells and Tedford 1995; Functional Morphology), both species were probably quite well adapted to browsing on the abundant shrubby or herbaceous vegetation of the interior. Indeed, almost every area from which *S. tindalei* is recorded, including northern South Australia, western New South Wales and the Nullarbor Plain, lies within the present range of chenopod shrublands (Leigh 1994). Cave assemblages of the southwest and Nullarbor Plain (with the notable exception of Leaena's Breath and Last Tree Caves) suggest better-developed forest and woodland (e.g., Lundelius and Turnbull 1989; Aplin et al. 1995; Gully 1997), habitats with which *S. tindalei* appears not to have otherwise been associated. Although the original collections from Balladonia (Glauert 1912) were made from a number of individual sites in the area (Merrilees 1968b), putative Pleistocene sediments have produced *Phascolonus gigas*, *Lasiorhinus latifrons* and *Diprotodon optatum*, which also suggest drier

Sthenurus tindalei

Sthenurus stirlingi

Figure 37. Middle to late Pleistocene distributions of *Sthenurus tindalei* and *Sthenurus stirlingi*.

conditions and a southward expansion of the arid zone. Balladonia lies in an intriguing paleobiogeographic position, rather analogous to the Wellington Caves (fig. 28). A detailed modern investigation of the paleontology, stratigraphy and chronology of the Balladonia sites and new assemblages from the Nullarbor caves may shed considerable light on both the expansion of the arid zone in the southwest, and the timing and dynamics of the spread of more mesic-adapted forms across the southern margin of the continent.

Geographic Size Variation

Recent advances in geochronology have slowly begun to transform our broad-brush view of Pleistocene mammal paleontology in Australia (e.g., Whitelaw 1991; Ayliffe et al. 1998; Grün et al. 2000; Prideaux et al. 2000; Roberts et al. 2001), but it will be some time before sufficient paleoenvironmental, biochronologic and zoogeographic data exist to allow detailed comparisons of faunal change dynamics between Australia and other continents. When they do, one facet that will deserve careful study is the basis for the spectrum of size differences in dentition, dentary proportions, cranial dimensions and presumably body mass seen between regions within some sthenurines. In *Si. occidentalis*, the only sthenurine well represented in Tasmania (fig. 29), dentary proportions and molar dimensions of the Tasmanian and southwestern samples are closer to each other than either is to mainland southeastern *Si. occidentalis*. Similarly, dental dimensions of *"Procoptodon" browneorum* from Kangaroo Island more closely match those of conspecifics from the west, despite the relative proximity of Kangaroo Island to the mainland of southeastern Australia. Are these patterns the result of temporal sampling biases, ecological differences between regions, or a combination of both?

It is significant that the populations with smaller craniodental dimensions are from Tasmania, Kangaroo Island and the southwest (fig. 28), relatively small isolated regions of mesic habitats compared with mainland southeastern Australia. This disparity was probably even more marked during cooler, drier phases of the Pleistocene. Indeed, paleobotanical data suggest that the High Rainfall Zone, which now takes in the extreme western edge of southwestern Australia, was lost completely during glacial maxima, severely restricting the ranges of mesic taxa (Hopper 1992; Archer 1996). Other wide-ranging Pleistocene marsupials (e.g., *Zygomaturus trilobus*, *Phascolarctos cinereus*, *Thylacoleo carnifex*, *Thylacinus cynocephalus*, *Sarcophilus harrisii*) are also smaller in the west, which suggests that an ecological phenomenon, namely island dwarfing, is the most apt explanation for the size patterns observed. However, this alone is insufficient to explain allometric differences between sthenurine species in size reduction. For example, dental dimensions of *Metasthenurus newtonae* from Western Australia (and Tasmania) fall well within the range of the mainland southeastern sample. The molars of southwestern *"Simosthenurus" pales* are 21% smaller on average than in the southeast, yet incisor and premolar dimensions fall within the lower end of the southeastern range. Conversely, premolar dimensions of western (Nullarbor Region) *Si. maddocki* are 21% smaller than in the southeast, while the molars are only 13% smaller. Two crania are known for southwestern *Si. occidentalis* and they lie only just outside of the southeastern range (6% smaller on average; no Tasmanian crania are known). Premolar dimensions fall within the southeastern range, but the molars are 17% smaller. Elucidation of the basis for these interspecific differences must await a marked refinement in our understanding of Pleistocene sthenurine paleoecology, zoogeography and chronology across southern Australia.

Extinction

One of the longest and most hotly discussed topics in Australian palaeontology concerns the reasons behind the demise of the Pleistocene megafauna, a group defined by Murray (1991b) as consisting of species that either exceeded 40 kg in weight or attained significantly larger body sizes than their closest living relatives. Sthenurines were the most diverse Pleistocene group to succumb, making up about 40% of the species tally. Explanations for megafaunal disappearance have focused on putative effects of climate change, human activities or a combination of both. A paucity of dated late Pleistocene sites, especially those showing clear stratigraphic presence then absence of taxa, has greatly impeded progress toward causal determination. Two recent studies have broken new ground by revealing a 50 ka extinction event for the giant flightless bird *Genyornis* at three well-stratified inland localities (Miller et al. 1999) and showing that very few articulated megafaunal remains date to younger than about 50 ka (Roberts et al. 2001). Humans were implicated as the most likely extinction cause, having apparently become established in Australia somewhere between 70 and 50 ka (Roberts et al. 1994; Thorne et al. 1999; Turney et al. 2001). This view, however, has been countered by a reiteration (Wroe et al. 2002) of the claim that increasing aridity in the lead-up to the Last Glacial Maximum probably sounded the death knell for the megafauna (e.g., Horton 1984; Field and Dodson 1999; see also Main 1978). Before outlining an extinction hypothesis that I believe best fits the data currently available, two scientifically unconvincing lines of reasoning recurrent within the literature need to be addressed, since interpretations of their importance have influenced perceptions of the likelihood of one extinction mechanism or another.

Based largely on superficial comparisons of craniodental morphology with closest living relatives or analogs, it is widely accepted that many of Australia's megafaunal taxa were browsers. None survived the Pleistocene, so why, it has often been asked (e.g., Wroe et al. 2002), were browsers so susceptible: human partiality or reduction in favorable habitats due to climate change? The idea of a bias in dietary guild extinction among the large herbivores has been reinforced by the fact that the five surviving large *Macropus* species (wherein only a subset of males exceed 40 kg) are principally grazers. Current paleodietary assertions are in dire need of confirmation with rigorous morphological studies as well as enamel microwear and stable isotope analyses, but it is critical to note that all large putative grazers also vanished. More large browsers became extinct not because their diet was necessarily disadvantageous, but because there were more of them to begin with. In one sense, the five smaller grazing kangaroos that survived are "'negative keys to the past', models for what their extinct relatives did not have or did not do" (Akersten 1996:3), but a preference for grass is an inadequate explanation by itself. Horton (1984) thought that the ecophysiological adaptations of *Macropus rufus* to aridity may explain its continued existence. On balance, however, while *M. robustus erubescens* is at least its equivalent in adaptive terms (Hume 1999 and references therein), it is very difficult to apply the same rationale to *M. antilopinus, M. fuliginosus* and especially *M. giganteus*. Moreover, the underlying premise that the extinct megafaunal inhabitants of arid and semi-arid regions were less well-adapted to their environment is a baseless assumption. Other factors, including aspects of social and defensive behavior, could have been important, but this remains an entirely speculative realm. Clearly, the loss of all species exceeding the four largest extant kangaroos and the emu in size shows that they were at a severe selective disadvantage. As Murray (1991b) and Johnson (2002) point out, low population sizes and slow reproductive rates were probably very significant. Such factors might also explain why non-megafaunal species (by any definition), like the putative grazers *Warendja wakefieldi* and *Vombatus hacketti*, and the putative browsers *"Procoptodon" gilli, Congruus congruus* and *Wallabia kitcheneri*, also became extinct in the late Pleistocene. Because smaller animals attract less attention and can be rarer in some deposits than their larger contemporaries, I suspect that the proportion of extinct non-megafaunal species is substantially underestimated.

Another persistent claim within the Australian extinction literature is that humans cannot be tangibly invoked as a likely cause until direct evidence of their predation of megafauna is found. This is a *non sequitur*: showing that one or more species within the diverse megafauna were human prey at some stage could never equate to evidence that humans drove the megafauna to extinction. If we accept that humans and megafauna overlapped in time and space, then we must acknowledge that "pursuit of at least the smaller megafauna was well within the capacity of Pleistocene hunters, given that people hunted animals as large as red kangaroo, emu, crocodile, seal and sea lion with equipment of very simple design, such as clubs, and a range of spears including thrusting javelins" (Mulvaney and Kamminga 1999:129). I suspect that ad hoc explanations for why it is unlikely that humans hunted the largest of the Pleistocene megafauna based on Holocene Aboriginal weapons and practices have their limitations. For example, it would be truly remarkable if *Diprotodon*-hunting technologies (if they ever existed) were retained for tens of millennia after they were last useful. Similarly, absence from the archeological record until 6 ka of "large stone spear-points typically associated with Ice Age big-mammal hunting elsewhere" (Wroe et al. 2002:58), and the fact that a range of American societies only occasionally used wooden-tipped projectiles to hunt big-game, does not preclude early Australians from having the technological enterprise to regularly do so. Perhaps most important of all is the fact that the majority of Australian megafaunal species weighed less than 100 kg (Murray 1984, 1991b). Absence of evidence for big-game hunting technologies in Australia is a red herring in the case against humans as an extinction cause.

One way to view Australia's megafauna is as an evolutionary end-product of 15 million years of adaptation to increasingly drier conditions. Sthenurines exemplified this as much as any other group. The more robust *Procoptodon* species likely originated and radiated in response to intense late Pliocene and Pleistocene aridity, while *Sthenurus* seems to have occupied savanna and open shrubland habitats for over three million years. *Diprotodon, Phascolonus, Genyornis* and several other large species were also adapted to relatively xeric conditions. That at least a proportion of them persisted into the Last Interstadial (59–24 ka; oxygen isotope stage 3) is not in dispute, yet all were apparently gone by the Last Glacial Maximum (22–18 ka; e.g., Field and Dodson 1999; Roberts et al. 2001). Non-reworked specimens of several mesic-adapted taxa from southwestern and eastern Australia also demonstrate survival through the prolonged and markedly arid end to the Penultimate Glacial (Grün et al. 2000; Prideaux et al. 2000; Roberts et al. 2001) and the many cycles of intense climatic fluctuations repeated since the late Pliocene. It is also clear that, while much of Australia became very dry during the 10,000 years leading up to the Last Glacial Maximum (e.g., Bowler et al. 1976; Wasson 1986; Chappell 1991; Harle 1997; Magee and Miller 1998), little evidence points to widespread marked aridity in the earlier half of the Last Interstadial (e.g., Bowler 1982; Chappell 1991; Croke et al. 1996; Ayliffe et al. 1998; Magee and Miller 1998; Harle et al. 1999). Given that many sthenurines and other megafaunal species were adapted to at least semi-arid conditions, and accepting that very few, if any, lasted until 30 ka, it is very difficult to envisage how climate change in the guise of extreme aridity could have accounted for their disappearance from every habitat (cf. Horton 1984; Field and Dodson 1999). The assertion that arid and semi-arid species were the first to go in response to greater aridity seems especially implausible (cf. Wroe et al. 2002). If anything, the reverse is more likely: ranges of more xeric-adapted taxa would have migrated toward the periphery of the continent, while mesic habitats became severely restricted or lost (e.g., Hopper 1992; Archer 1996).

Sound evidence (e.g., articulated remains, preserved anatomical features consistently broken during reworking, direct dating of fossils) of continued existence beyond the Penultimate Glacial Maximum has thus far been retrieved for 19 of the roughly 50 Pleistocene megafaunal species, including eight sthenurines (*Sthenurus andersoni, S. stirlingi, S. tindalei, Simosthenurus maddocki, Si. occidentalis, "Procoptodon" browneorum, "P." gilli, P. goliah* – Roberts et al. 2001; pers. obs.). Only three other Pleistocene sthenurines are known from articulated remains (*Metasthenurus newtonae, "Si." baileyi, "P." oreas*) and the relevant sites await dating. If we accept as reasonable the chance that a small proportion of non-articulation sites that postdate the Penultimate Glacial Maximum actually do contain megafaunal bones not reworked from older strata, and then consider the vast number of localities yet to be dated, we should accept 19 of 50 species as a patent underestimate of the number driven to extinction during this interval. What, then, was so distinctly different about the Last Glacial that this radiation of large marsupials could not endure? It is hard to overlook the arrival and spread of humans, despite the convenient lack of "emotive connotations" (Murray 1991b:1143) associated with invoking climate change as a cause. However, the blitzkrieg extinction model championed by Flannery (1990) does not bear up to the available evidence (e.g., Murray 1991b; Johnson 2002; Wroe et al. 2002). Given an overlap in error ranges for the youngest megafaunal dates, Roberts et al. (2001) could not bring themselves to wholly discount blitzkrieg, but a 1-in-30 probability is too low to consider it a realistic possibility. Moreover, not one piece of evidence for flagrant prey resource overuse exists (direct evidence should be essential to sustain such an extreme hypothesis), let alone a satisfactory explanation for how relatively low numbers of humans could have spread to all corners of the continent and blitzed both the megafauna and a range of smaller taxa from all habitats in no more than 1000 years.

With aridity and blitzkrieg discounted as possibilities, population attrition over a long duration, perhaps 20,000 years, due to a human-based elevation in predation pressure becomes a likely mechanism to account for the extinction of many Australian Pleistocene vertebrates across the body size spectrum. Habitat alteration or other indirect ecological factors might well have compounded their demise, but given the relatively low herbivore biomass (as reflected by the paucity of large carnivores in Australia's Quaternary ecosystems, e.g., Wroe 2002), the predatory effects of the first large terrestrial immigrant in the Cenozoic history of the continent could alone have been sufficient. A combination of comparatively low population sizes and slow reproductive rates may have rendered many species highly vulnerable to increased predation pressure (Murray 1991b). Recently, Johnson (2002) suggested that even sporadic predation of juveniles could have impacted the populations of slow-reproducing species. One implication of this is that direct evidence of human predation may never be found. If residents of habitats relatively inaccessible to humans stood a better chance of survival (Johnson 2002), then sthenurines that occupied relatively more uneven or denser terrain might have clung on for longer than those adapted to habitats easily accessible to humans. Others may have been behaviorally exapted to avoiding human predation. While the latter idea is probably impossible to test, the former should be more straightforward. Indeed, *Simosthenurus occidentalis* is the one sthenurine known to have persisted for at least 10,000 years after the arrival of humans in southwestern Australia (Prideaux et al. 2000; Ayliffe et al. in prep.), and aspects of its pedal anatomy suggest that it may have been adapted to denser or more uneven terrain compared with other sthenurines (Bishop 1997). Future work combining studies of locomotory adaptations with enamel microwear and stable isotope analyses of sthenurines in well-dated contexts will allow the late survival / habitat inaccessibility hypothesis to be tested.

That the magnitude of individual population or species extinctions might have been facilitated to some degree by climatic factors other than aridity should not, however, be discounted, even though I find myself drawn to the same overall conclusion as Murray (1991b:1142): "without the influence of aboriginal man, the megafauna would have survived until the arrival of the Europeans." Ironically, the main phase of extinction would seem to coincide with a relatively wetter period between 65 and 45 ka, during which the summer monsoon extended southward across a large part of Australia, favoring the proliferation of C_4 grasslands (Johnson et al. 1999). While C_4 grasses clearly spread at the expense of C_3 grasses, whether fragmentation or retraction of browse vegetation also correlated with this shift is unknown. If it did, this might have impinged upon a proportion of sthenurine and other browser populations, but it would not account for the concomitant loss of grazers. If humans crossed into northern Australia during oxygen isotope stage 4 (74–59 ka), at a time when the sea reached its lowest levels in almost 80,000 years, then the 65–45 ka wetter interval may well have paved the way for their dispersal into the far southern reaches of the continent. To bolster the extinction hypothesis outlined here, we would require examples of indubitable presence then absence of taxa in well-dated, unambiguous stratigraphic contexts from across the continent, with regional disappearances demonstrably overlapping the human record. Deciphering the precise behavioral and ecological reasons behind the variable vulnerability of Australia's Pleistocene vertebrate fauna seems likely to occupy palaeobiologists for decades to come.

SUMMARY AND CONCLUSIONS

This study provides the first systematic and evolutionary treatment of the subfamily Sthenurinae in more than three decades. It presents a detailed phylogenetic hypothesis regarding the origin of sthenurines and their interrelationships, and examines the dominant trends in the evolution of the sthenurine craniodental system from a functional standpoint. While the research strongly corroborates the long-held view that sthenurines evolved to exploit the more fibrous vegetation that spread across much of Australia during the late Cenozoic, it has been enhanced by the quantum improvements in the Neogene and Pleistocene records made over the past two decades. Its strengths lie in its testability and the congruence between the phylogeny, taxon distributions through time and space, and major directions of adaptive change within the group. The phylogeny may be tested with the discovery of more craniodental remains and a cladistic analysis of postcranial morphology. Functional or paleoecological inferences are testable with enamel isotope and microwear studies, and functional analyses of postcranial morphology. One third of the known sthenurine species are represented by at least a portion of the postcranial skeleton. Hypotheses concerning the spread of sthenurines through time and space will be tested as our knowledge of late Cenozoic geochronology and biostratigraphy continues to improve, and as fossils are retrieved from major geographic regions still largely unexplored.

Sthenurinae is divided into six genera: *Hadronomas* Woodburne, 1967, *Sthenurus* Owen, 1874a, *Metasthenurus* gen. nov., *Archaeosimos* gen. nov., *Simosthenurus* Tedford, 1966, and *Procoptodon* Owen, 1874b. Altogether, they contain 26 species, but *Hadronomas* and *Metasthenurus* are monotypic. Four new species, *S. murrayi, A. correlli, "Si." tirarensis* and *"P." williamsi*, are identified, while *Si. orientalis* Tedford, 1966, and *P. texasensis* Archer, 1978, are recognized as junior synonyms of *Si. occidentalis* Glauert, 1910, and *P. rapha* Owen, 1874b, respectively. Simosthenurini tribe nov. is raised to include the three brachycephalic genera, but no other groupings of equivalent rank are presently justified. Support for the proposed relationship between the extinct genus *Troposodon*, the banded hare-wallaby, *Lagostrophus*, and the sthenurines was found to be lacking following a detailed reexamination of the putative synapomorphies raised by Flannery (1983, 1989). This conclusion agrees with those drawn from previous investigations of the cranial and postcranial anatomy of *Hadronomas* (Murray 1991a, 1995). *Troposodon* is a macropodine; *Lagostrophus* is probably a macropodine, but it has several obscure features that suggest a distant relationship to most extant forms. Craniodental similarities shared between these genera and the sthenurines are likely to reflect dietary convergence in response to a browsing habit.

The cladistic analysis of the data matrix for 37 craniodental characters and 27 species (including 1 outgroup taxon) produced 15 equally most parsimonious trees of 115 steps. Consensus trees reveal a largely stable topology, with most uncertainty reflected in polytomies generated within *Sthenurus* and *Simosthenurus* due to poor character representation for several taxa. Sthenurinae is united by nine synapomorphies, and as such, it is currently the best-defined suprageneric taxon within the Macropodoidea. *Hadronomas* is the sister taxon of all other sthenurines, *Sthenurus* is the sister taxon of a *Metasthenurus*–Simosthenurini clade, and *Archaeosimos* is the most plesiomorphic of the simosthenurins. *Simosthenurus s. l.* includes the monophyletic *Simosthenurus s. s.*, and five inadequately understood simosthenurins: *"Si." brachyselenis*, *"Si." tirarensis*, *"Si." baileyi*, *"Si." antiquus* and *"Si." pales*. *Procoptodon* is expanded to include five species that arise pectinately between Owen's original three species (herein *Procoptodon s. s.*) and *Simosthenurus s. s.* The most plesiomorphic of these are *"P." browneorum* and *"P." gilli*.

An examination of the distribution of several molar crests that become hypertrophied in successively more derived simosthenurin species, but are absent in plesiomorphic sthenurines, reveals an excellent example of phylogenetic character reversal, because the crests are present in ancestral bulungamayines. One crest, the posthypocristid, has also returned in larger species of *Macropus*. Homoplasy is regularly equated with convergence and dismissed as noise hindering the exposé of evolutionary pattern. I show that within the framework of a solid phylogenetic hypothesis, which is supported by biochronologic data, the potential exists for revealing instances of the return of distantly ancestral features. I discuss underlying mechanisms that might facilitate the return of long-lost features, propose two possibly interrelated hypotheses involving differential enamel growth and developmental field expansion to explain the heterochronic postdisplacement and predisplacement patterns observed, and then suggest means by which these ideas might be tested. Phylogenetic character reversal may be an important, but underestimated mechanism of evolutionary change not only in kangaroos, but also in vertebrates as a whole.

Sthenurines evolved from a derived lineage of bulungamayines, phylogenetically close to the middle to ?late Miocene bulungamayine *Wanburoo*, as suggested by Cooke (1999). Additional features documented here, including upper incisor characteristics, support a view that *Wanburoo* is the sister taxon of the Sthenurinae. In all probability, sthenurines originated in the middle Miocene, an idea not only supported by the presence of *Hadronomas, Archaeosimos* and *Sthenurus* in the late Miocene, but also by the sthenurine-like postcranial remains present in the middle Miocene Bullock Creek LF (Murray and Megirian 1992). The origin and early evolution of the sthenurines took place during an especially portentous period in the evolution of Australian climate and environmental change, even though our understanding of the bulungamayine-sthenurine transition is severely hampered by the scant fossil record for the middle to late Miocene. During this time, Australia became increasingly drier and more seasonal, and by the end of the Miocene, a major transition was underway. Each of the terrestrial herbivore lineages (kangaroos, diprotodontoids, wombats) show adaptations to the consumption of more fibrous vegetation, including larger body sizes.

Wanburoo is a good phylogenetic link between derived bulungamayines and basal sthenurines, and an excellent ecomorphological intermediate as well. It is larger than any other bulungamayine known and better adapted to a more fibrous diet according to each craniodental attribute that allies it with the sthenurines. Sthenurines, however, took higher-fiber browsing to a new level. Body size increased markedly (*Hadronomas* is the oldest "kangaroo-sized" macropodoid) and the craniodental system was entirely restructured. The nine sthenurine synapomorphies recognized are integrated elements of an adaptive complex herein termed the sthenurine bauplan. As the most plesiomorphic sthenurine, *Hadronomas* exemplifies this bauplan in its most basic form.

Much of the craniodental restructuring appears to have centered on the more vertical orientation of the jaw adductor muscles and increased cranial rigidity. A more elevated and anteriorly rotated neurocranium, broadened frontals, deeper rostrum and more vertical ascending ramus were fundamental steps in the evolution of the sthenurine craniodental system. Lateral expansion of the frontals and development of the supraorbital crest were probably both geometrical and functional correlates of the more vertical temporalis musculature. An increased capacity for generating more orthal masticatory forces along the cheek tooth row was a major functional outcome of this reconfiguration. The simple, low-crowned molars of *Hadronomas* were not especially well adapted for grinding, but a shift in premolar morphology away from shearing to a dual shearing and grinding function is unmistakable. The widened condyle and greater structural support for the craniomandibular joint helped to counter large forces tending to disarticulate the condyle or drive it back against the postglenoid process. Enlargement of the digastric muscle reflects a need to ensure reseating of the condyle before the commencement of the next occlusal phase. Elongation of the I3 and reduction of the I1 and I2 reflect an efficient compromise between bite area and bite force transmission to resistant food items. Sthenurines were clearly consuming at least moderately tough browse vegetation before the end of the Miocene. Two macropodines, *Protemnodon* and *Troposodon*, had also attained larger body sizes by the beginning of the Pliocene and were clearly browsers, but their restricted diversity through the late Cenozoic probably reflects the constraints of their masticatory system. Functional convergence with sthenurines is seen in selected characters, but not in the wholesale restructuring of the cranium.

In combination with several postcranial adaptations, including a functionally monodactyl pes and an elongate, highly mobile forelimb, the eight craniodental synapomorphies that evolved in the sister lineage to *Hadronomas* represent key innovations that allowed sthenurines to become Australia's foremost radiation of medium to large browsers in the Pliocene and Pleistocene. Masticatory muscles, especially the masseter, became more anteriorly oriented, the orbits were projected anterolaterally, the mandibular symphysis was rigidified, grinding became the primary function of the premolars, the cheek tooth row became laterally curved, and molar complexity increased. These adaptations improved the capacity of the craniodental system to break down even more fibrous vegetation and paved the way for the cranial foreshortening characteristic of the Simosthenurini, which originated prior to the start of the Pliocene.

The presence of at least three *Sthenurus* species in the early late Pliocene, including *S. andersoni* and a form very close to the derived *S. tindalei*, indicates that most of the cladogenesis within the genus had occurred by that time. The persistence of *S. andersoni* and *S. tindalei* into the late Pleistocene infers at least three million years of adaptation to very similar habitats and provides indirect evidence for the presence of shrublands or savanna in inland Australia prior to 3.4 Ma. *Sthenurus* is characterized by a more gracile build, dolichocephaly, broader incisors and higher-crowned molars than the Simosthenurini, although the convergence of *S. stirlingi* toward a more simosthenurin-like craniodental form indicates a capacity for processing tougher vegetation than any other *Sthenurus* species. It also reflects a high level of craniodental integration and intimates the existence of inherent constraints that may canalize morphological responses to similar diets given a similar ancestral bauplan; seven of the nine craniodental autapomorphies for *S. stirlingi* originated independently in the Simosthenurini.

Sthenurus andersoni is the most widely distributed of all Pleistocene sthenurines. It is the only *Sthenurus* species found throughout the more mesic regions of southern Australia, although it is consistently uncommon in simosthenurin-dominated cave assemblages and probably occupied relatively more distal, open habitats. Overall, no *Sthenurus* species appear to have been particularly well equipped for the consumption of a high proportion of grasses, despite some superficial similarities in cranial and dental form to grazing macropodines. Most species probably subsisted largely on small-leafed, xeromorphic shrubs and dusty forbs widespread throughout the interior. Microwear and isotopic analyses of enamel hold great promise as independent avenues by which the dietary preferences of *Sthenurus* species may be tested, but it is pertinent to note that today the only large mammals that exploit this food resource are ungulates introduced by Eurasians.

The early Pliocene record for the Simosthenurini consists of one molar fragment from northeastern Queensland referred to *"Simosthenurus"* sp. cf. *baileyi*, and two molar fragments from southwestern Victoria referred to cf. *Archaeosimos* sp. In particular, the early Pliocene occurrence of a species close to *"Si." baileyi* implies that the simosthenurin radiation was well underway before the marked intensification of aridity in Australia in the late Pliocene. The quantum improvement in our understanding of this brachycephalic sthenurine group is largely a result of the past three decades of exploration into the Pleistocene cave deposits of southern and southeastern Australia. The number of species recognized has more than doubled and the plesiomorphic stage of evolution expressed by several of the taxa made them extremely useful in breaking up long branch lengths in the phylogenetic analysis. The major evolutionary trend within the Simosthenurini was, in essence, a continuation of the adaptive course set during earlier phases of sthenurine evolution: adaptation to increasingly fibrous browse vegetation.

Compared with the plesiomorphic *Metasthenurus*, simosthenurins are characterized by very robust, brachycephalic crania and robust, upturned lower incisors, as well as more complex molar surfaces designed to enhance their efficacy for grinding higher-fiber material. Initially, this might seem difficult to reconcile given that members of this tribe predominated in the relatively more mesic habitats of southern Australia. However, the overall trend from the late Miocene toward the retraction of wet forest and the expansion of drier, sclerophyll forest and woodland would have opened to the simosthenurins a range of higher-fiber browser niches that no other kangaroos were as well positioned to exploit. Simosthenurins may have gained their evolutionary foothold in the transitional areas between the more open sclerophyll habitats and wetter forest. Like the species of *Sthenurus*, their pes is functionally monodactyl (Tedford 1967; Adnams-Hodges 1988; Wells and Tedford 1995), which implies that the terrain to which they were originally adapted cannot have been particularly densely vegetated. This might be one factor that could explain their rarity in the known Pliocene deposits of eastern and inland Australia.

The phylogeny clearly reveals that *Procoptodon* is not an adaptively parallel lineage to *Simosthenurus*, but a derived clade arising from within that species assemblage. The almost entirely southern or southeastern distribution of the three most plesiomorphic species of *Procoptodon s. l.* suggests that it may have originated in this region. However, the concentration of the five most derived and robust *Procoptodon* species in the part of eastern Australia encompassing the Pleistocene deposits of central eastern New South Wales and southeastern Queensland intimates that this clade may have originated here. Two of these species (*"P." williamsi*, *P. goliah*), in addition to *"Si." pales*, *"Si." baileyi* and a form similar to *"Si." antiquus*, are the only simosthenurins known from the Pleistocene of inland Australia. Open sclerophyll woodlands, shrublands and grasslands expanded markedly in central eastern and northern New South Wales during the late Pliocene and early Pleistocene, and these habitats undoubtedly extended into the interior.

The most divergent sthenurines, indeed, the most specialized of all macropodoids, were *Procoptodon pusio*, *P. rapha* and their more completely known relative, *P. goliah.* These species were very large, markedly brachycephalic, and characterized by massive non-interlocking molars with multiple shearing crests, reduced premolars, very small incisors, and deep anteorbital and buccinator fossae suggestive of robust cheek muscles and manipulative lips. *P. goliah* probably weighed up to 200 kg and was more widespread throughout the open, drier regions of inland Australia than any other simosthenurin. Most aspects of its craniodental system indicate a capacity to generate and withstand larger masticatory forces than any other kangaroo, yet the elaborate pattern of high molar crests suggests a poly-shearing system at least as efficient as that seen in the large macropodine grazers. *P. goliah* and its two close relatives may have been consumers of a broad range of coarse, fibrous herbage, and in many respects, may have been macropodoid analogs of the elephants. Enamel isotope and microwear analyses promise to elucidate much more of the dietary ecologies of these remarkable kangaroos.

Sthenurines were widespread, diverse and abundant in the middle and early late Pleistocene, but were extinct by the Last Glacial Maximum along with the rest of Australia's megafauna. Irrefutable evidence exists for the survival of many sthenurines and other megafauna beyond the markedly arid end to the Penultimate Glacial, and at least a proportion persisted into the Last Interstadial. Because their adaptive radiation was fundamentally a response to the spread of sclerophyll and xeromorphic vegetation favored by the increasingly arid conditions of the late Cenozoic, and because they apparently vanished from a broad range of habitats during a relatively wetter interval, extreme aridity cannot account for their disappearance. The blitzkrieg model of human-driven megafaunal extinction is also irreconcilable with the available temporal, geographic and paleontological data, leaving attrition over an extended period due to a human-based elevation in predation pressure as the most defensible hypothesis. Even occasional removal of individuals would have impacted on populations of slow-reproducing species, gradually steering them down the path to extinction (e.g., Murray 1991b; Johnson 2002). In the end, we may find that the 65–45 ka extension of the summer monsoon across much of Australia facilitated the trans-continental migration of humans and, thus, the decline of the only radiation of large herbivorous marsupials ever to have evolved. Sthenurine kangaroos were by far the most speciose group within this radiation and, therefore, the only lineage of browsing marsupials comparable in diversity to the browsing artiodactyl guilds of other continents. Given that just under half of the Pleistocene sthenurines have only been recognized in the past decade, it is humbling to contemplate what awaits paleontologists prepared to get out into the field to actively explore Australia's vastly understudied mammal record.

Literature Cited

Adnams-Hodges, W. A. J. 1988. A partial reconstruction of the musculature and habits of an extinct browsing kangaroo (*Sthenurus occidentalis*). Unpublished B.Sc. (Hons) Thesis, Flinders University of South Australia, Adelaide.

Agenbroad, L. D. and J. I. Mead. 1996. Distribution and palaeoecology of central and western North American *Mammuthus.* Pp. 280-288 in J. Shoshani and P. Tassy (eds.), The Proboscidea: Evolution and Palaeoecology of Elephants and Their Relatives. Oxford: Oxford University Press.

Akersten, W. A. 1996. Diversity bottlenecks, oddball survivors, and negative keys. Pp. 3-15 in K. M. Stewart and K. L. Seymour (eds.), Palaeoecology and Palaeoenvironments of Late Cenozoic Mammals: Tributes to the Career of C. S. (Rufus) Churcher. Toronto: University of Toronto Press.

Alberch, P., S. J. Gould, G. F. Oster, and D. B. Wake. 1979. Size and shape in ontogeny and phylogeny. Paleobiology 5:296-317.

Anderson, C. 1932. The skull of *Sthenurus occidentalis* Glauert. Records of the Australian Museum 18:383-387.

Anonymous. 1910. Interesting fossil remain. Discovered in the Mammoth Cave. A link with the dead past. The West Australian Vol. 30, February 2, p. 4, cols 1-2.

Aplin, K., A. Baynes, L. M. Hatcher, and G. J. Prideaux. 1995. Lindsay Hall Cave: a recently discovered southern Nullarbor Pleistocene site. Abstracts of the 5th Conference on Australasian Vertebrate Evolution, Palaeontology and Systematics, Canberra, p. 27.

Archer, M. 1978. Quaternary vertebrate faunas from the Texas Caves of southeastern Queensland. Memoirs of the Queensland Museum 19:61-109.

_______. 1979. *Wabularoo naughtoni* gen. et. sp. nov., an enigmatic kangaroo (Marsupialia) from the middle Tertiary Carl Creek Limestone of northwestern Queensland. Results of the Ray E. Lemley Expeditions, part 4. Memoirs of the Queensland Museum 19:299-307.

_______. 1981. A review of the origins and radiations of Australian mammals. Pp. 1437-1488 in A. Keast (ed.), Ecological Biogeography in Australia. The Hague: W. Junk.

_______. 1984. The Australian marsupial radiation. Pp. 633-808 in M. Archer and G. Clayton (eds.), Vertebrate Zoogeography and Evolution in Australia (Animals in Space and Time). Perth: Hesperian Press.

_______. 1996. Refugial vertebrates from the Miocene rainforests of Western Australia: where are they? Pp. 100-103 in S. D. Hopper, J. A. Chappill, M. S. Harvey and A. S. George (eds.), Gondwanan Heritage: Past, Present and Future of the Western Australian Biota. Sydney: Surrey Beatty and Sons.

Archer, M. and A. Bartholomai. 1978. Tertiary mammals of Australia: a synoptic review. Alcheringa 2:1-19.

Archer, M. and L. Dawson. 1982. Revision of marsupial lions of the genus *Thylacoleo* Gervais (Thylacoleonidae, Marsupialia) and thylacoleonid evolution in the late Cainozoic. Pp. 477-494 in M. Archer (ed.), Carnivorous Marsupials. Sydney: Surrey Beatty and Sons and the Royal Zoological Society of New South Wales.

Archer, M. and T. F. Flannery. 1985. Revision of the extinct gigantic rat kangaroos (Potoroidae: Marsupialia), with description of a new Miocene genus and species and a new species of *Propleopus*. Journal of Paleontology 59:1331-1349.

Archer, M., S. J. Hand, and H. Godthelp. 1991. Riversleigh. The Story of Animals in Ancient Rainforests of Inland Australia. Sydney: Reed Books.

_______. 1994. Patterns in the history of Australia's mammals and inferences about palaeohabitats. Pp. 80-103 in R. S. Hill (ed.), History of the Australian Vegetation: Cretaceous to Recent. Melbourne: Cambridge University Press.

_______. 1995. Tertiary environmental and biotic change in Australia. Pp. 77-90 in E. S. Vrba, G. H. Denton, T. C. Partridge and L. H. Burckle (eds.), Paleoclimate and Evolution, with Emphasis on Human Origins. New Haven: Yale University Press.

Archer, M., S. J. Hand, H. Godthelp, and P. Creaser. 1997. Correlation of the Cainozoic sediments of the Riversleigh World Heritage Fossil Property, Queensland, Australia. Pp. 131-152 in J.-P. Aguilar, S. Legendre, and J.

Michaux (eds.), Actes du Congrès BiochroM'97. Memoires et Travaux de École Pratique des Hautes L'Institut de Montpelier.

Archer, M., S. J. Hand, H. Godthelp, and D. Megirian. 1989. Fossil mammals of Riversleigh, northwestern Queensland: preliminary overview of biostratigraphy, correlation and environmental change. The Australian Zoologist 25:29-65.

Archer, M. and M. Wade. 1976. Results of the Ray E. Lemley expeditions, part 1. The Allingham Formation and a new Pliocene vertebrate fauna from northern Queensland. Memoirs of the Queensland Museum 17:379-397.

Archie, J. W. 1989a. Homoplasy excess ratios: new indices for measuring levels of homoplasy in phylogenetic systematics and a critique of the consistency index. Systematic Zoology 38:253-269.

_______. 1989b. A randomization test for phylogenetic information in systematic data. Systematic Zoology 38:239-252.

_______. 1990. Homoplasy excess statistics and retention indices: a reply to Farris. Systematic Zoology 39:169-174.

Armand, L. K., W. D. L. Ride, and G. Taylor. 2000. The stratigraphy and palaeontology of Teapot Creek, MacLaughlin River, NSW. Proceedings of the Linnean Society of New South Wales 122:101-121.

Ayliffe, L. K., P. C. Marianelli, K. C. Moriarty, R. T. Wells, M. T. McCulloch, G. E. Mortimer, and J. C. Hellstrom. 1998. 500 ka precipitation record from southeastern Australia: evidence for interglacial relative aridity. Geology 26:147-150.

Bartholomai, A. 1963. Revision of the extinct macropodid genus *Sthenurus* Owen in Queensland. Memoirs of the Queensland Museum 14:51-76.

_______. 1967. *Troposodon*, a new genus of fossil Macropodinae (Marsupialia). Memoirs of the Queensland Museum 15:21-33.

_______. 1970. The extinct genus *Procoptodon* Owen (Marsupialia: Macropodidae) in Queensland. Memoirs of the Queensland Museum 15:213-233.

_______. 1972. Aspects of the evolution of the Australian marsupials. Proceedings of the Royal Society of Queensland 82:v-xviii.

______. 1973a. The genus *Protemnodon* Owen (Marsupialia: Macropodidae) in the Upper Cainozoic deposits of Queensland. Memoirs of the Queensland Museum 16:309-363.

______. 1973b. *Fissuridon pearsoni*, a new fossil macropodid (Marsupialia) from Queensland. Memoirs of the Queensland Museum 16:365-368.

______. 1975. The genus *Macropus* Shaw (Marsupialia: Macropodidae) in the Upper Cainozoic deposits of Queensland. Memoirs of the Queensland Museum 17:373-377.

______. 1977. The fossil vertebrate fauna from Pleistocene deposits at Cement Mills, Gore, southeastern Queensland. Memoirs of the Queensland Museum 18:41-51.

______. 1978a. The fossil kangaroos. Australian Mammalogy 2:15-22.

______. 1978b. The Macropodidae (Marsupialia) from the Allingham Formation, northern Queensland. Results of the Ray E. Lemley expeditions, part 2. Memoirs of the Queensland Museum 18:127-143.

Baverstock, P. R., B. J. Richardson, J. Birrell, and M. Krieg. 1989. Albumin immunologic relationships of the Macropodidae (Marsupialia). Systematic Zoology 38:38-50.

Baynes, A. 1997. The absolute last remake of *Beau Geste*: yet another review of the Australian megafaunal radiocarbon dates. Records of the Western Australian Museum Supplement No. 57:391.

Beecher, R. M. 1977. Function and fusion at the mandibular symphysis. American Journal of Physical Anthropology 47:325-336.

______. 1983. Evolution of the mandibular symphysis in Notharctinae (Adapidae, Primates). International Journal of Primatology 4:99-112.

Belperio, A. P. (comp.). 1995. Quaternary. Pp. 219-280 in J. F. Drexel and W. V. Preiss (eds.), The Geology of South Australia. Volume 2, The Phanerozoic. Geological Survey of South Australia, Bulletin No. 54.

Benbow, M. C., N. F. Alley, R. A. Callen, and D. R. Greenwood. 1995. Geological history and palaeoclimate. Pp. 208-217 in J. F. Drexel and W. V. Preiss (eds.), The Geology of South Australia. Volume 2, The Phanerozoic. Geological Survey of South Australia, Bulletin No. 54.

Berggren, W. A., F. J. Hilgen, C. G. Langereis, D. V. Kent, J. D. Obradovich, I. Raffi, M. E. Raymo, and N. J. Shackleton. 1995. Late Neogene chronology: new perspectives in high-resolution stratigraphy. Geological Society of America Bulletin 107:1272-1287. [Berggren et al. 1995a in text.]

Berggren, W. A., D. V. Kent, C. C. Swisher, and M.-P. Aubry. 1995. A revised Cenozoic geochronology and chronostratigraphy. Pp. 129-212 in W. A. Berggren, D. V. Kent, M.-P. Aubry, and J. Hardenbol (eds.), Geochronology, Time Scales, and Global Stratigraphic Correlation. Society for Sedimentary Geology Special Publication 54. [Berggren et al. 1995b in text.]

Berta, A. and A. R. Wyss. 1994. Pinniped phylogeny. Proceedings of the San Diego Society of Natural History 29:33-56.

Bishop, N. 1997. Functional anatomy of the macropodid pes. Proceedings of the Linnean Society of New South Wales 117:17-50.

Bodmer, R. E. 1990. Influence of digestive morphology on resource partitioning in Amazonian ungulates. Oecologia 85:361-365.

Boles, W. E. and B. S. Mackness. 1994. Birds from the Bluff Downs Local Fauna, Allingham Formation, Queensland. Records of the South Australian Museum 27:139-149.

Bowler, J. M. 1982. Aridity in the late Tertiary and Quaternary of Australia. Pp. 35-45 in W. R. Barker and P. J. M. Greenslade (eds.), Evolution of the Flora and Fauna of Arid Australia. Adelaide: Peacock Press.

Bowler, J. M., G. S. Hope, J. N. Jennings, G. Singh, and D. Walker. 1976. Late Quaternary climates of Australia and New Guinea. Quaternary Research 6:359-394.

Brown, H. Y. L. 1908. Bone breccia and rock phosphate at Brothers Islands. P. 343 in Records of the Mines of South Australia, 4th ed. Adelaide: Government Printer.

Brown, S. P. and R. T. Wells. 2000. A Middle Pleistocene vertebrate fossil assemblage from Cathedral Cave, Naracoorte, South Australia. Transactions of the Royal Society of South Australia 124:91-104.

Burk, A. and M. S. Springer. 2000. Intergeneric relationships among Macropodoidea (Metatheria: Diprotodontia) and the chronicle of kangaroo evolution. Journal of Mammalian Evolution 7:213-237.

Burk, A., M. Westerman, and M. S. Springer. 1998. The phylogenetic position of the Musky Rat-kangaroo and the evolution of bipedal hopping in kangaroos (Macropodidae: Diprotodontia). Systematic Biology 47:457-474.

Butler, P. M. 1956. The ontogeny of molar pattern. Biological Reviews 31:30-70.

Callen, R. A. and R. H. Tedford. 1976. New late Cainozoic rock units and depositional environments, Lake Frome area, South Australia. Transactions of the Royal Society of South Australia 100:125-167.

Campbell, C. R. 1973. A new species of *Troposodon* Bartholomai; from the early Pleistocene Kanunka Fauna, South Australia (Macropodinae; Marsupialia). Records of the South Australian Museum 16:1-18.

Case, J. A. 1984. A new genus of Potoroinae (Marsupialia: Macropodidae) from the Miocene Ngapakaldi Local Fauna, South Australia, and a definition of the Potoroinae. Journal of Paleontology 58:1074-1086.

Cerling, T. E., J. M. Harris, and M. G. Leakey. 1999. Browsing and grazing in elephants: the isotope record of modern and fossil proboscideans. Oecologia 120:364-374.

Chappell, J. 1991. Late Quaternary environmental changes in eastern and central Australia, and their climatic interpretation. Quaternary Science Reviews 10:377-390.

Chiarello, A. G. 1998. Diet of the Atlantic forest maned sloth *Bradypus torquatus* (Xenarthra: Bradypodidae). Journal of Zoology, London 246:11-19.

Cifelli, R. L., T. B. Rowe, W. P. Luckett, J. Banta, R. Reyes, and R. I. Howes. 1996. Fossil evidence for the origin of the marsupial pattern of tooth replacement. Nature 379:715-718.

Clemens, W. A., B. J. Richardson, and P. R. Baverstock. 1989. Biogeography and phylogeny of the Metatheria. Pp. 527-548 in D. W. Walton and B. J. Richardson (eds.), Fauna of Australia, Volume 1B. Mammalia. Canberra: Australian Government Publishing Service.

Colgan, D. J. 1999. Phylogenetic studies of marsupials based on phosphoglycerate kinase DNA sequences. Molecular Phylogenetics and Evolution 11:13-26.

Collett, R. 1884. On some apparently new marsupials from Queensland. Proceedings of the Zoological Society of London 26:381-388.

______. 1887. On a collection of mammals from central and northern Australia. Zoologisches Jahrbuch 2:829-940.

Cooke, B. N. 1992. Primitive macropodids from Riversleigh, northwestern Queensland. Alcheringa 16:201-217.

______. 1997a. Biostratigraphic implications of fossil kangaroos at Riversleigh, northwestern Queensland. Memoirs of the Queensland Museum 41:295-302.

______. 1997b. New Miocene bulungamayine kangaroos (Marsupialia: Potoroidae) from Riversleigh, northwestern Queensland. Memoirs of the Queensland Museum 41:281-294.

______. 1997c. Two new balbarine kangaroos and lower molar evolution within the subfamily. Memoirs of the Queensland Museum 41:269-280.

______. 1999. *Wanburoo hilarus* gen. et sp. nov., a lophodont bulungamayine kangaroo (Marsupialia: Macropodoidea: Bulungamayinae) from the Miocene deposits of Riversleigh, northwestern Queensland. Records of the Western Australian Museum Supplement No. 57:239-253.

______. 2000. Cranial remains of a new species of balbarine kangaroo (Marsupialia: Macropodoidea) from the Oligo-Miocene freshwater limestone deposits of Riversleigh World Heritage Area, northern Australia. Journal of Paleontology 74:317-326.

Cooke, B. N. and B. P. Kear. 1999. Evolution and diversity of kangaroos (Macropodoidea, Marsupialia). Australian Mammalogy 21:27-29.

Cooper, S. M. and N. Owen-Smith. 1985. Condensed tannins deter feeding by browsing ruminants in a South African savanna. Oecologia 67:142-146.

Crabb, P. L. 1982. Pleistocene dasyurids (Marsupialia) from southwestern New South Wales. Pp. 511-516 in M. Archer (ed.), Carnivorous Marsupials. Sydney: Surrey Beatty and Sons and the Royal Zoological Society of New South Wales.

Croke, J. C., J. W. Magee, and D. M. Price. 1996. Major episodes of Quaternary activity in the Neales River, north west Lake Eyre, central Australia. Palaeogeography, Palaeoclimatology, Palaeoecology 124:1-15.

Crosby, K. and M. Archer. 2000. Durudawirines, a new group of phalangeroid marsupials from the Miocene of Riversleigh, northwestern Queensland. Journal of Paleontology 74:327-335.

Davis, A. C. and W. D. L. Ride. 2000. *Procoptodon* Owen, 1874 (Mammalia, Marsupialia) and the specific names *P. rapha* Owen, 1874 and *P. pusio* Owen, 1874: proposed conservation. Bulletin of Zoological Nomenclature 57:103-107.

Davison, C. V. and W. G. Young. 1990. The muscles of mastication of *Phascolarctos cinereus* (Phascolarctidae: Marsupialia). Australian Journal of Zoology 38:227-240.

Dawson, L. 1985. Marsupial fossils from Wellington Caves, New South Wales; the historic and scientific significance of the collections in the Australia Museum, Sydney. Records of the Australian Museum 37:55-69.

_______. 1995. Biostratigraphy and biochronology of sediments from the Bone Cave, Wellington, NSW, based on vertebrate fossil remains. Unpublished abstracts of the Quaternary Symposium of the Linnean Society of New South Wales, Wellington, NSW, p. 4.

Dawson, L. and M. L. Augee. 1997. The Late Quaternary sediments and fossil vertebrate fauna from Cathedral Cave, Wellington, New South Wales. Proceedings of the Linnean Society of New South Wales 117:51-78.

Dawson, L. and T. F. Flannery. 1985. Taxonomic and phylogenetic status of living and fossil kangaroos and wallabies of the genus *Macropus* Shaw (Macropodidae: Marsupialia), with a new subgeneric name for the larger wallabies. Australian Journal of Zoology 33:473-498.

Dawson, L., J. Muirhead, and S. Wroe. 1999. The Big Sink Local Fauna: a lower Pliocene mammalian fauna from the Wellington Caves complex, Wellington, New South Wales. Records of the Western Australian Museum Supplement No. 57:265-290.

Demment, M. W. and P. J. Van Soest. 1985. A nutritional explanation for body-size patterns of ruminant and non-ruminant herbivores. The American Naturalist 125:641-672.

Dennis, A. J. and P. M. Johnson. 1995. Rufous Bettong, *Aepyprymnus rufescens* (Gray, 1837). Pp. 285-286 in R. Strahan (ed.), The Mammals of Australia. Sydney: Reed Books.

Desmarest, A. G. 1804. Noveau Dictionnaire d'Histoire Naturelle, applique aux arts, principalment à l'agriculture, à l'économie rurale et domestique: par un société de naturalistes et d'agriculteurs: Avec des figures tirées des trois règnes de la nature. Tableau Methode Mammifères. Paris: Deterville.

______. 1817. Noveau Dictionnaire d'Histoire Naturelle, applique aux arts . . . Vol. 25. Paris: Deterville.

______. 1822. Mammalogie ou description des especiès de Mammifères. In Encyclopédie Methodique . . . par un Société de Gens de Lettres, de Savans et d'Artistes. Paris. Tableau Methode Mammifères. Paris: Deterville.

De Vis, C. W. 1887. Notice of a probable new species of *Dendrolagus*. Proceedings of the Royal Zoological Society of Queensland 3:11-14.

______. 1895. A review of the fossil jaws of the Macropodidae in the Queensland Museum. Proceedings of the Linnean Society of New South Wales 10:75-133.

Dodson, J. R. 1989. Late Pleistocene vegetation and environmental shifts in Australia and their bearing on faunal extinctions. Journal of Archaeological Science 16:207-217.

Dodson, J. R., R. Fullager, J. H. Furby, R. Jones, and I. Prosser. 1993. Humans and megafauna in a late Pleistocene environment from Cuddie Springs, north western New South Wales. Archaeology in Oceania 28:94-99.

Domning, D. P. and L.-A. C. Hayek. 1984. Horizontal tooth replacement in the Amazonian manatee (*Trichechus inunguis*). Mammalia 48:105-127.

Dorland, W. A. N. and E. C. L. Miller. 1943. The American Illustrated Medical Dictionary, 19th ed. Philadelphia: W. B. Saunders.

Etheridge, R. 1878. A Catalogue of Australian Fossils (including Tasmania and the island of Timor). Cambridge: Cambridge University Press.

Farris, J. S. 1989. The retention index and homoplasy excess. Systematic Zoology 38:406-407.

Field, J. H. and J. R. Dodson. 1999. Late Pleistocene megafauna and archaeology from Cuddie Springs, southeastern Australia. Proceedings of the Prehistoric Society 65:275-301.

Flannery, T. F. 1983. Revision in the subfamily Sthenurinae (Marsupialia: Macropodoidea) and the relationships of the species of *Troposodon* and *Lagostrophus*. Australian Mammalogy 6:15-28.

_______. 1984. Kangaroos: 15 million years of Australian bounders. Pp. 817-835 in M. Archer and G. Clayton (eds.), Vertebrate Zoogeography and Evolution in Australia (Animals in Space and Time). Perth: Hesperian Press.

_______. 1989. Phylogeny of the Macropodoidea; a study in convergence. Pp. 1-46 in G. C. Grigg, P. J. Jarman and I. D. Hume (eds.), Kangaroos, Wallabies and Rat-kangaroos. Sydney: Surrey Beatty and Sons.

Flannery, T. F. and M. Archer. 1983. Revision of the genus *Troposodon* Bartholomai (Macropodidae: Marsupialia). Alcheringa 7:263-279.

_______. 1984. The macropodoids (Marsupialia) of the early Pliocene Bow Local Fauna, central eastern New South Wales. The Australian Zoologist 21:357-383.

_______. 1987a. *Bettongia moyesi*, a new and plesiomorphic kangaroo (Marsupialia: Potoroidae) from Miocene sediments of northwestern Queensland. Pp. 759-767 in M. Archer (ed.), Possums and Opossums: Studies in Evolution. Sydney: Surrey Beatty and Sons and the Royal Zoological Society of New South Wales.

_______. 1987b. *Hypsiprymnodon bartholomaii* (Potoroidae: Marsupialia), a new species from the Miocene Dwornamor Local Fauna and a reassessment of the phylogenetic position of *H. moschatus*. Pp. 749-758 in M. Archer (ed.), Possums and Opossums: Studies in Evolution. Sydney: Surrey Beatty and Sons and the Royal Zoological Society of New South Wales.

Flannery, T. F., M. Archer, and M. Plane. 1983. Middle Miocene kangaroos (Macropodoidea: Marsupialia) from three localities in northern Australia, with a description of two new subfamilies. Bureau of Mineral Resources Journal of Australian Geology and Geophysics 7:287-302.

Flannery, T. F. and B. Gott. 1984. The Spring Creek Locality, southwestern Victoria, a late surviving megafaunal assemblage. The Australian Zoologist 21:385-422.

Flannery, T. F. and J. H. Hope. 1983. Occurrences of the extinct macropodid marsupial *Simosthenurus maddocki* in New South Wales. Australian Mammalogy 6:37-39.

Flannery, T. F., M.-J. Mountain, and K. Aplin. 1983. Quaternary kangaroos (Macropodidae: Marsupialia) from Nombe Rock Shelter, Papua New Guinea, with comments on the nature of megafaunal extinctions in the New Guinea highlands. Proceedings of the Linnean Society of New South Wales 107:77-99.

Flannery, T. F. and T. H. Rich. 1986. Macropodoids from the middle Miocene Namba Formation, South Australia, and the homology of some dental structures in kangaroos. Journal of Paleontology 60:418-447.

Flannery, T. F., T. H. Rich, W. D. Turnbull, and E. L. Lundelius. 1992. The Macropodoidea (Marsupialia) of the early Pliocene Hamilton Local Fauna, Victoria, Australia. Fieldiana: Geology, new series 25:1-37.

Flannery, T. F. and F. S. Szalay. 1982. *Bohra paulae*, a new giant fossil tree kangaroo (Marsupialia: Macropodidae) from New South Wales, Australia. Australian Mammalogy 5:83-94.

Flood, J. M. 1974. Pleistocene man at Cloggs Cave: his tool kit and environment. Mankind 9:175-188.

Flower, B. P. and J. P. Kennett. 1994. The middle Miocene climatic transition: East Antarctic ice sheet development, deep ocean circulation and global carbon cycling. Palaeogeography, Palaeoclimatology, Palaeoecology 108:537-555.

Flower, W. H. 1867. On the development and succession of teeth in the Marsupialia. Philosophical Transactions of the Royal Society of London 157:631-641.

______. 1884. Catalogue of the specimens illustrating the osteology and dentition of vertebrated animals, living and extinct, contained in the museum of the Royal College of Surgeons of England. Part II, Class Mammalia other than man. London: Taylor and Francis.

Förster, F. and Lord Rothschild. 1907. Description of a new tree kangaroo. Novitates Zoologicae 14:506.

Fortelius, M. 1985. Ungulate cheek teeth: developmental, functional, and evolutionary interrelations. Acta Zoologica Fennica 180:1-76.

Frakes, L. A., B. McGowran, and J. M. Bowler. 1987. Evolution of Australian environments. Pp. 1-16 in G. R. Dyne and D. W. Walton (eds.), Fauna of Australia. Vol. 1A. General Articles. Canberra: Australia Government Publishing Service.

Freeland, W. J. 1991. Plant secondary metabolites: biochemical coevolution with herbivores. Pp. 61–81 in R. T. Palo and C. T. Robbins (eds.), Plant Defenses Against Mammalian Herbivores. CRC Press, Boca Raton, Florida.

Garrod, A. H. 1875. On the kangaroo called *Halmaturus luctuosus* by D'Albertis, and its affinities. Proceedings of the Zoological Society of London 1875:48-59.

Gill, E. D. 1957. The stratigraphical occurrence and palaeoecology of some Tertiary marsupials. Memoirs of the National Museum of Victoria 21:135-203.

Gillespie, A. 1997. *Priscileo roskellyae* sp. nov. (Thylacoleonidae, Marsupialia) from the Oligocene-Miocene of Riversleigh, northwestern Queensland. Memoirs of the Queensland Museum 41:321-327.

Gillette, D. D. and C. E. Ray. 1981. Glyptodonts of North America. Smithsonian Contributions to Paleobiology 40:1-255.

Glauert, L. 1909. A new species of *Sthenurus*. Quarterly Journal of the Geological Society of London 65:462.

_______. 1910a. The Mammoth Cave. Records of the Western Australian Museum and Art Gallery 1:11-36.

_______. 1910b. *Sthenurus occidentalis* (Glauert). Bulletin of the Geological Survey of Western Australia 36:53-69.

_______. 1912. Fossil marsupial remains from Balladonia in the Eucla Division. Records of the Western Australian Museum and Art Gallery 1:47-65.

_______. 1926. A list of Western Australian fossils. Geological Survey of Western Australia Bulletin 88:38-71.

_______. 1948. The cave fossils of the south-west. The Western Australian Naturalist 1:100-104.

Glauert, L. [In Anon.] 1914. Cave hunting. The discoverer at work. Mr Glauert's labours. The West Australian Vol. 26, October 1, p. 9, cols 5-6.

Goede, A. and P. F. Murray. 1979. Late Pleistocene bone deposits from a cave in the Florentine Valley, Tasmania. Papers and Proceedings of the Royal Society of Tasmania 113:39-52.

Gould, J. 1840. in Proceedings of the Zoological Society of London 1840:93.

______. 1841-1842. A Monograph of the Macropodidae or Family of Kangaroos. London: J. Gould.

Gould, S. J. 1980. Hen's teeth and horses' toes. Natural History 89:24-28.

Gould, S. J. and E. S. Vrba. 1982. Exaptation–a missing term in the science of form. Paleobiology 8:4-15.

Gray, J. E. 1821. On the natural arrangement of vertebrose animals. London Medical Repository 15:296-310.

______. 1837. Description of some new or little known Mammalia, principally in the British Museum collections. Magazine of Natural History 1: 583.

Graybeal, A. 1998. Is it better to add taxa or characters to a difficult phylogenetic problem? Systematic Biology 47:9-17.

Greaves, W. S. 1988. A functional consequence of an ossified mandibular symphysis. American Journal of Physical Anthropology 77:53-56.

Gresham, R. 2000. A palaeontological investigation of the uranium-series dated deposit in Grant Hall Chamber, Naracoorte Caves, South Australia. Unpublished B.Sc. (Hons) Thesis, Flinders University of South Australia, Adelaide.

Grün, R., K. C. Moriarty, and R. T. Wells. 2000. Electron spin resonance dating of the fossil deposits in the Naracoorte Caves, South Australia. Journal of Quaternary Science 16:49-59.

Gully, G. A. 1997. Tight Entrance Cave, southwestern Australia: a new megafaunal site and its palaeoclimatic and biogeographic implications. Unpublished B.Sc. (Hons) Thesis, Flinders University of South Australia: Adelaide.

Hall, B. K. 1984. Developmental mechanisms underlying the formation of atavisms. Biological Reviews 59:89-124.

Harle, K. J. 1997. Late Quaternary vegetation and climate change in southeastern Australia: palynological evidence from marine core E55-6. Palaeogeography, Palaeoclimatology, Palaeoecology 131:465-483.

Harle, K. J., A. P. Kershaw, and H. Heijnis. 1999. The contributions of uranium/thorium and marine palynology to the dating of the Lake Wangoom pollen record, western plains of Victoria, Australia. Quaternary International 57/58:25-34.

Harris, J. M. 1975. Evolution of feeding mechanisms in the family Deinotheriidae (Mammalia: Proboscidea). Zoological Journal of the Linnean Society 56:331-362.

Hayman, D. L. 1977. Chromosome number–constancy and variation. Pp. 27-48 in B. Stonehouse and D. Gilmore (eds.), The Biology of Marsupials. London: Macmillan Press.

_______. 1990. Marsupial cytogenetics. Australian Journal of Zoology 37:331-349.

Hayman, D. L. and P. G. Martin. 1974. Monotremata and Marsupialia, Chordata. Volume 4. Pp. 1-110 in B. John (ed.), Animal Cytogenetics. Berlin: Gebrüder Bornträger.

Hofmann, R. R. 1989. Evolutionary steps of ecophysiological adaptation and diversification of ruminants: a comparative view of their digestive system. Oecologia 78:443-457.

Hope, J. H. 1982. Late Cainozoic vertebrate faunas and the development of aridity in Australia. Pp. 85-100 in W. R. Barker and P. J. M. Greenslade (eds.), Evolution of the Flora and Fauna of Arid Australia. Adelaide: Peacock Press.

Hope, J. H., R. J. Lampert, E. Edmondson, M. J. Smith, and G. F. van Tets. 1977. Late Pleistocene faunal remains from Seton Rock Shelter, Kangaroo Island, South Australia. Journal of Biogeography 4:363-385.

Hopper, S. D. 1992. Patterns of plant diversity at the population and species levels in south-west Australian mediterranean ecosystems. Pp. 27-46 in R. J. Hobbs (ed.), Biodiversity of Mediterranean Ecosystems in Australia. Sydney: Surrey Beatty and Sons.

Horton, D. R. 1984. Red kangaroos: last of the Australian megafauna. Pp. 639-680 in P. S. Martin and R. G. Klein (eds.), Quaternary Extinctions: A Prehistoric Revolution. Tucson: University of Arizona Press.

Horton, D. R. and G. E. Connah. 1981. Man and megafauna at Reddestone Creek, near Glen Innes, northern New South Wales. Australian Archaeology 13:35-52.

Hume, I. D. 1999. Marsupial Nutrition. Cambridge: Cambridge University Press.

Hylander, W. L., M. J. Ravosa, C. F. Ross, and K. R. Johnson. 1998. Mandibular corpus strain in primates: further evidence for a functional link between symphyseal fusion and jaw-adductor muscle force. American Journal of Physical Anthropology 107:257-271.

International Commission on Zoological Nomenclature. 1999. International Code of Zoological Nomenclature, 4th ed. Adopted by the International Union of Biological Sciences. London: International Trust of Zoological Nomenclature.

Janis, C. M. 1976. The evolutionary strategy of the Equidae and the origins of rumen and cecal digestion. Evolution 30:757-774.

______. 1990. Why kangaroos (Marsupialia: Macropodidae) are not as hypsodont as ungulates (Eutheria). Australian Mammalogy 13:49-53.

Janis, C. M. and M. Fortelius. 1988. On the means whereby mammals achieve increased functional durability of their dentitions, with special reference to limiting factors. Biological Reviews 63:197-230.

Janis, C. M., K. M. Scott, and L. L. Jacobs (eds.). 1998. Evolution of Tertiary Mammals of North America. Vol. 1. Terrestrial Carnivores, Ungulates, and Ungulatelike Mammals. Cambridge: Cambridge University Press.

Jernvall, J. 1995. Mammalian molar cusp patterns: developmental mechanisms of diversity. Acta Zoologica Fennica 198:1-61.

______. 2000. Linking development with generation of novelty in mammalian teeth. Proceedings of the National Academy of Sciences USA 97:2641-2645.

Jernvall, J. and I. Thesleff. 2000. Reiterative signaling and patterning during mammalian tooth morphogenesis. Mechanisms of Development 92:19-29.

Johnson, B. J., G. H. Miller, M. L. Fogel, J. W. Magee, M. K. Gagan, and A. R. Chivas. 1999. 65,000 years of vegetation change in Central Australia and the Australian summer monsoon. Science 284:1150-1152.

Johnson, C. N. 2002. Determinants of loss of mammal species during the late Quaternary 'megafauna' extinctions: life history and ecology, but not body size. *Proceedings of the Royal Society of London, Series B* 269:2221-2227.

Kear B. P. and B. N. Cooke. 2001. A review of macropodoid (Marsupialia) systematics with the inclusion of a new family. Memoirs of the Association of Australasian Palaeontologists 25:83-101.

Keränen, S. V. E., T. Åberg, P. Kettunen, I. Thesleff, and J. Jernvall. 1998. Association of developmental regulatory genes with the development of different molar tooth shapes in two species of rodents. Development, Genes and Evolution 208:477-486.

Kershaw, A. P., H. A. Martin, and J. R. C. McEwen Mason. 1994. The Neogene: a period of transition. Pp. 299-327 in R. S. Hill (ed.), History of the Australian Vegetation: Cretaceous to Recent. Melbourne: Cambridge University Press.

Kierdorf, H. and U. Kierdorf. 1990. Erster Prämolar im Oberkiefer eines Rehbockes (*Capreolus capreolus* L.). Zeitschrift für Jagdwissenschaft 36:60-62.

King, D. R., A. J. Oliver, and R. J. Mead. 1978. The adaptation of some Western Australian mammals to food plants containing fluoroacetate. Australian Journal of Zoology 26:699-712.

King, J. E. 1972. Observations on phocid skulls. Pp. 81-115 in R. J. Harrison (ed.), Functional Anatomy of Marine Mammals. London: Academic Press.

Kirsch, J. A. W. 1977. The comparative serology of Marsupialia, and a classification of marsupials. Australian Journal of Zoology, Supplementary Series No. 52:1-152.

Kirsch, J. A. W., F.-J. Lapointe, and A. Foeste. 1995. Resolution of portions of the kangaroo phylogeny (Marsupialia: Macropodidae) using DNA hybridization. Biological Journal of the Linnean Society 55:309-328.

Kirsch, J. A. W., F.-J. Lapointe, and M. S. Springer. 1997. DNA-hybridisation studies of marsupials and their implications for metatherian classification. Australian Journal of Zoology 45:211-280.

Kluge, A. G. and J. S. Farris. 1969. Quantitative phyletics and the evolution of the anurans. Systematic Zoology 18:1-32.

Krefft, G. 1870. Wellington Caves (Correspondence relative to exploration of). Pp. 1-12 in New South Wales Parliamentary Paper. [Reproduced in Mahoney and Ride 1975.]

______. 1872. Natural history. The kangaroo tribe–(continued). The Sydney Mail and New South Wales Advertiser, no. 637, vol. 14, September 14, p. 327, cols 1-2. [Reproduced in Mahoney and Ride 1975.]

______. 1873. Natural history. Review of Professor Owen's papers on the fossil mammals of Australia. The Sydney Mail and New South Wales Advertiser, no. 686, vol. 16, September 23, p. 238, cols 1-4. [Reproduced in Mahoney and Ride 1975.]

______. 1874. Australian natural history. Mammals of Australia and their classification. Part 1–Ornithodelphia and Didelphia. Transactions of the Royal Society of New South Wales 7:135-147.

______. 1882. Exploration of the caves and rivers of New South Wales. Pp. 1-52 in New South Wales Parliamentary Paper.

Kurtén, B. 1963. Return of a lost structure in the evolution of the felid dentition. Commentationes Biologicae 26:1-12.

Lawler, I. R., W. J. Foley, B. M. Eschler, D. M. Pass, and K. Handasyde. 1998. Intraspecific variation in *Eucalyptus* secondary metabolites determines food intake by folivorous marsupials. Oecologia 116:160-169.

Leigh, J. H. 1994. Chenopod shrublands. Pp. 345-367 in R. H. Groves (ed.), Australian Vegetation, 2d ed. Melbourne: Cambridge University Press.

Le Souef, E. A. 1910. Letter to the Editor. The West Australian no. 7445, vol. 26, February 5, p. 5, col. 2.

Lesson, R. P. 1842. Nouveau Tableau du Règne Animal. Mammifères. Paris.

Levinton, J. S. 1986. Developmental constraints and evolutionary saltations: a discussion and critique. Pp. 253-288 in J. P. Gustafson, G. L. Stebbins and F. J. Ayala (eds.), Genetics, Development and Evolution. New York: Plenum Press.

Lieberman, D. E. and A. W. Crompton. 2000. Why fuse the mandibular symphysis? A comparative analysis. American Journal of Physical Anthropology 112:517-540.

Lister, A. M. 1996. Evolution and taxonomy of Eurasian mammoths. Pp. 203-213 in J. Shoshani and P. Tassy (eds.), The Proboscidea: Evolution and Palaeoecology of Elephants and Their Relatives. Oxford: Oxford University Press.

Locker, S. and E. Martini. 1986. Phytoliths from the southwest Pacific, Site 591. Pp. 1079-1084 in J. P. Kennett and C. C. von der Borch (eds.), Initial Reports of the Deep Sea Drilling Program No. 90.

_______. 1989. Phytoliths at DSDP Site 591 in the southwest Pacific and the aridification of Australia. Geologische Rundschau 78:1165-1172.

Longman, H. A. 1924. Some Queensland fossil vertebrates. Memoirs of the Queensland Museum 8:16-28.

Luckett, W. P. 1993. An ontogenetic assessment of dental homologies in therian mammals. Pp. 182-204 in F. S. Szalay, M. J. Novacek and M. C. McKenna (eds.), Mammal Phylogeny. New York: Springer-Verlag.

Luckett, W. P. and P. A. Woolley. 1996. Ontogeny and homology of the dentition in dasyurid marsupials: development in *Sminthopsis virginiae*. Journal of Mammalian Evolution 3:327-364.

Lundelius, E. L. 1963. Vertebrate remains from the Nullarbor Caves, Western Australia. Journal of the Royal Society of Western Australia 46:75-80.

Lundelius, E. L. and W. D. Turnbull. 1989. The mammalian fauna of Madura Cave, Western Australia. Part VII: Macropodidae: Sthenurinae, Macropodinae, with a review of the marsupial portion of the fauna. Fieldiana: Geology, new series 17:1-71.

Lydekker, R. 1887. Catalogue of the Fossil Mammalia in the British Museum (Natural History) Cromwell Road, S.W. Part V. Containing the Group Tillodontia, the Orders Sirenia, Cetacea, Edentata, Marsupialia,

Monotremata, and Supplement. London: British Museum (Natural History).

_______. 1891. On lower jaws of *Procoptodon*. Quarterly Journal of the Geological Society of London 47:571-574.

MacFadden, B. J. 1992. Fossil Horses: Systematics, Paleobiology, and Evolution of the Family Equidae. Cambridge: Cambridge University Press.

_______. 1998. Tale of two rhinos: isotopic ecology, paleodiet, and niche differentiation of *Aphelops* and *Teleoceras* from the Florida Neogene. Paleobiology 24:274-286.

MacFadden, B. J., N. Solounias, and T. E. Cerling. 1999. Ancient diets, ecology, and extinction of 5-million-year-old horses from Florida. Science 283:824-827.

Mackness, B. S. 1995a. *Palorchestes selestiae*, a new species of palorchestid marsupial from the early Pliocene Bluff Downs Local Fauna, northeastern Queensland. Memoirs of the Queensland Museum 38:603-609.

_______. 1995b. The Bluff Downs Local Fauna–a new synopsis. Abstracts of the 5th Conference on Australasian Vertebrate Evolution, Palaeontology and Systematics, Canberra, p. 15.

Mackness, B. S., P. W. Whitehead, and G. C. McNamara. 2000. New potassium-argon basalt date in relation to the Pliocene Bluff Downs Local Fauna, northern Australia. Australian Journal of Earth Sciences 47:807-811.

Macphail, M. K. 1996a. AGSO Wiluraratja Kapi ('Western Waters') Project Palynological Completion Report 2: A provisional palynostratigraphy for Late Cretaceous-Tertiary basins in the Alice Springs District. Australian Geological Survey Organisation Report No. 26.

_______. Neogene environments in Australia, 1: re-evaluation of microfloras associated with important Early Pliocene marsupial remains at Grange Burn, southwest Victoria. Review of Palaeobotany and Palynology 92:307-328.

_______. 1997. Late Neogene climates in Australia: fossil pollen- and spore-based estimates in retrospect and prospect. Australian Journal of Botany 45:425-464.

Macphail, M. K., N. F. Alley, E. M. Truswell, and I. R. K. Sluiter. 1994. Early Tertiary vegetation: evidence from spores and pollen. Pp. 189-261 in R. S. Hill (ed.), History of the Australian Vegetation: Cretaceous to Recent. Melbourne: Cambridge University Press.

Maddison, W. P. and D. R. Maddison. 1992. MacClade. Analysis of Phylogeny and Character Evolution, version 3.01. Sunderland, Mass.: Sinauer Associates.

Magee, J. W and G. H. Miller. 1998. Lake Eyre palaeohydrology from 60 ka to the present: beach ridges and glacial maximum aridity. Palaeogeography, Palaeoclimatology, Palaeoecology 144:307–329.

Maglio, V. J. 1972. Evolution of mastication in the Elephantidae. Evolution 26:638-658.

Mahoney, J. A. and W. D. L. Ride. 1975. Index to the genera and species of fossil Mammalia described from Australia and New Guinea between 1838 and 1968 (including citations of type species and primary type specimens). Special Publication of the Western Australian Museum No. 6:1-250.

Main, A. R. 1978. Ecophysiology: towards an understanding of late Pleistocene marsupial extinction. Pp. 169-183 in D. Walker and J. C. Guppy (eds.), Biology and Quaternary Environments. Canberra: Australian Academy of Science.

Marcus, L. F. 1962. A new species of *Sthenurus* (Marsupialia, Macropodidae) from the Pleistocene of New South Wales. Records of the Australian Museum 15:299-304.

_______. 1976. The Bingara Fauna. A Pleistocene vertebrate fauna from Murchison County, New South Wales, Australia. University of California Publications in Geological Sciences 114:1-145.

Marshall, C. R., E. C. Raff, and R. A. Raff. 1994. Dollo's law and the death and resurrection of genes. Proceedings of the National Academy of Sciences USA 91:12283-12287.

Marshall, L. G. 1973. Fossil vertebrate faunas from the Lake Victoria region, S.W. New South Wales, Australia. Memoirs of the National Museum of Victoria 34:151-171.

Marshall, L. G., J. A. Case, and M. O. Woodburne. 1990. Phylogenetic Relationships of Marsupials. Pp. 433-505 in H. H. Genoways (ed.), Current Mammalogy. New York: Plenum.

Marshall, L. G. and R. S. Corruccini. 1978. Variability, evolutionary rates, and allometry in dwarfing lineages. Paleobiology 4:101-119.

Martin, H. A. 1979. Stratigraphic palynology of the Mooki Valley, N.S.W. Journal and Proceedings, Royal Society of New South Wales 112:71-78.

_______. 1980. Stratigraphic palynology from shallow bores in the Namoi River and Gwydir River Valleys, north central New South Wales. Journal and Proceedings, Royal Society of New South Wales 113:81-87.

_______. 1981. Stratigraphic palynology of the Castlereagh River Valley, New South Wales. Journal and Proceedings, Royal Society of New South Wales 114:77-84.

_______. 1988. Stratigraphic palynology of the Lake Menindee region, New South Wales. Journal and Proceedings, Royal Society of New South Wales 121:1-9.

_______. 1990. The palynology of the Namba Formation in the Wooltana-1 bore, Callabonna Basin (Lake Frome), South Australia, and its relevance to the Miocene grasslands in central Australia. Alcheringa 14:247-255.

_______. 1993. The palaeovegetation of the Murray Basin, late Eocene to mid-Miocene. Australian Systematic Botany 6:491-531.

_______. 1994. Australian Tertiary phytogeography: evidence from palynology. Pp. 104-142 in R. S. Hill (ed.), History of the Australian Vegetation: Cretaceous to Recent. Melbourne: Cambridge University Press.

_______. 1997a. The use of ecological tolerances for the reconstruction of Tertiary palaeoclimates. Australian Journal of Botany 45:475-492.

_______. 1997b. The stratigraphic palynology of bores along the Darling River, downstream from Bourke, New South Wales. Proceedings of the Linnean Society of New South Wales 118:51-67.

_______. 1998a. Tertiary climatic evolution and the development of aridity in Australia. Proceedings of the Linnean Society of New South Wales 119:115-136.

______. 1998b. Late Cretaceous-Cainozoic palynology of the Poonarunna No. 1 well, central Australia. Transactions of the Royal Society of South Australia 122:89-138.

McCoy, F. 1879. *Procoptodon goliah*. P. 42 in Prodromus of the Palaeontology of Victoria, Decade VI. Geological Survey of Victoria. Melbourne: Government Printer.

McCune, A. R. 1990. Evolutionary novelty and atavism in the *Semionotus* complex: Relaxed selection during colonization of an expanded lake. Evolution 44:71-85.

McGowran, B. and Q. Li. 1994. The Miocene oscillation in southern Australia. Records of the South Australian Museum 27:197-212.

McEwan Mason, J. R. C. 1991. The late Cainozoic magnetostratigraphy and preliminary palynology of Lake George, New South Wales. Pp. 195-209 in M. A. J. Williams and A. P. Kershaw (eds.), The Cainozoic in Australia: A Re-appraisal of the Evidence. Geological Society of Australia Special Publication No. 18.

McNamara, J. A. 1994. A new fossil wallaby (Marsupialia; Macropodidae) from the south east of South Australia. Records of the South Australian Museum 27:111-115.

Megirian, D. 1992. Interpretation of the Carl Creek Limestone, northwestern Queensland. The Beagle, Records of the Northern Territory Museum of Arts and Sciences 9:219-248.

Megirian, D., P. F. Murray, and R. T. Wells. 1996. The late Miocene Ongeva Local Fauna of central Australia. The Beagle, Records of the Museums and Art Galleries of the Northern Territory 13:9-37.

Melloni, B. J. and G. M. Eisner. 1985. Illustrated Medical Dictionary, 2d ed. Baltimore: Williams and Wilkins.

Menzies, J. I. 1989. Observations on a captive forest wallaby (*Dorcopsis luctuosa*) colony. Pp. 629-631 in G. C. Grigg, P. J. Jarman and I. D. Hume (eds.), Kangaroos, Wallabies and Rat-kangaroos. Sydney: Surrey Beatty and Sons.

Menzies, J. I. and C. Ballard. 1994. Some new records of Pleistocene megafauna from New Guinea. Science in New Guinea 20:113-139.

Merrilees, D. 1965. Two species of the extinct genus *Sthenurus* Owen (Marsupialia, Macropodidae) from south-eastern South Australia, including *Sthenurus gilli* sp. nov. Journal of the Royal Society of Western Australia 48:22-32.

______. 1968a. South-western Australian occurrences of *Sthenurus* (Marsupialia, Macropodidae), including *Sthenurus brownei* sp. nov. Journal of the Royal Society of Western Australia 50:65-79.

______. 1968b. Man the destroyer: late Quaternary changes in the Australian marsupial fauna. Journal of the Royal Society of Western Australia 51:1-24.

______. 1969. A newly discovered bone-bearing deposit in Labyrinth Cave, near Augusta, Western Australia. The Western Australian Naturalist 11:86-87.

______. 1973. Fossiliferous deposits at Lake Tandou, New South Wales, Australia. Memoirs of the National Museum of Victoria 34:177-182.

______. 1979. The prehistoric environment in Western Australia. Journal of the Royal Society of Western Australia 62:109-128.

Merrilees, D. and W. D. L. Ride. 1965. *Procoptodon goliah* (Macropodidae, Marsupialia) from western Eyre Peninsula, South Australia. Transactions of the Royal Society of South Australia 89:139-142.

Miller, G. H., J. W. Magee, B. J. Johnson, M. L. Fogel, N. A. Spooner, M. T. McCulloch, and L. K. Ayliffe. 1999. Pleistocene extinction of *Genyornis newtoni*: human impact on Australian megafauna. Science 283:205-208.

Mitchell, T. L. 1838. Three Expeditions into the Interior of Eastern Australia, With Descriptions of the Recently Explored Region of Australia Felix, and of the Present Colony of New South Wales, 1st ed. London: T. and W. Boone.

Molnar, R. E. and C. Kurz. 1997. The distribution of Pleistocene vertebrates on the eastern Darling Downs, based on the Queensland Museum collections. Proceedings of the Linnean Society of New South Wales 117:107-134.

Müller, S. 1840. in C. J. Temminck (ed.), Natuurkundige Commissie in Oost-Indie. Verhandlungen over de natuurlijke Geschiedenis der Nederlandische Overzeesche Bezittingen, Door de Leden der Natuurkundige Commissie in

Indie en Andere Schrijvers. Zoologie. Leiden: S. and J. Luchtmans and C. C. van der Hoek.

Mulvaney, D. J. and J. Kamminga. 1999. Prehistory of Australia. Sydney: Allen and Unwin.

Murray, P. F. 1984. Extinctions downunder: a bestiary of extinct Australian late Pleistocene monotremes and marsupials. Pp. 600-628 in P. S. Martin and R. G. Klein (eds.), Quaternary Extinctions: A Prehistoric Revolution. Tucson: University of Arizona Press.

_______. 1989. The cranium of *Hadronomus* [sic] *puckridgi* Woodburne, 1967 (Macropodoidea: Macropodidae) a primitive macropodid kangaroo from the late Miocene Alcoota Fauna of the Northern Territory. The Beagle, Records of the Northern Territory Museum of Arts and Sciences 6:115-132.

_______. 1991a. The sthenurine affinity of the late Miocene kangaroo, *Hadronomas puckridgi* Woodburne (Marsupialia, Macropodidae). Alcheringa 15:255-283.

_______. 1991b. The Pleistocene megafauna of Australia. Pp. 1071-1164 in P. Vicker-Rich, J. M. Monaghan, R. F. Baird and T. H. Rich (eds.), Vertebrate Palaeontology of Australasia. Melbourne: Pioneer Design Studio.

_______. 1992. The smallest New Guinea zygomaturines–derived dwarfs or relict plesiomorphs? The Beagle, Records of the Northern Territory Museum of Arts and Sciences 9:89-110.

_______. 1995. The postcranial skeleton of the Miocene kangaroo, *Hadronomas puckridgi* Woodburne (Marsupialia, Macropodidae). Alcheringa 19:119-170.

Murray, P. F. and A. Goede. 1977. Pleistocene vertebrate remains from a cave near Montagu, N.W. Tasmania. Records of the Queen Victoria Museum 60:1-29.

Murray, P. F. and D. Megirian. 1992. Continuity and contrast in middle and late Miocene vertebrate communities from the Northern Territory. The Beagle, Records of the Northern Territory Museum of Arts and Sciences 9:195-218.

Murray, P. F., D. Megirian, and R. T. Wells. 1993. *Kolopsis yperus* sp. nov. (Zygomaturinae, Marsupialia) from the Ongeva Local Fauna: new evidence for the age of the Alcoota fossil beds of central Australia. The Beagle, Records of the Northern Territory Museum of Arts and Sciences 10:155-172.

Newman, D. M. R. 1983. One-humped camel, *Camelus dromedarius*. Pp. 497-499 in R. Strahan (ed.), The Australian Museum Complete Book of Australian Mammals. Sydney: Angus and Robertson.

Newton, C. A. 1988. A taphonomic and palaeoecological analysis of the Green Waterhole (5L81), a submerged late Pleistocene bone deposit in the lower southeast of South Australia. Unpublished B.Sc. (Hons) Thesis, Flinders University of South Australia, Adelaide.

Nix, H. 1982. Environmental determinants of biogeography and evolution in Terra Australis. Pp. 47-66 in W. R. Barker and P. J. M. Greenslade (eds.), Evolution of the Flora and Fauna of Arid Australia. Adelaide: Peacock Press.

Noble, V. E., E. M. Kowalski, and M. J. Ravosa. 1999. Orbital orientation and the function of the mammalian postorbital bar. Journal of Zoology, London 250:405-418.

Norrish, K. and J. G. Pickering. 1983. Clay minerals. Pp. 281-308 in Soils, an Australian Viewpoint. Melbourne: CSIRO/Academic Press.

Osborn, H. F. 1888. The evolution of mammalian molars to and from the tritubercular type. The American Naturalist 22:1067-1079.

Osborne, R. A. L. 1983. Cainozoic stratigraphy at Wellington Caves, New South Wales. Proceedings of the Linnean Society of New South Wales 107:129-145.

______. 1997. Rehabilitation of the Wellington Caves Phosphate Mine: implications for Cainozoic stratigraphy. Proceedings of the Linnean Society of New South Wales 117:175-180.

Owen, R. 1838. Fossil Marsupialia from the caves of Wellington Valley. Pp. 359-363 in T. L. Mitchell, Three Expeditions into the Interior of Eastern Australia, With Descriptions of the Recently Explored Region of Australia Felix, and of the Present Colony of New South Wales, 1st ed. London: T. and W. Boone.

_______. 1845a. *Macropus goliah* (fossil). P. 59 in G. R. Waterhouse, A Natural History of the Mammalia. Volume 1. Containing the Order Marsupiata, or Pouched Animals. London: Hippolyte Baillière.

_______. 1845b. Descriptive and illustrated catalogue of the fossil organic remains of Mammalia and Aves contained in the Museum of the Royal College of Surgeons of England. London: Richard and John E. Taylor.

_______. 1866. On the anatomy of vertebrates. Volume II. London: Longmans Green.

_______. 1868. On the anatomy of vertebrates. Volume III. London: Longmans Green.

_______. 1873a. On the fossil mammals of Australia. Part VIII. Family Macropodidae: Genera *Macropus, Osphranter, Phascolagus, Sthenurus,* and *Protemnodon*. Proceedings of the Royal Society of London 21:128.

_______. 1873b. On the fossil mammals of Australia. Family Macropodidae. Genera *Macropus, Pachysiagon, Leptosiagon, Procoptodon,* and *Palorchestes*. Part IX. Proceedings of the Royal Society of London 21:386-387.

_______. 1874a. On the fossil mammals of Australia. Part VIII. Family Macropodidae: Genera *Macropus*, *Osphranter*, *Phascolagus*, *Sthenurus* and *Protemnodon*. Philosophical Transactions of the Royal Society of London 164:245-288.

_______. 1874b. On the fossil mammals of Australia. Part IX. Family Macropodidae: Genera *Macropus*, *Pachysiagon*, *Leptosiagon*, *Procoptodon* and *Palorchestes*. Philosophical Transactions of the Royal Society of London 164:783-803.

_______. 1877a. On a new species of *Sthenurus*, with remarks on the relation of the genus to *Dorcopsis*. Proceedings of the Zoological Society of London 1877:352-361.

_______. 1877b. Researches on the fossil remains of the extinct mammals of Australia; with a notice of the extinct marsupials of England. London: J. Erxleben.

Patterson, B. 1956. Early Cretaceous mammals and the evolution of mammalian molar teeth. Fieldiana: Geology 13:1-105.

Patterson, C. and P. V. Rich. 1987. The fossil history of the emus, *Dromaius* (Aves: Dromaiinae). Records of the South Australian Museum 21:85-117.

Pearson, J. 1946. The affinities of the rat-kangaroos (Marsupialia) as revealed by a comparative study of the female urogenital system. Papers and Proceedings of the Royal Society of Tasmania 1945:13-25.

_______. 1950. The relationships of the Potoroidae to the Macropodidae (Marsupialia). Papers and Proceedings of the Royal Society of Tasmania 1949:211-229.

Peron, F. and C. A. Le Sueur. 1807. Voyage de découvertes aux Terres Australes . . . sur le Géographe, le Naturaliste, et la goëlette le Casuarina pendant les anées 1800, 1801, 1802, 1803 and 1804. Atlas.

Plane, M. D. 1967. Stratigraphy and vertebrate fauna of the Otibanda Formation, New Guinea. Bureau of Mineral Resources, Geology and Geophysics, Australia, Bulletin No. 86:1-64.

Pledge, N. S. 1980. Macropodid skeletons, including *Simosthenurus* Tedford, from an unusual "drowned cave" deposit in the south east of South Australia. Records of the South Australian Museum 18:131-141.

_______. 1984. A new Miocene vertebrate faunal assemblage from the Lake Eyre Basin: a preliminary report. The Australian Zoologist 21:345-355.

_______. 1990. The Upper Fossil Fauna of the Henschke Fossil Cave, Naracoorte, South Australia. Memoirs of the Queensland Museum 28:247-262.

_______. 1992. The Curramulka Local Fauna: a new late Tertiary fossil assemblage from Yorke Peninsula, South Australia. The Beagle, Records of the Northern Territory Museum of Arts and Sciences 9:115-142.

Pole, M. S., R. S. Hill, N. Green, and M. K. Macphail. 1993. The Oligocene Berwick Quarry Flora – rainforest in a drying environment. Australian Systematic Botany 6:399-427.

Prideaux, G. J. 1990. Behavioural patterns associated with the handling and mastication of certain food items in the potoroines. Unpublished third-year B.Sc. report, Flinders University of South Australia, Adelaide.

_______. 1994. A small sthenurine kangaroo from a Pleistocene cave deposit, Nullarbor Plain, Western Australia. Transactions of the Royal Society of South Australia 118:133-138.

_______. 1996. Origin and early evolution of the sthenurine kangaroos (Macropodidae, Diprotodontia). Journal of Vertebrate Paleontology 16, Supplement to No. 3:59A.

_______. 1997. Palaeontology of the Pliocene vertebrate deposit at Lake Kanunka in the Tirari Desert, South Australia. Unpublished report to the Royal Zoological Society of New South Wales, Sydney.

_______. 1999a. *Borungaboodie hatcheri* gen. et sp. nov., a very large bettong (Marsupialia: Macropodoidea) from the Pleistocene of southwestern Australia. Records of the Western Australian Museum Supplement No. 57:317-329.

_______. 1999b. Systematics and evolution of the extinct kangaroo subfamily, Sthenurinae. Journal of Vertebrate Paleontology 19, Supplement to No. 3:69A.

_______. 1999c. Systematics and evolution of the extinct kangaroo subfamily, Sthenurinae. Unpublished Ph.D. Thesis, Flinders University of South Australia, Adelaide.

_______. 2000. *Simosthenurus newtonae* sp. nov., a widespread sthenurine kangaroo (Diprotodontia: Macropodidae) from the Pleistocene of southern and eastern Australia. Records of the South Australian Museum 33:1-15.

Prideaux, G. J., G. A. Gully, L. K. Ayliffe, M. I. Bird, and R. G. Roberts. 2000. Tight Entrance Cave, southwestern Australia: a late Pleistocene vertebrate deposit spanning more than 180 ka. Journal of Vertebrate Paleontology 20, Supplement to No. 3:62A-63A.

Prideaux, G. J. and R. T. Wells. 1994. A new extinct sthenurine kangaroo (Marsupiala [*sic*], Macropodidae) from southeastern Australia. Records of the South Australian Museum 27:227.

_______. 1997. New *Sthenurus* species (Macropodidae, Diprotodontia) from Wellington Caves and Bingara, New South Wales. Proceedings of the Linnean Society of New South Wales 117:181-196.

_______. 1998. *Sthenurus baileyi* sp. nov., a new fossil kangaroo from the Pleistocene of southern Australia. Transactions of the Royal Society of South Australia 122:1-15.

Prince, R. I. T. 1995. Banded Hare-wallaby *Lagostrophus fasciatus* (Peron and Le Sueur, 1807). Pp. 406-408 in R. Strahan (ed.), Mammals of Australia. Sydney: Reed Books.

Quoy, J. R. C. and J. P. Gaimard. 1824. Zoologie. Part 1. In M. L. de Freycinet (ed.), Voyage autour du monde execute sur l'Uranie et la Physicienne pendant 1817-20. Paris.

Raff, R. A. 1996. The Shape of Life: Genes, Development, and the Evolution of Animal Form. Chicago: University of Chicago Press.

Ramsay, E. P. 1876. Description of a new genus and species of rat-kangaroo, allied to the genus *Hypsiprymnus*, proposed to be called *Hypsiprymnodon moschatus.* Proceedings of the Linnean Society of New South Wales 1:33-35.

Ramsay, E. P. 1883. Proceedings of the Linnean Society of New South Wales 8:17.

Raven, H. C. 1929. Kangaroo. Pp. 254-255 in Encyclopaedia Britannica, 14th ed., Vol. 13.

Raven, H. C. and W. K. Gregory. 1946. Adaptive branching of the kangaroo family in relation to habitat. American Museum Novitates 1309:1-33.

Ravosa, M. J. 1999. Anthropoid origins and the modern symphysis. Folia Primatologica 70:65-78.

Ravosa, M. J. and W. L. Hylander. 1994. Function and fusion of the mandibular symphysis in primates: stiffness or strength? Pp. 447-468 in J. G. Fleagle and R. F. Kay (eds.), Anthropoid Origins. New York: Plenum Press.

Ravosa, M. J. and E. L. Simons. 1994. Mandibular growth and function in *Archaeolemur.* American Journal of Physical Anthropology 95:63-76.

Reed, E. H. and S. J. Bourne. 2000. Pleistocene fossil vertebrate sites of the South East region of South Australia. Transactions of the Royal Society of South Australia 124:61-90.

Ride, W. D. L. 1957. *Protemnodon parma* (Waterhouse) and the classification of related wallabies (*Protemnodon*, *Thylogale*, and *Setonix*). Proceedings of the Zoological Society of London 128:327-346.

_______. 1959. Mastication and taxonomy in the macropodine skull. Pp. 33-59 in A. J. Cain (ed.), Function and Taxonomic Importance. Systematics Association Publication No. 3.

_______. 1971. On the fossil evidence of the evolution of the Macropodidae. The Australian Zoologist 16:6-16.

_______. 1993. *Jackmahoneya* gen. nov. and the genesis of the macropodiform molar. Memoirs of the Association of Australasian Palaeontologists 15:441-459.

Ride, W. D. L., P. A. Pridmore, R. E. Barwick, R. T. Wells, and R. D. Heady. 1997. Towards a biology of *Propleopus oscillans* (Marsupialia: Propleopinae, Hypsiprymnodontidae). Proceedings of the Linnean Society of New South Wales 117:243-328.

Roberts, R. G., T. F. Flannery, L. K. Ayliffe, H. Yoshida, J. Olley, G. J. Prideaux, G. Laslett, A. Baynes, M. Smith, R. Jones, and B. Smith. 2001. New ages for the last Australian megafauna: continent-wide extinction about 46,000 years ago. *Science* 292:1888-1892.

Roberts, R.G., R. Jones, N. A. Spooner, M. J. Head, A. S. Murray, and M. A. Smith. 1994. The human colonisation of Australia - optical dates of 53,000 and 60,000 years bracket human arrival at Deaf Adder Gorge, Northern Territory. Quaternary Science Reviews 13:575-583.

Rofe, R. H. 1979. G-banding and chromosome evolution in Australian marsupials. Unpublished Ph.D. Thesis, University of Adelaide, Adelaide. [Cited in Clemens et al. 1989.]

Sanson, G D. 1980. The morphology and occlusion of the molariform cheek teeth in some Macropodinae (Marsupialia: Macropodidae). Australian Journal of Zoology 28:341-365.

_______. 1989. Morphological adaptations of teeth to diets and feeding in the Macropodoidea. Pp. 151-168 in G. C. Grigg, P. J. Jarman and I. D. Hume (eds.), Kangaroos, Wallabies and Rat-kangaroos. Sydney: Surrey Beatty and Sons.

______. 1991. Predicting the diet of fossil mammals. Pp. 201-228 in P. Vickers-Rich, J. M. Monaghan, R. F. Baird and T. H. Rich (eds.), Vertebrate Palaeontology of Australasia. Melbourne: Pioneer Design Studio.

Sanson, G. D., S. J. Riley, and M. A. J. Williams. 1980. A late Quaternary *Procoptodon* fossil from Lake George, New South Wales. Search 11:39-40.

Scapino, R. P. 1981. Morphological investigations into functions of the jaw symphysis in carnivorans. Journal of Morphology 167:339-375.

Schlegel, H. and S. Müller. 1845. Over drie buideldieren uit de familie der kangoeroes. Pp. 129-148 in C. J. Temminck (ed.), Verhandlungen over de natuurlijke Geschiedenis der Nederlandische Overzeesche Bezittingen, Door de Leden der Natuurkundige Commissie in Indie en Andere Schrijvers. Zoologie. Leiden: S. and J. Luchtmans and C. C. van der Hoek.

Scott, H. H. 1906. Memoir on *Procoptodon rapha* (Owen) from King Island. Queen Victoria Museum, Launceston. Pamphlet (2 pp).

______. 1917. Some palaeontological notes (largely emendatory). Victoria Museum, Launceston, Brochure 6:1-8.

Shackleton, N. J. and J. P. Kennett. 1975. Paleotemperature history of the Cenozoic and initiation of Antarctic glaciation: oxygen and carbon isotopic analyses in DSDP Sites 277, 279, and 281. Pp. 743-755 in J. P. Kennett and R. E. Houtz (eds), Initial Reports of the Deep Sea Drilling Program No. 29.

Sharman, G. B. 1961. The mitotic chromosomes of marsupials and their bearing on taxonomy and phylogeny. Australian Journal of Zoology 9:38-60.

Shaw, G. 1790. The Naturalist's Miscellany. 1. London: Nodder and Co.

Simpson, G. G. 1930. Post-Mesozoic Marsupialia. Pp. 1-87 in J. F. Pompeckj (ed.), Fossilium Catalogus. 1: Animalia Pars 47. Berlin: W. Junk.

Skilbeck, C. G. 1980. A preliminary report on the late Cainozoic geology and fossil fauna of Bow, New South Wales. Proceedings of the Linnean Society of New South Wales 104:171-181.

Sluiter, I. R. K. and A. P. Kershaw. 1982. The nature of Late Tertiary vegetation in Australia. Alcheringa 6:211-222.

Solounias, N. and S. M. C. Moelleken. 1992. Tooth microwear analysis of *Eotragus sansaniensis* (Mammalia: Ruminantia), one of the oldest known bovids. Journal of Vertebrate Paleontology 12:113-121.

Solounias, N., M. Teaford, and A. Walker. 1988. Interpreting the diet of extinct ruminants: the case of a non-browsing giraffid. Paleobiology 14:287-300.

Specht, R. L. 1994. Heathlands. Pp. 321-344 in R. H. Groves (ed.), Australian Vegetation, 2d ed. Melbourne: Cambridge University Press.

Springer, M. S. and J. A. W. Kirsch. 1991. DNA hybridization, the compression effect, and the radiation of diprotodontian marsupials. Systematic Zoology 40:131-151.

Stein, R. and C. Robert. 1986. Siliciclastic sediments at Sites 588, 590, and 591: Neogene and Paleogene evolution in the southwest Pacific and Australian climate. Pp. 1437-1455 in J. P. Kennett and C. C. von der Borch (eds.), Initial Reports of the Deep Sea Drilling Program No. 90.

Stiassny, M. L. J. 1992. Atavisms, phylogenetic character reversals, and the origin of evolutionary novelties. Netherlands Journal of Zoology 42:260-276.

Stirton, R. A. 1955. Late Tertiary marsupials from South Australia. Records of the South Australian Museum 11:247-268.

_______. 1957. Tertiary marsupials from Victoria, Australia. Memoirs of the National Museum of Victoria 21:121-134.

_______. 1963. A review of the macropodid genus *Protemnodon*. University of California Publications in Geological Sciences 44:97-162.

_______. 1967. New species of *Zygomaturus* and additional observations on *Meniscolophus*, Pliocene Palankarinna Fauna, South Australia. Bureau of Mineral Resources, Geology and Geophysics, Bulletin 85:129-147.

Stirton, R. A. and L. F. Marcus. 1966. Generic and specific diagnoses in the gigantic macropodid genus *Procoptodon*. Records of the Australian Museum 26:349-359.

Stirton, R. A., R. H. Tedford, and M. O. Woodburne. 1961. Cenozoic stratigraphy and vertebrate paleontology of the Tirari Desert, South Australia. Records of the South Australian Museum 14:19-61.

Strickberger, M. W. 1976. Genetics, 2d ed. New York: Macmillan.

Swofford, D. L. 1993. Phylogenetic Analysis Using Parsimony (PAUP), version 3.1.1. Washington, DC: Smithsonian Institution.

Swofford, D. L. and W. P. Maddison. 1987. Reconstructing ancestral character states under Wagner parsimony. Mathematical Biosciences 87:199-229.

Szalay, F. S. 1994. The Evolutionary History of Marsupials and an Analysis of Osteological Characters. Cambridge: Cambridge University Press.

Tate, G. H. H. 1948. Results of the Archbold Expeditions. No. 59. Studies on the anatomy and phylogeny of the Macropodidae (Marsupialia). Bulletin of the American Museum of Natural History 91:233-351.

Tedford, R. H. 1955. Report on the extinct mammalian remains at Lake Menindee, New South Wales. Records of the South Australian Museum 11:299-305.

_______. 1966. A review of the macropodid genus *Sthenurus*. University of California Publications in Geological Sciences 57:1-72.

_______. 1967. The fossil Macropodidae from Lake Menindee, New South Wales. University of California Publications in Geological Sciences 64:1-165.

_______. 1994. Succession of Pliocene through medial Pleistocene mammal faunas of southeastern Australia. Records of the South Australian Museum 27:79-93.

Tedford, R. H. and R. T. Wells. 1990. Pleistocene deposits and fossil vertebrates from the "Dead Heart of Australia." Memoirs of the Queensland Museum 28:263-284.

Tedford, R. H., R. T. Wells, and S. F. Barghoorn. 1992. Tirari Formation and contained faunas, Pliocene of the Lake Eyre Basin, South Australia. The Beagle, Records of the Northern Territory Museum of Arts and Sciences 9:173-194.

Tedford, R. H. and M. O. Woodburne. 1987. The Ilariidae, a new family of vombatiform marsupials from Miocene strata of South Australia and an evaluation of the homology of molar cusps in the Diprotodontia. Pp. 401-418 in M. Archer (ed.), Possums and Opossums: Studies in Evolution. Sydney: Surrey Beatty and Sons and the Royal Zoological Society of New South Wales.

Thomas, O. 1887. On the wallaby commonly known as *Lagorchestes fasciatus*. Proceedings of the Zoological Society of London 1886:544-547.

_______. 1888. Catalogue of the Marsupialia and Monotremata in the Collection of the British Museum (Natural History). London: British Museum (Natural History).

_______. 1922. New mammals from New Guinea and neighbouring islands. Annual Magazine of Natural History 9:261-265.

Thorne, A., R. Grün, R., G. Mortimer, N. A. Spooner, J. J. Simpson, M. McCulloch, L. Taylor, and D. Curnoe. 1999. Australia's oldest human remains: age of the Lake Mungo 3 skeleton. Journal of Human Evolution 36:591-612.

Tindale, N. B. 1933. Tantanoola Caves, southeast of South Australia: geological and physiographical features. Transactions of the Royal Society of South Australia 57:130-142.

Todd, N. E. and V. L. Roth. 1996. Origin and radiation of the Elephantidae. Pp. 193-202 in J. Shoshani and P. Tassy (eds.), The Proboscidea: Evolution and Palaeoecology of Elephants and Their Relatives. Oxford: Oxford University Press.

Trouessart, E. L. 1905. Pp. 753-873 in Catalogus Mammalium tam viventium quam fossilium. Quinquenanale supplementum (1899-1904), Fasciculus IV, Cetacea, Edentata, Marsupialia, Allotheria, Monotremata. Berlin: R. Frielander and Son.

Truswell, E. M. and W. K. Harris. 1982. The Cainozoic palaeobotanical record in arid Australia: fossil evidence for the origins of an arid-adapted flora. Pp. 67-76 in W. R. Barker and P. J. M. Greenslade (eds.), Evolution of the Flora and Fauna of Arid Australia. Adelaide: Peacock Press.

Truswell, E. M., I. R. Sluiter, and W. K. Harris. 1985. Palynology of the Oligocene-Miocene sequence in the Oakvale-1 corehole, western Murray Basin, South Australia. Bureau of Mineral Resources Journal of Australian Geology and Geophysics 9:267-295.

Turnbull, W. D. 1970. Mammalian masticatory apparatus. Fieldiana: Geology 18:149-356.

Turnbull, W. D. and E. L. Lundelius. 1970. The Hamilton Fauna: a late Pliocene mammalian fauna from the Grange Burn, Victoria, Australia. Fieldiana: Geology 19:1-163.

Turnbull, W. D., E. L. Lundelius, and R. H. Tedford. 1992a. A Pleistocene marsupial fauna from Limeburner's Point, Victoria, Australia. The Beagle, Records of the Northern Territory Museum of Arts and Sciences 9:143-172.

_______. 1992b. The Limeburner's Point Local Fauna. The Beagle, Records of the Northern Territory Museum of Arts and Sciences 9:268-269.

Turney, C. S. M., M. I. Bird, L. K. Fifield, R. G. Roberts, M. A. Smith, C. E. Dortch, R. Grün, E. Lawson, G. H. Miller, J. Dortch, R. G. Cresswell, and L. K. Ayliffe. 2001. Breaking the radiocarbon barrier and early human occupation at Devil's Lair, southwestern Australia. Quaternary Research 55:3-13.

Tyndale-Biscoe, C. H. 1965. The female urogenital system and reproduction of the marsupial *Lagostrophus fasciatus*. Australian Journal of Zoology 13:255-267.

Van Huet, S. 1999. The taphonomy of the Lancefield swamp megafaunal accumulation, Lancefield, Victoria. Records of the Western Australian Museum Supplement No. 57:331-340.

Van Valen, L. M. 1964. Nature of the supernumerary molars of *Otocyon.* Journal of Mammalogy 45:284-286.

_______. 1966. Deltatheridia, a new order of mammals. Bulletin of the American Museum of Natural History 132:1-126.

_______. 1970. An analysis of developmental fields. Developmental Biology 23:456-477.

_______. 1978. Why not to be a cladist. Evolutionary Theory 3:285-299.

Veevers, J. J. 1984. Phanerozoic Earth History of Australia. Oxford Geological Sciences Series No. 2. Oxford: Clarendon Press.

Wang, Y., T. E. Cerling, and B. J. MacFadden. 1994. Fossil horses and carbon isotopes: new evidence for Cenozoic dietary, habitat, and ecosystem changes in North America. Palaeogeography, Palaeoclimatology, Palaeoecology 107:269-279.

Wasson, R. J. 1986. Geomorphology and Quaternary history of the Australian continental dune fields. Geographical Review of Japan 59:55-67.

Waterhouse, G. R. 1845. A Natural History of the Mammalia. Volume 1, containing the Order Marsupiata, or Pouched Animals. London: Baillière.

Wells, R. T. 1975. Reconstructing the past: excavations in fossil caves. Australian Natural History 18:208-211.

Wells, R. T. and R. A. Callen. 1986. The Lake Eyre Basin–Cainozoic sediments, fossil vertebrates and plants, landforms, silcretes and climatic implications. Australasian Sedimentologists Group Field Guide Series No. 4. Sydney: Geological Society of Australia.

Wells, R. T., K. C. Moriarty, and D. L. G. Williams. 1984. The fossil vertebrate deposits of Victoria Fossil Cave Naracoorte: an introduction to the geology and fauna. The Australian Zoologist 21:305-333.

Wells, R. T. and P. F. Murray. 1979. A new sthenurine kangaroo (Marsupialia, Macropodidae) from southeastern South Australia. Transactions of the Royal Society of South Australia 103:213-219.

Wells, R. T. and N. S. Pledge. 1983. Vertebrate fossils. Pp. 169-176 in M. J. Tyler, C. R. Twidale, J. K. Ling and J. W. Holmes (eds.), Natural History of the South East. Adelaide: Royal Society of South Australia.

Wells, R. T. and R. H. Tedford. 1995. *Sthenurus* (Macropodidae: Marsupialia) from the Pleistocene of Lake Callabonna, South Australia. Bulletin of the American Museum of Natural History 225:1-111.

Werdelin, L. 1987. Supernumerary teeth in *Lynx lynx* and the irreversibility of evolution. Journal of Zoology 211:259-266.

Whitelaw, M. J. 1989. Magnetic polarity stratigraphy and mammalian fauna of the late Pliocene (early Matuyama) section at Batesford (Victoria), Australia. Journal of Geology 97:624-631.

_______. 1991. Magnetic polarity stratigraphy of Pliocene and Pleistocene fossil vertebrate localities in southeastern Australia. Geological Society of America Bulletin 103:1493-1503.

_______. 1992. Magnetic polarity stratigraphy of three Pliocene sections and inferences for the ages of vertebrate fossil sites near Bacchus Marsh, Victoria, Australia. Australian Journal of Earth Sciences 39:521-528.

_______. 1993. Age constraints on the Duck Ponds and Limeburner's Point mammalian faunas based on magnetic polarity stratigraphy in the Geelong area (Victoria), Australia. Quaternary Research 39:120-124.

Williams, D. L. G. 1980. Catalogue of Pleistocene vertebrate fossils and sites in South Australia. Transactions of the Royal Society of South Australia 104:101-115.

Wilson, J. T. and J. P. Hill. 1897. Observations on the development and succession of the teeth in *Perameles*. Quarterly Journal of Microscopic Science 39:427-588.

Woodburne, M. O. 1967. The Alcoota Fauna, central Australia. An integrated palaeontological and geological study. Bureau of Mineral Resources, Geology and Geophysics, Bulletin 87:1-187.

_______. 1984a. Families of marsupials: relationships, evolution and biogeography. Pp. 48-71 in T. W. Broadhead (ed.), Mammals: Notes for a Short Course. University of Tennessee, Department of Geological Science, Studies in Geology 8.

_______. 1984b. *Wakiewakie lawsoni*, a new genus and species of Potoroinae (Marsupialia: Macropodidae) of middle Miocene age, South Australia. Journal of Paleontology 58:1062-1073.

Woodburne, M. O. and J. A. Case. 1996. Dispersal, vicariance, and the Late Cretaceous to Early Tertiary land mammal biogeography from South America to Australia. Journal of Mammalian Evolution 3:121-161.

Woodburne, M. O., B. J. MacFadden, J. A. Case, M. S. Springer, N. S. Pledge, J. D. Power, J. M. Woodburne, and K. B. Springer. 1994. Land mammal biostratigraphy and magnetostratigraphy of the Etadunna Formation (late Oligocene) of South Australia. Journal of Vertebrate Paleontology 13:483-515.

Woodburne, M. O., R. H. Tedford, and M. Archer. 1987. New Miocene ringtail possums (Marsupialia: Pseudocheiridae) from South Australia. Pp. 639-679 in M. Archer (ed.), Possums and Opossums: Studies in Evolution.

Sydney: Surrey Beatty and Sons and the Royal Zoological Society of New South Wales.

Woods, J. T. 1960. Fossiliferous fluviatile and cave deposits. Pp. 393-403 in D. Hill and A. K. Denmead (eds.), The Geology of Queensland. Journal of the Geological Society of Australia 7.

_______. 1962. Fossil marsupials and Cainozoic continental stratigraphy in Australia: a review. Memoirs of the Queensland Museum 14:41-49.

Woodward, B. H. 1909. Extinct marsupials of Western Australia. Geological Magazine 6:210-212.

_______. 1914. Further important discoveries in the Mammoth Cave. Records of the Western Australian Museum and Art Gallery 1:252.

Wopfner, H. 1997. Climatic and geodynamic significance of Cenozoic land surfaces and duricrusts of inland Australia. Proceedings of the 30th International Geological Congress 1:201-213.

Wroe, S. 2002. A review of terrestrial mammalian and reptilian carnivore ecology in Australian fossil faunas, and factors influencing their diversity: the myth of reptilian domination and its broader ramifications. Australian Journal of Zoology 50:1-24.

Wroe, S. and M. Archer. 1995. Extraordinary diphyodonty-related change in dental function for a tooth of the extinct marsupial *Ekaltadeta ima* (Propleopinae, Hypsiprymnodontidae). Archives of Oral Biology 40:597-603.

Wroe, S., J. Brammall, and B. N. Cooke. 1998. The skull of *Ekaltadeta ima* (Marsupialia, Hypsiprymnodontidae?): an analysis of some marsupial cranial features and a re-investigation of propleopine phylogeny, with notes on the inference of carnivory in mammals. Journal of Paleontology 72:738-751.

Wroe, S., J. H. Field, and R. Fullager. 2002. Lost giants. Nature Australia 27:54-61.

Wyss, A. R. 1988. On "retrogression" in the evolution of the Phocinae and phylogenetic affinities of the monk seals. American Museum Novitates 2924:1-38.

______. 1994. The evolution of body size in phocids: some ontogenetic and phylogenetic observations. Proceedings of the San Diego Society of Natural History 29:69-75.

Young, W. G., M. Stevens, and R. Jupp. 1990. Tooth wear and enamel structure in the mandibular incisors of six species of kangaroo (Marsupialia: Macropodidae). Memoirs of the Queensland Museum 28:337-347.

Plates

PLATE 1

Hadronomas puckridgi

Scale bars = 20 mm

A. NTM P98133, adult cranium, dorsal view. Alcoota, NT.

B. NTM P98133, palatal view.

A
B

PLATE 2

Hadronomas puckridgi

Scale bars = 20 mm

A. NTM P98140, partial adult cranium, lateral view. Alcoota, NT.

B. NTM P98140, mesial view.

C. NTM P98140, stereo occlusal view.

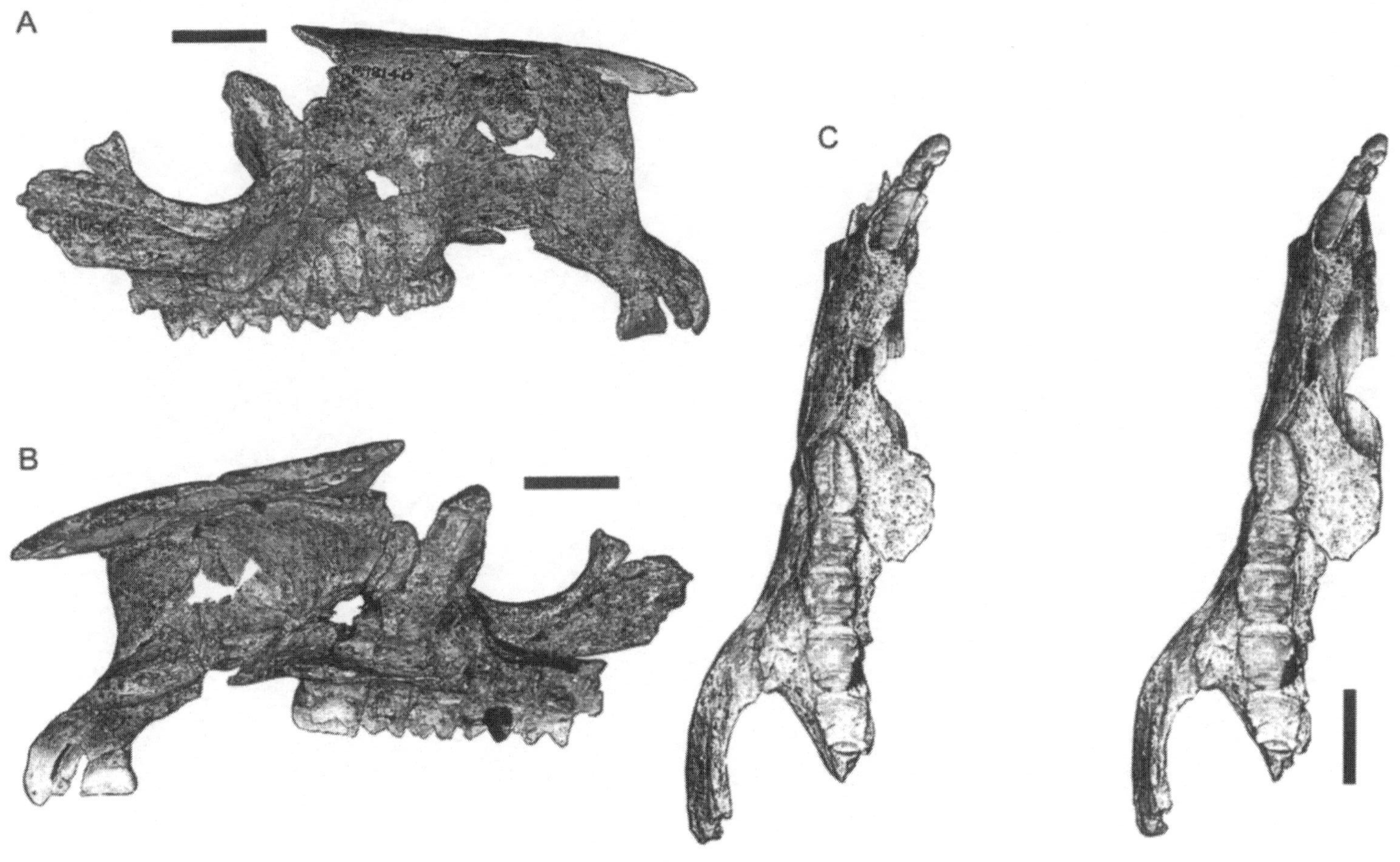
A
B
C

PLATE 3

Hadronomas puckridgi (A-C)

Hadronomas sp. (D)

Scale bars = 20 mm (A-B); 5 mm (C-D)

A. NTM P98132, adult cranium, lateral view. Alcoota, NT.
B. NTM P98132, palatal view.
C. NTM P98138, right P3, stereo occlusal view. Alcoota, NT.
D. SAM P32556, left P3, stereo occlusal view. Lake Kanunka, SA.

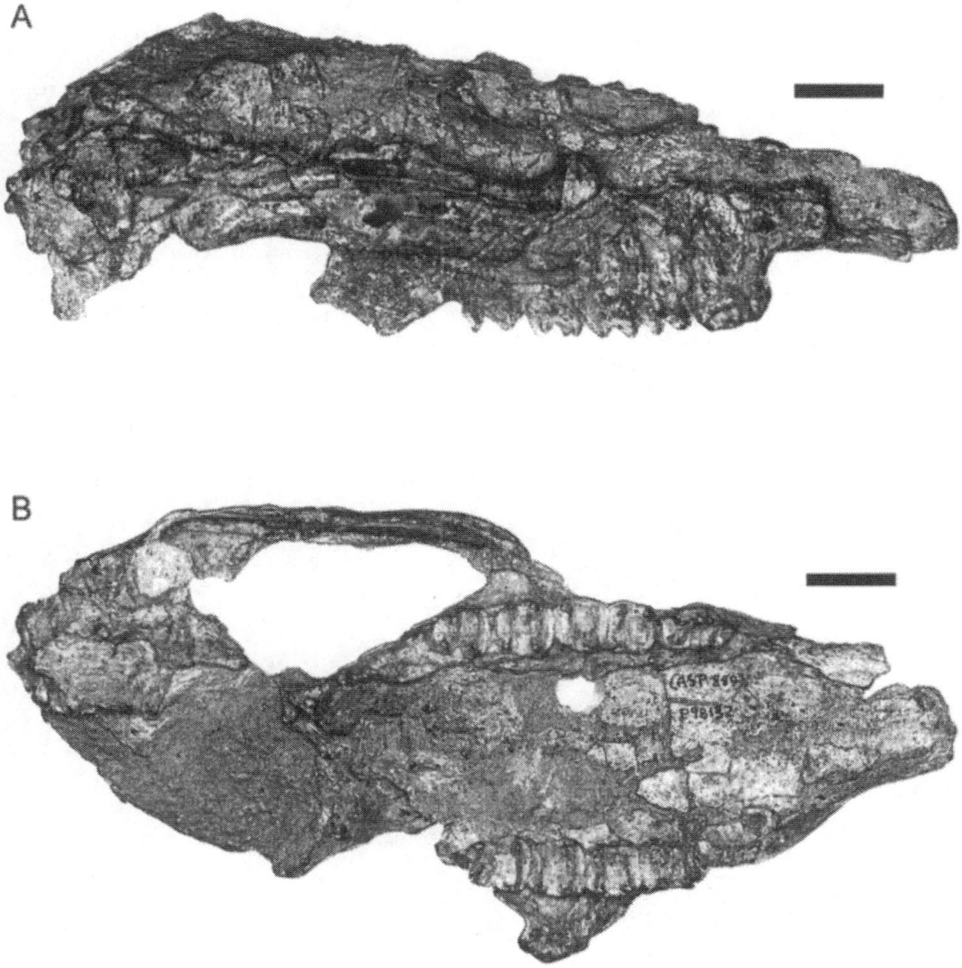
A
B
C
D

PLATE 4

Hadronomas puckridgi

Scale bars = 20 mm

A. NTM P98135, left adult dentary, lateral view. Alcoota, NT.

B. NTM P98135, mesial view.

C. NTM P98139, right adult dentary, stereo occlusal view. Alcoota, NT.

D. NTM P98134, right adult dentary, mesial view (photo reversed). Alcoota, NT.

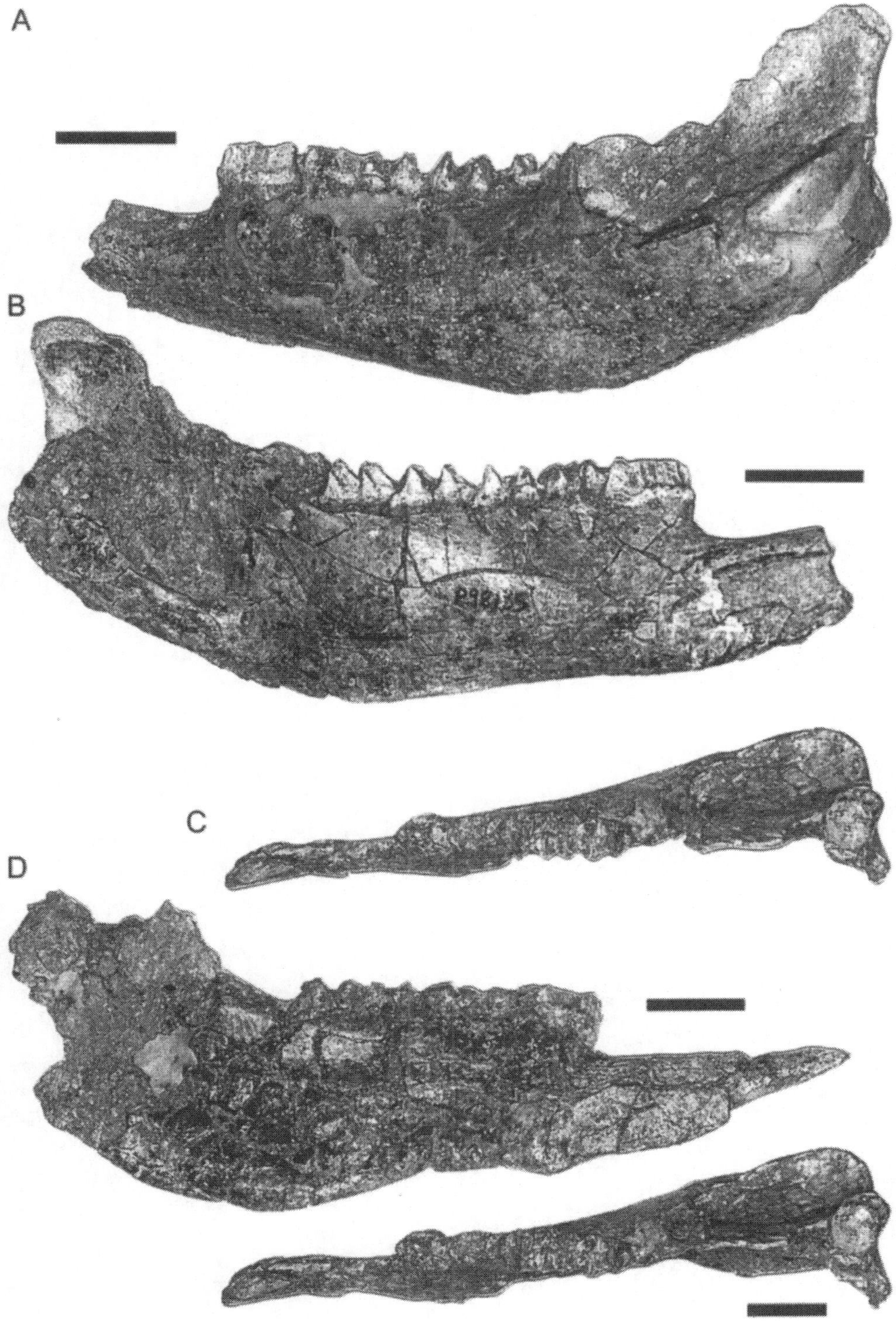
A
B
C
D

PLATE 5

Hadronomas puckridgi

Scale bars = 20 mm (A-B); 10 mm (C)

A. NTM P98142, right adult dentary, lateral view. Alcoota, NT.

B. NTM P98142, mesial view.

C. NTM P98142, stereo occlusal view.

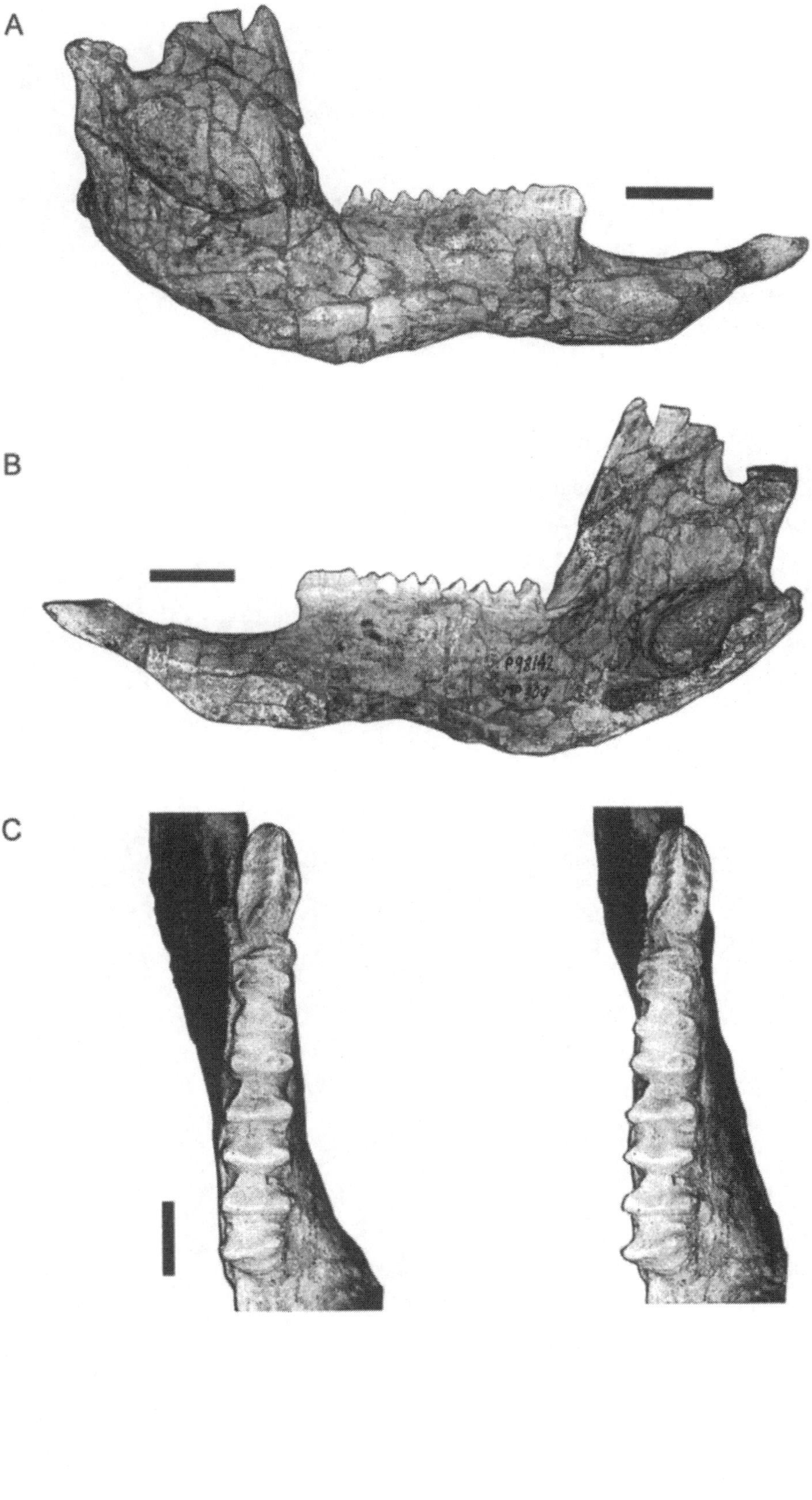
A
B
C

PLATE 6

Hadronomas puckridgi

Scale bars = 10 mm (A-B, D-E); 5 mm (C)

A. NTM P98146, right juvenile dentary, lateral view. Alcoota, NT.

B. NTM P98146, stereo occlusal view.

C. NTM P98146, right p3, stereo occlusal view. Alcoota, NT.

D. SAM P38771, right i1, buccal view. Alcoota, NT.

E. SAM P38771, lingual view.

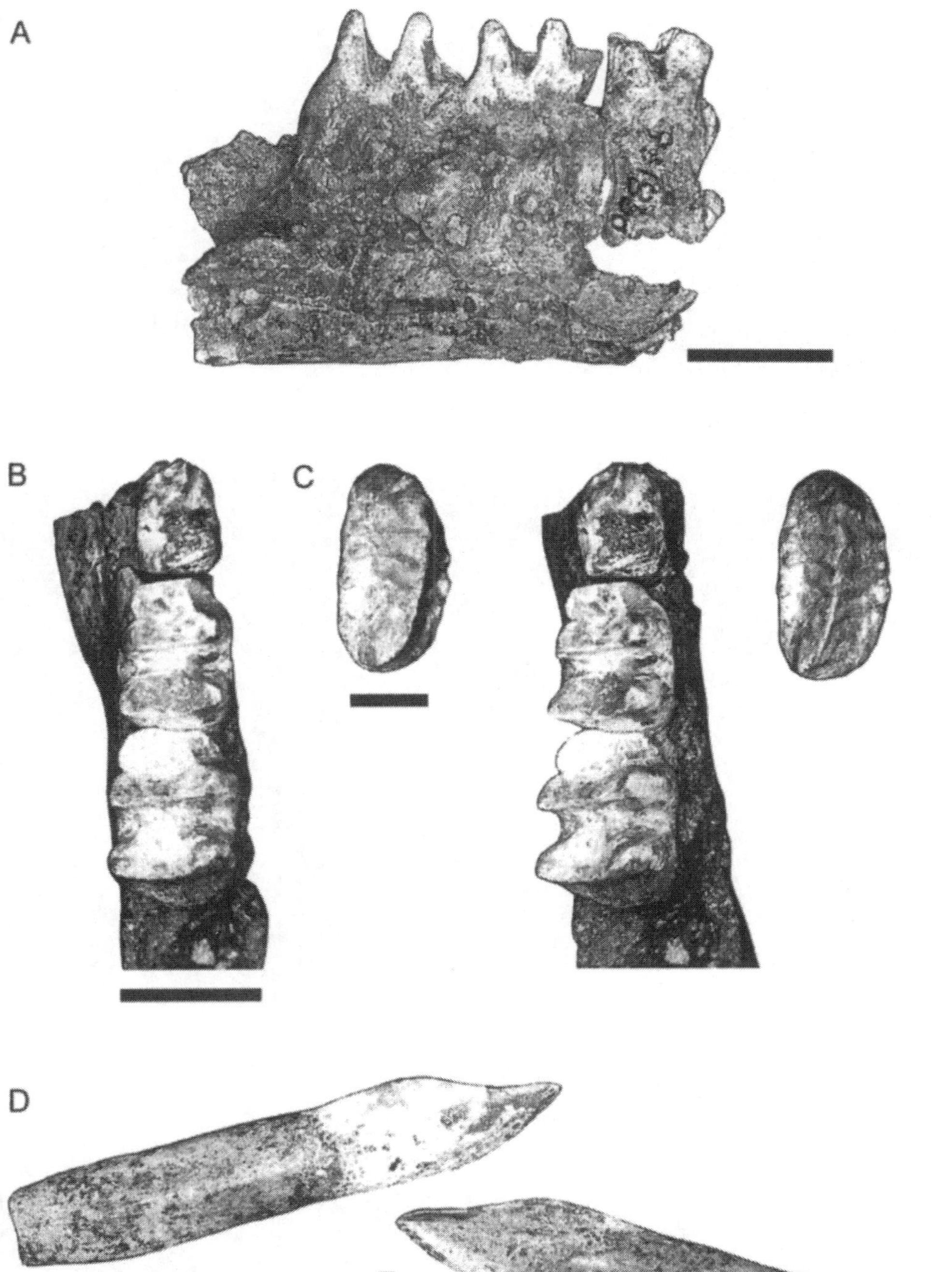
A
B
C
D
E

PLATE 7

Sthenurus andersoni

Scale bars = 20 mm

A. SAM P29570, adult cranium, lateral view. Henschke's Fossil Cave, SA.

B. SAM P29570, dorsal view.

C. SAM P29570, palatal view.

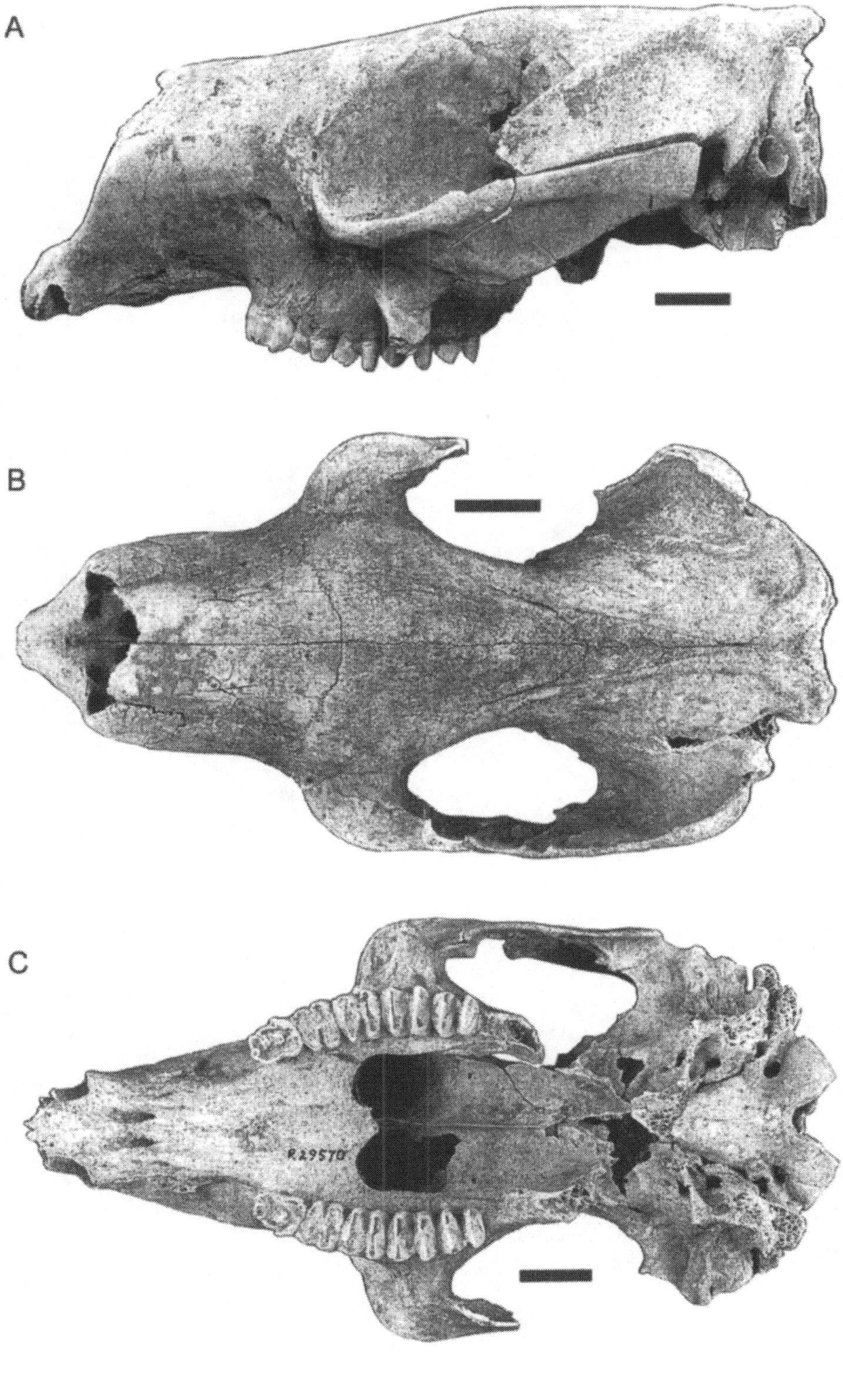
A
B
C
R29570

PLATE 8

Sthenurus andersoni

Scale bars = 20 mm

A. SAM P29570, adult cranium, anterior view. Henschke's Fossil Cave, SA.
B. SAM P31877, partial juvenile cranium, anterior view. Henschke's Fossil Cave, SA.
C. SAM P16547, juvenile cranium (M3 erupting), lateral view. Victoria Fossil Cave, SA.
D. SAM P20499, juvenile cranium (M4 erupting), lateral view. Victoria Fossil Cave, SA.
E. SAM P31877, partial juvenile cranium, lateral view (photo reversed).
F. SAM P20461, adult cranium, lateral view. Victoria Fossil Cave, SA.

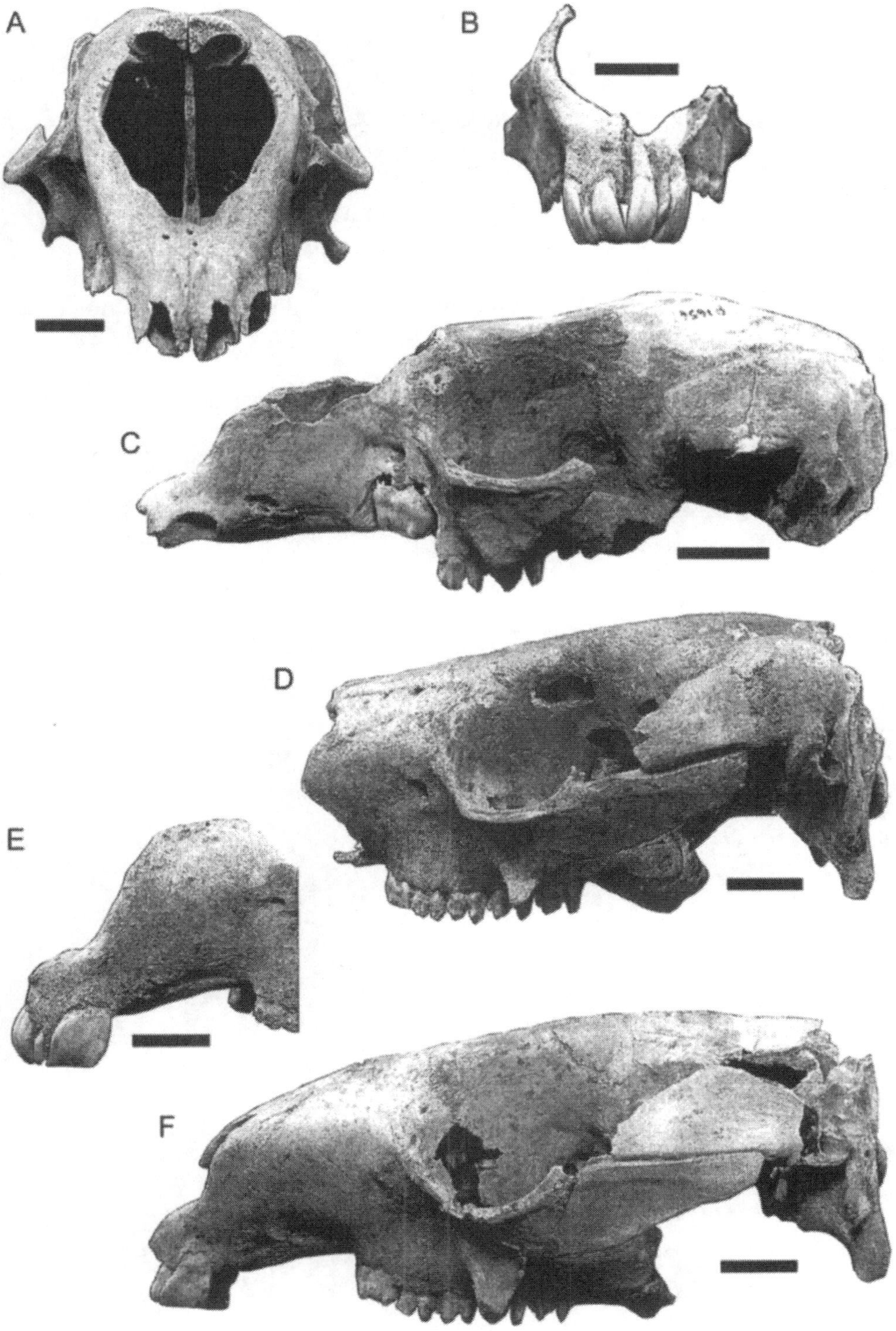
A
B
C
D
E
F

PLATE 9

Sthenurus andersoni

Scale bars = 20 mm

A. SAM P20792, juvenile cranium (M3 erupting), dorsal view. Victoria Fossil Cave, SA.
B. SAM P20499, juvenile cranium (M4 erupting), dorsal view. Victoria Fossil Cave, SA.
C. SAM P20461, adult cranium, dorsal view. Victoria Fossil Cave, SA.

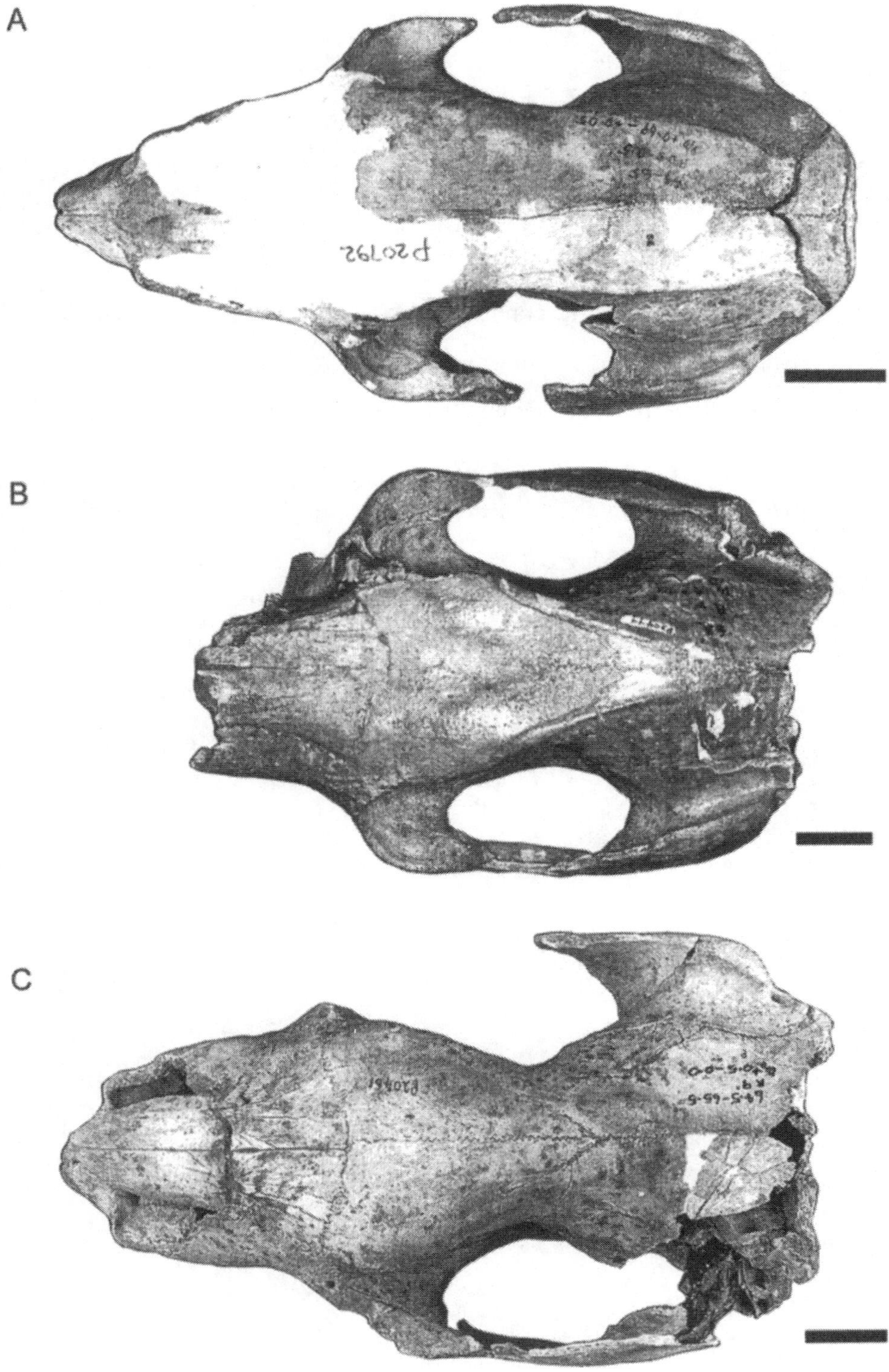
A
B
C

PLATE 10

Sthenurus andersoni

Scale bars = 20 mm

A. SAM P20792, juvenile cranium (M3 erupting), with excavated left P3, palatal view. Victoria Fossil Cave, SA.
B. SAM P20499 juvenile cranium (M4 erupting), palatal view. Victoria Fossil Cave, SA.
C. SAM P31877, partial juvenile cranium, palatal view. Henschke's Fossil Cave, SA.
D. SAM P20461, adult cranium, palatal view. Victoria Fossil Cave, SA.

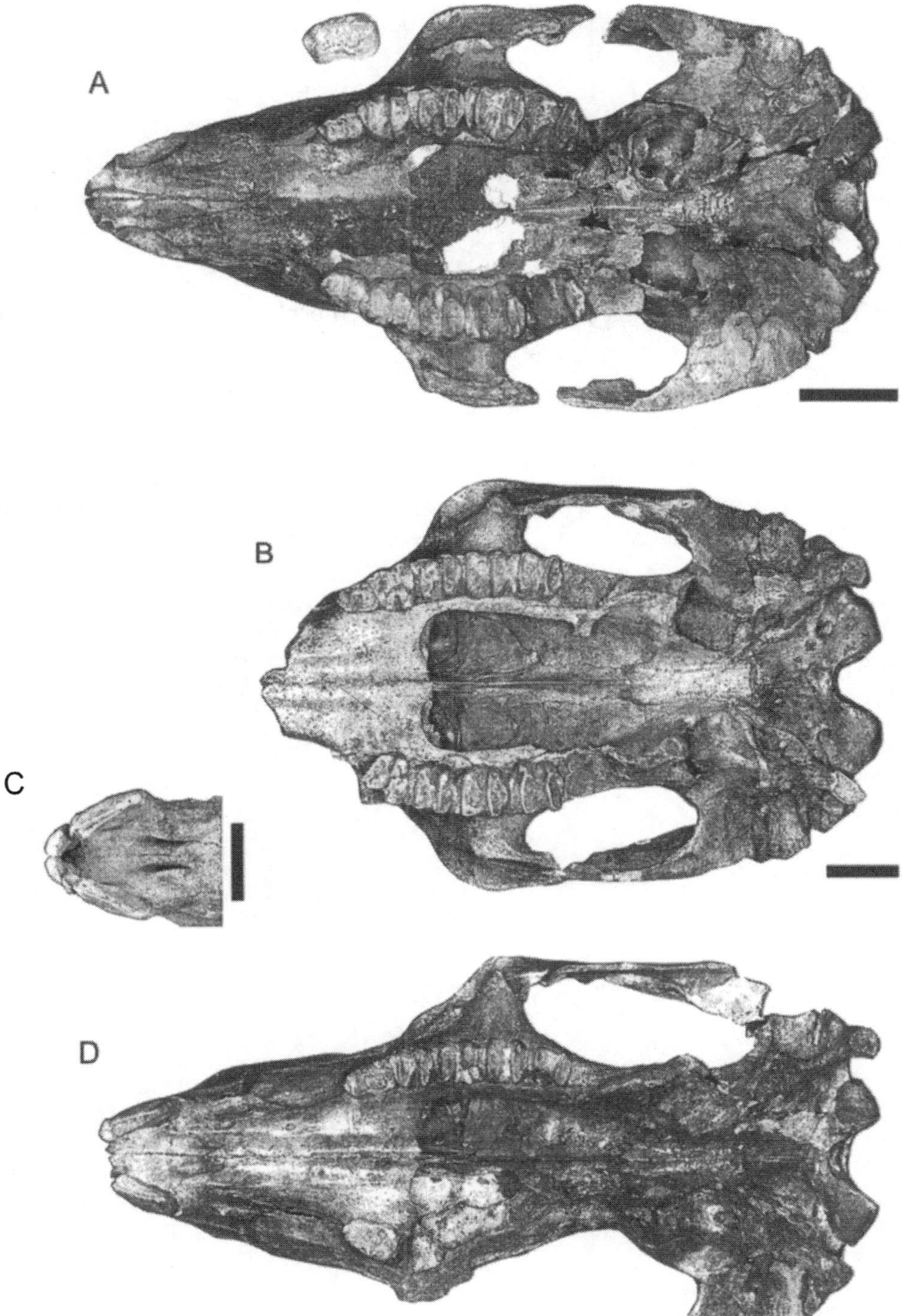
A
B
C
D

PLATE 11

Sthenurus andersoni

Scale bars = 20 mm (A, C, F); 10 mm (B, D-E)

A. SAM P16671, right juvenile maxilla, stereo occlusal view. Victoria Fossil Cave, SA.
B. AM F29590, right P3, occlusal view. Wellington Caves, NSW.
C. AM F88610, right adult maxilla, occlusal view. Wellington Caves, NSW.
D. WAM 03.11.08, excavated right P3, stereo occlusal view. Lindsay Hall Cave, WA.
E. WAM 03.11.08, right juvenile maxilla, stereo occlusal view.
F. SAM P40121, right adult maxilla, stereo occlusal view. Henschke's Fossil Cave, SA.

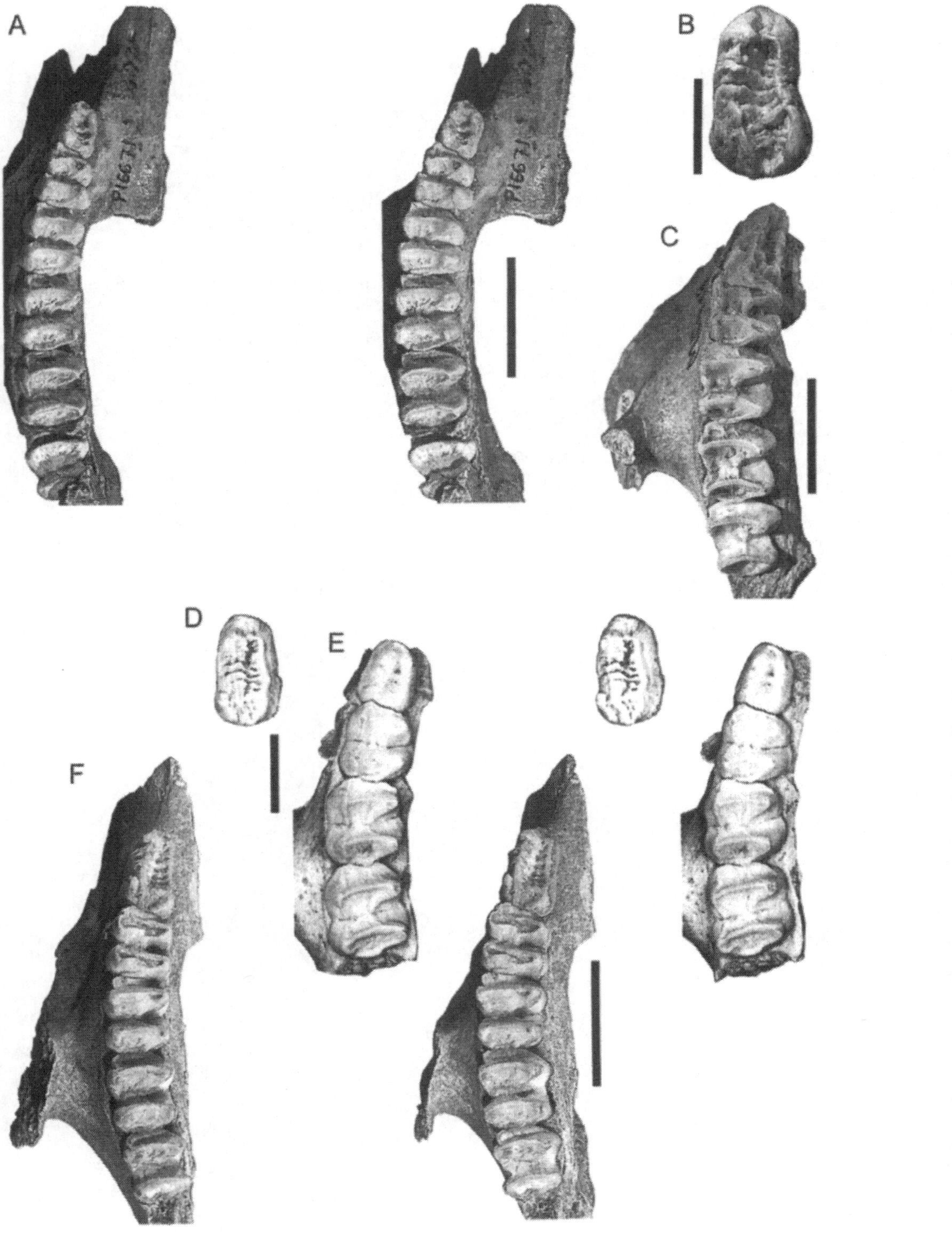
A
B
C
D
E
F

PLATE 12

Sthenurus andersoni

Scale bars = 20 mm

A. SAM P40120, left juvenile dentary, mesial view. Henschke's Fossil Cave, SA.
B. SAM P40120, lateral view.
C. SAM P16628, left juvenile dentary, mesial view. Victoria Fossil Cave, SA.
D. SAM P16628, lateral view.
E. WAM 03.11.02, left dentary, mesial view. Tight Entrance Cave, WA.
F. WAM 03.11.02, lateral view.
G. AM F29578, right dentary, lateral view. Wellington Caves, NSW.
H. AM F29578, mesial view.

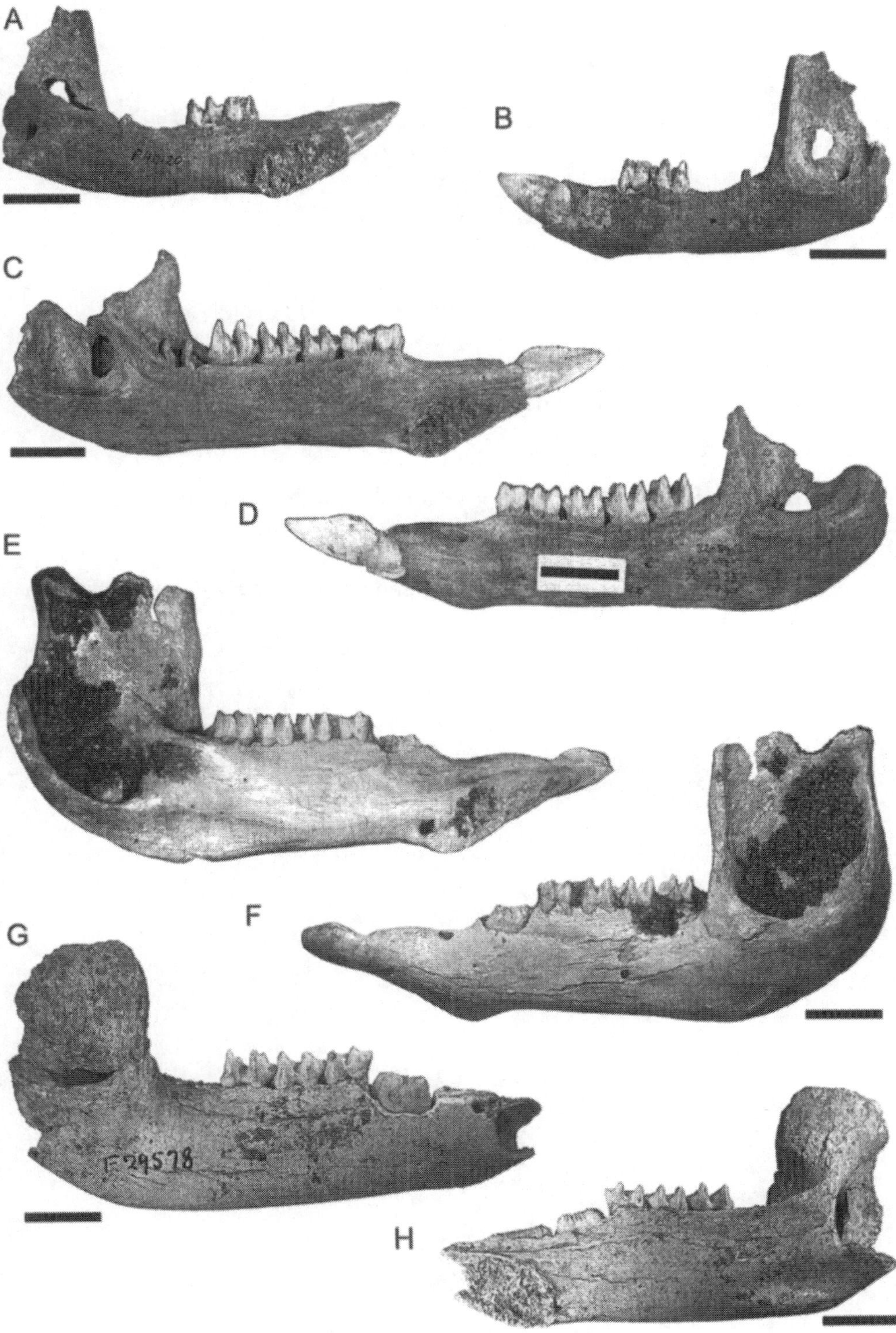
A
B
C
D
E
F
G
H
F29578

PLATE 13

Sthenurus andersoni

Scale bars = 20 mm

A. AM F88545, right adult dentary, lateral view. Bingara, NSW.

B. AM F88545, mesial view.

C. SAM P18501, left adult dentary, mesial view. Henschke's Fossil Cave, SA.

D. SAM F18501, lateral view.

C. SAM P34820, right adult dentary, lateral view. Normanville, SA.

F. SAM P34820, mesial view.

G. SAM P20496, right adult dentary, lateral view. Victoria Fossil Cave, SA.

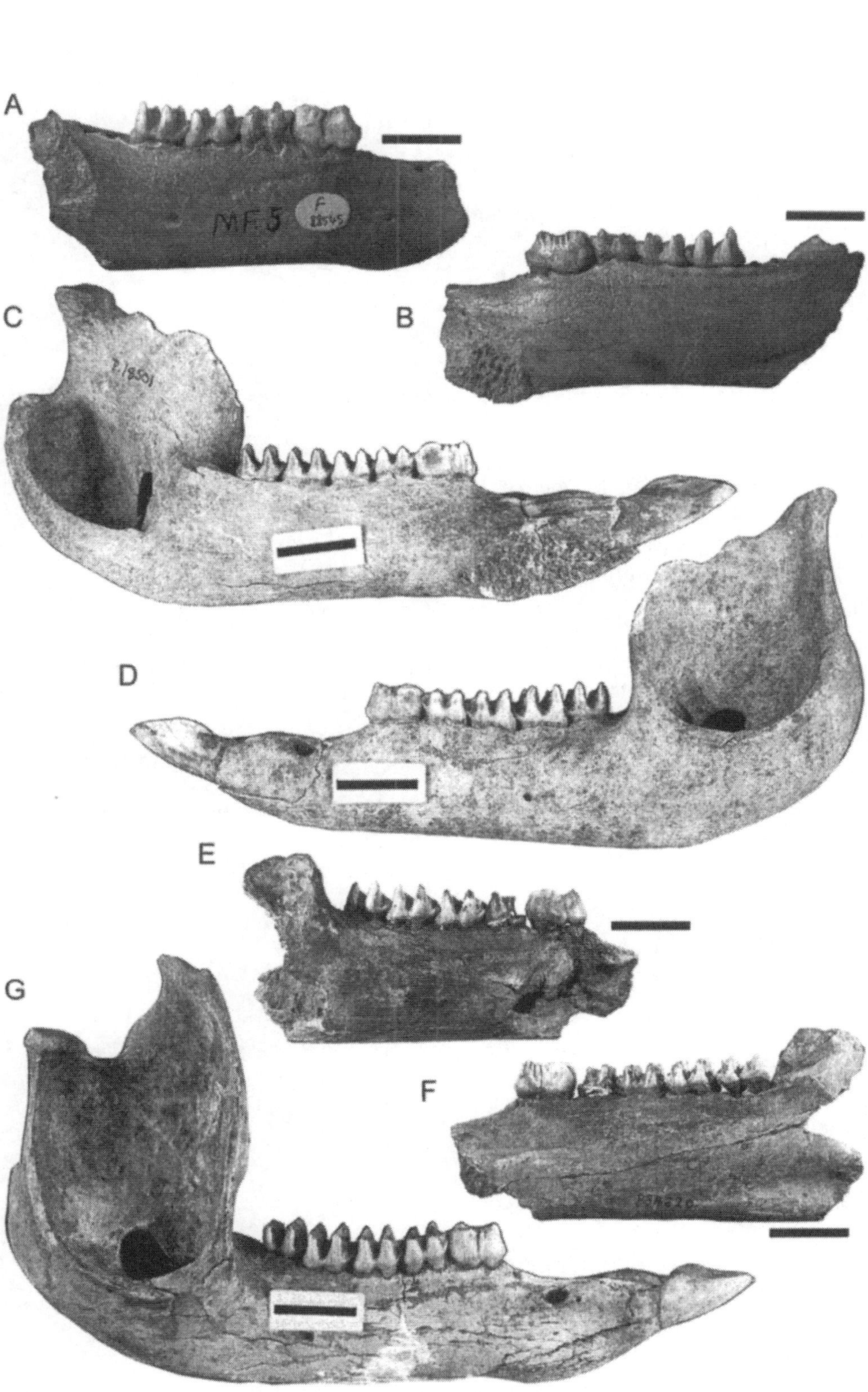
A
MF.5
B
C
D
E
F
G

PLATE 14

Sthenurus andersoni

Scale bars = 20 mm

A. FU 0463, right juvenile dentary, stereo occlusal view. Victoria Fossil Cave, SA.
B. WAM 03.11.02, left dentary, stereo occlusal view. Tight Entrance Cave, WA.
C. AM F29578, right p3-m1, occlusal view. Wellington Caves, NSW.
D. AM F88545, right adult dentary, occlusal view. Bingara, NSW.
E. SAM P18501, left adult dentary, stereo occlusal view. Henschke's Fossil Cave, SA.
F. AM F88605, right adult dentary, occlusal view. Wellington Caves, NSW.
G. SAM P34820, right adult dentary, occlusal view. Normanville, SA.

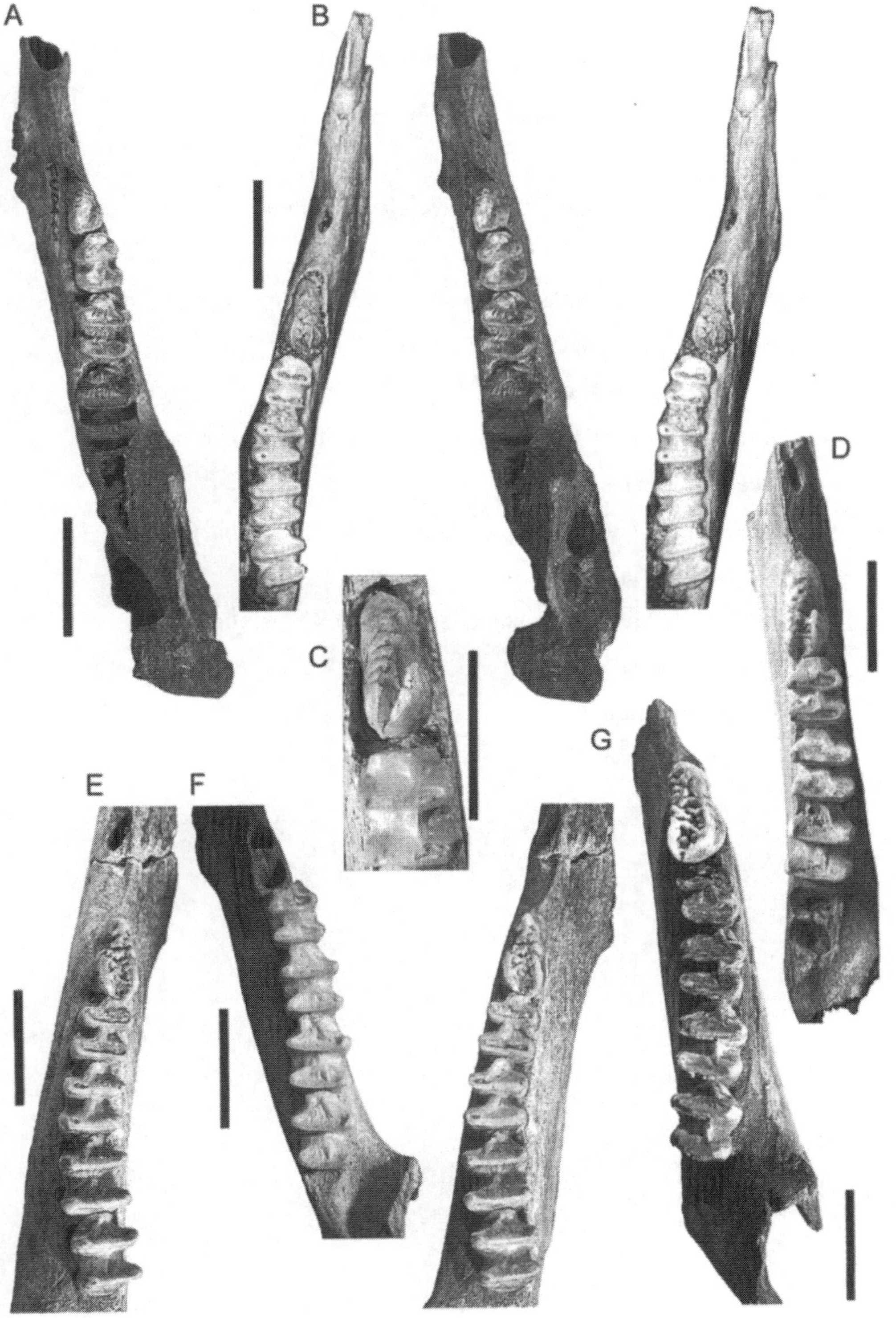
A
B
C
D
E
F
G

PLATE 15

Sthenurus andersoni

Scale bars = 20 mm (A, C-K); 10 mm (B)

A. QM F3813, left juvenile dentary, lateral view. King Creek, Pilton, QLD.

B. QM F3813, excavated p3, occlusal view.

C. QM F3813, occlusal view.

D. QM F3811, right adult dentary, lateral view. Darling Downs, QLD.

E. QM F3811, occlusal view.

F. QM F814, left juvenile dentary, lateral view. Chinchilla, QLD.

G. QM F814, mesial view.

H. QM F813, left juvenile dentary, lateral view. Chinchilla, QLD.

I. QM F813, mesial view.

J. QM F814, occlusal view.

K. QM F813, occlusal view.

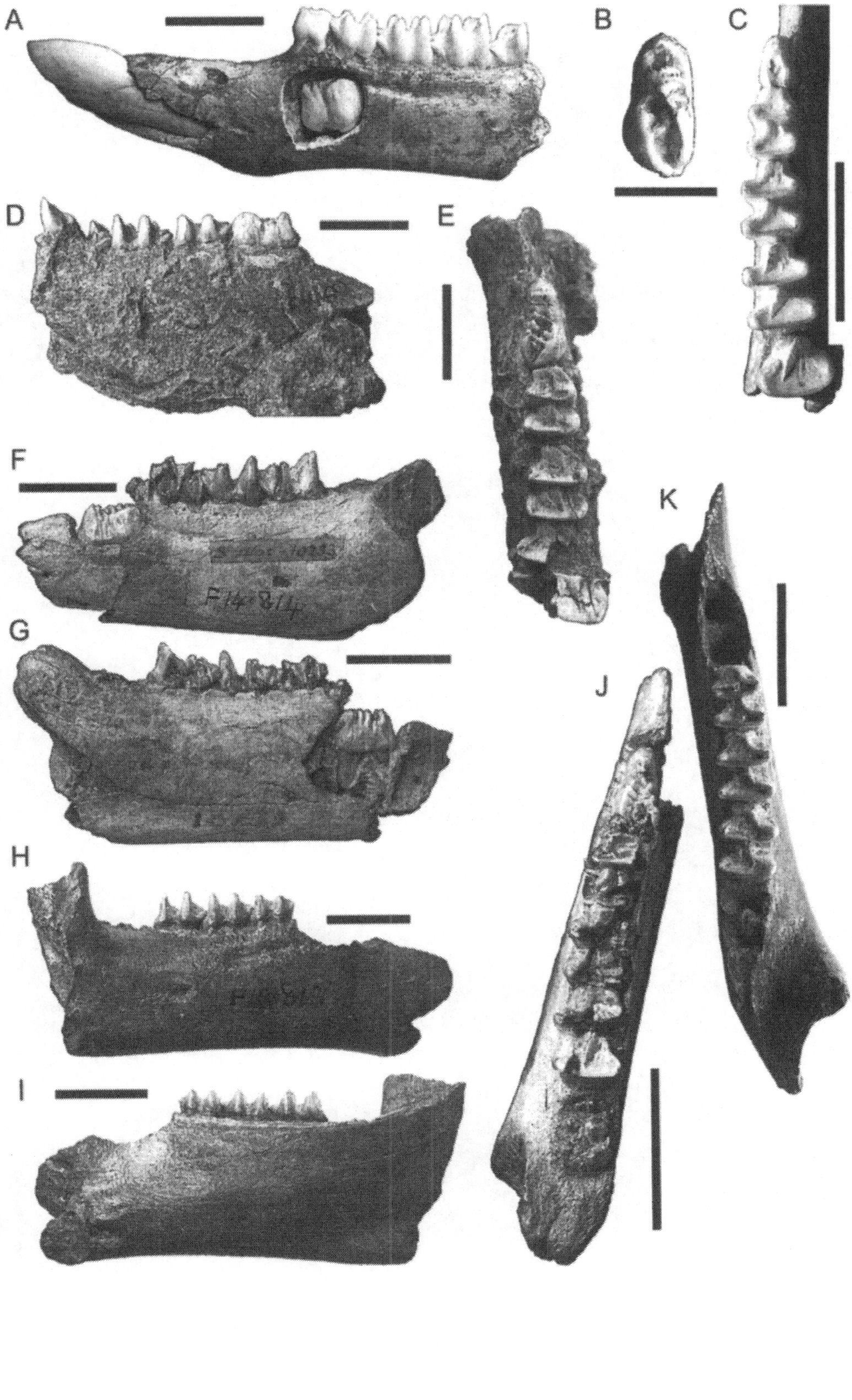
A
B
C
D
E
F
G
H
I
J
K

PLATE 16

Sthenurus atlas

Scale bars = 20 mm

A. AM F121396, partial adult cranium, lateral view. Wellington Caves, NSW.

B. AM F121396, dorsal view.

C. AM F121396, palatal view.

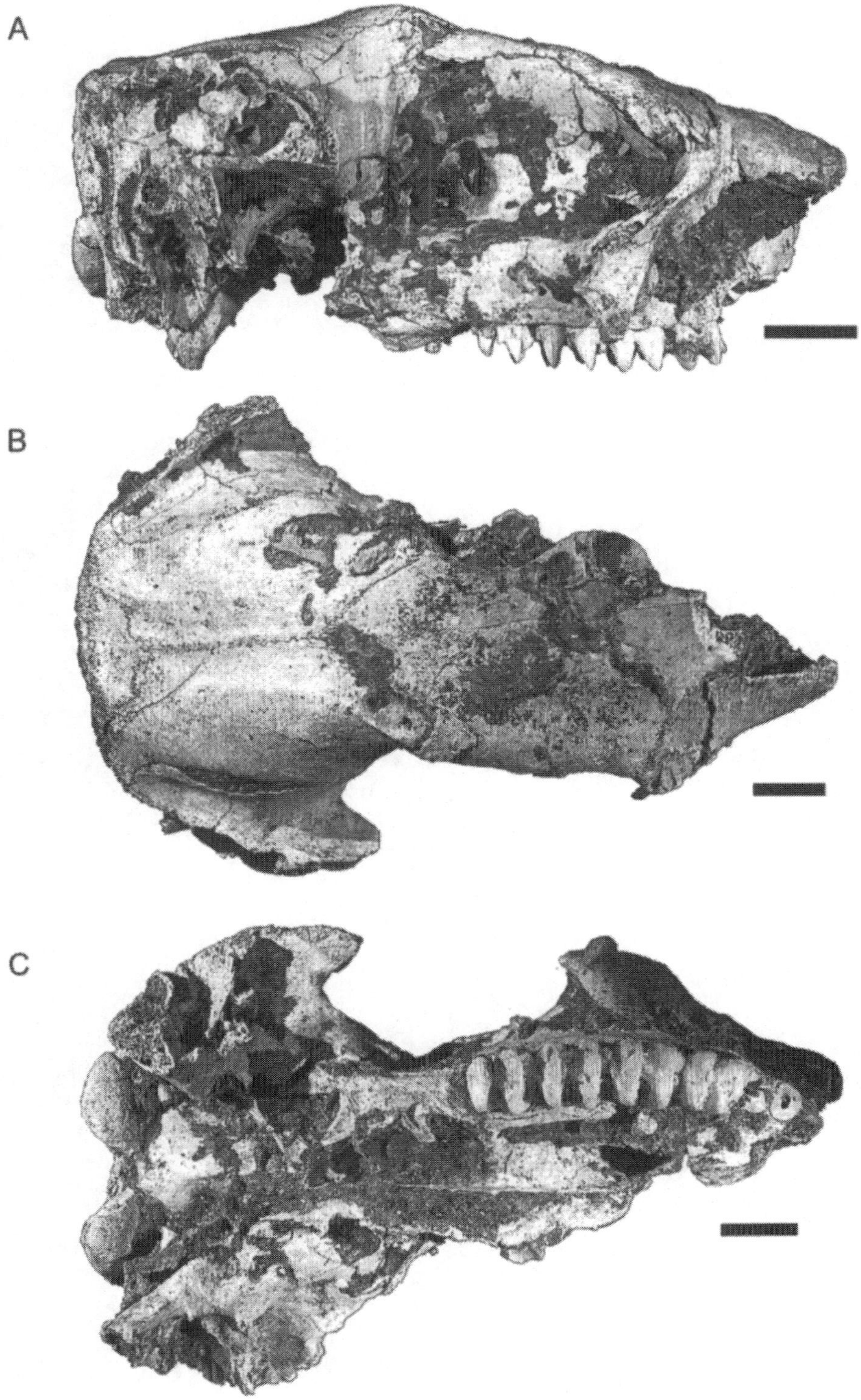
A
B
C

PLATE 17

Sthenurus atlas

Scale bars = 20 mm

A. AM F88550, partial juvenile cranium, lateral view. Wellington Caves, NSW.

B. AM F88550, dorsal view.

C. AM F88550, palatal view.

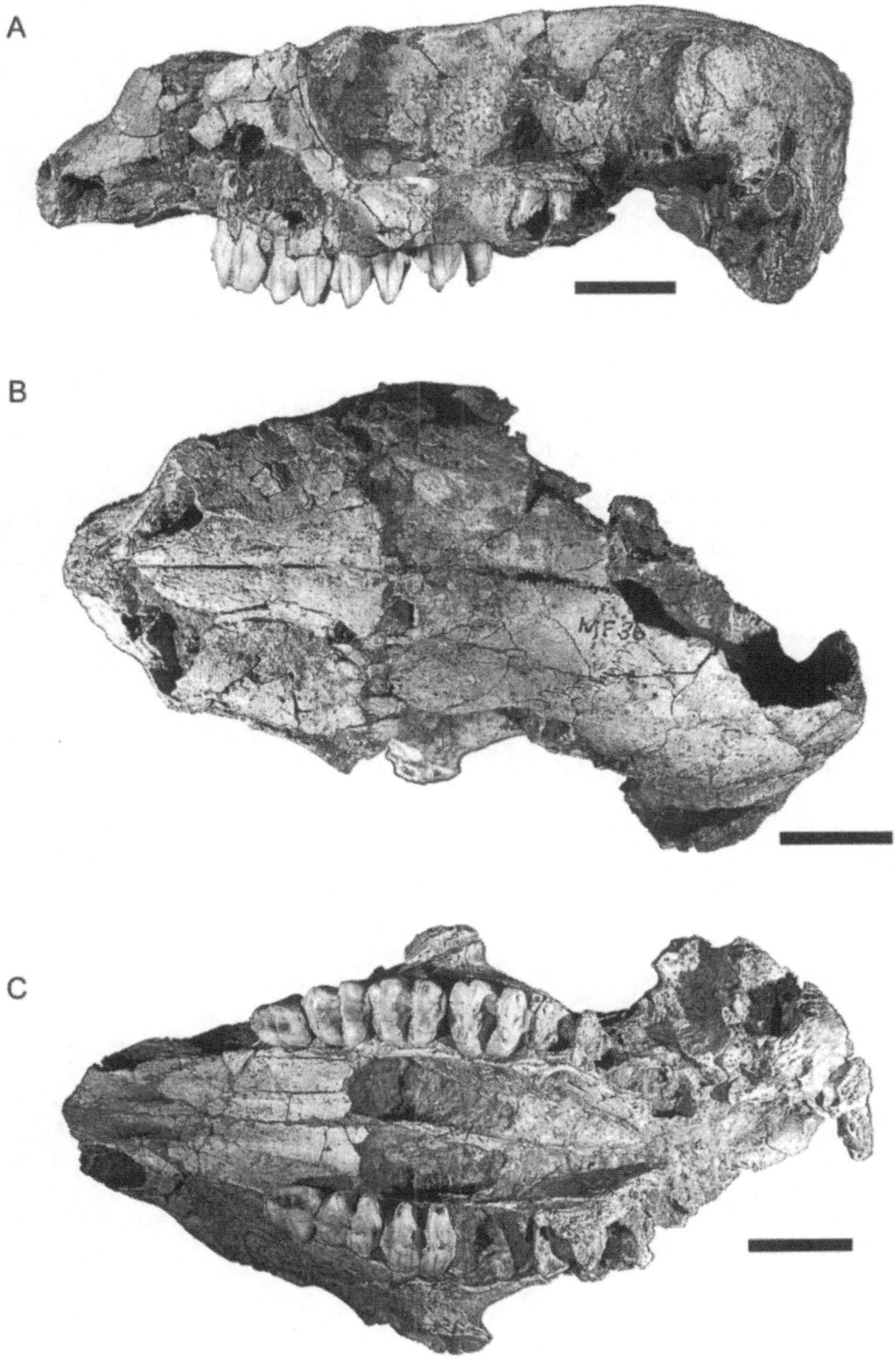
A
B
MF36
C

PLATE 18

Sthenurus atlas

Scale bars = 20 mm (A-B, F-G); 10 mm (C-E)

A. AM F47086, partial adult cranium, lateral view. Wellington Caves, NSW.

B. AM F47086, palatal view.

C. AM F30623, right P3, occlusal view. Wellington Caves, NSW.

D. AM F29591, left P3, occlusal view. Wellington Caves, NSW.

E. AM F29593, left P3, occlusal view. Wellington Caves, NSW.

F. AM F106044, left juvenile maxilla, occlusal view. Bone Cave, NSW.

G. UCMP 57394, right adult maxilla, stereo occlusal view. Wellington Caves, NSW.

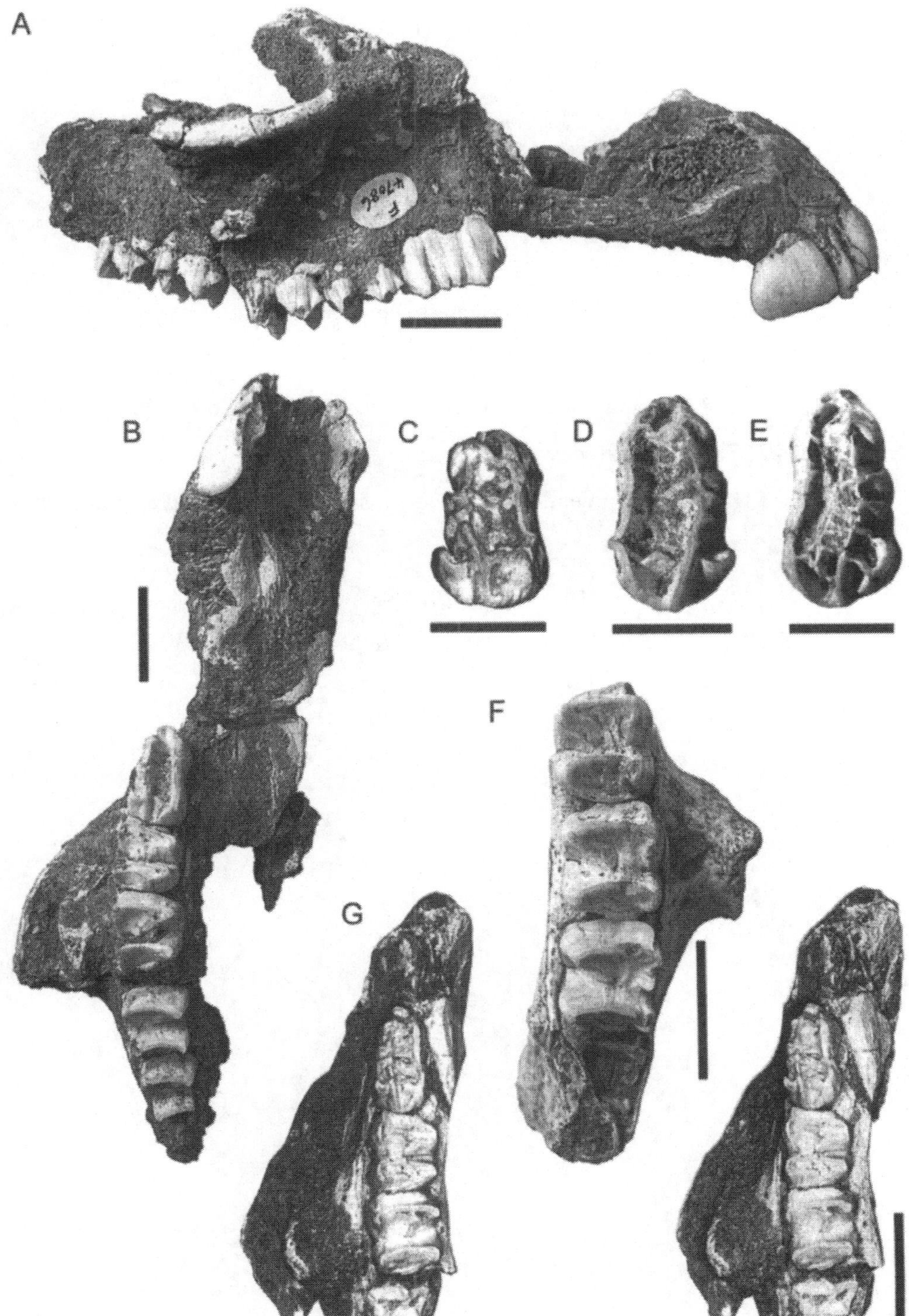
A
B
C
D
E
F
G

PLATE 19

Sthenurus atlas

Scale bars = 20 mm (A-F); 10 mm (G-H)

A. AM F47052, left adult dentary, lateral view. Wellington Caves, NSW.
B. AM F47052, mesial view.
C. QM F5767, left adult dentary, lateral view. Freestone Creek, QLD.
D. QM F5767, occlusal view.
E. AM F106049, left juvenile dentary, occlusal view. Bone Cave, Wellington, NSW.
F. AM F47052, left and right adult dentaries, occlusal view.
G. AM F88553, right p3, buccal view. Wellington Caves, NSW.
H. AM F88553, occlusal view.

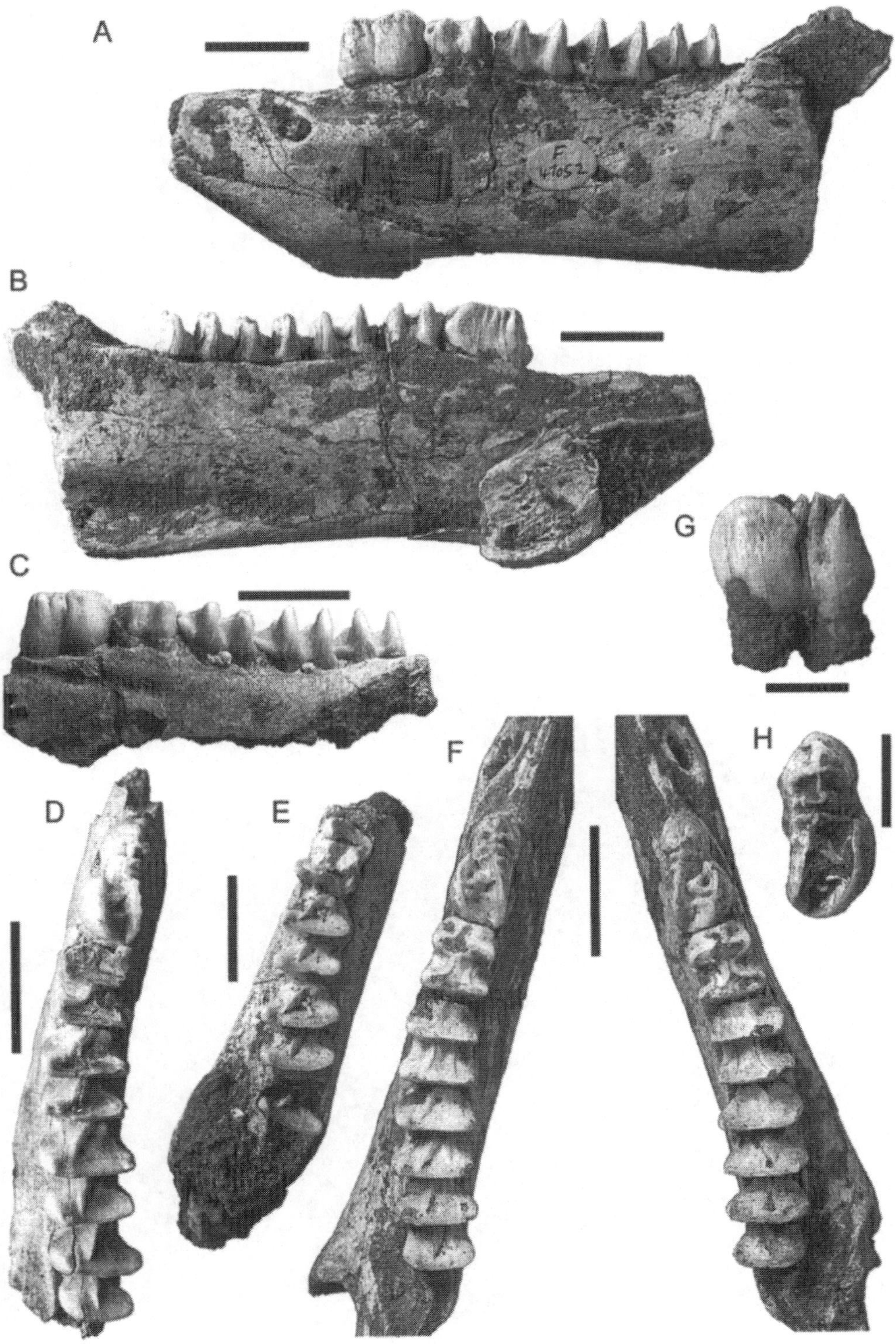
A
F
41052
B
C
G
D
E
F
H

PLATE 20

Sthenurus atlas

Scale bars = 20 mm

A. AM F106047, right adult dentary, lateral view. Bone Cave, NSW.
B. AM F106047, mesial view.
C. SAM P25453, right adult dentary, lateral view. Warburton River, SA.
D. SAM P25453, left and right adult dentaries, stereo occlusal view.
E. AM F29579, left adult dentary, occlusal view. Wellington Caves, NSW.

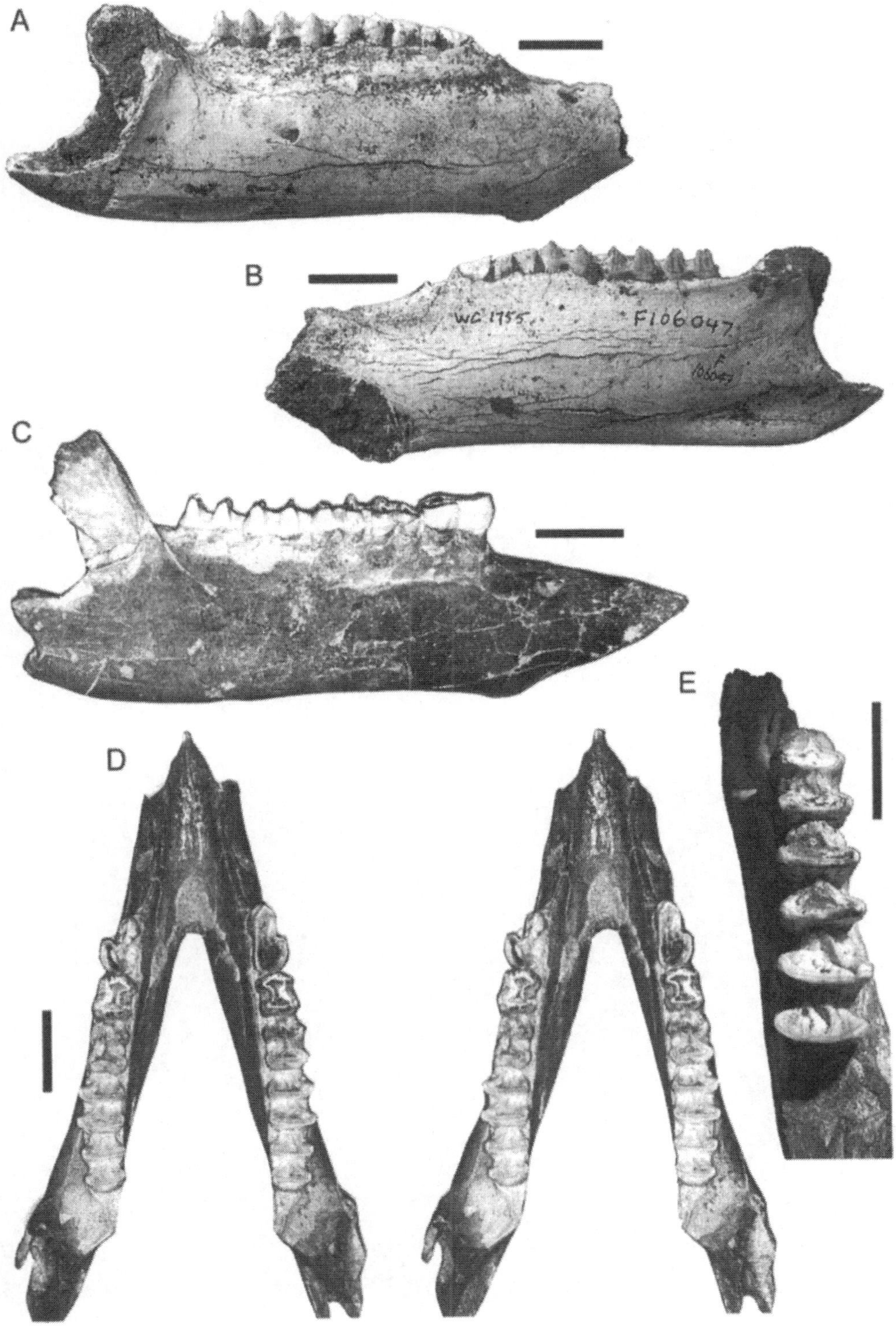
A
B
WC 1755
F106047
C
D
E

PLATE 21

Sthenurus atlas

Scale bars = 20 mm

A. AM F29573, excavated left p3, buccal view. Wellington Caves, NSW.

B. AM F29573, excavated left p3, occlusal view.

C. AM F29573, left juvenile dentary, lateral view.

D. AM F29573, mesial view.

E. AM F29573, occlusal view.

F. UCMP 52429, left juvenile dentary, lateral view. Wellington Caves, NSW.

G. UCMP 52429, mesial view.

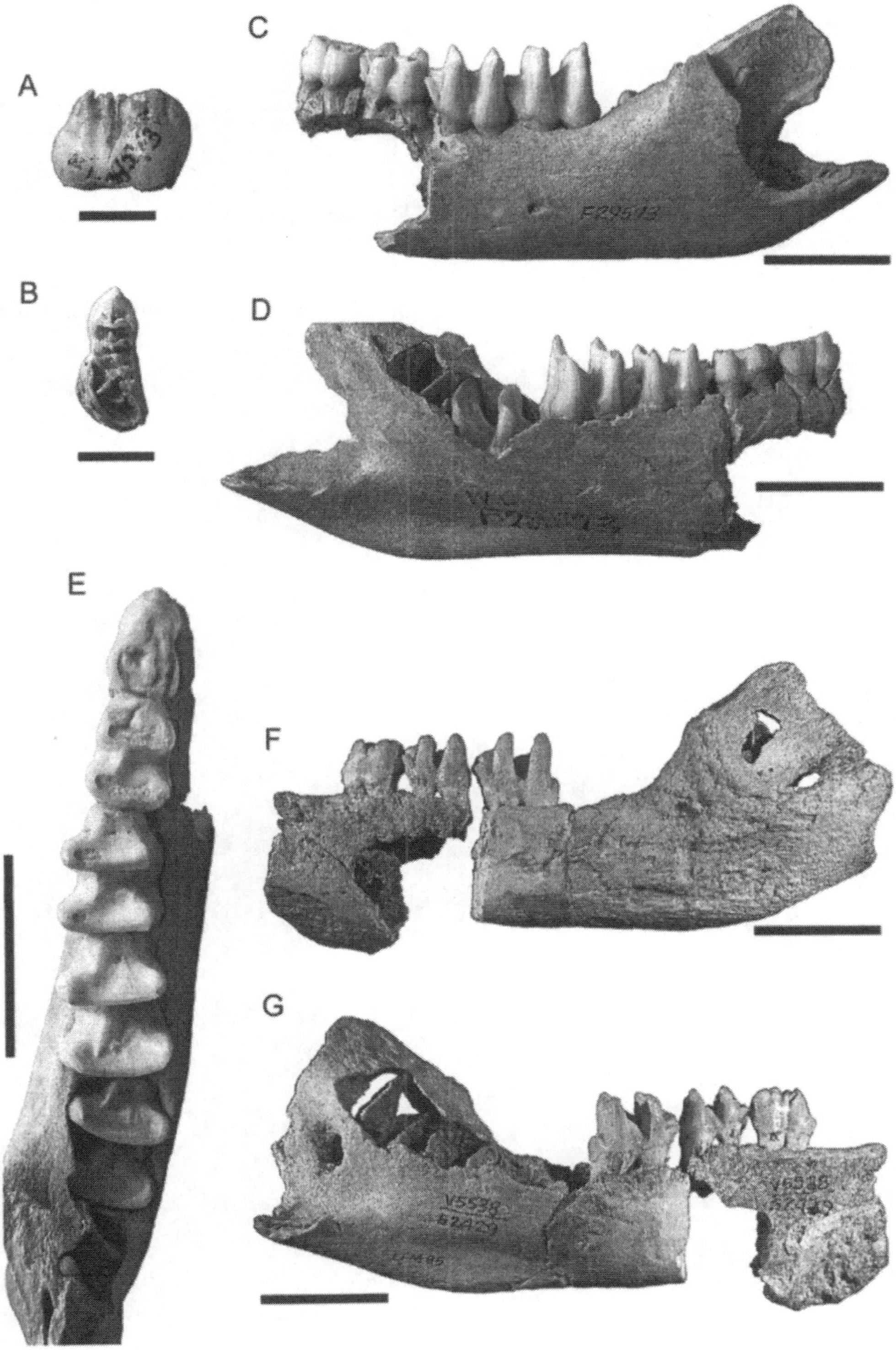
A
B
C
D
E
F
G

PLATE 22

Sthenurus murrayi

Scale bars = 20 mm (A, D); 10 mm (B-C)

A. NMV P28267 (holotype), partial juvenile cranium, lateral view. Lake Victoria, NSW.

B. NMV P28267, excavated left P3, buccal view.

C. NMV P28267, excavated left P3, stereo occlusal view.

D. NMV P28267, stereo palatal view.

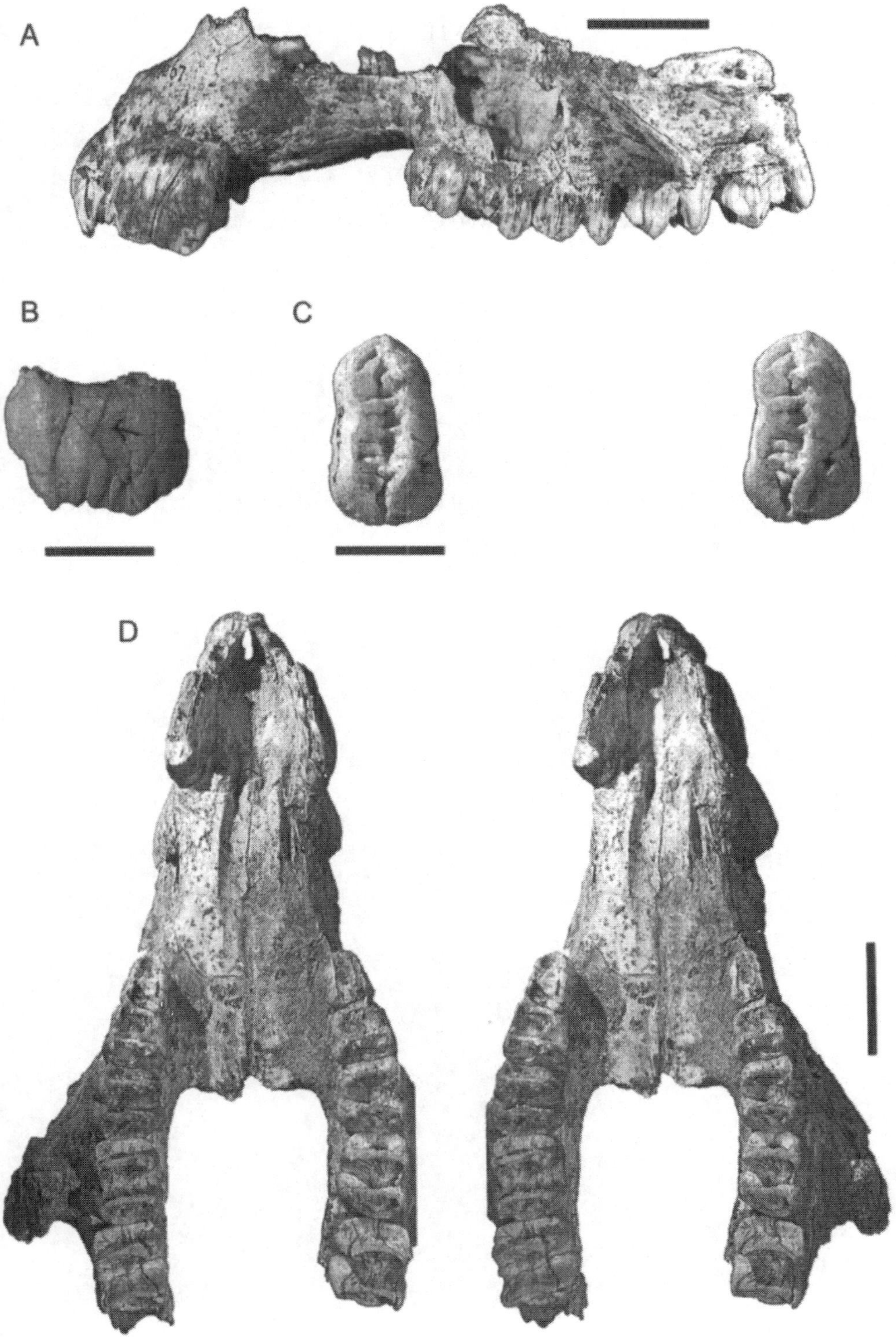
A
B
C
D

PLATE 23

Sthenurus notabilis

Scale bars = 20 mm (A-C); 10 mm (D-F)

A. QM F3817 (holotype), right juvenile dentary, lateral view. Chinchilla, QLD.
B. QM F3817, mesial view.
C. QM F3817, occlusal view.
D. QM F3817, excavated p3, buccal view.
E. QM F3817, excavated p3, lingual view.
F. QM F3817, excavated p3, occlusal view.

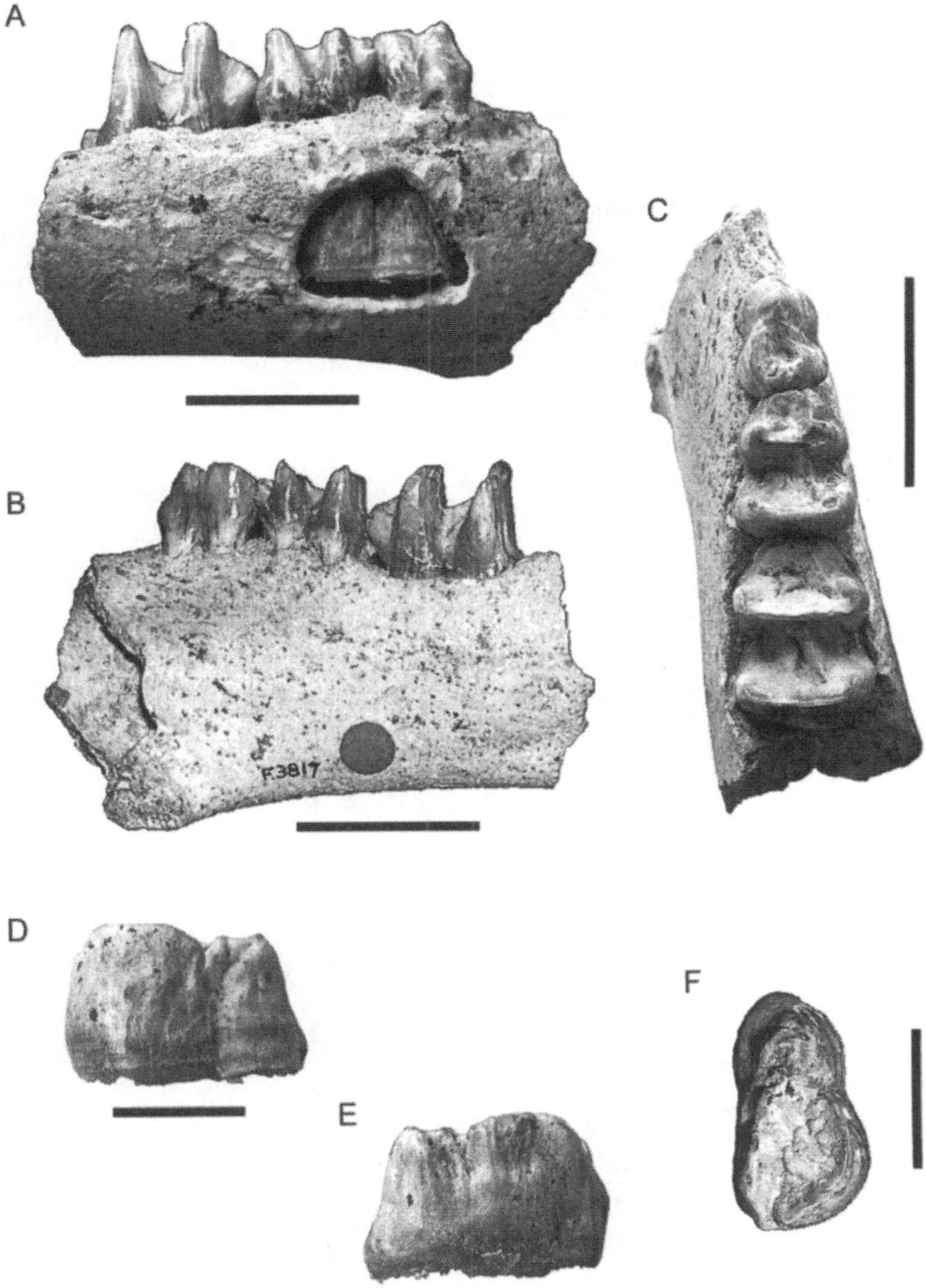
A
B
C
D
E
F
F.3817

PLATE 24

Sthenurus stirlingi

Scale bars = 20 mm

A. SAM P22533 (holotype), adult cranium, lateral view. Lake Callabonna, SA.
B. SAM P22533, dorsal view.
C. SAM P22533, palatal view.

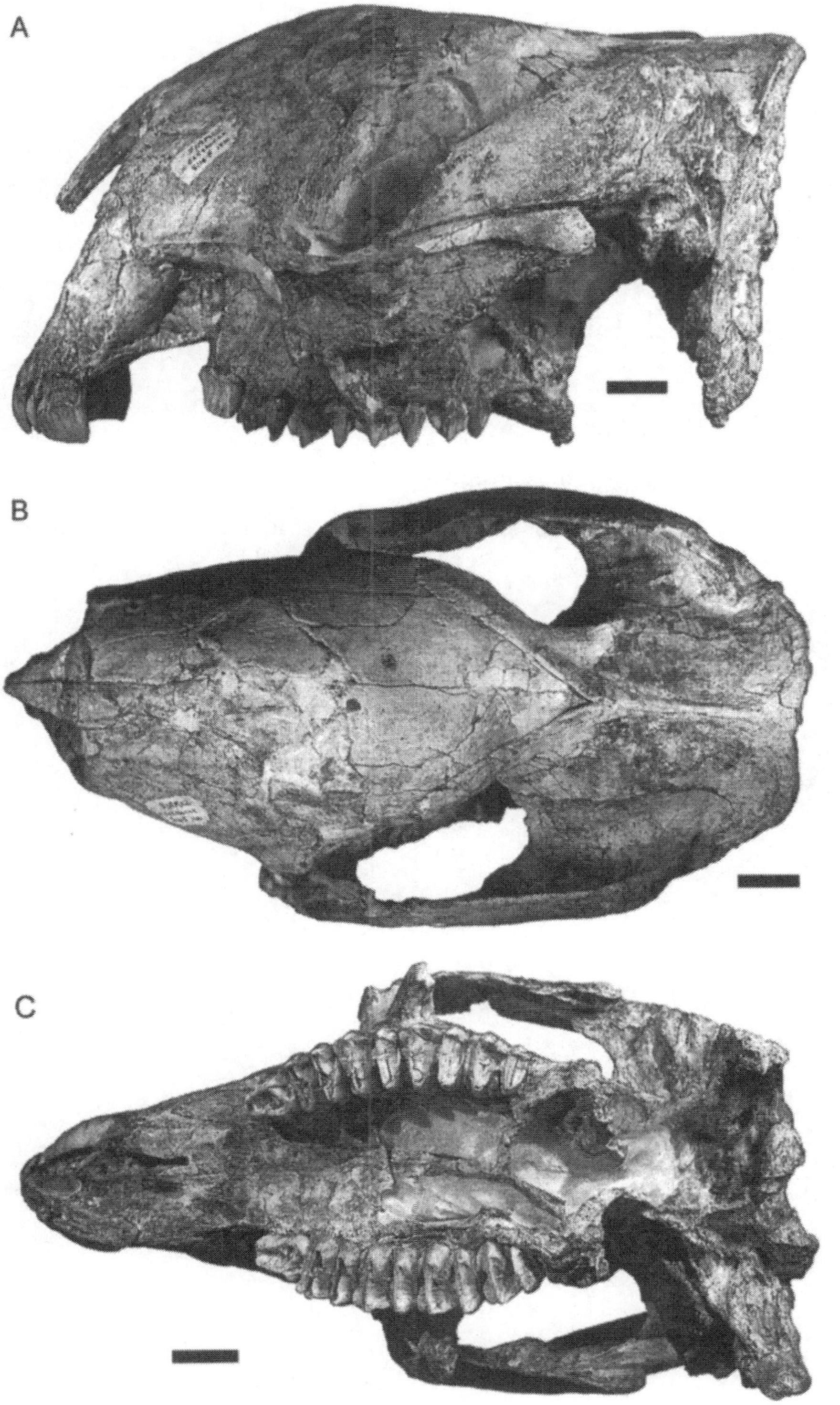
A
B
C

PLATE 25

Sthenurus stirlingi

Scale bars = 20 mm

A. SAM P22533 (holotype), left adult dentary, mesial view. Lake Callabonna, SA.
B. SAM P22533, lateral view.
C. SAM P22533, stereo occlusal view.
D. SAM P22533, upper left cheek tooth row (cast), stereo occlusal view.

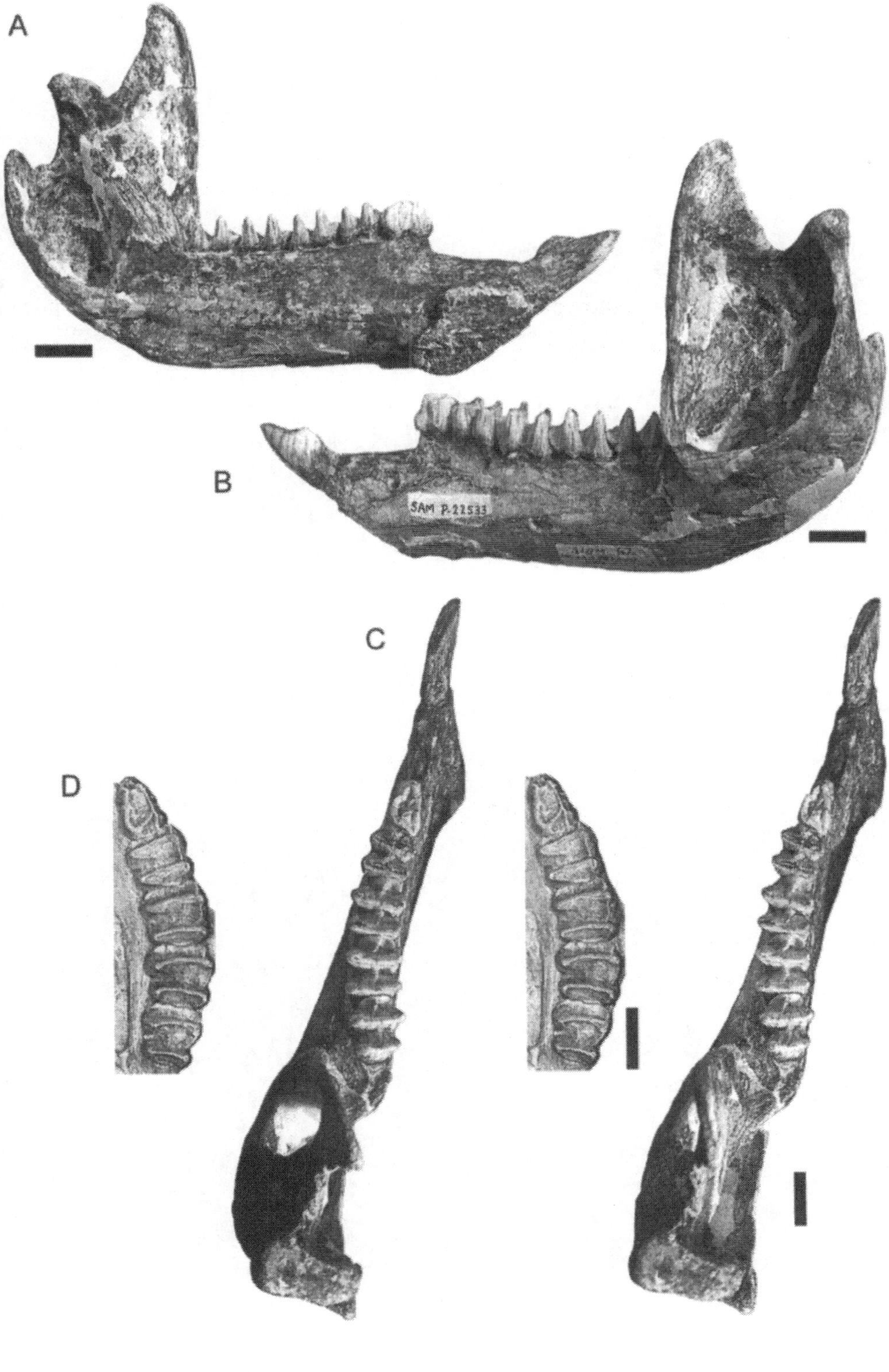
A
B
SAM P.22533
C
D

PLATE 26

Sthenurus tindalei

Scale bars = 20 mm

A. SAM P13820 (holotype), adult cranium, lateral view. Lake Menindee, NSW.
B. SAM P13820, dorsal view (dorsal surface removed by post-depositional erosion).
C. SAM P13820, palatal view.

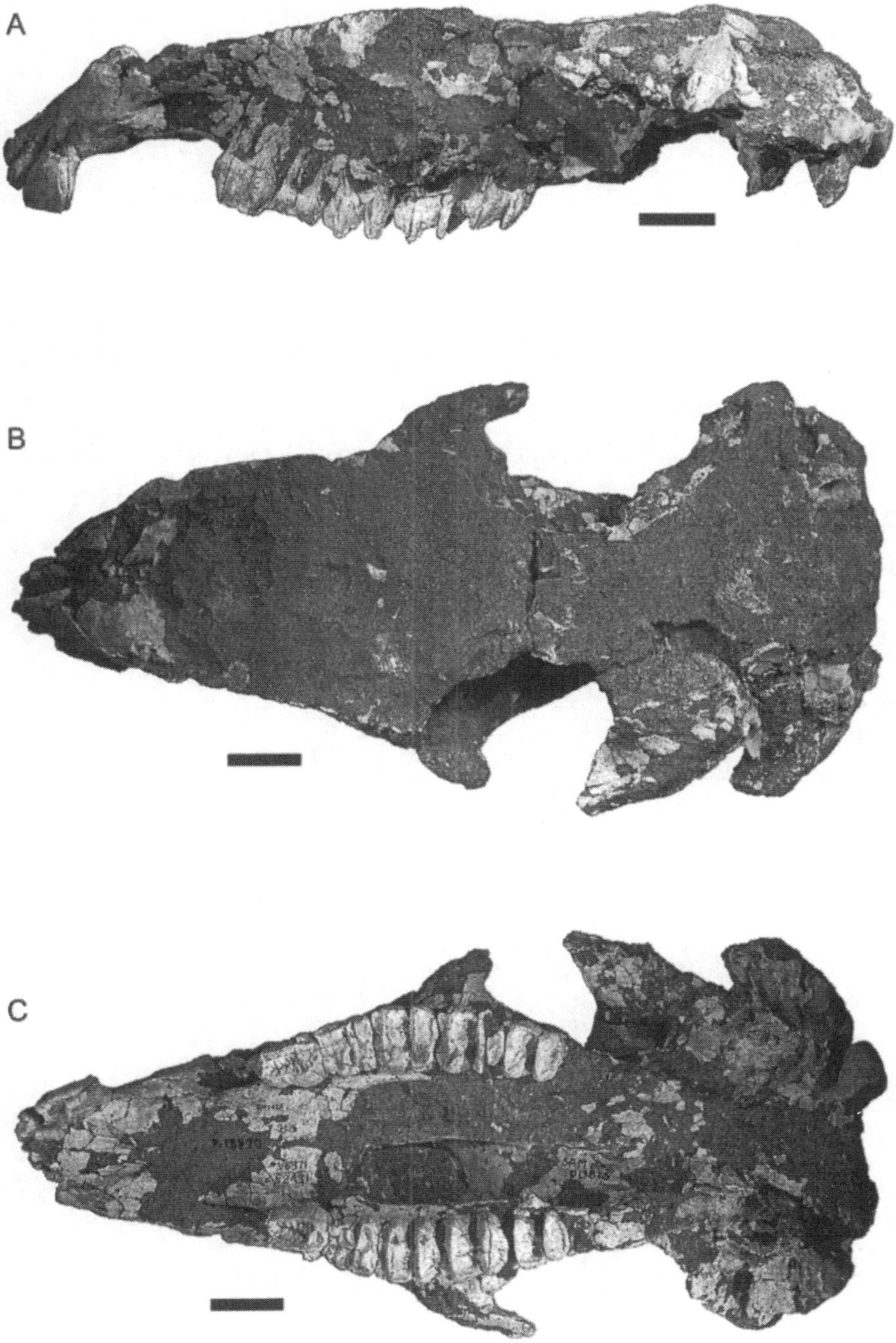
A
B
C

PLATE 27

Sthenurus tindalei

Sthenurus sp. cf. *tindalei*

Scale bars = 20 mm

A. AMNH 117499, left adult maxilla (cast), stereo occlusal view. Lake Callabonna, SA.
B. UCMP 56928, right juvenile maxilla, stereo occlusal view. Lake Kanunka, SA.

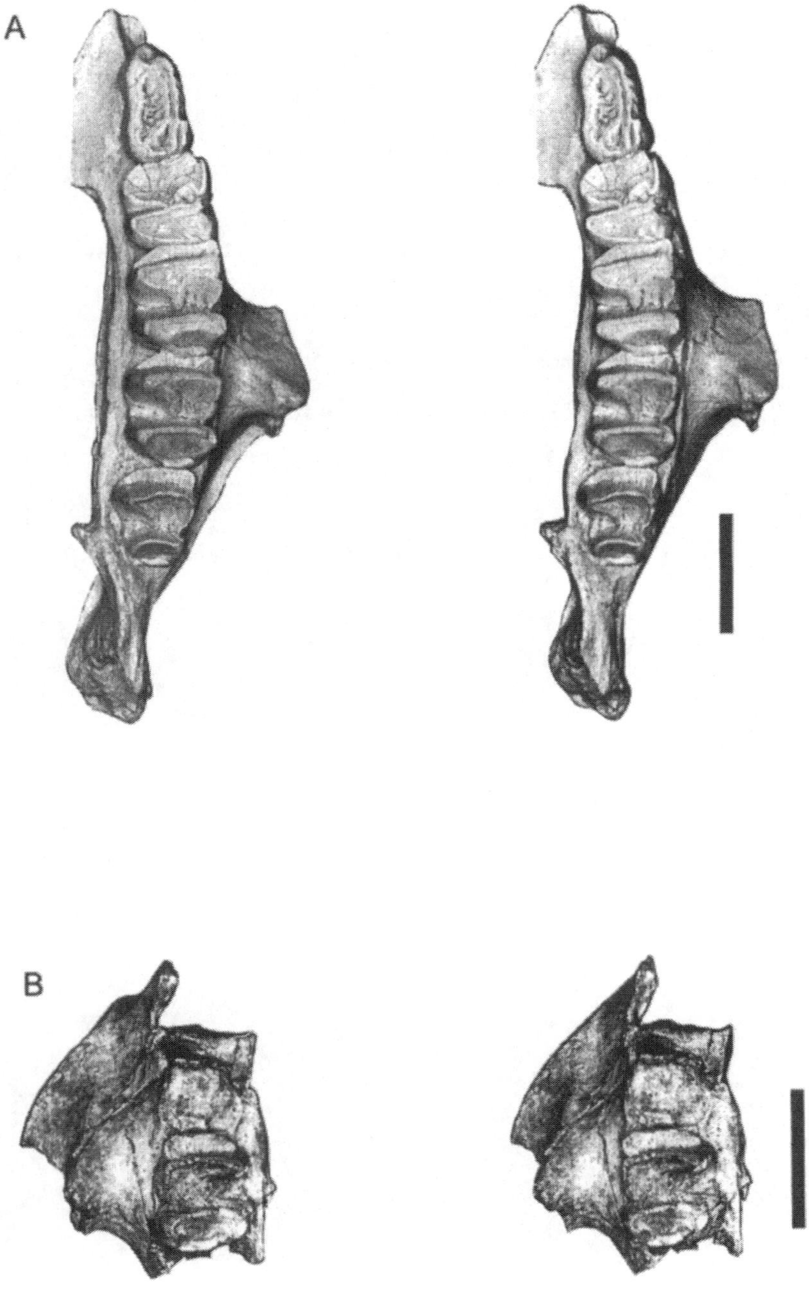
A
B

PLATE 28

Sthenurus tindalei

Scale bars = 20 mm

A. UCMP 56471, right juvenile dentary, lateral view. Cooper Creek, SA.

B. UCMP 56471, mesial view.

C. UCMP 56471, stereo occlusal view.

D. UCMP 56471, close-up of cheek tooth row, occlusal view.

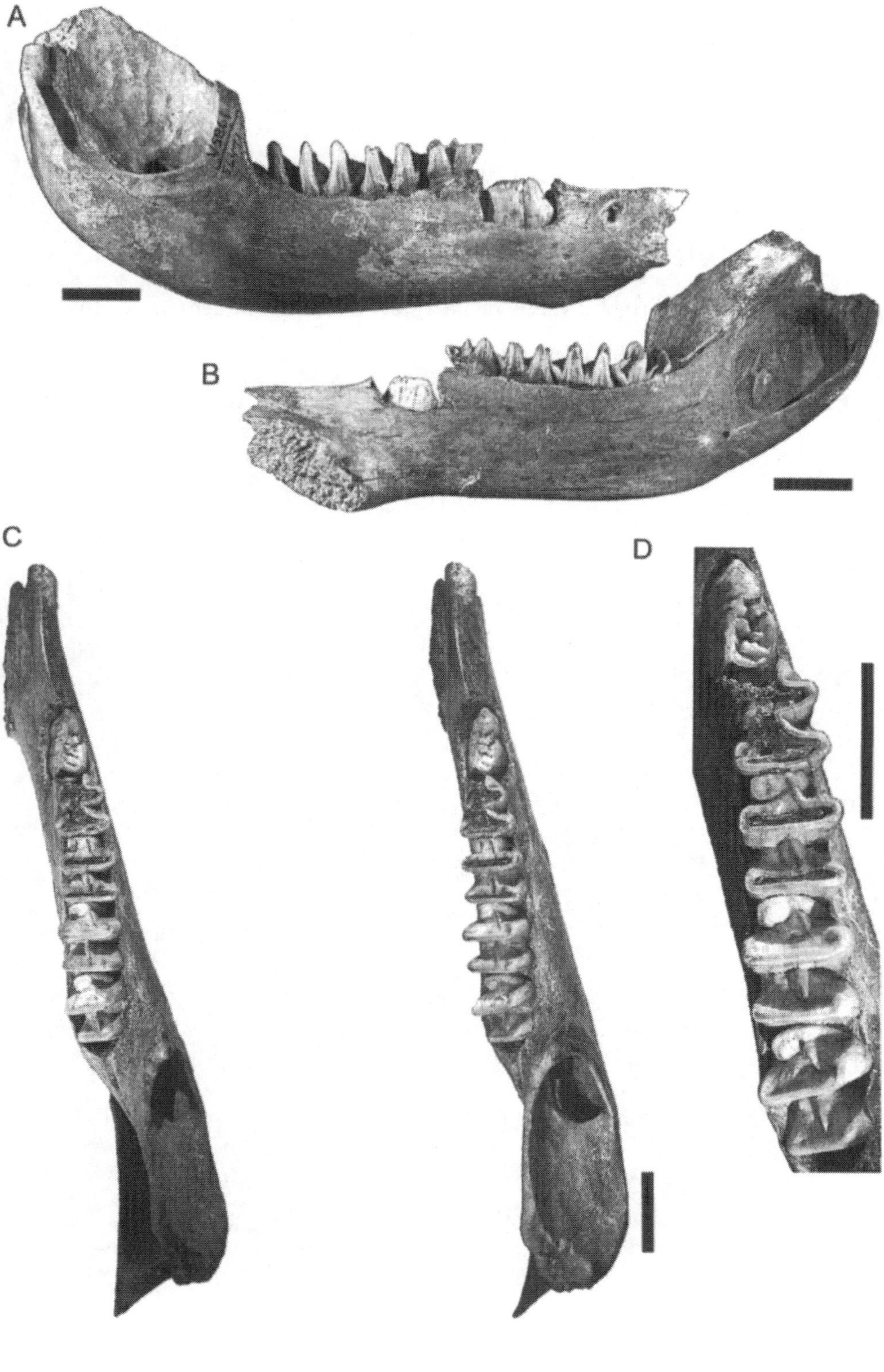
A
B
C
D

PLATE 29

Sthenurus tindalei

Scale bars = 20 mm

A. SAM P25172, right juvenile dentary, lateral view. Cooper Creek, SA.
B. SAM P25172, stereo occlusal view.
C. WAM 63.11.1/-.4/-.9/-.10, left juvenile dentary, stereo occlusal view. Balladonia, WA.

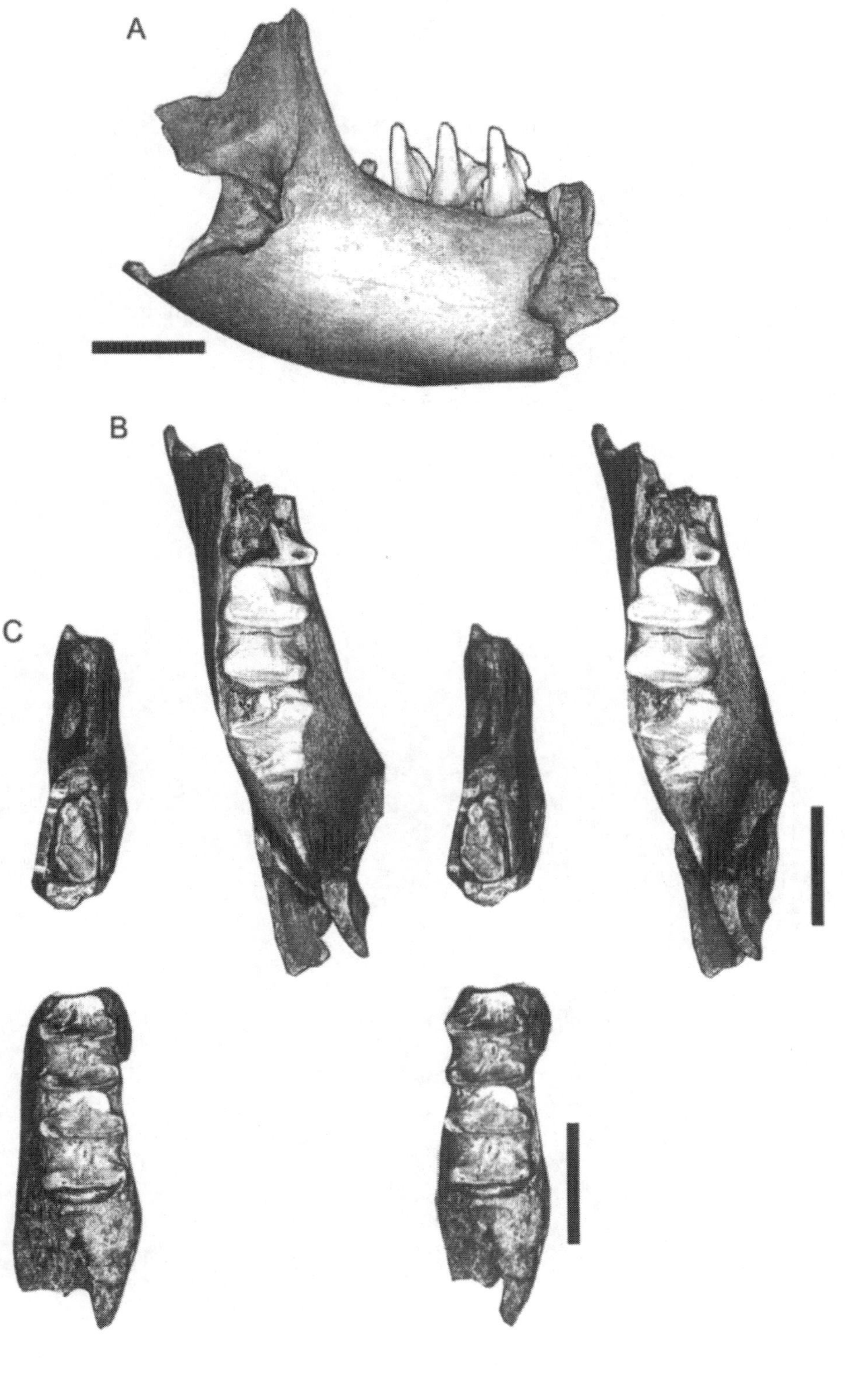
A
B
C

PLATE 30

Metasthenurus newtonae

Scale bars = 20 mm

A. SAM P17249 / P17250 (holotype), adult cranium and right dentary, lateral view. Green Waterhole Cave, SA.
B. SAM P17249, adult cranium, stereo palatal view.
C. SAM P17250, right adult dentary, stereo occlusal view.

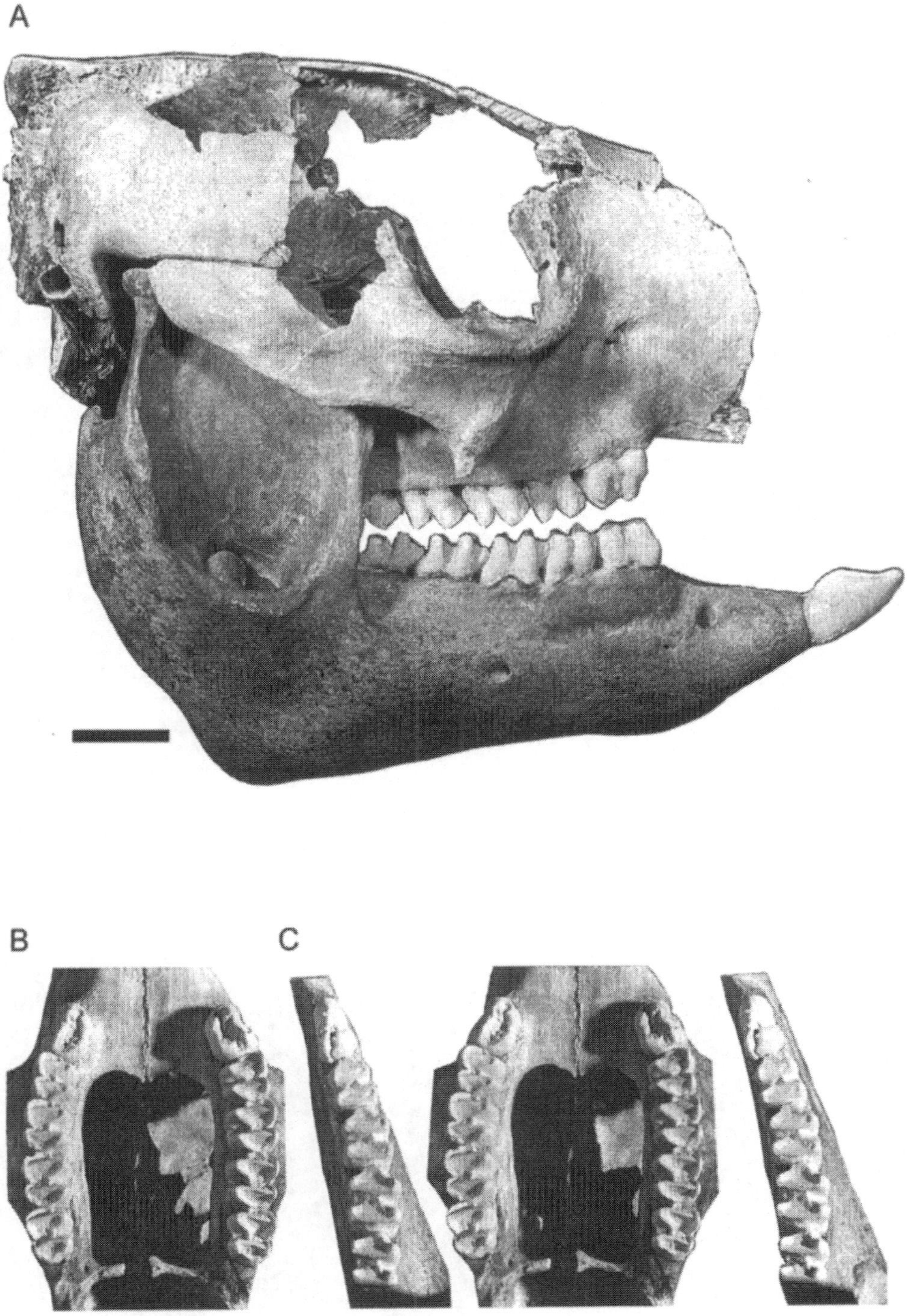
A
B
C

PLATE 31

Metasthenurus newtonae

Scale bars = 20 mm

A. SAM P28969, adult cranium, dorsal view. Victoria Fossil Cave, SA.
B. SAM P28969, adult cranium, palatal view.
C. SAM P20255, partial adult cranium, lateral view.

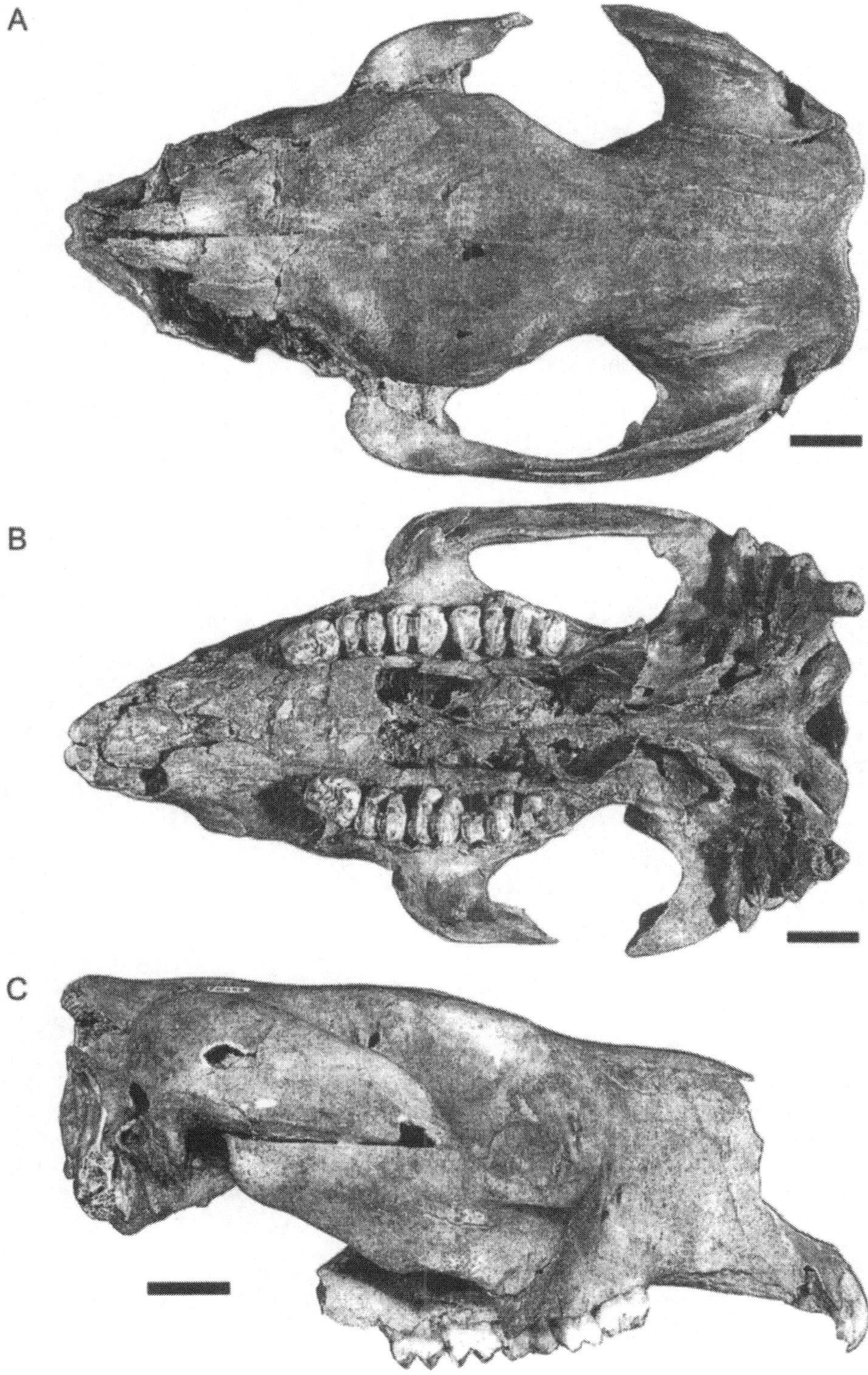
A
B
C

PLATE 32

Metasthenurus newtonae

Scale bars = 20 mm

A. FU 0227, adult cranium, dorsal view. Victoria Fossil Cave, SA.
B. FU 0227, adult cranium, lateral view.
C. FU 0227, adult cranium, stereo palatal view.

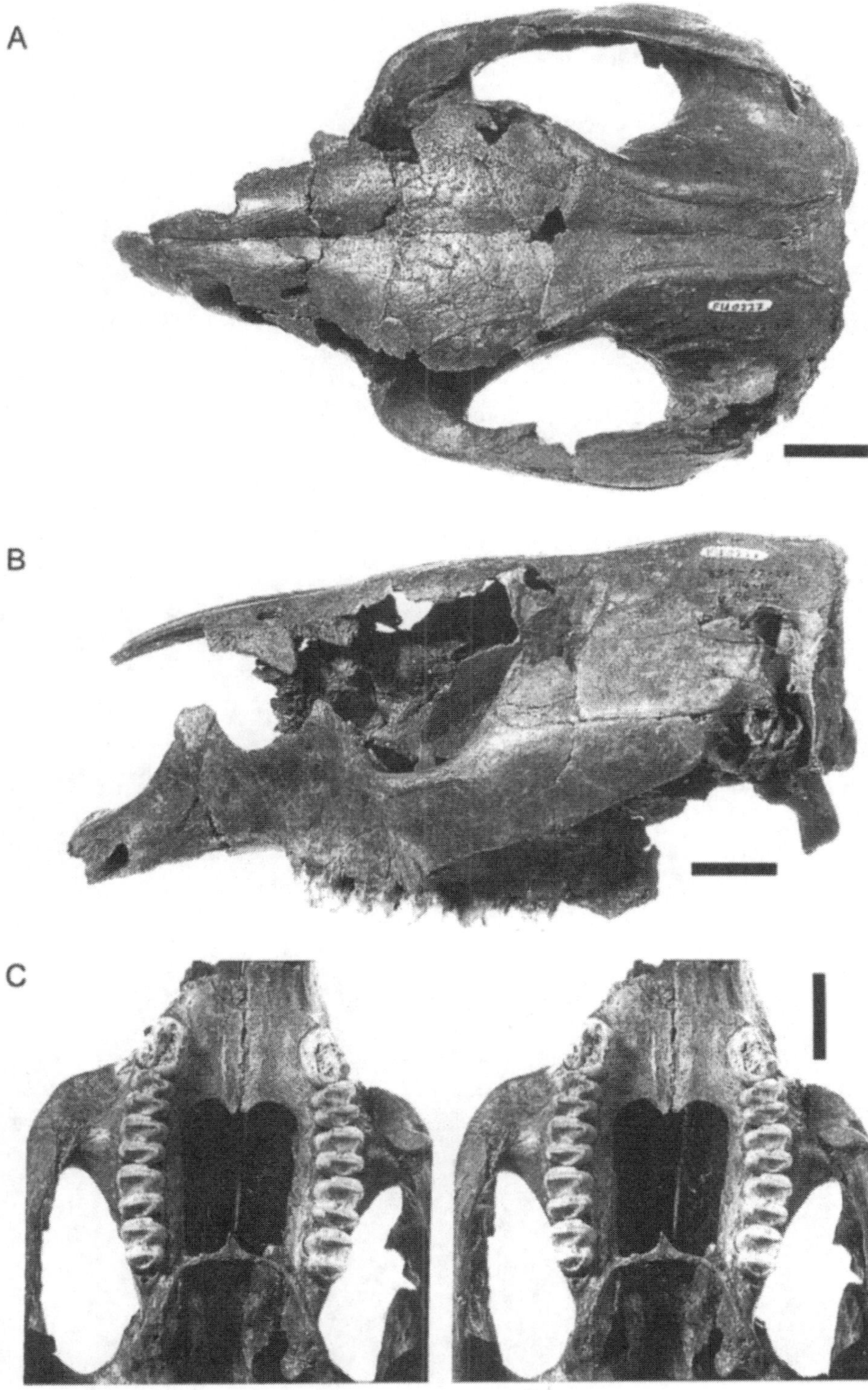
A
B
C

PLATE 33

Metasthenurus newtonae

Scale bars = 20 mm (A-B, D); 10 mm (C)

A. FU 0252, juvenile cranium, dorsal view. Victoria Fossil Cave, SA.

B. FU 0252, lateral view.

C. FU 0252, excavated left P3, stereo occlusal view.

D. FU 0252, stereo palatal view.

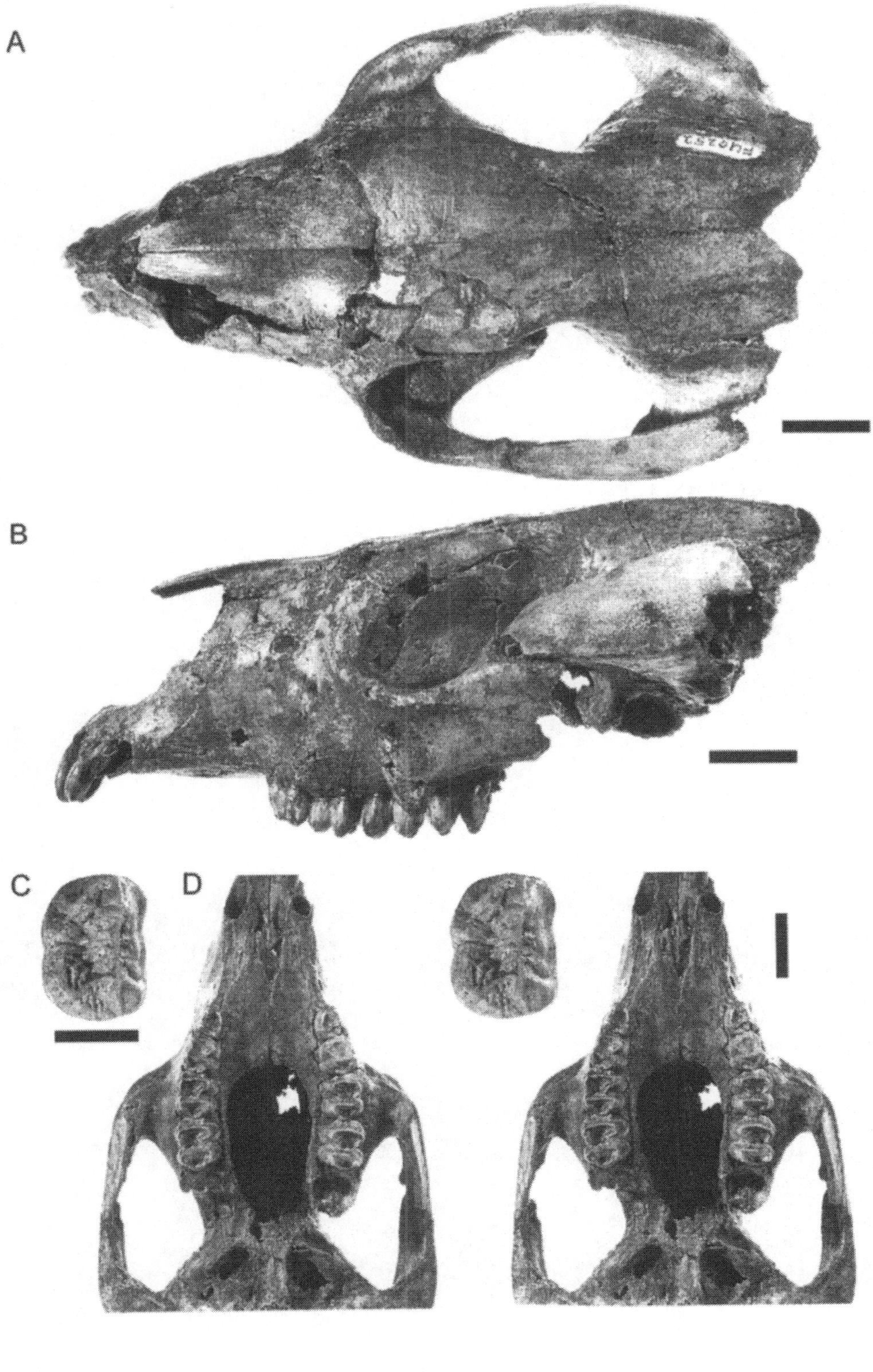
A
B
C
D

PLATE 34

Metasthenurus newtonae

Scale bars = 20 mm (A-D); 10 mm (E)

A. FU 0179, right juvenile dentary, lateral view. Victoria Fossil Cave, SA.

B. FU 0179, mesial view.

C. SAM P28996, left juvenile dentary, stereo occlusal view.

D. FU 0179, stereo occlusal view.

E. FU 0179, excavated right p3, stereo occlusal view.

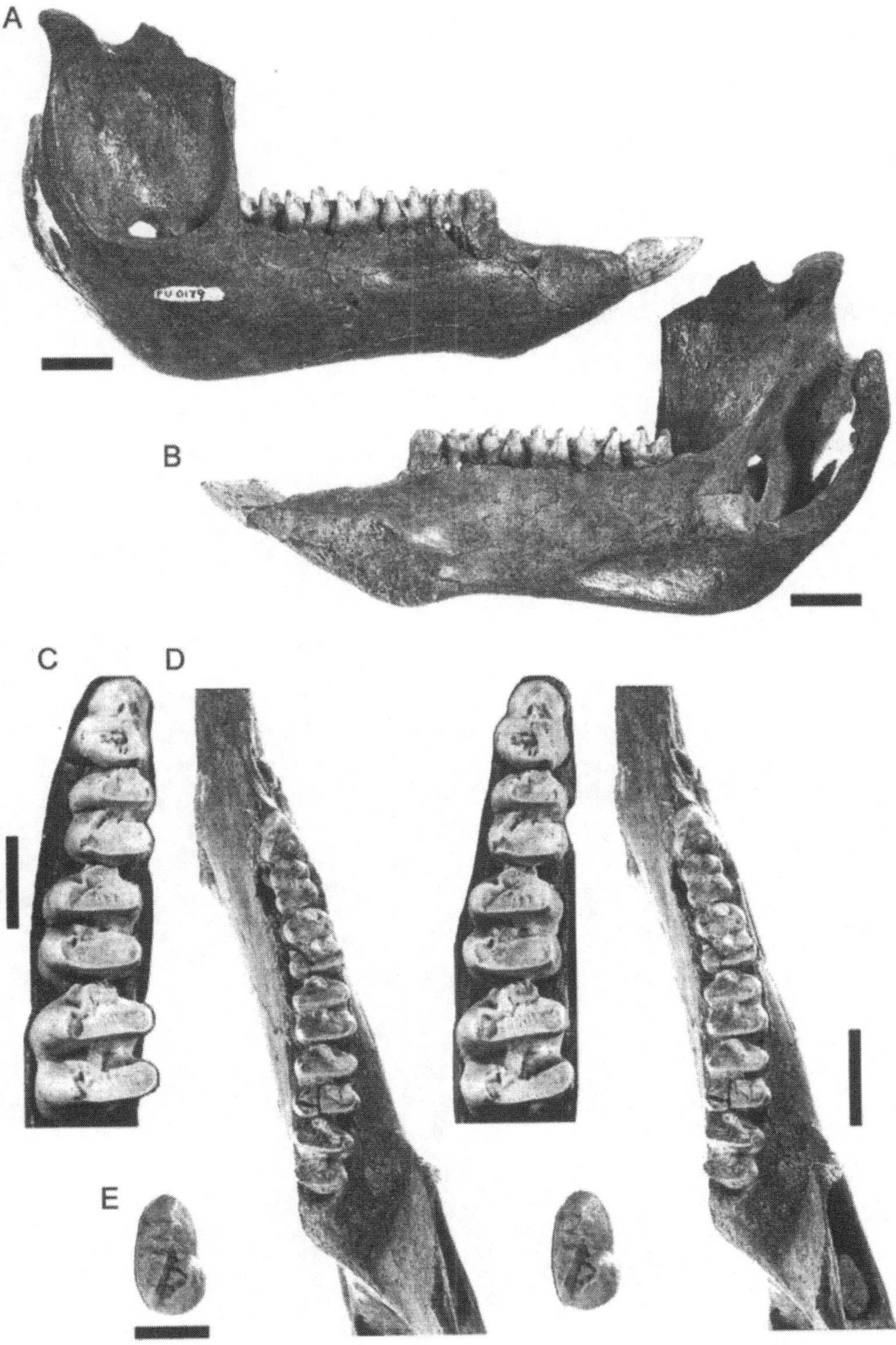

A
FU 0179
B
C
D
E

PLATE 35

Metasthenurus newtonae

Scale bars = 10 mm

A. FU 0887, left juvenile maxilla, stereo occlusal view. Victoria Fossil Cave, SA.
B. AM F18872, left P3, stereo occlusal view. Wellington Caves, NSW.
C. SAM, three right P3s, two left P3s, stereo occlusal view. Victoria and Henschke's Fossil Caves, SA.
D. WAM 00.1.1, right juvenile maxilla, occlusal view. Lindsay Hall Cave, WA.
E. SAM, four left p3s, stereo occlusal view. Victoria and Henschke's Fossil Caves, SA.
F. QM F2978, right adult maxilla, occlusal view. Darling Downs, QLD.

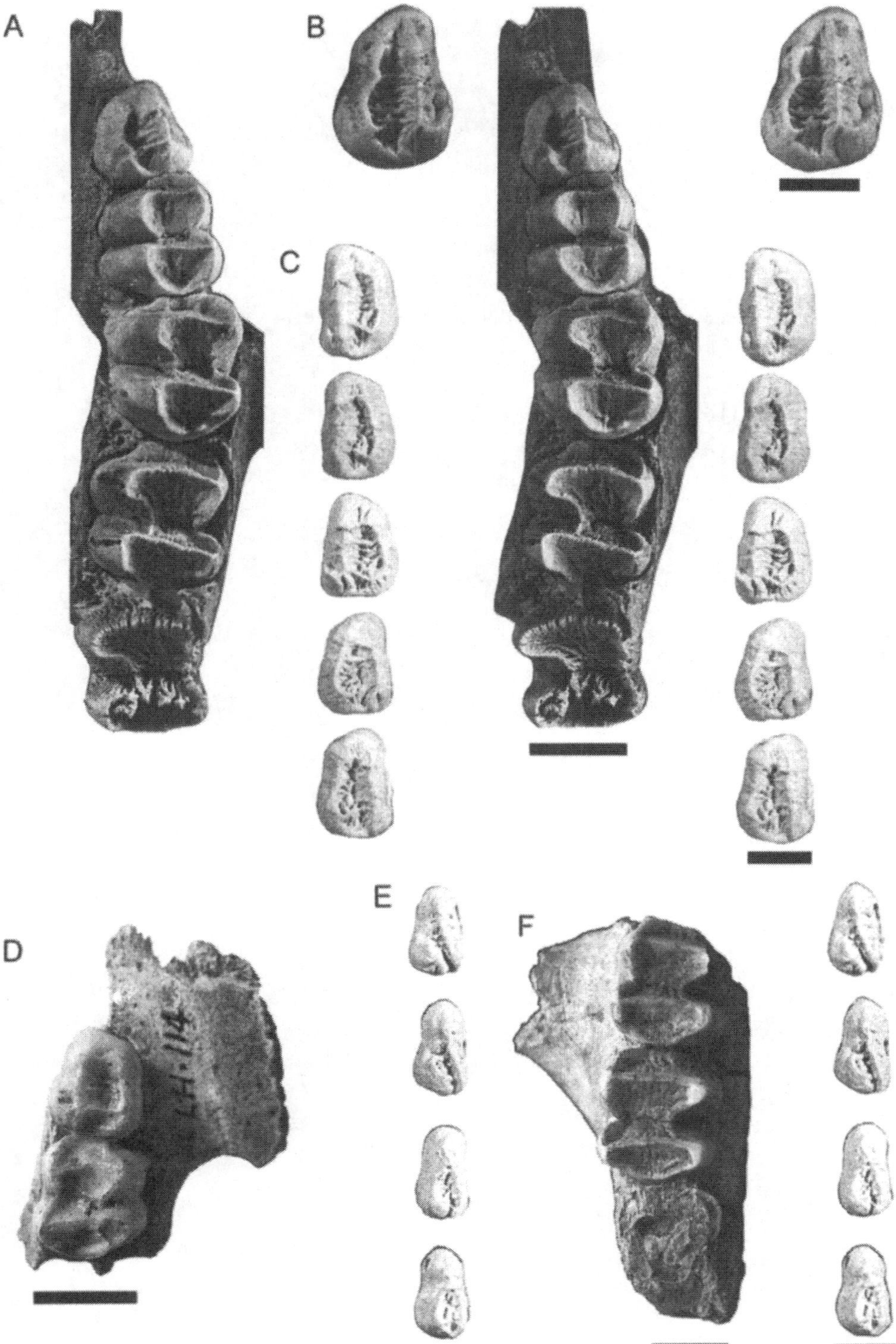
A
B
C
D
E
F

PLATE 36

Archaeosimos cegsai

Scale bars = 20 mm

A. SAM P31800 (holotype), left adult dentary, lateral view. Corra-Lynn Cave, SA.
B. SAM P31800, mesial view.
C. SAM P31800, stereo occlusal view.

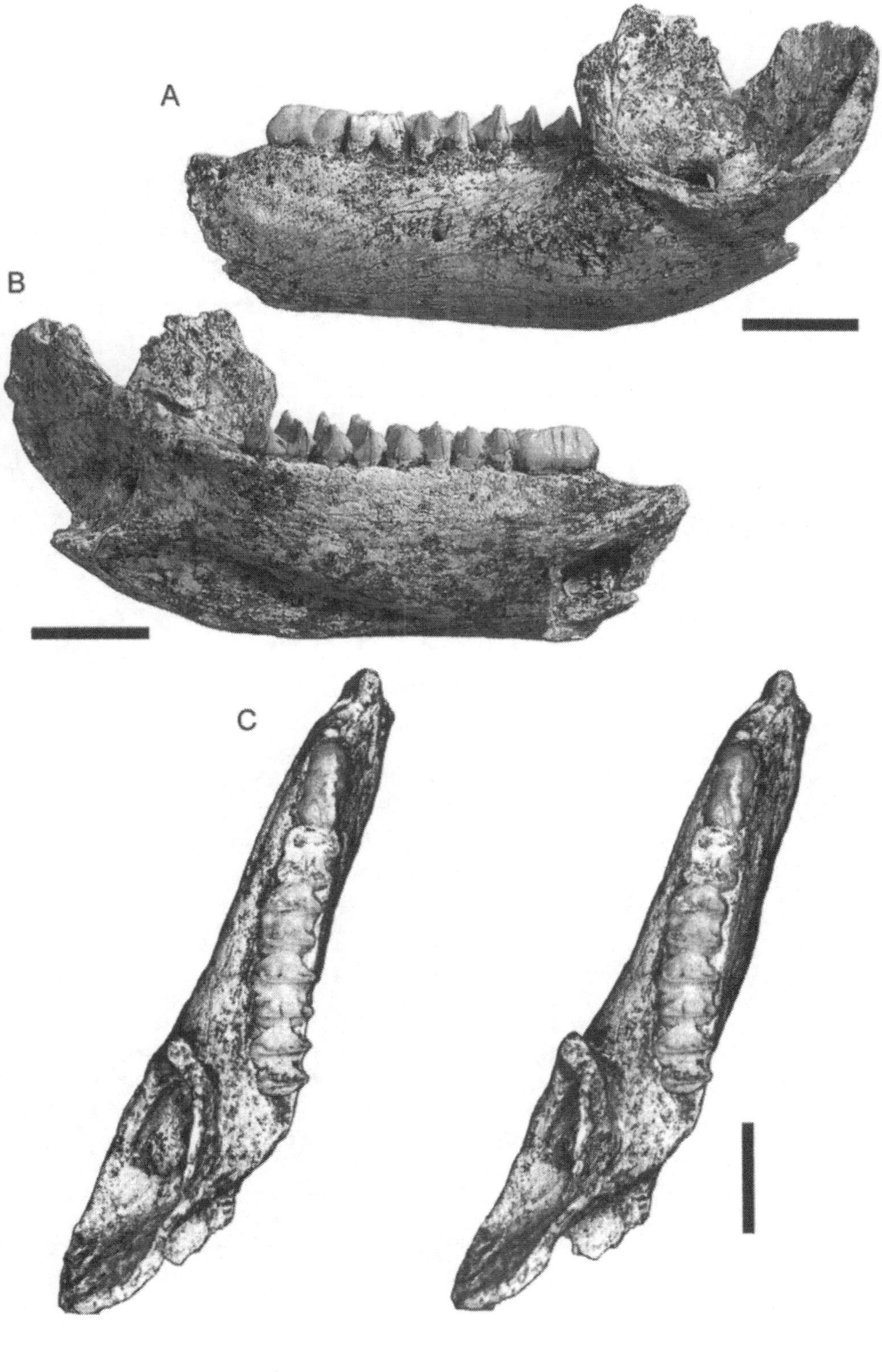
A
B
C

PLATE 37

Archaeosimos cegsai

Scale bars = 10 mm (A-B); 5 mm (C-F)

A. SAM P29917, left adult maxilla, stereo occlusal view. Corra-Lynn Cave, SA.
B. SAM P31335, right dentary with m3, stereo occlusal view. Corra-Lynn Cave, SA.
C. SAM P29930, right I3, buccal view. Corra-Lynn Cave, SA.
D. SAM P29930, right I3, lingual view.
E. SAM P29872, left i1, lingual view. Corra-Lynn Cave, SA.
F. SAM P29872, left i1, buccal view.

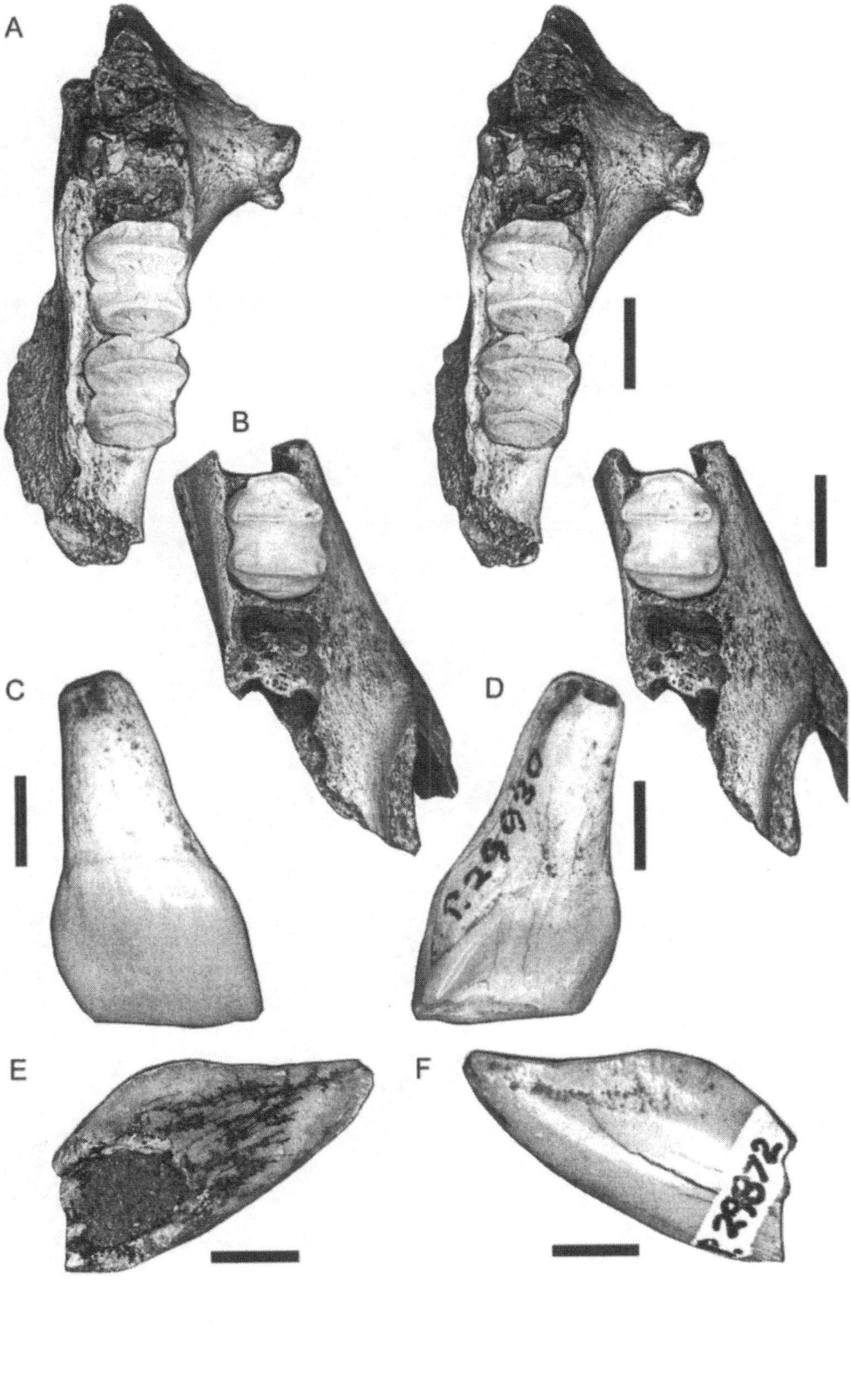
A
B
C
D
E
F
P.29930

PLATE 38

Archaeosimos correlli

Scale bars = 20 mm (A-C); 10 mm (D-F)

A. SAM P38801 (holotype), left juvenile dentary, lateral view. Corra-Lynn Cave, SA.
B. SAM P38801, mesial view.
C. SAM P38801, stereo occlusal view.
D. SAM P38801, excavated p3, stereo occlusal view.
E. SAM P38801, excavated p3, buccal view.
F. SAM P38801, excavated p3, lingual view.

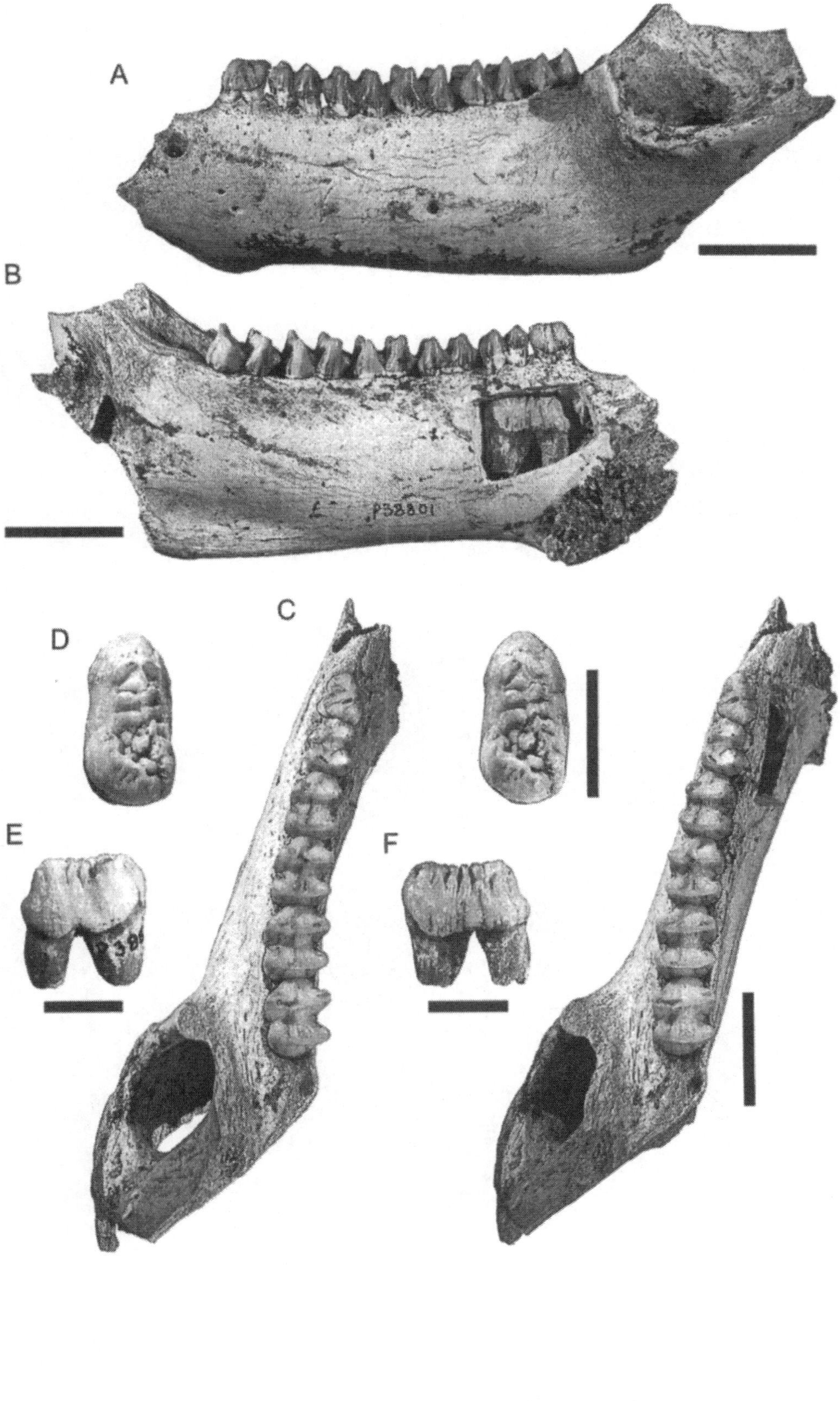
A
B
P38801
C
D
E
F

PLATE 39

Archaeosimos correlli

cf. *Archaeosimos* sp.

Scale bars = 20 mm (A-C); 10 mm (D-H)

A. SAM P29900, left adult dentary, lateral view. Corra-Lynn Cave, SA.
B. SAM P29900, mesial view.
C. SAM P29900, occlusal view.
D. SAM P31340, left I3, buccal view.
E. SAM P31340, left I3, lingual view.
F. UCR 22656, right ?m2, buccal view. Lake Kanunka, SA.
G. UCR 22656, lingual view.
H. UCR 22656, occlusal view.

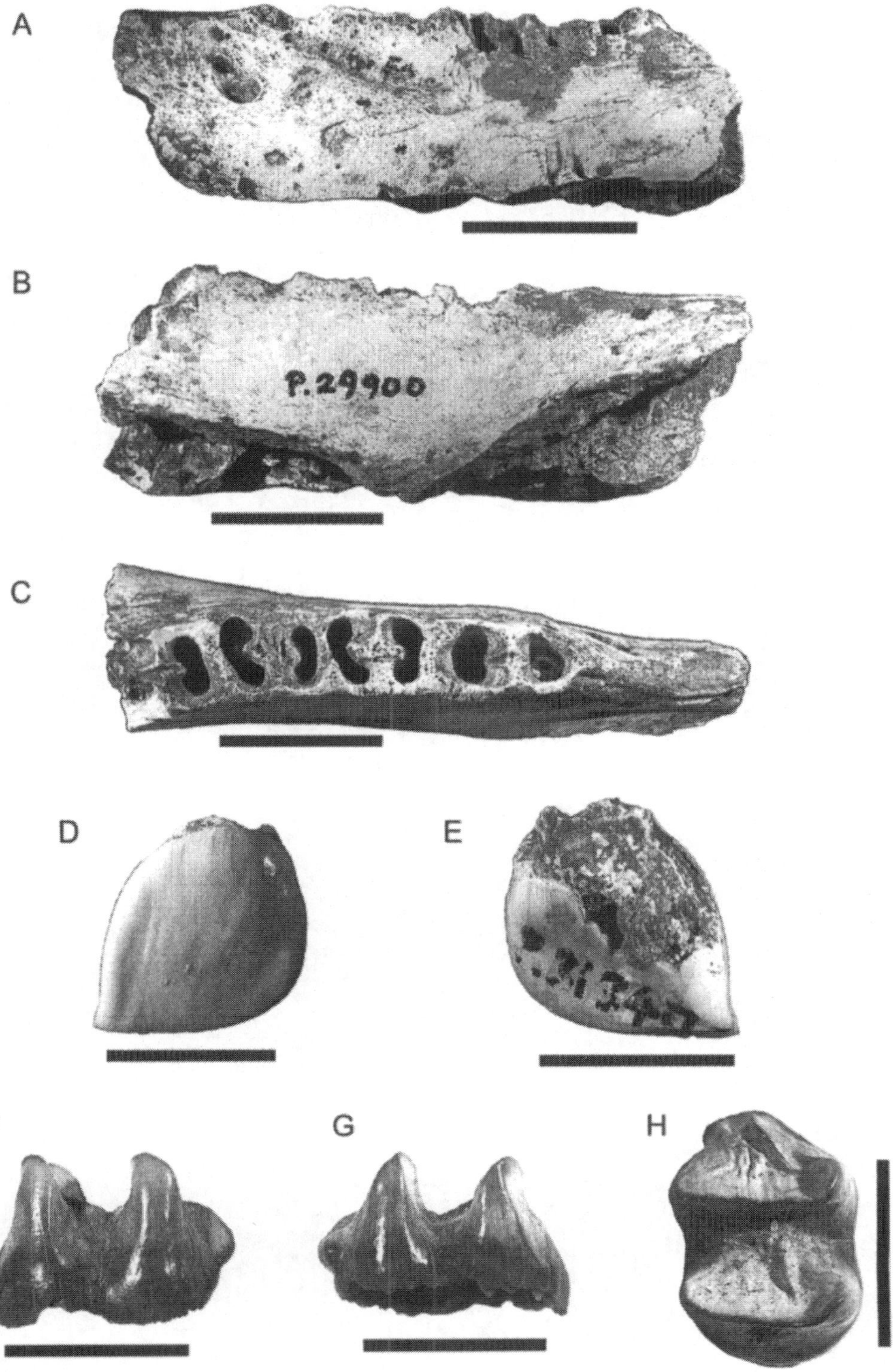
A
B
P.29900
C
D
E
F
G
H

PLATE 40

Simosthenurus euryskaphus

Scale bars = 20 mm (A-B, E-G); 10 mm (C-D)

A. AM F113131 (holotype), left juvenile dentary, lateral view. Bingara, NSW.

B. AM F113131, mesial view.

C. AM F113131, stereo occlusal view.

D. AM F113131, excavated p3, stereo occlusal view.

E. AM F73723, left juvenile dentary, lateral view. Bingara, NSW.

F. AM F73723, mesial view.

G. AM F73723, occlusal view.

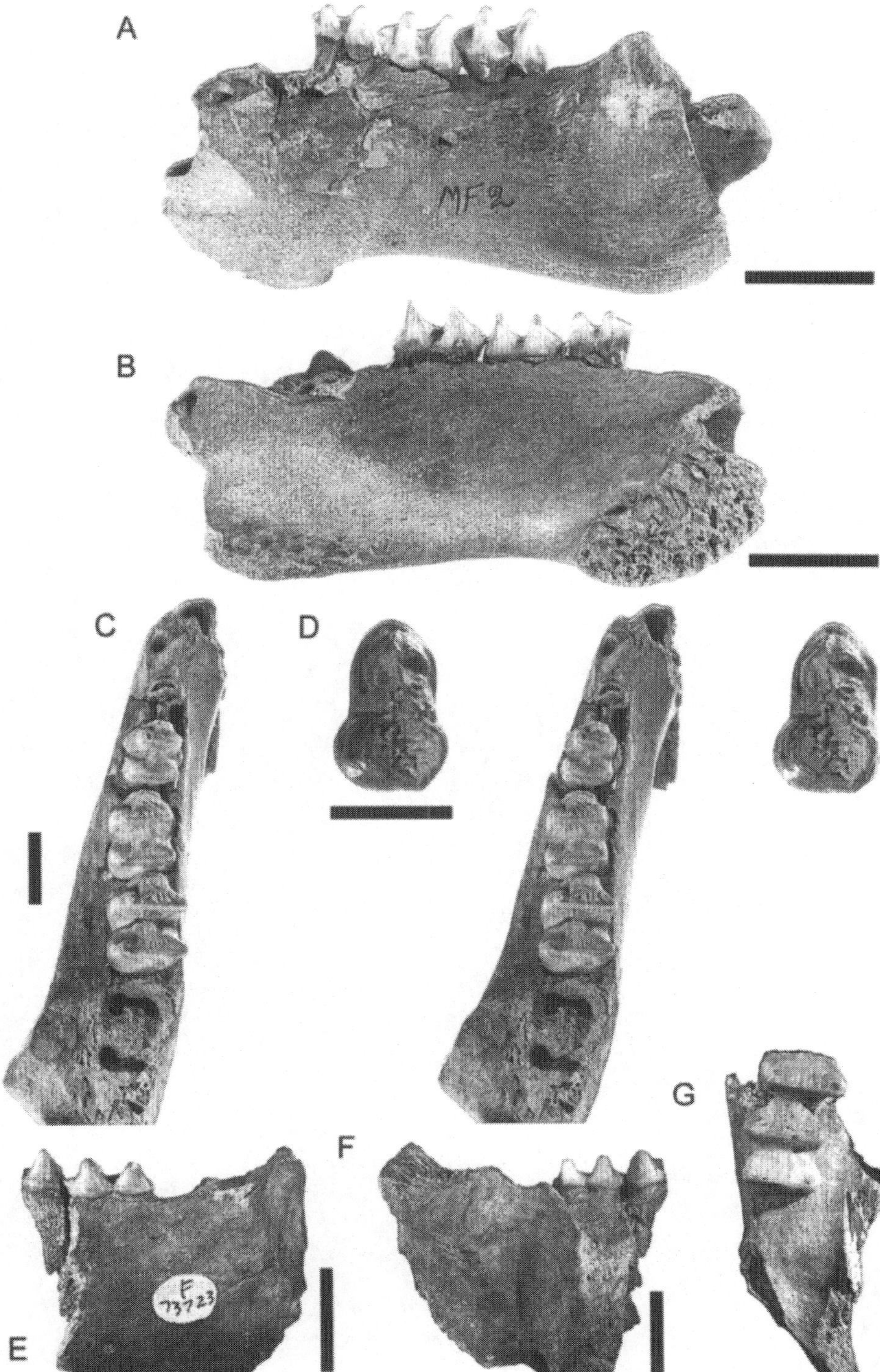

A
MF 2
B
C
D
G
F
E
F 73723

PLATE 41

Simosthenurus maddocki

Scale bars = 20 mm

A. SAM P17247, adult cranium and dentaries, lateral view. Green Waterhole Cave, SA.
B. SAM P17247, dorsal view.

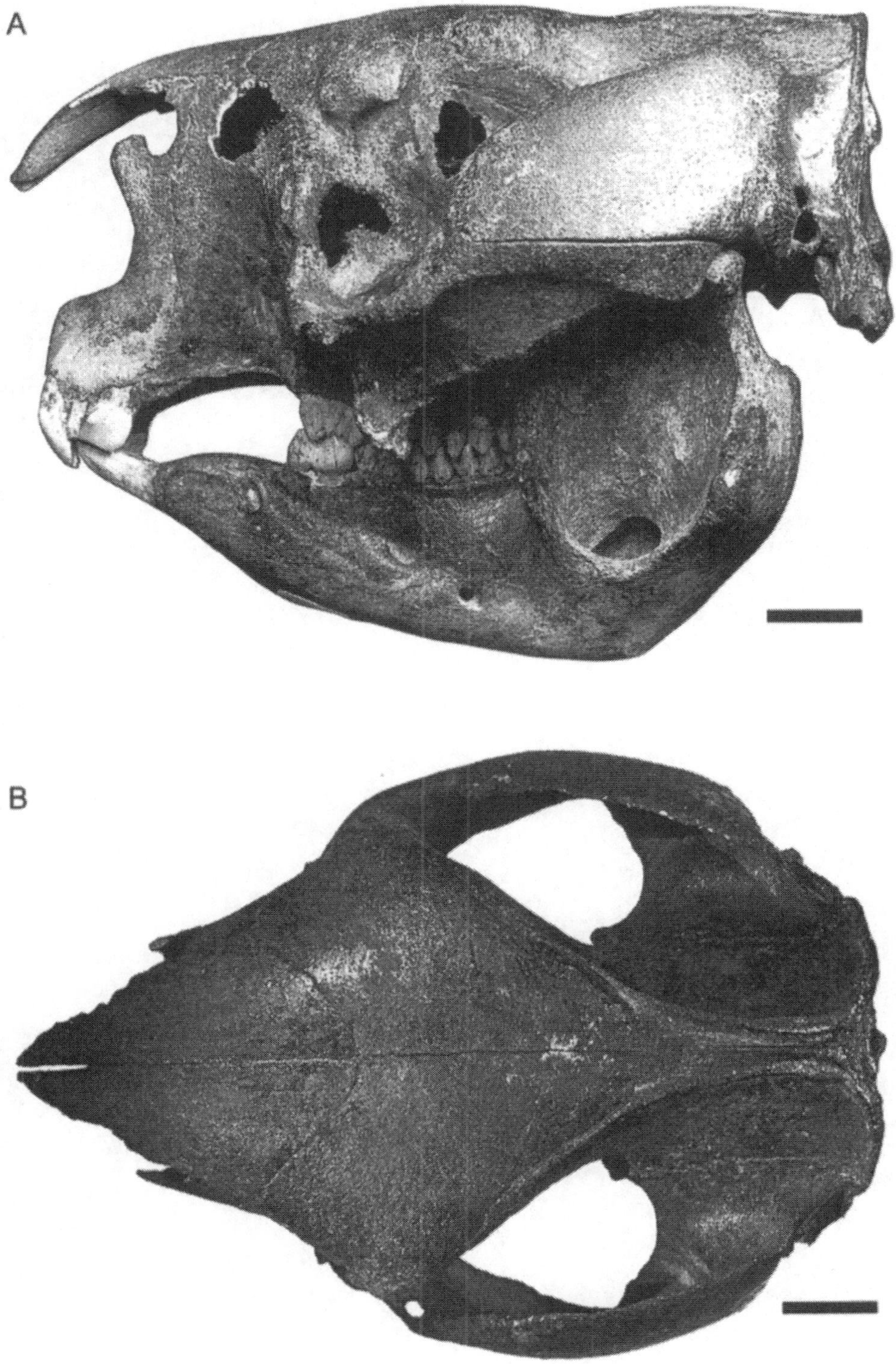
A
B

PLATE 42

Simosthenurus maddocki

Scale bars = 20 mm

A. SAM P17247, adult cranium, palatal view. Green Waterhole Cave, SA.

B. SAM P17247, adult cranium and dentaries, ventral view.

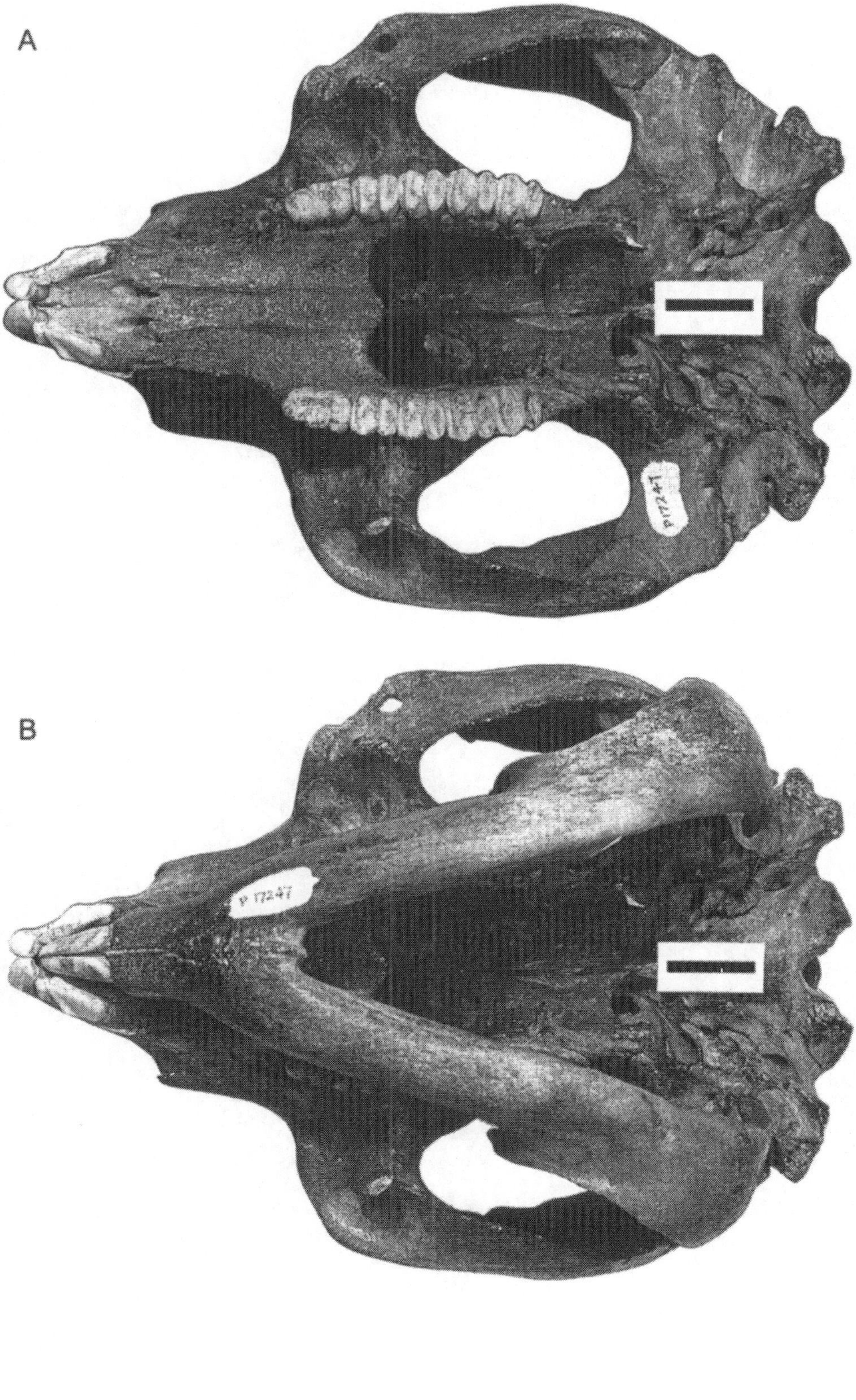
A
P17247
B
P 17247

PLATE 43

Simosthenurus maddocki

Scale bars = 20 mm

A. SAM P16999 (holotype), cranium, lateral view. Victoria Fossil Cave, SA.
B. SAM P16999, dorsal view.
C. WAM 03.11.09, right adult maxilla, lateral view. Lindsay Hall Cave, WA.

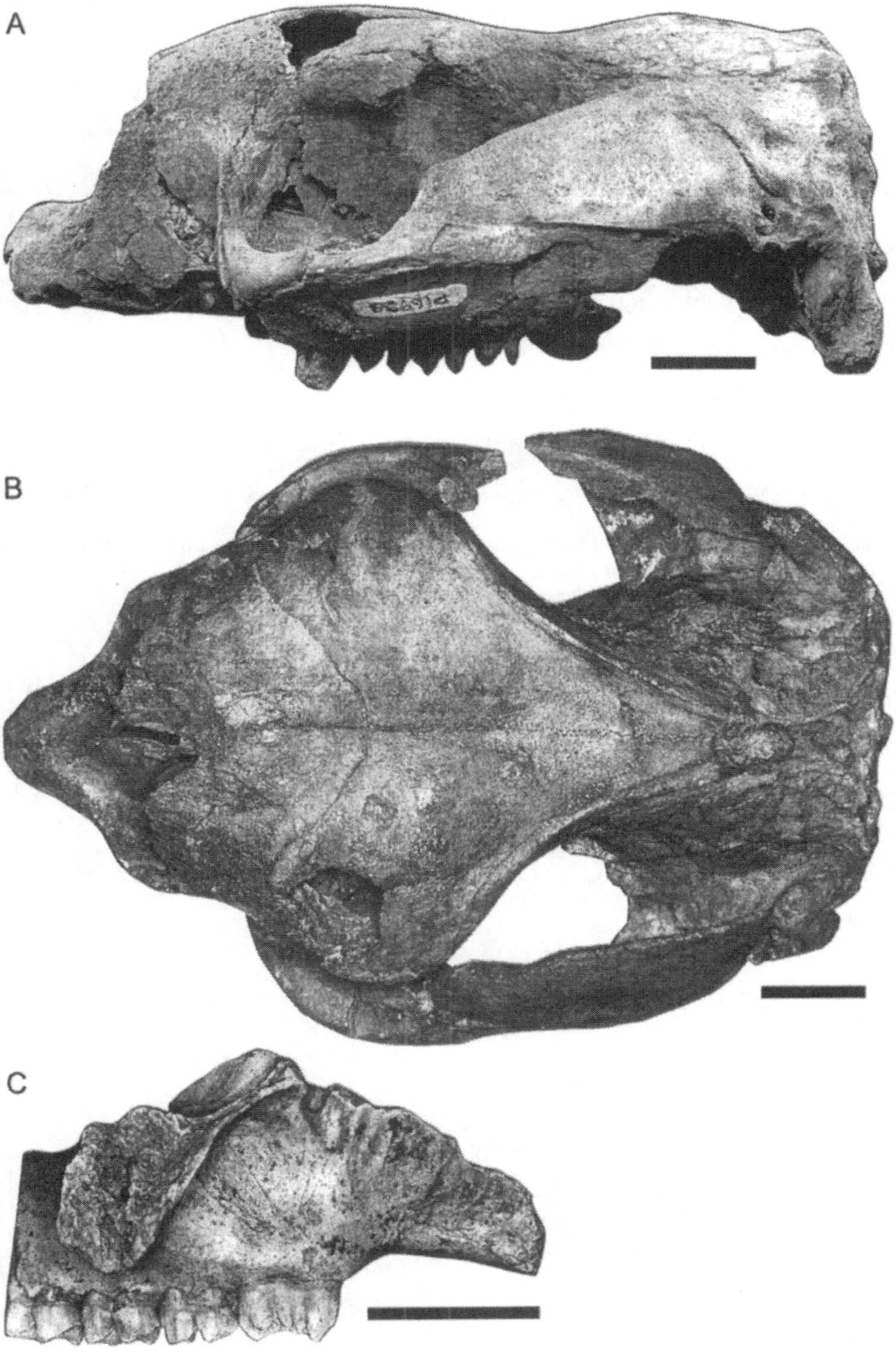
A
B
C

PLATE 44

Simosthenurus maddocki

Scale bars = 10 mm

A. WAM 92.9.8, left juvenile maxilla, stereo occlusal view. Lindsay Hall Cave, WA.
B. WAM 92.9.8, left dP2, stereo occlusal view.
C. WAM 92.9.8, excavated left P3, stereo occlusal view.
D. WAM 92.12.7, left P3, stereo occlusal view. Lindsay Hall Cave, WA.
E. QM F8529, right juvenile maxilla fragment, stereo occlusal view. Texas Caves, QLD.
F. SAM P27752, left P3, stereo occlusal view. Victoria Fossil Cave, SA.
G. WAM 03.11.09, right adult maxilla, stereo occlusal view. Lindsay Hall Cave, WA.
H. FU 1471, left P3, stereo occlusal view. Victoria Fossil Cave, SA.
I. SAM P20004, right P3, stereo occlusal view. Victoria Fossil Cave, SA.
J. SAM P20011, excavated right P3, stereo occlusal view. Sand Funnel Cave, SA.

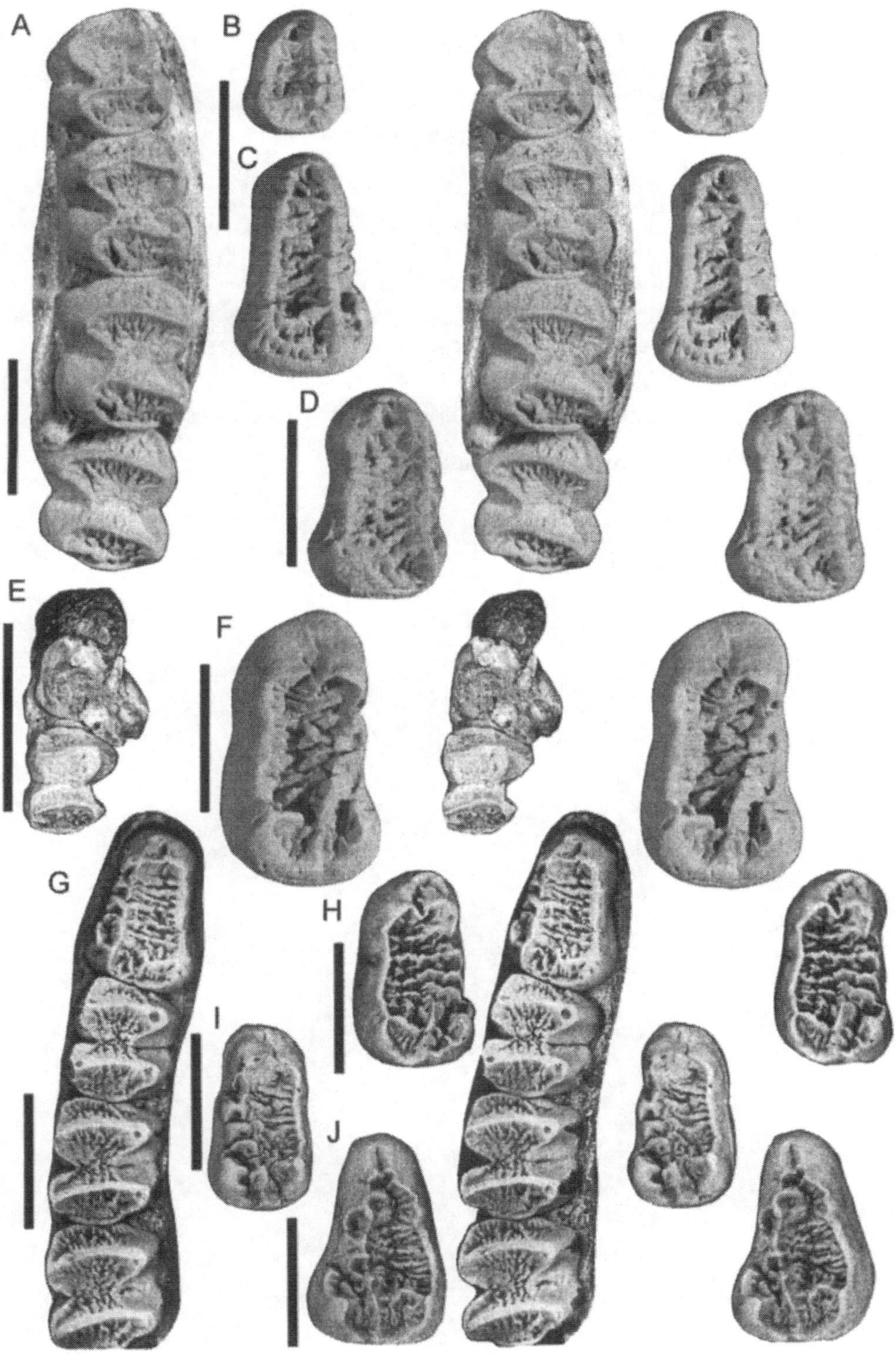
A
B
C
D
E
F
G
H
I
J

PLATE 45

Simosthenurus maddocki

Scale bars = 20 mm (A-B); 10 mm (C)

A. SAM P17247, left and right adult dentaries, occlusal view. Green Waterhole Cave, SA.
B. SAM P16627, left adult dentary, lateral view. Victoria Fossil Cave, SA.
C. SAM P16627, stereo occlusal view.

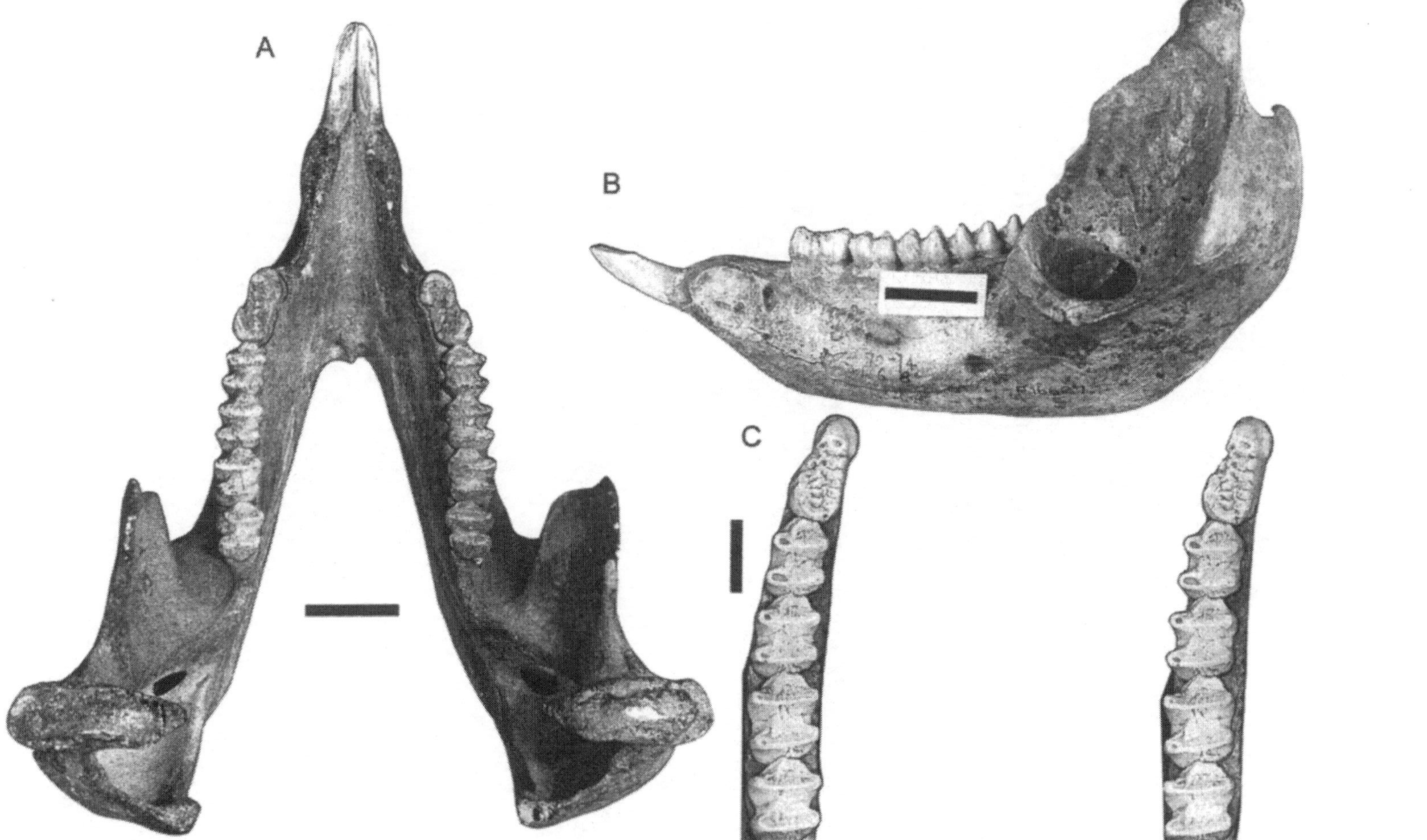
A
B
C

PLATE 46

Simosthenurus maddocki

Scale bars = 20 mm (A-D); 10 mm (E)

A. SAM P27357, left juvenile dentary, mesial view. Victoria Fossil Cave, SA.
B. SAM P27357, lateral view.
C. SAM P20624, right juvenile dentary, lateral view. Victoria Fossil Cave, SA.
D. SAM P20624, mesial view.
E. SAM P20508, left juvenile dentary, stereo occlusal view. Victoria Fossil Cave, SA.

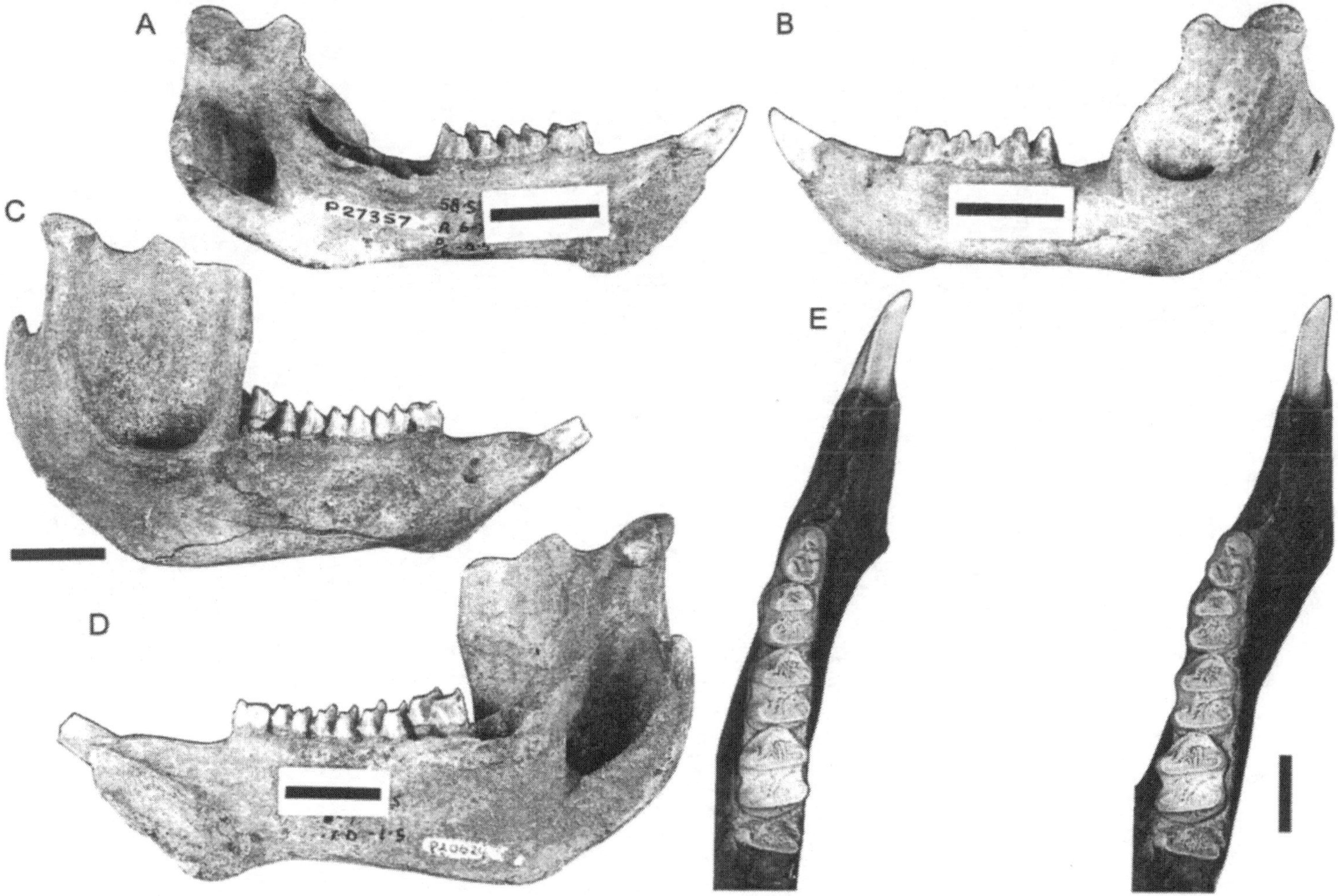

A
B
C
D
E
P27357

PLATE 47

Simosthenurus sp. cf. *maddocki*

Scale bars = 20 mm (A-B); 10 mm (C-G)

A. AM F73720, right juvenile dentary, lateral view. Wellington Caves, NSW.
B. AM F73720, mesial view.
C. AM F73720, stereo occlusal view.
D. AM F73720, excavated p3, stereo occlusal view.
E. SAM P35054, right juvenile maxilla, lateral view. Town Well Cave, Curramulka, SA.
F. SAM P35054, occlusal view.
G. SAM P35054, excavated P3, occlusal view.

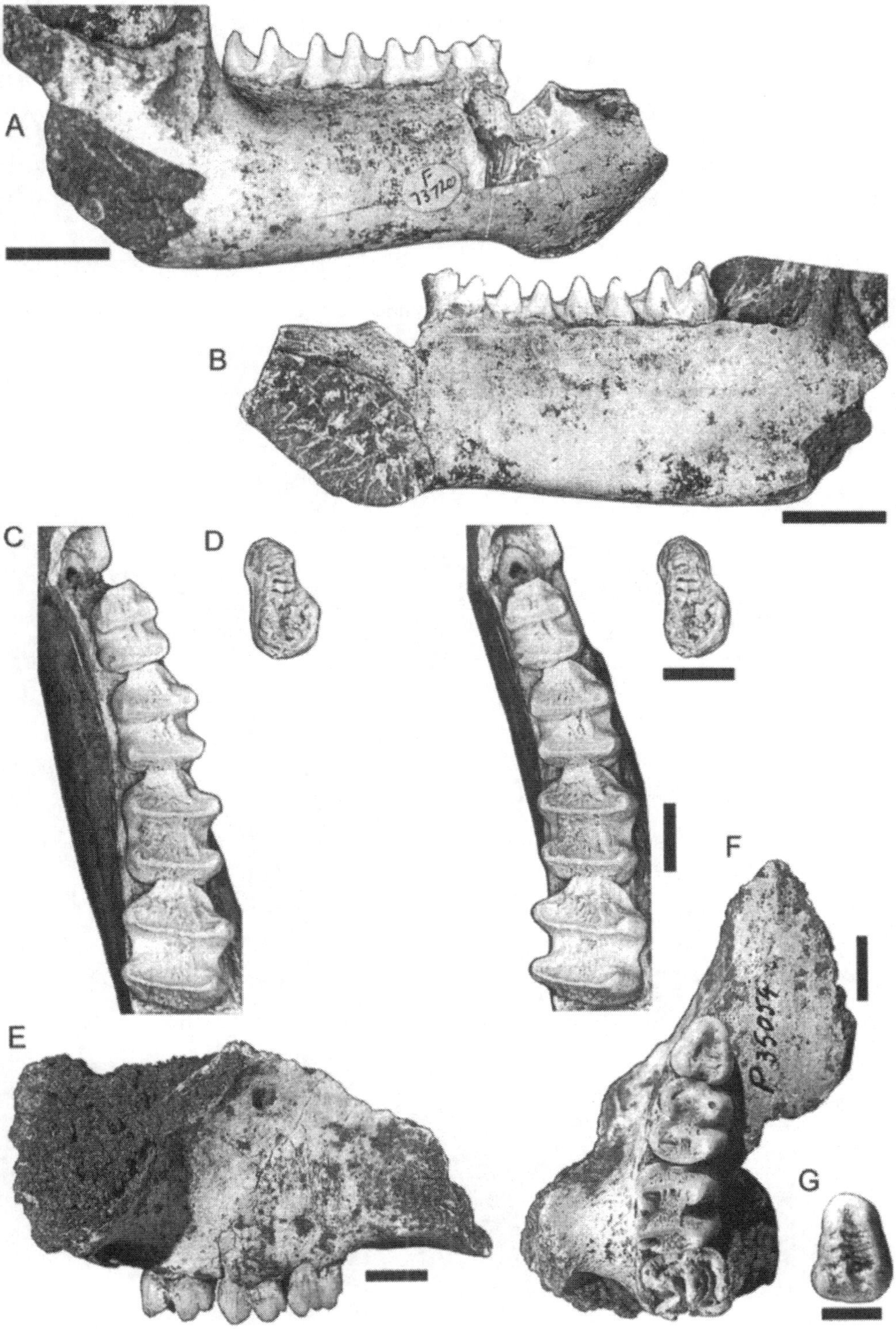
A
F
73720
B
C
D
E
F
P35054
G

PLATE 48

Simosthenurus occidentalis

Scale bars = 20 mm

A. SAM P20820, adult cranium and dentaries, lateral view. Green Waterhole Cave, SA.
B. SAM P20820, dorsal view.
C. SAM P20820, palatal view (cast).

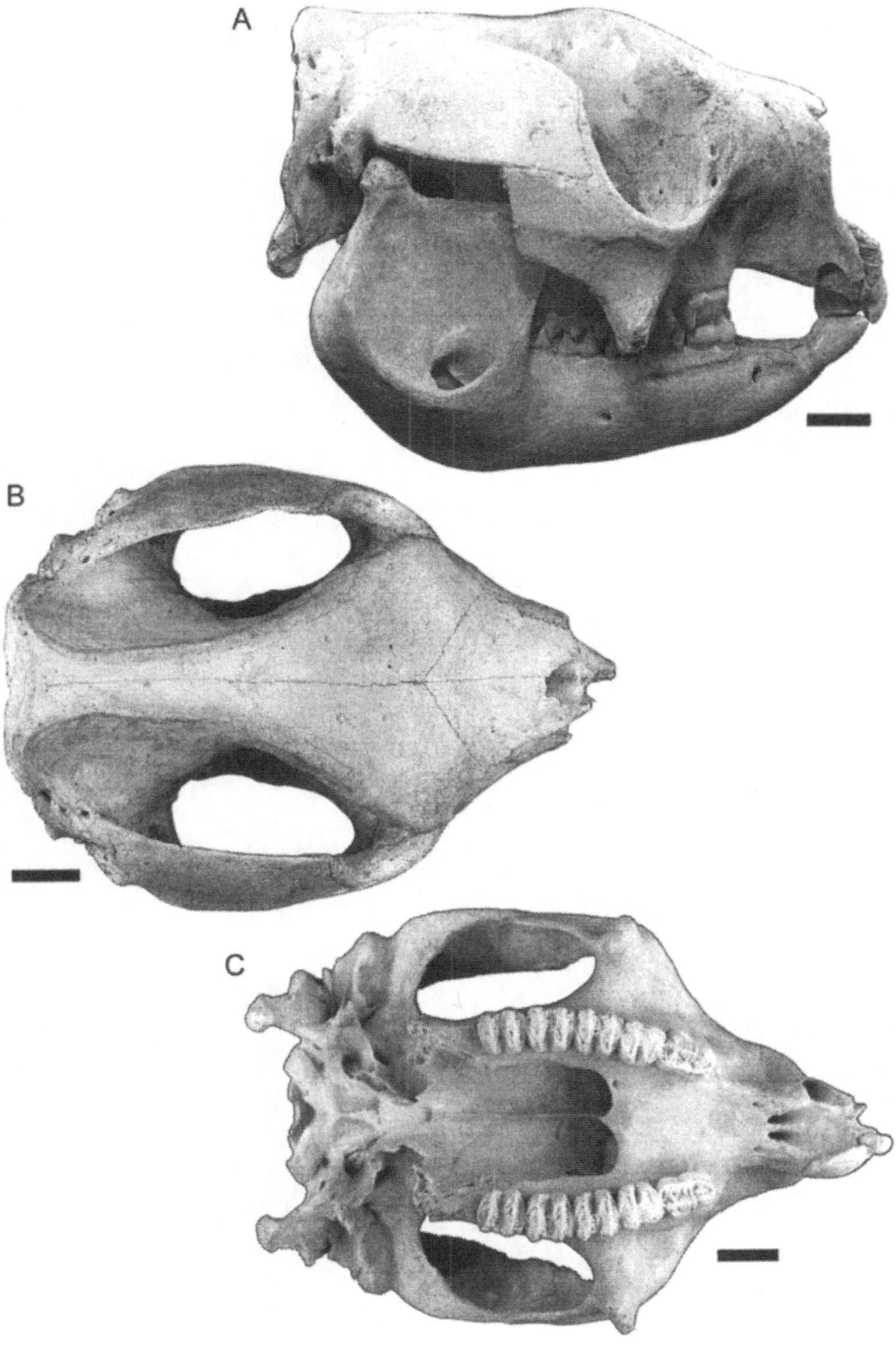
A
B
C

PLATE 49

Simosthenurus occidentalis

Scale bars = 20 mm

A. FU 1662, adult cranium, lateral view. Green Waterhole Cave, SA.

B. SAM P16648, adult cranium, lateral view. Victoria Fossil Cave, SA.

C. AM F53801, partial cranium, lateral view. Cow Flat Quarries, Bathurst, NSW.

A
B
C
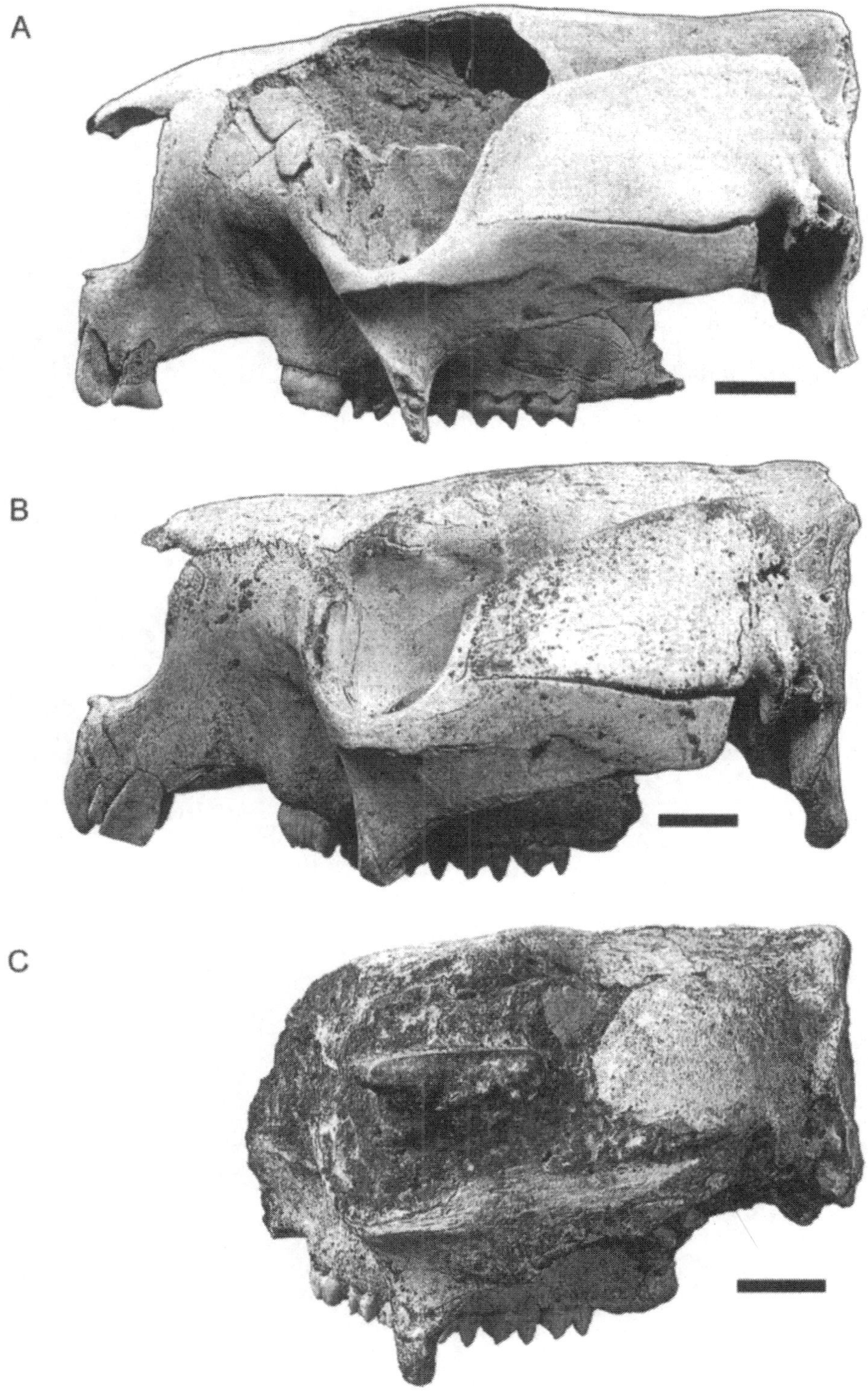

PLATE 50

Simosthenurus occidentalis

Scale bars = 20 mm

A. SAM P17245, left half of adult cranium and left dentary, anterior view. Green Waterhole Cave, SA.
B. SAM P20820, adult cranium and dentaries, anterior view. Green Waterhole Cave, SA.
C. SAM P17498, partial juvenile cranium, anterior view. Green Waterhole Cave, SA.
D. SAM P17245, left half of adult cranium and left dentary, posterior view.
E. SAM P16648, adult cranium, posterior view. Victoria Fossil Cave, SA.

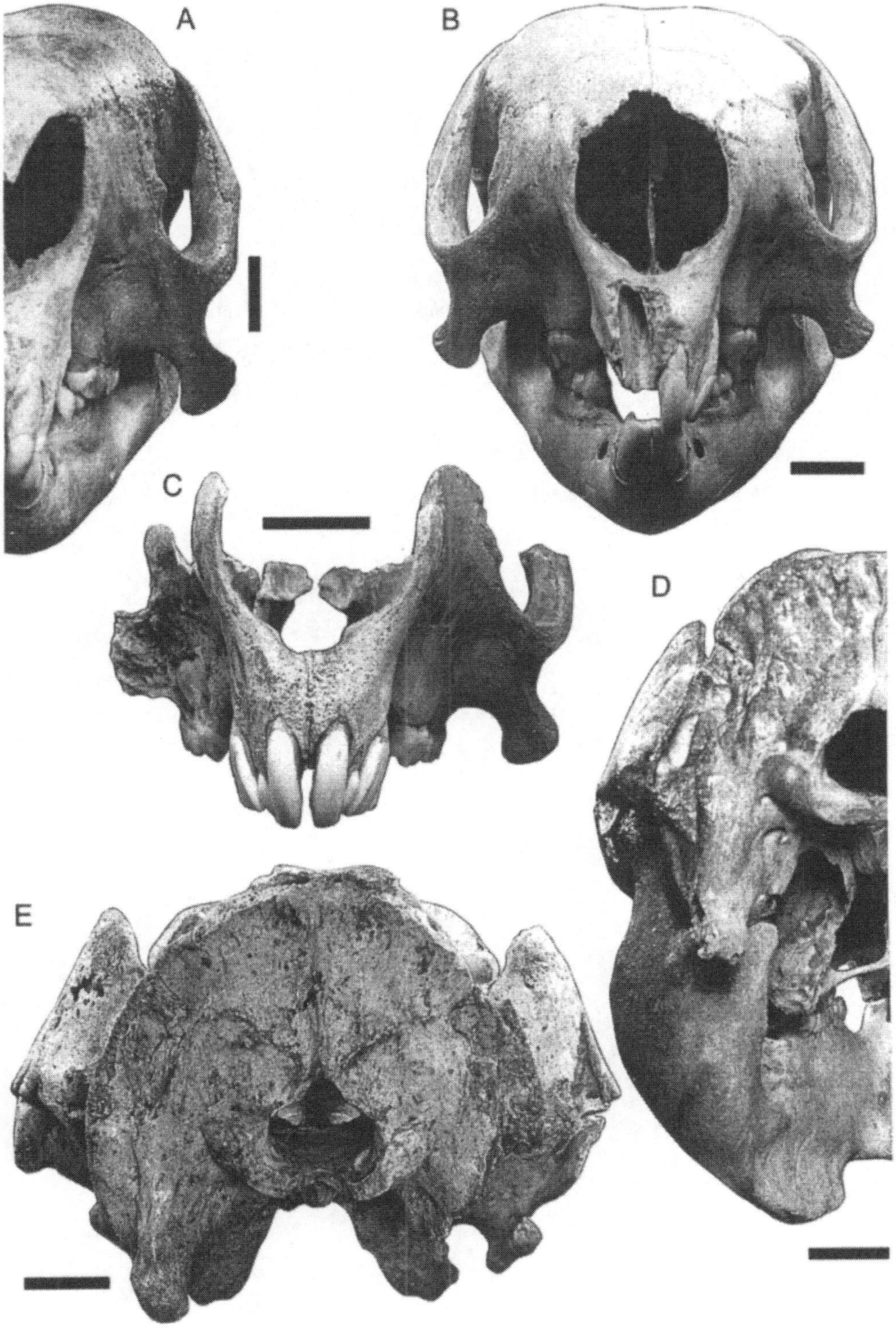
A
B
C
D
E

PLATE 51

Simosthenurus occidentalis

Scale bars = 20 mm

A. WAM 99.10.6, adult cranium, dorsal view. Tight Entrance Cave, WA.

B. WAM 62.8.31, adult cranium, dorsal view. Mammoth Cave, WA.

C. SAM P20803, adult cranium, dorsal view. Victoria Fossil Cave, SA. (Note zygomatic arches slightly distorted mesially.)

D. SAM P16648, adult cranium, dorsal view. Victoria Fossil Cave, SA.

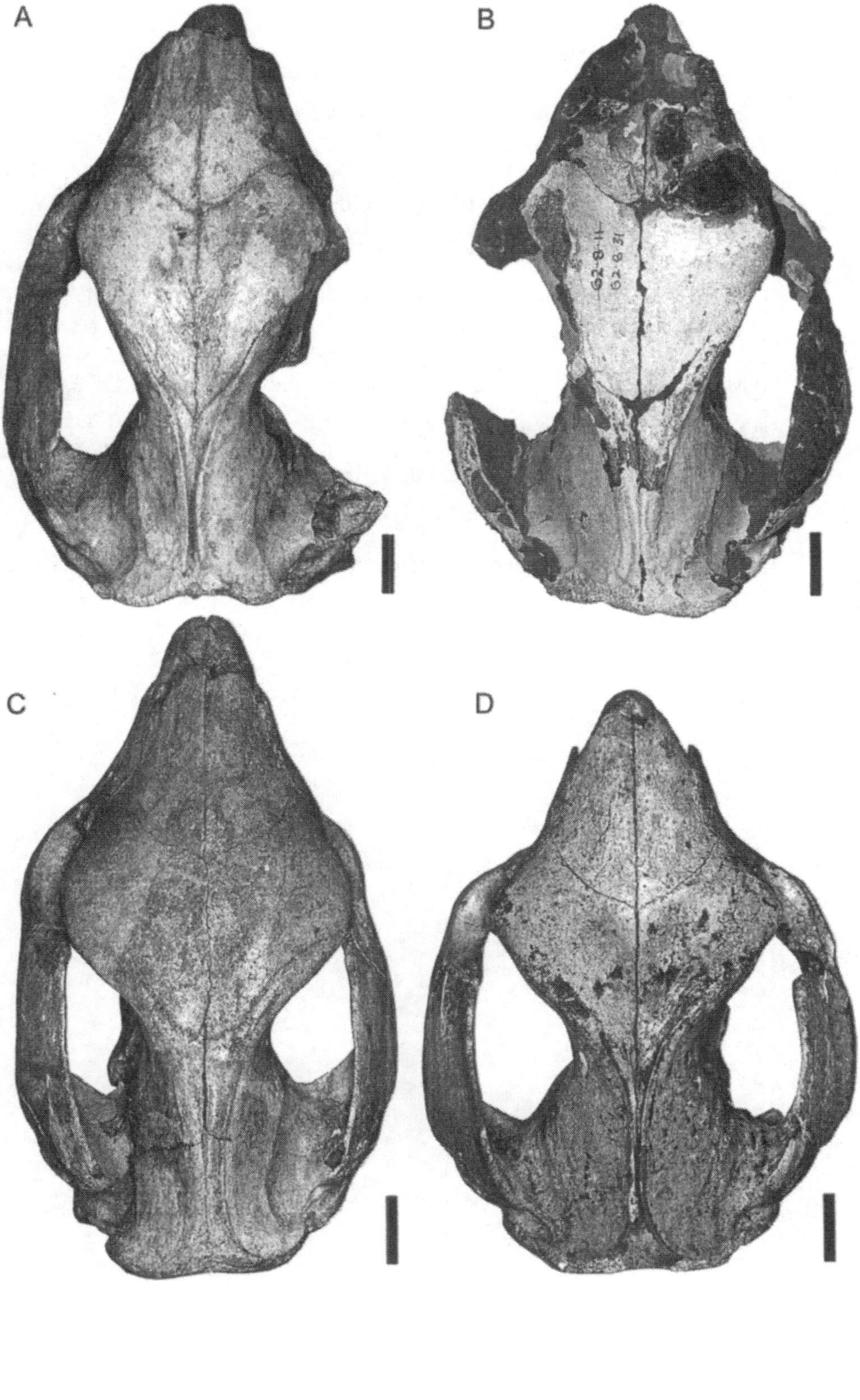
A
B
C
D

PLATE 52

Simosthenurus occidentalis

Scale bars = 20 mm

A. QVM:1971:39:1, partial juvenile cranium, dorsal view. King Island, TAS.
B. QVM:1971:39:1, lateral view.
C. SAM P27296, partial juvenile cranium, dorsal view. Alexandra Cave, SA.
D. SAM P27296, lateral view.
E. SAM P20489, partial juvenile cranium, lateral view. Victoria Fossil Cave, SA.
F. SAM P17498, partial juvenile cranium, lateral view. Green Waterhole Cave, SA.

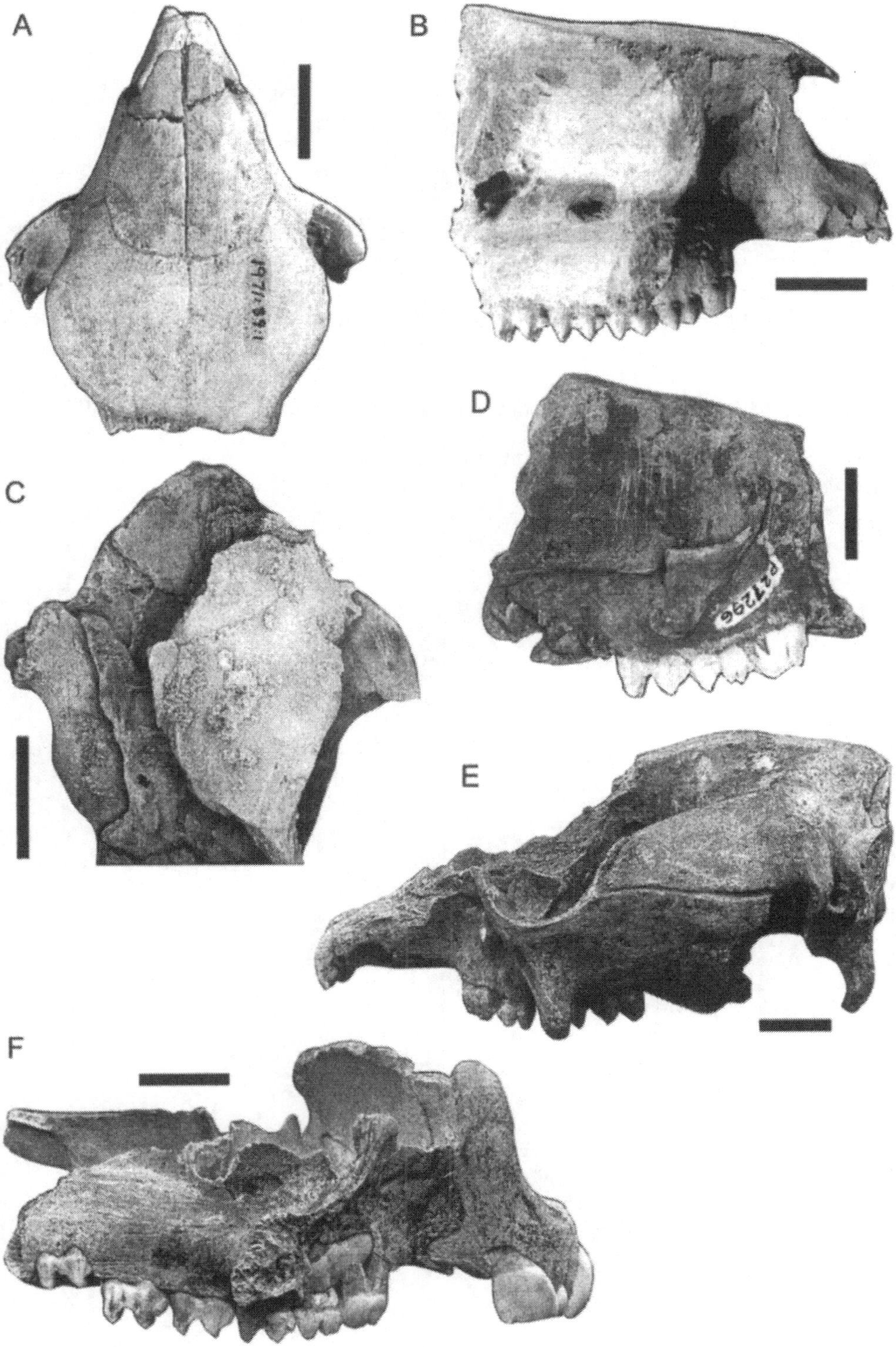
A
B
C
D
E
F

PLATE 53

Simosthenurus occidentalis

Scale bars = 20 mm

A. SAM P20489, partial juvenile cranium, palatal view. Victoria Fossil Cave, SA.
B. WAM 99.10.6, adult cranium, palatal view. Tight Entrance Cave, WA.
C. SAM P17245, adult cranium and dentaries, ventral view. Green Waterhole Cave, SA.

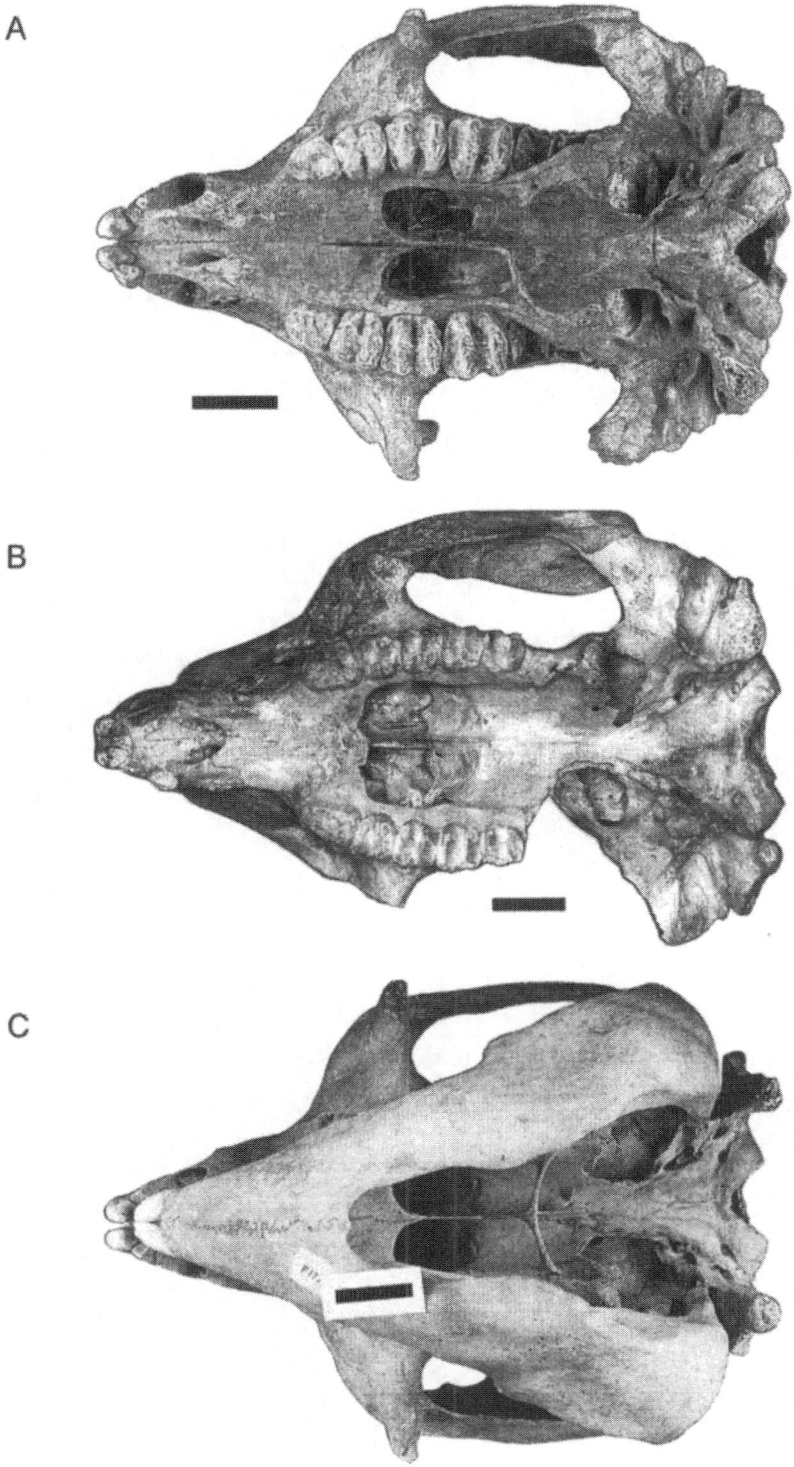
A
B
C

PLATE 54

Simosthenurus occidentalis

Scale bars = 10 mm

A. SAM P27296, right maxilla of partial juvenile cranium, stereo occlusal view. Alexandra Cave, SA.
B. WAM 97.5.78, right adult maxilla, stereo occlusal view. Tight Entrance Cave, WA.
C. SAM P16648, right maxilla of adult cranium, stereo occlusal view. Victoria Fossil Cave, SA.
D. SAM P20803, left P3, stereo occlusal view. Victoria Fossil Cave, SA.

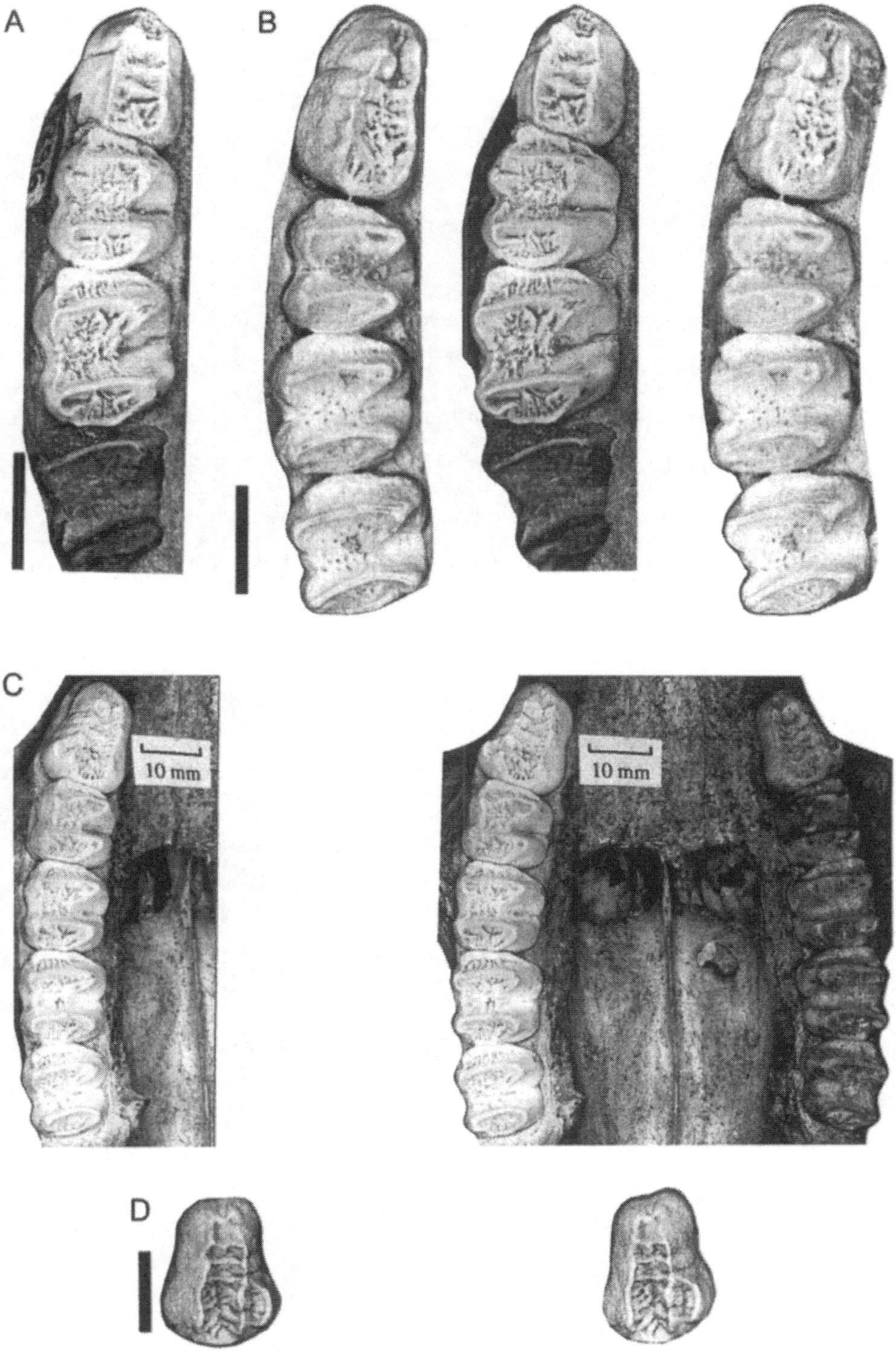

A
B
C
10 mm
10 mm
D

PLATE 55

Simosthenurus occidentalis

Scale bars = 20 mm

A. SAM P16536, left juvenile dentary, mesial view. Victoria Fossil Cave, SA.

B. SAM P16536, lateral view.

C. SAM P16664, left juvenile dentary, mesial view. Victoria Fossil Cave, SA.

D. SAM P16664, lateral view.

E. FU 1662, right adult dentary, lateral view. Green Waterhole Cave, SA.

F. SAM P20803, left dentary, lateral view. Victoria Fossil Cave, SA.

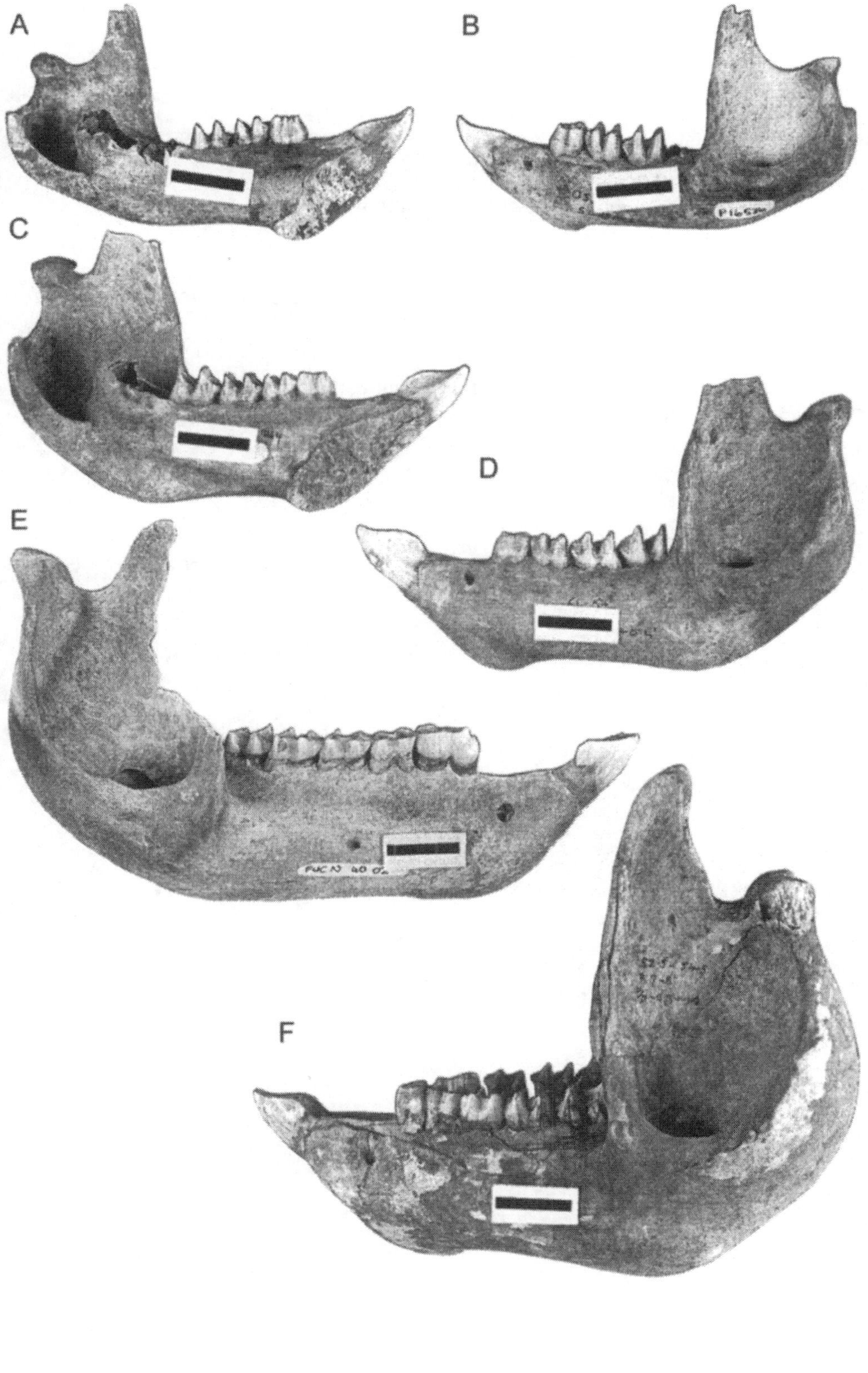
A
B
C
D
E
F

PLATE 56

Simosthenurus occidentalis

Scale bars = 20 mm

A. SAM P20820, left and right adult dentaries, occlusal view. Green Waterhole Cave, SA.
B. FU unregistered, left and right juvenile dentaries, occlusal view. Green Waterhole Cave, SA.
C. FU 1662, left and right adult dentaries, occlusal view. Green Waterhole Cave, SA.
D. SAM P20803, left and right dentaries, occlusal view. Victoria Fossil Cave, SA.
E. SAM P17245, anterior portion of left and right adult dentaries, occlusal view. Green Waterhole Cave, SA.

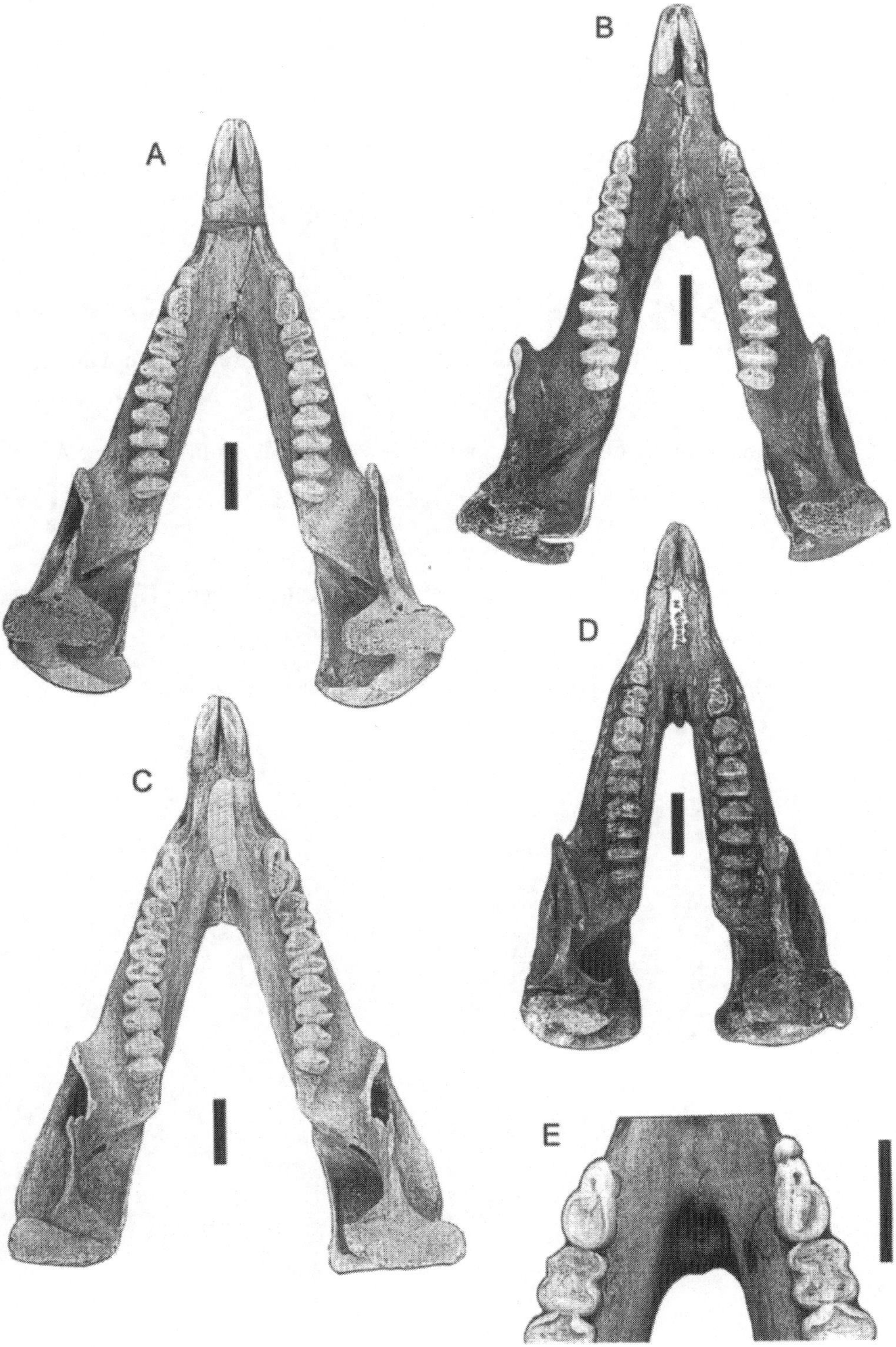
A
B
C
D
E

PLATE 57

Simosthenurus occidentalis

Scale bars = 20 mm (A-C); 10 mm (D-F)

A. WAM 97.5.74, right adult dentary, mesial view. Tight Entrance Cave, WA.
B. SAM P20820, right adult dentary, mesial view. Green Waterhole Cave, SA.
C. FU 1662, right adult dentary, mesial view. Green Waterhole Cave, SA.
D. WAM 63.2.78, left adult dentary, stereo occlusal view. Mammoth Cave, WA.
E. WAM 97.5.74, right adult dentary, stereo occlusal view. Tight Entrance Cave, WA.
F. SAM P16664, left juvenile dentary, stereo occlusal view. Victoria Fossil Cave, SA.

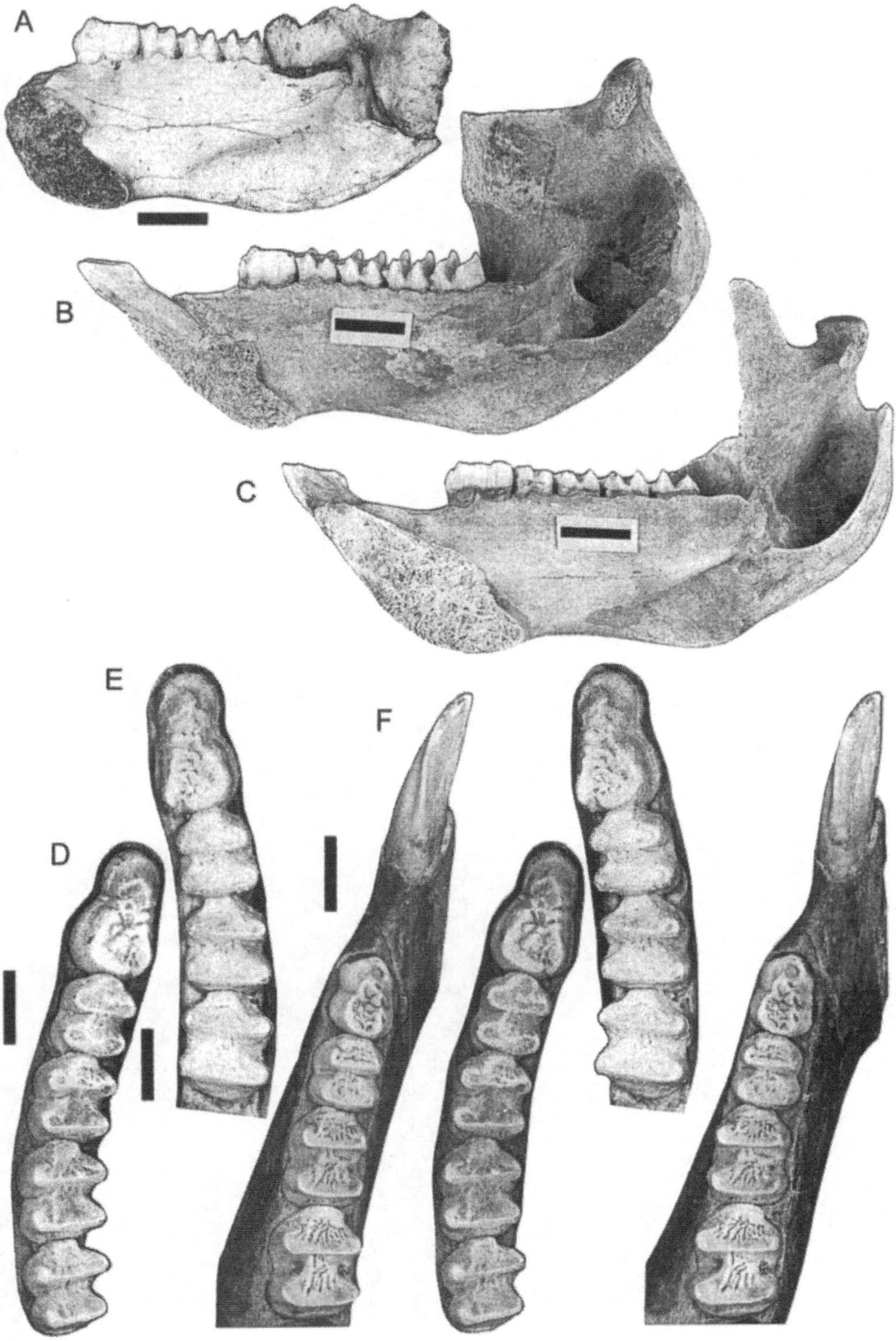
A
B
C
D
E
F

PLATE 58

"Simosthenurus" antiquus

Scale bars = 20 mm

A. QM F2975 (holotype), left adult maxilla, lateral view. Chinchilla, QLD.
B. QM F2975, mesial view.
C. QM F2975, occlusal view.
D. QM F2975, sketch of left P3, occlusal view (posterolingual corner reconstructed).

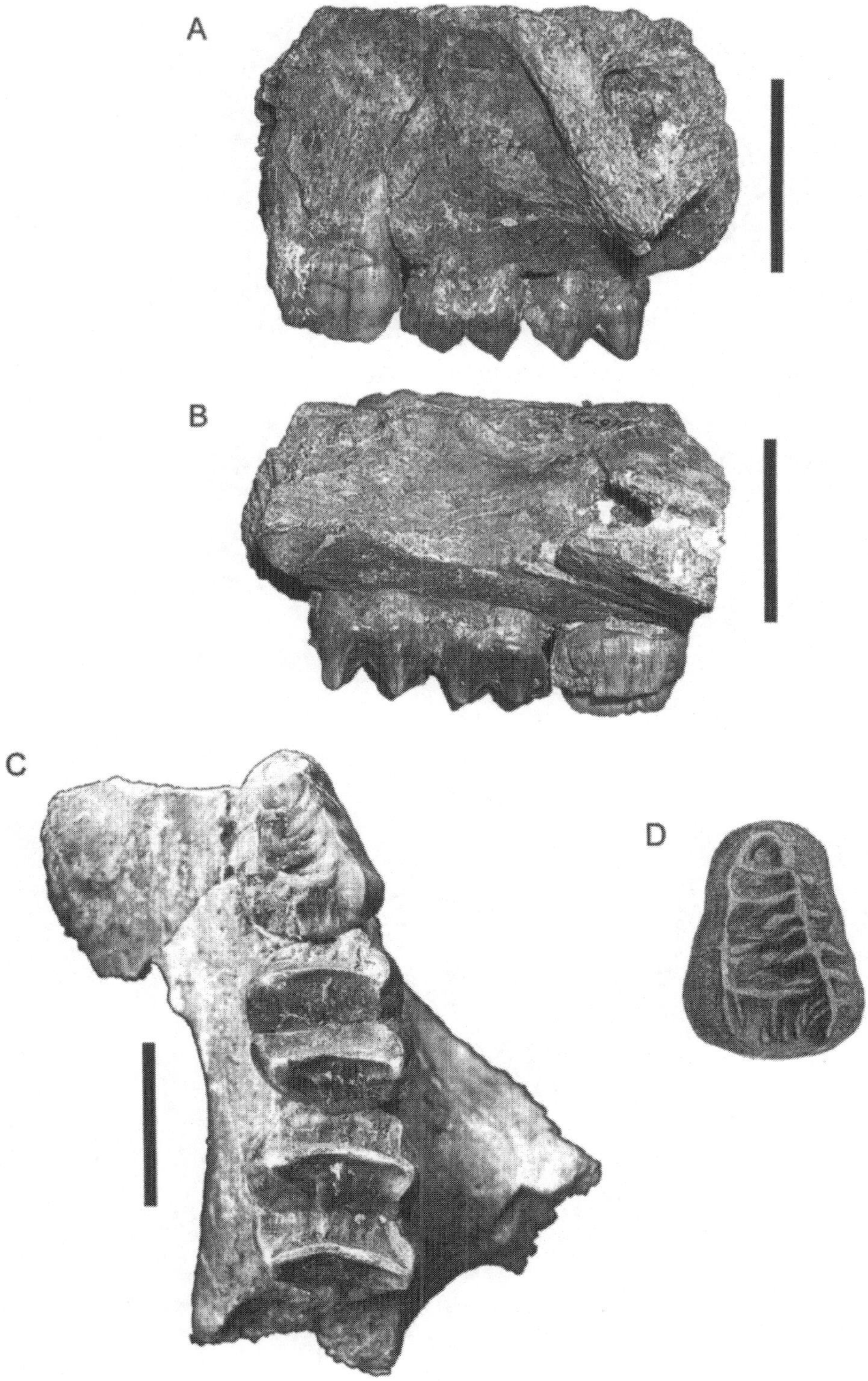
A
B
C
D

PLATE 59

"Simosthenurus" antiquus

Scale bars = 20 mm

A. QM F2931, left adult dentary, lateral view. Darling Downs, QLD.
B. QM F2931, left adult dentary, mesial view.
C. QM F2931, close-up of m2-4, occlusal view.
D. QM F2931, left adult dentary (cast), stereo occlusal view.

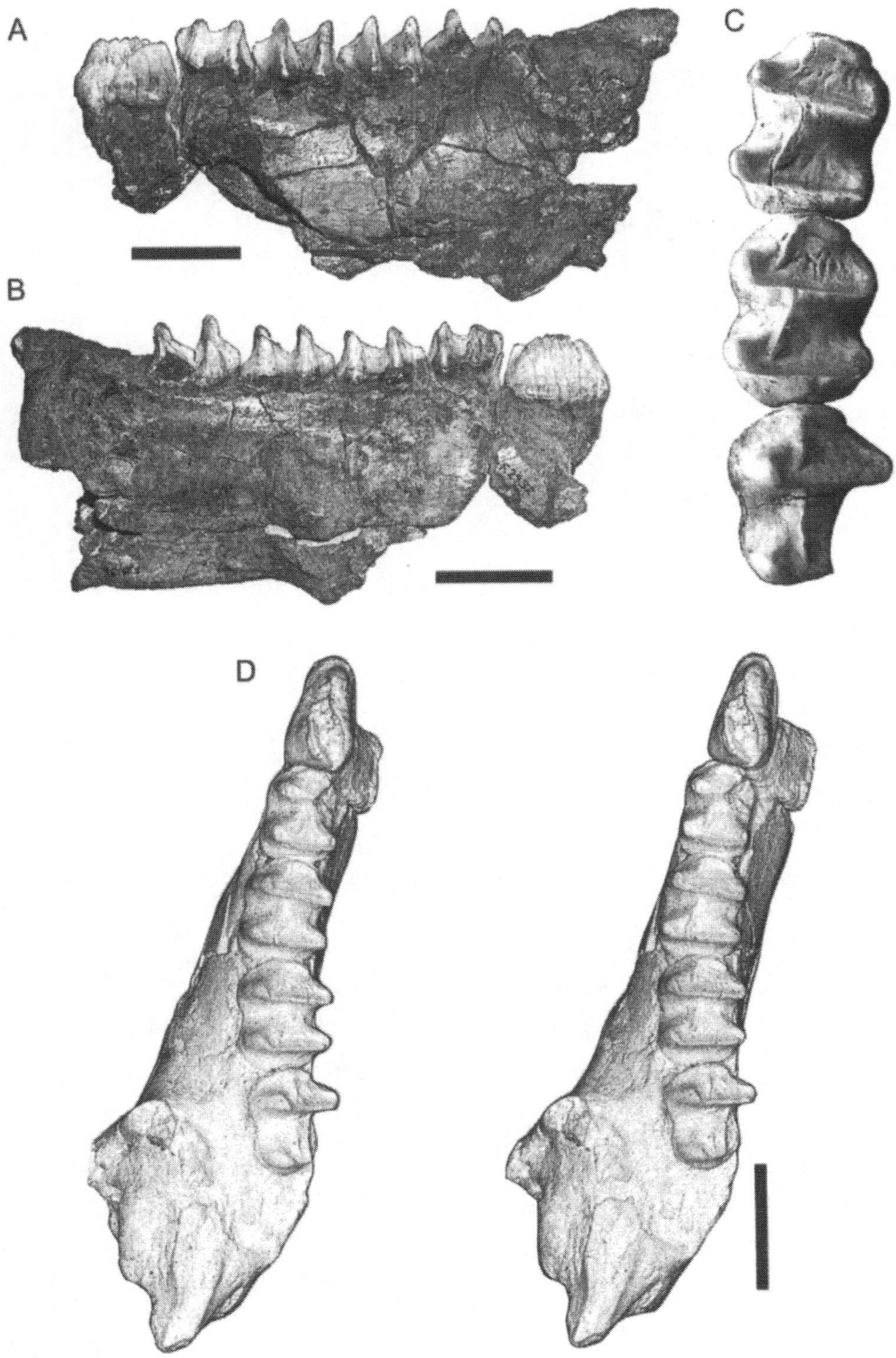
A
B
C
D

PLATE 60

"Simosthenurus" sp. cf. *antiquus*

Scale bars = 20 mm (A-C); 10 mm (D)

A. UCMP 47947, right adult dentary, lateral view. Warburton River, SA.
B. UCMP 47947, mesial view.
C. UCMP 47947, stereo occlusal view.
D. UCMP 56901, left adult maxilla, stereo occlusal view. Warburton River, SA.

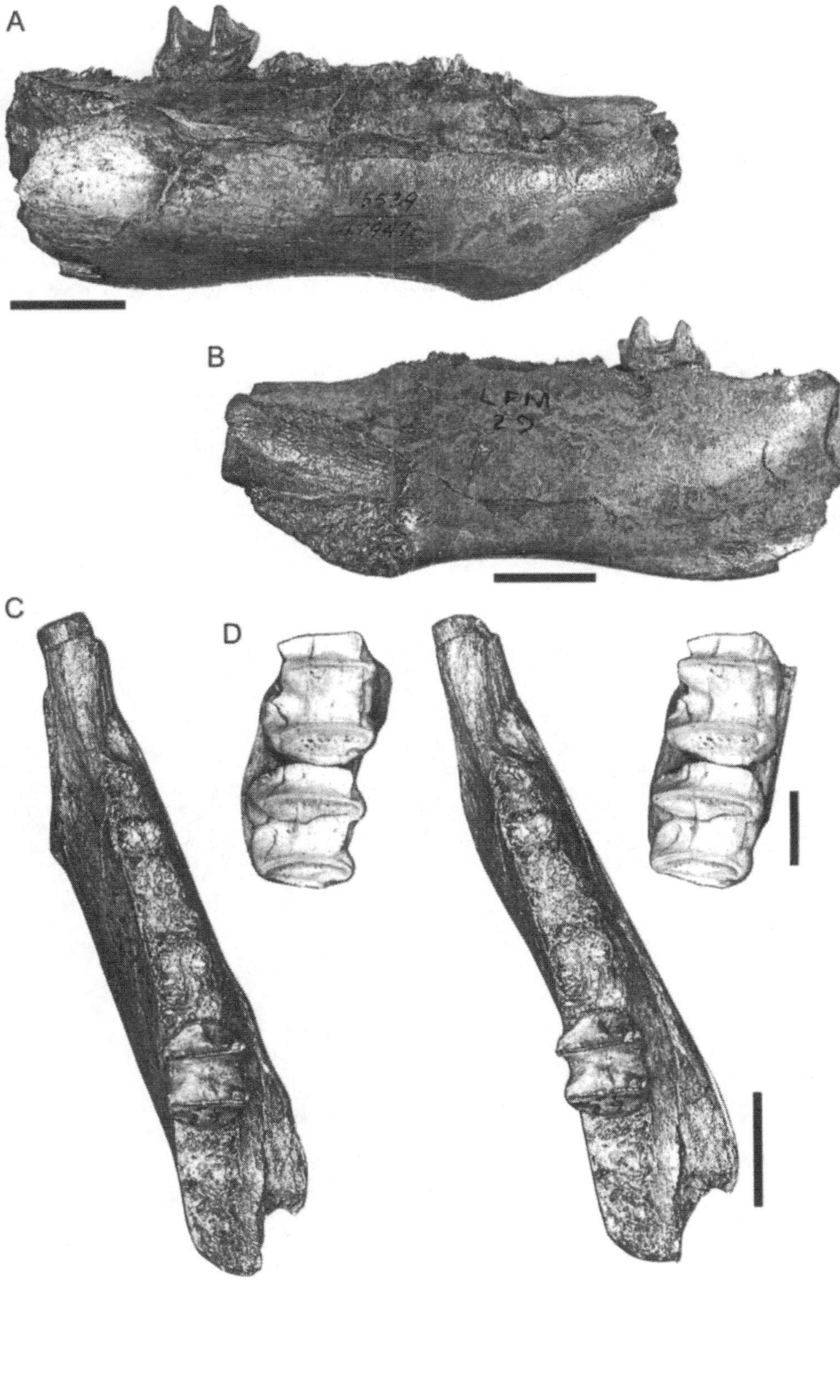
A
B
LFM
29
C
D

PLATE 61

"Simosthenurus" baileyi

Scale bars = 20 mm

A. SAM P13670 (holotype), partial cranium, lateral view. Western Brothers Island, SA.
B. SAM P13670, dorsal view.
C. SAM P13670, stereo palatal view.

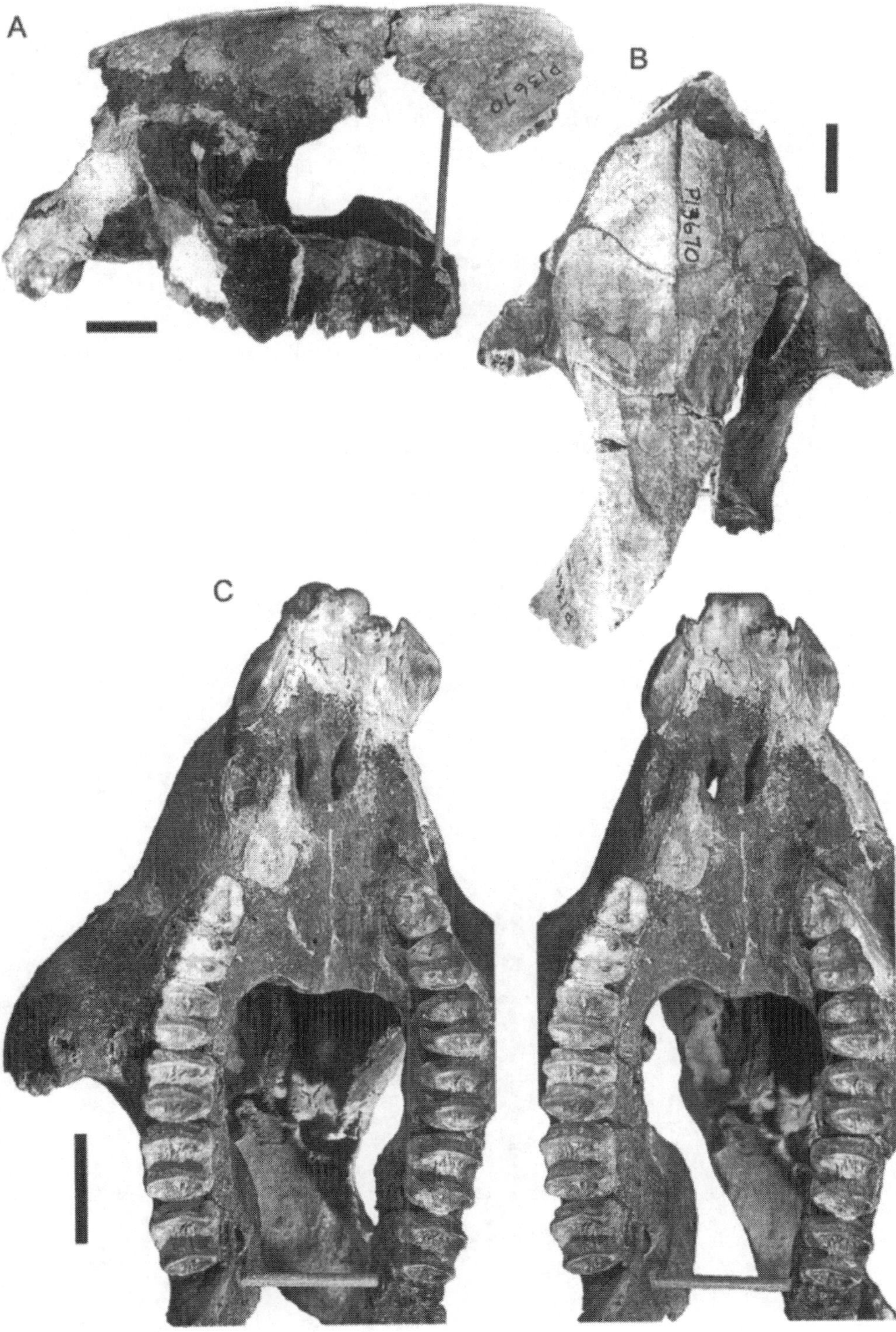
A
B
C
P13670
P13670

PLATE 62

"Simosthenurus" baileyi (A-F)

"Simosthenurus" sp. cf. *baileyi* (G)

Scale bars = 10 mm

A. SAM P13670 (holotype), close-up of left maxilla, occlusal view. Western Brothers Island, SA.
B. SAM P13670, excavated left P3, stereo occlusal view.
C. FU 0167, left P3, stereo occlusal view.
D. UCMP 60674, left adult maxilla, lateral view. Cooper Creek, SA.
E. UCMP 60674, mesial view.
F. UCMP 60674, occlusal view.
G. QM F9104, right M1 metaloph, stereo posterior view (inverted). Bluff Downs, QLD.

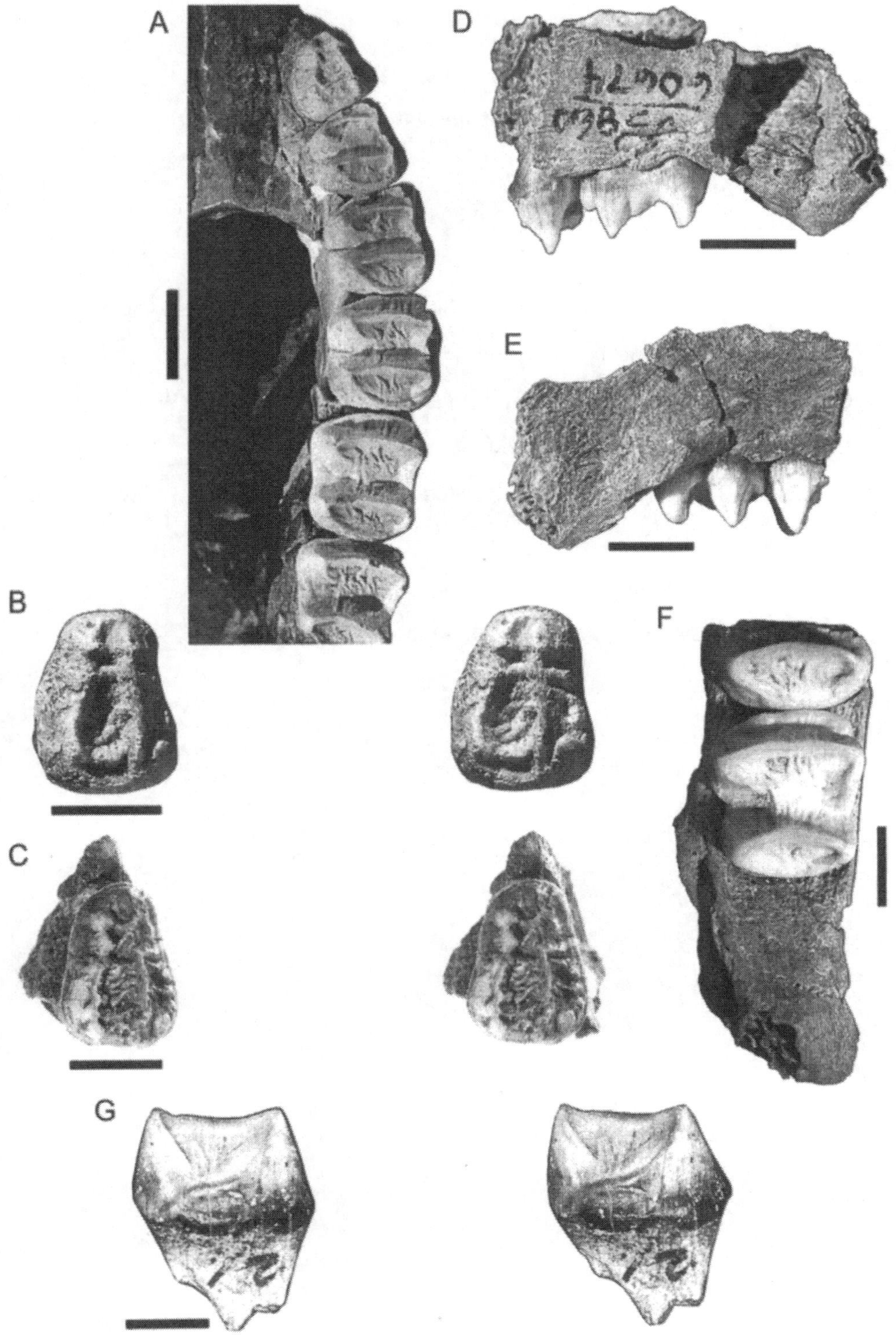
A
B
C
D
E
F
G

PLATE 63

"Simosthenurus" baileyi

Scale bars = 20 mm (A-B, F); 10 mm (C-E)

A. SAM P16558, left adult dentary, lateral view. Victoria Fossil Cave, SA.
B. SAM P16558, mesial view.
C. WAM 03.11.01, right p3, buccal view. Lindsay Hall Cave, WA.
D. WAM 03.11.01, occlusal view.
E. SAM P13670 (holotype), excavated left p3, stereo occlusal view. Western Brothers Island, SA.
F. SAM P16558, stereo occlusal view.
G. SAM P13670, right lower cheek tooth row, occlusal view.

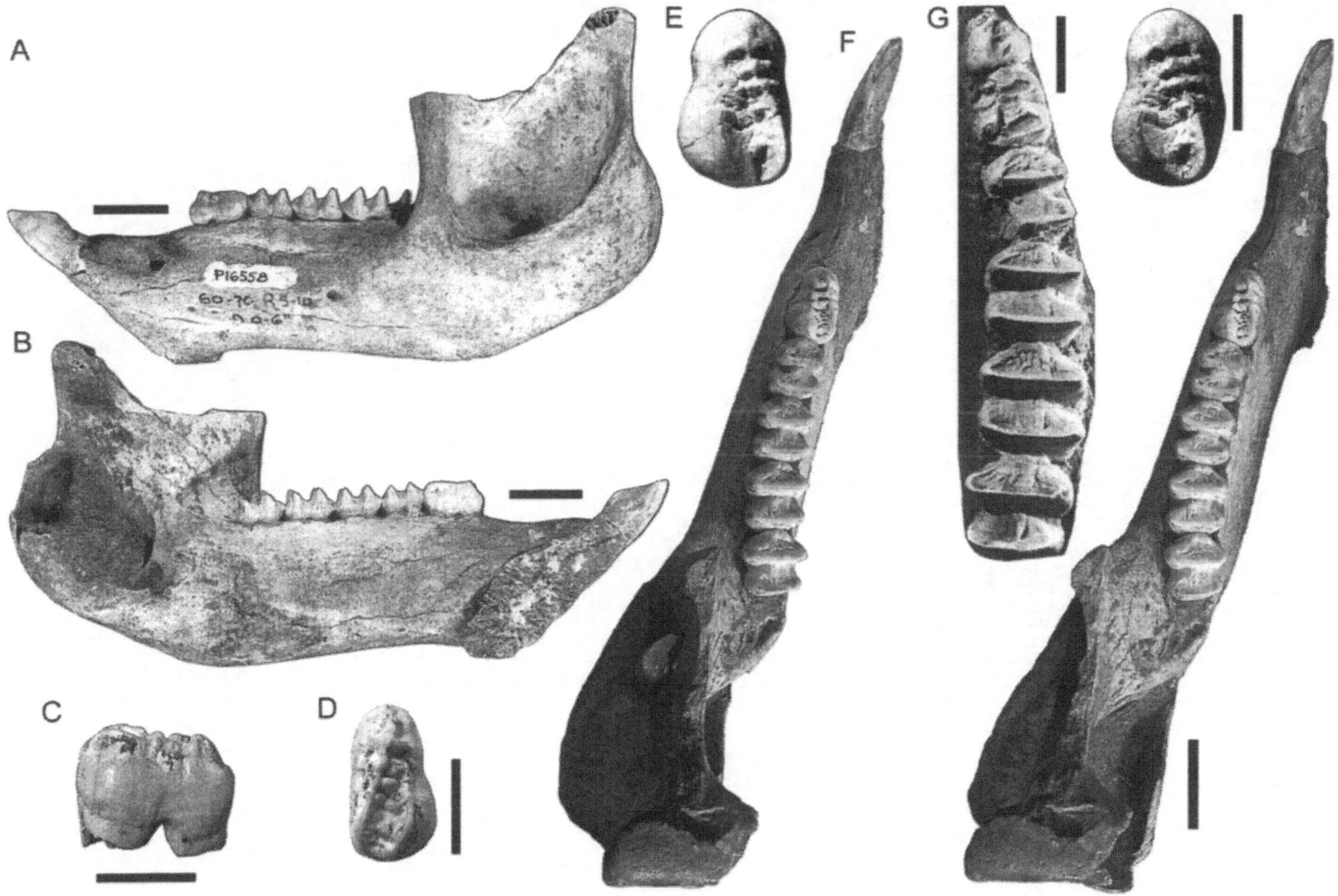
A
B
C
D
E
F
G
P16558
60-70 R5-10
0-6"

PLATE 64

"Simosthenurus" brachyselenis

Scale bars = 20 mm (A-B); 10 mm (C-D)

A. AM F31026 (holotype), right juvenile dentary, lateral view. Phosphate Mine, Wellington Caves, NSW.
B. AM F31026, mesial view.
C. AM F31026, stereo occlusal view.
D. AM F31026, excavated right p3, stereo occlusal view.

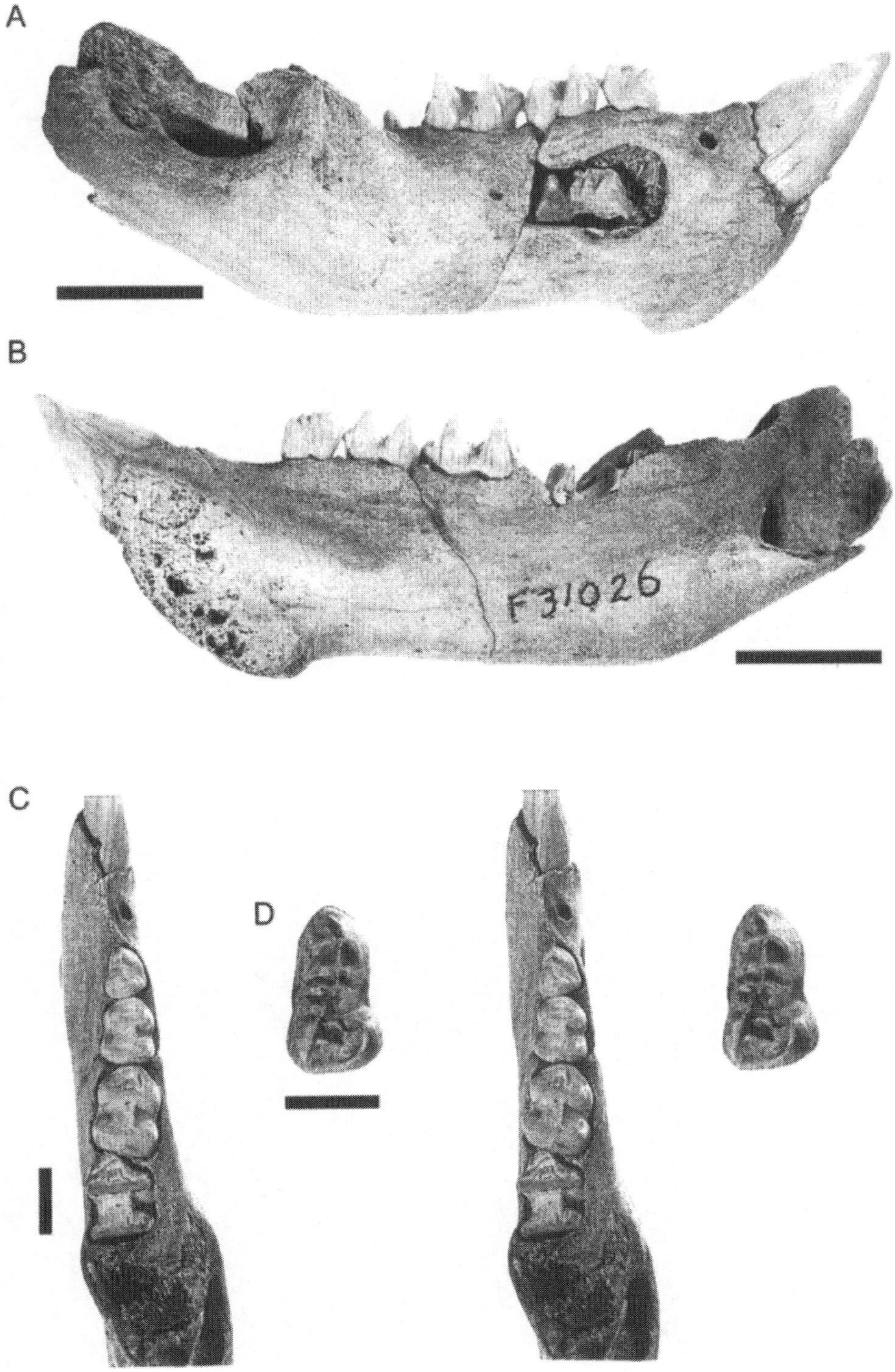
A
B
F31026
C
D

PLATE 65

"Simosthenurus" brachyselenis

Scale bars = 20 mm

A. AM F106043, left adult dentary, lateral view. Bone Cave, NSW.
B. AM F106043, mesial view.
C. AM F106043, stereo occlusal view.

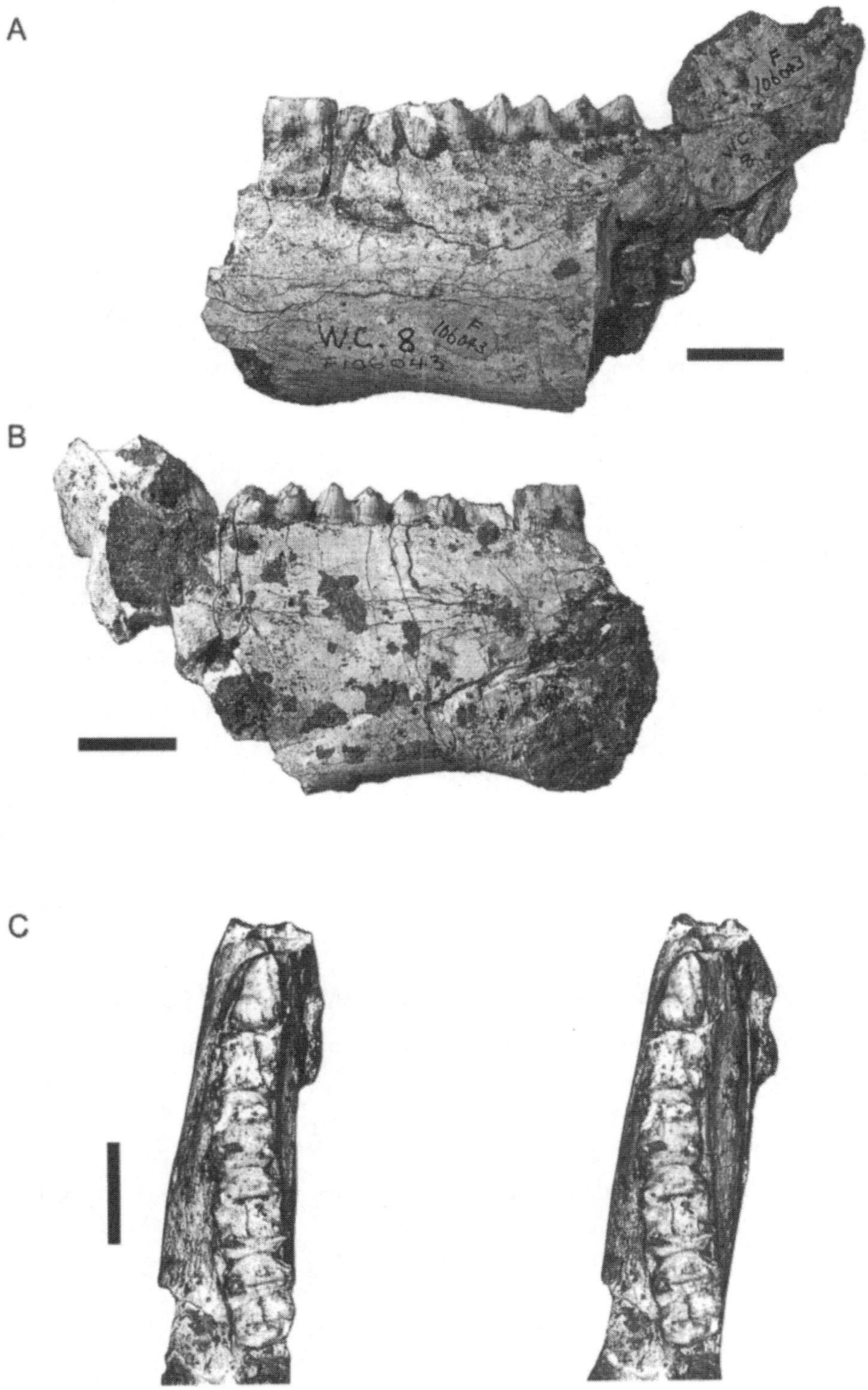
A
W.C. 8
F106043
B
C

PLATE 66

"Simosthenurus" pales

Scale bars = 10 mm

A. QM F815 (holotype), right P3, buccal view. Darling Downs, QLD.
B. QM F815, lingual view.
C. QM F815, occlusal view.
D. SAM P unregistered a, right P3, occlusal view. Henschke's Fossil Cave, SA.
E. SAM P unregistered b, right P3, occlusal view. Henschke's Fossil Cave, SA.
F. SAM P unregistered c, right P3, occlusal view. Henschke's Fossil Cave, SA.
G. WAM 99.11.270, right juvenile maxilla, stereo occlusal view. Tight Entrance Cave, WA.
H. SAM P unregistered d, left P3, occlusal view. Henschke's Fossil Cave, SA.
I. SAM P32320, right adult maxilla, occlusal view. Henschke's Fossil Cave, SA.
J. FU 0377, left adult maxilla, occlusal view. Victoria Fossil Cave, SA.
K. WAM 03.11.03, left juvenile premaxilla / maxilla, occlusal view. Tight Entrance Cave, WA.
L. AM F41534, left adult maxilla, occlusal view. Geurie, NSW.
M. SAM P32318, left adult maxilla, occlusal view. Henschke's Fossil Cave, SA.

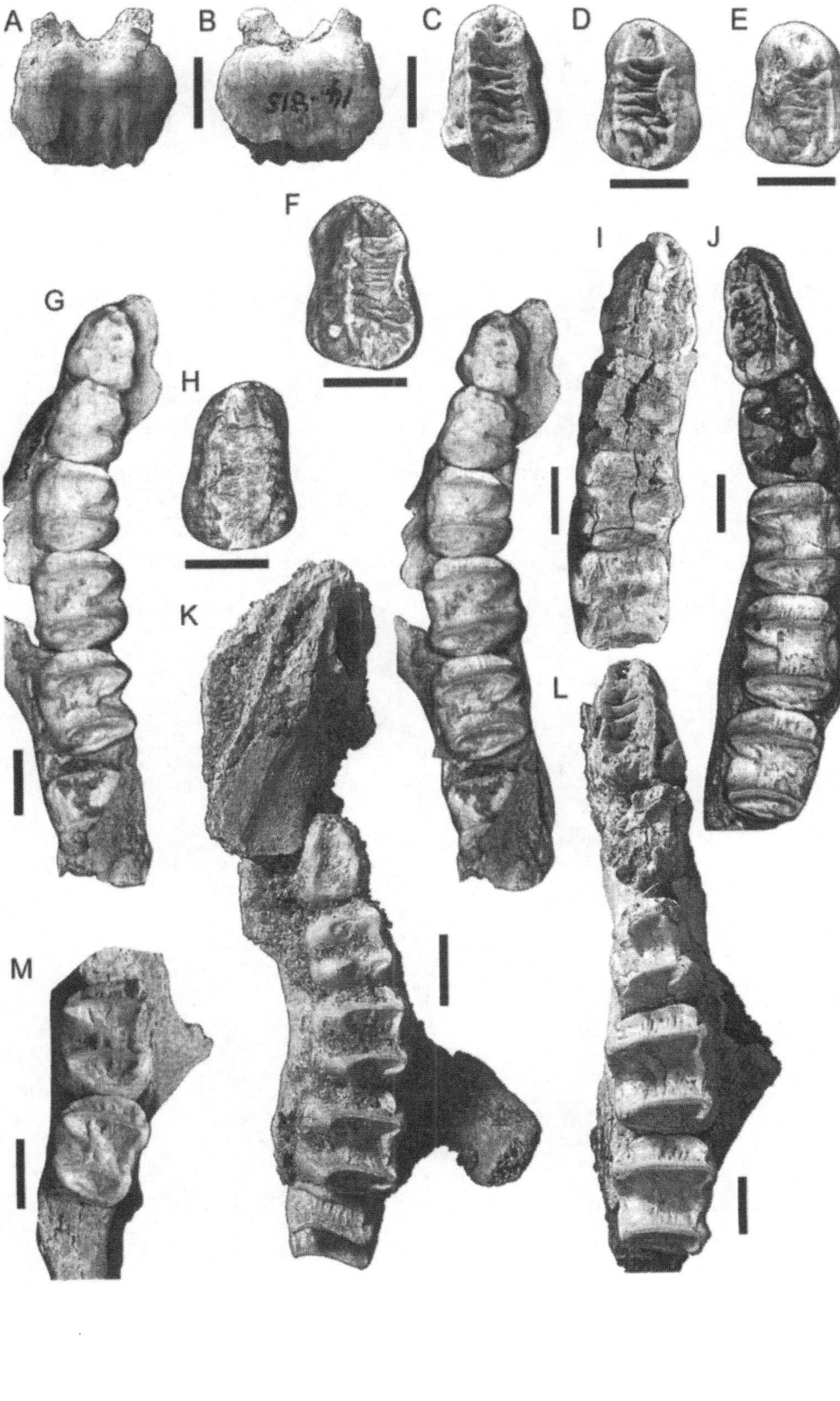
A
B
C
D
E
F
G
H
I
J
K
L
M

PLATE 67

"Simosthenurus" pales

Scale bars = 20 mm (A-C); 10 mm (D-F)

A. WAM 03.11.04, right adult dentary, lateral view. Tight Entrance Cave, WA.
B. WAM 03.11.04, mesial view.
C. WAM 03.11.04, stereo occlusal view.
D. SAM P unregistered, left and right i1s, anterior view. Henschke's Fossil Cave, SA.
E. SAM P unregistered, right I3, buccal view. Henschke's Fossil Cave, SA.
F. SAM P unregistered, left I3, buccal view. Henschke's Fossil Cave, SA.

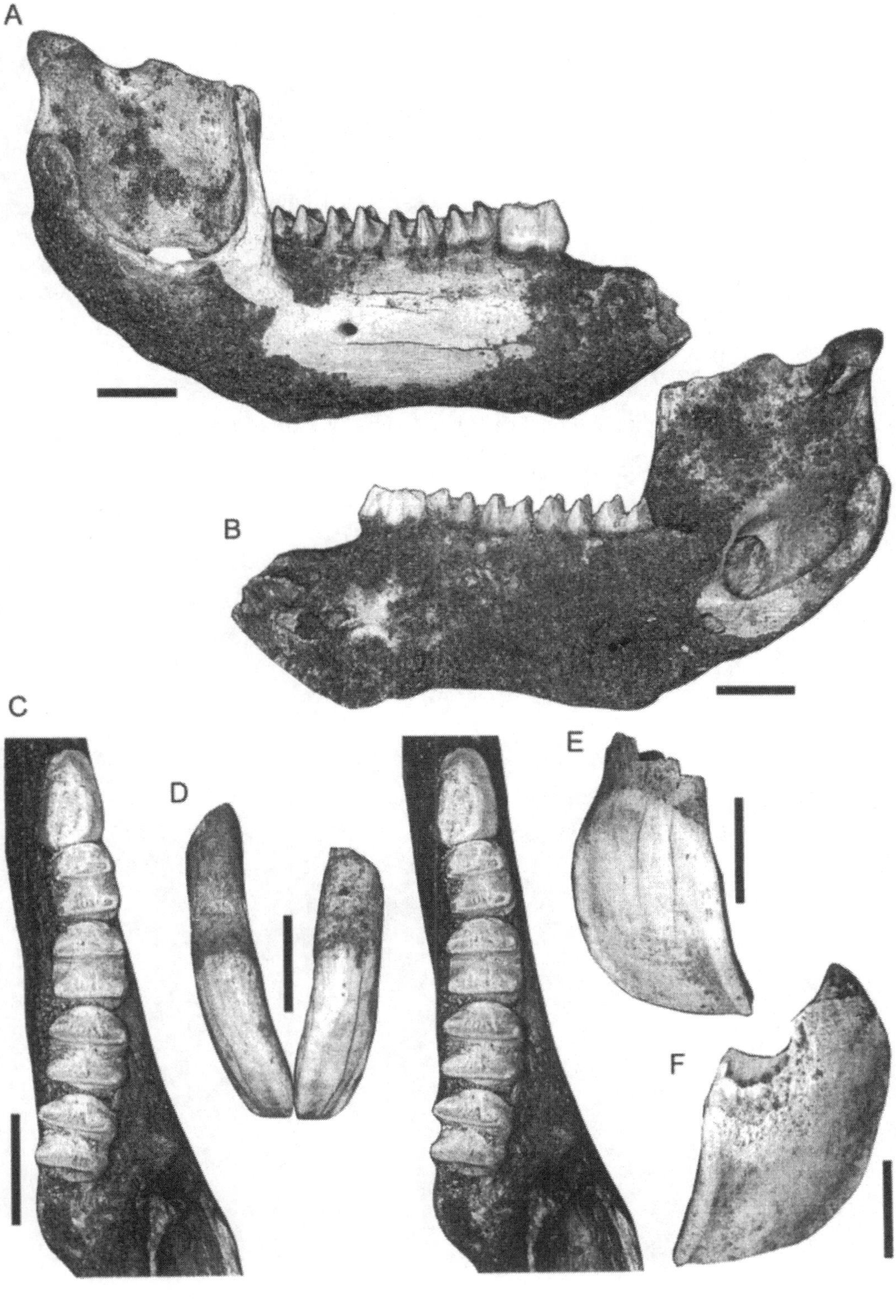
A
B
C
D
E
F

PLATE 68

"Simosthenurus" pales

Scale bars = 20 mm (A-E, G); 10 mm (F)

A. SAM P27797, right juvenile dentary, lateral view. Victoria Fossil Cave, SA.
B. SAM P27797, mesial view.
C. FU 1083, right adult dentary, lateral view. Victoria Fossil Cave, SA.
D. FU 1083, mesial view.
E. SAM P27797, stereo occlusal view.
F. SAM P27797, excavated p3, stereo occlusal view.
G. FU 1083, stereo occlusal view.

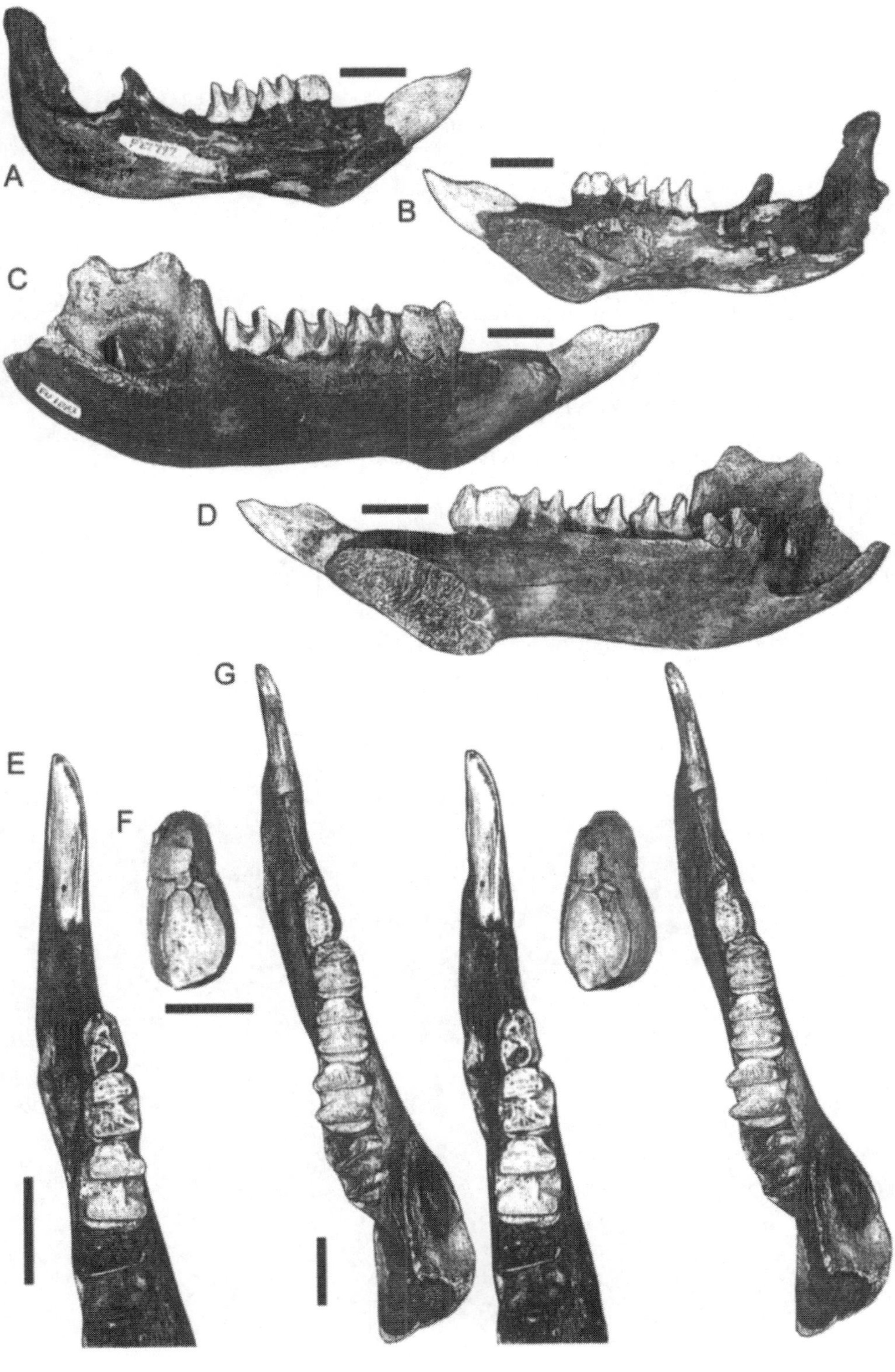
A
B
C
D
E
F
G

PLATE 69

"Simosthenurus" pales

Scale bars = 20 mm (A-B, D-F); 10 mm (C, G-H)

A. QM F9428, left and right juvenile dentaries, lateral view. Brisbane River, QLD.
B. QM F9428, occlusal view.
C. QM F9428, excavated p3, stereo occlusal view.
D. UCMP 56807, left juvenile dentary, lateral view. Warburton River, SA.
E. UCMP 56807, mesial view.
F. UCMP 56807, occlusal view.
G. AM F29594, right p3, buccal view. Wellington Caves, NSW.
H. AM F29594, occlusal view.

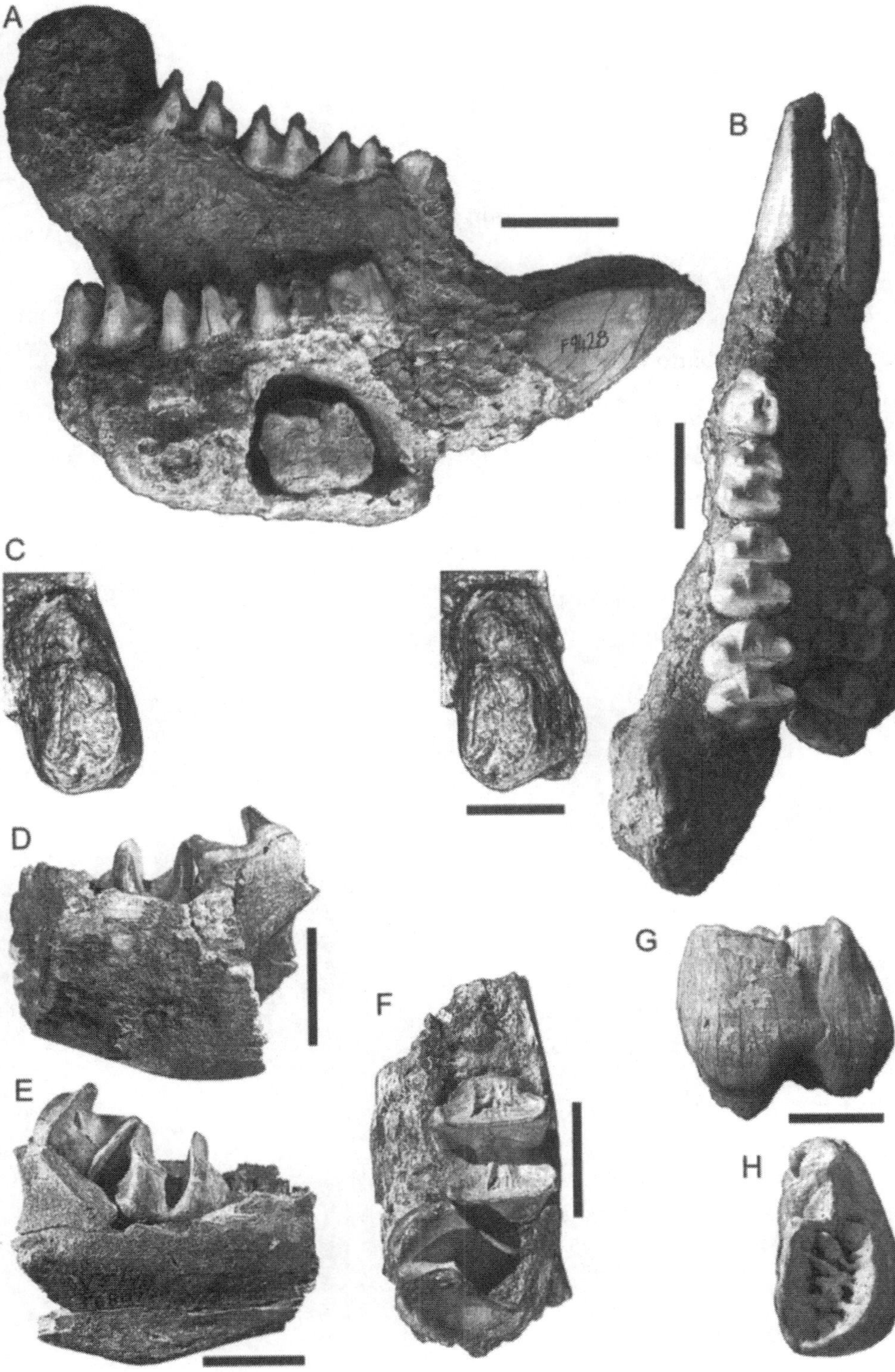
A
B
C
D
E
F
G
H

PLATE 70

"Simosthenurus" tirarensis

Scale bars = 20 mm (A-B); 10 mm (C-F)

A. SAM P40128 (holotype), left juvenile dentary, lateral view. Camel Swamp Yard, Warburton River, SA.
B. SAM P40128, mesial view.
C. SAM P40128, excavated p3, buccal view.
D. SAM P40128, excavated p3, lingual view.
E. SAM P40128, excavated p3, stereo occlusal view.
F. SAM P40128, stereo occlusal view.

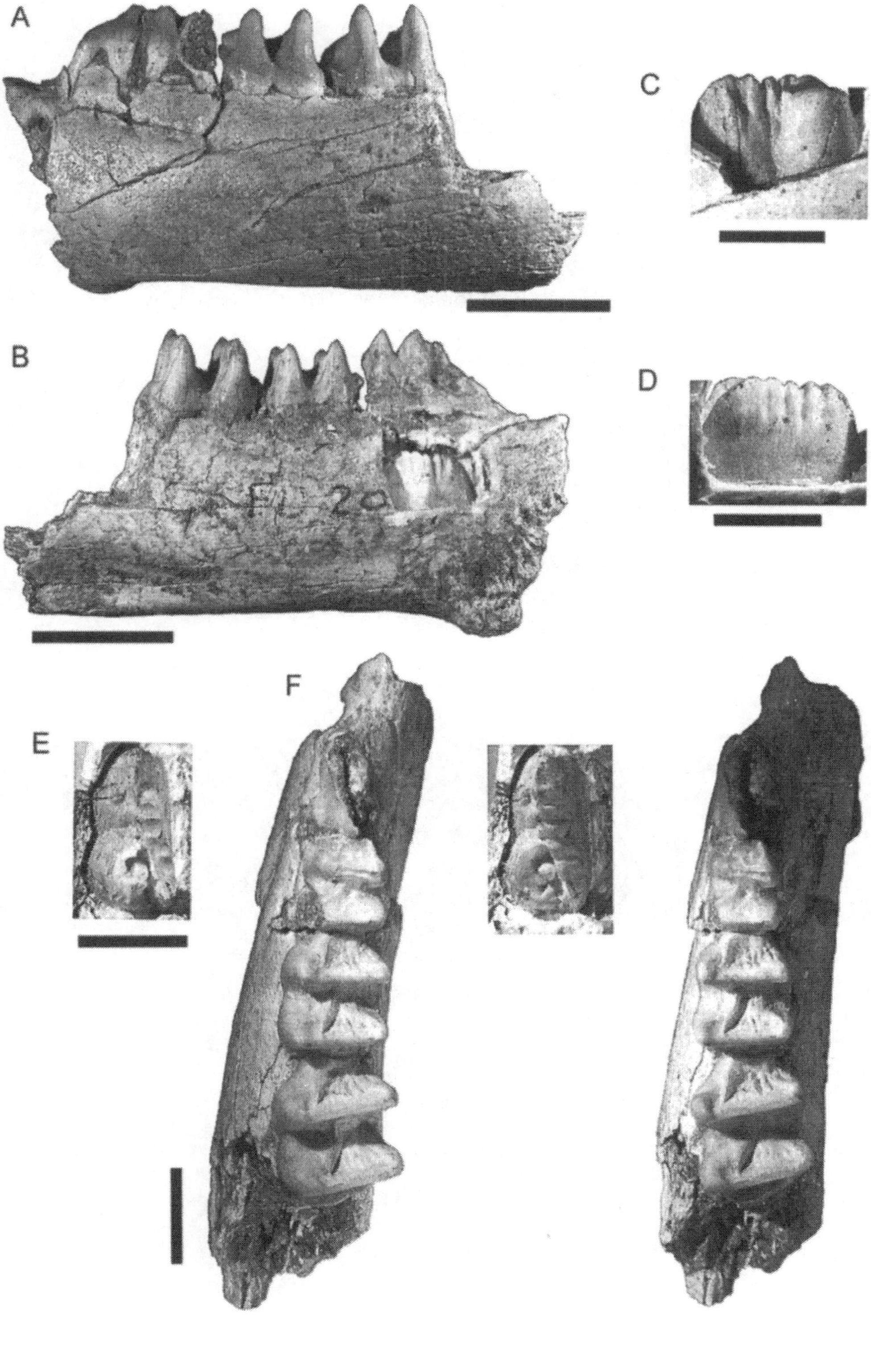
A
B
C
D
E
F

PLATE 71

Procoptodon goliah

Scale bars = 20 mm

A. SAM P16695, adult cranium, dorsal view. Victoria Fossil Cave, SA.

B. SAM P16695, palatal view.

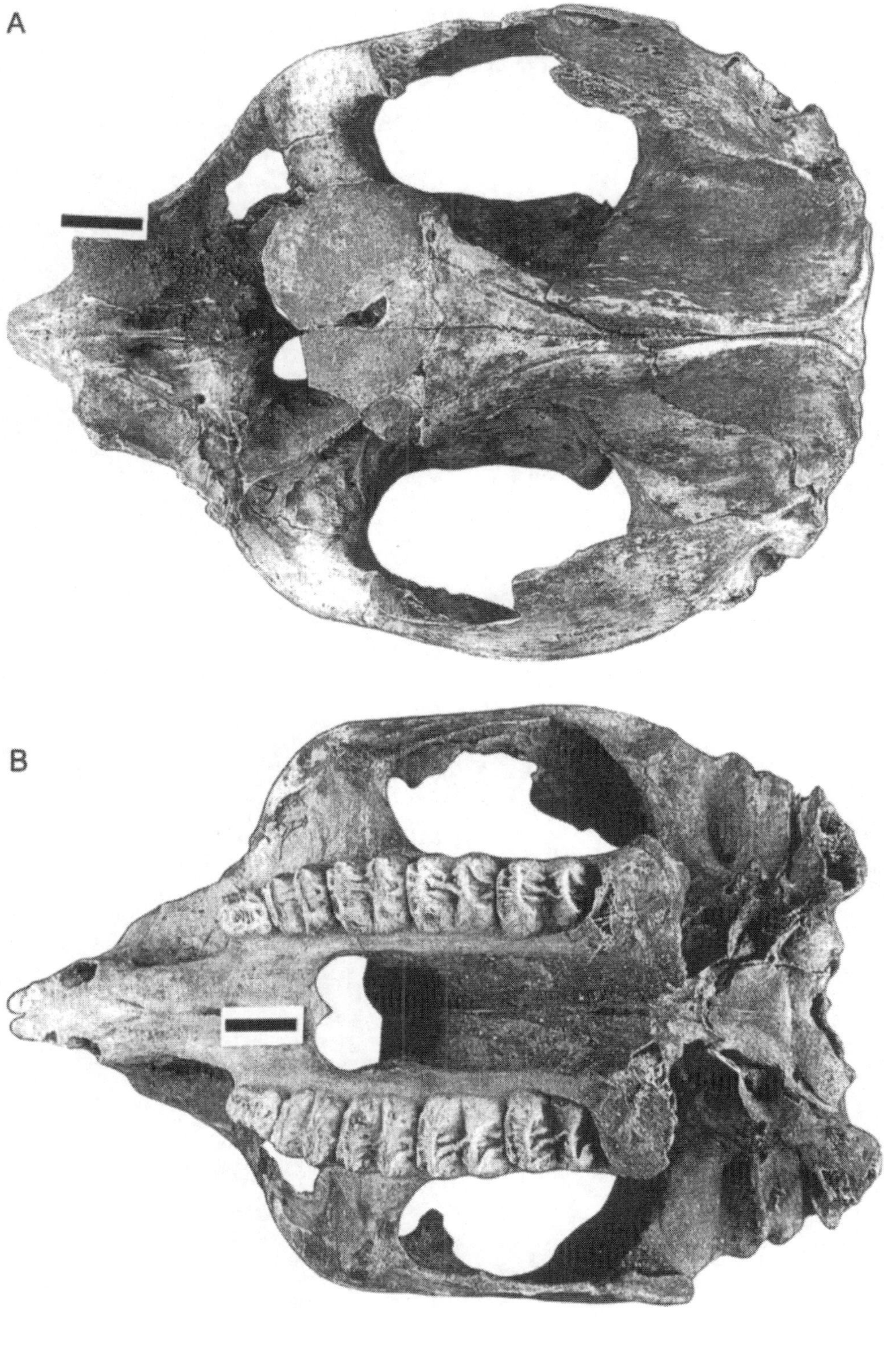

PLATE 72

Procoptodon goliah

Scale bars = 20 mm

A. SAM P16696, partial juvenile cranium, dorsal view. Victoria Fossil Cave, SA.
B. SAM P16696, palatal view.

A

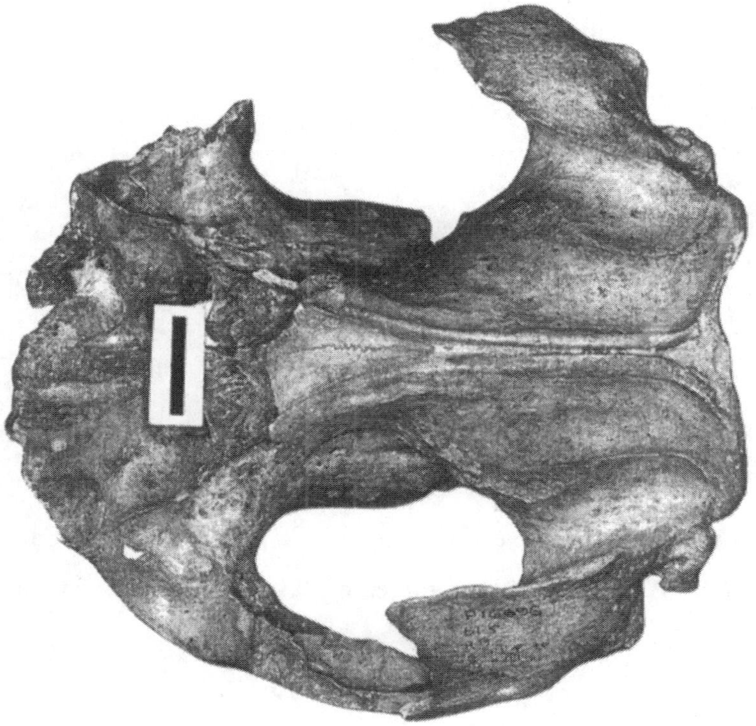

B

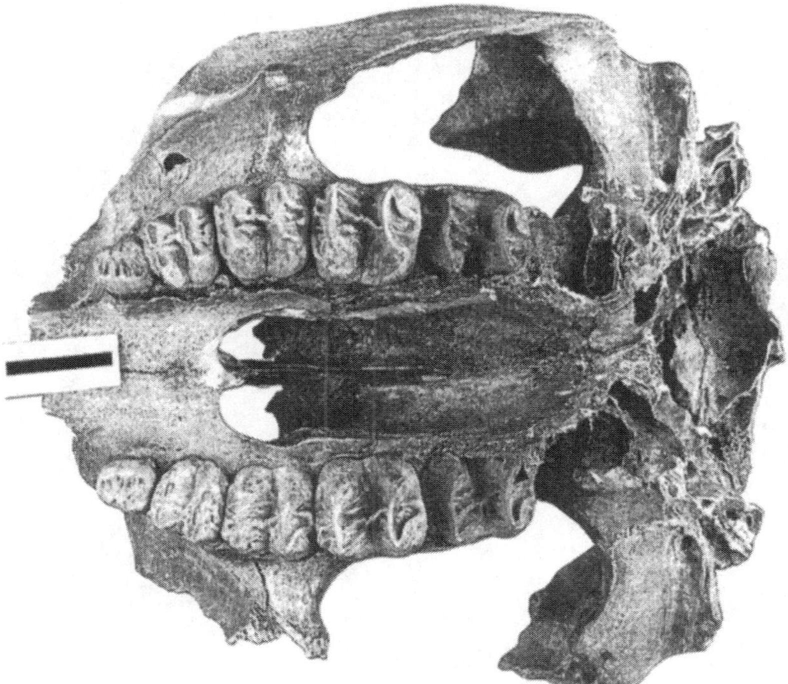

PLATE 73

Procoptodon goliah

Scale bars = 20 mm

A. SAM P27293, partial juvenile cranium, lateral view. Victoria Fossil Cave, SA.
B. SAM P16696, partial juvenile cranium, lateral view. Victoria Fossil Cave, SA.
C. SAM P16695, adult cranium, lateral view. Victoria Fossil Cave, SA.
D. SAM P27886, partial adult cranium and dentaries, lateral view. Lake Victoria, NSW.

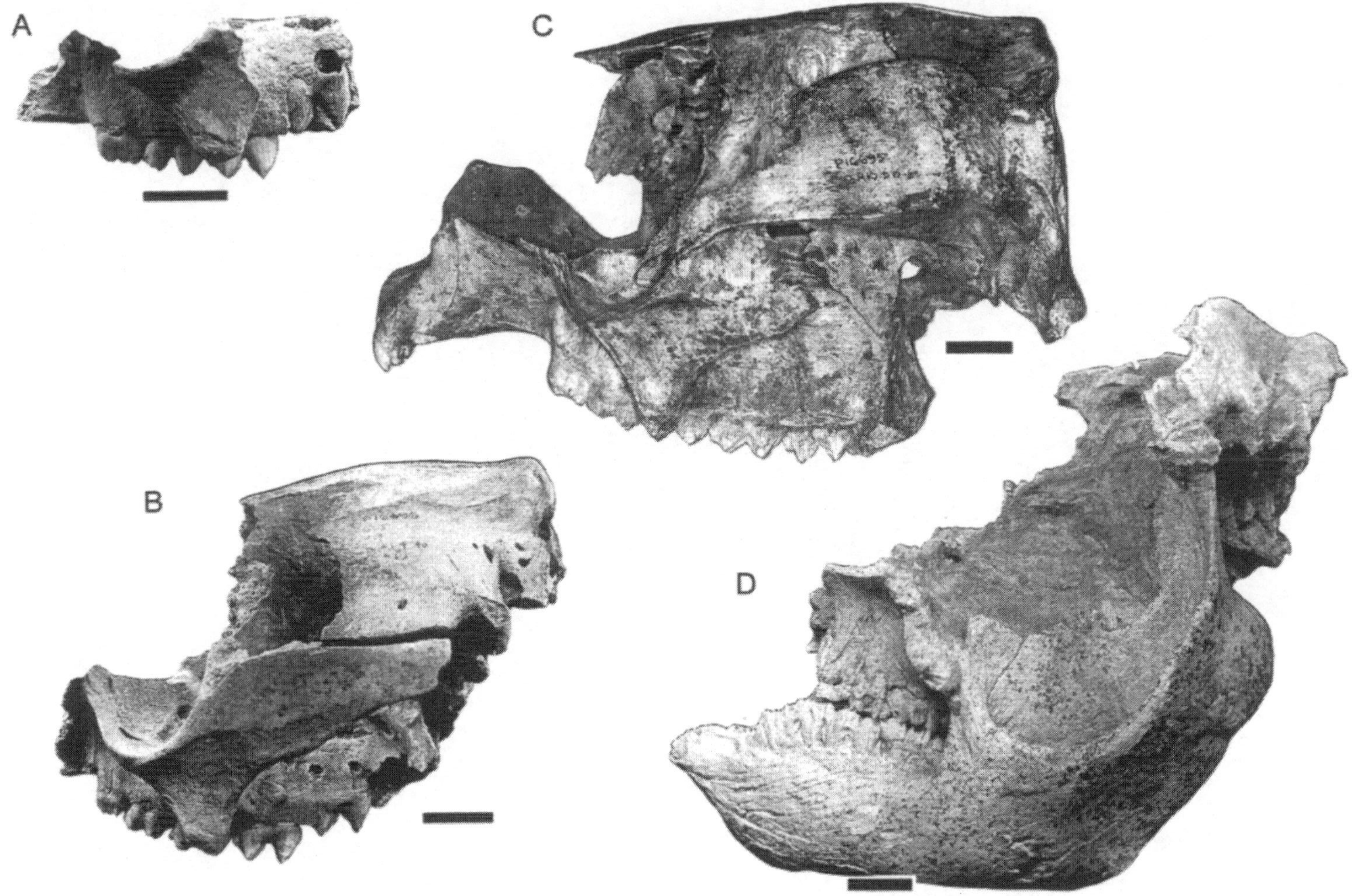
A
B
C
D

PLATE 74

Procoptodon goliah

Scale bars = 20 mm

A. SAM P16696, partial juvenile cranium, anterior view. Victoria Fossil Cave, SA.
B. SAM P16696, posterior view.
C. SAM P16695, adult cranium, anterior view. Victoria Fossil Cave, SA.
D. SAM P16695, posterior view.

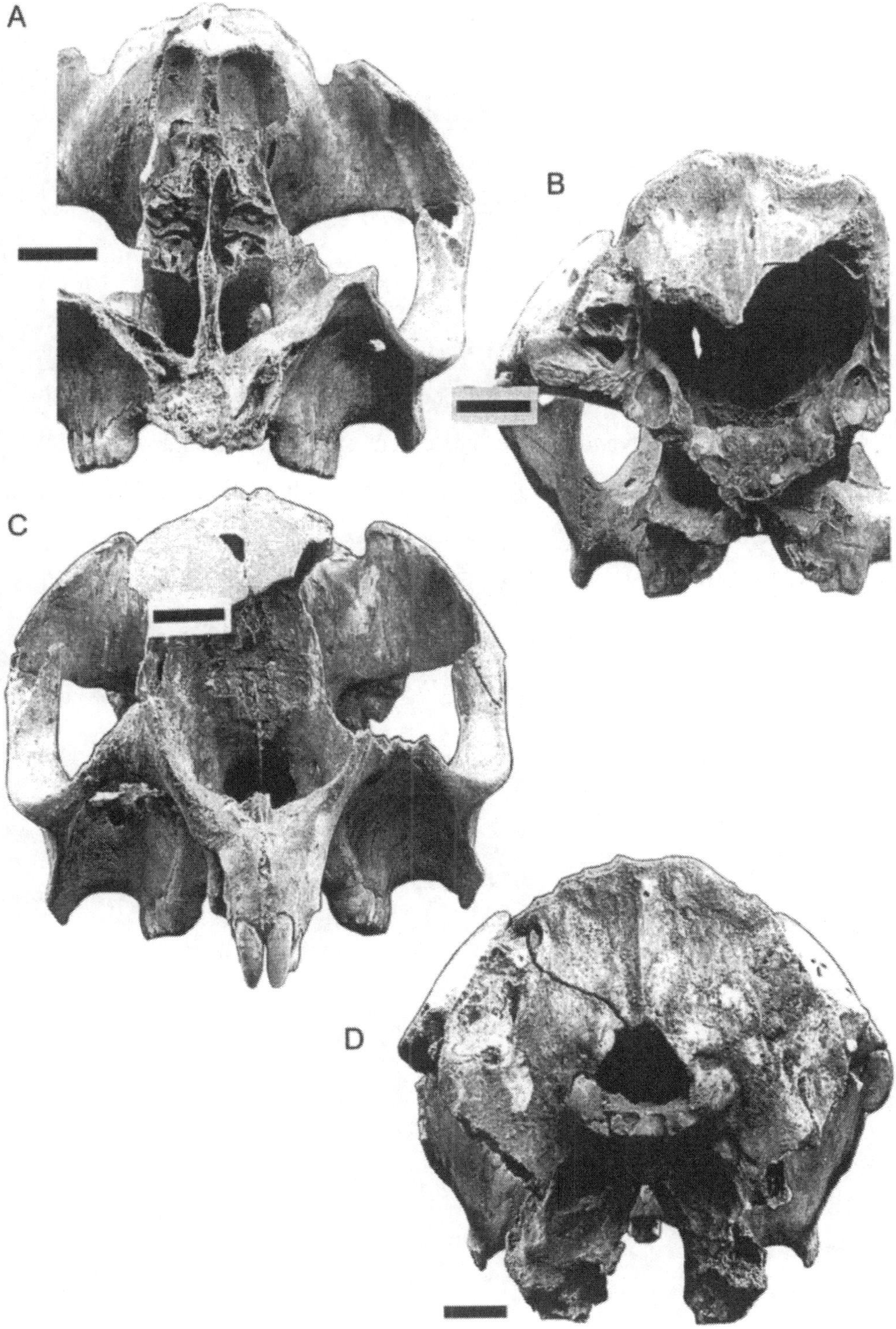
A
B
C
D

PLATE 75

Procoptodon goliah

Scale bars = 10 mm (A-F); 5 mm (G-I)

A. AM F102010, left juvenile maxilla, occlusal view. Bingara, NSW.
B. AM F106574, left juvenile maxilla, occlusal view. Bingara, NSW.
C. AM F106574, excavated left P3, occlusal view.
D. SAM P27293, juvenile palate, stereo occlusal view. Alexandra Cave, SA.
E. SAM P16696, excavated right P3, stereo occlusal view. Victoria Fossil Cave, SA.
F. SAM P16696, right maxilla of partial juvenile cranium, stereo occlusal view.
G. FU 0147, left I3, stereo buccal view. Victoria Fossil Cave, SA.
H. FU 0147, stereo lingual view.
I. FU 0147, occlusal view.

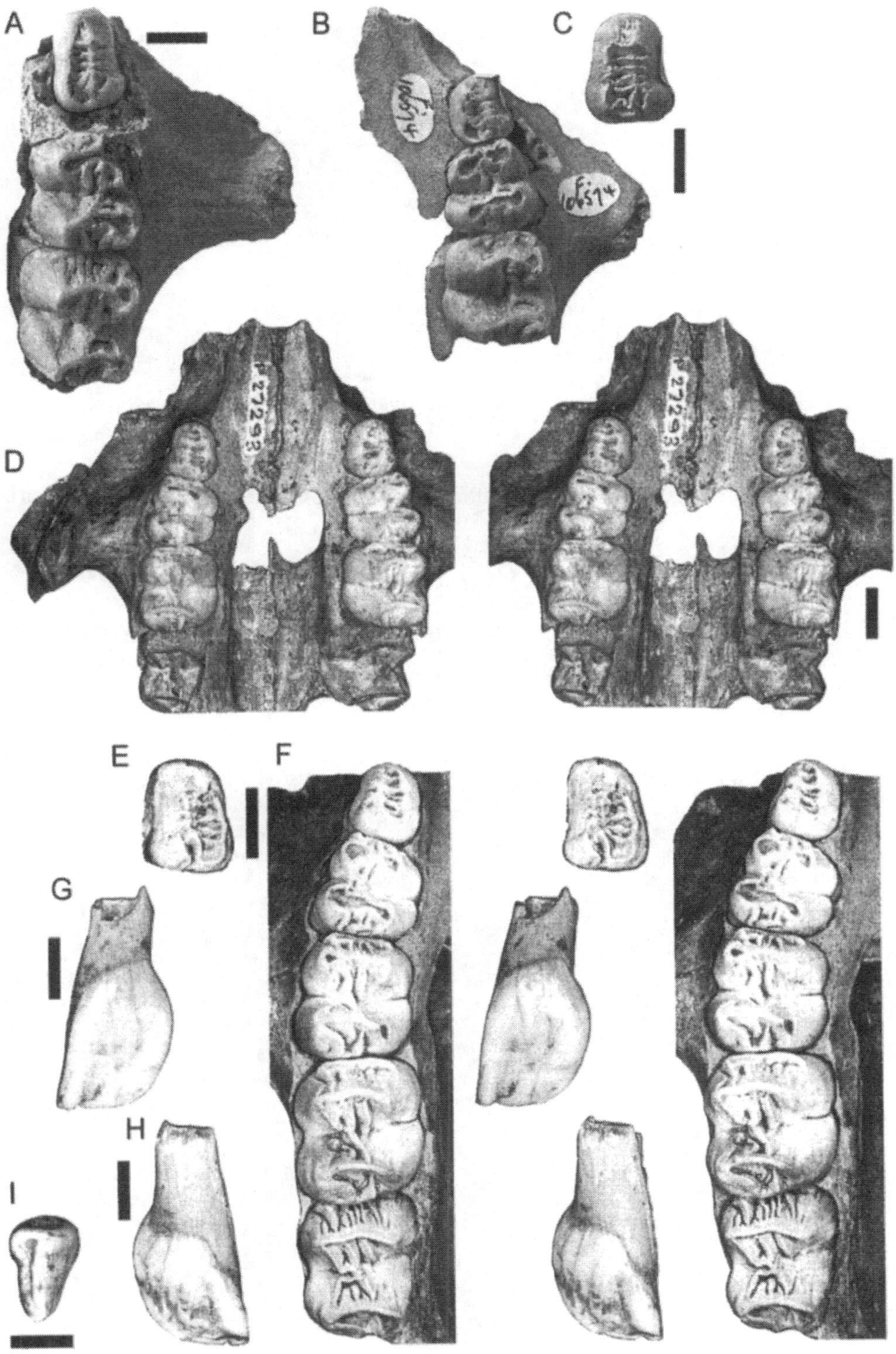
A
B
C
D
E
F
G
H
I

PLATE 76

Procoptodon goliah

Scale bars = 20 mm

A. SAM P20666, left juvenile dentary, mesial view. Victoria Fossil Cave, SA.
B. SAM P20666, lateral view.
C. SAM P16704, left juvenile dentary, mesial view. Victoria Fossil Cave, SA.
D. SAM P16704, lateral view.
E. SAM P16690, right adult dentary, mesial view (photo reversed). Victoria Fossil Cave, SA.
F. SAM P16690, lateral view (photo reversed).
G. AM F101638, right juvenile dentary, mesial view (photo reversed). Bingara, NSW.
H. AM F102004, left adult dentary, lateral view. Bingara, NSW.
I. QM F4457, left adult dentary, lateral view. Locality unknown.

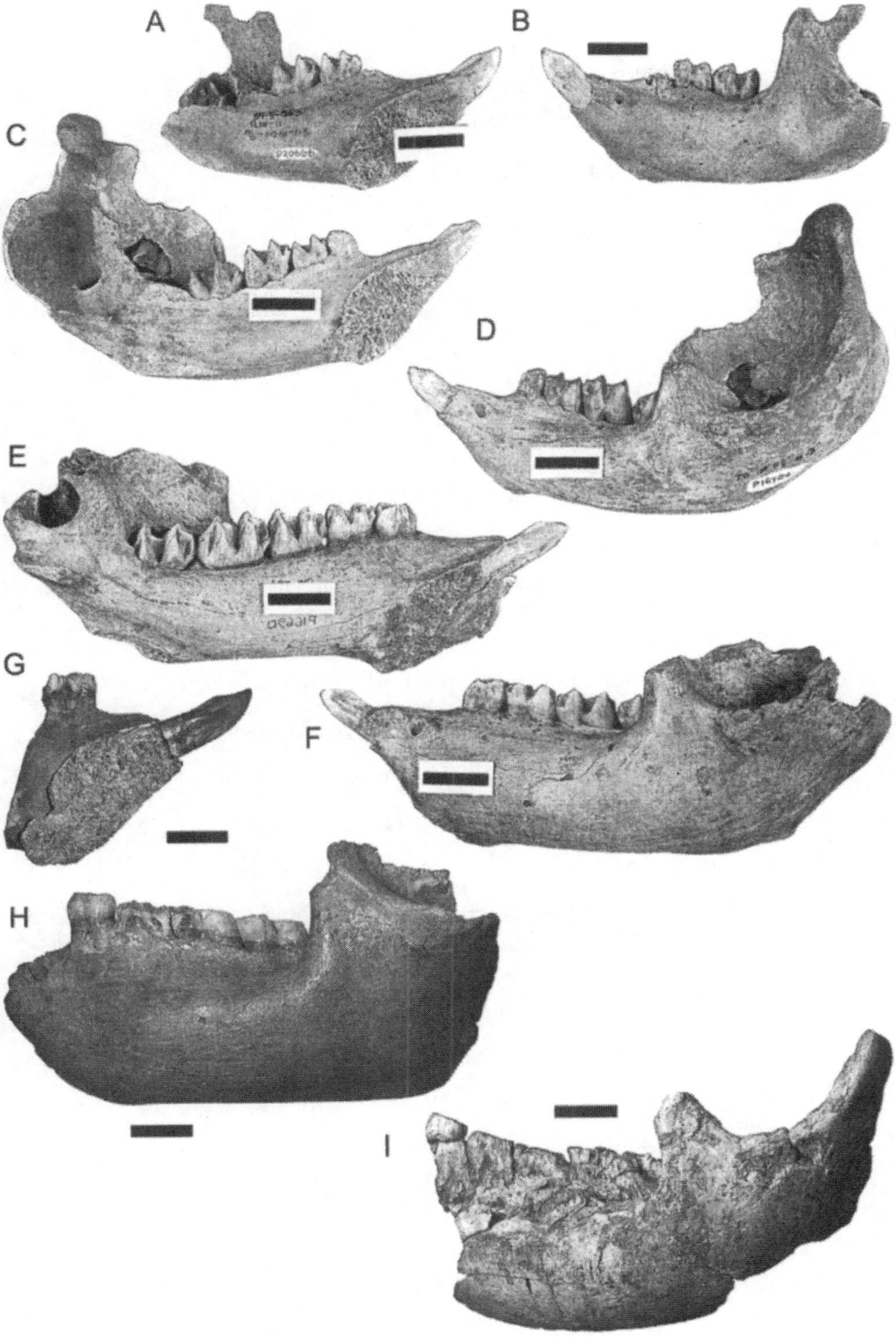
A
B
C
D
E
G
F
H
I

PLATE 77

Procoptodon goliah

Scale bars = 20 mm (A-F, H-J); 10 mm (G)

A. SAM P20666, left juvenile dentary, occlusal view. Victoria Fossil Cave, SA.
B. SAM P16704, left juvenile dentary, occlusal view. Victoria Fossil Cave, SA.
C. SAM P16694, left and right juvenile dentaries, occlusal view. Victoria Fossil Cave, SA.
D. SAM P16690, right adult dentary, stereo occlusal view. Victoria Fossil Cave, SA.
E. AM F102005, left adult dentary, occlusal view. Bingara, NSW.
F. QM F797, right adult dentary, occlusal view. Darling Downs, QLD.
G. AM F101638, excavated right p3, occlusal view. Bingara, NSW.
H. AM F101638, excavated right p3, buccal view.
I. SAM P16690, ventral view.
J. SAM P27886, partial adult cranium and dentaries, ventral view. Lake Victoria, NSW.

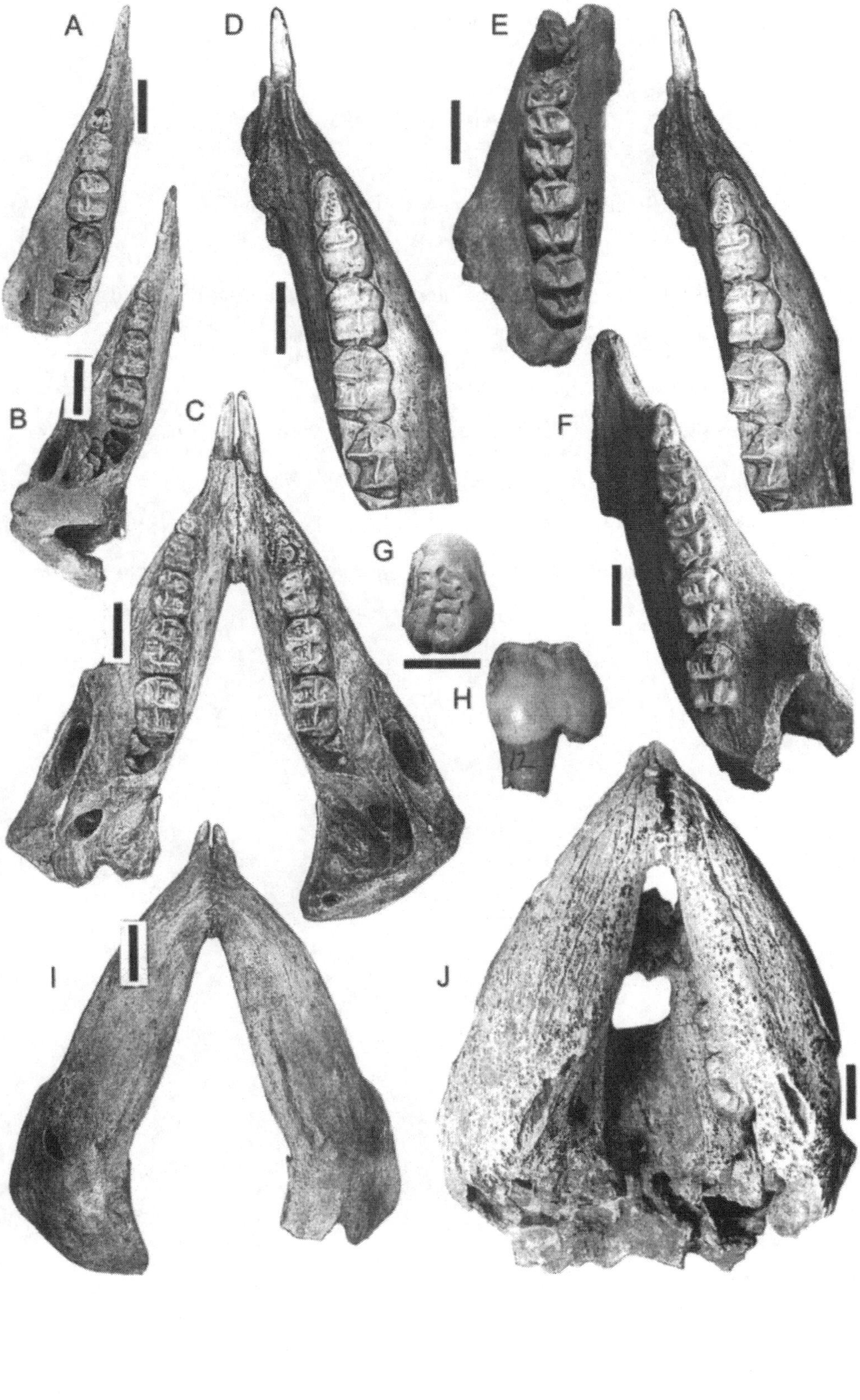
A
D
E
B
C
F
G
H
I
J

PLATE 78

Procoptodon pusio

Scale bars = 20 mm (A-B, D-G); 10 mm (C)

A. QM F2989, left adult maxilla, lateral view. Condamine River, QLD.

B. QM F2989, occlusal view.

C. AM F89068, left P3, occlusal view.

D. QM F33018, partial adult cranium, occlusal view. King's Creek, QLD.

E. AM F89061, right maxilla, occlusal view.

F. AM F89072, right maxilla, stereo occlusal view. Bingara, NSW.

G. AM F88579, right juvenile maxilla, stereo occlusal view. Bingara, NSW.

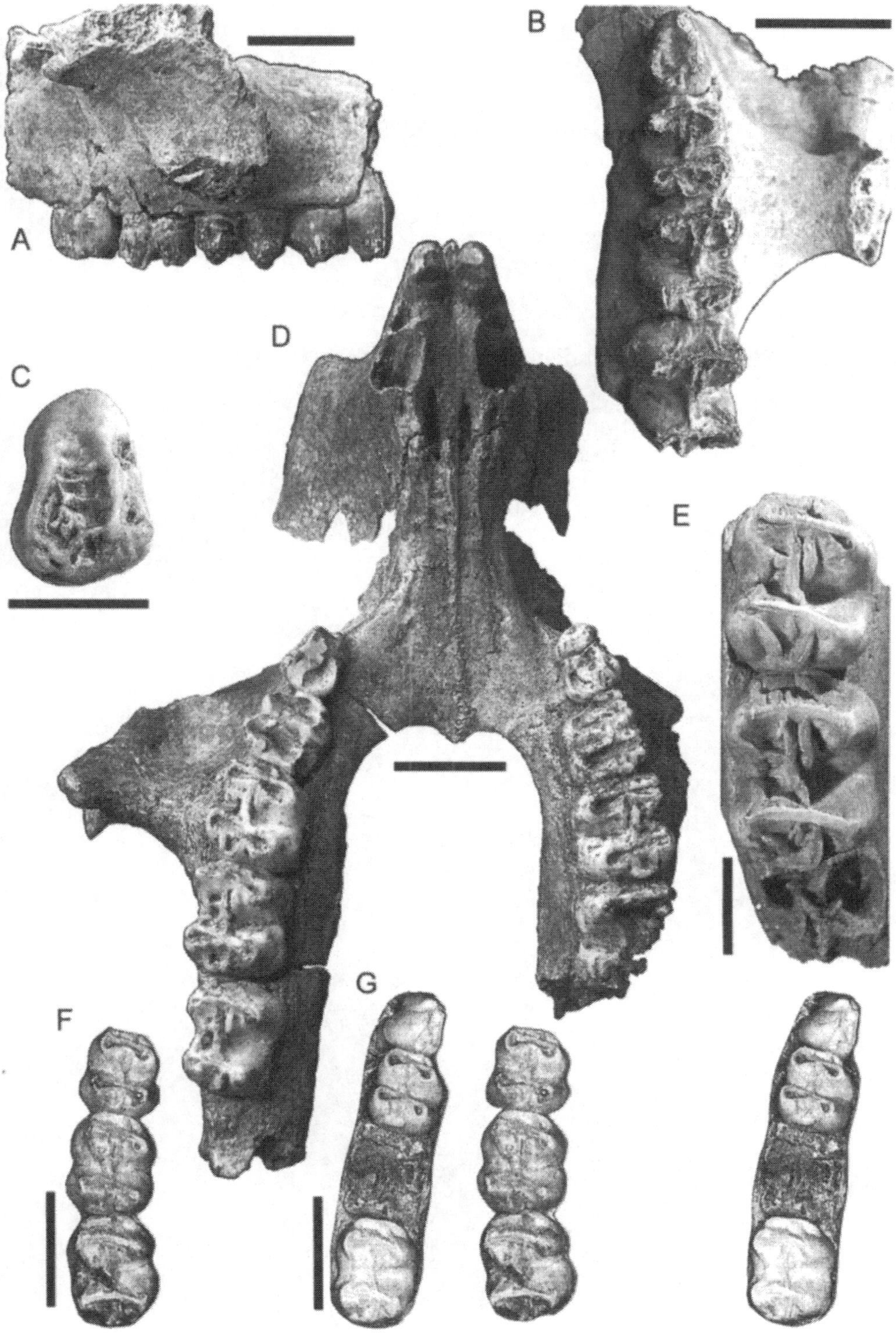
A
B
C
D
E
F
G

PLATE 79

Procoptodon pusio

Scale bars = 20 mm

A. AM F89073, right juvenile dentary, lateral view. Bingara, NSW.

B. AM F89073, left and right juvenile dentaries, ventral view.

C. AM F89073, stereo occlusal view.

D. AM F89073, excavated left p3, occlusal view.

E. AM F89073, close-up of left m2-3, occlusal view.

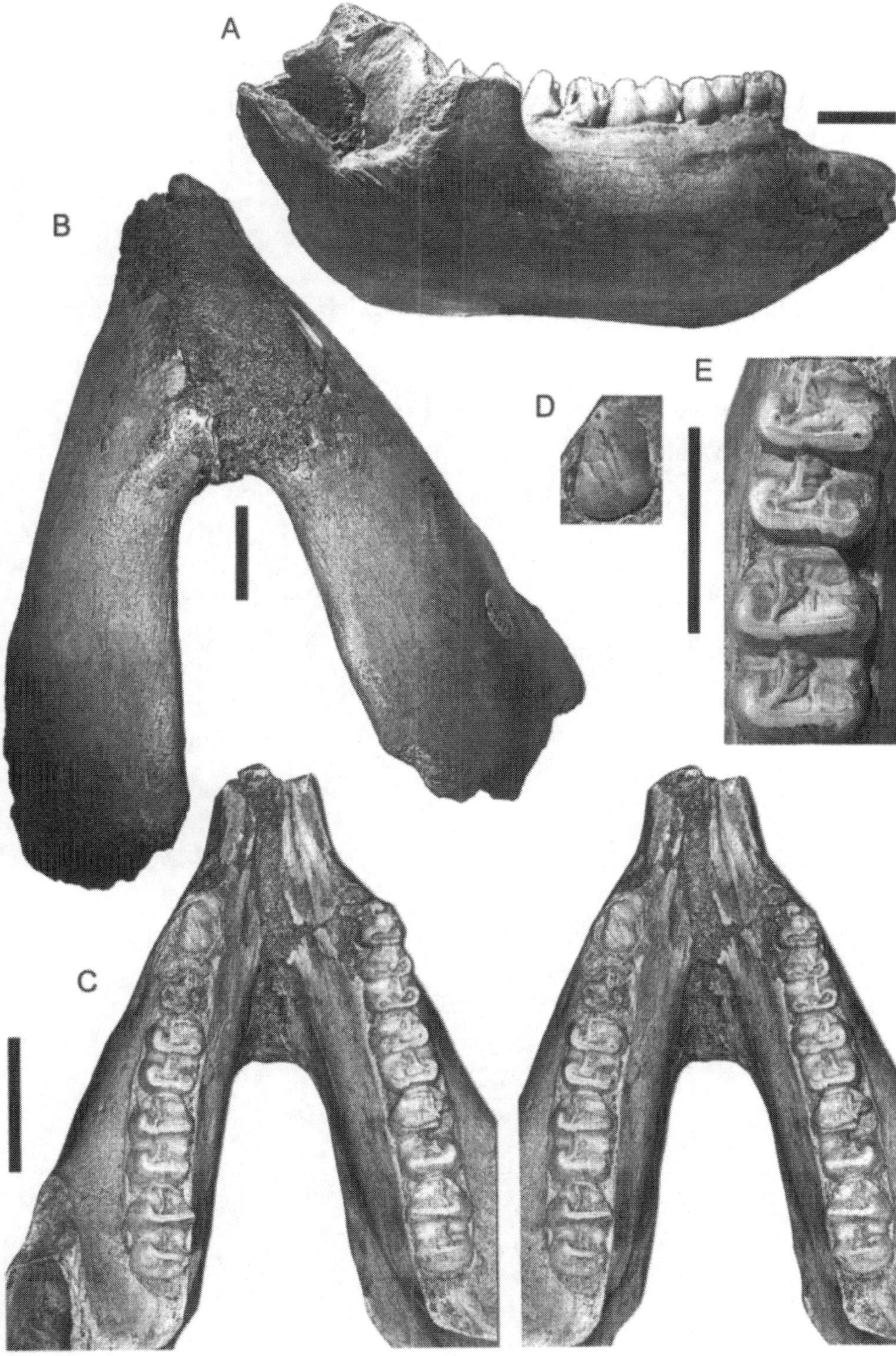
A
B
C
D
E

PLATE 80

Procoptodon pusio

Scale bars = 20 mm

A. QM F2983, right juvenile dentary, lateral view. Locality unknown.
B. QM F2983, mesial view.
C. QM F4467, left adult dentary, lateral view. Darling Downs, QLD.
D. QM F4467, mesial view.
E. QM F808, right adult dentary, mesial view. Darling Downs, QLD.
F. QM F808, lateral view.
G. QM F4468, right adult dentary, lateral view. Locality unknown.
H. AM F88560, right adult dentary, mesial view. Bingara, NSW.

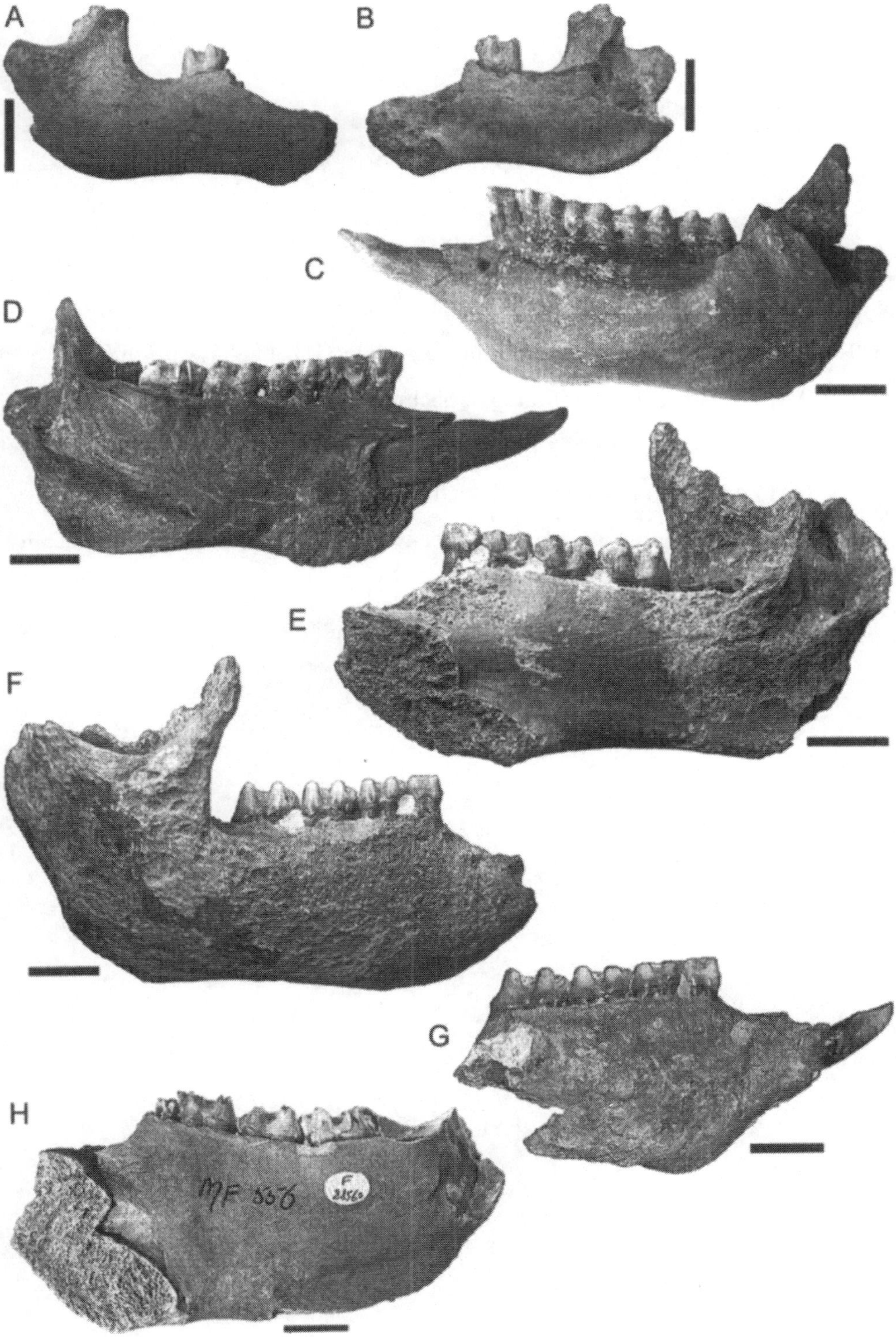
A
B
C
D
E
F
G
H

PLATE 81

Procoptodon pusio

Scale bars = 20 mm (A-B, E-G); 10 mm (C-D)

A. QM F2983, right juvenile dentary, occlusal view. Locality unknown.
B. QM F808, right adult dentary, occlusal view. Darling Downs, QLD.
C. AM F88574, excavated right p3, occlusal view. Bingara, NSW.
D. AM F30583, excavated right p3, occlusal view. Bingara, NSW.
E. QM F4467, left adult dentary, occlusal view. Darling Downs, QLD.
F. QM F4468, close-up of right cheek tooth row, occlusal view. Locality unknown.
G. QM F4468, left and right adult dentaries, occlusal view.

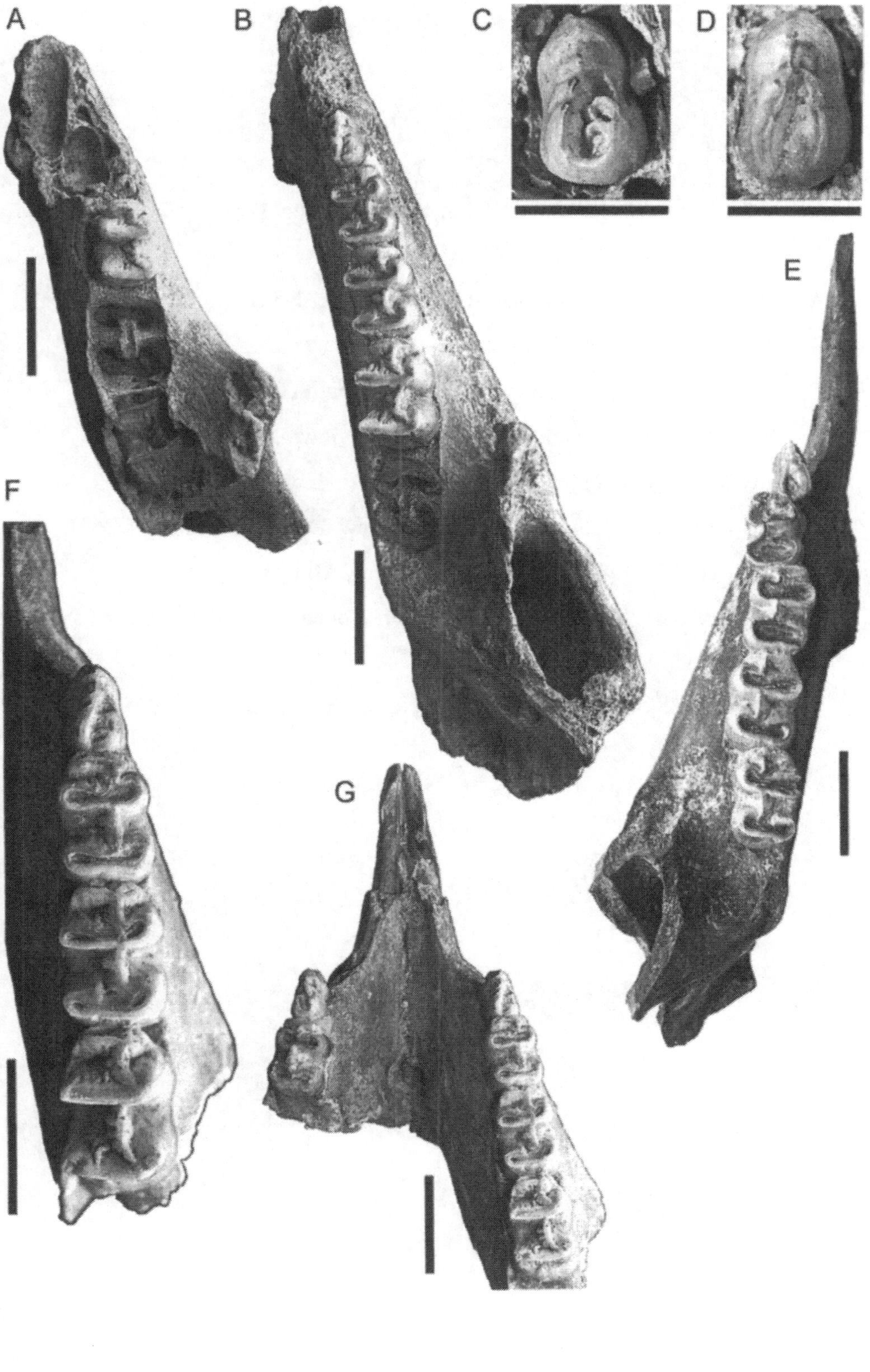
A
B
C
D
E
F
G

PLATE 82

Procoptodon rapha

Scale bars = 10 mm (A-C); 20 mm (D-I)

A. UCMP 60048, right P3, lateral view. Bingara, NSW.

B. UCMP 60048, occlusal view.

C. AM F60967, left P3, occlusal view. Wellington Caves, NSW.

D. AM F102255, left adult maxilla, occlusal view. Bingara, NSW.

E. AM F102261, right adult maxilla, occlusal view. Bingara, NSW.

F. QM F799, left adult maxilla, occlusal view. Darling Downs, QLD.

G. AM F102259, right maxilla, occlusal view. Bingara, NSW.

H. SAM P31045, right adult maxilla, stereo occlusal view. Henschke's Fossil Cave, SA.

I. AM F88565, left maxilla, stereo occlusal view. Bingara, NSW.

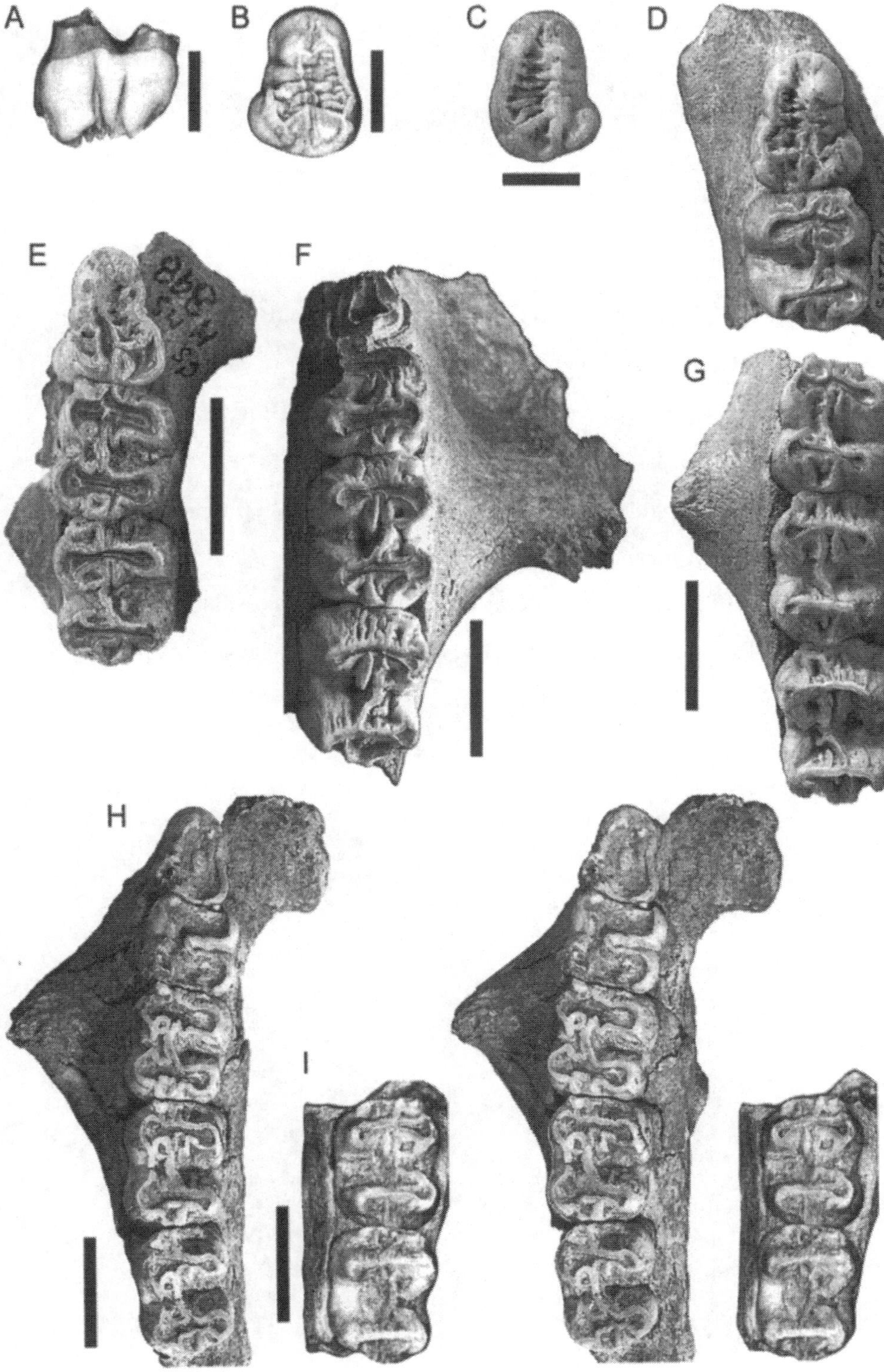

A
B
C
D
E
F
G
H
I

PLATE 83

Procoptodon rapha

Scale bars = 20 mm

A. AM F102247, right juvenile dentary, lateral view. Bingara, NSW.

B. AM F102247, mesial view.

C. AM F102233, left juvenile dentary, mesial view. Bingara, NSW.

D. AM F102233, lateral view.

E. AM F19757, right juvenile dentary, lateral view. Wellington Caves, NSW.

F. AM F19757, mesial view.

G. AM F3198, left dentary, lateral view. Attunga Springs, NSW.

H. NMV P200725, left adult dentary, mesial view. Lancefield, VIC.

I. NMV P200725, lateral view.

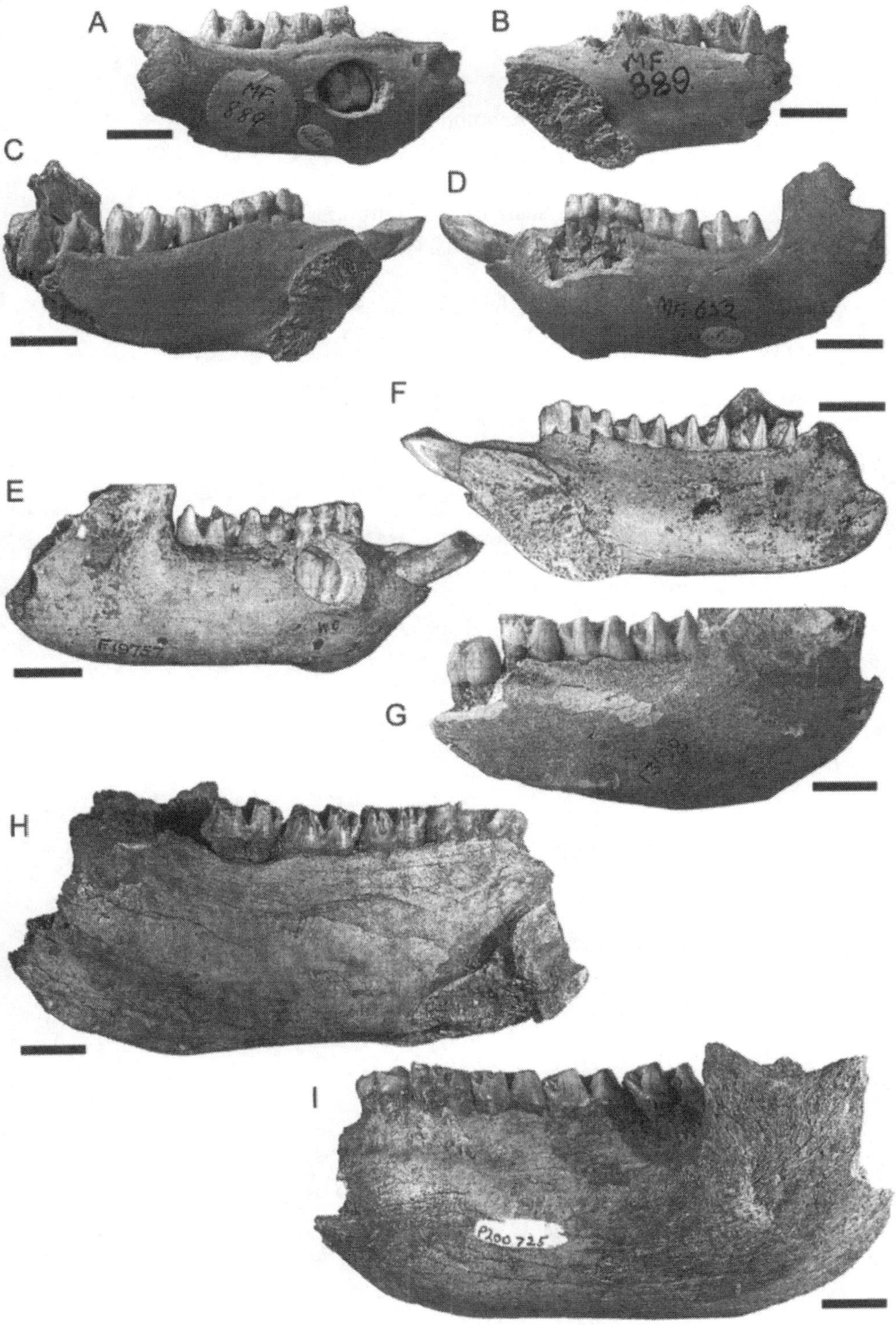

A
B
MF 889
C
D
E
F
G
H
I
P200725

PLATE 84

Procoptodon rapha

Scale bars = 20 mm

A. AM F102233, left juvenile dentary, occlusal view. Bingara, NSW.
B. QM F9156, left adult dentary, occlusal view. Darling Downs, QLD.
C. AM F102246, left adult dentary (close-up of p3), occlusal view. Bingara, NSW.
D. NMV P200725, left adult dentary, occlusal view. Lancefield, VIC.
E. AM F102242, left adult dentary, occlusal view. Bingara, NSW.
F. AM F102246, close-up of m2-4, occlusal view.
G. AM F19757, right juvenile dentary, stereo occlusal view. Wellington Caves, NSW.
H. SAM P31045, right adult dentary, stereo occlusal view. Henschke's Fossil Cave, SA.

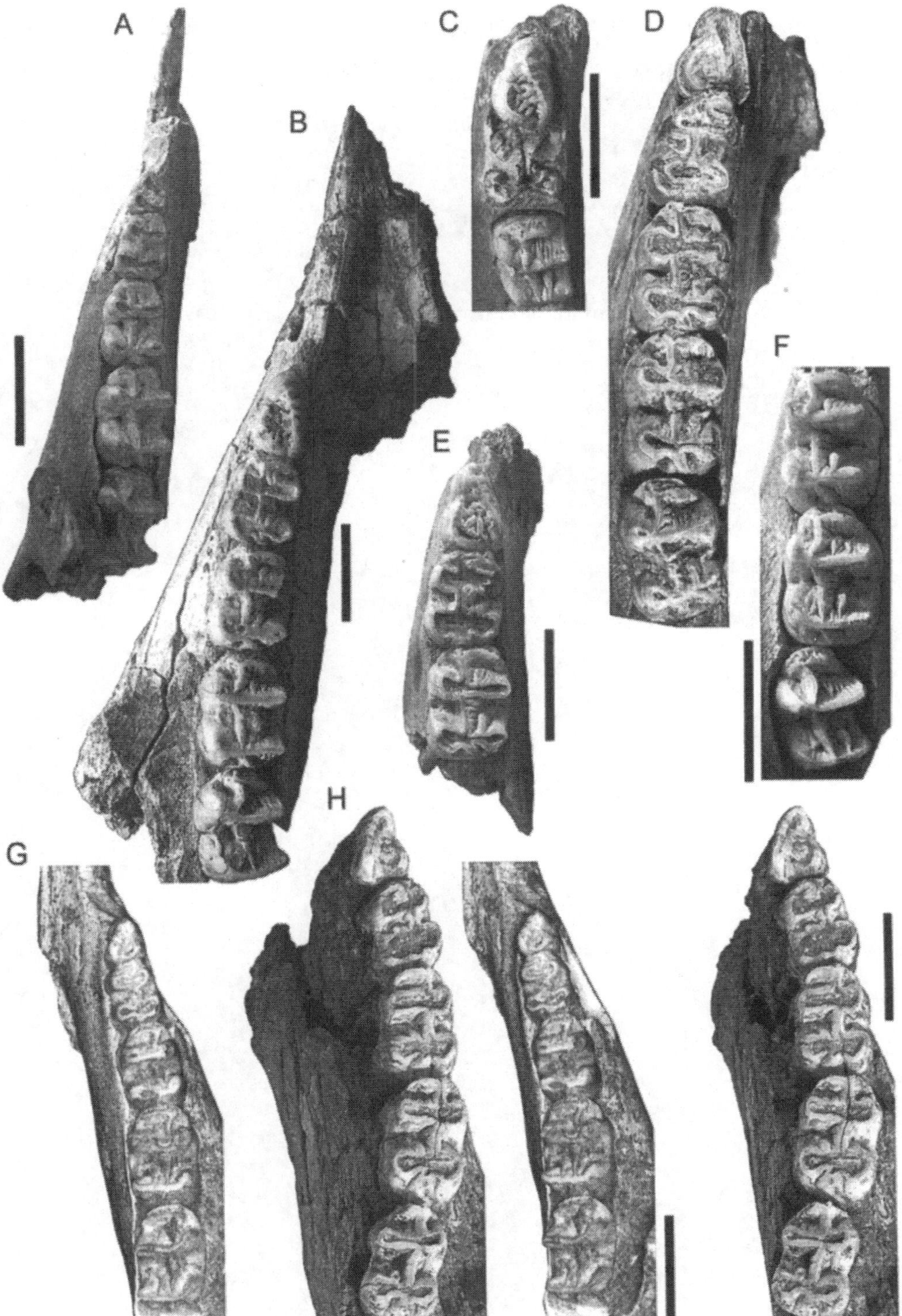
A
B
C
D
E
F
G
H

PLATE 85

"Procoptodon" browneorum

Scale bars = 20 mm

A. FU 1011, adult cranium, lateral view. Cathedral Cave, SA.

B. FU 1011, dorsal view.

C. FU 1011, palatal view.

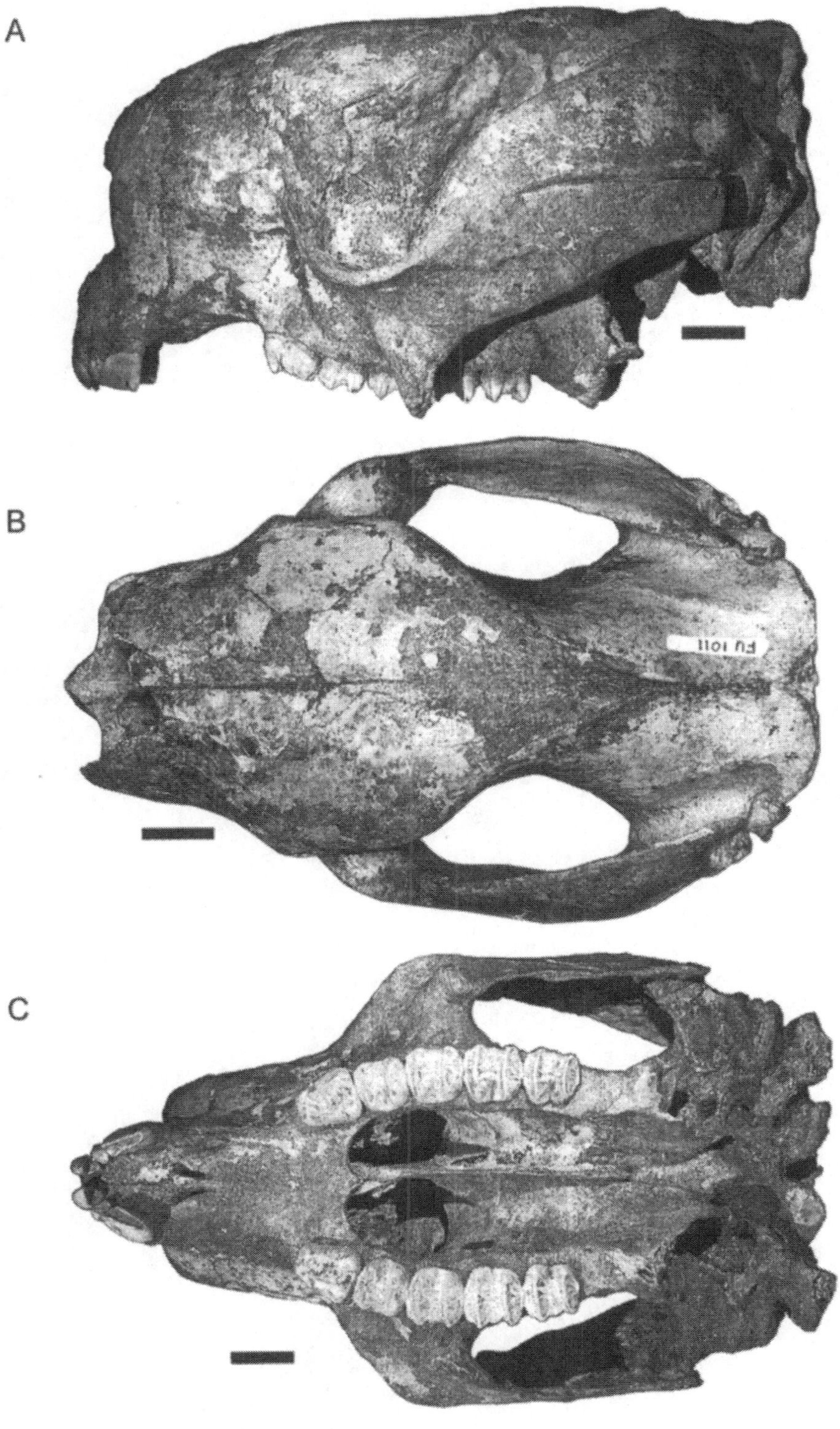
A
B
FU 1011
C

PLATE 86

"Procoptodon" browneorum

Scale bars = 20 mm

A. WAM 63.2.198, adult cranium, lateral view. Mammoth Cave, WA.

B. WAM 63.2.198, dorsal view.

C. WAM 63.2.198, palatal view.

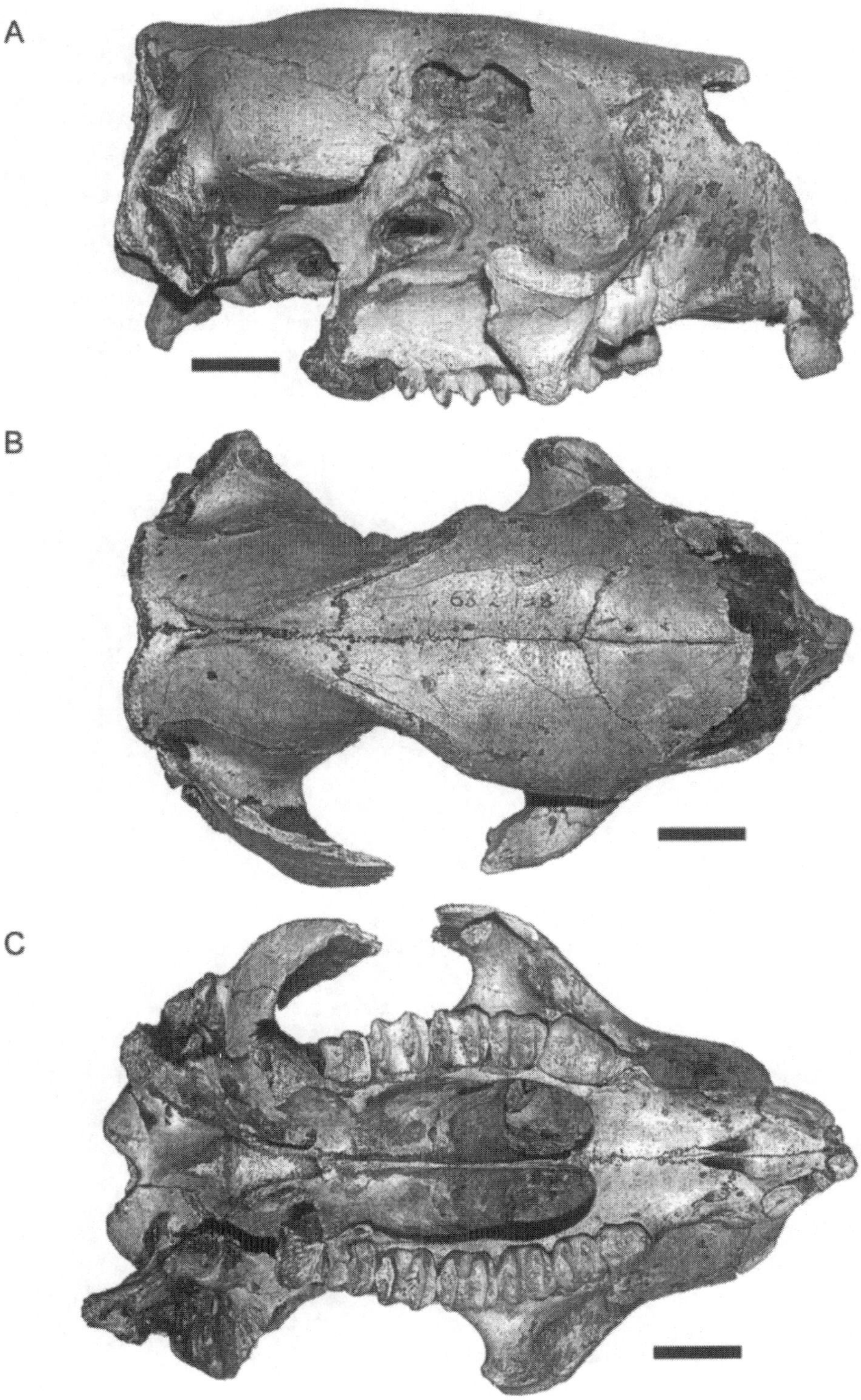
A
B
C

PLATE 87

"Procoptodon" browneorum

Scale bars = 20 mm

A. SAM P20677, juvenile cranium, lateral view. Victoria Fossil Cave, SA.
B. SAM P20677, anterior view.
C. SAM P20677, palatal view.
D. SAM P20677, dorsal view.
E. SAM P20677, posterior view.

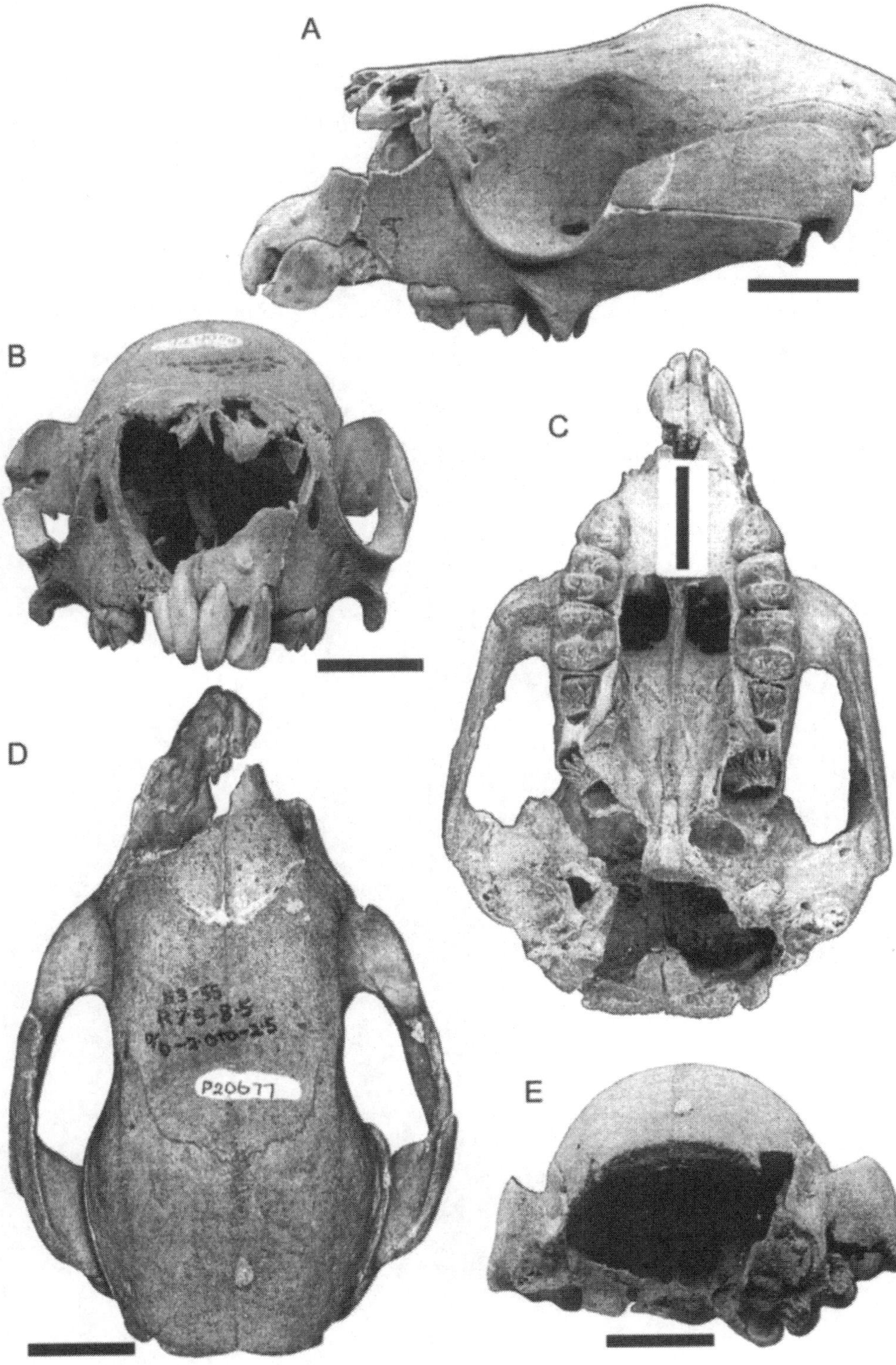
A
B
C
D
E
P20677

PLATE 88

"Procoptodon" browneorum

Scale bars = 20 mm

A. SAM P16646, juvenile cranium, lateral view. Victoria Fossil Cave, SA.

B. SAM P16649, adult cranium, lateral view. Victoria Fossil Cave, SA.

C. SAM P28677, adult cranium, lateral view. Victoria Fossil Cave, SA.

D. SAM P25525, adult cranium, lateral view. Victoria Fossil Cave, SA.

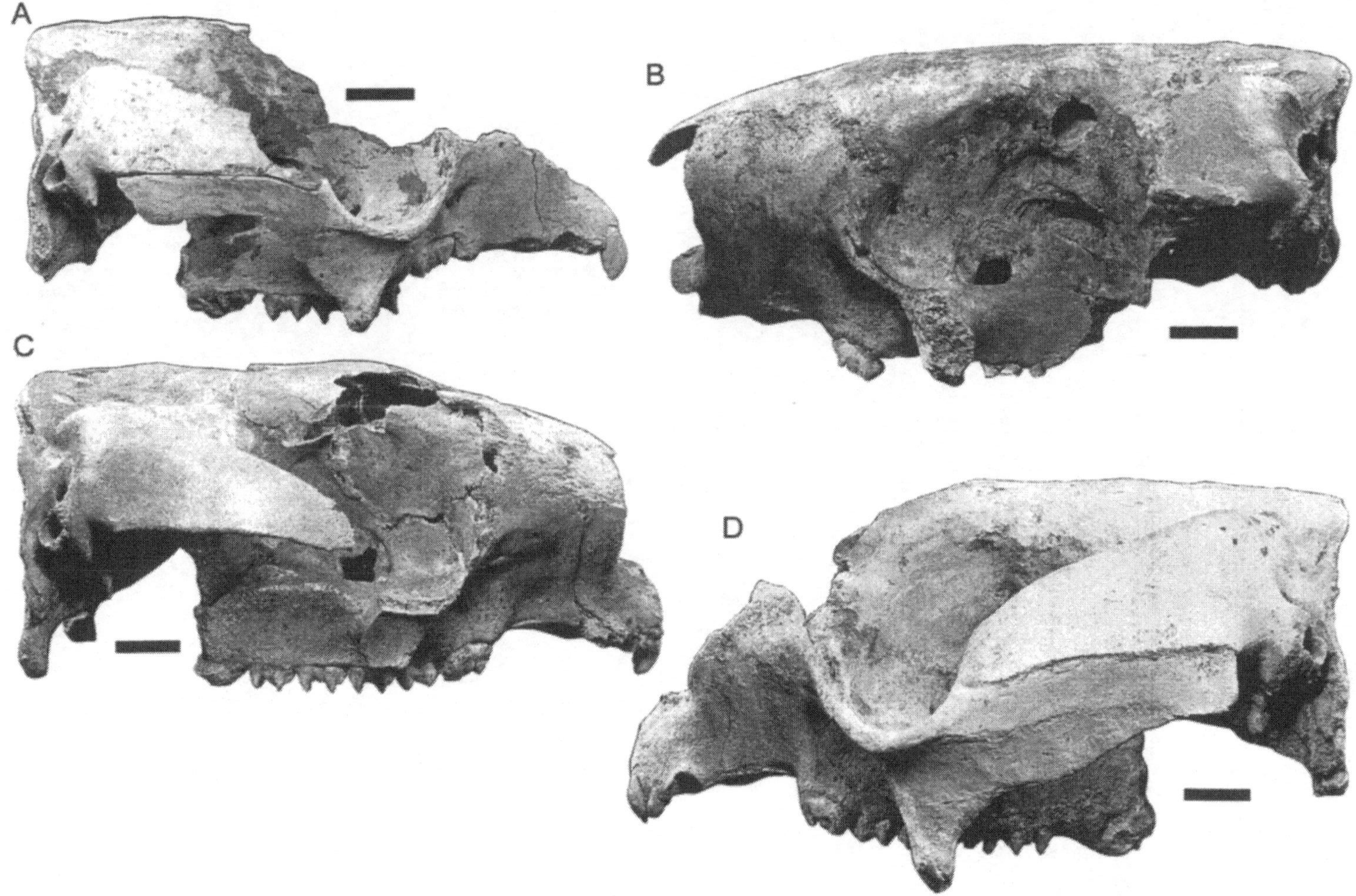
A
B
C
D

PLATE 89

"Procoptodon" browneorum

Scale bars = 20 mm

A. SAM P16542, juvenile cranium, dorsal view. Victoria Fossil Cave, SA.

B. SAM P16646, juvenile cranium, dorsal view. Victoria Fossil Cave, SA.

C. SAM P16649, adult cranium, dorsal view. Victoria Fossil Cave, SA.

D. SAM P25525, adult cranium, dorsal view. Victoria Fossil Cave, SA.

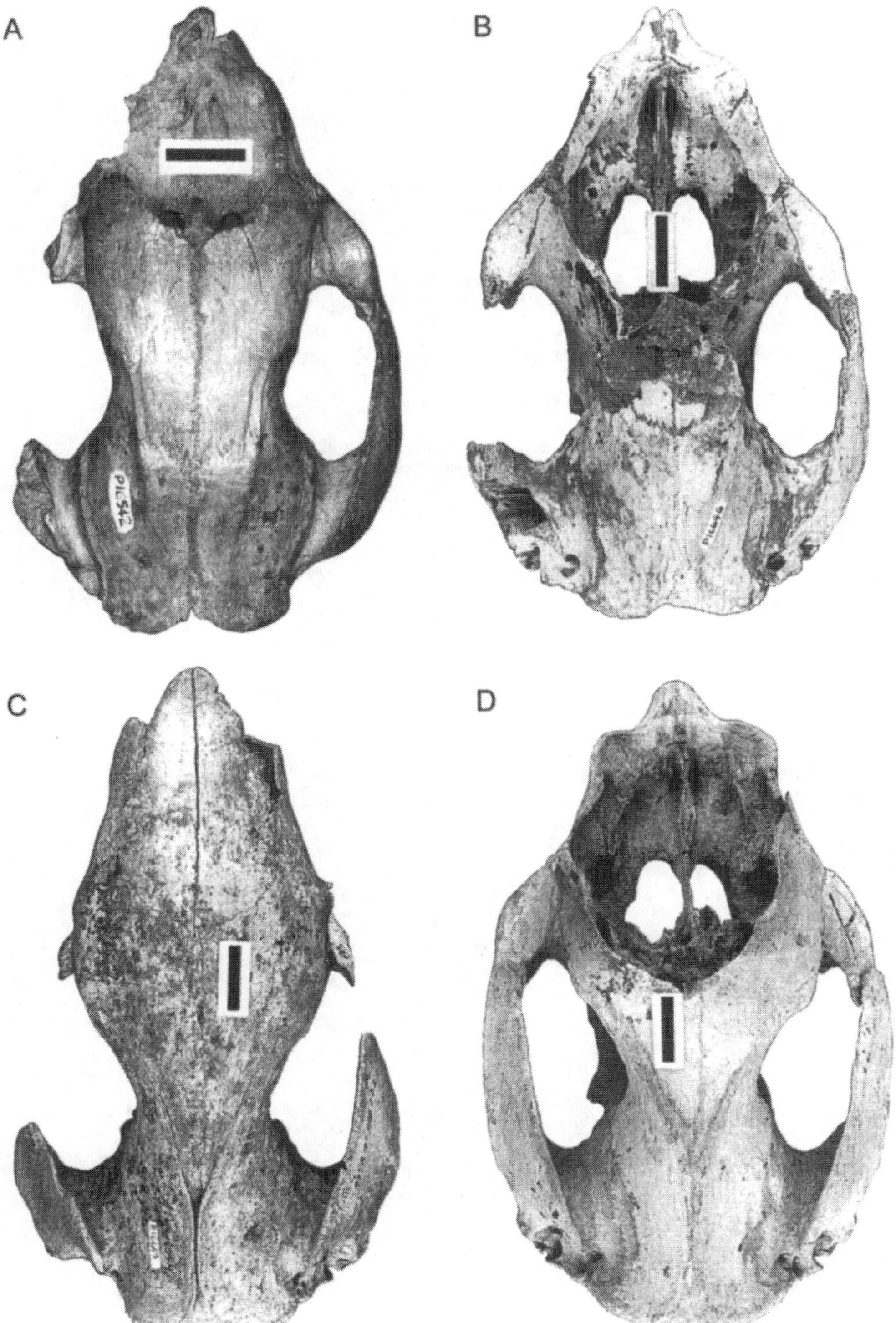
A
B
C
D

PLATE 90

"Procoptodon" browneorum

Scale bars = 20 mm

A. SAM P16646, juvenile cranium, palatal view. Victoria Fossil Cave, SA.
B. SAM P25525, adult cranium, palatal view. Victoria Fossil Cave, SA.
C. SAM P16652, adult cranium, palatal view. Victoria Fossil Cave, SA.

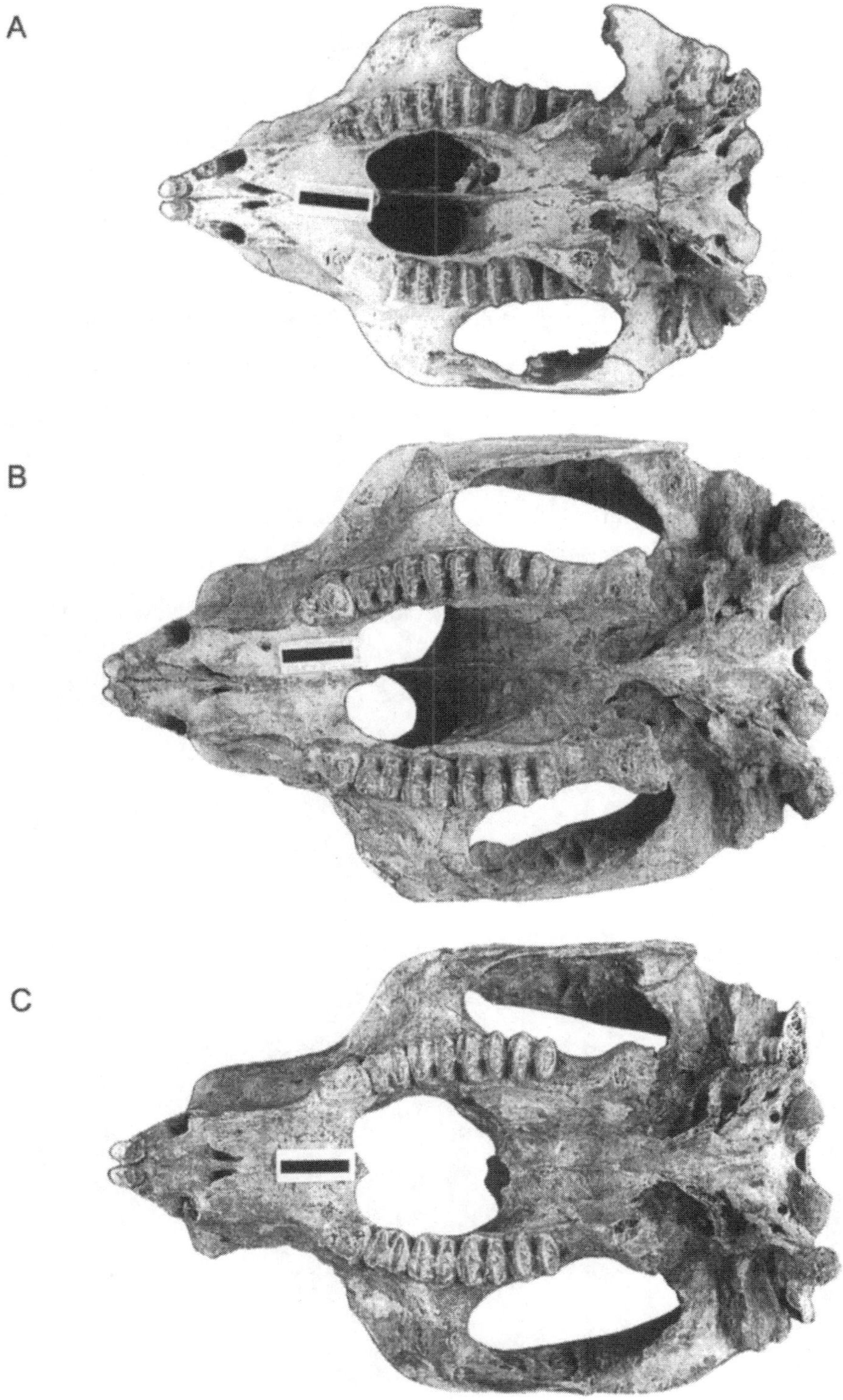
A
B
C

PLATE 91

"Procoptodon" browneorum

Scale bars = 20 mm

A. SAM P25525, adult cranium, anterior view. Victoria Fossil Cave, SA.
B. SAM P16649, adult cranium, anterior view. Victoria Fossil Cave, SA.
C. SAM P20257, right juvenile maxilla, stereo occlusal view. Victoria Fossil Cave, SA.
D. WAM HC LH12, left adult maxilla, stereo occlusal view. Lindsay Hall Cave, WA.
E. SAM P20471, left and right adult dentaries, ventral view. Victoria Fossil Cave, SA.
F. SAM P20471, posterior view.

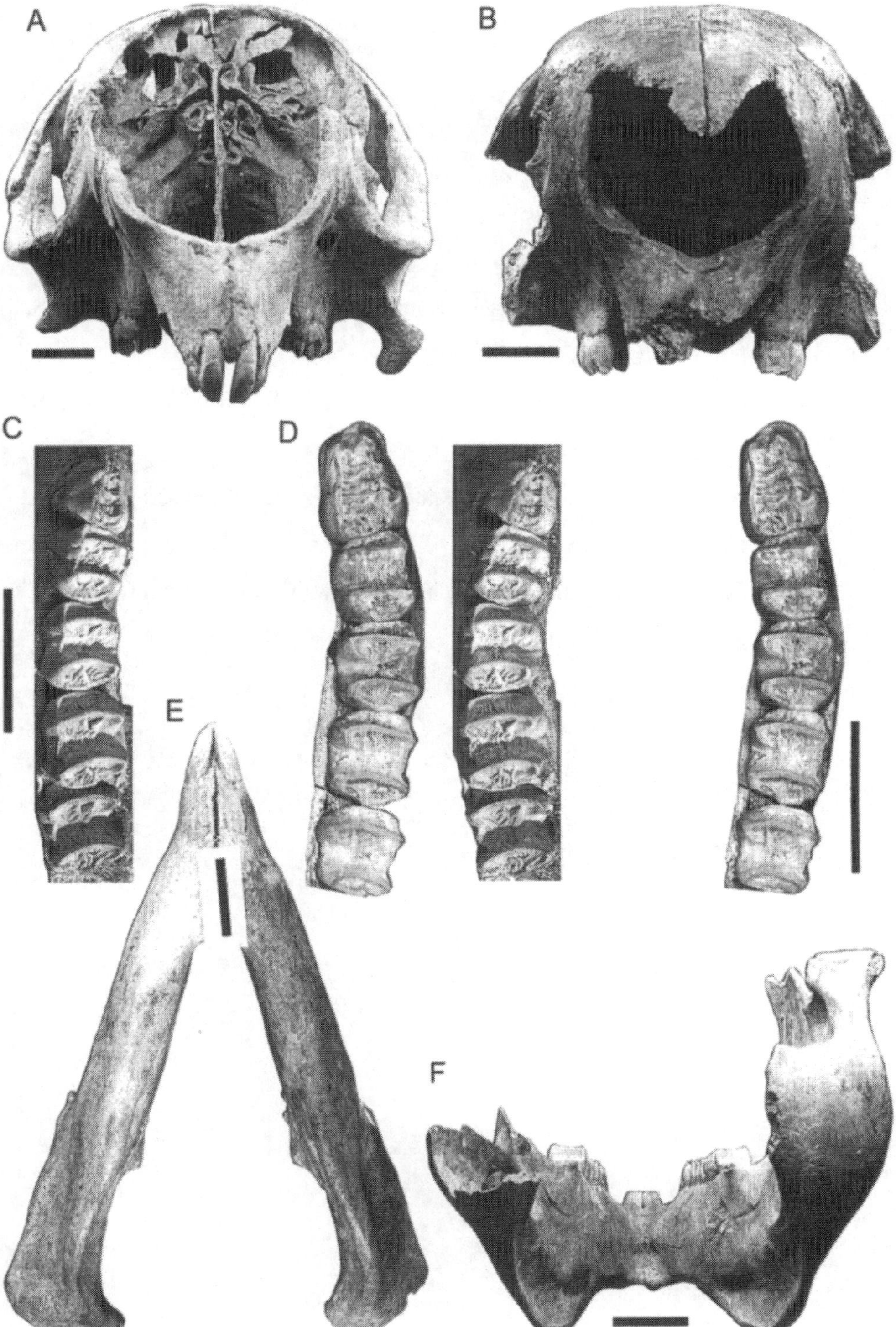
A
B
C
D
E
F

PLATE 92

"Procoptodon" browneorum

Scale bars = 20 mm

A. SAM P16537, left juvenile dentary, mesial view. Victoria Fossil Cave, SA.
B. SAM P16537, lateral view.
C. SAM P20470, left juvenile dentary, mesial view. Victoria Fossil Cave, SA.
D. SAM P20470, lateral view.
E. SAM P16532, right adult dentary, lateral view. Victoria Fossil Cave, SA.
F. SAM P16532, mesial view.
G. SAM P20465, right adult dentary, lateral view. Victoria Fossil Cave, SA.
H. SAM P20465, mesial view.

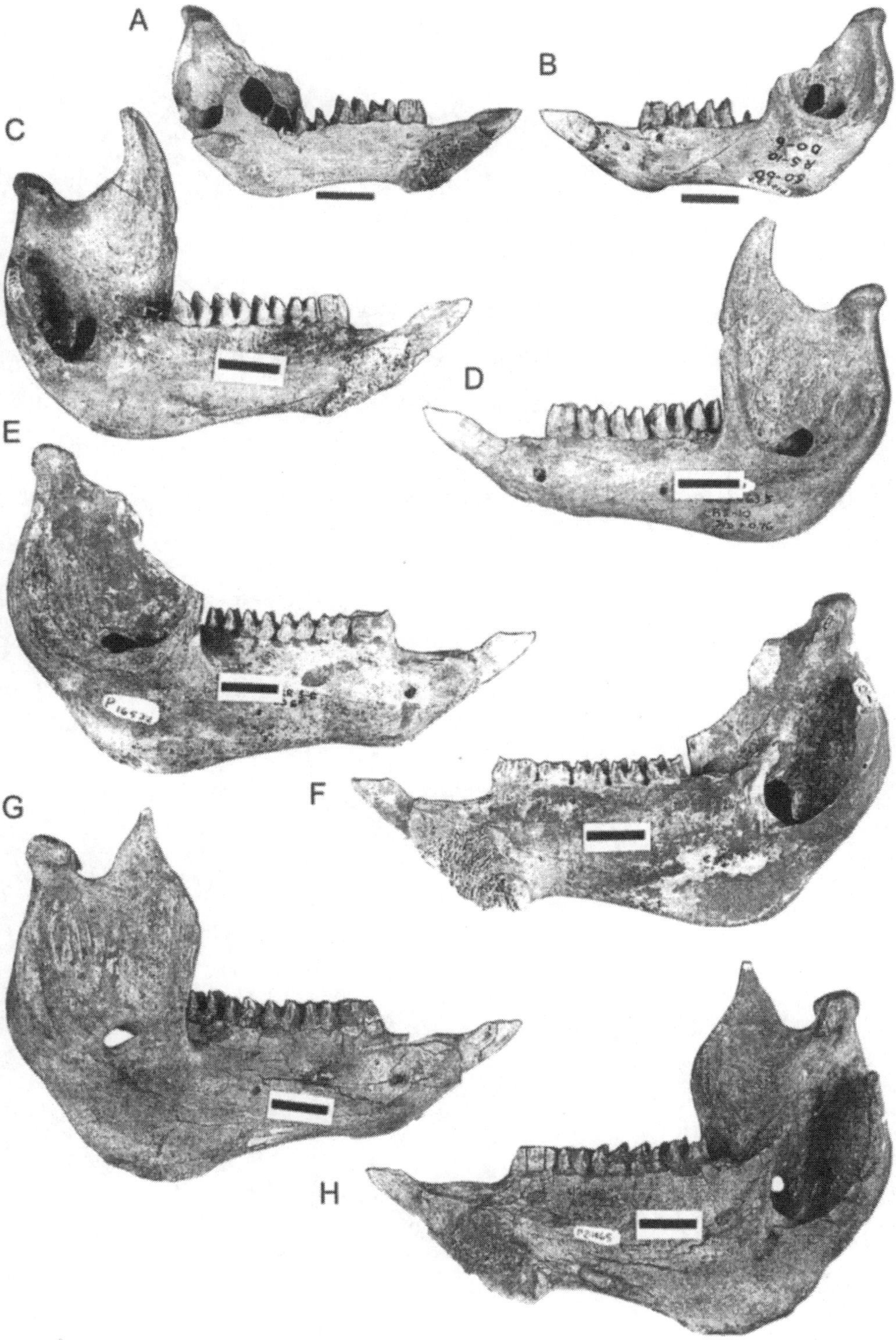
A
B
C
D
E
F
G
H

PLATE 93

"Procoptodon" browneorum

Scale bars = 20 mm

A. FU 1061, left and right adult dentaries, occlusal view. Cathedral Cave, SA.
B. SAM P16532, right adult dentary, occlusal view. Victoria Fossil Cave, SA.
C. FU 1089, right juvenile dentary, stereo occlusal view. Victoria Fossil Cave, SA.
D. SAM P16638, right adult dentary, stereo occlusal view. Victoria Fossil Cave, SA.

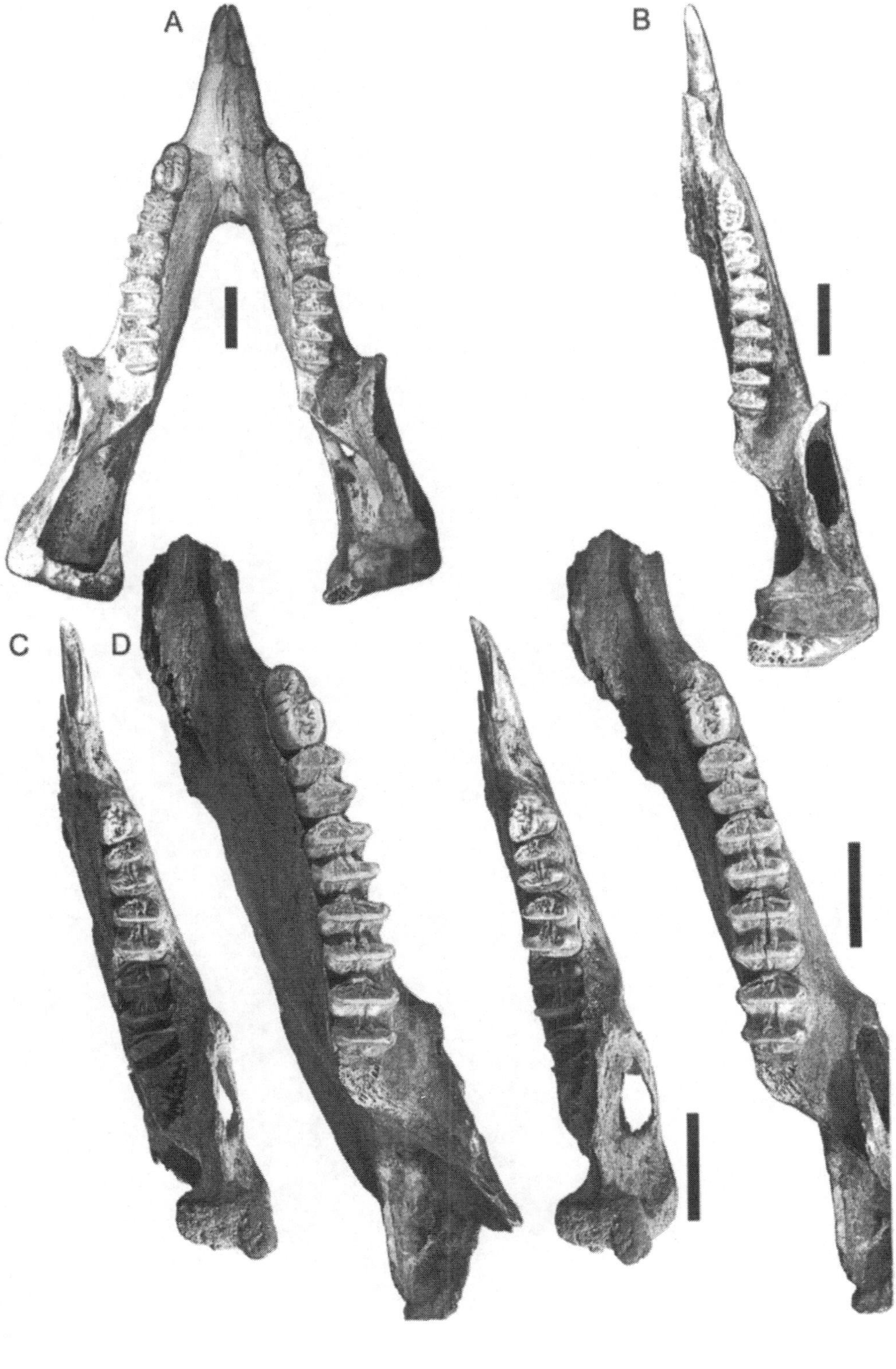
A
B
C
D

PLATE 94

"Procoptodon" gilli

Scale bars = 20 mm

A. FU 1653, adult cranium, dorsal view. Green Waterhole Cave, SA.

B. FU 1653, adult cranium and dentaries, lateral view.

C. FU 1653, palatal view.

D. FU 1653, anterior view.

E. FU 1653, right dentary, mesial view.

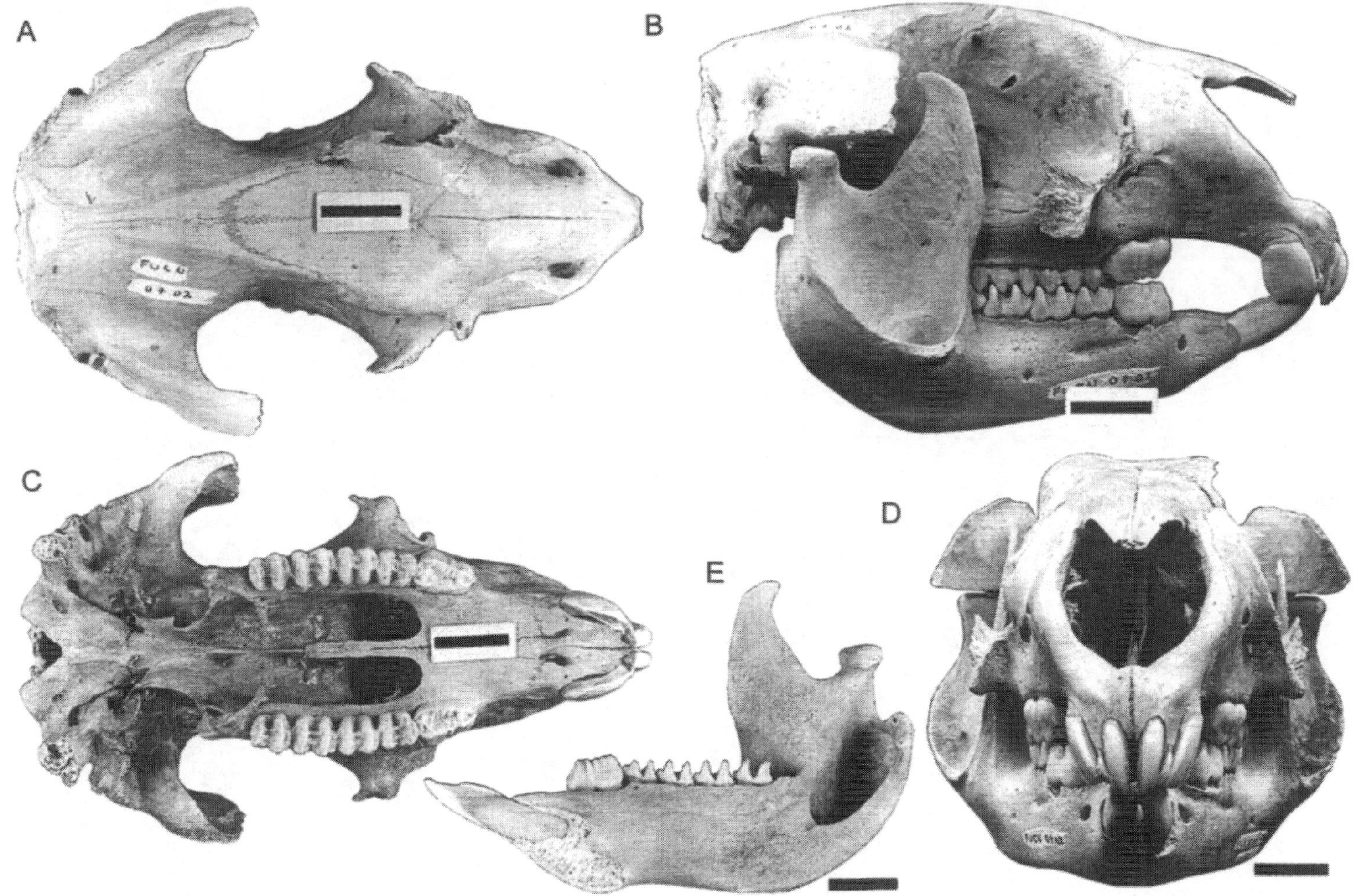
A
B
C
D
E

PLATE 95

"Procoptodon" gilli

Scale bars = 20 mm

A. SAM P20568, juvenile cranium, lateral view. Victoria Fossil Cave, SA.
B. SAM P16506, juvenile cranium, lateral view (photo reversed). Victoria Fossil Cave, SA.
C. SAM P27297, left juvenile maxilla and dentary, lateral view. Victoria Fossil Cave, SA.
D. SAM P16647, adult cranium, lateral view. Victoria Fossil Cave, SA.
E. SAM P17248, adult cranium and dentaries, lateral view (photo reversed). Green Waterhole Cave, SA.

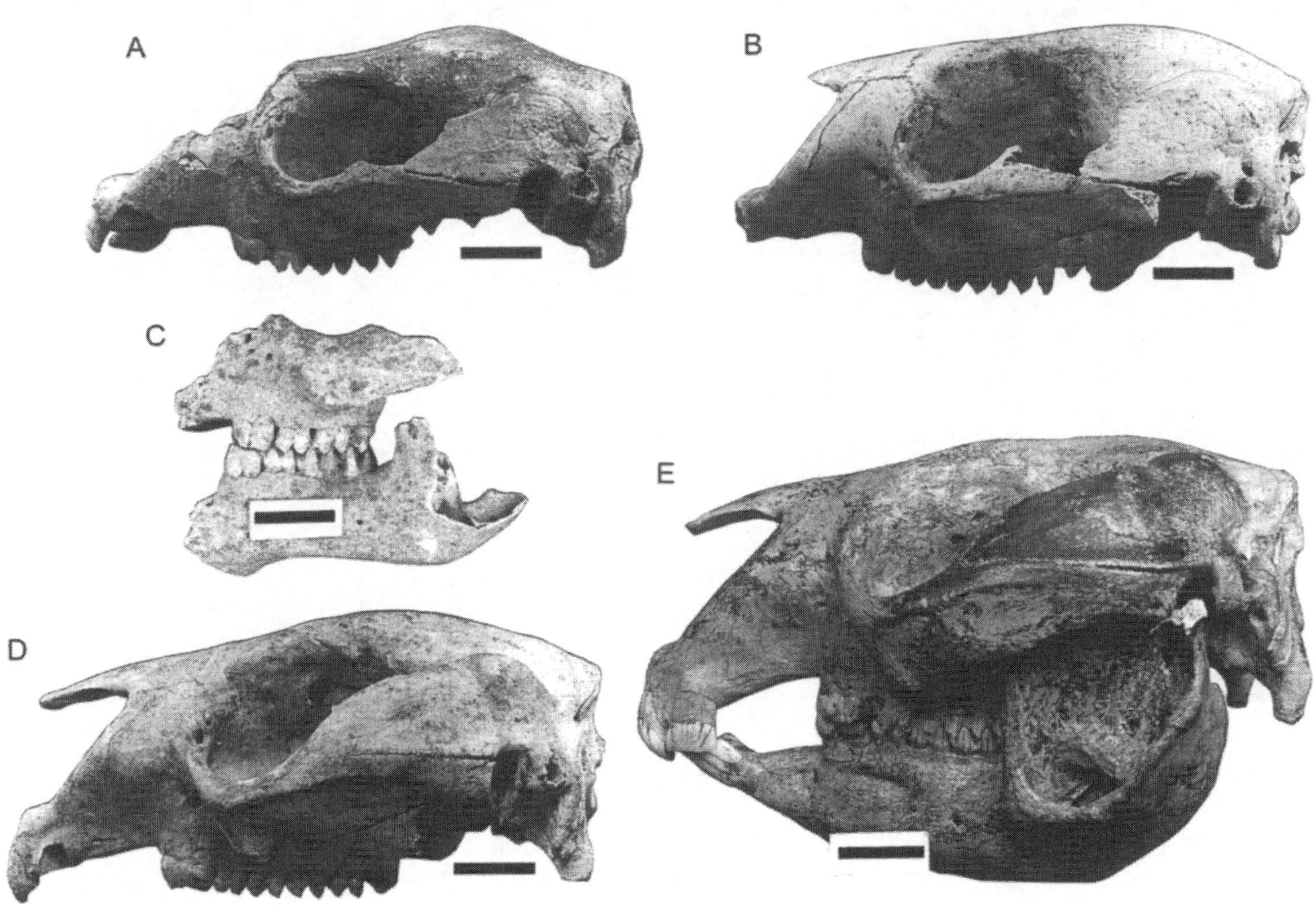
A
B
C
D
E

PLATE 96

"Procoptodon" gilli

Scale bars = 20 mm

A. SAM P20568, juvenile cranium, anterior view. Victoria Fossil Cave, SA.

B. SAM P20568, posterior view.

C. SAM P16506, juvenile cranium, anterior view. Victoria Fossil Cave, SA.

D. SAM P16506, posterior view.

E. FU 1653, adult cranium, anterior view. Green Waterhole Cave, SA.

F. FU 1653, adult cranium and dentaries, posterior view.

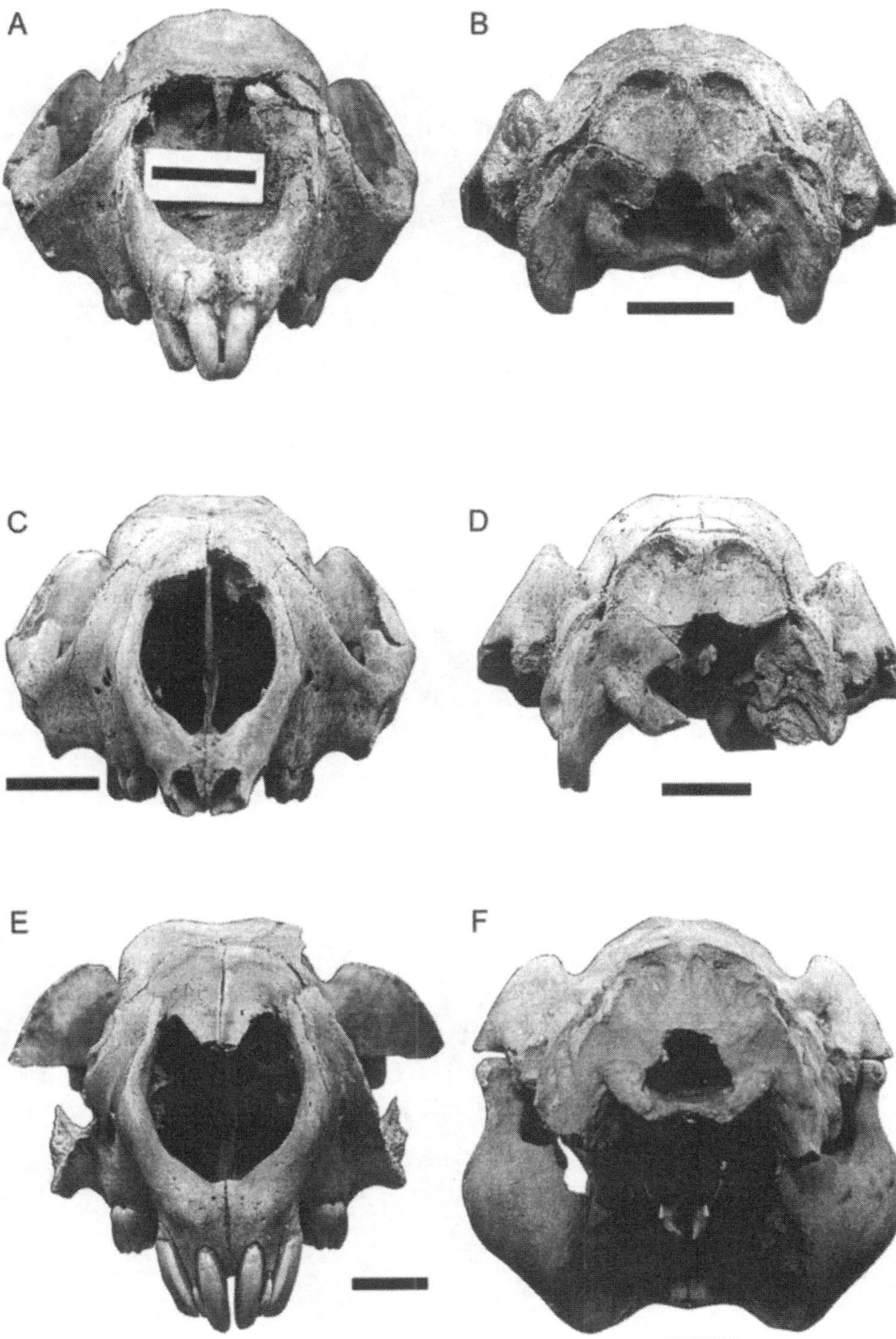
A
B
C
D
E
F

PLATE 97

"Procoptodon" gilli

Scale bars = 20 mm

A. SAM P20568, juvenile cranium, dorsal view. Victoria Fossil Cave, SA.
B. SAM P16506, juvenile cranium, dorsal view. Victoria Fossil Cave, SA.
C. SAM P16647, adult cranium, dorsal view. Victoria Fossil Cave, SA.

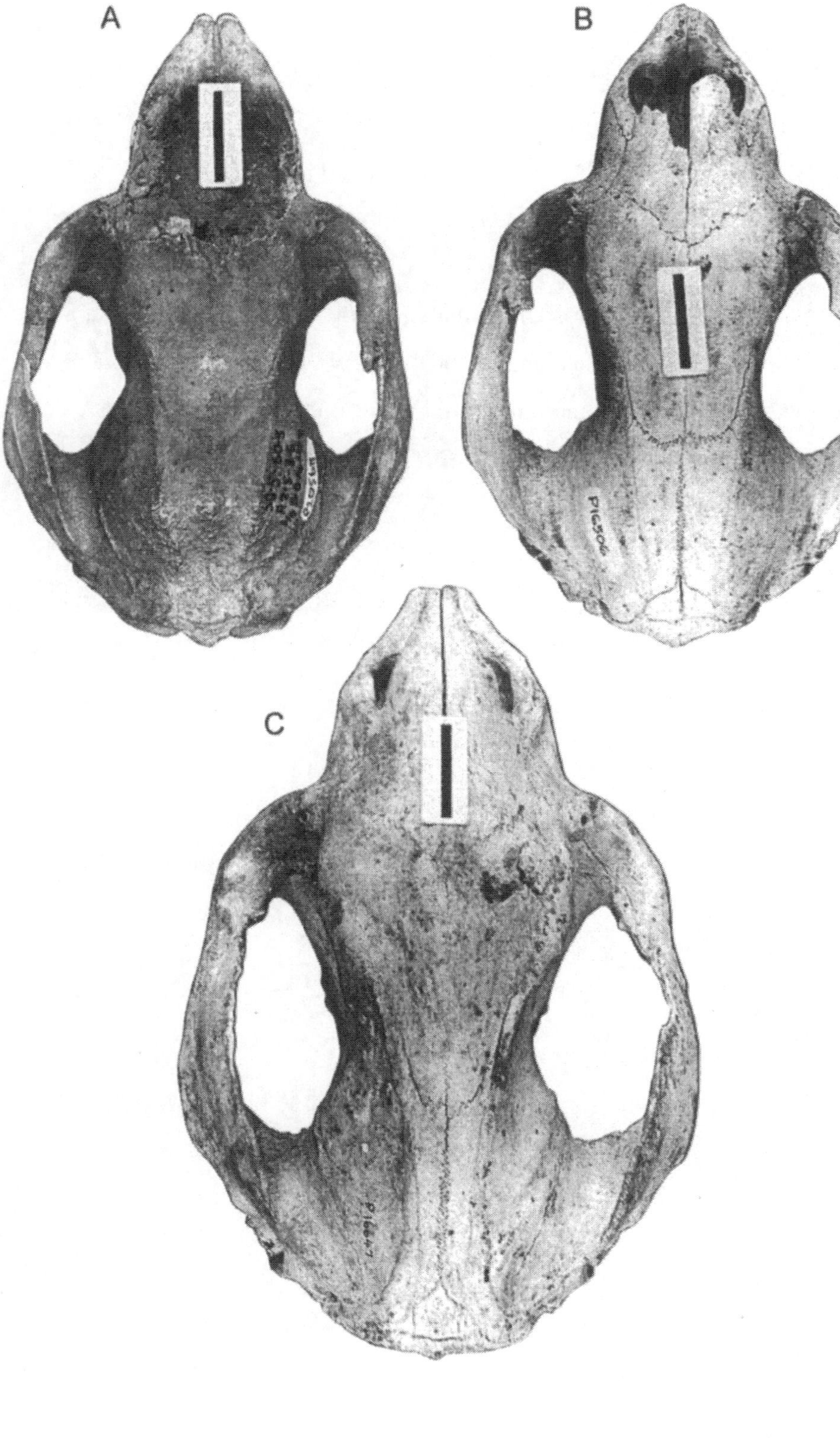
A
B
C

PLATE 98

"Procoptodon" gilli

Scale bars = 20 mm

A. SAM P20568, juvenile cranium, palatal view. Victoria Fossil Cave, SA.

B. SAM P16506, juvenile cranium, palatal view. Victoria Fossil Cave, SA.

C. SAM P16647, adult cranium, palatal view. Victoria Fossil Cave, SA.

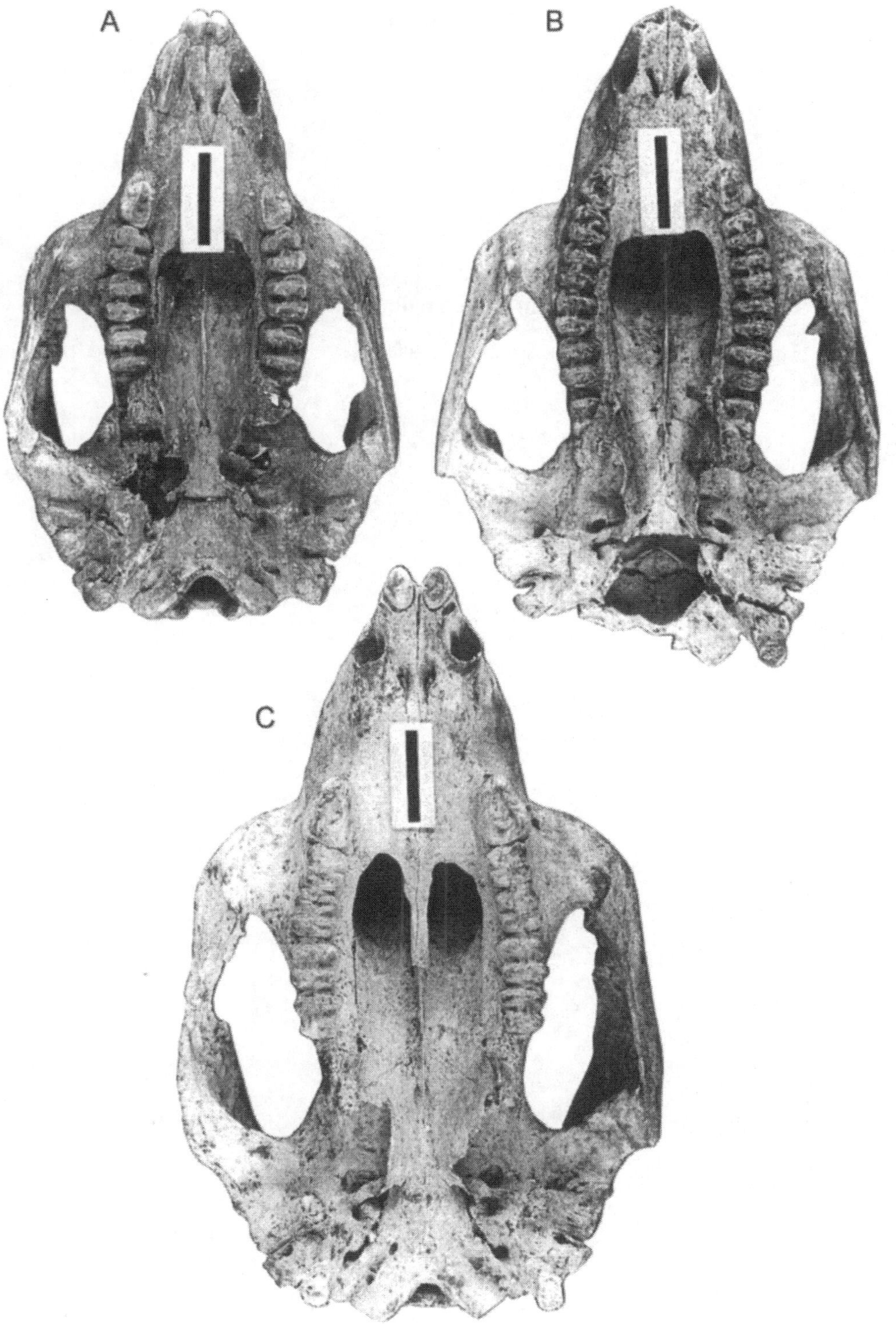
A
B
C

PLATE 99

"Procoptodon" gilli

Scale bars = 20 mm

A. FU 0676, juvenile cranium, stereo palatal view. Victoria Fossil Cave, SA.

B. FU 0003, adult cranium, stereo occlusal view. Victoria Fossil Cave, SA.

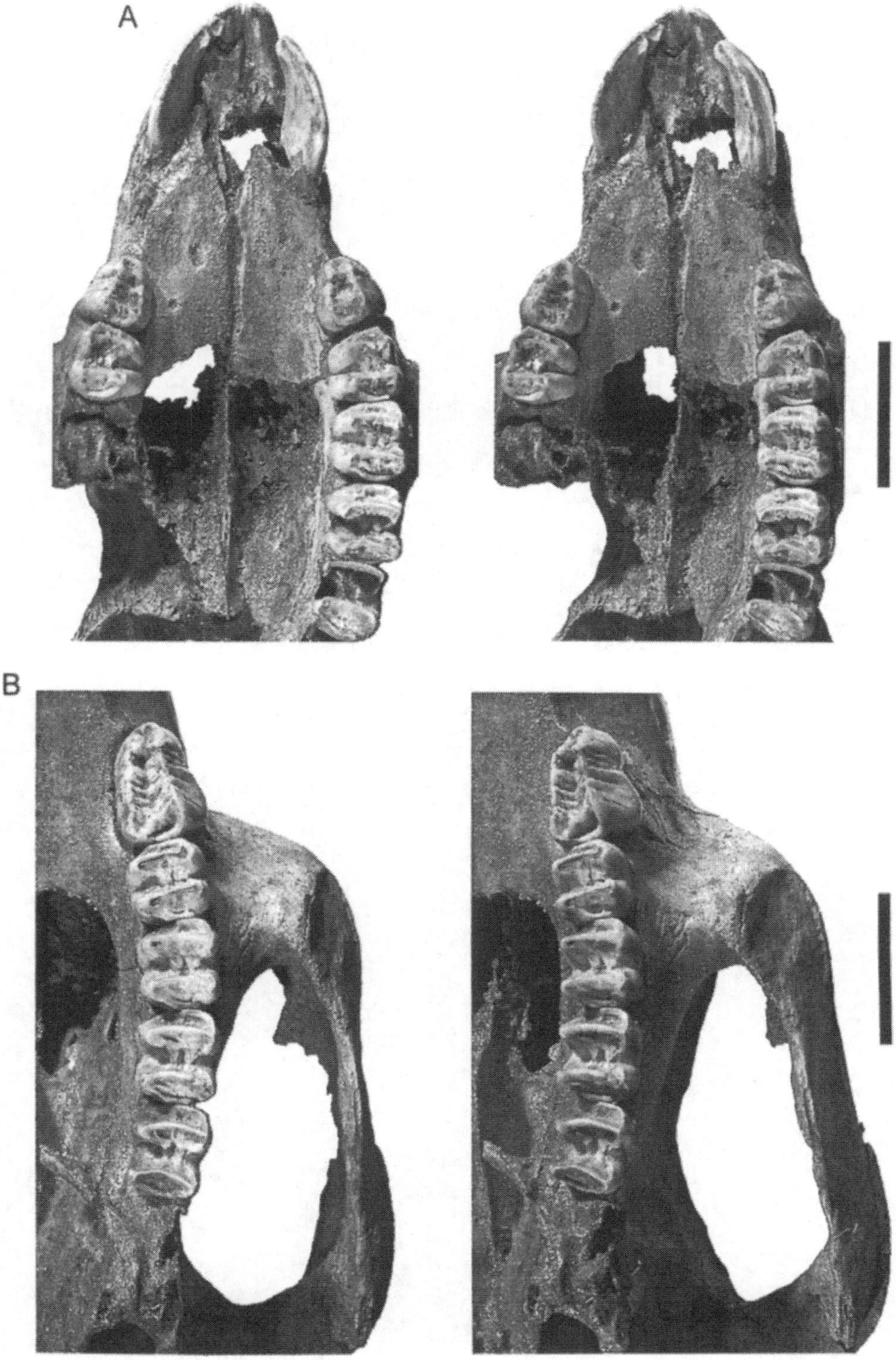
A
B

PLATE 100

"Procoptodon" gilli

Scale bars = 20 mm

A. SAM P20727, left juvenile dentary, mesial view. Victoria Fossil Cave, SA.
B. SAM P20727, lateral view.
C. SAM P20670, right juvenile dentary, mesial view (photo reversed). Victoria Fossil Cave, SA.
D. SAM P16510, left juvenile dentary, lateral view. Victoria Fossil Cave, SA.
E. SAM P20634, right juvenile dentary, mesial view (photo reversed). Victoria Fossil Cave, SA.
F. SAM P20634, lateral view (photo reversed).
G. SAM P16528, left dentary, mesial view. Victoria Fossil Cave, SA.
H. SAM P16528, lateral view.
I. SAM P20520, left adult dentary, mesial view. Victoria Fossil Cave, SA.
J. SAM P20520, lateral view.

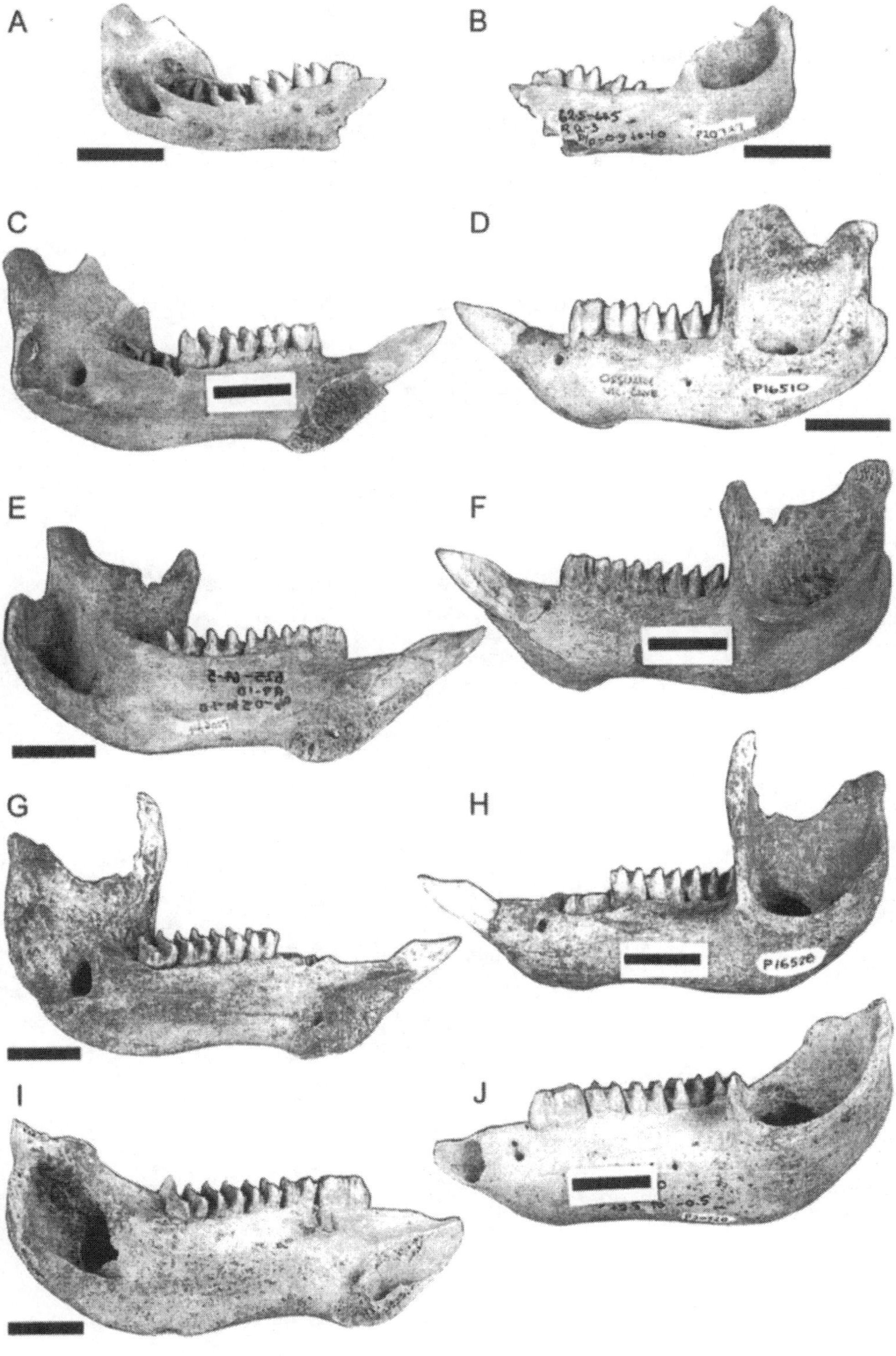
A
B
C
D
P16510
E
F
G
H
P16578
I
J

PLATE 101

"Procoptodon" gilli

Scale bars = 20 mm

A. SAM P16510, left and right juvenile dentaries, occlusal view. Victoria Fossil Cave, SA.
B. SAM P20797, left and right juvenile dentaries, occlusal view. Victoria Fossil Cave, SA.
C. SAM P28982, left juvenile dentary, stereo occlusal view. Victoria Fossil Cave, SA.
D. SAM P20520, left adult dentary, stereo occlusal view. Victoria Fossil Cave, SA.

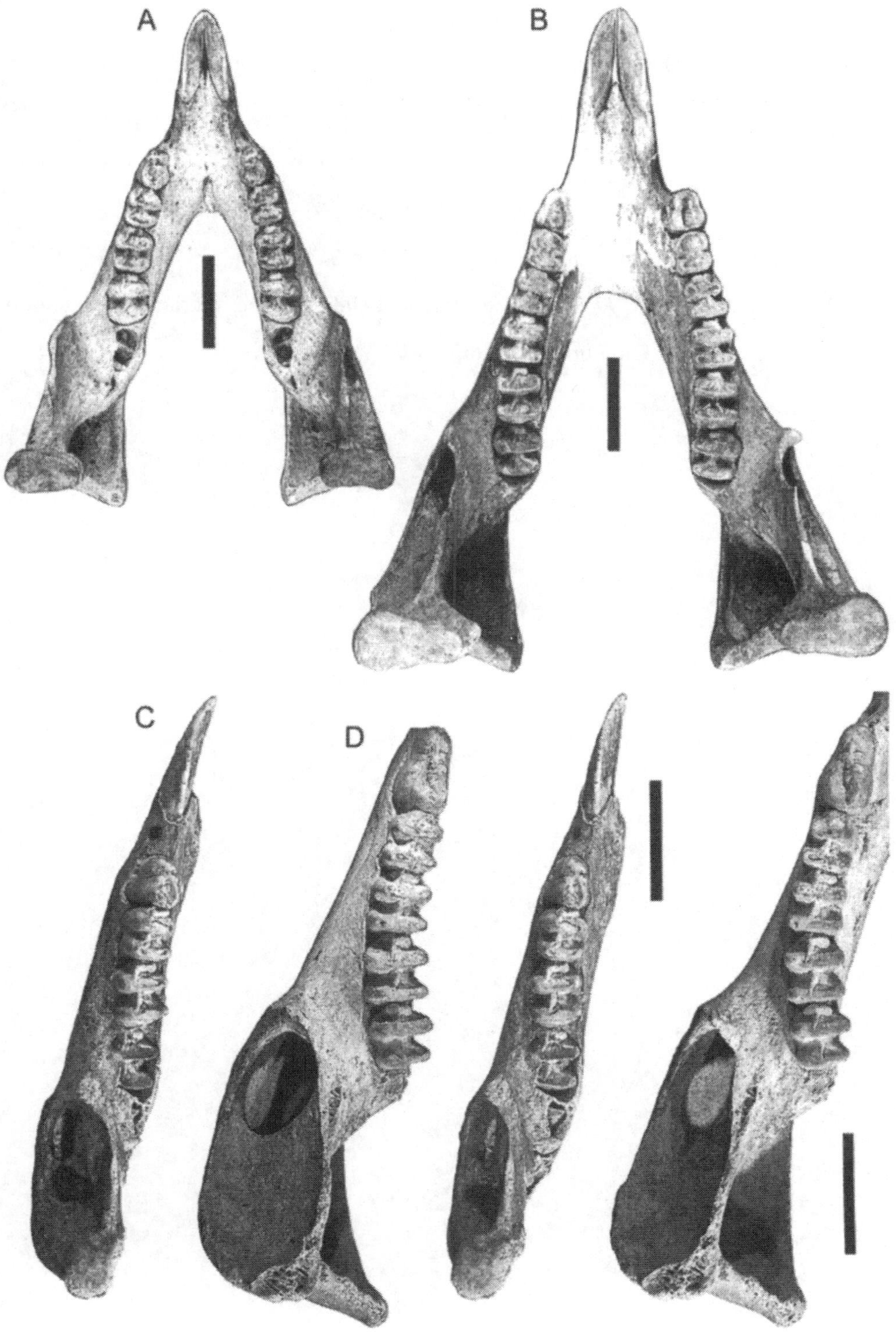
A
B
C
D

PLATE 102

"Procoptodon" mccoyi

Scale bar = 20 mm

NMV P23271 / P23272 (holotype), left and right adult dentaries (cast), stereo occlusal view. Limeburner's Point, VIC.

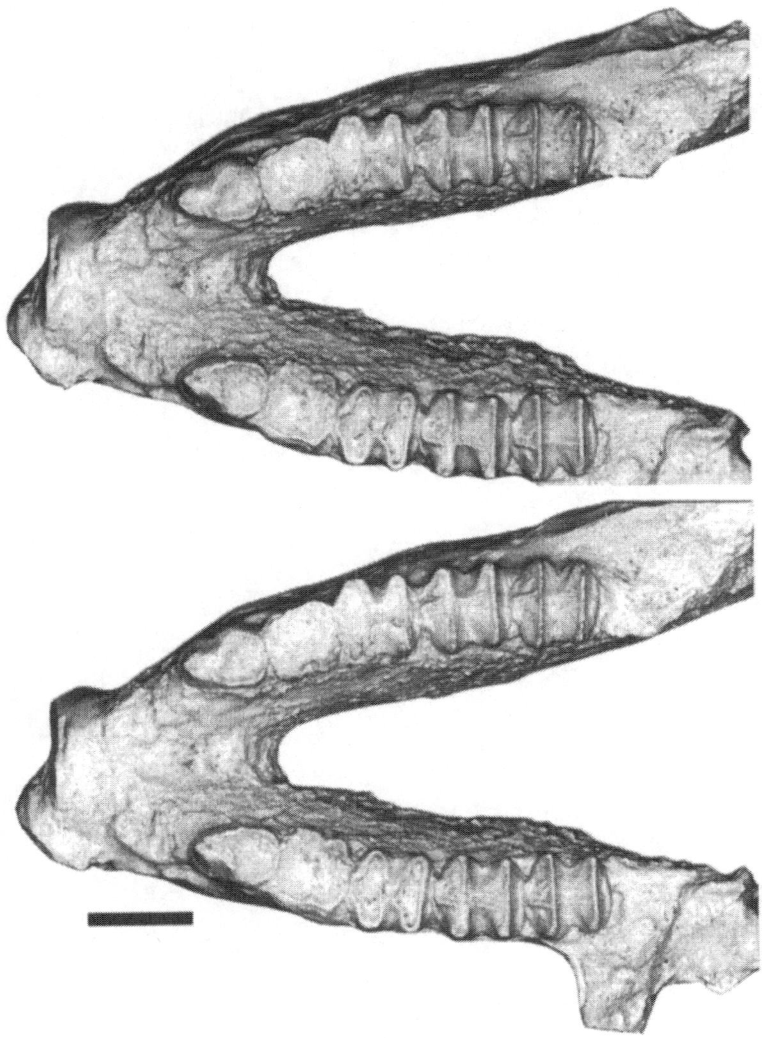

PLATE 103

"Procoptodon" oreas

Scale bars = 20 mm

A. QM F2923 (holotype), left and right adult dentaries, stereo occlusal view. Darling Downs, QLD.
B. QM F2923, mesial view.
C. QM F2923, lateral view.
D. QM F2923, close-up of right cheek tooth row, occlusal view.

A
B
C
D

PLATE 104

"Procoptodon" oreas

Scale bars = 20 mm

A. QM F3814, left maxilla, stereo occlusal view. Gore, QLD.

B. AM F88541, left adult dentary, lateral view. Wellington Caves, NSW.

C. AM F88541, mesial view.

D. AM F88541, occlusal view.

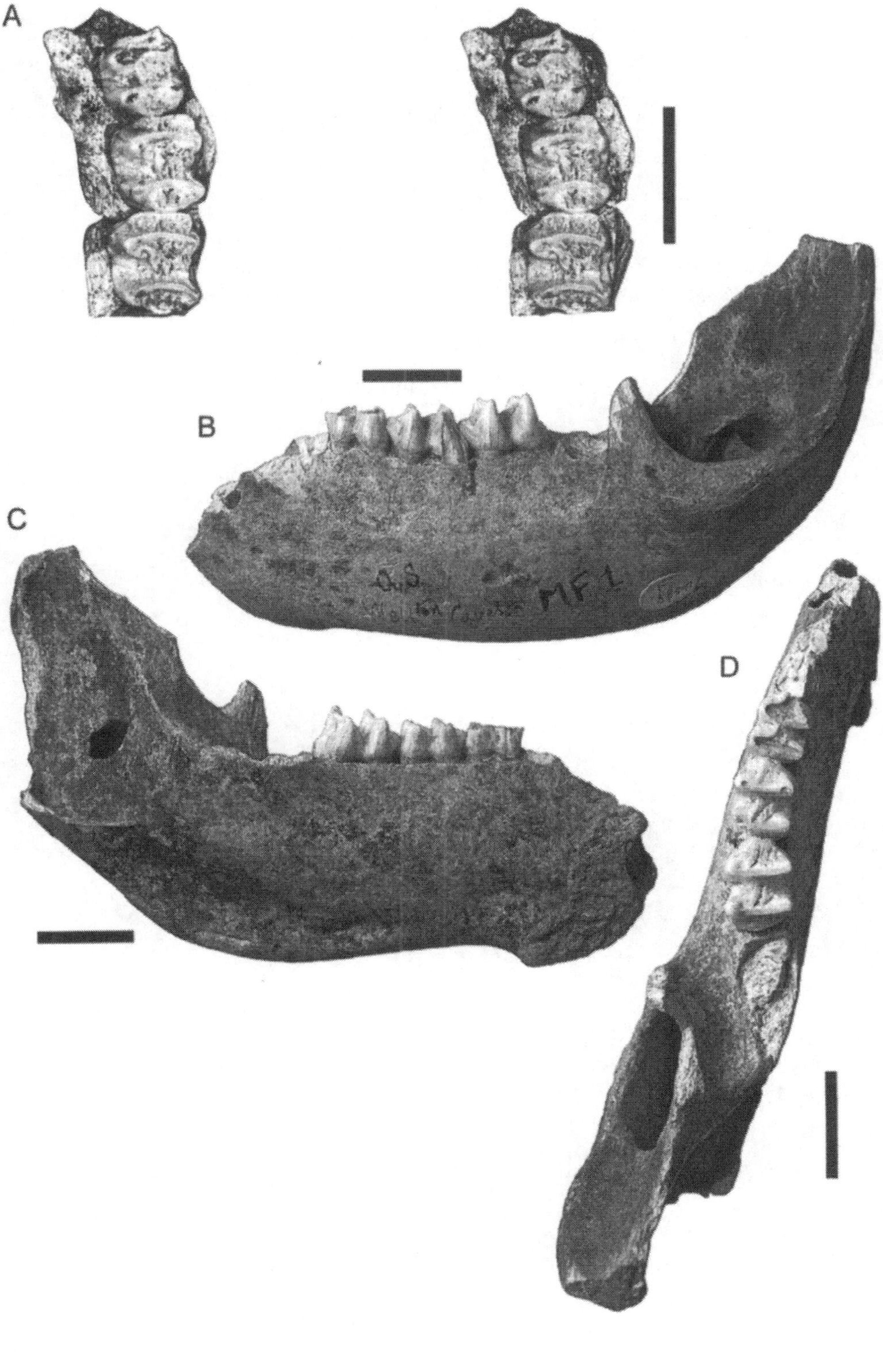
A
B
C
D

PLATE 105

"Procoptodon" oreas

Scale bars = 20 mm

A. AM F69880, left juvenile dentary, lateral view. Cathedral Cave, NSW.

B. AM F69881, right juvenile dentary, lateral view. Cathedral Cave, NSW.

C. AM F69881, right juvenile dentary, mesial view.

D. AM F69880, left juvenile dentary, mesial view.

E. AM F69880 / F69881, left and right juvenile dentaries, occlusal view.

F. AM F69881, right juvenile dentary, close-up of cheek tooth row, occlusal view.

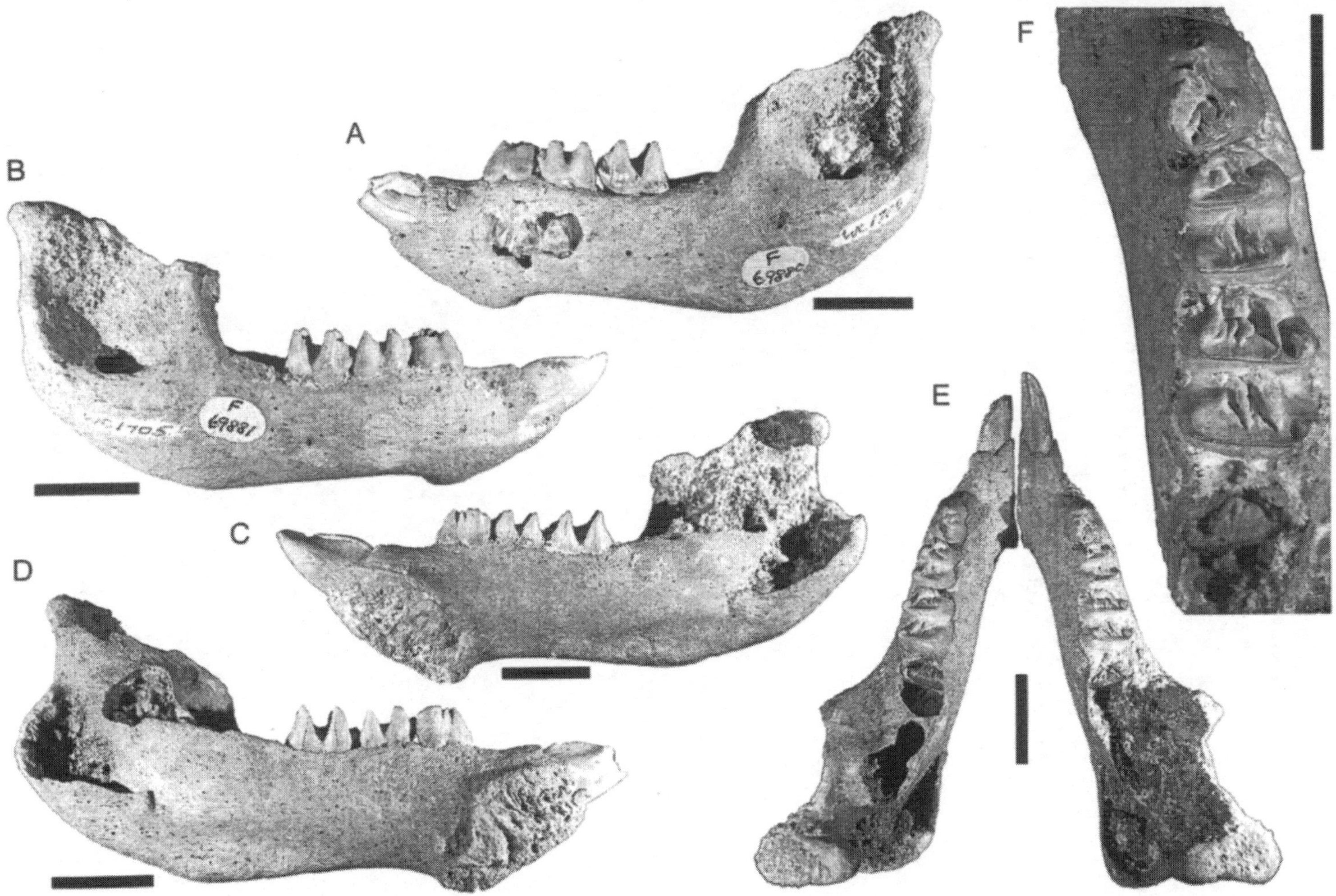
A
B
C
D
E
F

PLATE 106

"Procoptodon" oreas

Scale bars = 20 mm (A-C); 10 mm (D)

A. AM F31922, left juvenile dentary, lateral view. Molong, NSW.

B. AM F31922, mesial view.

C. AM F31922, stereo occlusal view.

D. AM F31922, excavated p3, occlusal view.

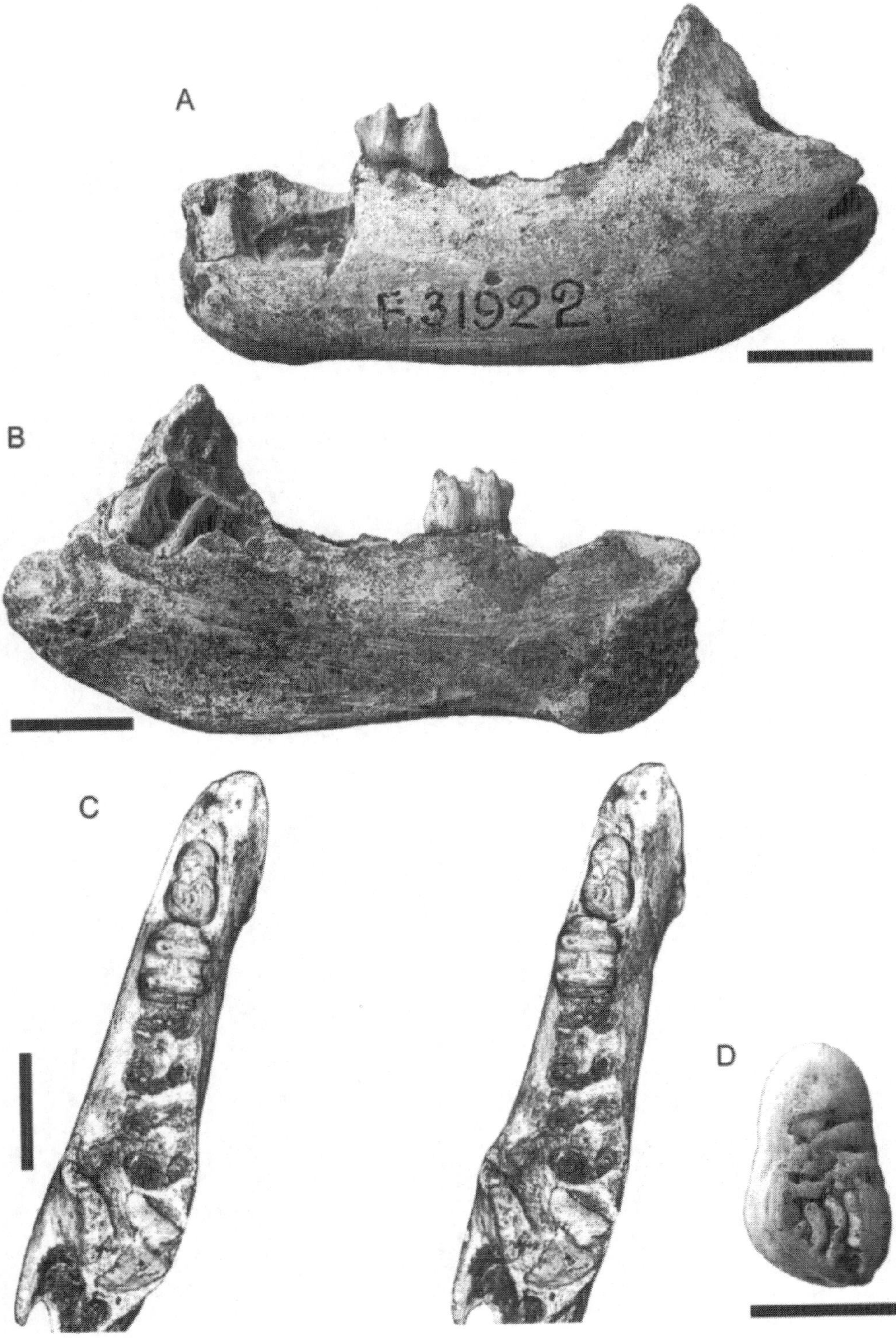
A
F.31922
B
C
D

PLATE 107

"Procoptodon" williamsi

Scale bars = 20 mm (A-C); 10 mm (D-H)

A. SAM P25059, left adult maxilla, lateral view. Malkuni Waterhole, SA.
B. SAM P25059, mesial view.
C. SAM P25059, stereo occlusal view.
D. AM F18903, right M1 in maxilla fragment, anterior view. Wellington Caves, NSW.
E. AM F18903, buccal view.
F. AM F18903, lingual view.
G. AM F18903, posterior view.
H. AM F18903, stereo occlusal view.

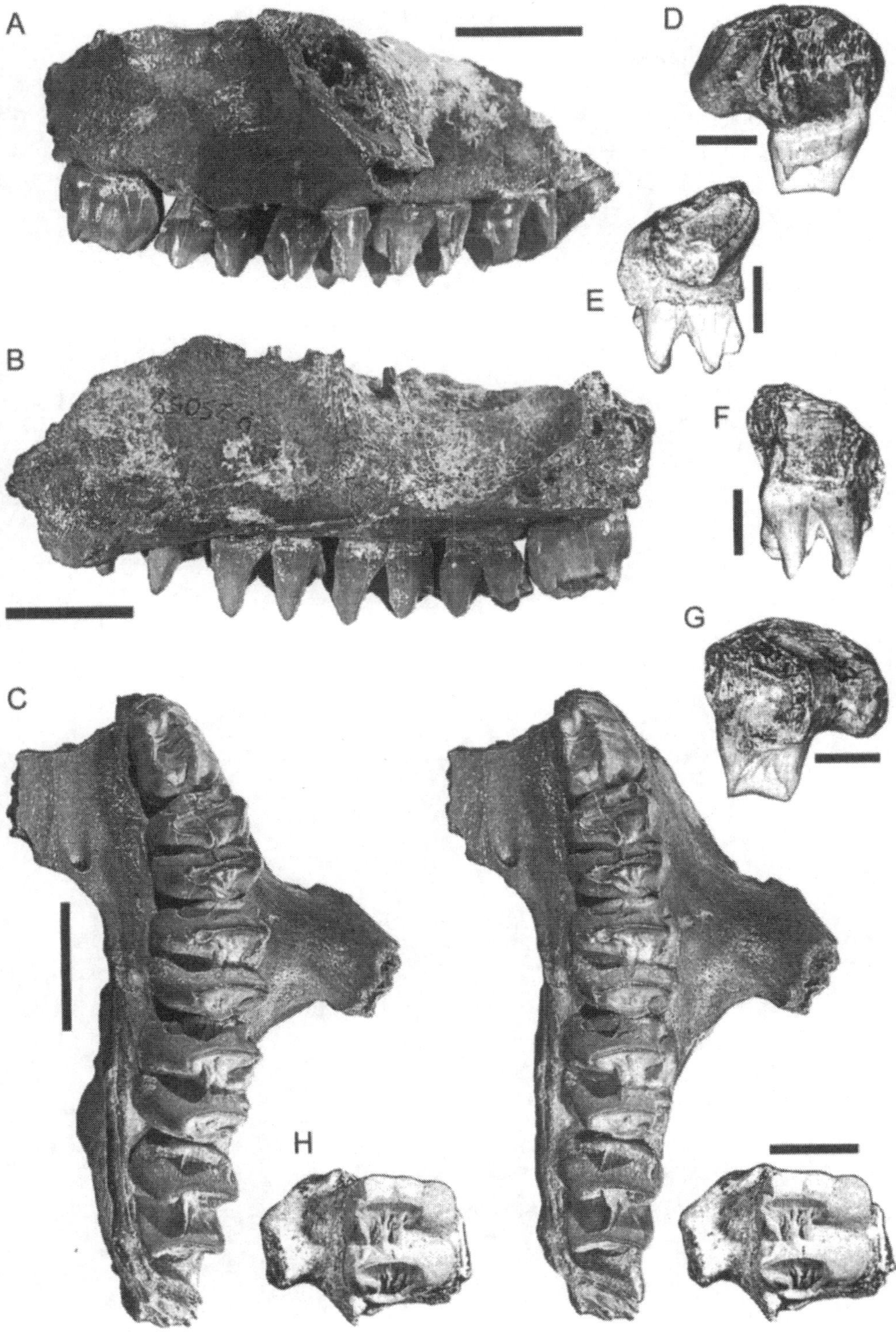
A
B
C
D
E
F
G
H

PLATE 108

Sketch reconstruction of the late Miocene sthenurine *Hadronomas puckridgi* by Peter F. Murray.

PLATE 109

Sketch reconstruction of the middle to late Pleistocene sthenurine *Sthenurus stirlingi* by Peter F. Murray.

PLATE 110

Full reconstruction of the middle to late Pleistocene sthenurine *Procoptodon goliah* by Stephen Hayter. Photographer: Steven Bourne, Naracoorte Caves National Park.